INTERNATIONAL ENCYCLOPEDIA OF
PHANRMACOLOGY AND THERAPEUTICS

Executive Editor: A. C. SARTORELLI, *New Haven*

Section 126

MECHANISMS OF CELLULAR TRANSFORMATION
BY CARCINOGENIC AGENTS

EDITORIAL BOARD

D. BOVET, *Rome*

W. C. BOWMAN, *Glasgow*

A. M. BREKENRIDGE, *Liverpool*

A. S. V. BURGEN, *London*

J. CHEYMOL, *Paris*

G. B. KOELLE, *Philadelphia*

P. LECHAT, *Paris*

H. RASKOVA, *Prague*

A. C. SARTORELLI, *New Haven*

V. V. ZAKUSOV, ✠ *Moscow*

Some Recent and Forthcoming Volumes

113 MATSUMURA
Differential Toxicities of Insecticides and Halogenated Aromatics

114 BALFOUR
Nicotine and the Tobacco Smoking Habit

115 MITCHELL
The Modulation of Immunity

116 SHUGAR
Viral Chemotherapy, Volume 2

117 De WEID, GISPEN and van WIMERSMA GREIDANUS
Neuropeptides and Behaviour, Volume 1

118 GOLDMAN
Membrane Transport of Antineoplastic Agents

119 DORNER and DREWS
Pharmacology of Bacterial Toxins

120 IHLER
Methods of Drug Delivery

121 DETHLEFSEN
Cell Cycle Effects

122 ROWLAND and TUCKER
Pharmacokinetics: Theory and Methodology

123 De WEID, GISPEN and van WIMERSMA GREIDANUS
Neuropeptides and Behaviour, Volume 2

124 DENBOROUGH
The Role of Calcium Ions in Drug Actions

125 WEBBE
The Toxicology of Molluscicides

126 GRUNBERGER & GOFF
Mechanisms of Cellular Transformation by Carcinogenic Agents

127 PESTKA
Interferon

NOTICE TO READERS

Dear Reader

If your library is not already a standing order customer to this series, may we recommend that you place a standing order to receive immediately on publication all new volumes published in this valuable series. Should you find that these volumes no longer serve your needs your order can be cancelled at any time without notice.

The Editors and Publisher will be glad to receive suggestions or outlines of suitable titles for consideration for rapid publication in this series.

ROBERT MAXWELL
Publisher at Pergamon Press

INTERNATIONAL ENCYCLOPEDIA OF
PHARMACOLOGY AND THERAPEUTICS

Section 126

MECHANISMS OF CELLULAR TRANSFORMATION BY CARCINOGENIC AGENTS

SECTION EDITORS

D. GRUNBERGER

and

S. P. GOFF

*College of Physicians and Surgeons of Columbia University,
Cancer Center, New York, USA*

RC268.5
M426
1987

PERGAMON PRESS

OXFORD · NEW YORK · BEIJING · FRANKFURT
SÃO PAULO · SYDNEY · TOKYO · TORONTO

U.K.	Pergamon Press, Headington Hill Hall, Oxford OX3 0BW, England
U.S.A.	Pergamon Press, Maxwell House, Fairview Park, Elmsford, New York 10523, U.S.A.
PEOPLE'S REPUBLIC OF CHINA	Pergamon Press, Room 4037, Qianmen Hotel, Beijing, People's Republic of China
FEDERAL REPUBLIC OF GERMANY	Pergamon Press, Hammerweg 6, D-6242 Kronberg, Federal Republic of Germany
BRAZIL	Pergamon Editora, Rua Eça de Queiros, 346, CEP 04011, Paraiso, São Paulo, Brazil
AUSTRALIA	Pergamon Press Australia, P.O. Box 544, Potts Point, N.S.W. 2011, Australia
JAPAN	Pergamon Press, 8th Floor, Matsuoka Central Building, 1-7-1 Nishishinjuku, Shinjuku-ku, Tokyo 160, Japan
CANADA	Pergamon Press Canada, Suite No 271, 253 College Street, Toronto, Ontario, Canada M5T 1R5

Copyright © 1987 Pergamon Books Ltd.

All Rights reserved. No part of this publication may be reproduced, stored in a retrieval system or transmitted in any form or by any means: electronic, electrostatic, magnetic tape, mechanical, photocopying, recording or otherwise, without permission in writing from the publishers.

First edition 1987

Library of Congress Cataloging-in-Publication Data

Mechanisms of cellular transformation by
carcinogenic agents.
(International encyclopedia of pharmacology
and therapeutics; section 126)
Includes index.
1. Carcinogenesis. 2. Cell transformation.
3. Oncogenes. I. Grunberger, D. (Dezider),
1922–   . II. Goff, S. III. Series. [DNLM:
1. Cell Transformation, Neoplastic—chemically
induced. 2. Oncogenic Viruses.
QV 4 158 section 126]
RC268.5.M426   1987   616.99'4071   87–2355

British Library Cataloguing in Publication Data

Mechanisms of cellular transformation by
carcinogenic agents.—(International
encyclopedia of pharmcology and therapeutics;
section 126).
1. Carcinogenesis
I. Grunberger, D.   II. Goff, S.   III. Series
616.99'407   RC268.5
ISBN 0-08-034204-3

Published as Supplement 26 (1987) to the review journal
Pharmacology and Therapeutics

PREFACE

THE study of cellular transformation and tumorigenesis has historically been divided into two major areas: the study of carcinogenesis by chemical and physical agents on the one hand, and the investigation of tumor viruses on the other. At times these two camps have been at odds with each other, debating both the methodologies of the opposing field and the relative significance of these two mechanisms in human cancers. In the last five years an extraordinary unification of these two previously diverse fields of research has occurred. The union is largely attributable to an experimental finding that these two seemingly unconnected mechanisms of transformation are in fact closely related, in that both agents apparently act *via* the alteration of a limited number of target genes in the host, the infamous proto-oncogenes. The critical alteration can be of many sorts. The target gene may suffer point mutations induced by chemical mutagens; it may be translocated to new chromosomal locations and joined to novel DNA after physical damage; it may be activated by viral transcriptional elements. The wide role of these genes in transformation has focused the attention of both camps on these sequences. Perhaps even more importantly it has provided a comprehensible framework within which both camps can proceed to study how chemicals or radiation, and how viruses, act to alter cellular DNA. This volume is a result of that union of fields.

Investigators at the leading edge of both expanding areas of research have been asked to contribute reviews to this volume. The collected chapters are here grouped into a total of five clusters. The first section, including five contributions, deals with the mechanisms of transformation by chemical carcinogens; the second section, a single chapter, describes transformation by physical agents. The third and fourth sections describe transformation by various selected members of the DNA and RNA virus families, including a total of eight contributions. The last section, containing a single chapter, summarizes the role of cellular oncogenes in carcinogenesis. A brief outline of each of the chapters follows in the remaining pages of this preface.

An important finding from a decade of studies on chemical carcinogenesis is the fact that such carcinogenesis is usually a multistage process that can extend over a considerable fraction of the total life span of the individual. In such studies, two qualitatively different stages in the transformation process were clearly defined, each mediated by separate compounds: tumor initiation and tumor promotion. The critical target in the initiation phase is the cellular DNA, the introduction of mutations into cellular genes. The primary effect in the promotion phase is less clear but probably results in alterations in the properties of the cell membrane and membrane proteins. The first chapter describes the metabolism of chemical carcinogens that are involved in the initiating step of transformation.

A wide range of different chemicals associated with cancer induction require metabolic activation. Depending on the routes of activation all carcinogenic compounds can be assigned to one of three broad classes. The first class is comprised of carcogens for which known metabolites have been shown to exhibit pronounced carcinogenic activity. Most the substances in this class require more than one enzymatic step for activation and thus are converted initially to proximate carcinogens. The second class of carcinogens includes those for which strong indirect evidence points to a particular route of metabolic activation but for which no metabolite of comparable carcinogenic potency than the parent compound can be directly demonstrated. The third class of carcinogens contains all those chemicals for which a route of metabolic activation has not yet been established. It should be stressed that metabolic activation plays an important role in determining the site of cancer development by the availability of appropriate enzymes in various tissues in which metabolic activation occur.

In the second chapter modification of DNA by chemical carcinogens is reviewed. Only covalent interactions of the carcingens with DNA are described. Different techniques used for the detection of DNA-carcinogen adducts is discussed thoroughly. The classification of chemical carcinogens is based here mainly upon their chemical structure. Some simple alkylating agents bind directly to the DNA but others require the sorts of activation described in the previous chapter. A large series of compounds, polycyclic aromatic hydrocarbons, react with DNA only after such events and their major target is usually the N2 position of guanosine residues. The class of N-substituted aromatic amines, after activati, have as their main target the C-8 position of guanosine residues; often these adducts induce significant distortions in the DNA helix. A class of naturally occurring carcinogens comprises the aflatoxins, which bind to the N-7 position of guanosine residues. Finally, compounds like psoralen which require photochemical rather than metabolic activation for reactivity are described

The following chapter reviews the recent developments in the role of tumor promoters in cell transformation. The concept of two-stage carcinogenesis is discussed, focussing on the mouse skin system and on other *in vivo* and *in vitro* model systems. After a presentation of the chemistry of phorbol diesters and other skin tumor promoters, the biological and biochemical effects of promoters on the cell surface and plasma membrane of cells in culture is presented with emphasis on the binding of phorbol diesters to specific receptor sites and activation of the Ca^{++}-sensitive phospholipid-dependent protein kinase C. The role of tumor promoters in cell proliferation, gene amplification and cell differentiation is further discussed. In the final part of the chapter, it is suggested that tumor promoters might enhance the efficiency of transformation by tumor viruses.

In the last chapter of this section the role of DNA lesions and repair in the transformation of human cells is discussed. The critical factor which can influence the final outcome of the DNA damage is whether the modifications can be repaired. The data indicate that excision repair in human fibroblasts, an error-prone process, is important; cells deficient in excision repair are abnormally sensitive to the transforming effects of carcinogens. It is further suggested that the critical event for transformation is DNA replication on a template containing unexcised lesions.

The second section of this volume, containing only a single chapter (5), deals with the cellular and molecular mechanisms involved in the induction and control of radiogenic transformation *in vitro*. Culture systems currently used in transformation by ionizing as well as by non-ionizing radiation are described. The role of carcinogens and modulators of radiogenic transformation in enhancing or inhibiting the process is further discussed. Finally, consideration is given to the possible involvement of specific cellular oncogenes in the radiogenic transformation of cells in culture.

In the remaining portion of the book, we move into the camp of those investigators who study the process of transformation by viral agents. The third section specifically describes transformation by a number of DNA viruses. These viruses were among the first biological agents to be clearly associated with transformation, and the genes of the viruses responsible were among the first and most intensively studied of the transforming oncogenes. The first chapter in this section, the sixth of the book, describes the properties of the papova (for Papilloma, Polyoma, and Vacuolating) viruses known as SV40 and polyoma in causing transformation. These viruses can be lytic viruses in an appropriately permissive host, leaving few viable cells able to exhibit any growth, normal or abnormal; but in non-permissive host species these viruses can lead to potent tumors. The key product of these viruses is the famous T (for Tumor) antigen; this nuclear protein product of the early region is essential for the ability of the viruses to replicate in permissive hosts, and to cause tumors in other hosts. The situation has become more complicated by the finding of multiple, alternately spliced, mRNA products from a single region of the genome. The current state of knowledge of the role of these several proteins in eliciting alterations in growth are summarized in this chapter.

The next chapter (7) describes another, equally popular, and more complex object of study: the Adenoviruses. Here too multiple genes and gene products are involved, and the dissection of the interactions of these proteins has required efforts in both molecular biology and genetics. The early region, denoted El, is the essential region, but numerous overlapping transcripts and proteins are encoded there; simple mutations have multiple effects on these gene products and therefore often induce multiple phenotypes. Surgical attacks on the region have been made by site-directed *in vitro* mutagenesis to alter only one or a few of the gene products, and the analysis of these mutations has allowed at least partial assignment of function. This chapter describes the structure of these mutations and the current status of the role of the gene products in eliciting the transformed state.

The herpes viruses do not seem to contain one essential gene for transformation; tumor cells seem not always to retain a particular portion of the genome. Our understanding of transformation by these important agents is much less detailed than for the other DNA viruses. Many of the advances in therapy in recent years have stemmed from efforts to block replication of the viruses rather than the actual transformation event itself. The chapter on the herpes viruses discusses the life cycle and summarizes recent progress in the development of antiviral agents.

The hepadna (hepatitis B-like) viruses bridge the world of DNA and RNA tumor viruses; they are DNA viruses in the sense that the virion-associated genome is DNA, but they are most likely related to the RNA tumor viruses in their replication. In the infected cell the circular DNA genome is transcribed into a more-than-full length RNA, and this RNA is reverse transcribed into DNA before packaging into the progency virions. The viruses can apparently replicate without integration into cellular DNA, but many hepatic tumors contain viral sequences. As for the herpes viruses, it is not clear whether particular viral genes are essential for tumorigenesis or whether alterations in cellular genes are responsible. A chapter on transformation by these viruses discusses their role in induction of human carcinomas.

A fourth section of this volume addresses the important topic of transformation by the RNA tumor viruses, now more generally named retroviruses. These viruses have the unique capacity to transduce cellular sequences by the acquisition of such sequences onto the viral genome. When such acquired genes have the property of inducing alterations in the parameters of growth of the host cell, we term them cellular oncogenes: genes which when altered by their incorporation into the virus can cause tumors. Analysis of the genes transduced by the acutely-transforming retroviruses has led to the identification of the most extensive collection of potential transforming genes of any of the available routes toward such genes. The feline viruses have been a particularly rich source of new oncogenes; the first chapter in this section discusses the properties of several feline sarcoma viruses, and the important relationship of their transduced oncogenes to the cellular genes encoding cell-surface receptors for polypeptide growth factors. A second chapter in this series describes the *fos* gene, a gene transduced by a mouse osteosarcoma virus, encoding a nuclear protein factor likely to be involved in induction of expression of other cellular genes. These chapters highlight the diverse properties of the proteins encoded by the various classes of cellular oncogenes.

Two other chapters in this section describe transformation by replication-competent retroviruses. These slowly-transforming viruses, such as the avian leukosis viruses, do not carry oncogenes of their own; the major initiating event is apparently the activation of expression of selected genes in the host by a process termed promoter insertion or enhancer insertion. Here the insertion of the viral genome is the key: the proviral DNA is found integrated in the vicinity of a target gene in tumor cells. In some cases the target genes are identical to those identified by transduction events that led to the formation of one or more acute transforming viruses. The mechanisms of activation are many. For the human T-lymphotropic viruses, the integration site seems not be critical, and if activation events are involved in transformation they must be mediated by trans-acting factors encoded by the virus. The discovery of autostimulatory factors for the virus's own transcription has strengthened the case for such a mechanism of transformation in the rare human tumors induced by these agents.

The final section ties together the processes of transformation by chemical and environmental carcinogens and transformation by viruses. The single chapter (14) describes the mutations identified in various human tumors: translocations, amplifications, and point mutations. In many such cases the site of the critical mutation has been identified, and the astonishing result is that the target genes altered in these tumors by various chemical and physical abuses are the very same cellular oncogenes activated by the intervention of viruses. They include representatives of all the classes of the known viral oncogenes: those encoding cell-surface proteins, cytoplasmic proteins, and nuclear proteins. Yet it is clear that these tumors are of non-viral etiology, and it is only at the underlying gene level that the similarity of the mechanisms of transformation becomes apparent. It seems appropriate to close on this note. The intense efforts of biological scientists in their two-pronged attack on the mechanism of transformation has resulted in a true Grand Unified Theory of transformation. The unification has great explanatory power for the results of the past, and it has given us a new framework on which to hang the results of future explorations. The chapters in this volume herald that future work.

CONTENTS

LIST OF CONTRIBUTORS — xiii

1. **METABOLISM OF CHEMICAL CARCINOGENS** — 1
   A. DIPPLE, C. J. MICHEJDA, *NCI-Frederick Cancer Research Faculty, Maryland, USA* and E. K. WEISBURGER, *National Cancer Institute, Bethesda, Maryland, USA*
   1. Introduction — 1
   2. Chemical carcinogens — 2
   3. Mechanism of action of chemical carcinogens — 5
   4. Metabolic activation of chemical carcinogens — 7
   5. Conclusions — 24

2. **DNA MODIFICATION BY CHEMICAL CARCINOGENS** — 33
   A. M. JEFFERY, *Cancer Center/Institute of Cancer Research, Columbia University, New York, USA*
   1. Introduction — 33
   2. Types of interaction with DNA — 34
   3. Detection of DNA adducts — 34
   4. Classes of chemical carcinogens — 39
   5. Summary — 56

3. **TUMOR PROMOTERS AND CELL TRANSFORMATION** — 73
   L. DIAMOND, *The Wistar Institute of Anatomy and Biology, Philadelphia, USA*
   1. Introduction — 73
   2. Tumor promotion *in vivo* — 75
   3. Chemistry of phorbol diesters and other skin tumor promoters — 80
   4. Biological and biochemical effects of phorbol diesters on cells in culture — 84
   5. Models of two-stage transformation *in vitro* — 98
   6. Interactions between tumor promoters and tumor viruses — 108
   7. Conclusions — 112

4. **ROLE OF DNA LESIONS AND REPAIR IN THE TRANSFORMATION OF HUMAN CELLS** — 135
   V. M. MAHER and J. J. McCORMICK, *Michigan State University, USA*
   1. Introduction — 135
   2. Effect of repair on the cytotoxicity of carcinogens in human cells — 137
   3. Effect of excision repair on the frequency of mutations induced in human cells by carcinogens — 139
   4. Effect of allowing time for excision before DNA synthesis — 141
   5. Studies on neoplastic transformation of diploid human fibroblasts — 143
   6. Role of DNA excision repair in the transformation process — 145

7.	Effect on transformation of allowing time for excision repair before onset of DNA synthesis	145
8.	Conclusion	146

5. THE INDUCTION AND REGULATION OF RADIOGENIC TRANSFORMATION *IN VITRO*: CELLULAR AND MOLECULAR MECHANISMS — 151
C. BOREK, *Columbia University College of Physicians and Surgeons, New York, USA*

1.	Introduction	152
2.	Cell cultures	152
3.	Cell transformation *in vitro*	152
4.	Radiogenic transformation *in vitro*	164
5.	Nonionizing radiation	173
6.	Cocarcinogens and modulators of radiogenic transformation	174
7.	Oncogenes in radiation transformation	182
8.	Discussion	184
9.	A final word	188

6. THE MECHANISM OF CELL TRANSFORMATION BY SV40 AND POLYOMA VIRUS — 197
C. BASILICO, *New York University School of Medicine, USA*

1.	Introduction	197
2.	Genome organization and expression	199
3.	Effects of viral infection on the host cell metabolism	204
4.	Cell transformation by polyoma and SV40	205
5.	The state of the viral genome in polyoma or SV40 transformed cells	208
6.	Expression of integrated genomes	213
7.	Evolution of transformed cell lines	217
8.	Role of the viral proteins in transformation	218
9.	Host control of viral transformation	224

7. CELLULAR TRANSFORMATION BY ADENOVIRUSES — 237
S. J. FLINT, *Princeton University, New Jersey, USA*

1.	Introduction	237
2.	Identification of adenovirus transforming genes	238
3.	The adenovirus E1A and E1B regions and their products	241
4.	The roles of E1A and E1B gene products in transformation	246
5.	Tumorigenicity of adenoviruses and adenovirus transformed cells	255
6.	Properties of E1A and E1B polypeptides	258

8. CELLULAR TRANSFORMATION BY THE HERPESVIRUSES AND ANTIVIRAL DRUGS — 271
J.-C. LIN and J. S. PAGANO, *University of North Carolina at Chapel Hill, USA*

1.	Introduction	271
2.	The herpesviruses	274
3.	The antiviral drugs	279
4.	Mechanism of action of drugs	283
5.	A perspective on treatment of EBV infection states	285
6.	General discussion and summary: herpesviruses and human neoplasia	289

9. **THE ROLE OF HEPADNA VIRUSES IN HEPATOCELLULAR CARCINOMA** — 299
   W. S. ROBINSON, R. H. MILLER and P. L. MARION, *Stanford University School of Medicine, California, USA*
   1. Introduction — 299
   2. Virus morphology and antigenic structure — 300
   3. Virion polypeptides — 301
   4. Physical and genetic structure of the viral genome — 302
   5. Genome homology of hepadnaviruses and retroviruses — 305
   6. Codon preferences in hepadnavirus genes — 307
   7. The mechanism of hepadnavirus replication — 308
   8. Relationship of hepadnaviruses and retroviruses — 311
   9. The association of hepadnavirus infection and hepatocellular carcinoma (HCC) — 313
   10. The state of hepadnaviruses in HCC — 315

10. **TRANSFORMATION BY FELINE RETROVIRUSES** — 325
    C. J. SHERR, *St Jude Children's Research Hospital, Memphis, Tennessee, USA*
    1. Introduction — 325
    2. Genetic structure and replication of FeLV — 326
    3. Exposure to feline leukemia virus — 326
    4. FeLV-induced disease — 327
    5. Isolation of feline sarcoma viruses — 328
    6. Structure and replication of feline sarcomaa viruses — 329
    7. Regulation of FeSV transcription — 330
    8. Oncogenes encoding tyrosine – specific protein kinases v-*fes*, v-*abl* and v-*fgr* — 331
    9. An unusual member of the kinase gene family v-*fms* — 332
    10. The v-*fms* gene product is related to the CSF-1 receptor — 334
    11. Transforming genes and growth factors (v-*sis*) — 335
    12. Conclusions — 335

11. **THE *fos* GENE** — 341
    I. M. VERMA and W. R. GRAHAM, *The Salk Institute, San Diego, California, USA*
    1. Historical background — 341
    2. Biology of *fos* viruses — 341
    3. Structure of the *fos* gene and protein — 343
    4. Transformation by *fos* gene — 345
    5. Expression of the c-*fos* gene — 347
    6. Transcription of the c-*fos* gene — 350
    7. Regulation of *fos* expression — 350

12. **CELLULAR TRANSFORMATION BY AVIAN VIRUSES** — 355
    D. R. MAKOWSKI, P. G. ROTHBERG and S. M. ASTRIN, *Institute for Cancer Research, Philadelphia, Pennsylvania, USA*
    1. Introduction — 355
    2. Rous sarcoma virus — 356
    3. Defective avian retroviruses — 365
    4. Lymphoid leukosis virus — 374

13. **CELLULAR TRANSFORMATION BY HUMAN T LYMPHOTROPIC RETROVIRUSES** — 391
    P. S. SARIN, *National Cancer Institute, Bethesda, Maryland, USA*
    1. Introduction — 391
    2. Isolation of HTLV — 391
    3. Other retroviruses — 394
    4. Epidemiology — 394
    5. *In vitro* transmission — 395
    6. T-cell tropism — 396
    7. Cell transformation — 397
    8. Phenotype of HTLV 1-transformed T-cells — 399
    9. Release of lymphokines — 400
    10. Mechanism of transformation — 401

14. **THE ROLE OF CELLULAR ONCOGENES IN CANCERS OF NON-VIRAL ETIOLOGY** — 407
    M. P. GOLDFARB, *Columbia University, New York, USA*
    1. Human and mouse B-cell malignancies have altered c-*myc* expression — 407
    2. Altered c-*abl* gene in human chronic myelogenous leukemia — 410
    3. Certain tumors have amplified c-*onc* genes — 411
    4. Activation of c-*mos* in murine myeloma by retrovirus-related element — 412
    5. The transforming genes in many types of tumors are c-*ras* — 413
    6. Directions for future research — 416

INDEX — 423

LIST OF CONTRIBUTORS

SUSAN M. ASTRIN
Institute for Cancer Research
Fox Chase Cancer Center
7701 Burholme Avenue
Philadelphia, PA 19111
USA

CLAUDIO BASILICO
Department of Pathology
NYU School of Medicine
550 First Avenue
New York, NY 10016
USA

CARMIA BOREK
Radiological Research Laboratory
Departments of Radiology and Pathology
Columbia University
College of Physicians & Surgeons
New York, NY 10032
USA

LEILA DIAMOND
The Wistar Institute of Anatomy and Biology
36th Street at Spruce
Philadelphia, PA 19104
USA

ANTHONY DIPPLE
Laboratory of Chemical and Physical
Carcinogenesis
LBI-Basic Research Program
NCI-Frederick Cancer Research Facility
PO Box B
Frederick, MD 21701
USA

S. FLINT
Department of Molecular Biology
Biochemical Sciences Laboratory
Princeton University
Princeton, NJ 08544
USA

MITCHELL P. GOLDFARB
Department of Biochemistry and Molecular
Biophysics
Columbia University
630 West 168th Street
New York, NY 10032
USA

CHRISTOPHER J. MICHEJDA
Laboratory of Chemical and Physical
Carcinogenesis
LBI-Basic Research Program
NCI-Frederick Cancer Research Facility
PO Box B
Frederick, MD 21701
USA

R. H. MILLER
Department of Medicine
Stanford University School of Medicine
Stanford, CA 94305
USA

JOSEPH S. PAGANO
Departments of Microbiology and Immunology
Lineberger Cancer Research Center
School of Medicine
University of North Carolina
Chapel Hill, NC 27514
USA

W. S. ROBINSON
Department of Medicine
Stanford University School of Medicine
Stanford, CA 94305
USA

PAUL G. ROTHBERG
Department of Human Genetics
Roswell Park Memorial Institute
666 Elm Street
Buffalo, NY 14263
USA

PREM S. SARIN
Laboratory of Tumor Cell Biology
National Cancer Institute
Bethesda, MD 20892
USA

CHARLES J. SHERR
Department of Tumor Cell Biology
St Jude Children's Research Hospital
332 North Lauderdale
PO Box 318
Memphis, TN 38101
USA

INDER M. VERMA
Molecular Biology and Virology Laboratory
The Salk Institute
PO Box 85800
San Diego, CA 92138
USA

ELIZABETH K. WEISBURGER
Division of Cancer Etiology
National Cancer Institute
Bethesda, MD 20892
USA

CHAPTER 1

METABOLISM OF CHEMICAL CARCINOGENS

Anthony Dipple,* Christopher J. Michejda* and Elizabeth K. Weisburger†

*Laboratory of Chemical and Physical Carcinogenesis, LBI-Basic Research Program,
NCI-Frederick Cancer Research Faculty, Frederick, MD, U.S.A.
†Division of Cancer Etiology, National Cancer Institute, Bethesda, MD, U.S.A.

Abstract—An overview of the *metabolism* of *chemical carcinogens* focuses on the *activation* of those substances to their *proximate* and *ultimate carcinogenic forms*. Less emphasis is placed on detoxification pathways. The carcinogens considered in this review include *aromatic amines, azo dyes, polycyclic aromatic hydrocarbons, N-nitrosamines, triazenes, hydrazines* and selected *carcinogenic natural products*. It is shown that in practically all cases, the evidence suggests that the active metabolites are electrophiles which are capable of interacting with cellular macromolecules. However, in many cases the presumed reactive metabolites have not been isolated and tested for their carcinogenic activity. Their involvement in the carcinogenic process is based on indirect evidence.

1. INTRODUCTION

Over the last fifty years, a wide variety of chemical substances have proven capable of eliciting a carcinogenic response in experimental animals. This property defines these chemicals as potential human carcinogens and, in the last 15 years in particular, public concern about the presence of such chemicals in the human environment has grown considerably. The problem is not specifically of recent origin, however, because sources of chemical carcinogens include plants and fungi (reviewed by Weisburger, 1977) and combustion products (reviewed by Dipple, 1983) which have been present in the environment for thousands of years.

The origins of the modern study of chemical carcinogenesis are closely associated with carcinogenic combustion and pyrolysis products because, in 1775, Percival Pott, surgeon to St. Bartholomew's Hospital in London, identified the first occupational carcinogen by attributing the high incidence of scrotal cancer in chimney sweepers to their occupational exposure to soot. Later, another major discovery was the first successful demonstration of chemical carcinogenesis in experimental animals by Japanese workers (Yamagiwa and Ichikawa, 1915) in which coal-tar was used as the carcinogenic agent. The final development of this early phase of research was the recognition, in 1930 by Kennaway and Hieger, that a pure chemical compound, dibenz[a,h]anthracene [1] was capable of eliciting tumors in animals. This last discovery was made in the course of studies designed to identify the major carcinogenic component of coal-tar, which later was shown to be benzo[a]pyrene [2] (Cook *et al.*, 1933).

Recognition of carcinogenic activity in relatively simple chemical structures stimulated a flow of research which widely extended the number of known chemical carcinogens. New carcinogens are still being discovered today, even after the passage of some fifty years. Thus, by 1962, Clayson could report that over five hundred chemical carcinogens had been identified. This number has certainly increased substantially since then. A more contemporary listing can be found in Searle's monograph (1976), and in the updated second edition (1984), but not all of the noxious substances described therein are to be found in our environment. Many carcinogens have been synthesized on a research scale for investigations of various structure-activity relationships. It should be obvious from the foregoing, however, that a comprehensive discussion of every known chemical carcinogen within the confines of a single article is not possible. Moreover, since the metabolism of chemical carcinogens is known to be influenced by experimental variables such as the species, strain, sex, age or endocrine status of the animals used, as well as by their diets or various drugs which can be administered, it is difficult to present even an adequate summary of this vast topic of metabolism of chemical carcinogens. For these reasons, we have cited many reviews of the literature and have focused this article on the metabolic reactions associated with the activation rather than the detoxification of chemical carcinogens.

2. CHEMICAL CARCINOGENS

A wide range of different chemical structures has been associated with cancer induction in animals and, apart from their carcinogenic properties, many of these structures bear no particularly obvious relationship to one another. As noted above, concern about the high incidence of skin cancers in workers exposed to soot, coal-tar, and pitch ultimately led to the discovery of the polycyclic aromatic hydrocarbon carcinogens (recounted by Kennaway, 1955) in the early 1930s. This discovery was rapidly followed by demonstrations that many, but by no means all, polycyclic aromatic hydrocarbons could be shown to be carcinogenic in animal studies. For the unsubstituted benzenoid hydrocarbons containing up to six rings, at least 16 structures have been found to be carcinogenic and many alkyl derivatives of the hydrocarbons also exhibit activity (reviewed by Dipple, 1976). For example, 7,12-dimethylbenz[a]anthracene [3] is one of the most potent of the hydrocarbon carcinogens, even though the unsubstituted benz[a]anthracene parent structure is one of the weaker carcinogens in this class. The hydrocarbons are widely distributed in the environment because of their origin in the incomplete combustion of fossil fuels and organic matter, and they also represent a very large group of carcinogens.

[1] Dibenz[a,h]anthracene
[2] Benzo[a]pyrene
[3] 7,12-Dimethylbenz[a]anthracene
[4] Dibenz[a,h]acridine
[5] 2',3-Dimethyl-4-aminoazobenzene
[6] 4-Dimethylaminoazobenzene
[7] 2-Naphthylamine
[8] 2-Fluorenylacetamide
[9] Carbon tetrachloride
[10] Ethylene dibromide
[11] Vinyl chloride
[12] Urethan
[13] N-Nitrosodimethylamine
[14] N-Nitrosomethyl-n-propylamine
[15] N-Nitrosomorpholine
[16] N-Nitroso-N-methylurea
[17] 1,2-Dimethylhydrazine
[18] Azoethane
[19] Azoxymethane
[20] 1-Phenyl-3,3-dimethyltriazene
[21] Aflatoxin B_1
[22] Safrole
[23] Cycasin
[24] 4-Nitroquinoline-N-oxidel
[25] N-[4-(5-nitro-2-furyl)-2-thiazolyl]formamide
[26] 1,3-Diphenyltriazene
[27] 1,3,3-Trimethyltriazene
[28] [1-Methyl-3-(2,4,6-trichlorophenyl)-2-triazeno]methyl-β-D-glucopyranoside uronic acid
[29] Estragole

A closely related group of carcinogens are the heterocyclic aromatics which contain structures such as dibenz[a,h]acridine [4]. These were discovered through synthetic manipulations of the hydrocarbons in which one or more carbon atoms in the skeleton were replaced by nitrogen (Barry et al., 1935). Such compounds are also formed through combustion and Lacassagne, Buu-Hoi and their colleagues synthesized and examined the carcinogenic activities of many heterocyclic aromatics (Lacassagne et al., 1956). The heterocyclic aromatics are generally similar to the hydrocarbons in their carcinogenic actions and may perhaps work through similar mechanisms of metabolic activation.

Another substantial group of carcinogens is the aromatic amines and azo dyes which are principally products of the synthetic organic chemist. The initial observation which ultimately led to the discovery of this group of carcinogens was due to the concern of Rehn (1895) over the urinary bladder cancers he noted in three workers in an 'aniline dye' factory in Germany. This led to the subsequent testing in animals of various chemicals to which such workers were exposed and the consequent discovery of carcinogenic activity for the liver of rats and mice in the azo dye, 2′,3-dimethyl-4-aminoazobenzene [5] (Yoshida, 1933). An isomeric compound, N,N-dimethyl-4-aminoazobenzene [6] was subsequently found to be a liver carcinogen also (Kinosita, 1936), and a large number of aminoazo dyes have since been synthesized and found to be carcinogenic (Miller and Miller, 1953; Clayson and Garner, 1976). The cause of the bladder tumors observed in workers in the dye industry was established by the epidemiological investigations of Case et al. in 1954 to be 2-naphthylamine [7]. This aromatic amine induced bladder cancer in dogs (Hueper et al., 1938) but not in rats. Another carcinogen of major importance experimentally which belongs in this group of carcinogens is N-2-fluorenylacetamide [8]. This compound was intended to be used as an insecticide but, despite its lack of acute toxicity, was found to be carcinogenic for the rat after chronic exposures (Wilson et al., 1941). As for the azo dyes, many derivatives of the aromatic amines and amides have been studied and found to be carcinogenic (reviewed by Weisburger and Weisburger, 1958; Clayson and Garner, 1976; Garner et al., 1984).

Apart from obvious structural differences between the aromatic hydrocarbons on the one hand and the azo dyes and aromatic amines on the other, these two groups of carcinogens are also different in their tumor-inducing properties. Thus, quite small doses of the hydrocarbons administered subcutaneously or topically yield tumors at the site of administration. With the azo dyes and aromatic amines, such local tumor induction does not occur and relatively large doses have to be administered parenterally over a substantial time in order to obtain tumors. Early on in research into their mechanisms of action, these properties of the dyes and aromatic amines suggested that a metabolite rather than the administered carcinogen was responsible for tumor induction. It may surprise relative newcomers to the field of chemical carcinogenesis to learn that it was not until the late 1960s and early 1970s that it was widely accepted that metabolism may also play a significant role in the carcinogenic actions of the polycyclic aromatic hydrocarbons.

The 1940s saw the discovery of other carcinogens in addition to 2-fluorenylacetamide. In 1941, Edwards reported carcinogenic properties for carbon tetrachloride [9] and, though discovered much later, a number of other simple aliphatic halides, such as ethylene dibromide [10] and vinyl chloride [11], have also been identified as carcinogens (reviewed by Lawley, 1976). Thus, halogenated aliphatic hydrocarbons comprise another, albeit small, group of carcinogens. The carcinogenic properties of urethane [12] for the mouse lung were recognized when this agent was being used as an anesthetic for mice (Nettleship et al., 1943). A number of related compounds have been studied (reviewed by Mirvish, 1968) but, again, this agent is not representative of a large group of carcinogens. The carcinogenic properties of thiourea and thioacetamide, and of inorganic chemicals such as beryllium oxide were found in the forties (reviewed in Miller, 1978), but it was not until the mid-fifties that another really substantial group of carcinogens, the N-nitroso compounds, was discovered.

In 1956, Magee and Barnes reported that N-nitrosodimethylamine [13] was carcinogenic in rat liver and, since that time, this relatively simple structure which contains only eleven atoms has been found to be carcinogenic in every species in which it has been tested (reviewed by Magee and Barnes, 1967; Magee et al., 1976). Many other nitrosamines were shown to be carcinogenic by Druckrey and co-workers (1967) and by 1976, Magee et al. could report that about 100 N-nitroso compounds were known to be carcinogenic. By 1984 (Preussman and Stewart), this number is now over 300. Among these carcinogens are symmetrical nitrosodialkylamines such as N-nitrosodimethylamine itself [13], asymmetrical nitrosodialkylamines such as N-nitrosomethyl-n-propylamine [14], a whole range of cyclic nitrosamines, such as N-nitrosomorpholine [15] and many other nitrosamines which are derivatives of these basic types. These carcinogens are frequently administered in the drinking water or by gavage, and they affect an even wider range of internal organs than do the aromatic amines.

A large sub-group of nitroso compounds represent a separate group of carcinogens. These are the alkyl nitrosoureas, nitrosourethanes, and nitrosoguanidines such as N-methyl-N-nitrosourea [16] which in contrast to other nitroso compounds, give rise to reactive intermediates without the intervention of cellular metabolism. Such compounds can cause tumors to arise at the site of subcutaneous injections and this property sets them apart from the other nitrosamines. Without doubt, the N-nitroso compounds represent the most versatile class of chemical carcinogens yet discovered with respect to their ability to produce tumors in various organs. This is one factor which suggests that they could be causative agents for some human cancers.

Many other kinds of chemical carcinogen have been discovered. These include the hydrazines whose carcinogenic potency was first recognized in 1959 (reviewed by Balo, 1979; Zedeck, 1984). One of the simplest structures exhibiting carcinogenic activity is 1,2-dimethylhydrazine [17], which has been extensively studied in model experiments for colon carcinogenesis. Azoethane [18] and azoxyalkanes, such as azoxymethane [19] are known carcinogens for the rat as are the 1-aryl-3,3-dialkyltriazenes, such as 1-phenyl-3,3-dimethyltriazene [20] (reviewed in Magee et al., 1976; Kolar, 1984). While the compounds summarized so far have been combustion products or synthetic compounds, carcinogenic properties are also associated with some natural products.

The best known of the carcinogenic natural products is aflatoxin B_1 [21]. The discovery of this carcinogen arose from investigations of a disease which killed some 100,000 turkey poults in Great Britain in 1960. The disease was eventually traced to certain batches of Brazilian ground-nut meal and then to toxins formed by certain strains of *Aspergillus flavus* with which this meal was infected (reviewed in Roe and Lancaster, 1964; Selkirk, 1980). Some other naturally occurring carcinogens are safrole [22] (Homburger et al., 1961) which has been used in root beer flavoring, cycasin [23] (Laqueur et al., 1963) a constituent of cycad nuts, and the pyrrolizidine alkaloids found in certain plants (reviewed by Schoental, 1976).

This outline of the types of structure which have been associated with carcinogenic activity does not do justice to the effort and perspicacity of the many investigators involved in these discoveries and the reader is encouraged to examine some of the excellent reviews of chemical carcinogens that have been cited above. However, the few structures of carcinogens that have been presented [1]–[23] should illustrate the wide diversity of structure associated with carcinogenic activity. We also hope to have communicated the fact that while some carcinogens were discovered by probing occupational environments associated with high tumor incidences in man, others were discovered serendipitously, e.g. as a by-product of investigations of the cause of a turkey disease. Clearly, our understanding of chemical carcinogenesis is far from complete and we cannot yet assign structural limits to chemicals which are associated with carcinogenic activity.

3. MECHANISM OF ACTION OF CHEMICAL CARCINOGENS

Although the last 20 years has led to considerable advances in our understanding of the way in which chemical carcinogens exert their noxious effects, it would be wrong to suggest that there is any complete understanding of the mechanism of action of any given carcinogen. Most of the progress that has been made is in the realization that, for many classes of chemical carcinogen, the carcinogenic process is initiated by a metabolite of the chemical regarded as a carcinogen and, in a few cases, identification of those metabolites that are carcinogenic. Although it is widely believed that these carcinogenic metabolites are mutagens, the mechanism through which mutagenesis might lead to carcinogenesis is far from clear. To some extent, the reason that our understanding of overall mechanisms is rather limited is because it is not obvious how such mechanistic studies should be approached. Following the recent discovery that the DNA of chemically transformed cells is capable of transforming other cells (Shih et al., 1979), fascinating discoveries with respect to the role of oncogenes in chemical carcinogenesis are beginning to emerge (Sukumar et al., 1983). One of the most challenging aspects of research into chemical carcinogenesis mechanisms at present lies in studies of the linkage between chemical carcinogens and their metabolites and the changes in oncogene expression or oncogene structure that are presently being associated with the cancerous state.

Albeit limited from the perspective of the overall mechanism of carcinogenesis, the understanding of mechanism of action gained over the past 20–30 years is impressive. A major approach to mechanistic investigations of chemical carcinogenesis has been through structure-activity studies. Herein, the alteration in biological activity wrought by structural change is supposed to lead to a description of the structural feature responsible for carcinogenic activity. Through the 1930s, 40s and 50s, this approach created a great wealth of data (for reviews see Arcos and Argus, 1974; Searle, 1984) and many correlations of structure and activity were uncovered. Unfortunately, these correlations usually applied only to a limited series of closely related compounds, such that no single structural feature that could be generally related to carcinogenic activity was unearthed. This could be interpreted to indicate that there are many discrete mechanisms for chemical carcinogenesis and that different classes of carcinogens work through different mechanisms. However, it now seems that the lack of a simple structural correlate with biological activity arose from the attention given in these studies to the chemical carcinogen rather than to its metabolite which is the real carcinogenic agent (defined as the ultimate carcinogen). Thus, when the structures of ultimate carcinogens are considered in relation to carcinogenic activity, it becomes clear that there is a common factor amongst all ultimate carcinogens and that this is their chemical reactivity towards cellular constituents (Miller and Miller, 1966, 1971; Dipple et al., 1968). For the potent experimental organic chemical carcinogens, the essence of carcinogenic activity lies in the chemical reactivity of their ultimate carcinogens towards cellular macromolecules.

Important observations which led to this surprisingly simple link between such widely diverse structures accumulated over a period of many years. Amongst these observations was the early report by Miller and Miller (1947) that carcinogenic aminoazo dyes become bound to rat liver constituents *in vivo*. Such binding requires that the relatively unreactive dye be converted to a metabolite that is reactive toward tissue constituents *in vivo*. Another important factor was the recognition that alkylating agents, which are reactive without metabolic intervention and had been known to be mutagens for some time, are carcinogenic (reviewed by Walpole, 1958). The carcinogenic effect of alkylating agents constituted the first direct linkage between chemical reactivity and carcinogenic activity. Perhaps the most significant observation of all was the demonstration that the N-hydroxy metabolite of the carcinogen, 2-acetylaminofluorene [8], was a more potent carcinogen than the parent compound, and that this proximate carcinogen was also active in tissues where the carcinogen itself was inactive (reviewed by Miller and Miller, 1966). While the role of carcinogen metabolism in chemical carcinogenesis had been debated for some time, this finding of increased carcinogenic potency in a metabolite represented the first solid proof that carcinogen metabolism could be an activation as well as a detoxification process. Other key developments leading to the present-day views of chemical carcinogenesis were the report by Brookes and Lawley (1964) showing that a correlation existed between the relative carcinogenic potencies of a series of polycyclic aromatic hydrocarbons and the extents to which they become covalently bound to mouse skin DNA *in vivo*; and the combination of this finding of the importance of DNA binding, along with the Millers' finding of the importance of metabolic activation, to yield an assay for mutagenic activity in which many compounds could be examined in a short period of time (Ames et al., 1975). The Ames assay, despite some limitations of the use of subcellular fractions for metabolic activation (Bigger et al., 1980), has played a major role in showing that many chemical carcinogens probably act through a DNA-reactive metabolite.

For carcinogens with intrinsic reactivity towards cellular macromolecules i.e. alkylating agents such as N-nitroso-N-methylurea [16], cellular metabolism is not necessary. However, for the vast majority of organic chemical carcinogens the reactivity currently believed to be essential for carcinogenic activity is generated through one or several metabolic reactions. The metabolic reactions involved are not unusual ones and they can be carried out on many substrates without risk to the cell. Chemical carcinogens are characterized by the presence of certain structural features which result in chemically reactive metabolites being created and being allowed to damage important informational macromolecules of the cell.

4. METABOLIC ACTIVATION OF CHEMICAL CARCINOGENS

For the majority of chemical carcinogens, mechanisms of metabolic activation have not been intensively investigated. However, for a range of carcinogens belonging to different chemical classifications, the common theme of formation of a DNA-reactive ultimate carcinogen, as discussed above, has emerged consistently. It is likely, therefore, that the general types of metabolic reaction discussed herein will be applicable to other carcinogens of the same chemical class.

4.1. Aromatic Amines and Aminoazo Dyes

As mentioned previously, the first direct demonstration that a metabolite was a more potent carcinogen than the administered chemical arose from the Millers' work with 2-fluorenylacetamide [8]. Cramer et al. (1960) noted that, after feeding 2-fluorenylacetamide to rats for a prolonged period (approximately 1 week), a new metabolite, identified as a glucuronide of the N-hydroxy derivative, appeared in the urine. Miller et al. (1961) subsequently showed that the N-hydroxy compound was a more potent carcinogen than 2-fluorenylacetamide, in contrast to findings for the ring-hydroxylated metabolites (Weisburger et al., 1956; Morris et al., 1960). This carcinogenic activity, together with the fact that feeding 3-methylcholanthrene in the diet drastically reduced both the carcinogenicity of N-2-fluorenylacetamide as well as the excretion of the N-hydroxy metabolite, led to the conclusion that the N-hydroxylation reaction represented an activation step in the carcinogenic action of this aromatic amide (Fig. 1). There is now ample evidence to support this conclusion as well as to indicate that the generation of N-hydroxy derivatives of a wide range of carcinogenic N-substituted aromatic compounds is a generally required step for metabolic activation.

This evidence (reviewed by Garner et al., 1984) includes the findings that guinea pigs are refractory to the hepatocarcinogenic action of 2-fluorenylacetamide and that the N-hydroxy derivative is not a major metabolite in this species though it is carcinogenic in this species. In addition, the N-hydroxy derivatives of several aromatic amines have been found to have equal or greater carcinogenic potencies than the parent amines themselves. A particularly convincing argument can be made from the work of Gutmann et al. (1967) who showed that two aromatic amides which were essentially noncarcinogenic, i.e. N-(7-hydroxy-2-fluorenyl) acetamide and N-2-fluorenylbenzamide, could be converted to highly carcinogenic compounds by synthetically converting them into the N-hydroxy derivatives.

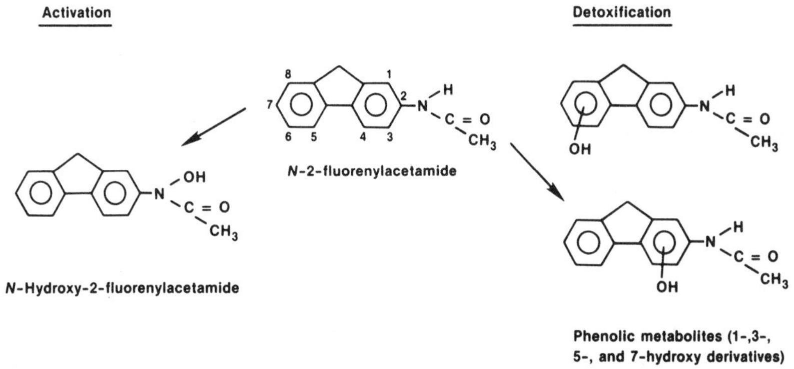

FIG. 1. Activation and detoxification reactions for 2-fluorenylacetamide.

These workers also showed that N-benzoylphenylhydroxylamine was not carcinogenic indicating that, although N-hydroxylation may be necessary for the expression of carcinogenic activity, not all arylhydroxylamines are carcinogenic.

There are two major routes, oxidation and reduction, through which N-hydroxy compounds can arise metabolically *in vivo* (Fig. 2) (reviewed in King, 1982; Garner *et al.*, 1984). Thus, nitro compounds such as the carcinogen 4-nitroquinoline-1-oxide [24] can give rise to an hydroxylamine by a reductive process and this hydroxylamine has been found to be a more potent carcinogen than the nitro compound itself (reviewed in Tada, 1981). Further reduction to the amino compound also occurs *in vivo* but this metabolite lacks carcinogenic activity indicating that, for this compound, oxidation of the amine to the hydroxylamine does not occur to any substantial extent. Another class of carcinogens for which the reduction of a nitro group may be an important activating step (Swaminathan and Lower, 1978) is the 5-nitrofurans. These include compounds such as N-[4-(5-nitro-2-furyl)-2-thiazolyl]formamide [25] which is a useful experimental bladder carcinogen (reviewed in Cohen, 1978) but for which some evidence indicates that activation may involve prostaglandin H synthase rather than a nitroreductase (Murasaki *et al.*, 1984).

Carcinogens presently believed to be activated through the oxidative route to N-hydroxy compounds include 2-fluorenylacetamide, discussed above, 4-biphenylamine, benzidine, 2-naphthylamine, 4-stilbenylacetamide, 4-biphenylacetamide and 2-phenanthrenyl-acetamide. The aminoazo dyes are also activated through this route but, for 4-dimethylaminoazobenzene [6] for example, an initial N-demethylation reaction precedes the oxidation step which, in turn, yields an N-hydroxy-N-methyl-4-aminoazobenzene (Kadlubar *et al.*, 1976). In an earlier attempt to prepare this N-hydroxy-N-methyl compound, Poirier *et al.* (1967) had obtained its benzoyloxy ester and had found it to be a potent carcinogen and to be reactive towards cellular nucleophiles *in vitro*. This latter property of chemical reactivity is necessary for covalent reaction with cellular constituents and was considered an essential property of ultimate carcinogens. Thus, the idea was developed that the N-hydroxy compounds which were reactive only under acidic (i.e. non-physiological) conditions (Kriek, 1965) were proximate carcinogens and that a further metabolic step, probably an esterification, was necessary to generate the metabolite which is the ultimate carcinogen (Miller, 1970). This idea has endured and has been most extensively investigated for 2-fluorenylacetamide.

It was shown that a synthetic N-acetoxy-2-fluorenylacetamide was reactive towards nucleosides *in vitro* (Kriek *et al.*, 1967) and, shortly thereafter, that soluble fraction from rat liver could catalyze the formation of a sulfate ester of N-hydroxy-2-fluorenylacetamide and that such esters were reactive toward nucleic acids (Fig. 3) (King and Phillips, 1968; DeBaun *et al.*, 1968). Since male rats, which are more sensitive to the hepatocarcinogenic action of 2-fluorenylacetamide than females, have higher levels of sulfotransferases than females (DeBaun *et al.*, 1970a), the carcinogenic action of this agent in rat liver may well involve such sulfate esters. Furthermore, Weisburger *et al.* (1972) found that the administration of acetanilide to rats to deplete their sulfate levels rendered them less susceptible to

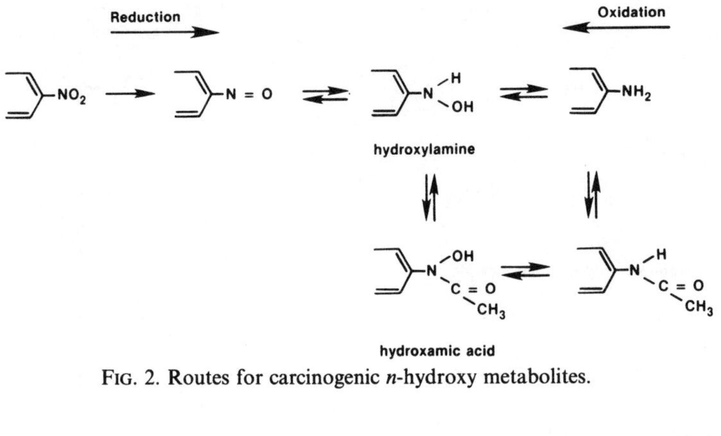

FIG. 2. Routes for carcinogenic n-hydroxy metabolites.

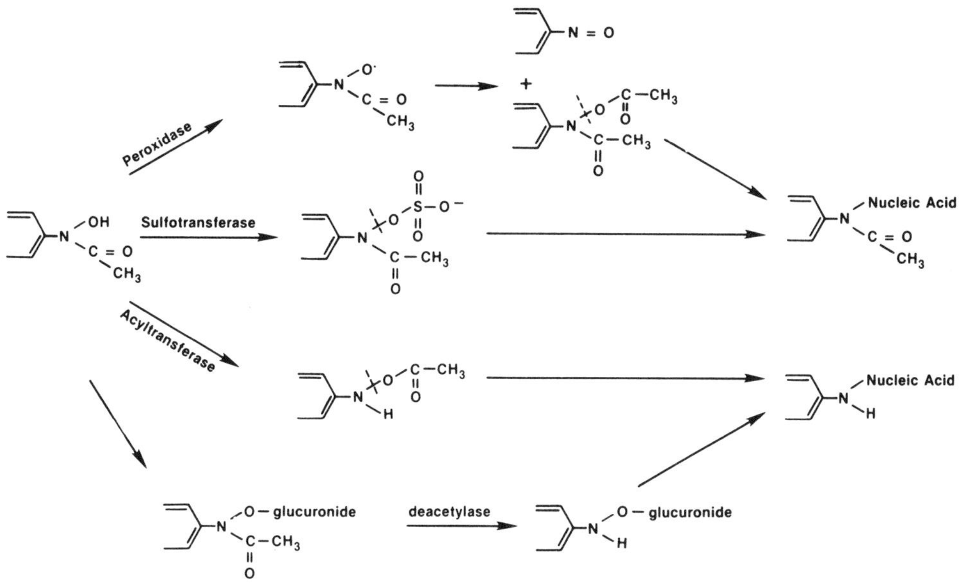

FIG. 3. Potential ultimate carcinogens derived from N-hydroxy-2-fluorenylacetamide.

the carcinogenic effect of N-hydroxy-2-fluorenylacetamide and DeBaun et al. (1970b) showed that large amounts of sodium sulfate increased the acute hepatotoxicity of this carcinogen. Sulfotransferase activity has not been detected in all organs susceptible in 2-fluorenylacetamide carcinogenesis (Irving et al., 1971) however, and even in the liver, substantial reaction with DNA occurs with loss of the N-acetyl group (Kriek, 1969; Irving and Veazey, 1969; Beland et al., 1982). These latter observations require that additional routes of conversion of the N-hydroxy-derivative to an ultimate carcinogen be available.

One such route involves a soluble acyltransferase activity (Bartsch et al., 1972) that transfers the acetyl group from the nitrogen to the oxygen (Fig. 3) to yield an N-acetoxyarylamine which reacts with nucleic acids to yield non-acetylated adducts. Another route to such adducts which may occur in some species (Cardona and King, 1976) involves deacetylation of the glucuronide of N-hydroxy-2-fluorenylacetamide. Removal of the acetyl group substantially increases the reactivity of this metabolite towards cellular constituents (Irving and Russell, 1970). Other potential pathways to reactive ultimate carcinogens have been proposed wherein N-hydroxy-2-fluorenylacetamide is converted by a peroxidase into nitroxyl radicals. Two of these radicals can dismutate to yield 2-nitrosofluorene and N-acetoxy-2-fluorenylacetamide (Fig. 3) (Bartsch et al., 1971; Floyd et al., 1976). Besides peroxidases carcinogenic aromatic amines are also activated through a prostaglandin endoperoxide synthetase system. This may account in part for their activity as bladder carcinogens (Robertson et al., 1983; Wise et al., 1984).

There is no absolute method, at present, for determining which of the possible ultimate carcinogens of 2-fluorenylacetamide are responsible for carcinogenic activity in specific organs and species. In fact, it is conceivable that for species that have slightly acidic urine, the N-hydroxy derivative of some aromatic amines (2-naphthylamine and 4-aminobiphenyl) may be directly responsible for urinary bladder carcinogenesis (reviewed in Miller and Miller, 1981). It is quite clear, however, that the formation of the N-hydroxylated metabolite is a prerequisite for carcinogenic activity. The metabolic formation of the somewhat unstable N–O bond is, in fact, the real source of the carcinogenic activity of the aromatic amines. This bond can be cleaved by mild acid conditions or through various esterification reactions which serve to generate a better leaving group. Thus, once formed, this N–O bond has the potential thereafter to undergo cleavage under favorable conditions, thereby yielding a potent arylamidating agent. Another pathway of aromatic amides to arylating intermediates involves metabolic oxidation to N-acetylquinone imines which may be responsible for some of the toxic effects of the parent compounds (Home et al., 1984; Streeter and Baillie, 1985).

Hydroxylations also occur on the aromatic rings of 2-fluorenylacetamide (Fig. 1). The phenolic carbon-oxygen bonds so formed are stable, however, under physiological conditions even after conjugation to form sulfates. In this situation, these metabolic reactions achieve what is presumed to be their objective of solubilizing the xenobiotic and making it more readily excretable. In general, such metabolic detoxification greatly exceeds metabolic activation and it is the balance between these two general types of metabolism which will determine carcinogenic potency. This balance can vary with different tissues and species and can also be deliberately altered by the administration of chemicals which interfere with either route of metabolism.

The above discussion has focused primarily on the metabolic activation of 2-fluorenylacetamide but structures [5]–[8] and [24] and [25] are all activated through the mechanisms discussed herein. There are of course very many other aromatic amines which behave similarly and these have been reviewed (Garner et al., 1984).

4.2. POLYCYCLIC AROMATIC HYDROCARBONS

Although the aromatic hydrocarbons were the first pure chemical carcinogens identified, understanding of the mechanisms through which they are metabolically activated developed only within the last decade. For the aromatic amines, the key discovery was finding a metabolite, N-hydroxy-2-fluorenylacetamide, of greater carcinogenic activity than the parent compound but, for the polycyclic aromatic hydrocarbons, such discoveries were not made until the active metabolites had been recognized by other means. In the seventies when the metabolic activation of the hydrocarbons was being elucidated, it was fairly widely believed that chemical carcinogens were active through DNA-reactive metabolites. The ultimate carcinogens were sought, therefore, by looking for reactive metabolites which, with DNA *in vitro*, yielded the same adducts as those obtained when the parent hydrocarbon was bound to DNA (through metabolic activation) in cellular systems (reviewed in Dipple et al., 1984).

The sequence of events which led to the discovery of the bay region dihydrodiol-epoxide route of metabolic activation for the hydrocarbons really began back in 1950 when Boyland suggested that the vicinal *trans* dihydrodiol metabolites that had been identified in urine had probably arisen from an initially formed epoxide. He further suggested that such epoxides might be the active metabolites of the hydrocarbons responsible for tumor initiation. Since theoretical chemists had concluded that the K-regions (i.e. phenanthrene-like double bonds) were the critical sites for interaction with cellular constituents (Pullman and Pullman, 1955), it became popular to ascribe the carcinogenic activity of the hydrocarbons to metabolically-formed K-region epoxides. The first synthesis of such compounds by Newman and Blum in 1964 allowed carcinogenicity testing of some K-region epoxides but these were found to be much less active than the hydrocarbons from which they were derived (Miller and Miller, 1967). Contrasting results were reported using *in vitro* systems to monitor biological activity. The K-region epoxides were found to be cytotoxic and mutagenic and to be very effective in cellular transformation assays (reviewed in Heidelberger, 1973; Sims and Grover, 1974). Nevertheless, when the DNA adducts formed from the K-region epoxide of 7-methylbenz[a]anthracene were compared with those formed by the metabolic activation of this hydrocarbon in cultured cells, it became clear that this K-region epoxide was not the metabolically-formed DNA-reactive metabolite of this carcinogen (Baird et al., 1973).

In the same year, Borgen et al. (1973) reported their findings on the microsome-catalyzed binding of benzo[a]pyrene and various *trans* dihydrodiol metabolites of benzo[a]pyrene to DNA. These authors found that the binding of the 7,8-dihydrodiol was more than ten-fold greater than that of benzo[a]pyrene itself and concluded that the 7,8-dihydrodiol was an intermediate in the activation of benzo[a]pyrene for DNA binding and that this diol must be further metabolized to yield an active alkylating agent. Based on this report, Sims and his colleagues (1974) conceived the idea that this active alkylating agent was a dihydrodiol epoxide formed by further oxidation of the dihydrodiol (Fig. 4). Also, they were able to

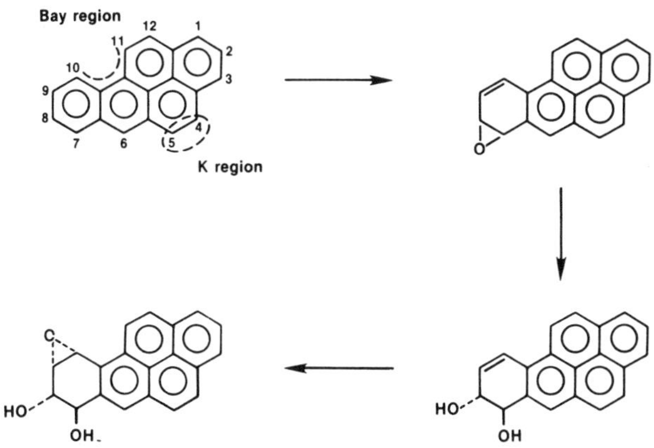

FIG. 4. Bay region dihydrodiol epoxide route of metabolic activation of hydrocarbons.

show that a synthetic 7,8-dihydrodiol-9,10-epoxide of benzo[a]pyrene reacted with DNA *in vitro* to yield products which exhibited the same chromatographic properties as did the adducts formed when rodent embryo cells were exposed to the parent hydrocarbon, benzo[a]pyrene.

While there are two structurally isomeric vicinal dihydrodiol epoxides that could be formed from benzo[a]pyrene (i.e., the 7,8-dihydrodiol 9,10-epoxide and the 9,10-dihydrodiol 7,8-epoxide), there are four that could be formed from 7,12-dimethylbenz[a]anthracene [3] and several other carcinogenic hydrocarbons. In these latter cases, there was no immediate means of knowing which of the possible isomeric dihydrodiol epoxides was likely to be the ultimate carcinogen. This difficulty was resolved by a generalization introduced by Jerina and Daly in 1976. Based on structure-activity relationships, they successfully predicted that the active dihydrodiol epoxide metabolites of the hydrocarbons would be those in which the epoxide ring was adjacent to a bay region (Fig. 4). Subsequent investigations of other hydrocarbon-DNA interactions and of the carcinogenic activities of dihydrodiols and dihydrodiol epoxides of various hydrocarbons have been broadly supportive of this bay region dihydrodiol epoxide mechanism for hydrocarbon activation (reviewed in Sims and Grover, 1981; Harvey, 1981; Conney, 1982; Dipple *et al.*, 1984).

Since the dihydrodiol epoxides contain chiral carbon centers, four stereoisomers for each dihydrodiol epoxide are possible (Fig. 5). These four isomers consist of an enantiomeric pair of structures where the benzylic hydroxyl group and the epoxide oxygen are on opposite sides of the ring system (these have been called *anti* or *trans* dihydrodiol epoxides) and a second enantiomeric pair where this hydroxy and the epoxide oxygen are on the same side of the ring system (*syn* or *cis* dihydrodiol epoxides) (Fig. 5). For a number of hydrocarbons, racemic *syn* and *anti* bay region dihydrodiol epoxides have been synthesized and their tumor initiating activities examined (reviewed in Dipple *et al.*, 1984). These studies have shown that the most potent tumor initiating activity resides in the *anti* dihydrodiol epoxides, except in the case of benzo[c]phenanthrene where both *syn* and *anti* dihydrodiol epoxides are equally potent (Levin *et al.*, 1980). The dihydrodiol epoxides of benzo[c]phenanthrene are also unusual in that the *trans* hydroxyl groups prefer the quasi-diequatorial conformation in both the *syn* and *anti* dihydrodiol epoxides while, in most other cases, only the *anti* dihydrodiol epoxides preferentially adopt this conformation (Sayer *et al.*, 1981). Thus, for the synthetic dihydrodiol epoxides, tumor initiating activity is usually associated with the diastereomer(s) in which the hydroxyl groups adopt the quasi-diequatorial conformation.

For some hydrocarbons, both optically active enantiomers of each diastereomer have been separately synthesized and tested for carcinogenic activity. Substantial carcinogenic activity is usually associated with only one optically active enantiomeric dihydrodiol epoxide, as demonstrated initially for the case of benzo[a]pyrene where the (+)*anti* derivative (Fig. 5) is clearly the most carcinogenic (Buening *et al.*, 1978; Slaga *et al.*, 1979). In Fig. 5, the heavy and light arrows indicate the relatively major and minor metabolic

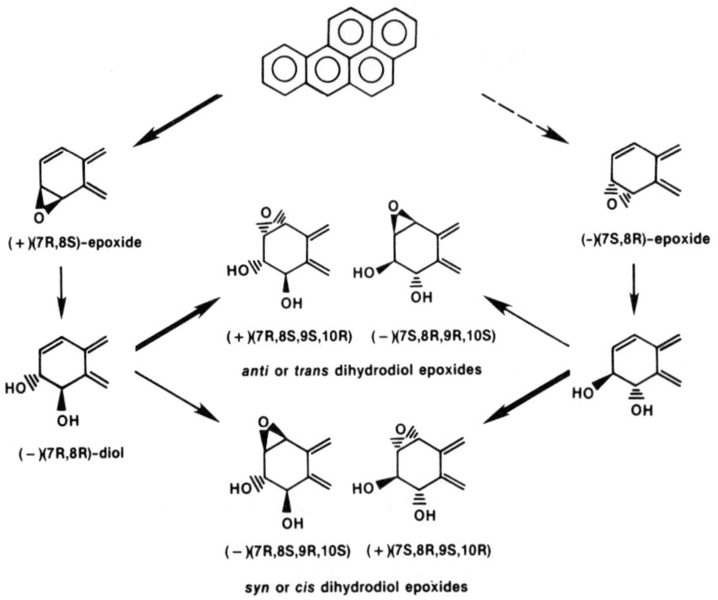

Fig. 5. Stereoisomeric benzo[a]pyrene bay region dihydrodiol epoxides.

pathways for benzo[a]pyrene with microsomal preparations *in vitro*. These relationships have been worked out principally through the researches of Yang and Gelboin and of Jerina and their colleagues (reviewed in Cooper *et al.*, 1983; Conney, 1982; Dipple *et al.*, 1984) who have shown that a remarkable degree of stereospecificity or stereoselectivity is apparent in these reactions. For the case of benzo[a]pyrene and liver microsomes from rats exposed to 3-methylcholanthrene (i.e. that illustrated in Fig. 5), the stereoselectivities are such that the metabolic formation of the most tumorigenic stereoisomeric dihydrodiol epoxide is favored (Conney, 1982). However, it can be shown that these stereoselectivities vary with the microsomal preparation used *in vitro* and that, while the (+) *anti* dihydrodiol epoxide of benzo[a]pyrene is responsible for the vast majority of DNA binding by benzo[a]pyrene in most cellular systems, this is not the case in cells of rat origin in particular (Autrup *et al.*, 1980; Pruess-Schwartz *et al.*, 1984). There is, then, no single description for the distribution of metabolism along the various alternate pathways in Fig. 5: there are only descriptions which are valid for specific systems under specific conditions.

As with benzo[a]pyrene, initial information on the ultimate carcinogens for other hydrocarbons came, in many cases, from studies of their DNA-bound forms (reviewed in Sims and Grover, 1981). This was followed by the synthesis of the appropriate dihydrodiols and dihydrodiol epoxides, mostly in Jerina's and Harvey's laboratories, and by the testing of these synthetic products for tumorigenic activity. It is generally true that, of all the metabolites of hydrocarbons that have been tested for carcinogenic activity, those that are not associated with the bay region dihydrodiol epoxide pathway exhibit little if any carcinogenic potency. It is also generally true that the dihydrodiol precursors of the bay region dihydrodiol epoxides are either of equal or of greater potency as tumor initiators than the hydrocarbons from which they were derived; this is consistent with the idea that these metabolites are proximate carcinogens. Tumor initiation data for the dihydrodiol epoxides are not so readily interpreted however. For benz[a]anthracene (Levin *et al.*, 1978) and benzo[c]phenanthrene (Levin *et al.*, 1980), the bay region dihydrodiol epoxides are clearly more potent tumor initiators than the parent hydrocarbon but this is not the case for some other hydrocarbons, such as 7-methylbenz[a]anthracene, chrysene (Slaga *et al.*, 1980) and benzo[a]pyrene (Slaga *et al.*, 1979). This reduced activity of the supposed ultimate carcinogen can be attributed to rapid hydrolysis and inactivation of these chemically reactive metabolites but it is also possible that tumor initiation requires some other metabolite in addition to the dihydrodiol epoxides. These concerns are allayed, somewhat, but not entirely, by the demonstration that, in a newborn-mouse lung-adenoma induction assay, the bay region dihydrodiol epoxide of benzo[*a*]pyrene is clearly a far more potent carcinogen than is benzo[*a*]pyrene (Buening *et al.*, 1978)

The metabolic activation of the polycyclic aromatic hydrocarbon carcinogens is viewed, therefore, as a sequence of three metabolic reactions leading to the formation of bay region dihydrodiol epoxides (Fig. 4), while metabolic reactions at other double bonds in the molecule are generally inactivating with respect to carcinogenesis. For several years now, it has been clear that the initial oxidation of most polycyclic hydrocarbon double bonds gives rise to an epoxide (reviewed by Daly *et al.*, 1972; Jerina and Daly, 1974), so it is interesting to consider why some of these oxidations lead on to a carcinogenic metabolite while others do not. For any initially formed epoxide, there are apparently three possible dispositions (Fig. 6), i.e., rearrangement to a phenol, hydration through the action of the enzyme epoxide hydrolase to yield a *trans* dihydrodiol and glutathione conjugate formation (catalyzed by the glutathione-*S*-transferases) which is the first step in the generation of mercapturic acid excretion products. For any given epoxide, any one of these three reactions, if it occurs rapidly enough, would effectively destroy the inherent reactivity of the epoxide sytem for informational cellular macromolecules. In fact, this is the fate of most of the epoxide metabolites formed from the polycyclic aromatic hydrocarbon carcinogens. Interestingly, these fates vary with the nature of the aromatic double bond that has been epoxidized. Thus, if the bond localization energy for this bond is high, i.e. it tends towards a truly aromatic benzenoid bond, the major product tends to be the phenol. This is the only way in which aromaticity is preserved. At the other extreme of a bond of low bond localization energy, which tends toward the properties of an isolated double bond, i.e. usually a K-region bond, glutathione conjugates seem to predominate. Somewhere between these extremes, dihydrodiol formation is the preferred route of detoxification for these epoxide metabolites. These generalizations are based primarily on Sim's metabolism studies and are discussed more fully elsewhere (Dipple, 1976; Dipple *et al.*, 1984). In the light of studies presently in the scientific literature, it appears that all primary epoxide metabolites of the polycyclic aromatic hydrocarbons are so efficiently inactivated by one of these three routes that they have no opportunity for chemical reaction with cellular DNA or initiation of carcinogenesis.

If, however, an epoxide formed in an angular benzene ring has been hydrated to a dihydrodiol in a secondary metabolic event, the isolated adjacent double bond can be oxidized to an epoxide in a tertiary metabolic step to yield a vicinal dihydrodiol epoxide. This does not occur readily when the initially formed dihydrodiol is in the quasidiaxial conformation (Thakker *et al.*, 1978) because of its proximity to a bay region or a methyl group and this tends to select for the formation of bay region dihydrodiol epoxides (i.e. where the epoxide is adjacent to a bay region) rather than a dihydrodiol epoxide where the dihydrodiol grouping is adjacent to the bay region. Since bay region dihydrodiol epoxides have been found to be the metabolites responsible for the majority of hydrocarbon-DNA binding which occurs, it is clear that these metabolites are not effectively inactivated by further metabolism. *In vitro* studies have shown that bay region dihydrodiol epoxides can

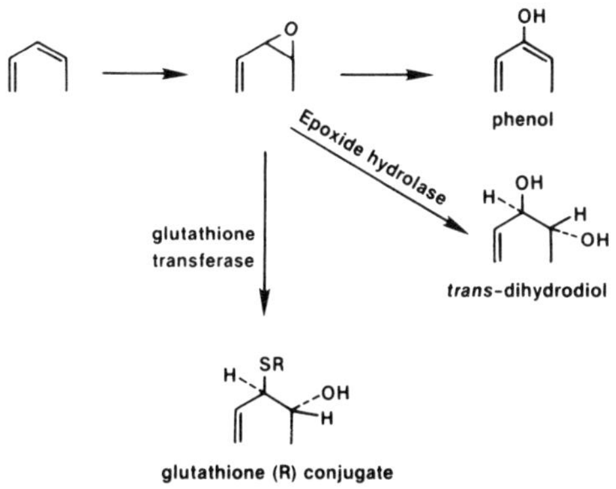

FIG. 6. Disposition of metabolically generated epoxides.

be converted to glutathione conjugates (Cooper et al., 1980), but these metabolites do not serve as substrates for epoxide hydrolase (Sims et al., 1974; Bentley et al., 1977) which may be an important factor in allowing them to be tumorigenic.

The metabolic activation of the polycyclic aromatic hydrocarbons bears some similarities to that of the aromatic amines in that more than one metabolic reaction is usually required for activation, and a proximate carcinogen (the N-hydroxy derivatives for the aromatic amines and the dihydrodiol precursors of the dihydrodiol epoxides for the hydrocarbons) can be isolated in each case and can be shown to exhibit comparable or greater carcinogenic potency than the parent compounds. In each case, the source of cell-damaging and tumor initiating activity lies in a carcinogen-oxygen bond introduced by metabolism i.e. the N–O bond for the aromatic amines and the strained three membered-ring C–O bond for the hydrocarbons.

4.3. N-Nitrosamines

It was realized very soon after the discovery of the carcinogenicity in rats of N-nitrosodimethylamine (Magee and Barnes, 1956) that the compounds of this class needed to be metabolized before their carcinogenic potential could be expressed. Since that time, hundreds of papers have dealt with various aspects of metabolism of nitrosamines by whole animals, organ and tissue cultures, subcellular fractions and various enzyme preparations. The interested reader is referred to several excellent reviews for a full account of this work (Magee and Barnes, 1967; Magee et al., 1976; Lai and Arcos, 1980; Preussmann and Stewart, 1984). This discussion will be divided into subsections, each dealing with a specific class of nitrosamines. These subsections are: (1) nitrosodialkylamines, (2) cyclic nitrosamines, (3) β-oxidized nitrosamines, and (4) nitrosoarylalkylamines.

4.3.1. Nitrosodialkylamines

The simplest member of this group, N-nitrosodimethylamine [13], has been studied most extensively and could be regarded as a prototype of the others. It will be apparent however, that N-nitrosodimethylamine may be rather special in terms of metabolism, although it shares some general features with other nitrosodialkylamines. As early as 1956 it was demonstrated that [^{14}C]-N-nitrosodimethylamine was metabolized in vivo to ^{14}CO$_2$ (Dutton and Heath, 1956). This suggested that the chemical was oxidatively demethylated during in vivo metabolism. Experiments utilizing postmitochondrial liver homogenate fractions (designated S9 fractions) or microsomes demonstrated that the metabolism of N-nitrosodimethylamine in vitro required oxygen and NADPH (Magee and Vandekar, 1958). In the presence of trapping agents, such as semicarbazide, it was demonstrated that formaldehyde was formed from N-nitrosodimethylamine (Mizrahi and Emmelot, 1962), acetaldehyde from nitrosodiethylamine (Arcos et al., 1976), propionaldehyde from nitrosodi-n-propylamine (Park and Archer, 1978), and both formaldehyde and acetaldehyde from nitrosomethylethylamine (Chau et al., 1978). Carbonyl products from other nitrosamines have been detected by trapping with 2,4-dinitrophenylhydrazine (Farrelly, 1980). These experiments indicated that, at least, simple nitrosodialkylamines are metabolized by the monooxygenase enzymes belonging to the cytochrome P-450 group. The early experiments suggested furthermore that the principal reaction leading to N-dealkylation of the nitrosamines and the ultimate formation of aldehydes is an oxidation at the α-carbon of one of the alkyl groups. This is believed to result in the formation of the unstable α-hydroxylated derivatives, which then fragment down spontaneously to the aldehydes and the corresponding nitrosomonoalkylamines. The latter are unstable and break down to an alkyldiazonium ion or directly to the carbocation, depending on their relative stabilities. It might be pointed out here that, in the case of N-nitrosodimethylamine, the methyl cation is *not* an intermediate because it is much too unstable to be formed in aqueous environment.

The α-hydroxylation mechanism is now considered to be the principal activation pathway for many nitrosamines. The general reaction scheme is shown in Fig. 7, together

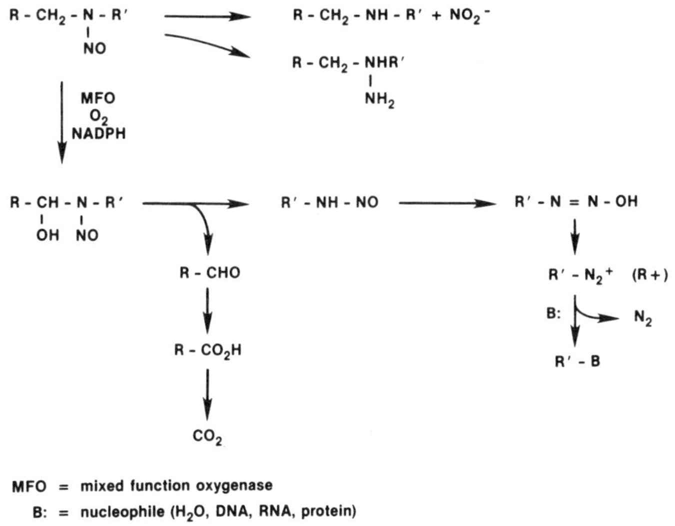

FIG. 7. α-Hydroxylation pathway of N-nitrosamine metabolism.

with some other metabolic reactions, most of which are believed to result in detoxification. Although it is now reasonably certain that the ultimate carcinogenic metabolite from nitrosodialkylamines is an alkyldiazonium ion or the corresponding carbocation, it was necessary to exclude the possibility that nitrosodialkylamines were metabolized to the highly electrophilic diazoalkanes. This was done by treating rats with perdeuterated N-nitrosodimethylamine and N-nitrosodiethylamine. The DNA and RNA isolated from these rats was shown to contain only completely deuterated methyl and ethyl groups, respectively (Lijinksy et al., 1968). Had diazomethane or diazoethane been the reactive metabolites, the methyl and ethyl groups would have contained one deuterium less, since these groups would have incorporated a proton from the body water. Keefer and co-workers (1973) also used α-deuterium labeled nitrosamines to show that in most cases a pronounced isotope effect existed in the carcinogenicity of these compounds. The α-deuterated compounds were found to be generally less carcinogenic than the corresponding undeuterated compounds. These results suggested that the α-hydroxylation step was the rate limiting step in tumor initiation by nitrosamines.

The α-hydroxylation reaction sequence shown in Fig. 7 indicates that a molecule of nitrogen is an obligatory product of the reaction. This fact provided a method to quantitate the role of the α-hydroxylation pathway for several nitrosamines. Thus, Magee, Holsman and Haliday (Magee, 1976) used N-nitrosodimethylamine, doubly-labeled with ^{15}N, in an *in vivo* experiment to show that approximately 90% of the nitrogen could be detected in the expired air. More recently Milstein and Guttenplan (1979), using unlabeled N-nitrosodimethylamine found a high yield of molecular nitrogen being formed during the *in vitro* metabolism of N-nitrosodimethylamine by mouse and rat microsomes. These results, however, do not agree with the data of Cottrell and co-workers (1977) who found that < 5% of ^{15}N$_2$ was released *in vitro* from N-nitrosodimethylamine, under conditions where N-nitroso-α-acetoxymethylmethylamine (see below) released N$_2$ quantitatively. Recently, Michejda and co-workers studied the release of ^{15}N$_2$ from doubly-labeled N-nitrosodimethylamine, and obtained results which generally corroborate the findings of the earlier workers. The α-hydroxylation reaction was found to be somewhat less efficient *in vitro* [33% using rat liver S9 (Kroeger-Koepke et al., 1981) and 47% using isolated hepatocytes (Koepke et al., 1984)], than *in vivo* [67% (Michejda et al., 1982)]. The nitrogen-release experiments indicate that the α-hydroxylation pathway is an important but not an exclusive route of nitrosamine metabolism.

Nitrosamines which have a hydroxyl group in the α-position are not stable. The esters of the alcohols however, can be prepared by indirect methods. Thus, N-nitroso-α-acetoxymethylmethylamine (Roller et al., 1975; Wiessler, 1975) was found to be a potent carcinogen (Rice *et al.*, 1975; Habs *et al.*, 1978). The ester is readily hydrolyzed by esterases (Roller *et al.*, 1975) and also undergoes apparent dissociation to the nitrosamino cation (Wiessler, 1979), as indicated in Fig. 8. Recent preparative methods for *N*-nitroso-α-acetoxydialkylamines have greatly improved the convenience and the yields of the original preparations (Saavedra, 1978; Kupper and Michejda, 1980). The chemistry of the α-esters of nitrosamines has been reviewed (Wiessler, 1979). Recently, the preparation of the first authentic α-hydroxylated nitrosamines has been described (Mochizuki *et al.*, 1980). The behavior of these substances fully supports the conclusions obtained from the metabolic studies. The synthetic putative proximate carcinogens appear to be more powerful carcinogens than their parent compounds. It must be realized however, that a precise comparison cannot be made because administration of the synthetic activated compounds to animals creates an abnormal distribution of the active materials, different from that when the activated compounds are formed by metabolism.

The enzymology of activation of nitrosamines to carcinogenic metabolites is still not completely understood. As was mentioned previously, there is good evidence that α-hydroxylation is catalyzed, at least in part, by cytochrome P-450. However, considerable confusion exists in the literature regarding the effects of various inducers or inhibitors of nitrosamine bioactivation, particularly in the case of *N*-nitrosodimethylamine. This unsettled state of affairs can be resolved by assuming that there are at least two forms of *N*-nitrosodimethylamine-demethylase, one operating at low substrate concentrations and the other at high substrate concentrations. These two forms can be demonstrated kinetically *in vitro* and it can be shown that various modifiers of metabolism affect the two forms differently (Arcos *et al.*, 1977; Mostafa *et al.*, 1981). While these two forms of *N*-nitrosodimethylamine-demethylase appear to be the important enzymes *in vitro* and have been shown to be isozymes of the cytochrome P-450 monooxygenase family, there have been reports of *N*-nitrosodimethylamine-demethylases which are not hemoproteins. Thus, amine oxidases have been implicated in the metabolism of *N*-nitrosodimethylamine (Lake *et al.*, 1978) and a soluble liver enzyme referred to as the pH 5 enzyme, has been shown to have demethylase activity (Grilli and Prodi, 1975). The latter was also investigated by Kroeger-Koepke and Michejda (1979), who provided evidence that this preparation possesses a very low K_M value, consistent with the apparent K_M of the enzyme responsible for *N*-nitrosodimethylamine metabolism *in vivo* (Skipper *et al.*, 1983). Although the metabolism of *N*-nitrosodimethylamine appears to be similar to that of other nitrosamines, there are differences which suggest that enzymes responsible for *N*-nitrosodimethylamine oxidation may be selective for that substrate. For example, it has been shown that ethanol is an inducer of hepatic nitrosamine oxidases when rats are pretreated with the chemical. On the other hand, ethanol administered concurrently is an inhibitor of *in vitro* metabolism of low concentrations of nitrosamines, but especially *N*-nitrosodimethylamine (Peng *et al.*, 1982; Tomera *et al.*, 1984). Thus, ethanol is an inhibitor of the low K_M nitrosodimethylamine-demethylase but has little or no effect on the high K_M form. It has been demonstrated that ethanol inhibits *N*-nitrosodimethylamine metabolism *in vivo* in the rat (Skipper *et al.*, 1983) and in man (Spiegelhalder *et al.*, 1982).

From the foregoing, it is apparent that while the details of bioactivation of simple nitrosodialkylamines are still imperfectly known, the overall result is an oxidative transformation which leads to reactive electrophilic agents. The metabolic picture for nitrosamines

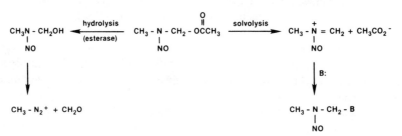

FIG. 8. Reactions of *N*-nitroso-(acetoxymethyl)methylamine.

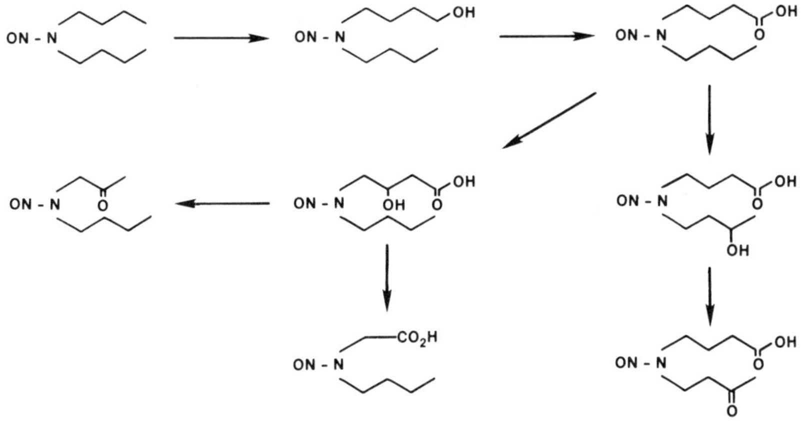

FIG. 9. Metabolic reaction following ω-oxidation of N-nitrosodi-n-butylamine.

becomes more complicated when longer chain N-nitrosomethylalkylamines or nitrosodialkylamines are considered. Much of our knowledge of the metabolic processes in these compounds comes from examination of urinary metabolites. One of the best studied nitrosamines in this regard is the urinary bladder carcinogen N-nitrosodi-n-butylamine (Suzuki and Okada, 1980; Blattmann and Preussmann, 1974). The oxidation of that chemical occurs not only at the α-position but also at all the other carbons. Oxidation of the terminal methyl group leads to the formation of the 4-hydroxy compound, a bladder carcinogen itself, which can then be oxidized further to the carboxylic acid, and so on. The sequence in Fig. 9 illustrates the complexities arising from oxidation at the terminal carbon (ω-oxidation). Even though a great deal is known about the metabolism of this chemical in rats and other species, it is still difficult to determine which of the many metabolic products is responsible for bladder tumor initiation. It appears that some of these metabolic reactions, perhaps in addition to α-hydroxylation, could be responsible for the critical metabolites. It has been suggested that ω-oxidation is important in bladder carcinogenesis by this chemical (c.f. Preussmann et al., 1982). The metabolism of nitrosomethyl-(long chain alkyl)amines appears to follow a pattern similar to that for normal fatty acid degradation (Okada et al., 1976; Singer et al., 1981). Singer et al., investigated a series of N-nitrosomethylalkylamines with chain lengths varying from C_4 to C_{14}. The even-numbered chains were degraded by two carbon fragments to a number of products, but principally to N-nitrososarcosine, N-nitrosomethyl-(3-carboxypropyl)amine and N-nitrosomethyl-(2-oxopropyl)amine. The odd numbered alkyl-nitrosamines were metabolized to N-nitrosomethyl-(2-carboxyethyl)amine (Fig. 10). It is interesting to note that the even numbered N-nitrosomethylalkylamines ($C_n > C_6$) are mainly bladder carcinogens, while the odd-membered derivatives attack other organs, principally the liver (Lijinsky et al., 1981). Thus it appears that N-nitrosomethyl-N-(3-carboxypropyl)amine (Okada et al., 1976; Singer et al., 1981), is necessary for bladder carcinogenesis by these compounds.

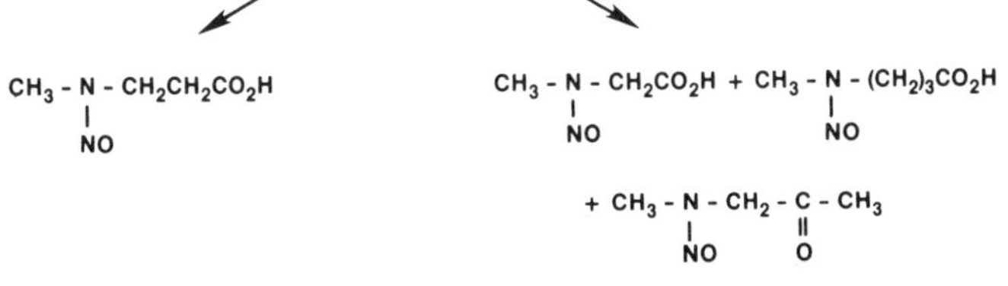

FIG. 10. Metabolic products of odd- and even-numbered N-nitrosomethylalkylamines.

4.3.2. Cyclic Nitrosamines

This group of chemicals is related to nitrosodialkylamines but differs from them in that oxidative metabolism usually results in retention of the entire carbon framework in the same molecule. In contrast with the nitrosodialkylamines, the cyclic compounds do not give metabolites which bind extensively to cellular nucleic acids. Thus, while there is considerable information about metabolic pathways, relatively little is known about the mechanisms by which these chemicals initiate the carcinogenic process.

N-Nitrosopyrrolidine, a liver carcinogen in rats, nitrosohexamethyleneimine, a liver and esophageal carcinogen in rats, and nitrosoheptamethyleneimine, a lung and esophageal carcinogen, were shown to be metabolized *in vivo* to CO_2 and various urinary metabolites (Snyder *et al.*, 1977). N-Nitrosopyrrolidine, N-nitrosopiperidine and di-N-nitrosopiperazine as well as N-nitrosodimethylamine and N-nitrosodiethylamine were found to be metabolized by cultured human bronci (Harris *et al.*, 1977). The metabolism of N-nitrosopyrrolidine by rat liver microsomes (Chen *et al.*, 1978a) and by rat liver microsomes plus cytosol (Hecker *et al.*, 1979) was examined in detail. The principal activation pathway was found to be a-hydroxylation, mediated by a polychlorobiphenyl-inducible form of cytochrome P-450. The reaction sequence is shown in Fig. 11. Some β-hydroxylation, resulting in the retention of the nitrosamine moiety, has also been observed *in vitro*. Presumably, the carcinogenic action of N-nitrosopyrrolidine is due to the interaction of the intermediate electrophilic diazonium ion species with a critical cellular target. The nature of that reaction, however, is not known. Other cyclic nitrosamines, such as N-nitrosopiperidine have also been shown to be oxidized by a-hydroxylation (Leung *et al.*, 1978).

The tobacco-specific nitrosamines, N-nitrosonornicotine and N-nitrosoanabasine are important environmental carcinogens. The metabolism of N-nitrosonornicotine was investigated in detail by Hecht, Hoffman and their collaborators (Chen *et al.*, 1978b; Hecht *et al.*, 1981, 1982). Since this chemical is related to N-nitrosopyrrolidine, it is not surprising that the general features are similar. However, since N-nitrosonornicotine is not symmetrical, there are two a-positions, each giving rise to a different set of metabolites. Beside a-hydroxylation, β-hydroxylation was also observed, which leads to metabolites with the retained nitrosamine structure (Fig. 12). In terms of activation to an ultimate carcinogen, both of the a-hydroxylation pathways lead to an electrophilic intermediate. Oxidation at the 5'-position, however, leads to a secondary arylalkyl cation which would not be expected to be as reactive as the primary diazonium ion derived from the 2'-oxidation. Thus, it would be reasonable to assume that the 2'-oxidation is the critical metabolic reaction for carcinogenic activation (Hecht and Young, 1982).

N-Nitrosomorpholine is a potent liver carcinogen in rats, as is its methylated derivative, 2,6-dimethyl-N-nitrosomorpholine. The latter, however, has been found to be a pancreatic carcinogen in hamsters. Incorporation of deuterium into the a-positions of N-nitrosomor-

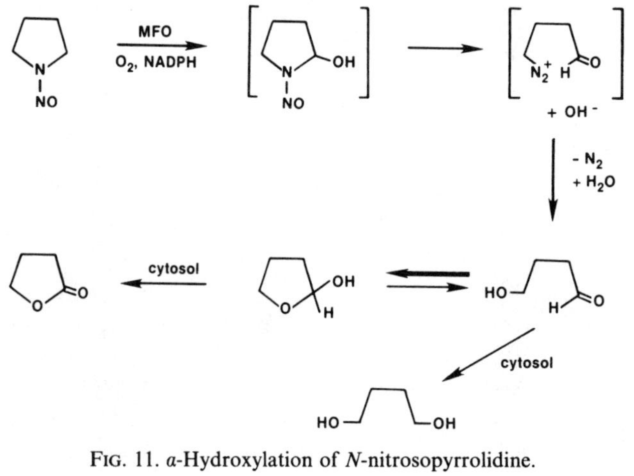

FIG. 11. a-Hydroxylation of N-nitrosopyrrolidine.

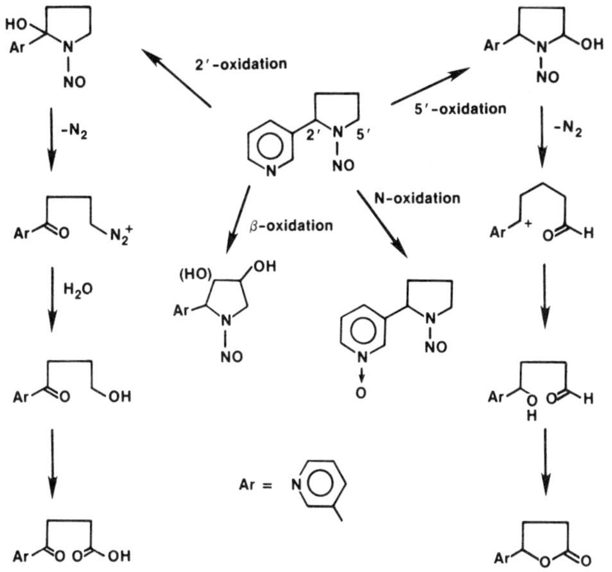

FIG. 12. Metabolic pathways for the tobacco-specific carcinogen, N'-nitrosonornicotine (NNN).

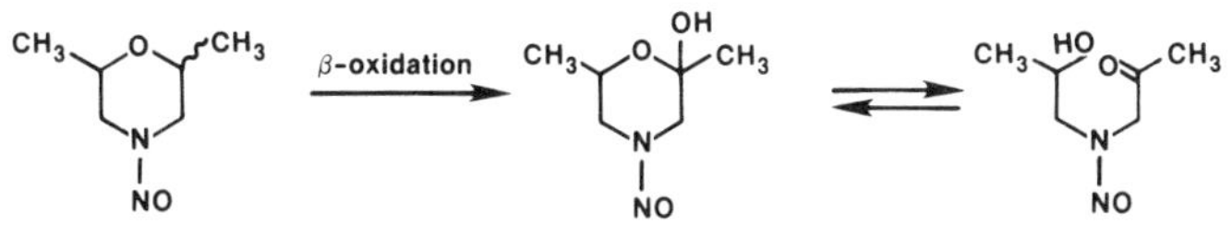

FIG. 13. Hydroxylation of N-nitroso-2,6-dimethylmorpholine.

pholine decreased the carcinogenic potency of the chemical in rats, relative to the undeuterated compound (Lijinsky et al., 1976). The same was true for 2,6-dimethyl-3,3,5,5-tetradeutero-N-nitrosomorpholine in rats. In hamsters, however, deuterium in the β-positions (next to the oxygen) decreased carcinogenic potency of 2,6-dimethyl-N-nitrosomorpholine while deuteration of the positions alpha to the nitrogen actually enhanced carcinogenicity (Rao et al., 1981; Lijinsky et al., 1980c). These data suggest strongly that β-hydroxylation is the important activation pathway for that nitrosamine in hamsters. The reason for that becomes clear when one considers the consequences of β-hydroxylation. These are shown in the reaction sequence of Fig. 13. Hydroxylation of the carbon next to the oxygen leads to the formation of a hemiacetal which is in equilibrium with its open form N-nitroso-2-hydroxypropyl-2-oxopropylamine. The latter compound is a potent pancreatic carcinogen in hamsters (Pour et al., 1979). The metabolism of 2,6-dimethyl-N-nitrosomorpholine demonstrates that activation of N-nitrosamines does not necessarily involve α-hydroxylation, at least in hamsters. The important conclusion to draw from these data is that activation of procarcinogens is a consequence of xenobiotic metabolism, the purpose of which is detoxification and elimination, which sometimes gives rise to highly reactive products. Thus, carcinogen activation can be looked at as an aberration of normal detoxification pathways.

4.3.3. β-Oxidized Nitrosamines

It was stated above that 2,6-dimethyl-N-nitrosomorpholine was transformed metabolically to the open-chain nitrosamine, N-nitroso-2-hydroxypropyl-2-oxopropylamine. The latter can be oxidized metabolically to N-nitrosobis-2-oxopropylamine, or reduced to N-nitrosobis-2-hydroxypropylamine (Whalley et al., 1981), as shown in Fig. 14. These three compounds are all pancreatic carcinogens in the hamster, but have different organs as targets in other animals. Analysis of the DNA of rats treated with a related compound, N-

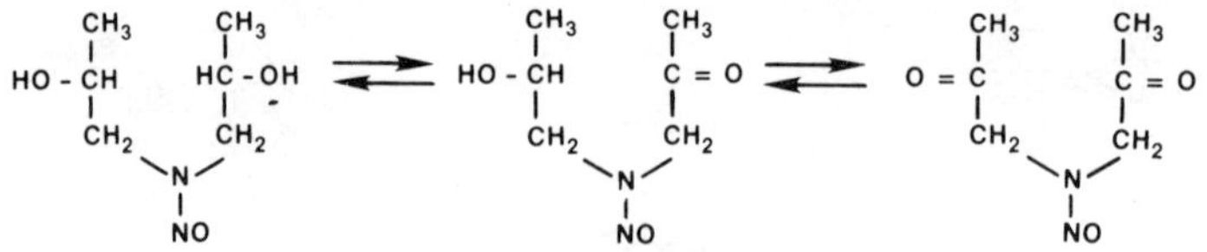

FIG. 14. Redox interconversion of β-oxidized N-nitrosodipropyl system.

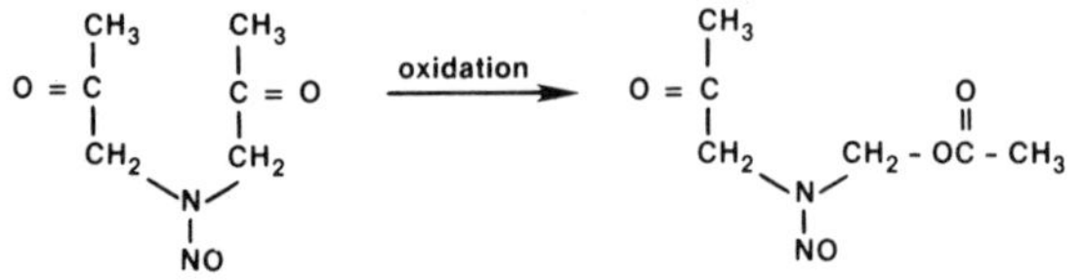

FIG. 15. Baeyer–Villiger oxidation of N-nitrosobis-(2-oxopropyl)amine.

nitroso-2-oxopropylpropylamine revealed high levels of methylation of the guanine residues (Leung et al., 1980). The mechanism of formation of the methylating agent is still not completely clear, but there is a possibility that the nitrosamine undergoes an enzymatically catalyzed version of the well-known organic oxidation, the Baeyer-Villiger oxidation depicted in Fig. 15 (Kupper, Farrelly and Michejda, unpublished data). The product of the oxidation, N-nitrosoacetoxymethyl-2-oxopropylamine, readily breaks down to diazomethane (Leung and Archer, 1984), which, in an aqueous environment forms the methyldiazonium ion.

One of the most significant environmental nitrosamines, N-nitrosodiethanolamine, has been found to be a strong liver carcinogen in rats (Lijinsky et al., 1980b). Most of the chemical, however, is excreted unchanged in the urine (Preussmann et al., 1978). Moreover, very little metabolism is seen in vitro (Farrelly et al., 1984), although oxidation of the chemical to N-(2-hydroxyethyl)-N-(formylmethyl) nitrosamine and then to the corresponding carboxylic acid by rat S9 has been observed (Airoldi et al., 1984). The chemical does not appear to be subject to α-hydroxylation. A related compound, N-nitrosomethyl-2-hydroxyethylamine, a strong liver carcinogen in female Fischer rats (Koepke and Michejda, unpublished data), is not metabolized by liver S9, but is metabolized by isolated hepatocytes. Experiments with doubly ^{15}N-labeled nitrosamine indicated that the metabolism did not involve α-hydroxylation. Thus, it appears that β-hydroxylated nitrosamines must utilize different pathways of bioactivation than those found for simple dialkyl and cyclic nitrosamines. Michejda and co-workers (1979) proposed a mechanism involving sulfate conjugation of the hydroxyl group, followed by intramolecular nucleophilic displacement to form an oxadiazolium ion, which has been shown to be a directly acting mutagen. These reactions are shown in Fig. 16. Although the chemical evidence for this sequence is strong, the biological data are more ambiguous, and hence this mechanism is still speculative. Loeppky and co-workers (1979) have proposed that β-hydroxylated nitrosamines can undergo a retroaldol fragmentation. Thus, N-nitrosomethyl-2-hydroxyethylamine could, in principle, fragment to N-nitrosodimethylamine and formaldehyde. Although there is some evidence for an enzymatically induced retroaldol reaction of β-N-nitrosohydroxyamines (Loeppky et al., 1981), the importance of the reaction in carcinogen bioactivation is difficult to assess at the present time.

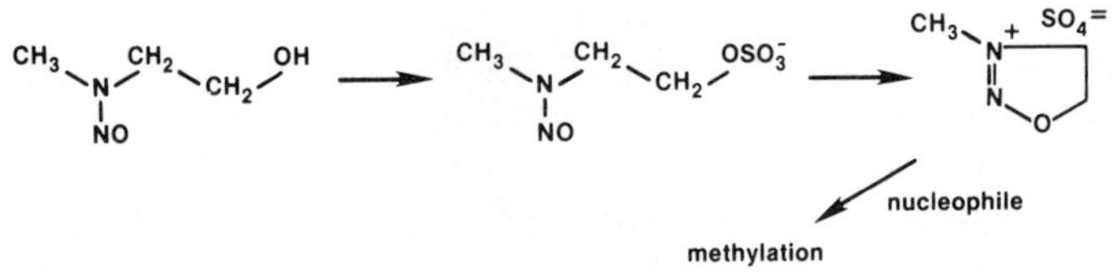

FIG. 16. Hypothetical activation of β-hydroxynitrosamines through sulfate conjugation.

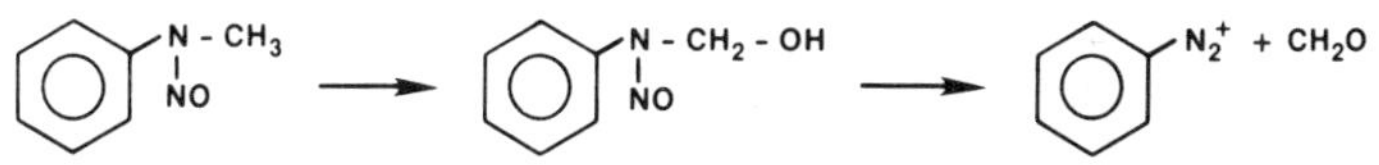

FIG. 17. Formation of benzene diazonium ion by oxidative demethylation of N-nitrosomethylaniline.

4.3.4. N-nitrosoarylalkylamines

This group is represented by a relatively small number of chemicals, which nevertheless, are important for mechanistic reasons. The prototype of these compounds is N-nitrosomethylaniline, an esophageal carcinogen in rats (Kroeger-Koepke et al., 1983). It is readily metabolized in vitro and in vivo (Kroeger-Koepke et al., 1981; Koepke et al., 1984; Michejda et al., 1982). The only significant metabolism, besides denitrosation, appears to be α-hydroxylation at the methyl group. Release of formaldehyde leads to the formation of a benzenediazonium ion (Fig. 17).

It is interesting to note that of the three isomeric N'-nitrosomethyl aminopyridines, only N'-nitroso-N'-methyl-2-aminopyridine is carcinogenic in the rat (Preussmann et al., 1979). The oxidative demethylation of the carcinogenic isomer was much more rapid in vitro than that of the noncarcinogenic isomers (Heydt et al., 1982). It is also significant that detoxification pathways for the carcinogenic isomer (denitrosation, N-oxide formation) were considerably slower than demethylation. The subsequent fate of the presumed carcinogen, pyridine-2-diazonium ion, is unknown at present.

Arenediazonium ions are very different chemically from alkyldiazonium ions since they are much more stable and less reactive. Their reactions are frequently the result of azo-coupling, displacement of nitrogen by nucleophiles or induction of free radical processes. Which of these reactions is important in initiation of carcinogenesis is unknown at the present time.

4.4. HYDRAZINES

Various aspects of carcinogenicity, toxicity and biochemistry of hydrazines and their oxidation products, azoalkanes and azoxyalkanes have been reviewed recently (Zedeck, 1984). Hydrazine itself, N_2H_4, is a carcinogen in rats and mice, but little is known about its mode of action, which must be very different from its alkyl-substituted derivatives. Most of the simple alkylated hydrazines are also carcinogens (Zedeck, 1984). Thus, 1,2-dimethylhydrazine induces tumors in the rat colon, rectum, duodenum, jejunum, ileum, liver and kidneys (Druckrey, 1970). The next higher homolog, 1,2-diethylhydrazine induces tumors of the olfactory system, brain, mammary gland and the liver, but not in the gastrointestinal system (Druckrey, 1970). Naturally occurring carcinogenic hydrazine derivatives include N-methyl-N-formylhydrazine, which is found in the false morel mushroom, and agaritine, β-N-[λ-L-(+)-glutamyl]-4-hydroxymethylphenylhydrazine, which is found in the common mushroom, Agaricus bisporus. The naturally occurring hydrazines have been reviewed by Toth (1980).

The metabolism of 1,2-dimethylhydrazine, which is a prototype of other 1,2-dialkylhydrazines, has been elucidated largely through the work of Fiala (Fiala, 1975; Fiala et al., 1977). The metabolic sequence is shown in Fig. 18. Several points should be noted about the sequence. Dimethylhydrazine is converted by microsomal oxidation to azomethane, which is subsequently oxidized to azoxymethane. All three compounds are stable and all are carcinogens. The third oxidation step converts the azoxymethane to the proximate carcinogen, methylazoxymethanol. This step is equivalent to the α-hydroxylation step of nitrosamines. Methylazoxymethanol is relatively unstable and fragments to formaldehyde and the methyldiazonium ion, precisely the same ultimate carcinogen which is produced by the metabolism of N-nitrosodimethylamine. The methyldiazonium ion can also be formed from the hydrolytic breakdown of methylazoxyformamide, which can be formed by the alcohol dehydrogenase oxidation of methylazoxymethanol (Zedeck, 1984). Other hydra

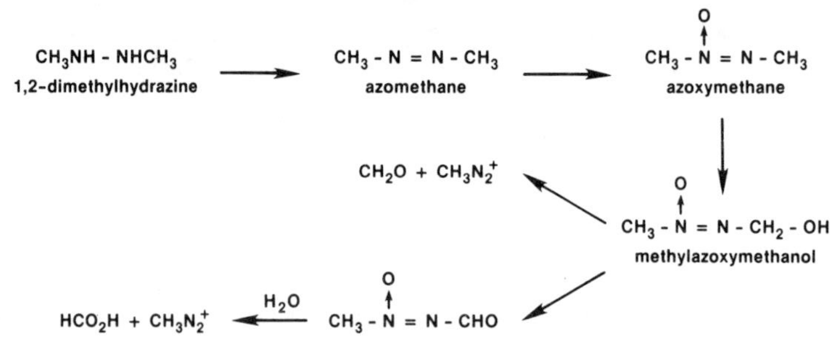

FIG. 18. Metabolic scheme for 1,2-dimethylhydrazine and its oxidation products, azomethane and azoxymethane.

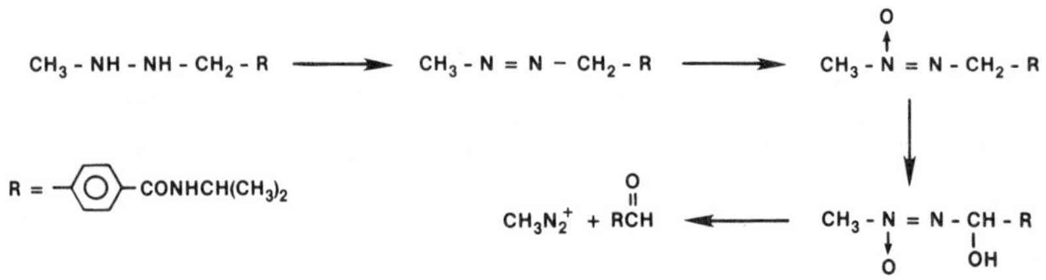

FIG. 19. Metabolism of the antitumor drug, procarbazine, to the methyldiazonium ion.

zine derivatives are metabolized by similar pathways. Especially interesting in this regard are some therapeutically useful hydrazine drugs. For example, procarbazine is an antitumor agent, which has also been found to be a carcinogen in several species (IARC, 1981). Metabolic transformation to a methyldiazonium ion is outlined in the simplified scheme shown in Fig. 19 (Weinkamp and Shiba, 1978).

Procarbazine as well as 1,2-dimethylhydrazine and other hydrazine derivatives are metabolized to ultimate carcinogens which are very similar to those derived from nitrosamines. Since azo- and azoxycompounds are intermediates in the activation of hydrazines, their ultimate carcinogenic forms are also closely related to those arising from nitrosamines.

4.5. TRIAZENES

The chemistry of triazenes has been reviewed recently (Benson, 1984) and their biological properties were also reviewed (Kolar, 1984). Although triazenes have been known for many years, recent interest has developed because of their chemotherapeutic properties (Comis, 1976). However, they also form an important class of chemical carcinogenic agents. Triazenes can be divided into three broad classes: 1,3-diaryltriazenes, 1-aryl-3,3-dialkyltriazenes and related 1-aryl-3-alkyltriazenes, and 1,3,3-trialkyltriazenes. Representative members are shown [26], [20] and [27] respectively. Since there have been relatively few studies of the biological activity of diaryltriazenes, these substances will not be discussed here. The trialkyltriazenes have considerable biological activity but do not require metabolic activation and, consequently, are beyond the scope of this article (Sieh et al., 1980). Aryldialkyltriazenes, on the other hand, are biologically active and their metabolism has been studied. The simplest member of the series is 1-phenyl-3,3-dimethyltriazene. It is a potent carcinogen in rats, giving mainly tumors of the central nervous system and of the kidneys (Preussmann et al., 1974). It is known to alkylate the DNA and RNA of target and nontarget tissue (Kleihues et al., 1976). It was shown that rat liver microsomes dealkylate triazenes, as evidenced by the formation of formaldehyde from 1-phenyl-3,3-dimethyltriazene, and higher aldehydes from alkyl groups longer than methyl (Preussmann et al., 1969). The in vivo metabolism of 1-[^{14}C]phenyl-3,3-dimethyltriazene was studied by Kolar and Schlesiger (1976). The principal metabolites were aniline and various hydroxylated anilines. About 1% of the dose, however, was a metabolite with an intact triazene skeleton. The metabolism of 1-(2,4,6-trichlorophenyl)-3,3-dimethyltriazene in rats gave only one isolable metabolite which proved to be the glucuronide of the methylhydroxylated triazene [28]

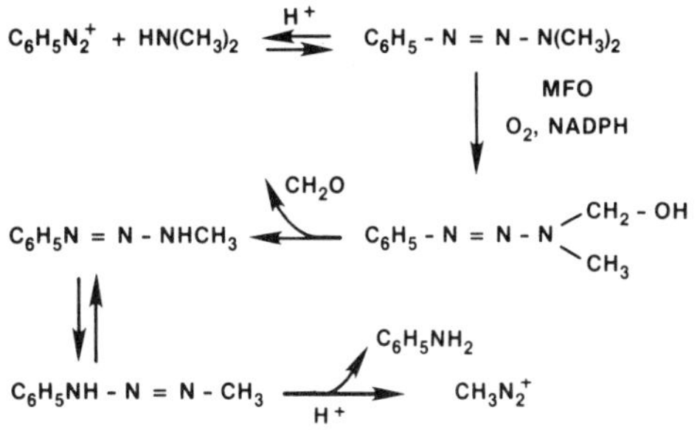

$$C_6H_5N_2^+ + HN(CH_3)_2 \underset{}{\overset{H^+}{\rightleftharpoons}} C_6H_5-N=N-N(CH_3)_2$$

$$\downarrow \text{MFO, } O_2, \text{NADPH}$$

$$C_6H_5N=N-NHCH_3 \underset{CH_2O}{\overset{}{\rightleftharpoons}} C_6H_5-N=N-N\underset{CH_3}{\overset{CH_2-OH}{}}$$

$$\updownarrow$$

$$C_6H_5NH-N=N-CH_3 \xrightarrow[H^+]{C_6H_5NH_2} CH_3N_2^+$$

FIG. 20. Metabolic pathways for 1-phenyl-3,3-dimethyltriazene.

(Kolar and Carubelli, 1979). This compound could be cleaved by a glucuronidase, whereupon it decomposed further to the monoalkyltriazene by loss of formaldehyde. The arylmonoalkyltriazenes are unstable in an aqueous environment and decompose to the corresponding aniline and the alkyldiazonium ion (Vaughan and Stevens, 1978). Thus, the picture which emerges for the metabolism of aryldialkyltriazenes is very similar to the ones described earlier for dialkylnitrosamines and hydrazines. However, an additional pathway must be considered. Aryldialkyltriazenes are usually prepared by mixing the aryldiazonium salt with an appropriate dialkylamine. Thus, it is conceivable that the reverse reaction might be important in some cases. This was indeed found to be when the aryldiazonium ion was stabilized by electron donating groups on the aromatic ring (e.g. 1-(4-methoxyphenyl)-3,3-dimethyltriazene) (Kolar and Preussmann, 1971). Thus the metabolic scheme for aryldialkyltriazenes, as exemplified by 1-phenyl-3,3-dimethyltriazene, is shown in Fig. 20.

It is interesting to note that this triazene produces apparently the same ultimate carcinogen, the methyldiazonium ion, as N-nitrosodimethylamine and 1,2-dimethylhydrazine. Each of these carcinogens produces a very different tumor spectrum. It is clear then, that the identification of the ultimate carcinogenic form of a procarcinogen does not give much information about the type of tumor and the site where it will be found. It is equally clear that pharmacokinetics and the pharmacology of the procarcinogens and of some of their intermediate metabolites must play an important role in tumor induction at specific sites.

4.6. CARCINOGENIC NATURAL PRODUCTS

As indicated earlier, aflotoxin B_1 [21] (reviewed by Busby and Wogan, 1984) is a naturally occurring carcinogen. It is distinguished by being the most potent chemical carcinogen known for rat liver although it is essentially noncarcinogenic for mouse liver. At this time, no metabolite of aflatoxin B_1 which exhibits carcinogenic potency comparable with that of this mycotoxin has been isolated. However, as with the polycyclic hydro+carbons, studies of the nucleic acid bound forms of aflatoxin B_1 have suggested that it is activated through an epoxide generated, in this case, at the 2,3-double bond in the terminal furan ring.

This conclusion arose from a series of investigations beginning with demonstrations that the 2,3-double bond in aflatoxin B_1 is essential for toxicity and carcinogenicity (Wogan et al., 1971). Thereafter, Swenson et al. (1973) reported that aflatoxin 2,3-dihydrodiol could be obtained by mild acid hydrolysis of RNA to which aflatoxin B_1 had been bound in the presence of microsomes. Several laboratories then independently characterized the major DNA and RNA adduct formed from aflatoxin B_1 (Essigman et al., 1977; Martin and Garner, 1977; Lin et al., 1977) showing that it was 2,3-dihydro-2-(guan-7-yl)-3-hydroxy-aflatoxin B_1 which presumably arose from the 2,3-epoxide. This constitutes a one-step metabolic activation sequence with no relatively stable proximate carcinogen being formed as an intermediate. No success in the synthesis of this epoxide has been achieved so far and the identification of the carcinogenic metabolite cannot be tested directly. As with the hydrocarbons, the metabolically generated strained ring C–O bond is again the source of the biological activity of this carcinogen.

Safrole [22] and estragole [29] are products of green plants and both produce liver tumors in mice (reviewed by Miller and Miller, 1981). In both cases, a 1'-hydroxy metabolite has been found to exhibit greater hepatocarcinogenicity than the parent alkenylbenzenes (Borchert et al., 1973; Drinkwater et al., 1976) and investigations of the structures of DNA adducts formed suggests that an ester (possibly a sulfate ester) of these secondary benzylic alcohols mediates their nucleic acid binding and probably their carcinogenicity (Phillips et al., 1981a,b). Again, a metabolic oxidation seems to have created a potentially unstable C–O bond in forming the benzylic alcohol, an esterification to create a better leaving group than OH^- is then sufficient to create a reactive and carcinogenic metabolite.

This very brief description of the metabolic activation of just three naturally occurring carcinogens can be augmented by reference to detailed discussions of aflatoxins (Busby and Wogan, 1984), fusarial mycotoxins (Schoental, 1984), and carcinogens in plants (Schoental, 1976). However, this brief discussion serves to show that naturally occurring chemical carcinogens are subject to the same general type of activation processes found to apply to synthetic chemicals.

5. CONCLUSIONS

Although a limited number of known chemical carcinogens have been discussed, sufficient information has been presented to indicate the general routes of metabolic activation that have been discovered and to familiarize the reader with the kind of evidence from which such activation routes were devised. Thus, experimental approaches to the determination of which specific metabolites are involved in the expression of carcinogenic activity are varied. The most compelling data, which is nevertheless not definitive, arises from assays for carcinogenic activity of carcinogen metabolites. If a metabolite exhibits carcinogenic activity which is comparable to or greater than that of the parent carcinogen and/or exhibits activity in species or organs where the parent carcinogen is inactive, this must be considered to be strong evidence indicating that the carcinogenic activity of the parent carcinogen is expressed through this metabolite. Negative findings in such assays are more difficult to interpret. For example, the sulfate ester of N-hydroxy-2-fluorenylacetamide, which is believed to be the ultimate carcinogenic metabolite of N-2-fluorenylacetamide, does not display any significant carcinogenic activity (Garner et al., 1984). This can be attributed to solvolytic destruction of the metabolite during administration, or to the difficulty cells experience in taking up anions. However, these rationalizations of negative data are never totally satisfying. Another example can be drawn from the polycyclic aromatic hydrocarbon field where the 7,8-epoxide of benzo[a]pyrene has not been found to be a particularly potent carcinogen, even though it is believed to be a proximate carcinogenic metabolite of benzo[a]pyrene (Figs 4 and 5).

A second type of data which has been used extensively in recent years involves the presumption that the metabolite which initiates the carcinogenic process is also responsible for the reactions of the carcinogen with cellular DNA. Indeed, many people believe that initiation of carcinogenesis and reaction with DNA are one and the same event but this grossly oversimplifies the operational definition of tumor initiation. Nevertheless, the carcinogen residue that binds to DNA clearly arises from a reactive metabolite of the carcinogen which has not been inactivated by other metabolic reactions. Thus, the structures of these nucleic acid adducts are valuable indicators of the structure of the ultimate carcinogens.

Other criteria which can be applied to the identification of carcinogenic metabolites include syntheses of model putative proximate or ultimate carcinogens for evaluation of biological activity in cases where the proximate or ultimate carcinogens themselves are synthetically inaccessible. Additionally, various structure-activity approaches have been used to identify sites of metabolic activation of carcinogens. For example, the replacement of hydrogen atoms with deuterium in various nitrosamines, and azo- and azoxy-compounds can modify carcinogenic potency and indicate which C–H or C–D bond needs to be broken in the course of metabolic activation (Section 4.3).

Carcinogenic compounds can be assigned to one of three broad classes depending on the kind of evidence used to establish their route of metabolic activation. The first class would be comprised of carcinogens for which known metabolites have been shown to exhibit pronounced carcinogenic activity and for which considerable indirect evidence about the structure of nucleic acid bound metabolites etc. also exists. This class includes the aromatic amines, the N-hydroxy metabolites of which are potent carcinogens, and are intermediates in DNA binding. Polycyclic aromatic hydrocarbons also fall into this category since most of them are activated through mechanisms similar to that described for benzo[a]pyrene, and all the dihydrodiol precursors of the bay region dihydrodiol epoxide metabolites are potent carcinogens. Similar criteria, i.e. the demonstration of the formation of carcinogenic metabolites, have been obtained for dialkylhydrazines, azo- and azoxy-compounds, triazenes and the naturally occurring alkenylbenzenes, estragole and safrole. Long alkyl-chain nitrosamines fall into this category because it has been shown that chain-shortened metabolites are at least as active in the induction of cancer as the parent compounds. Most of the substances grouped together in this class require more than one enzymatic step for activation and thus are converted initially to proximate carcinogens and subsequently, by further metabolism, to ultimate carcinogens. Proximate carcinogens are relatively stable, chemically, in comparison with ultimate carcinogens, and they can be detected as metabolites and be administered for carcinogenicity testing much more readily for this reason.

The second class of carcinogens would include those for which strong indirect evidence points to a particular route of metabolic activation but for which no metabolite of comparable or greater carcinogenic potency than the parent compound can be directly demonstrated. In this class are the naturally occurring toxin, aflatoxin B_1 and some related compounds, as well as most of the simple N-nitrosodialkylamines. In these cases, only a single metabolic reaction is required for activation such that the ultimate carcinogen is formed directly. These metabolites are therefore very reactive and difficult to obtain or to subject to carcinogenicity testing. In the case of the N-nitrosodialkylamines, metabolic conversion to the unstable α-hydroxy derivative occurs. Although some α-hydroxylated nitrosamines have been made synthetically and many esters of the α-hydroxylated nitrosamines have been prepared and studied, the involvement of α-hydroxylation in carcinogenic activation is still based on indirect evidence.

The third class of carcinogens contains all those chemicals for which a route of metabolic activation has not yet been established. This is the most perplexing class and it contains compounds such as N-nitrosodiethanolamine and the antihistaminic, methapyrilene (Lijinsky et al., 1980a). In general, these chemicals are not reactive towards cellular macromolecules in the sense that the nitrosoalkylureas are, and hence it is difficult to see how they could be directly acting carcinogens. Nevertheless, it has not been possible to detect metabolic processes which would activate these clearly carcinogenic compounds. Cyclic nitrosamines also fall into this category. Although it has been shown that these substances are metabolized by α-hydroxylation, little or no binding to nucleic acids has been detected and no metabolites of notable carcinogenic potency have been isolated. Carcinogenic potency has been reduced by α-deuterium substitution in some cases (see Section 4.3) but the direct involvement of metabolites of cyclic nitrosamines in tumor initiation remains to be established. Further investigation will undoubtedly reveal whether the compounds presently in this third class exert their carcinogenic potential through a distinctly different mechanism or whether they are indeed activated through metabolites yet to be defined.

From the work on those compounds in the first two classes, above, where metabolic activation is well established, some generalizations can be made. One important point is that metabolic activation plays a key role in determining the site of cancer development. Thus, in addition to pharmacologic and pharmacokinetic factors, the availability of appropriate metabolizing enzymes in various tissues determines those organs in which metabolic activation can occur. Another interesting point is that the metabolic activation reactions that have been recognized are not unique or special reactions. These reactions are routinely employed in the metabolism of foreign substances but certain structural peculiarities in chemical carcinogens lead, in these cases, to the generation of highly reactive electrophilic metabolites. Lastly, the nature of these reactive metabolites is specific to each class of carcinogens. Thus, the source of reactivity of the ultimate carcinogen from the aromatic amines i.e. a reactive N–O bond does not allow one to predict the ultimate carcinogen for polycyclic hydrocarbons which contain a reactive C–O bond in a strained three membered ring. It is quite obvious, therefore, that our knowledge of metabolic activation at present is far from complete. The combination of the complex array of enzyme systems available in mammalian systems with the enormous range of chemical structure to which these systems can be exposed can surely result in many new subtle mechanisms of generating reactive and carcinogenic metabolites.

Acknowledgement—The authors' work has been supported by the NCI both directly (EKW) and through Contract No. N01-CO-23909 with Litton Bionetics, Inc.

REFERENCES

AIROLDI, L., BONFANTI, M., FANELLI, R., BOVE, B., BENFENATI, E. and GARIBOLDI, P. (1984) Identification of a nitrosamino aldehyde and a nitrosamino acid resulting from β-oxidation of N-nitrosodiethanolamine. *Chem.-Biol. Interactions*, **51**: 103–113.

AMES, B. N., MCCANN, J. and YAMASAKI, E. (1975) Methods for detecting carcinogens and mutagens with the *Salmonella*/mammalian-microsome mutagenicity test. *Mutat. Res.* **31**: 347–364.

ARCOS, J. C. and ARGUS, M. F. (1974) *Chemical Induction of Cancer Volume IIa*, Academic Press, New York.

ARCOS, J. C., BRYANT, G. M., PASTOR, K. M. and ARGUS, M. F. (1976) Structural limits of specificity of methylcholanthrene-repressible nitrosamine *N*-dealkylase inhibition of analog substrates. *Z. Krebsforsch.* **86**: 171–183.

ARCOS, J. C., DAVIES, D. L., BROWN, CH. E. L. and ARGUS, M. F. (1977) Repressible and inducible enzymic forms of dimethylnitrosamine-demethylase, *Z. Krebsforsch.* **89**: 181–199.

AUTRUP, H., WEFOLD, F. C., JEFFREY, A. M., TATE, H., SCHWARTZ, R. D., TRUMP, B. F. and HARRIS, C. C. (1980) Metabolism of benz[*a*]pyrene by cultured tracheobronchial tissues from mice, rats, hamsters, bovines and humans. *Int. J. Cancer* **25**: 293–300.

BAIRD, W. M., DIPPLE, A., GROVER, P. L., SIMS, P. and BROOKES, P. (1973) Studies on the formation of hydrocarbon-deoxyribonucleoside products by the binding of derivatives of 7-methylbenz[*a*]anthracene to DNA in aqueous solution and in mouse embryo cells in culture. *Cancer Res.* **33**: 2386–2392.

BALO, J. (1979) Role of hydrazine in carcinogenesis. *Adv. Cancer Res.* **30**: 151–164.

BARRY, G., COOK, J. W., HASLEWOOD, G. A. D., HEWETT, C. L., HIEGER, I. and KENNAWAY, E. L. (1935) Production of cancer by pure hydrocarbons. *Proc. R. Soc.* **B117**: 318–351.

BARTSCH, H., DWORKIN, M., MILLER, J. A. and MILLER, E. C. (1972) Electrophilic *N*-acetoxyaminoarenes derived from carcinogenic *N*-hydroxy-*N*-acetylaminoarenes by enzymatic deacetylation and transacetylation in liver. *Biochim. biophys. Acta* **286**: 272–298.

BARTSCH, H., TROUT, M. and HECKER, E. (1971) On the metabolic activation of *N*-hydroxy-*N*-2-acetylaminofluorene. II. Simultaneous formation of 2-nitrosofluorene and *N*-hydroxy-*N*-2-acetylaminofluorene *via* a free radical intermediate. *Biochim. biophys. Acta* **237**: 556–566.

BELAND, F. A., DOOLEY, K. L. and JACKSON, C. D. (1982) Persistence of DNA adducts in rat liver and kidney after multiple doses of the carcinogen *N*-hydroxy-2-acetylaminofluorene. *Cancer Res.* **42**: 1348–1354.

BENSON, F. R. (1984) *The High Nitrogen Compounds*, John Wiley and Sons, New York.

BENTLEY, P., OESCH, F. and GLATT, H. (1977) Dual role of epoxide hydratase in both activation and inactivation of benzo[*a*]pyrene. *Arch. Toxicol.* **39**: 65–75.

BIGGER, C. A. H., TOMASZEWSKI, J. E., DIPPLE, A. and LAKE, R. S. (1980) Limitations of metabolic activation systems used with *in vitro* tests for carcinogens. *Science* **209**: 503–505.

BLATTMANN, L. and PREUSSMAN, R. (1974) Oxidative biotransformation of di-*n*-butylnitrosamine. *Z. Krebsforsch.* **88**: 311–314.

BORCHERT. P., MILLER, J. A., MILLER, E. C. and SHIRES, T. K. (1973) 1'-Hydroxysafrole: a proximate carcinogenic metabolite of safrole in the rat and mouse. *Cancer Res.* **33**: 590–600.

BORGEN, A., DARVEY, H., CASTAGNOLI, N., CROCKER, T. T., RASMUSSEN, R. E. and WANG, I. Y. (1973) Metabolic conversion of benzo[*a*]pyrene by Syrian hamster liver microsomes and binding of metabolites to deoxyribonucleic acid. *J. med. Chem.* **16**: 502–506.

BOYLAND, E. (1950) The biological significance of metabolism of polycyclic compounds. *Biochem. Soc. Symp.* **5**: 40–54.

BROOKES, P. and LAWLEY, P. D. (1964) Evidence for the binding of polynuclear aromatic hydrocarbons to the nucleic acids of mouse skin: relation between carcinogenic power of hydrocarbons and their binding to deoxyribonucleic acid. *Nature, Lond.* **202**: 781–784.

BUENING, M. K., WISLOCKI, P. G., LEVIN, W., YAGI, H., THAKKER, D. R., ARKAGI, H., KOREEDA, M., JERINA, D. M. and CONNEY, A. H. (1978) Tumorigenicity of the optical enantiomers of the diastereomeric benzo[*a*]pyrene 7,8-diol-9,10-epoxides in newborn mice: exceptional activity of (+)-7β,8a-dihydroxy-9a,10a-epoxy-7,8,9,10-tetrahydrobenzo[a]pyrene. *Proc. natn. Acad. Sci. U.S.A.* **75**: 5358–5361.

BUSBY, W. F. and WOGAN, G. N. (1984) Aflatoxins. In: *Chemical Carcinogens* (2nd Edn), pp. 945–1136, SEARLE, C. E. (ed.) Am. chem. Soc. Monogr. 182, Am. chem. Soc., Washington, D.C.

CARDONA, R. A. and KING, C. M. (1976) Activation of the O-glucuronide of the carcinogen N-hydroxy-2-fluorenylacetamide by enzymatic deacetylation *in vitro*: formation of fluorenylamine/tRNA adducts. *Biochem. Pharmac.* **25**: 1051–1056.

CASE, R. A. M., HOSKER, M. E., MCDONALD, D. B. and PEARSON, J. T. (1954) Tumours of the urinary bladder of workmen engaged in the manufacture and use of certain dye stuff intermediate in the British chemical industry. *Br. J. Ind. Med.* **11**: 213–216.

CHAU, I. Y., DAGANI, D. and ARCHER, M. C. (1978) Kinetic studies on the hepatic microsomal metabolism of dimethylnitrosamine, diethylnitrosamine and methylethylnitrosamine in the rat. *J. natn. Cancer Inst.* **61**: 517–521.

CHEN, C. B., MCCOY, G. D., HECHT, S. S., HOFFMANN, D. and WYNDER E. L. (1978a) High-pressure liquid chromatographic assay for α-hydroxylation of N-nitrosopyrrolidine by isolated rat liver microsomes. *Cancer Res.* **38**: 3812–3816.

CHEN, C. B., HECHT, S. S. and HOFFMANN, D. (1978b) Metabolic α-hydroxylation of the tobacco-specific nitrosamine, N'-nitrosonornicotine. *Cancer Res.* **38**: 3639–3645.

CLAYSON, D. B. (1962) *Chemical Carcinogenesis,* Little Brown and Co., Boston, Massachusetts.

CLAYSON, D. B. and GARNER, R. C. (1976) Carcinogenic aromatic amines and related compounds. In: *Chemical Carcinogens,* pp. 366–461, SEARLE, C. E. (ed.) Am. chem. Soc. Monogr. 173, Am. chem. Soc., Washington, D. C.

COHEN, S. M. (1978) Toxicity and carcinogenicity of nitrofurans. In: *Nitrofurans,* pp. 171–231. BRYAN, G. T. (ed.) Raven Press, New York.

COMIS, R. L. (1976) DTIC (NSC-45388) in malignant melanoma: a perspective, *Cancer Treat. Rep.* **42**: 4875–4917.

COOK, J. W., HEWITT, C. L. and HIEGER, I. (1933) The isolation of a cancer producing hydrocarbon from coal tar. *J. chem. Soc.* 395–405.

COOPER, C. S., GROVER, P. L. and SIMS, P. (1983) The metabolism and activation of benzo[a]pyrene. *Prog. Drug. Metab.* **7**: 295–396.

COOPER, C. S., HEWER, A., RIBEIRO, O., GROVER, P. L. and SIMS, P. (1980) The enzyme-catalysed conversion of *anti* benzo[a]pyrene-7,8-diol 9,10-oxide into a glutathione conjugate. *Carcinogenesis* **1**: 1075–1080.

COTTRELL, R. C., LAKE, B. G., PHILLIPS, J. C. and GANGOLLI, S. D. (1977) The hepatic metabolism of ^{15}N-labelled dimethylnitrosamine in the rat. *Biochem. Pharmac.* **26**: 809–813.

CRAMER, W., MILLER, J. A. and MILLER, E. C. (1960) N-Hydroxylation: a new metabolic reaction observed in the rat with the carcinogen 2-acetylaminofluorene. *J. biol. Chem.* **235**: 885–888.

DALY, J. W., JERINA, D. M. and WITKOP, B. (1972) Arene oxides and the NIH shift: the metabolism, toxicity and carcinogenicity of aromatic compounds *Experientia* **28**: 1129–1264.

DEBAUN, J. R., MILLER, E. C. and MILLER, J. A. (1970a) N-Hydroxy-2-acetylaminofluorene sulfotransferase: its probable role in carcinogenesis and protein-(methion-S-yl)binding in rat liver. *Cancer Res.* **30**: 577–595.

DEBAUN, J. R., ROWLEY, J. Y., MILLER E. C. and MILLER, J. A. (1968) Sulfotransferase activation of N-hydroxy-2-acetylaminofluorene in rodent livers susceptible and resistant to this carcinogen. *Proc. Soc. exp. Biol. Med.* **129**: 268–273.

DEBAUN, J. R., SMITH, J. Y. R., MILLER, E. C. and MILLER, J. A. (1970b) Reactivity *in vivo* of the carcinogen N-hydroxy-2-acetylaminofluorene: increase by sulfate ion. *Science* **167**: 184–186.

DIPPLE, A. (1976) Polynuclear aromatic carcinogens. In: *Chemical Carcinogens,* pp. 245–314, SEARLE, C. E. (ed.) Am. chem. Soc. Mongr. 173. Am. chem. Soc., Washington, D.C.

DIPPLE, A. (1983) Formation metabolism and mechanism of action of polycyclic aromatic hydrocarbons. *Cancer Res. (Suppl.)* **43**: 2422s–2425s.

DIPPLE, A., LAWLEY, P. D. and BROOKES, P. (1968) Theory of tumor initiation by chemical carcinogens: dependence of activity on structure of ultimate carcinogen. *Eur. J. Cancer* **4**: 493–506.

DIPPLE. A., MOSCHEL, R. C. and BIGGER, C. A. H. (1984) Polynuclear aromatic carcinogens. In: *Chemical Carcinogens* (2nd Edn), pp. 41–163, SEARLE, C. E. (ed.) Am. chem. Soc. Monogr. 182, Am. chem. Soc., Washington, D.C.

DRINKWATER, N. R., MILLER, E. C., MILLER, J. A. and PITOT, H. C. (1976) Hepatocarcinogenicity of estragole (1-allyl-4-methoxybenzene) and 1'-hydroxyestragole in the mouse and mutagenicity of 1'-acetoxyestragole in bacteria. *J. natn. Cancer Inst.* **57**: 1323–1331.

DRUCKREY, H. (1970) Production of colonic carcinomas by 1,2-dialkylhydazines and azoxyalkanes. In: *Carcinoma of the Colon and Antecedent Epithelium,* pp. 267–279, BURDETT, W. J. (ed.) Charles C. Thomas, Springfield, Illinois.

DRUCKREY, H., PREUSSMANN, R., IVANKOVIC, S. and SCHMÄHL, D. (1967) Organotrope carcinogene Wirkungen bei 65 verschiedenen N-nitrosoverbindungen an BD Ratten, *Z. Krebsforsch.* **69**: 103–201.

DUTTON, A. H. and HEATH, D. F. (1956) Demethylation of dimethylnitrosamine in rats and mice. *Nature* **178**: 644.

EDWARDS, J. E. (1941) Hepatomas in mice induced with carbon tetrachloride. *J. natn. Cancer Inst.* **2**: 197–199.

ESSIGMAN, J. M., CROY, R. G. NADZAN, A. M., BUSBY, W. F., JR., REINHOLD, V. N., BÜCHI, G. and WOGAN, G. N. (1977) Structural identification of the major DNA adduct formed by aflatoxin B_1 *in vitro. Proc. natn. Acad. Sci. U.S.A.* **74**: 1870–1874.

FARRELLY, J. G. (1980) A new essay for the microsomal metabolism of nitrosamines. *Cancer Res.* **40**: 3241–3244.

FARRELLY, J. G. STEWART, M. L. and LIJINSKY, W. (1984) The metabolism of nitrosodi-n-propylamine, nitrosodiallylamine and nitrosodiethanolamine. *Carcinogenesis* **5**: 1015–1019.

FIALA, E. S. (1975) Investigations into the metabolism and mode of action of the colon carcinogen, 1,2-dimethylhydrazine. *Cancer* **36**: 2407.

FIALA, E. S., BOBOTAS, G., KULAKIS, C., WATTENBERG, L. W. and WEISBURGER, J. H. (1977) Effects of disulfiram and related compounds on the metabolism *in vivo* of the colon carcinogen, 1,2-dimethylhydrazine. *Biochem. Pharmac.* **26**: 1763–1768.

FLOYD, R. A., SOONG, L. M., WALKER, R. N. and STUART, M. (1976) Lipid hydroperoxide activation of N-hydroxy-N-acetylaminofluorene via a free radical route. *Cancer Res.* **36**: 2761–2767.

GARNER, R. C., MARTIN, C. N. and CLAYSON, D. B. (1984) Carcinogen aromatic amines and related compounds. In: *Chemical Carcinogens* (2nd Edn), pp. 175–276, SEARLE, C. E. (ed.) Am. chem. Soc. Monogr. 182, Am. chem. Soc., Washington, D.C.

GRILL, S. and PRODI, G. (1975) Identification of dimethylnitrosamine metabolites *in vitro*. *Gann* **66**: 473–480.

GUTMANN, H. R., GALITSKI, S. B. and FOLEY, W. A. (1967) The conversion of noncarcinogenic aromatic amides to carcinogenic arylhydroxamic acids by synthetic N-hydroxylation. *Cancer Res.* **27**: 1443–1455.

HABS, M., SCHMÄHL, D. and WIESSLER, M. (1978) Carcinogenicity of acetoxymethyl-methyl-nitrosamine after subcutaneous, intravenous and intrarectal application in rats. *Z. Krebsforsch.* **91**: 217–221.

HARRIS, C. C., AUTRUP, H., STONER, G. D., MCDOWELL, E. M., TRUMP, B. F. and SCHAFER, P. (1977) Metabolism of acyclic and cyclic N-nitrosamines in cultured human bronchi. *J. natn. Cancer Inst.* **59**: 1401–1406.

HARVEY, R. G. (1981) Activated metabolites of carcinogenic hydrocarbons. *Acc. Chem. Res.* **14**: 218–226.

HECHT, S. S. and YOUNG, R. (1982) Regiospecificity in the metabolism of the homologous cyclic nitrosamines, N'-nitrosonornicotine and N'-nitrosoanabasine. *Carcinogenesis* **3**: 1195–1199.

HECHT, S. S., LIN, D. and CHEN, C. B. (1981) Comprehensive analysis of urinary metabolites of N'-nitrosonornicotine. *Carcinogenesis* **2**: 833–838.

HECHT, S. S., REISS, B., LINN, D. and WILLIAMS, G. M. (1982) Metabolism of N'-nitrosonornicotine by cultured rat esophageus. *Carcinogenesis* **3**: 453–456.

HECKER, L. I. FARRELLY, J. G., SMITH, J. H., SAAVEDRA, J. E. and LYON, P. A. (1979) Metabolism of the liver carcinogen N-nitrosopyrrolidine by rat liver microsomes. *Cancer Res.* **39**: 3679–3686.

HEIDELBERGER, C. (1973) Chemical oncogenesis in culture. *Adv. Cancer Res.* **18**: 317–366.

HEYDT, G., EISENBRAND, G. and PREUSSMANN, R. (1982) Metabolism of carcinogenic and non-carcinogenic N-nitroso-N-methylaminopyridines I. Investigations *in vitro*. *Carcinogenesis* **3**: 455–458.

HOLME, J. A., DAHLIN, D. C., NELSON, S. D. and DYBING, E. (1984) Cytotoxic effects of N-acetyl-p-benzoquinone imine, a common arylating intermediate of paracetamol and N-hydroxyparacetamol. *Biochem. Pharmac.* **33**: 401–406.

HOMBURGER, F., KELLEY, T. and FRIEDLER, G. (1961) Nutritional factors modifying hepatic adenomatosis induced by safrole (4-allyl-1,2-methylenedioxybenzene). *Proc. Am. Ass. Cancer Res.* **3**: 236.

HUEPER, W. C., WILEY, F. H. and WOLFE, H. D. (1938) Experimental production of bladder tumors in dogs by administration of beta-naphthylamine, *J. ind. Hyg. Toxicol.* **20**: 46–84.

IARC (1981) Evaluation of the carcinogenic risk of chemicals to humans. *IARC Monograph, Lyon* **26**: 311–334.

IRVING, C. C. and RUSSELL, L. T. (1970) Synthesis of the O-glucuronide of N-2-fluorenylhydroxylamine. Reaction with nucleic acids and with guanosine 5'-monophosphate. *Biochemistry* **9**: 2471–2476.

IRVING, C. C. and VEAZEY, R. A. (1969) Persistent binding of 2-acetylaminofluorene to rat liver DNA *in vivo* and consideration of the mechanism of binding of N-hydroxy-2-acetylaminofluorene to rat liver nucleic acids. *Cancer Res.* **29**: 1799–1804.

IRVING, C. C., JANSS, D. H. and RUSSELL, L. T. (1971) Lack of N-hydroxy-2-acetylaminofluorene sulfotransferase activity in the mammary gland and Zymbal's gland of the rat. *Cancer Res.* **31**: 387–391.

JERINA, D. M. and DALY, J. W. (1974) Arene oxides: a new aspect of drug metabolism. *Science* **185**: 573–582.

JERINA, D. M. and DALY, J. W. (1976) Oxidation at carbon. In: *Drug Metabolism*, pp. 13–32, PARKE, D. V. and SMITH, R. L. (eds) Taylor and Francis, London.

KADLUBAR, F. F., MILLER, J. A. and MILLER, E. C. (1976) Hepatic metabolism of N-hydroxy-N-methyl-4-aminoazobenzene and other N-hydroxyarylamines to reactive sulfuric acid esters. *Cancer Res.* **36**: 2350–2359.

KEEFER, L. K., LIJINSKY, W. and GARCIA, H. (1973) Deuterium isotope effect on the carcinogenicity of dimethylnitrosamine in rat liver. *J. natn. Cancer Inst.* **51**: 299–302.

KENNAWAY, E. L. (1955) The identification of a carcinogenic compound in coal-tar. *Br. med. J.* **2**: 749–752.

KENNAWAY, E. L. and HIEGER, I. (1930) Carcinogenic substances and their fluorescence spectra. *Br. med. J.* **1**: 1044–1046.

KING, C. M. (1982) N. Substituted aromatic compounds. In: *Chemical Carcinogenesis*, pp. 35–46, NICOLINI, C. (ed.) Plenum Press, New York.

KING, C. M. and PHILLIPS, B. (1968) Enzyme-catalyzed reactions of the carcinogen N-hydroxy-2-fluorenylacetamide with nucleic acid. *Science* **159**: 1351–1353.

KINOSITA, R. (1936) Researches on the carcinogenesis of the various chemical substances. *Gann* **30**: 423–426.

KLEIHUES, P., KOLAR, G. F. and MARGISON, G. P. (1976) Interaction of the carcinogen 3,3-dimethyl-1-phenyltriazene with nucleic acids of various rat tissues and the effects of a protein-free diet. *Cancer Res.* **36**: 2184–2193.

KOEPKE, S. R., TONDEUR, Y., FARRELLY, J. G., STEWART, M. L., MICHEJDA, C. J. and KROEGER-KOEPKE, M. B. (1984) Metabolism of ^{15}N-labeled N-nitrosodimethylamine and N-nitroso-N-methylaniline by isolated rat hepatocytes. *Biochem. Pharmac.* **33**: 1509–1513.

KOLAR, G. F (1984) Triazenes. In: *Chemical Carcinogens* (2nd Edn), pp. 869–914, SEARLE, C. E. (ed.) Am. chem. Soc. Monogr. 184, Am. chem. Soc., Washington, D.C.

KOLAR, G. F. and CARUBELLI, R. (1979) Urinary metabolite of 1-(2,4,6-trichlorophenyl)-3,3-dimethyltriazene with an intact diazoamino structure. *Cancer Lett.* **7**: 209–214.

KOLAR, G. F. and PREUSSMANN, R. (1971) Validity of a linear Hammett plot for the stability of some carcinogenic 1-aryl-3,3-dimethyltriazenes in an aqueous system. *Z. Naturforsch.* **256**: 950–953.

KOLAR, G. F. and SCHLESIGER, J. (1976) Urinary metabolites of 3,3-dimethyl-1-phenyltriazenes. *Chem. biol. Interact.* **14**: 301–311.

KRIEK, E. (1965) On the interaction of N-2-fluorenylhydroxylamine with nucleic acids *in vitro*. *Biochem. biophys. Res. Commun.* **20**: 793–799.

KRIEK, E. (1969) On the mechanism of action of carcinogenic aromatic amines I. Binding of 2-acetylaminofluorene and N-hydroxy-2-acetylaminofluorene to rat liver acids *in vivo*. *Chem. biol. Interact.* **1**: 3–17.

KRIEK, E., MILLER, J. A., JUHL, U. and MILLER, E. C. (1967) 8-(N-2-Fluorenylacetamido)guanosine, an arylamidation reaction product of guanosine and the carcinogen N-acetoxy-N-2-fluorenylacetamide in neutral aqueous solution. *Biochemistry* **6**: 177–182.

KROEGER-KOEPKE, M. B. and MICHEJDA, C. J. (1979) Oxidation of dimethylnitrosamine and phenylmethylnitrosamine by rat liver fractions. Evidence for several dimethylnitrosamine demethylase enzymes. *Cancer Res.* **39**: 1587–1591.

KROEGER-KOEPKE, M. B., KOEPKE, S. R., MCCLUSKY, G. A., MAGEE, P. N. and MICHEJDA C. J. (1981) α-Hydroxylation pathway in the *in vitro* metabolism of carcinogenic nitrosamines: N-nitrosodimethylamine and N-nitroso-N-methylaniline. *Proc. natn. Acad. Sci. U.S.A.* **79**: 6489–6493.

KROEGER-KOEPKE, M. B., REUBER, M. D., IYPE, P. T., LIJINSKY, W. and MICHEJDA, C. J. (1983) The effect of substituents in the aromatic ring on carcinogenicity of N-nitrosomethylaniline in F344 rats. *Carcinogenesis* **2**: 157–160.

KUPPER, R. and MICHEJDA, C. J. (1980) N-Nitrosoenamines; versatile new synthesis intermediates. *J. org. Chem.* **45**: 2119–2921.

LACASSAGNE, A., BUU-HOI, N. P., DAUDEL, R. and ZAJDELA, F. (1956) The relation between carcinogenic activity and the physical and chemical properties of angular benzacridines. *Adv. Cancer Res.* **4**: 315–369.

LAI, D. Y. and ARCOS, J. C. (1980) Mini review. Dialkylnitrosamine bioactivation and carcinogeneses. *Life Sci.* **27**: 2149–2165.

LAKE, B. G., PHILLIPS, J. C., COTTRELL, R. C. and GANGOLLI, S. D. (1978) The possible involvement of a microsomal amine oxidase enzyme in hepatic dimethylnitrosamine degradation *in vitro*. In: *Biological Oxidation of Nitrogen*, pp. 131–135, GORROD, J. E. (ed.) Elsevier/North Holland Biomedical Press, Amsterdam.

LAQUEUR, G. L., MICKELSEN, O., WHITING, M. G. and KURLAND, L. T. (1963) Carcinogenic properties of nuts from Cycas circinalis indigenous to Guam. *J. natn. Cancer Inst.* **31**: 919–951.

LAWLEY, P. D. (1976) Carcinogenesis by alkylating agents. In: *Chemical Carcinogens*, pp. 83–244, SEARLE, C. E. (ed.) Am. chem. Soc. Monogr. 173, Am. chem. Soc., Washington, D.C.

LEUNG, K. H. and ARCHER, M. C. (1984) Studies on the metabolic activation of β-ketonitrosamines: mechanisms of DNA methylation by N-(2-oxopropyl)-N-nitrosourea and N-nitroso-N-acetoxymethyl-N-(2-oxopropyl)-amine. *Chem. biol. Interact.* **48**: 169–179.

LEUNG, K. H., PARK, K. K. and ARCHER, M. C. (1978) Alpha-hydroxylation in the metabolism of N-nitrosopiperidine by rat liver microsomes: formation of 5-hydroxypentanol. *Res. Commun. chem. Path. Pharmac.* **19**: 201–211.

LEUNG, K. H., PARK, K. K. and ARCHER, M. C. (1980) Methylation of DNA by N-nitroso-2-oxopropyla-mine: formation of O^6 and 7-methylguanine and studies on the methylation mechanism. *Toxicol. appl. Pharmac.* **53**: 29.

LEVIN, W., THAKKER, D. R., WOOD, A. W., CHANG, R. L. LEHR, R. E. JERINA, D. M. and CONNEY, A. H. (1978) Evidence that benzo[a]anthracene 3,4-diol-1,2-epoxide is an ultimate carcinogen on mouse skin. *Cancer Res.* **38**: 1705–1710.

LEVIN, W., WOOD, A. W., CHANG, R. L., ITTAH, Y., CROISY-DELCEY, M., YAGI, H., JERINA, D. M. and CONNEY, A. H. (1980) Exceptionally high tumor-initiating activity of benzo[c]phenanthrene bay region diolepoxides on mouse skin. *Cancer Res.* **40**: 3910–3914.

LIJINSKY, W., LOO, J. and ROSS, A. E. (1968) Mechanism of alkylation of nucleic acids by nitrosomethylamine. *Nature* **218**: 1174–1175.

LIJINSKY, W., REUBER, M. D. and BLACKWELL, B.-N. (1980a) Liver tumors induced in rats by oral administration of the antihistaminic methapyrilene hydrochloride. *Science* **209**: 817–819.

LIJINSKY, W., REUBER, M. D. and MANNING, W. B. (1980b) Potent carcinogenicity of nitrosodiethanolamine in rats. *Nature* **288**: 589–590.

LIJINSKY, W., SAAVEDRA, J. E., REUBER, M. D. and BLACKWELL, B.-N. (1980c) The effect of deuterium labeling on the carcinogenicity of N-nitroso-2,6-dimethylmorpholine in rats. *Cancer Lett.* **10**: 325–331.

LIJINSKY, W., SAAVEDRA, J. E. and REUBER, M. D. (1981) Induction of carcinogenesis in Fischer rats by methylalkylnitrosamines. *Cancer Res.* **41**: 1288–1292.

LIJINSKY, W., TAYLOR, H. W. and KEEFER, L. K. (1976) Reduction of rat liver carcinogenicity by N-nitrosomorpholine by alpha-deuterium substitution, *J. natl Cancer Inst.* **57**: 1311–1313.

LIN, J. K., MILLER, J. A. and MILLER, E. C. (1977) 2,3-Dihydro-2-(guan-7-yl)-3-hydroxyaflatoxin B_1, a major acid-hydrolysis product of aflatoxin B_1-DNA or -ribosomal RNA adducts formed in hepatic microsome-mediated reactions and in rat liver *in vivo*. *Cancer Res.* **37**: 4430–4438.

LOEPPKY, R. N., GNEWUCH, C. T., HAZLITT, L. G. and MCKINLEY, W. A. (1979) N-Nitrosamine fragmentation and N-nitrosamine transformation. In: *N-Nitrosamines*, pp. 109–123, ANSELME, J.-P. (ed.) Am. chem. Soc., Washington, D.C.

LOEPPKY, R. N., OUTRAM, J. R., TOMASIK, W. and McKinley, W. (1981) Chemical and biochemical transformation of β-oxidized nitrosamines. In: *N-Nitroso Compounds*, pp. 21–37, SCANLAN, R. A. and TANNENBAUM, S. R. (eds) Am. chem. Soc., Washington, D.C.

MAGEE, P. N. (1976) Nitrosamine activation. In: In Vitro *Metabolic Activation in Mutagenesis Testing*, pp. 213–216, DE SERRES, F. J., FOUTS, J. R., BEND, J. R. and PHILPOT, R. M. (eds). Elsevier/North Holland Biomedical Press, Amsterdam.

MAGEE, P. N. and BARNES, J. M. (1956) The production of malignant primary hepatic tumours in the rat by feeding dimethylnitrosamine. *Br. J. Cancer* **10**: 114–122.

MAGEE, P. N. and BARNES, J. M. (1967) Carcinogenic nitroso compounds. *Adv. Cancer Res.* **10**: 163–246.

MAGEE, P. N. and VANDEKAR, M. (1958) Metabolism of dimethylnitrosamine *in vitro*. *Biochem. J.* **70**: 600–605.

MAGEE, P. N., MONTESANO, R. and PREUSSMANN, R. (1976) N-Nitroso compounds and related carcinogens. In: *Chemical Carcinogens*, pp. 491–625, SEARLE, C. E. (ed.) Am. chem. Soc. Monogr. 173, Am. chem. Soc., Washington, DC.

MARTIN, C. N. and GARNER, R. C. (1977) Aflatoxin B_1-oxide generated by chemical or enzymic oxidation of aflatoxin B_1 causes guanine substitution in nucleic acids. *Nature* **267**: 863–865.

MICHEJDA, C. J., ANDREWS, A. W. and KOEPKE, S. R. (1979) Derivatives of side-chain hydroxylated nitrosamines: direct acting mutagens in *Salmonella typhimurium. Mutat. Res.* **67**: 301–308.

MICHEJDA, C. J., KROEGER-KOEPKE, M. B., KOEPKE, S. R., MAGEE, P. N. and CHU, C. (1982) Nitrogen formation during *in vivo* and *in vitro* metabolism of *N*-nitrosamines. In: *Banbury Report 12: Nitrosamines And Human Cancer*, pp. 69–85, MAGEE, P. N. (ed.) Cold Spring Harbor Laboratory, New York.

MILLER, E. C. (1978) Some current perspectives on chemical carcinogenesis in humans and experimental animals: Presidential address. *Cancer Res.* **38**: 1479–1496.

MILLER, E. C. and MILLER, J. A. (1947) The presence and significance of bound aminoazo dyes in the livers of rats fed *p*-dimethylaminoazobenzene. *Cancer Res.* **7**: 468–480.

MILLER, E. C. and MILLER, J. A. (1966) Mechanisms of chemical carcinogenesis: nature of proximate carcinogens and interactions with macromolecules. *Pharmac. Rev.* **18**: 805–838.

MILLER, E. C. and MILLER, J. A. (1967) Low carcinogenicity of the K-region epoxides of 7-methylbenz[*a*]anthracene and benz[*a*]anthracene in the mouse and rat. *Proc. Soc. exp. Biol. Med.* **124**: 915–919.

MILLER, E. C. and MILLER, J. A. (1981) Searches for ultimate chemical carcinogens and their reactions with cellular macromolecules. *Cancer* **47**: 2327–2345.

MILLER, E. C., MILLER, J. A. and HARTMANN, H. (1961) *N*-Hydroxy-2-acetylaminofluorene: a metabolite of 2-acetylaminofluorene with increased carcinogenic activity in the rat. *Cancer Res.* **21**: 815–824.

MILLER, J. A. (1970) Carcinogenesis by chemicals: an overview-G.H.A. Clowes Memorial Lecture. *Cancer Res.* **30**: 559–576.

MILLER, J. A. and MILLER, E. C. (1953) The carcinogenic aminoazo dyes. *Adv. Cancer Res.* **1**: 339–396.

MILLER, J. A. and MILLER, E. C. (1971) Chemical Carcinogenesis: mechanisms and approaches to its control. *J. natn. Cancer Inst.* **47**: V–XIV.

MILSTEIN, S. and GUTTENPLAN, J. B. (1979) Near quantitative production of molecular nitrogen from metabolism of dimethylnitrosamine. *Biochem. biophys. Res. Commun.* **87**: 337–342.

MIRVISH, S. S. (1968) The carcinogenic action and metabolism of urethane and *N*-hydroxyurethane. *Adv. Cancer Res.* **11**: 1–42.

MIZRAHI, I. J. and EMMELOT, P. (1962) The effect of cysteine on the metabolic changes produced by two carcinogenic *N*-nitrosodialkylamines in rat liver. *Cancer Res.* **22**: 339–351.

MOCHIZUKI, M., ANJO, T. and OKADA, M. (1980) Isolation and characterization of *N*-alkyl-(*N*-hydroxymethyl)-nitrosamines from *N*-alkyl-*N*-(hydroperoxymethyl)nitrosamines by deoxygenation. *Tetrahedron Lett.* 3693–3696.

MORRIS, H. P., VELAT, C. A., WAGNER, B. P., DAHLGARD, M. and RAY, F. E. (1960) Studies of carcinogenicity in the rat of derivatives of aromatic amines related to *N*-2-fluorenylacetamide. *J. nat. Cancer Inst.* **24**: 149–180.

MOSTAFA, M. H., RUCHIRAWAT, M. and WEISBURGER, E. K. (1981) Comparative studies on the effects of various microsomal enzyme inducers on the *N*-demethylation of dimethylnitrosamine. *Biochem Pharmac.* **30**: 2007–2011.

MURASAKI, G., ZENSER, T. V., DAVIS, B. B. and COHEN, S. M. (1984) Inhibition by aspirin of N-[4-(5-nitro-2-furyl)-2-thiazolyl]formamide-induced bladder carcinogenesis and enhancement of forestomach carcinogenesis. *Carcinogenesis* **5**: 53–55.

NETTLESHEP, A., HENSHAW, P. S. and MEYER, H. L. (1943) Induction of pulmonary tumors in mice with ethyl carbamate (urethane). *J natn. Cancer Inst.* **4**: 309–319.

NEWMAN, M. S. and BLUM, S. J. (1964) A new cyclization reaction leading to epoxides of aromatic hydrocarbons. *J. Am. chem. Soc.* **86**: 5598–5600.

OKADA, M., SUZUKI, E. and MOCHIZUKI, M. (1976) Possible important role of urinary *N*-methyl-*N*-(3-carboxypropyl)nitrosamine in the induction of bladder tumors in rats by *N*-methyl-*N*-dodecylnitrosamine. *Gann* **67**: 771–772.

PARK, K. K. and ARCHER, M. C. (1978) Microsomal metabolism of di-*n*-propylnitrosamine: formation of products resulting from α- and β-oxidation. *Chem. biol. Interact.* **22**: 83–90.

PENG, R., TU, Y. Y. and YANG, C. L. (1982) The induction and competitive inhibition of a high affinity microsomal nitrosodimethylamine demethylase by ethanol. *Carcinogenesis* **3**: 1457–1461.

PHILLIPS, D. H. MILLER, J. A., MILLER, E. C. and ADAMS, B. (1981a) Structures of the DNA adducts found in mouse liver after administration of the proximate hepatocarcinogen 1'-hydroxyestragole. *Cancer Res.* **41**: 176–186.

PHILLIPS, D. H., MILLER, J. A., MILLER, E. C. and ADAMS, B. (1981b) N^2-Atom of guanine and N^6-atom of adenine residues as sites for covalent binding of metabolically activated 1'-hydroxysafrole to mouse liver DNA *in vivo*, *Cancer Res.* **41**: 2664–2671.

POIRIER, L. A., MILLER, J. A., MILLER, E. C. and SATO, K. (1967) *N*-Benzoyloxy-*N*-methyl-4-aminoazobenzene: its carcinogenic activity in the rat and its reactions with proteins and nucleic acids and their constituents *in vitro*. *Cancer Res.* **27**: 1600–1613.

POTT, P. (1775) Chirurgical observations relative to the cancer of the scrotum. Reprinted in *Natn. Cancer Inst. Monograph* **10**: 7–13 (1963).

POUR, P., WALLCAVE, L., GINGELL, R., NAGEL, D., LAWSON, T., SALMASI, S., and TINES, S. (1979) Carcinogenic effect of *N*-nitroso(2-hydroxypropyl)(2-oxopropyl)amine, a postulated proximate pancreatic carcinogen in Syrian hamsters. *Cancer Res.* **39**: 3828–3833.

PREUSSMANN, R. and STEWART B. W. (1984) *N*-Nitroso carcinogens, In: *Chemical Carcinogens* (2nd Edn), pp. 643–828, SEARLE, C. E. (ed.) Am. chem. Soc. Monogr. 182, Am. chem. Soc., Washington, D.C.

PREUSSMANN, R., HABS, M. and POOL, B. L. (1979) Carcinogenicity and mutagenicity testing of three isomeric *N*-nitroso-*N*-methylaminopyridines in rats. *J. natn. Cancer Inst.* **62**: 153–156.

PREUSSMANN, R., HABS, M., HABS, H. and STUMMEYER, D. (1982) Fluoro-substituted *N*-nitrosamines. 5. Carcinogenicity of *N*-nitroso-bis-(4,4,4-trifluoro-*n*-butyl)amine in rats. *Carcinogenesis* **3**: 1219–1222.

PREUSSMANN, R., IVANKOVIC, S., LANDSCHÜTZ, C., GIMMY, J., FLOHR, E. and GRIESBACH, U. (1974) Carcinogene Wirkung von 13 Aryldialkyltriazene an BD-Ratten. *Z. Krebsforsch.* **81**: 285–310.

PREUSSMANN, R., VON HODENBERG, A. and HENGY, H. (1969) Mechanism of carcinogenesis with 1-aryl-3,3-dialkyltriazenes. Enzymatic dealkylation by rat liver microsomal fraction *in vitro. Biochem. Pharmac.* **18**: 1–13.

PREUSSMANN, R., WURTELE, G., EISENBRAND, G. and SPIEGELHALDER, B. (1978) Urinary excretion of N-nitrosodiethanolamine administered orally to rats. *Cancer Lett.* **4**: 207–209.

PRUESS-SCHWARTZ, D., SEBTI, S. M., GILHAM, P. T. and BAIRD, W. M. (1984) Analysis of benzo[*a*]pyrene: DNA adducts formed in cells in culture by immobilized boronate chromatography. *Cancer Res.* **44**: 4104–4110.

PULLMAN, A. and PULLMAN, B. (1955) Electronic structure and carcinogenic activity and aromatic molecules: new developments. *Adv. Cancer Res* **3**: 117–169.

RAO, M. S., SCARPELLI, D. G. and LIJINSKY, W. (1981) Carcinogenesis in Syrian hamsters by N-nitroso-2,6-dimethylmorpholine, its *cis* and *trans* isomers, and the effects of deuterium labeling. *Carcinogenesis* **2**: 731–735.

REHN, L. (1895) Blasengeschwülste bei Fuchsin-Arbeitern. *Arch. Klin. Chir.* **50**: 588–600.

RICE, J. M., JOSHI, S. R., ROLLER, P. P. and WENK, M. L. (1975) Methyl(acetoxymethyl)nitrosamine: a new carcinogen highly specific for colon and small intestine. *Proc. Am. Ass. Cancer Res.* **16**: 32.

ROBERTSON, I. G. C., SIVARAJAH, K., ELING, T. E. and ZEIGER, E. (1983) Activation of some aromatic amines to mutagenic products by prostaglandin endoperoxide synthetase. *Cancer Res.* **43**: 476–480.

ROE, F. J. C. and LANCASTER, M. C. (1964) Natural, metallic and other substances as carcinogens. *Brit. med. Bull.* **20**: 127–133.

ROLLER, P. P., SHIMP, D. R. and KEEFER, L. K. (1975) Synthesis and solvolyis of methyl-(acetoxymethyl)nitrosamine. Solution chemistry of the presumed carcinogenic metabolite of dimethylnitrosamine. *Tetrahedron Lett.* 2065–2068.

SAAVEDRA, J. E. (1978) Oxidative decarboxylation of nitrosamino acids: a synthetic approach to cyclic α-acetoxynitrosamines. *Tetrahedron Lett.* 1923–1926.

SAYER, J. M., YAGI, H., CROISY-DELCEY, M. and JERINA, D. M. (1981) Novel bay-region diol epoxides from benzo[*c*]phenanthrene, *J. Am. chem. Soc.* **103**: 4970–4972.

SCHOENTAL, R. (1976) Carcinogens in plants and microorganisms. In: *Chemical Carcinogens,* pp. 626–689, SEARLE, C. E. (ed.) Am. chem. Soc. Monogr. 176, Am. chem. Soc., Washington, DC.

SCHOENTAL, R. (1984) Fusarial mycotoxins and cancer. In: *Chemical Carcinogens* (2nd Edn), pp. 1137–1169, Am. chem. Soc. Monogr. 182, Am. chem. Soc., Washington, DC.

SEARLE, C. E. (1976) *Chemical Carcinogens,* Am. chem. Soc. Monogr. 173, Am. chem. Soc., Washington, DC.

SEARLE, C. E. (1984) *Chemical Carcinogens* (2nd Edn), Am. chem. Soc. Monogr. 182, Am. chem. Soc., Washington, DC.

SELKIRK, J. K. (1980) Chemical carcinogenesis: a brief overview of the mechanism of action of polycyclic hydrocarbons, aromatic amines, nitrosamines and aflatoxins. In: *Modifiers of Carcinogenesis,* pp. 1–31, SLAGA, T. J. (ed.) Raven Press, New York.

SHIH, C., SHILO, B.-Z., GOLDFARB, M. P., DANNENBERG, A. and WEINBERG, R. A. (1979) Passage of phenotypes of chemically transformed cells via transfection of DNA and chromatin. *Proc. natl. Acad. Sci. U.S.A.* **76**: 5714–5718.

SIEH, D. H., ANDREWS, A. W. and MICHEJDA, C. J. (1980) Mutagenicity of trialkyltriazenes: mutagenic potency of alkyldiazonium ions, the putative ultimate carcinogens from dialkylnitrosamines. *Mutat. Res.* **73**: 227–235.

SIMS, P. and GROVER, P. L. (1974) Epoxides in polycyclic aromatic hydrocarbon metabolism and carcinogenesis. *Adv. Cancer Res.* **20**: 166–274.

SIMS, P. and GROVER, P. L. (1981) Involvement of dihydrodiols and diol epoxides in the metabolic activation of polycyclic hydrocarbons other than benzo[*a*]pyrene. In: *Polycyclic Hydrocarbons and Cancer,* pp. 117–181. GELBOIN, H. C. and TS'O, P. O. P. (eds) Academic Press, New York.

SIMS, P., GROVER, P. L., SWAISLAND, A., PAL, K. and HEWER, A. (1974) Metabolic activation of benzo[*a*]pyrene proceeds by a diol-epoxide. *Nature* **252**: 326–328.

SINGER, G. M., LIJINSKY, W., BUETTNER, L. and MCCLUSKEY, G. A. (1981) Relationship of rat urinary metabolites of N-nitrosomethyl-N-alkylamine to bladder carcinogenesis. *Cancer Res.* **41**: 4942–4946.

SKIPPER, P. L., TOMERA, J. F., WISHNOK, J. S., BRUNENGRABER, H. and TANNENBAUM, S. R. (1983) Pharmacokinetic model for N-nitrosodimethylamine based on Michaelis–Menten constants determined with the isolated perfused rat liver. *Cancer Res.* **43**: 4786–4790.

SLAGA, T. J., BRACKEN, W. J., GLEASON, G., LEVIN, W., YAGI, H., JERINA, D. M. and CONNEY, A. H. (1979) Marked differences in the skin tumor-initiating activities of the optical enantiomers of the diastereomeric benzo[*a*]pyrene 7,8-diol-9,10-epoxides. *Cancer Res.* **39**: 67–71.

SLAGA, T. J., GLEASON, G. L., MILLS, G., EWALD, L., FU, P. P., LEE, H. M. and HARVEY, R. G. (1980) Comparison of the skin tumor-initiating activities of dihydrodiols and diol-epoxides of various polycyclic aromatic hydrocarbons. *Cancer Res.* **40**: 1981–1984.

SNYDER, C. M., FARRELLY, J. G. and LIJINSKY, W. (1977) Metabolism of three cyclic nitrosamines in Sprague-Dawley rats. *Cancer Res.* **37**: 3530–3532.

SPIEGELHALDER, B., EISENBRAND, G. and PREUSSMANN R. (1982) Urinary excretion of N-nitrosamines in rats and man. In: N-*Nitroso Compounds: Occurrence and Biological Effects,* pp. 443–449, BARTSCH, H., O'NEILL, I. K., CASTEGNARO, M. and OKADA, M. (eds) International Agency for Research on Cancer, Lyon.

STREETER, A. J. and BAILLIE, T. A. (1985) 2-Acetamido-p-benzoquinone: a reactive arylating metabolite of 3'-hydroxyacetanilide. *Biochem. Pharmac.* **34**: 2871–2876.

SUKUMAR, S., NOTARIO, V., MARTIN-ZANCA, D. and BARBACID, M. (1983) Induction of mammary carcinomas in rats by nitroso-methylurea involves malignant activation of H-*ras*-1 locus by single point mutations. *Nature* **306**: 658–661.

SUZUKI, E. and OKADA, M. (1980) Metabolic fate of N,N-dibutylnitrosamine in the rat. *Gann* **71**: 863–870.

SWAMINATHAN, S. and LOWER, G. M., JR. (1978) Biotransformations and excretion of nitrofurans. In: *Nitrofurans,* pp. 59–97, BRYAN, G. T. (ed.) Raven Press, New York.

SWENSON, D. H., MILLER, J. A. and MILLER, E. C. (1973) 2,3-Dihydro-2,3-dihydroxy-aflatoxin B_1: An acid hydrolysis product of an RNA-aflatoxin B_1 adduct formed by hamster and rat liver microsomes *in vitro*. *Biochem. biophys. Res. Commun.* **53**: 1260–1267.

TADA, M. (1981) Metabolism of 4-nitroquinoline 1-oxide and related compounds. In: *The Nitroquinolines*, pp. 24–25, SUGIMURA, T. (ed.) Raven Press, New York.

THAKKER, D. R., YAGI, H., LEHR, R. E., LEVIN, W., BUENING, M., LU, A. Y. H., CHANG, R. L., WOOD, A. W., CONNEY, A. H. and JERINA, D. M. (1978) Metabolism of *trans*-9,10-dihydroxy-9,10-dihydrobenzo[*a*]pyrene occurs primarily by arylhydroxylation rather than formation of a diol epoxide. *Molec. Pharmac.* **14**: 502–513.

TOMERA, J. F., SKIPPER, P. L., WISHNOK, J. S., TANNENBAUM, S. R. and BRUNENGRABER, H. (1984) Inhibition of *N*-nitrosodimethylamine metabolism by ethanol and other inhibitors in the isolated perfused rat liver. *Carcinogenesis* **5**: 113–116.

TOTH, B. (1980) Actual new cancer causing hydrazines, hydrazides and hydrazones. *J. Cancer Res. clin. Oncol.* **97**: 97–108.

VAUGHAN, K. and STEVENS, M. F. G. (1978) Monoalkyltriazenes. *Chem. Soc. Rev.* **7**: 377–397.

WALPOLE, A. L. (1958) Carcinogenic action of alkylating agents. *Ann. N.Y. Acad. Sci.* **68**: 750–761.

WEINKAMP, R. J. and SHIMBA, D. A. (1978) Metabolic activation of procarbazine. *Life Sci.* **22**: 937–946.

WEISBURGER, E. K. (1977) Carcinogenic natural products. In: *Structural Correlates of Carcinogenesis and Mutagenesis*, pp. 184–192, ASHER, I. M. and ZERVOS, C. (eds) Proceedings of the 2nd FDA Office of Science Summer Symposium, FDA, Rockville, Maryland.

WEISBURGER E. K. AND WEISBURGER, J. H. (1958) Chemistry, carcinogenicity and metabolism of 2-fluorenamine and related compounds. *Adv. Cancer Res.* **5**: 331–431.

WEISBURGER, J. H., WEISBURGER, E. K. and MORRIS, H. P. (1956) Urinary metabolites of the carcinogen *N*-2-fluorenylacetamide, *J. natn. Cancer Inst.* **17**: 345–361.

WEISBURGER, J. H., YAMAMOTO, R. S., WILLIAMS, G. M., GRANTHAM, P. H., MATSUSHIMA, T. and WEISBURGER, E. K. (1972) On the sulfate ester of *N*-hydroxy-*N*-2-fluorenylacetamide as a key ultimate hepatocarcinogen in the rat. *Cancer Res.* **32**: 491–500.

WHALLEY, C. E., IQBAL, Z. M. and EPSTEIN, S. S. (1981) *In vivo* and microsomal metabolism of the pancreatic carcinogen *N*-nitroso-bis-(2-oxopropyl)amine by the Syrian golden hamster. *Cancer Res.* **41**: 482–486.

WIESSLER, M. (1975) Chemie der Nitrosamine. II. Synthese α-funktioneller Dimethylnitrosamine. *Tetrahedron Lett.* 2575–2578.

WIESSLER, M. (1979) Chemistry of α-substituted *N*-nitrosamines. In: *N-Nitrosamines*, pp. 57–75, ANSELME, J.-P. (ed.) Am. chem. Soc., Washington, DC.

WILSON, R. H., DEEDS, F. and COX, A. J., JR. (1941) The toxicity and carcinogenic activity of 2-acetaminofluorene. *Cancer Res.* **1**: 595–608.

WISE, R. W., ZENSER, T. V., KADLUBAR, F. F. and DAVIS B. B. (1984) Metabolic activation of carcinogenic aromatic amines by dog bladder and kidney prostaglandin H synthetase. *Cancer Res.* **44**: 1893–1897.

WOGAN, G. N., EDWARDS, G. S. and NEWBERNE, P. M. (1971) Structure–activity relationships in toxicity and carcinogenicity of aflatoxins and analogs. *Cancer Res.* **31**: 1936–1942.

YAMAGIWA, K. and ICHIKAWA, K. (1915) Experimentelle Studie über die Pathogenese der Epithelialgeschwulste. *Mitt. Med. Fak. Tokio* **15**: 295–344.

YOSHIDA, T. (1933) Uber die serienweise Verfolgung der Veränderungen der Leber der experimentellen Hepatomerzeugung durch *o*-Aminazotoluol. *Trans. japan path. Soc.* **23**: 636–638.

ZEDECK M.S. (1984) Hydrazine derivatives, azo and azoxy compounds, and methylazoxymethanol and cycasin. In: *Chemical carcinogens* (2nd Edn), pp. 915–944, SEARLE, C. E. (ed.) Am. chem. Soc. Monog. 182, Am. chem. Soc., Washington, D.C.

CHAPTER 2

DNA MODIFICATION BY CHEMICAL CARCINOGENS

ALAN M. JEFFERY

Cancer Center/Institute of Cancer Research, Division of Environmental Sciences and Department of Pharmacology, Columbia University, New York, NY 10032, U.S.A.

Abstract—A review of the types of *interactions* which can occur between *xenobiotics* and *DNA* is presented. Consideration is given to the *methods* available for the *detection* of these DNA adducts in biological systems. A number of specific groups of carcinogen-DNA adducts are considered including *simple alkylating agents, polycyclic aromatic hydrocarbons, aromatic amines,* and a variety of *natural products,* as well as some *photochemically* activated compounds.

1. INTRODUCTION

There are to date about six million compounds which have been identified and catalogued. Of these, only a small percent have ever been investigated with respect to their general toxicity and only about 30 have been clearly shown to be associated with increased risk of human cancer (IARC Monograph, 1982). Despite the limited number of compounds which are recognized as human carcinogens the list can be expanded to several hundred if animal test data is included. The exact value depends upon how stringent are the criteria used for evaluation of the animal test data. Many more have been shown to transform or mutate cells in culture. Such evidence provides concern that such compounds pose health hazards to man although accurate risk assessment is often very difficult (National Academy of Sciences, 1983).

The process of chemical carcinogenesis has been divided into two stages, initiation and promotion (Berenblum, 1944), although additional steps are now recognized (Slaga *et al.*, 1980). One of the major distinctions between initiation and promotion is that chemicals in the former group often undergo chemical reaction with DNA while those in the latter do not. It is chemicals of the former class which will be considered in this review.

Of compounds classified as initiating chemical carcinogens most have also been shown to be mutagenic (Sugimura *et al.*, 1976). Although initiating carcinogens and mutagens do not necessarily function through the same mechanism, a common point for both types of compounds is almost always their ability to react with DNA (Miller and Miller, 1974, 1981). Treatment of hamster embryo cells with 5-bromodeoxyuridine allows incorporation of this modified base into DNA. Subsequent treatment with near u.v. light results in neoplastic transformation of these cells (Barrett *et al.*, 1978). This strongly suggests, since the u.v. light alone at the doses used was insufficient to induce transformation, that the photochemical damage induced in the DNA was critical for transformation. For these reasons, the study of DNA adduct formation seems justified. Other evidence that DNA can be the critical target for chemical carcinogens comes from the recent data on the activation of C-Ha-*ras*-1-proto-oncogene by treatment with nitrosomethylurea (Sukumaran *et al.*, 1983) or $7\beta,8\alpha$-dihydroxy-7,8,9,10-tetrahydrobenzo[a]pyrene $9\alpha,10\alpha$-epoxide (*anti*BPDE) (Marshall *et al.*, 1984). In this case it is clear that only DNA was damaged but it is not yet certain that oncogene mutation is the only pathway for initiation by chemical carcinogens even if DNA is always the target.

2. TYPES OF INTERACTIONS WITH DNA

Two types of interactions of chemical carcinogens have been recognized: non-covalent and covalent. The former is a reversible reaction and, although this can have profound effects upon DNA structure effecting gene expression, it is not the subject of this review. The latter type of interaction may result either from the intrinsic reactivity of a chemical with DNA or by an activation process involving one or more enzyme or photochemical reactions. The distinction between these directly and indirectly acting carcinogens has led to a subclassification of initiating carcinogens as precarcinogens—those requiring activation, and ultimate carcinogens—those compounds capable of direct reaction with DNA. Metabolism of precarcinogens is therefore a critical aspect of the carcinogenic process and is reviewed elsewhere in this book (Dipple *et al.*, 1987).

In general, natural products and environmental contaminants will be expected to fall into the class of compounds which require some form of host metabolic or photochemical activation since, if they were sufficiently reactive to bind to DNA, they would be expected to react with other molecules in the environment before reaching the DNA of the target cell. Direct acting or ultimate carcinogens are, therefore, normally the product of the chemical or pharmaceutical industries or metabolites formed within an animal susceptible to a particular precarcinogen.

Factors governing which adducts might be formed, and to what extent, from a particular carcinogen and DNA are not fully understood. Several parameters have, however, been recognized. These include the reaction mechanism itself (S_{N^1} or S_{N^2}), Swain and Scott parameters (Swain and Scott, 1953), hardness of the nucleophile (Pearson and Songstad, 1967) and the steric constraints of the nucleophile, as well as the possibility of non-covalent transient interactions or activation by adjacent groups near the nucleophilic centers (Burfield *et al.*, 1981). The presence in cells of other proteins such as the P450 enzymes, which may impose metabolic specificity, histones or other specific proteins such as the *lac* repressor, resulting in steric or electronic constraints, which may alter the reaction of carcinogens with DNA are recognized. No evidence exists, however, for a direct enzymic transfer of a carcinogen to DNA. The types of DNA adducts formed *in vivo*, therefore, often match those formed *in vitro*. Exceptions to this generally occur because of the much higher concentrations of reagents used *in vitro*, resulting in higher levels of DNA modification, or the fact that homopolymers, or even the deoxyribonucleosides themselves, are used under unusual conditions of pH or the presence of non-aqueous solvents. Such reactions are of considerable importance in preparation of synthetic modified bases to aid in structural elucidation of adducts formed *in vivo* (see Section 3).

The strongest nucleophilic sites in DNA include the ring-nitrogens, particularly the N^7 of guanine, while the hydroxyl and amino groups are weaker. Thus, carcinogens which react predominantly by an S_{N^2} mechanism will, without other constraints, favor the strongest nucleophilic centers. Those reacting by an S_{N^1} mechanism are less selective and react with other sites such as the N^2 or O^6 of guanine residues. In some instances the dominant factor may be steric effects since different enantiomers can show major differences in reactivity. These effects, while not adequately understood to make definite predictions regarding the sites at which reaction may occur, have been extensively reviewed by Lawley (1984). With simple compounds calculations on sites of reactivity have been made (e.g. Kikuchi and Hopfinger, 1980).

3. DETECTION OF DNA ADDUCTS

A variety of approaches have been developed with various objectives in mind. Because of the very low levels of modification of DNA which occurs in biological systems, normally below one adduct per 10^5 bases, all methods require an intrinsically high sensitivity or some form of amplification of the damage to a measurable phenomenon. Some assays do not directly attempt to measure the adducts themselves but rather the consequence of such damage. For example, the Ames assay (Ames et al., 1973, 1975; McCann et al., 1975b; Yamasaki and Ames, 1977) depends upon the fact that strains of Salmonella were developed which could no longer grow in the absence of histidine. Damage to the DNA may induce a back mutation in a very few bacteria such that they can once again grow in plates lacking adequate histidine and therefore form easily visible colonies. Some bacterial strains were developed which contained error prone repair enzymes to further increase the chance of a back mutation being induced and thereby increase the sensitivity of the assay (McCann et al., 1975a). Indeed, some mutational events seem to depend entirely upon an error-prone repair mechanism for their expression (Ivanovic and Weinstein, 1980). Similar approaches have been made using mammalian systems in which the end point can be either mutation or transformation (for review see National Academy of Sciences, 1983).

Such approaches become difficult to evaluate when extrapolation to human exposures is needed. For these reasons, it would be desirable to measure the levels of DNA adducts formed in exposed human populations in attempts to correlate this with the risk to such groups and a number of such methods have been developed. In general, good correlations have been obtained between the ability of a carcinogen to act as an initiating carcinogen and its ability to bind to DNA (Ashurst et al., 1983; Brookes and Lawley, 1964; Wigley et al., 1979), especially within closely related groups of compounds. Metabolites of the procarcinogen which are more carcinogenic generally bind better to DNA (Borgen et al., 1973; Brookes, 1979; Kapitulnik et al., 1977) while those which are less carcinogenic do not (Cohen et al., 1980; Levin et al., 1980; Slaga et al., 1978; Kapitulnik et al., 1976). Exceptions have been noted (Goshman and Heidelberger, 1967; Phillips et al., 1979) but this is perhaps not surprising depending upon the linearity of dose response curves (Dunn, 1983; Lutz et al., 1978), if adducts vary with respect to time after exposure (Baird et al., 1983) or differ in their ability to induce mutations (Brookes and Osborne, 1982) and cells vary in their sensitivity to transformation with a particular carcinogen (Gehly and Heidelberger, 1982; Lo and Kakunaga, 1982; Tejwan et al., 1982). Correlations of a particular base are sometimes better than overall binding (DiGiovanni et al., 1979). Such considerations are important when extrapolating to human risk (Ehling et al., 1983).

3.1. DIRECT ANALYSIS OF DNA ADDUCTS

As mentioned above, the levels of modification *in vivo* are low. While high levels of modification of DNA can be obtained by chemical reactions of the ultimate carcinogens with DNA *in vitro* to allow direct isolation and chemical identification of adducts, this has not been generally possible with *in vivo* samples. 3-Methyladenine, which is not a normal urinary product, has been proposed as an index of exposure to methylating agents (Shaikh et al., 1980) and the highly fluorescent aflatoxin adducts have also been detected in human urine (Autrup et al., 1983).

Most methods involve some purification of the DNA prior to adduct analysis in order that other macromolecules to which the carcinogen may have become bound, or residual free carcinogen and its metabolites, do not interfere with the assays. The exact methods used will depend upon the source of the DNA but will often involve a proteolytic digestion in order to simplify the subsequent solvent extractions with phenol or phenol chloroform mixtures which separate the proteins and nucleic acids. Final ethanol precipitation not only removes the phenol and any residual carcinogen but also concentrates the DNA. If appropriate, a ribonuclease treatment of the DNA may also be undertaken together with further purification by cesium chloride centrifugation (Harris et al., 1976) or hydroxyapatite chromatography (Beland et al., 1979; Shoyab, 1979).

3.1.1. *Radio Labelling of the Carcinogen*

Most work has involved the use of ^{3}H and ^{14}C although, with a few compounds, other radioisotopes are potentially available. The advantage to the use of ^{3}H as the isotope is that it is relatively easily introduced into the carcinogens and can be obtained at higher specific activities than ^{14}C because of the former's shorter half-life. However, it is also very susceptible to loss during metabolism or by exchange reactions. This can result in the formation of tritiated water which in turn may become incorporated into non-exchangeable positions in newly synthesized DNA. Although now at very low specific activity, this tritium can contribute significantly to the overall level of apparent binding (Wigley et al., 1979). The use of tritium labelled compounds allow detection down to a level of a few fentomoles or, depending upon the amount of DNA which is available, about one adduct per 10^9 bases. This is about 200 times more sensitive than can be obtained with ^{14}C analogs although the latter generally circumvents the metabolic incorporation of problems described above. Adduct identification is made by chromatographic comparison with synthetic standards. Such comparisons should be carefully undertaken by analysis on different columns or with derivatization to ensure chromatographic identity since many of these adducts are difficult to separate. With the improved resolution of HPLC systems isotopic separations may become significant (Jeffrey and Fu, 1977).

3.1.2. Post Labelling Technique of DNA

Post labelling of DNA is a more recently introduced method (Gupta et al., 1982; Phillips et al., 1984; Randerath et al., 1984; Reddy et al., 1984) which is particularly suited to the evaluation of complex mixtures or human exposures where the administration of labelled carcinogen is not possible. DNA, isolated from the source of interest, is digested enzymatically to 3'-deoxyribonucleoside phosphates which are then phosphorylated with [γ-^{32}P]ATP in their 5' positions. ^{32}P can be obtained at about 100 times higher specific activity than ^{3}H and can be counted with higher efficiency because the β particle which it emits upon decay is of higher energy. This allows the method to be intrinsically much more sensitive. However, the major difficulty is the separation of the overwhelming excess of radioactivity associated with residual ATP, inorganic phosphate, the normal nucleotides which also become phosphorylated, and any radiochemical decomposition products which might have been formed. Clearly, depending upon the level of modification of the DNA by the carcinogen, which may be in the range of one adduct per 10^{6-9} bases, only about that fraction of the disintegrations will be those of interest. This assumes that the kinase will work equally on modified and unmodified 3'-phosphates which appears to be only partially true.

An elaborate separation technique was therefore developed, in which the mixture of products is placed in the center of a PEI cellulose tlc plate and the plate washed first in two directions, each time cutting away the edge of the plate into which the bulk of the unwanted radioactivity moves. The final two directions of chromatography allows for separation of the carcinogen modified bases. This approach works quite well providing the carcinogen is a hydrophobic structure such as the polycyclic aromatic hydrocarbons but is much less sensitive when simple alkylating agents are used.

Alternative approaches involve the separation of the unmodified bases prior to the kinase reaction (Gupta, 1985) or subsequent pre-separation on reverse phase tlc plates of the modified and unmodified bases before final transfer and separation of the adducts by PEI cellulose tlc. This leads to increased sensitivity but also effort per assay and a loss in accuracy in quantitation. In general, for PAH detection, levels of one adduct per 10^7 can be obtained although Dr Randerath has been able to obtain positive results down to one adduct per 10^{10} bases (Reddy and Randwarth, 1986). A recent report demonstrated that adducts could be isolated from placental DNA of women who smoked but was absent from those who did not (Everson et al., 1986).

3.2. Immunological Techniques, DNA Adduct Detection

This approach relies upon the preparation of antisera which recognize specifically the modified bases in DNA. Such antibodies were prepared initially against methylated bases

(Muller and Rajewsky, 1978, 1980; Rajewsky *et al.*, 1980; Wild *et al.*, 1983), but are becoming increasingly available for more complex modifications (Groopman *et al.*, 1982; Hertzog *et al.*, 1982; Hsu *et al.*, 1981; Leng *et al.*, 1978; Perera *et al.*, 1982; Poirier, 1981; Poirier *et al.*, 1982; Santella *et al.*, 1984; Strickland and Boyle, 1981; Van der Laken *et al.*, 1982; Wallin *et al.*, 1984). (For reviews see Muller *et al.*, 1982; Rajewsky *et al.*, 1980.) Many variations exist on the method by which the antisera may be used but one simple approach is described below.

In the enzyme-linked immunosorbent assay (ELISA) DNA to be tested is coated onto the surface of a suitable plate. Antisera, specific for the adduct of interest, is added and after incubation any non-complexed antibody is removed. If this antisera was prepared, for example, in a rabbit, a second anti-rabbit antibody which has been previously linked to a suitable marker enzyme, for example alkaline phosphatase, is now added. This will bind to the plate only if there was an interaction between the modified DNA and the primary antibody. The plate is again washed and the enzyme's substrate, typically *p*-nitrophenyl-phosphate, is added. As hydrolysis proceeds, which will be proportional to the extent of modification originally present on the DNA, the colour formed from the released nitrophenyl anion can be measured. By using specifically designed plates, 96 samples can be assayed at one time and the plates automatically washed and the color development measured.

This assay is clearly attractive because of its simplicity of automation once established and should be applicable equally to a wide variety of DNA-adducts. The sensitivity of the assay can be further increased by using competitive versions of the assay, substrates which yield fluorescent products (Poirer *et al.*, 1985), by biotin-avidin-enzyme complexes (Shamauddin and Harris, 1983; Fioles *et al.*, 1985), following the hydrolysis of $[^{32}P]ATP$ (Hsu *et al.*, 1981) or other modifications of the basic technique (Nehls *et al.*, 1984). The main restrictions relate to preparation of DNA adequately modified to act as immunogen and the relative affinity constants of those antibodies produced for very low levels of modified bases compared to the normal sequences of DNA present in large excess. The former may be solved by further development of *in vitro* immunization of spleen cells. The latter has both probable and theoretical limits which we may already be approaching. Current limits of detection are about 0.1 fmole/μg DNA or one adduct/10^8 bases.

The antibodies prepared to date have shown varying specificity towards different adducts. For example, it may be possible to distinguish between different alkyl substitutions at a particular site (Rajewsky *et al.*, 1980) or to use the antibodies to detect adducts released into the urine by repair enzymes or spontaneous depurination. Such specific probes also allow for the location of adducts within cells and tissues or along the DNA chain (Slor *et al.*, 1981). Image enhanced immunofluorescence microscopic approaches have allowed detection of as few as 700 O^6-ethylguanine residues per diploid genome (Adamkiewicz *et al.*, 1983). Although such experiments are still in a very early stage, these antibodies clearly provide a tool for investigating not only the presence but also the location at which a particular adduct exists (Kurth and Bustin, 1985; Paules *et al.*, 1985; Muysken-Schoen *et al.*, 1985).

3.3. Fluorescence Techniques, DNA Adduct Detection

Guanosine shows significant fluorescence and this often increases with alkylation. Similarly some highly fluorescent adenosine adducts have been prepared. However, the short wavelengths at which they absorb generally precludes the use of this technique to study such adducts unless the DNA is digested and the adducts first separated by HPLC (Hemminki, 1980; Herron and Shank, 1979; Lindamood *et al.*, 1982). By contrast, studies on the polycyclic aromatic hydrocarbons and aflatoxins have made great use of fluorescence since these compounds absorb beyond 310 nm, where DNA gives little interference. In addition, the quantum yields of adducts may be increased by cooling to liquid nitrogen temperatures (Ivanovic *et al.*, 1976) and the sensitivity of the instruments increased by photon counting (Daudel *et al.*, 1975) or laser excitation of the sample (Dovichi *et al.*, 1983) or by the simultaneous scanning of both the excitation and emission monochomoters with a fixed difference in their wavelengths appropriate to the particular carcinogen under study (Vahakangas et al., 1985). The advantage of this approach is that it requires no prior knowledge or preparation of the DNA adduct. Success depends, however, considerably upon the type of metabolic activation which occurs. For example, in the case of DMBA, several groups have reported fluorescence spectra for DNA isolated from a variety of sources after exposure to the hydrocarbon (Ivanovic et al., 1978; Moschel et al., 1983; Vigny et al., 1981). All describe metabolic activation having occurred in the 1-4 positions. However, dimethylanthracene absorbs strongly at long wavelengths to give a very characteristic spectrum. Had metabolism occurred in the 8-11 positions, the residual phenanthrene chromophore would have been much harder to detect. Only after more thorough HPLC purification of the DNA adducts was it possible to be certain that the majority of adduct resulted from 'Bay Region' activation (Cooper et al., 1980c; Moschel et al., 1983; Vigny et al., 1981).

Another limitation of this approach is that, because of the relatively broad excitation and emission spectra, it is severely limited when attempts are made to analyze exposures to complex mixtures of PAHs. In the case of PAHs themselves, significant improvement in the quality of spectra have been obtained by matrix isolation techniques (Stroupe et al., 1977), working at liquid helium temperatures to remove the effects of thermal broadening of the spectral lines. Under such conditions, it has been possible to analyze directly quite complex mixtures of hydrocarbons. However, the materials generally used for the matrix, such as nitrogen or hexane, are quite unsuited to PAH-modified DNA samples. Mixtures of water, ethylene glycol and ethanol do, however, form suitable glasses in which DNA samples can be embedded. In such glasses, the bound hydrocarbon moieties will be in any one of a large variety of micro-environments and, while the individual molecules may themselves have sharp excitation and emission spectra, the result of these overlapping spectra is again a broad fluorescent spectrum. These subpopulations, or isochromats, may be individually excited with very narrow band pass light such as produced by a laser. This approach, called 'fluorescence line narrowing', has been applied to a study of B[a]P bound to DNA (Heisig et al., 1984). The spectra are much more characteristic than the broad band spectra of equivalent samples (Ivanovic et al., 1976). Subtle differences even allow distinction between 7,8,9,10-tetrahydroB[a]P and its corresponding tetraol. At the resolution so far obtained, distinction between the tetraol, BPDE modified DNA and the BPDE-deoxyguanosine moity isolated from such modified DNA could not be made. The emission spectra are highly characteristic and complex mixtures containing up to nine PAH metabolites, have recently been resolved (Sanders et al., 1986). In addition, it has been possible to detect the B[a]PDE-deoxyguanosine moiety present in DNA isolated from mouse fibroblast 10T1/2 cells exposed for 24 hr to 1 μg of B[a]P where the level of adduct was one per 10^6 bases (Brown et al., 1979) (unpublished results). This approach provides an opportunity to begin to investigate directly the types of DNA adducts which are formed when cells are exposed to complex mixtures of PAHs.

An alternative approach to measure covalent binding of B[a]P to mouse skin has been to release from the isolated DNA the hydrocarbon-DNA adducts by acid hydrolysis (Rahn et al., 1980, 1982; Shugart et al., 1983). Since, in the case of BPDE-DNA adducts, they are acid labile and the tetraols produced are not only more fluorescent than the adducts themselves but also more easily extracted from the large excess of unmodified bases, this provides a convenient approach. Detection levels of ~ 1 adduct/10^5 bases using 40 μg of DNA have been obtained (Rahn et al., 1980). The method does, however, require the certainty that hydrolysis of the adducts will occur and that the products will be stable or, if degradation occurs, that they can still be recognized.

In the case of aflatoxin, the major N^7 adduct undergoes, in addition to stabilization by ring opening, depurination (Section 4.6.1) (Essigmann et al., 1977). It was this adduct which, after extensive purification, was tentatively identified in the human urine of individuals who had high exposures to aflatoxins (Autrup et al., 1983).

3.4. OTHER INDIRECT TECHNIQUES

These include measurements of unscheduled DNA synthesis (Probst et al., 1981; Williams, 1977), or single or double strand breaks induced by chemicals by sedimentation, alkaline elution (Kohn et al., 1976) or other techniques to determine the molecular weight of the modified DNA such as base sequencing methods (Haseltine et al., 1980).

4. CLASSES OF CHEMICAL CARCINOGENS

Any classification scheme has limitations in that certain carcinogens may appropriately belong to more than one group. In addition, perhaps the most useful form of classification, based on the mechanism by which the resulting modified DNA exerts an effect upon the cell, cannot be made because of our lack of understanding of the biochemical events underlying cell transformation. The approach used here is to divide them based roughly upon their chemical structure.

4.1. SIMPLE ALKYLATING AGENTS

These compounds are classified based upon the alkyl group, which is normally methyl or ethyl, that they transfer to DNA rather than an overall simple structure. They comprise of a larger group and include diazoalkanes (Farmer et al., 1973), alkylmethane sulfonates, dialkylsulfates, alkyliodides, trialkylphosphates, hydrazines, certain ring strained lactones and nitroso derivatives, as well as more complex molecules such as S-adenosylethionine.

An analysis of the DNA from these simple alkylating agents is aided, as described in Section 3, by the availability of good reference standards and chromatographic techniques for the separation of the various alkylated derivatives (Beranek et al., 1980; Herron and Shank, 1979) as well as antibodies (Rajewsky et al., 1980) which will recognize specific alkylated bases. Thus, although the amount of work required is considerable, the potential exists to estimate essentially all possible alkylation sites. Such a comprehensive analysis is, however, not often undertaken. In general, at equimolar doses, methylating agents are about 10–20 times more effective than their corresponding ethyl derivatives. The ethyl and higher alkyl derivatives react with more S_{N^1} character producing higher O^6- to N^7-guanine ratios. [For example, dimethyl- (O'Connor et al., 1973) and diethyl- (Pegg and Balog, 1979; Scherer et al., 1977) nitrosamines (Singer and Grunberger, 1983).] In the case of N-nitroso-N-benzylurea model reactions have produced ratios greater than unity and the reactions have sufficient S_{N^1} character that modification of the exocyclic amino groups of guanine and adenine have been reported (Moschel et al., 1979, 1980).

Dimethyl (Lawley and Warren, 1976; Swann and Magee, 1968), diethyl (Sun and Singer, 1975), sulfates and methyl (Beranek et al., 1980; Frei and Lawley, 1976; Lawley and Brookes, 1963; Swann and Magee, 1968) and ethyl (Beranek et al., 1980; Frei et al., 1978; Sun and Singer, 1975) methanesulfonate, azaserine (Zurlo et al., 1982), propanesultone (Hemminki, 1983) and ethyleneimine (Hemminki, 1984) all react primarily with the N^7 of guanine (>65%), the next most predominant site being the N^3 position of adenosine (4–15%). It was this extensive reaction with the N^7 of guanine and subsequent instability of the adduct which led into the first methods for DNA sequencing (Maxam and Gilbert, 1977).

These compounds are all relatively weak carcinogens. Variation between data reported for in vivo and in vitro reactions for direct acting carcinogens reflect a combination of the effects of carcinogen concentration, chromatin structure and the action of repair enzymes. In most instances in vivo and in vitro results are similar, e.g. dimethyl sulfate, methyl methanesulfonate or methyl nitrosourea. In some instances the in vivo results have been found to be highly dose and time dependent (Pegg and Balog, 1979; Swenberg et al., 1984).

The dialkyl nitrosamines and alkylnitrosoureas show similar patterns of DNA adduct formation. The main differences between the two series of alkylating agents result from the fact that the former require metabolic conversion to the latter and tissues lacking appropriate enzymic abilities will not therefore be targets for these compounds. Of the bases modified, relatively high levels of O^6 occurs although in some cases as much as 70% of the reaction is on the phosphate groups (Sun and Singer, 1975). Repair rates for adducts are not always equal (O'Connor et al., 1973; Scherer et al., 1977; Swenberg et al., 1984) and the brain is generally poor in its ability to repair O^6 adducts (Kleihues and Rajewsky, 1984).

An additional observation has been made with the higher homologue of alkyl nitrosamines (Kruger and Bertram, 1973a,b). When using 1-[^{14}C]di-n-propyl- or β-hydroxypropyl-propyl-nitrosamines, but not when labelled in the 2-positions, 7-[^{14}C]methylguanine was formed from these compounds. Since the corresponding methyl analogues are more potent alkylating agents any metabolic oxidation of higher alkyl derivatives to methylnitrosamines could account for the preferential identification of 7-methylguanine. The hydroxy derivatives gave about 50 times the amount of 7-methylguanine suggesting that the β-oxidation was the rate limiting step in the reactions. More recent experiments with N-(2-oxopropyl)-N-nitrosourea, which reacts with calf thymus DNA to yield O^6- and 7-methylguanine and 3-methyladenine, suggest that this compound decomposes to yield 2-oxopropyldiazoate (Leung and Archer, 1984). Guanine sites in DNA depurinate very slowly and have a half-life of $\sim 1.5 \times 10^7$ hr. By contrast 7-methylguanine has, under physiological conditions, a half-life of approximately 155 hr which may be further reduced by repair enzymes. Consequently persistent adducts are usually, therefore, O^6 or imidazole ring opened derivatives at the 8,9 bond (Beranek et al., 1983; Chetsanga et al., 1982; Chetsanga and Makaroff, 1982; Kadlubar et al., 1984; Yagi et al., 1984). Some of the chemistry of their rearrangement products with respect to their ribose moiety have recently been investigated (Tomasz et al., 1985).

β-Propiolactone is highly reactive with water because of ring strain and, although carcinogenic, is only weakly so. The major site of alkylation in DNA is the N^7 of guanine both in vivo (Colburn and Boutwell, 1966) and in vitro (Roberts and Warwick, 1963). The N^1 of adenine (Chen et al., 1981; Mate et al., 1977) and the 3 position of cytosine (Segal et al., 1981) also yield carboxethyl derivatives. In addition, fluorescent cyclic adducts have been identified (Chen et al., 1981) analogous to those formed with chloroacetaldehyde.

Simple aldehydes, such as formaldehyde, which has been shown to be carcinogenic at high doses, form Schiff base derivatives with nucleic acid and proteins. These are generally reversible reactions and may not be critical in the carcinogenicity of these compounds, although cross links have been observed (Chaw et al., 1980).

4.1.1. Bifunctional Alkylating Agents

Several compounds of this class are of chemotherapeutic interest, e.g. bis(chloroethyl)nitrosourea (BCNU), bis(fluoroethyl)nitrosourea (BFNU) and N-(2-chloroethyl)-N'-cyclohexyl-N-nitrosourea (CCNU). Interstrand cross links appear to be important with respect to this therapeutic value and the major cross linked adduct has been tentatively identified (Tong et al., 1982). A scheme (Gombar et al., 1980) has been proposed for the decomposition of BCNU during its reaction with DNA to yield, as the major product, 7-(β-hydroxyethyl)guanine, together with 7-(β-chloroethyl)guanine. Minor products include O^6-(β-hydroxyethyl)guanine, 7-aminoethylguanine, 3-hydroxyethylcytidine, two cyclic adducts 3,N^4-ethenocytidine and 1,N^6-ethenoadenine, as well as an example of N^7-guanine cross link: 1,2-(diguan-7-yl)ethane (Gombar et al., 1980; Tong et al., 1982; Tong and Ludlum, 1981) and C-G cross links. The hydroxy ethyl derivatives do not arise from hydrolysis of the corresponding chloro compound since that reaction is too slow and an oxadiazoline intermediate has been proposed. The production of the aminoethylguanine is unexpected and it was suggested (Gombar et al., 1980) that it may have arisen from the reaction of chloroethylamine, formed from BCNU, with DNA. Similar adducts have been reported by BFNU (Tong and Ludlum, 1978).

Bifunctional derivatives such as glyoxal (Broude and Budowsky, 1971; Czarnik and Leonard, 1980), malondialdehyde (Basu and Marnett, 1984; Moschel and Leonard, 1976)

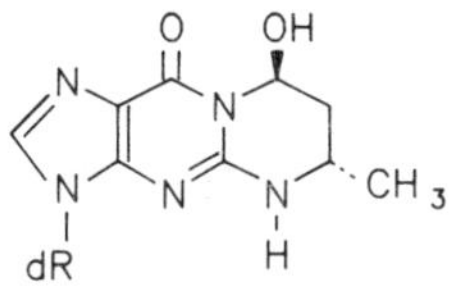

FIG. 1. Structure of deoxyguanosine adduct formed by the reaction of α-acetoxy-N-nitrosopyrrolidine or 4-(carbethoxynitrosamino)butanal in the presence of esterases or directly with crotonaldehyde. Two products were formed in about equal amounts which were enantiomeric about the chiral centers formed during guanine modification. Interestingly, the hydroxyl and methyl groups were *trans* to each other and the *cis* adducts were not detected. dR indicates the deoxyribose moiety.

chloroacetaldehyde (Section 4.1.2), etc. (Shapiro and Hachmann, 1966) can form cyclic structures which are, depending upon the pH, more stable than normal Schiff base complexes. Studies have shown malondialdehyde lacks carcinogenic activity on mouse skin (Fischer *et al.*, 1983).

Crotonaldehyde (butenal) is also capable of forming $1,N^2$ cyclic adducts with guanine (Fig.1) (Chung and Hecht, 1983; Chung *et al.*, 1984) equivalent to those produced by the reaction of α-acetoxy-N-nitrosopyrrolidine with DNA.

Acrylonitrile undergoes a very slow chemical reaction with DNA and a number of adducts have been isolated (Solomon *et al.*, 1984). The role of these adducts in the carcinogenicity of acrylonitrile has, however, yet to be determined.

Ethyl carbamate (urethan) has been suggested to undergo reactions with DNA via first conversion to vinyl carbamate, which is more carcinogenic than ethylcarbamate and then to an epoxide. A 7-(2-oxoethyl)-guanine adduct was identified (Scherer *et al.*, 1980) in hepatic DNA of the rat. Two cyclic adducts, $1,N^6$-ethenoadenosine and $3,N^4$-ethenocytidine, have also been isolated from mouse liver RNA (Ribovich *et al.*, 1982).

4.1.2. *Simple Alkylating Agents Requiring Metabolic Activation*

The nitrosamines have already been discussed above in relation to the nitrosoureas. A large number of hydrazine derivatives occur both as synthetic and natural products including sources of human food. This has been reviewed recently (Toth, 1980). These compounds would seem, therefore, to deserve further study.

Hydrazine, which does not have any carbon atoms, has been shown to stimulate DNA methylation, giving both N^7- and O^6-guanine derivatives (Becker *et al.*, 1981). Several other hepatotoxic compounds have also been shown to cause methylation of DNA in rats, which may occur via S-adenosyl-methionine (Barrows and Shank, 1981).

1,2-Dimethyl hydrazine, a potent colon carcinogen in rodents, requires metabolic activation which probably results in a methyldiazonium ion. High levels of N^7-guanine alkylation have been observed together with about 5% O^6-guanine adducts (Beranek *et al.*, 1983; Likhachev *et al.*, 1977; Rogers and Pegg, 1977). Strand scission can result from exposure to hydrazine derivatives (Augusto *et al.*, 1984).

A large number of triazene derivatives have been investigated as cancer chemotherapeutic agents. Studies with 1-phenyl-3,3-dimethyltriazine showed that in microsomal system it could be oxidized, via its monomethyl derivative, to products, possibly methyldiazohydroxide, which are capable of methylating DNA (Kruger *et al.*, 1971; Preussmann and Hodenberg, 1969), to yield N^7-methylguanine. Only minor amounts of O^6-methylguanine were isolated (Margison *et al.*, 1979), although they were repaired more slowly in the rats' brain, one of the principal target organs, and accumulate with time (Cooper *et al.*, 1978).

Ethionine, a hepato carcinogen, forms 7-ethylguanine with liver DNA (Swann *et al.*, 1971). It is metabolized to S-adenosylethionine although it is not certain whether this metabolite is responsible for DNA alkylation. S-vinylhomocysteine binds more strongly to DNA (Leopold *et al.*, 1982) but it is unclear that it is an intermediate in the activation of ethionine. The normal analog, S-adenosylmethionine has been shown to react *in vitro* with DNA to yield N^7-methylguanine as the major product (Barrows and Magee, 1982).

Vinyl chloride produces angiosarcomas of the liver in humans. Since it is a relatively rare tumor its epidemiologic association with vinyl chloride exposure was correspondingly easy to demonstrate. Subsequently it was shown in addition to produce tumors at other sites. Vinyl chloride is metabolized to its corresponding epoxide which is a highly reactive compound having a half-life of less than a minute under physiological conditions yielding chloracetaldehyde. Both the epoxide and aldehyde react with DNA, however, some differences in the products have been observed which suggest that the oxide rather than the aldehyde may be critical *in vivo*. The oxide reacts with deoxyguanine to give 7-(2-oxoethyl)-deoxyguanosine (Scherer *et al.*, 1981). The corresponding base has been isolated from the DNA of animals given vinyl chloride (Laib *et al.*, 1981; Osterman-Golkar *et al.*, 1977). This compound is interesting based upon its possible equilibrium existence with the cyclic hemiacetal O^6, 7-(1'-hydroxyetheno)guanine. In contrast, chloracetaldehyde has long been known to react with the amino groups on guanine, adenosine or cytidine to give a Schiff base which can then rearrange with the elimination of HCl to hydroxyetheno derivatives (Kusmierek and Singer, 1982). These in turn undergo an acid catalysed dehydration, with a half-life of about 1.4 hr for the adenosine derivatives under physiological conditions, to the corresponding $1,N^6$-, $3,N^4$-, and $1,N^2$- or $N^2,3$- adenine, cytidine or guanine etheno derivatives respectively. These adducts have also been observed by some investigators *in vivo* (Green and Hathaway, 1978). The possible reasons for variations in the *in vivo* results has been discussed (Laib *et al.*, 1981; Kusmierek and Singer, 1982). Reaction of chloracetaldehyde with DNA produces only small amounts of $N^2,3$-ethenoguanine (Oesch and Doerjer, 1982).

Many polyhalogenated compounds have been shown to be carcinogenic. In general, as the extent of halogenation increases, the extent to which they can be metabolized decreases. The extremely high toxicity of tetrachlorodibenzodioxin, for example, is unrelated to its ability to bind to DNA which occurs at the limits of current technology to detect. By contrast an unusual activation of 1,2-dibromoethane through a reduced glutathione intermediate has recently been reported (Inskeep and Gengerich, 1984). In general, such GSH conjugates would be considered to be on the pathway of detoxification but, because of the vicinyl bromine, addition reactions occur to yield S-[2-(N^7guanyl)ethyl] glutathione adducts. The generality of this type of reaction has not yet been fully explored.

4.2. Simple Oxides

4.2.1. *Alkene Oxides*

Many are genotoxic (Ehrenberg and Hussain, 1981). The simplest homologue is ethylene oxide. This compound is both formed metabolically from ethylene and is widely used industrially as a sterilant. It has been associated with increased leukemia incidence and increases in sister chromatid exchanges (Stolley *et al.*, 1984). Reaction of both ethylene oxide (Segerback, 1983) and propylene oxides (Lawley and Jarman, 1972) with DNA have been studied, the major products being the hydroxyalkyl N^7-guanine derivatives (Ehrenberg and Hussain, 1981; Lawley and Jarman, 1972).

Other simple epoxides which have been studied include butadiene diepoxide, 2,3-epoxybutane and styrene oxide (Hemminki and Hesso, 1984; Hemminki *et al.*, 1980b; Hemminki and Vainio, 1984). Although the N^7-guanine adduct has often been identified by reactions *in vitro* with deoxyribonucleosides the products which are formed with DNA *in vivo* have been less well documented. Introduction of two epoxide functions as in butadiene diepoxide, produces marked differences in reactivity based upon which isomer is used. The L isomer reacts best with DNA to form cross links, the D is slightly weaker, while the *meso* is strongly inhibited (Matagne *et al.*, 1969; Verly *et al.*, 1971). More recent studies (Castleman *et al.*, 1983) have shown that DNA in the Z form can react to form cross linked derivatives which stabilize the Z conformations. Conversely, if the reaction is carried out in the B form monoadducts predominate which, if the DNA is then converted to the Z form, then form cross links (Kang *et al.*, 1985). Interest in butadiene may be renewed based upon recent studies showing its carcinogenicity at low doses (Gratton and Wise, 1985).

1,2:5,6-Dianhydrogalactitol reacts as expected preferentially to yield N^7-guanine derivatives of which about 30% are cross linked (Institoris, 1981; Institoris and Tamas, 1980).

Trichloropropylene oxide has been studied extensively as an inhibitor of epoxide hydratase. In addition, it has been found to react with several deoxyribonucleosides (Hemminki *et al.*, 1980).

4.3. Complex Epoxides

Glycidaldehyde is bifunctional, in terms of its reaction with DNA yielding a cyclic adduct by reaction at the N^2 and N^1 position of guanine. The product, which is highly fluorescent, has been characterized (Goldschmidt *et al.*, 1978).

4.4. Polycyclic Aromatic Hydrocarbons

This series of compounds represents an extensively studied group of compounds from which considerable information has been derived over the past few years regarding factors which influence their carcinogenic potential. One major determining factor, reviewed elsewhere (Dipple *et al.*, 1985), is their metabolic activation. Many of these compounds will react strongly with DNA in aqueous solution to form non-covalent complexes. However, these complexes appear biologically inert with respect to mutation or transformation in systems which are unable to metabolically activate these hydrocarbons. Studies both with crude and purified enzymes have demonstrated that many of the carcinogenic PAHs require several enzymes for their metabolic activation.

In general, Brookes and Lawley (1964) found a good correlation between the carcinogenicity of PAHs and their ability to bind covalently to DNA or RNA but not to proteins. Evidence has been obtained which is suggestive that individuals with a presumed disposition to lung cancer may have an increased ability to bind BP to the DNA of their monocytes (Rudiger *et al.*, 1985).

4.4.1. *Benzo[a]pyrene*

The most extensively studied PAH is benzo[a]pyrene. This probably arises for a number of reasons. It was isolated and its structure determined many years ago by Cook and Kennaway. Its strong u.v. absorption and fluorescence spectra have made it simple to monitor and, at least in comparison with some other PAH derivatives, it proved relatively easy chemically to synthesize many of its metabolites, which has been critical in the study of DNA adduct formation. Although the compound is clearly a potent animal carcinogen and is formed ubiquitously by incomplete combustion processes, it is not clear that it is a critical compound in the etiology of any human cancer. As studies have progressed with other PAHs, however, it has been found to present a good model for the general process of PAH carcinogenesis.

Suggestive evidence for multiple steps in the activation of BP to compounds which could bind to DNA come from studies (Borgen *et al.*, 1973) which showed that, comparing BP with its metabolites, the highest level of DNA binding was obtained with *trans*-7,8-dihydro-7,8-dihydroxy-BP. This metabolite is formed by the action of the P450 enzyme complex on BP via an intermediary 7,8-oxide which undergoes enzymic hydrolysis to the dihydrodiol. Sims and co-workers (Sims *et al.*, 1974) proposed that the dihydrodiol was again oxidized by the P450 complex to yield BPDE. This metabolite is highly unstable and, as with many arene oxides and diol epoxides, has not been isolated directly although its non-enzymic (or enzymic) hydrolysis products, tetraols, have.

There are four possible optical or stereochemical isomers of this compound, all of which have been prepared chemically (Fu and Harvey, 1977; Yagi *et al.*, 1977). Using these compounds it has become clear that, both their metabolic formation and their reaction with DNA is highly stereoselective. From a combination of studies on: (1) the direct carcinogenicity of these diol epoxides; (2) their reaction *in vitro* with DNA; and (3) their reaction with DNA *in vivo*, either directly or as a result of metabolic activation of BP, the most carcinogenic isomer is also the one which is predominantly formed metabolically.

The major DNA adduct isolated from the metabolic activation of BP or direct reaction of BPDE, *in vivo* or *in vitro*, and most other PAH diol epoxides so far analyzed, result from attack of the 10 position of the oxide, or its corresponding position in other diol epoxides, at the N^2 position of guanine (Fig. 2) (Ashurst *et al.*, 1983; Baird and Diamond, 1977; Baird *et al.*, 1983; Brown *et al.*, 1979; Daniel *et al.*, 1983; Jeffrey *et al.*, 1976c, 1977; Koreeda *et al.*, 1978; Meehan and Straub, 1979; Meehan *et al.*, 1976a,b). Although other homopolymers, for example polydA or dC, react well *in vitro* they normally comprise of even less than 10% of the adducts formed *in vivo* (Brown *et al.*, 1979; Jeffrey *et al.*, 1977; Weinstein *et al.*, 1976) or when BPDE is reacted *in vitro* (Meehan *et al.*, 1977; Straub *et al.*, 1977) with DNA. This further supports the ideas of critical stereochemical interactions between BPDE and DNA. Other sites of modification such as the N^7 of guanine (King *et al.*, 1979; Osborne *et al.*, 1978, 1981; Sage and Haseltine, 1984) are, as would be predicted, unstable and have not been demonstrated *in vivo*. In addition, adenosine (Jeffrey *et al.*, 1979; Sage and Haseltine, 1984; Straub *et al.*, 1977) and cytosine adducts have been described (Jennette *et al.*, 1977; Sage and Haseltine, 1984; Straub *et al.*, 1977). Both the N^7 and O^6 adducts were derived from the (−) BPDE enantiomer, the minor metabolite (Osborne *et al.*, 1981). Phosphate ester formation has been reported although it appears to be very minor (Gamper *et al.*, 1977, 1980; Koreeda *et al.*, 1976).

Direct chemical analysis of the major DNA adduct formed from BPDEI and DNA showed that the N^2 amino group of guanine reacted with the 10 position of the original oxide and that it is the 7-R enantiomer which reacts preferentially (Meehan and Straub, 1979; Nakanishi *et al.*, 1977). Based on the high stability of the carbonium ion formed at the 10 position (Section 4.4.1.1) a mixture of *cis* and *trans* adducts would be expected. This is indeed what was found with homo polymers of guanine (Jeffrey *et al.*, 1976c). However, in the case of DNA the *trans* isomer is formed to greater than 90% of the adducts. This probably reflects stereochemical constraints on the reaction of the carbonium ion with DNA in competition with its reaction water rather than a mechanistic change. Evidence has been presented that BPDE can intercalate with DNA prior to its hydrolysis (Geacintov *et al.*, 1980, 1981, 1984) and models have been calculated for potential complexes (Klopman *et al.*, 1979; Poulos *et al.*, 1982; Subbiah *et al.*, 1983) and reactivity (Yeh *et al.*, 1978). No direct evidence exists, however, to indicate an essential role of such complexes in the reaction with DNA.

The role of the adducts in the carcinogenic process itself is not clear. Studies on the conformation of the major BPDE-DNA adduct show it lies in the minor groove of the DNA helix (Geacintov *et al.*, 1981). Such a model would require only slight distortion of B form of DNA to accommodate the hydrocarbon (Jeffrey *et al.*, 1980; Taylor *et al.*, 1983; Miller *et al.*, 1985). Thus, in contrast to the results obtained with AAF (Section 4.5.1) only slight effects were observed with respect to sensitivity to single strand specific agents such as S_1 nucleases, or reaction with formaldehyde, and the modification had little effect on the

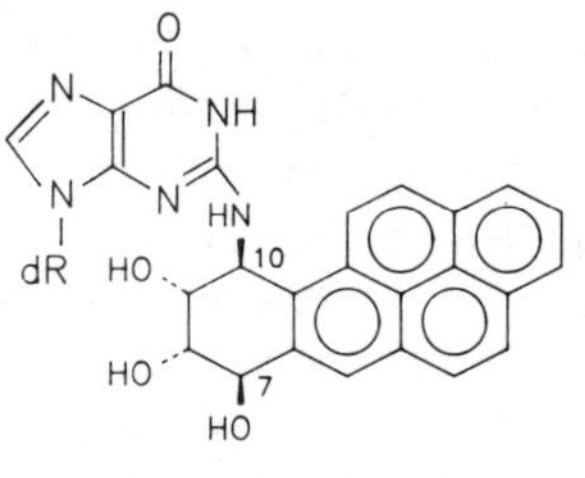

FIG. 2. Structure of the major adduct formed by the reaction of racemic *anti*-BPDE with DNA. The corresponding adduct formed from the enantiomeric diol epoxide is a minor product in DNA although it is formed during reactions with polydeoxyguanoside. dR indicates the deoxyguanosine moiety.

thermal melting of DNA samples (Jeffrey et al., 1980). The consequences of such a conformation may be that the normal repair processes for DNA damage may have greater difficulty in detecting such lesions which may, therefore, be abnormally persistent (Dipple and Schultz, 1979; Feldman et al., 1980). The repair of this type of DNA damage, especially prior to cell division, has been shown dramatically to reduce the potency of toxic effects and has been recently reviewed (Maher and McCormick, 1987). Exceptions to this general situation were noted in some bacterial strains where the SOS error prone repair system appeared critical for a biological effect (Ivanovic and Weinstein, 1980). This adduct inhibits polymerase 1 (Moore and Strauss, 1979) and induces point mutations (King and Brookes, 1984). Other models for the conformation of the BPDE residue in which it is more intercalated into the DNA helix have been proposed (Drinkwater et al., 1978; Hogan et al., 1981; MacLeod and Tang, 1985). Additional work will be required, however, before an unambiguous and detailed answer can be presented.

4.4.1.1. *Bay region theory.*

Prior to the elucidation of the structure of BP-DNA adducts K-region oxides had been favored as candidates for the reactive metabolites. This was reasonable from the electronic point of view, in that metabolism might well have been expected to occur at these positions, as indeed it does, to produce K-region arene oxides. However, it appears that many of these are good substrates for hydrolysis by the epoxide hydratases (Oesch et al., 1978) and react too slowly with DNA. The diol epoxide, despite the fact that two additional enzyme steps are required for their formation, appear to be relatively poor substrates for epoxide hydratase although not to dehydrogenases (Glatt et al., 1982) or GSH transferases (Glatt et al., 1983; Hesse et al., 1982). In addition, they react well by an S_{N1} mechanism having a well delocalized carbonium ion and the greater extent of reaction at the relatively harder N^2 position of guanine may be expected.

Developing from these observations were calculations (Jerina et al., 1980) showing that the more potent PAHs all had sites at which, with appropriate metabolic activation, oxides could be formed which, in turn, would open to give particularly stable carbonium ions. Analogous to the visual identification of K-regions in PAH systems as single double bonds which can be removed without disturbing the resonance of the other ring systems, these positions at which stable carbonium ions can be formed are recognizable as 'bay regions'. Additional calculations can at least partially account for differences resulting from methylation (Silverman, 1981), but even so, certain exceptions have been noted to this theory. For example, based on this type of calculation, benzo[a] and benzo[e]pyrenes should be about equally carcinogenic. However, the latter is much less active. Comparative studies on the metabolism of B[a]P and B[e]P show that the formation of the appropriate dihydrodiols leading to diol epoxides in the case of B[e]P is considerably inhibited by the presence of the adjacent extra aromatic ring (MacLeod et al., 1979; Wood et al., 1979b). In addition, even if the dihydrodiols are prepared chemically the B[e]P derivative is much less carcinogenic than the B[a]P analog. This is in part due to the subsequent metabolism of these two compounds: in the case of B[a]P the diol epoxide is formed, in the case of B[e]P the conformation of the hydroxyl groups which are quasi axial, rather than quasi equatorial, as is the case for B[a]P, result in metabolites other than B[e]P diol epoxide being formed. Even when B[e]P diol epoxide is prepared chemically it reacts very poorly with DNA and the adducts have not been characterized. This again probably results from the conformation of the hydroxyl groups of the molecule rather than the relocation of the aromatic ring. Analogs lacking the hydroxyl groups, B[a]P and B[e]P tetrahydro oxides, both react readily with DNA (Kinoshita et al., 1982) and are mutagenic (Wood et al., 1980). In this case, however, the enantiomeric specificity and *trans* stereoselective of addition seen with B[a]PDE is largely lost and both enantiomers of both B[a]P and B[e]P tetrahydro epoxides react giving mixtures of *cis* and *trans* adducts.

An alternate type of exception to the bay region theory can be seen with BA and DMBA (Fig. 3). Both calculate to give equally stable carbonium ions using simple models and yet

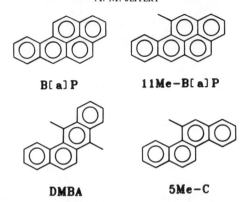

FIG. 3. Comparison of carcinogenic polycyclic aromatic hydrocarbons with respect to their bay regions. Each compound undergoes metabolic activation and binding to DNA through the benzo ring analogous in location to that of the benzo ring of B[a]P. Similarly, substitution with methyl groups is particularly efficient at enhancing the carcinogenicity of these compounds when it occurs on the ring adjacent to the benzo ring involved in metabolic activation.

the former is significantly less carcinogenic. The reasons for these differences have been less well documented than those between B[a]P and B[e]P, but it appears likely that since BA is mainly metabolized to a non-bay region diol epoxide it will not yield metabolites which can bind as efficiently to DNA. In contrast, methyl groups in the 7 or 8 positions of BA may inhibit metabolism in that ring and thus increase the amount of bay region diol epoxide formed. A similar effect is achieved with 3-methylcholanthrene where again metabolism in the non-bay region is inhibited.

The effect of methyl substitution, therefore, is generally that when in the ring required for metabolic activation to the bay region diol epoxide is substituted, or the group is positioned such that the hydroxyl groups of the intermediate diol formed would be quasi-axial, then carcinogenicity will be reduced because of at least partial inhibition of metabolism in those positions. When substitution occurs in the bay region but on the ring adjacent to that activated, e.g. 11-methyl BP, 5-chrysene, 12-methyl BA (Hecht *et al.*, 1979) or 11-methyl 15,16-dihydrocyclopenta[a]phenanthrene-17-one, there is an increase in carcinogenicity (DiGiovanni *et al.*, 1983a). The reasons for this are not clear. It may be a combination of factors: reduction of metabolism to give the diol in the bay region itself as a primary metabolite, e.g. 11-methylB[a]P 9,10-dihydrodiol; increased stability with respect to epoxide hydration of the diol epoxide once formed; increased stability of the carbonium ion; or difficulty in repair of the DNA adducts.

Another apparent exception of the bay region theory is with carcinogenic PAHs which do not have bay regions. Cyclopenta[cd]pyrene is very carcinogenic and yet does not have a bay region (Eisenstadt *et al.*, 1982). Calculation shows, however, that despite the lack of a formal bay region it does generate a particularly stable carbonium ion which is the fundamental aspect of this theory.

4.4.1.2. *Non-bay region adducts of benzo[a]pyrene.*

The first intermediate considered as the ultimate carcinogenic metabolite of BP was its 4,5-oxide. This compound and its DNA-adducts have been prepared. Although it has been shown to be formed in microsomal systems (Selkirk *et al.*, 1975) and is mutagenic and evidence for its formation of DNA *in vitro* has been presented (King *et al.*, 1976; Santella *et al.*, 1979), *in vivo* evidence is limited (Baer-Dubowska *et al.*, 1981; Baird *et al.*, 1975). The oxide is probably too rapidly hydrolyzed by epoxide hydrolase.

9-Hydroxy-BP has been proposed as an intermediate in DNA binding (Ashurst and Cohen, 1981; Robertson *et al.*, 1984; Vigny *et al.*, 1980a). Because of the similarities in the fluorescence spectra of these two compounds they are not easily distinguished except by chromatographic separation. To date, it has not been possible to prepare 9-hydroxy-BP

4,5-oxide chemically and so confirmation of the presence of these adducts has been difficult (King *et al.*, 1976; Robertson *et al.*, 1984; Santella *et al.*, 1979). Recently, the 9-methoxy analog was synthesized and it has been used to demonstrate that at least using phenobarbital induced microsomal systems this DNA adduct is formed (Heisig *et al.*, 1983).

Other DNA adducts may be formed via radical mechanisms. Several studies have shown that such oxidations can lead to DNA adducts formation (Cavalieri *et al.*, 1982; Rogan *et al.*, 1976, 1978). However, the chemical structure of these adducts has not yet been determined in detail. Analysis of *in vivo* samples of BP modified DNA shows that many of the adducts are not retained by the chromatographic systems used: either Sephadex LH20 or reverse phase HPLC columns (Eastman *et al.*, 1978). These adducts can vary from a few percent of the total, in the case of human bronchial explants (Jeffrey *et al.*, 1977), to being the major components in hepatocytes (Boroujerdi *et al.*, 1981). Extensive investigations (Eastman *et al.*, 1978) into the nature of the adducts has shown they do not result from, (1) tritium released from the BP during phenol and quinone formation being incorporated into the normal deoxyribonucleosides, (2) contamination by proteins, (3) or simply to incomplete digestion of the DNA. Radical oxidation products are still, therefore, possible candidates to explain some of these unidentified adducts.

4.4.2. Benz[a]anthracene Derivatives

These compounds have been the subject of considerable investigation partly owing to the dramatic increase in carcinogenicity upon methyl-substitution at particularly the 7-, 12-, and 6,8-positions (Huggins *et al.*, 1967; Levin *et al.*, 1983). BA itself is weakly carcinogenic and binds poorly to DNA (Phillips *et al.*, 1979). It has been shown to undergo metabolic activation, primarily to 8,9-dihydrodiol. When BA-modified DNA was investigated it appeared that both the non-bay and to a lesser extent the bay region diol epoxides are involved (Cary *et al.*, 1980; Cooper *et al.*, 1980a, b, d; MacNicoll *et al.*, 1979, 1981; Vigny *et al.*, 1980b). These results are consistent with metabolic studies in which only a minor amount of the bay region diol epoxide is formed compared to that at the non-bay region. Direct experimentation with the synthetic diol epoxides showed the former, as expected, is more capable of reacting with DNA (Hemminki *et al.*, 1980a). They react with guanine (Cooper *et al.*, 1980b) and induce single strand breaks (Cooper *et al.*, 1983). Using DNA from mouse skin or hamster embryo cells, previously exposed to BA, the adducts which were isolated from Sephadex LH20 columns had phenanthrene-like fluorescence spectra consistent with activation through a non-bay region diol epoxide (Vigny *et al.*, 1980b). In subsequent studies by Cooper and co-workers (Cooper *et al.*, 1980d) using more detailed HPLC approaches it appears that both the non-bay and to a lesser extent the bay region diol epoxides are involved.

The 7-methyl groups of 7-methylbenzanthracene and 7,12-dimethyl-BA inhibit metabolism at the 8,9-positions and, therefore, the amounts of non-bay region diols which can be formed are reduced. In addition, the conformation of these diols are quasi-axial which suppresses further epoxidation. The bay region diol epoxide adducts may also be difficult to repair (Dipple and Hayes, 1979). In both cases, the fluorescence spectra of the DNA isolated from a variety of sources, their light sensitivity, formation from the corresponding 3,4-dihydrodiol and additional evidence based upon HPLC analysis of the DNA adducts all agree with the conclusion that the major metabolite responsible for binding is the bay region diol epoxide (Fig. 4) (Baird and Dipple, 1977; Bigger *et al.*, 1983; Cooper *et al.*, 1980c; Dipple *et al.*, 1984a, b; Ivanovic *et al.*, 1978; Jeffrey *et al.*, 1980; Malaveille *et al.*, 1977; Moschel *et al.*, 1983; Slaga *et al.*, 1979a, b; Tierny *et al.*, 1978; Vigny *et al.*, 1981). Hydroxylation of the methyl groups, although often a significant metabolic pathway, appears not to be a major route to adduct formation although such adducts have been identified (Daniel and Joyce, 1983; DiGiovanni *et al.*, 1983b; Dipple *et al.*, 1979; Joyce and Daniel, 1982; Watabe *et al.*, 1982). Recent evidence with DMBA suggests that, in addition to the *anti* diol epoxide, which is the major metabolite formed in the case of BP, significant

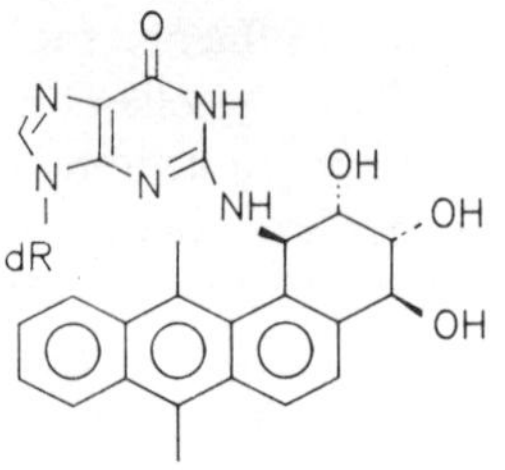

FIG. 4. The structure of the major adduct formed by metabolic activation of DMBA in 10T1/2 cells and human bronchus or by the reaction of *anti*-DMBA diol epoxide with DNA *in vitro*. The absolute stereochemistry of the adduct is not known but the relative stereochemistry at the 1 position (point of attachment of guanine) is *cis* with respect to the 4-hydroxy group and *trans* with respect to the 2- and 3-hydroxyl groups.

amounts of adducts arise from the *syn* diol epoxide and adenosine (Dipple *et al.*, 1983; Moschel *et al.*, 1983; Sawicki *et al.*, 1983). This is based upon the chromatographic properties of the adducts in the presence of borate which complex with *cis* hydroxyl groups found only in adducts derived from *anti* diol epoxides. Repair of these adducts is relatively slow in mouse embryo cells (Dipple and Hayes, 1979).

Several DMBA K-region oxide DNA adducts have been isolated (Jeffrey *et al.*, 1976a,b). Some have been detected as a result of *in vivo* exposure to DMBA when high concentrations were used.

3-Methylcholanthrene yields multiple DNA adducts which can perhaps be explained in part by the additional hydroxylation of the methyl group (Cooper *et al.*, 1980e; Eastman and Bresnick, 1979; King *et al.*, 1978; Levin *et al.*, 1979; Vigny *et al.*, 1977). When these adducts have been purified most seem to have a fluorescence which is again characteristically anthracene-like suggesting the DNA adducts are derived from bay region diol epoxides.

4.4.3. *Chrysene Derivatives*

Chrysene is only weakly carcinogenic despite the fact it has two bay regions. The reason for this is not clear since the explanations used with B[e]P, which similarly has two bay regions, do not apply in this case. The size of the ring system is also probably adequate since the addition of a methyl group at the 5 position results in a significant increase in carcinogenicity (Hecht *et al.*, 1974). Determining the key factor(s) to this enhanced effect is clearly a matter of significant interest since the principle appears to be quite general. In the case of chrysene itself, both direct and indirect evidence has been presented for the formation of bay region diol epoxide adducts (Hodgson *et al.*, 1983a; Vigny *et al.*, 1982; Wood *et al.*, 1979a). An interesting alternative pathway, analogous to that proposed for 9-hydroxy BP 4,5-oxide and which may account for as much as 50% of the adducts formed in some systems, involves a triol epoxide (Hodgson *et al.*, 1983b; Hulbert and Grover, 1983; Hodgson *et al.*, 1985). The general role of such phenol epoxides needs further study.

Evidence from both animal testing of halogenated 5-methylchrysene derivatives and direct analysis of the PAH adducts suggest that activation occurs through the bay region involving the methyl group (Hecht *et al.*, 1979; Melikian *et al.*, 1982, 1983, 1984, 1985a,b). The N^2-guanine adducts were characterized as major products.

4.4.4. *Related Compounds*

15,16-Dihydro-11-methylcyclopenta[a]phenanthrene-17-one has been studied extensively by Coombs and co-workers (Abbott and Coombs, 1981; Wiebers *et al.*, 1981). The major DNA adduct (80%) formed in mouse skin is derived from guanine and, by mass spectral analysis, appeared to be attached through exocyclic amino group to the C^1 of a 1,2,3,4-tetrahydro-2,3,4-trihydroxy derivative of the original carcinogen. These results have been compared to 15,16-dihydro-1,11-methanocyclopenta[a]phenanthrene-17-one in which the bay region was obstructed (Hadfield et al., 1984). Despite this substitution a similar pattern of metabolic activation was seen and the major DNA adduct isolated was thought to be derived from a bay region diol epoxide.

4.5. N-Substituted Aromatic Compounds

These compounds have recently been reviewed (Schut, 1984). Many years ago it was recognized that aromatic amines may be carcinogenic. Several examples were discovered as a result of increased incidence of bladder cancer in dye workers. Studies by the Millers in the 1960s demonstrated that metabolic activation was necessary *in vivo* but that synthetic analogs could be prepared which, upon reaction with DNA, would give the same spectra of DNA adducts. In particular, aminofluorene has been extensively studied.

4.5.1. *Aminofluorene*

N-Acetoxyacetylaminofluorene reacts *in vitro* with DNA to give one major product: N-(deoxyguanos-8-yl)-AAF (Kriek et al., 1967). This represented about 80% of the total DNA adducts, most of the remainder was shown to be 3-(deoxyguanos-N^2-yl)-AAF (Westra et al., 1976). This latter adduct could only be formed in native DNA but not in denatured DNA or RNA. *In vivo* in rats or using hepatocyte cells in culture, additional adducts are formed resulting from either deacetylation of the AAF adduct or an N,O-acyltransferase catalyzed formation of N-acetoxy-AF. In mice, evidence has been presented for the importance of the sulfotransferase (Lai et al., 1985). These may represent essentially all of the total adducts in some systems (Irving et al., 1969; Allaben et al., 1983). The C^8 deacetylated derivatives and N^2 adducts appear to be more stable *in vivo* with respect to repair, possibly because they induce less distortion in the DNA helix than the acetylated C^8 derivative (Lipkowitz et al., 1982; Yamasaki et al., 1977). In addition to a model in which the bulky groups cause a base displacement of the guanine residue (Grunberger et al., 1970; Jeffrey et al., 1980), an alternative model has been proposed (Santella et al., 1981) in which substitution on poly(dG-dC)·poly(dG-dC) enhances formation of the Z-conformation of this polymer. Under slightly basic conditions ring opening of this product occurs to yield two stereoisomeric 1-[6-(2,5-diamino-4-oxo-pyrimidimyl-N^6-deoxyribose)]-3-(2-fluoryl)ureas (Kriek and Westra, 1980). These ring opened products appear to be minor components *in vivo*. In addition, some rapidly depurinating derivatives, probably by substitution at the N^7 position of guanine, have been reported (Tarpley et al., 1982). More recently studies using prostaglandin H synthase and other non P-450 type one electron oxidation systems have shown that they are capable of activating a wide spectrum of carcinogens including a number of aromatic amines (Krauss and Eling, 1985; Tsuruta et al., 1985) to products which bind to DNA. A full evaluation of the relative importance of these pathways has not yet been completed.

4.5.2. *Naphthylamines*

2-Naphthylamine has long been recognized for its association with human bladder cancer as a result of occupational exposure, which has now been substantially reduced. Studies showed that [^{3}H]-2-naphthylamine forms mainly an N^6-adenine and, to a lesser extent, C-8 and N^2-guanine adducts when given to male dogs (Kadlubar et al., 1981a). Similar spectra of adducts were obtained in both liver and urothelium. The major differences appeared to be the ratio of binding in the urothelium, which is the target organ, to the non-target liver. This ratio increased with respect to time indicating the greater repair capability of the liver for these adducts, especially the C^8-guanine adduct.

To investigate the structure of the adducts they reacted N-hydroxy-2-naphthylamine, the metabolite responsible for DNA binding, with DNA *in vitro* at pH 5.0 (Kadlubar et al.,

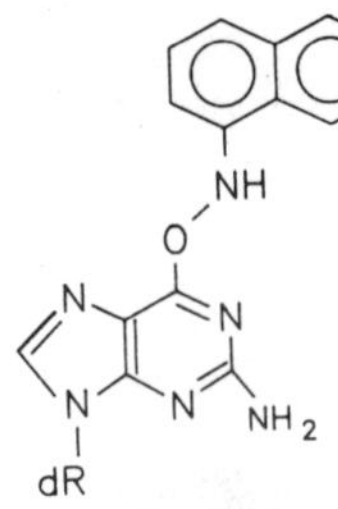

FIG. 5. Structure of the major adduct formed by the reaction of N-hydroxy-1-naphthylamine with DNA. Of particular interest is the reaction having occurred at the O^6 position.

1980, 1981c). Under these slightly acidic conditions, which might be expected to occur in the bladder, the metabolite is unstable and decomposition to an arylnitrenium ion carbocation occurs which can then react with nucleophilic sites in DNA. At pH 7.0 the C^8 adduct can undergo imidazole ring opening at the 8,9 bond (Kadlubar et al., 1981c). The structures of adducts were then determined by standard approaches.

1-Naphthylamine, although considered a potential human carcinogen, is clearly much less active than its 2-isomer. The reason for this appears to be the relative inability for N-hydroxylation at the 1-position since the corresponding N-hydroxy-1-naphthylamine is both highly carcinogenic and highly reactive with DNA. The two products, N-(deoxyguanosin-O^6-yl)-1-naphthylamine (Fig. 5) and 2-(deoxyguanosine-C^1-yl)-1-naphthylamine, isolated, which accounted for essentially all the adducts, are, however, different from those of the 2-isomer and other aryl amines in that reaction occurs at O^6-guanine (Dooley et al., 1984; Kadlubar et al., 1981b). The conformation of these adducts in DNA has also been studied (Kadlubar et al., 1981b).

4.5.3. Benzidine

Benzidine has also been shown to be a human and animal carcinogen. The metabolic activation of this compound is complicated because of the additional amino group. However, studies with [acetyl ^{3}H]-acetyl benzidine showed that the radioactive compound isolated from the DNA was identical to that obtained when animals were treated with benzidine itself. Martin et al. (1982, 1983) showed that the major DNA adduct formed when benzidine was given in the drinking water to mice was N-(deoxyguanosine-8-yl)-N'-acetylbenzidine. The order in which the metabolic steps occur to yield N-hydroxy-N'-acetylbenzidine and the ultimate carcinogen are not clear (Kennelly et al., 1984). Further activation of this metabolite appeared to be dependent upon N,O-acyltransferases and O-acetyltransferases in the rat and mouse livers respectively (Frederick et al., 1985). Ring opened forms of this adduct have not been described.

4.5.4. Other Aromatic Amines

In comparison with AAF the in vivo stability of the deacetylated adducts appears similar and much larger than the corresponding acetylated derivatives.

4-Acetylaminobiphenyl reacts with DNA although more weakly than 2-acetylaminofluorene. Two adducts have been characterized: N-(deoxyguanosin-8-yl)-acetylaminobiphenyl and 3-(deoxyguanosin-N^2yl)-acetylaminobiphenyl.

N-Hydroxy-4-acetylamino-4'-fluorobiphenyl shows similar characteristics to AAF in that three products have been detected in vivo with rates of repair comparable to AAF. Thus, the N-(deoxyguanosin-8-yl)-4-acetylamino-4'-fluorobiphenyl had a half-life of about two days while the deacetylated adduct was about ten days, and the 3-(deoxyguanosin-N^2yl)-4-acetylamino-4'-fluorobiphenyl was the most persistent (Kreik and Hengeveld, 1978). 3,2'-Dimethyl-4-aminobiphenyl (DMABP) induces tumors of both the colon and bladder probably via an N-hydroxy intermediate. Reaction of the latter compound with at pH 4.5 or at pH 7.4 in the presence of S-acetyl coenzyme A in the presence of hepatic or intestinal cytosol gives the same pattern of products bound to DNA, being mainly N-(deoxyguanosin-8-yl)-DMABP (Flammang et al., 1985), while N,O-acyltransferases and sulfotransferases appeared not to be involved.

N-Hydroxy-2-acetylamidophenathracene esters form at least two adducts with DNA. In this case modification of both guanine and adenine occurred although the ratio depended upon the ester. Using the acetate a ratio of N^6-1-(2-acetamido-phenathryl) adenosine to 8-(N^2-phenanthrylacetamido) guanine of 8:2 was obtained while this changed to 4:6 with the sulfate ester. This would suggest that either a non-covalent complex was formed first, which may have a direct effect on the reaction, or the leaving group directly influenced the site of reaction (Scribner and Naimy, 1975).

4-Dimethylaminoazobenzene was the first compound studied with respect to its in vivo binding to macromolecules (Miller and Miller, 1947). N-Dimethyl-4-aminoazobenzene is again analogous to other aromatic amines with respect to adduct formed (Beland et al., 1980; Tarpley et al., 1980, 1981). The C^8 adduct, although initially predominant, is removed more rapidly by the rat liver than the persistent N^2 derivative (Beland et al., 1980; Tarpley et al., 1982). The C^8 derivative was also found to be the major persistent adduct in mice which, unlike the situation with Sprague-Dawley rats, could also be formed when animals were exposed to 4-aminoazobenzene (Delclos et al., 1984). Another very persistent adduct has been identified as 3-(deoxyadenosin-N^6yl)-methylazobenzene. The in vitro analog of the ultimate carcinogen of this azo dye used in the preparation of synthetic DNA adducts for chemical characterization and chromatographic markers was its N-benzoyloxy derivative (Beland et al., 1981; Tarpley et al., 1980). As with BPDE high levels of modification of DNA can be obtained (1–2% of the bases). At least six adducts were detected in significant quantities. Most were also detectable in vivo although the ratios were often substantially different (Tillis et al., 1981).

Most of the aromatic amines discussed so far are likely to be human health hazards as a result of industrial exposure although some, e.g. the naphthylamines, are found in cigarette smoke. It is important to know whether industrial production of carcinogens is a major source of our exposure to such hazardous compounds: a common preconception. It was of particular interest, therefore, that a number of pyrolysate products from the charred surfaces of meat and fish have been isolated (Hirota and Sugimura, 1983). About a dozen such compounds have been isolated as pyrolysis products of tryptophane, glutamic acid, phenylalanine, lysine, soy bean globulin and fried beef. In particular, two of the tryptophane and glutamic acid products have been shown to be carcinogenic.

The binding of two of these products, 3-amino-1-methyl-5H-pyrido[4,3-b]indole (Trp-P-2) (Hashimoto et al., 1979) and 2-amino-6-methyldipyrido[1,2-a:3′,2′d]imidazole (Glu-P-1) (Hashimoto et al., 1980) involve N-hydroxylation for activation and the C^8 position of guanine as the major binding site in DNA. Another closely related glutamic pyrolysis product appears to undergo similar activation (Loukakou et al., 1985). Further studies on these types of products will be of particular interest with respect to the possible etiology of human cancers.

4.5.5. Nitroaromatics

Interest was generated in this class of compounds when it was discovered that these were particularly mutagenic and carcinogenic (El-Bayoumy et al., 1984; Lofroth et al., 1980; Oghaki et al., 1984; Rosenkranz, 1984) and components of diesel engine emissions and photocopiers. The 1-nitropyrene is less potent than the 1,6- and 1,8-dinitro derivatives but is more abundant in the extracts analyzed. Studies on the 1-nitropyrene-DNA adduct formed in S. typhimurium (Howard et al., 1983b; Messier et al., 1981) show that the major product results from the reduction of the nitro group to a probable hydroxylamine which then reacts at the C^8 position of guanine residues in DNA. The adduct, as expected, is moderately unstable to both acid and base and a number of decomposition products have also been detected although not fully characterized. Activation of the dinitropyrene appears to be additionally dependent upon acetyl coenzyme A (Djuric et al., 1985). Alternative oxidative pathway may also exist for some nitroaromatics (El-Bayoumy and Hecht, 1984) but the DNA adducts have yet to be identified. *In vitro* transformation of human fibroblast has also been achieved with the nitroaromatics (Howard et al., 1983a).

Many acridine derivatives bind non-covalently to DNA. Recently (Pawlak et al., 1983) found that some 1-nitroacridine derivatives required metabolic activation and can yield quite high levels of DNA modification. Tentative identification of the adducts, based on mass spectral analysis of the isolated modified bases, suggests that both mono and cross linked adducts can be formed involving deoxy-guanosine, adenosine and cytosine residues. A more detailed study of these adducts would be of interest.

4.5.6. *4-Nitroquinoline Oxide* (4NQO)

The DNA adducts found with 4NQO have been characterized by acid hydrolysis of the DNA isolated from rat, HeLa, AH 130 or *E. coli* cells exposed to [^{3}H] 4NQO (Tada and Tada, 1971, 1976). The major adduct was N-(deoxyguanosin-8-yl)-4-aminoquinoline 1-oxide. One of the minor adducts formed with adenine has been partially characterized (Kawazoe et al., 1975) as being most probably 3(N^6adenyl)-4-aminoquinoline 1-oxide. A C^8-guanine adduct has also been characterized although it appears not to be present at more than about 20% of the total adducts found *in vivo* (Bailleul et al., 1981). 4-Acetoxyaminoquinoline oxide has been used as a model for the ultimate carcinogen (Galiegue-Zouitina et al., 1983).

4.6. NATURAL PRODUCTS

As mentioned above a major concern is the identification of carcinogens which may naturally occur in our environment. They may be ubiquitous and present in sufficiently low concentrations that they are difficult to detect epidemiologically or even chemically. With the enormous diversity of metabolism which exists in plants and micro-organisms it is not surprising that potent carcinogens have been isolated from these sources.

4.6.1. *Aflatoxins*

Aflatoxins are perhaps the best characterized of these carcinogens. This occurred for two reasons: (1) they are highly fluorescent and therefore can be detected at very low concentrations and (2) they are extremely potent carcinogens. A wide number of these derivatives have been isolated, not all of which are carcinogenic. Analysis of the structure of these compounds and the fact that those which are carcinogenic require metabolic activation suggested that the 2,3-double bond was a critical position. Chemical or enzymic epoxidation of this bond, for example aflatoxin B_1 or G_1, leads to DNA binding (Gurtoo et al., 1978). The proposed intermediate oxides have been too unstable to isolate.

The major adduct which is formed, 2,3-dihydro-2-(N^7-guanyl)-3-hydroxyaflatoxin B_1, results from reaction at the N^7 of guanine through the 2 position of the oxide (Autrup et al., 1979; Croy et al., 1978; Croy and Wogan, 1981; Essigmann et al., 1977, 1982; Garner et al., 1979; Lin et al., 1977; Martin and Garner, 1977; Stoner et al., 1982). As with other N^7 modified guanine the positive charge on the imidazole ring results in labile glycosidic bonds. Non-enzymic depurination or ring opening of the modified bases occur. Depurination may be aided *in vivo* by repair enzymes and urinary excretion of the adducts is observed (Autrup et al., 1983; Bennett et al., 1981). Depurination reactions may not be critical to the carcinogenicity of aflatoxins since repair of such sites appears quite efficient (Bose et al., 1978).

However, the imidazole ring is sensitive to opening of the 7,8-bond by base to yield an aflatoxin formamidopyridine derivative (Essigmann et al., 1977). The N^9-nitrogen of the original guanine no longer carries the charge which was associated with the imidazole ring and the glycoside bond is stabilized. At longer time intervals this adduct is therefore the major species bound to DNA (Essigmann *et al.*, 1982).

Sterigmatocystin, which is less potent a carcinogen than AFB_1, is structurally related and appears to give similar types of DNA adducts (Essigmann *et al.*, 1979, 1980).

4.6.2. *Safrole*

Safrole (4-alkyl-1,2-methylenedioxybenzene) is a hepatocarcinogen in rats and was widely used in the food industry as a major component in sassafras oil. The compound requires metabolic activation via 1'-hydroxysafrole and probably a sulfate ester. Synthetically the 1'-acetoxy derivative reacts with deoxyguanosine and adenosine to give derivatives chromatographically identical with those obtained from liver DNA of mice treated with 1'-hydroxysafrole. Most of the adducts formed *in vivo* were deoxyguanosine derivatives and the major one was identified as N^2-(*trans*-isosafrol-3-yl)deoxyguanosine (Phillips *et al.*, 1981b). The other deoxyguanosine derivative was not fully characterized. The deoxyadenosine derivative was a minor *in vivo* product.

A closely related compound, 1-alkyl-4-methoxybenzene, is also quite widespread occurring in several herbs including sweet basil, tarragon and chervil. Not unsurprisingly it is also a hepatocarcinogen and appears to undergo a very similar series of metabolic steps and DNA binding. The major *in vivo* DNA adduct has been identified as N^2-(*trans*-isoestrangol-3'-yl)deoxyguanosine with lesser amounts of N^2-(estragol-1'-yl)deoxyguanosine, and traces of N^6-(*trans*-isoestrangol-3'-yl)deoxyguanosine, and what was probably N^2-(*cis*-isoestragol-3'-yl)deoxyguanosine (Phillips *et al.*, 1981a). Most of these DNA adducts appear to be repaired relatively quickly. Detailed analysis of the structures of adducts formed from a synthetic analog, 1'-hydroxy-2',3'-dehydroestragole, in mice showed that again the N^2 position of guanine accounted for over 85% of the adducts formed and that metabolic activation involved hepatic sulfotransferase(s) (Fennell *et al.*, 1985).

4.6.3. *Cycacin*

Cycacin, the β-glycoside of methylazoxymethanol, occurs in the cycad nut and is only carcinogenic after bacterial hydrolysis to methylazoxymethanol (Laqueur and Spatz, 1975). The aglycone is carcinogenic in both normal and germ-free animals and has been shown to react directly with DNA, probably via a methyldiazonium ion, which in turn loses nitrogen to yield a carbonium ion. The major product is N^7-methylguanine (Matsumoto and Higa, 1966). This compound is therefore related to other hydrazine derivatives discussed in Section 4.1.3.

4.6.4. *Mitomycin C*

Mitomycin C is an antibiotic antineoplastic agent which binds to DNA to give both mono adducts and cross links. Recently, evidence has been presented that the major mono adduct is derived from addition between the 1 position of mitomycin C and the O^6 of guanine as the major product (Fig. 6), from microsomal or chemical reductions of mitomycin C in the presence of dinucleoside monophosphates (Tomasz *et al.*, 1983), or DNA (Hashimoto *et al.*, 1982), although N^2-guanine and N^6-adenine adducts were also identified. The reactivity of the O^6 position to such an extent is surprising and may relate to the possible steric constraints of the reaction with DNA since the addition appears, as noted above with active metabolites of the PAHs, very stereoselective (Tomasz *et al.*, 1983). Further examination of the products formed under mildly acidic conditions showed that N-7 modification of guanine residues could occur (Tomasz *et al.*, 1985; Verdine and Nakanishi, 1985). The chemistry of the cross linked adducts remains to be elucidated.

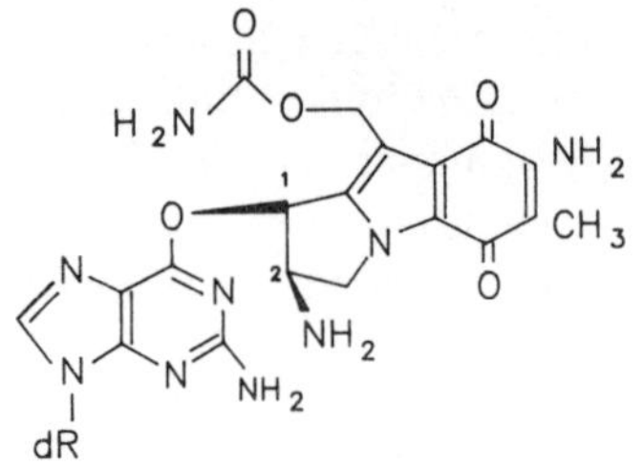

FIG. 6. Structure of the major DNA adduct formed with mitomycin C. These have eluded full chemical characterization for a number of years and the elucidation of this structure is therefore particularly significant. Reaction at O^6 was unexpected.

4.6.5. Streptozotocin

Streptozotocin is a 2-deoxyglucose derivative of methylnitrourea and as such gives preferentially N^7-guanine adducts both *in vivo* and *in vitro* (Bennet and Pegg, 1981).

4.6.6. Pyrrolo[1,4]benzodiazepine Antibiotics

These compounds represent an interesting group which present a transition between those compounds which are covalently bound to DNA and examples of which have been described here, and those which only form complexes with DNA such as BP itself of ethidium bromide. These derivatives form very tight complexes in the minor groove of double stranded DNA (Petrusek *et al.*, 1981) and yet it seems that, without the stabilization of the complex by additional non-covalent hydrogen bonding, the bond between the carbinolamine carbon and the N^2 of guanine is extremely labile and the adduct is unstable to even slightly acidic conditions. By using a deoxyhexanucleotide it has been possible to isolate the complex directly and analyze its NMR spectra to confirm earlier models for the structure of the adduct (Graves *et al.*, 1984, 1985).

4.6.7. Pyrrolizidine Alkaloids

This group of compounds is widespread in nature and are a source of significant exposure of humans to toxic and potentially carcinogenic compounds. Their chemistry and occurrence has been reviewed (Robins, 1982; Schoental, 1982; Mattocks, 1986). In order to be toxic they require 1,2-unsaturation and must be esterified in the 9 position. They are metabolically oxidized to the corresponding pyrrole. These oxidation products react with DNA and some of the adducts have been characterized, for example, that from monocrotaline (Robertson, 1982) which reacts through the 7 position to bind to the N^2 position of deoxyguanosine (Fig. 7). The reaction is thought to be S_{N^1} based upon the production of two adducts in approximately equal yields and enantiomeric at the 7 position.

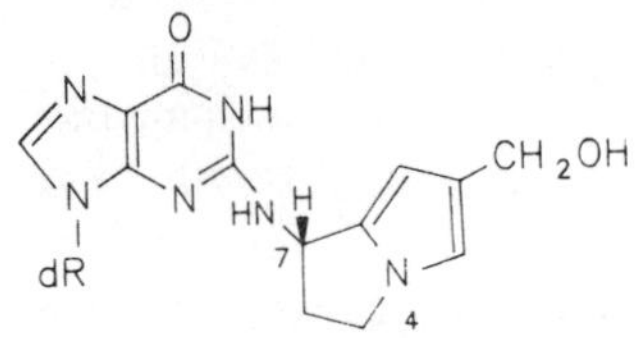

FIG. 7. Structure of major deoxyguanosine adduct formed during reaction of monocrotaline with DNA. dR is the deoxyguanosine moiety.

4.6.8. CC-1065

CC-1065 is an unusual antibiotic for *Streptomyces* which bind to the N^3 position of adenine (Svenson *et al.*, 1982). It is thought to lie in the minor groove and, because of its extended structure, causes considerable helix stabilization (Hurley *et al.*, 1984). Unlike, for example bleomycin, it does not cause strand breaks, but inhibited susceptibility of DNA to S_1 digestion. It is extremely cytotoxic and may function through inhibition of DNA synthesis.

4.7. PHOTOCHEMICALLY PRODUCED DNA ADDUCTS

The toxicity of u.v. light has long been recognized (Giese, 1971). Its association with skin neoplasia is most clearly seen in Xeroderma pigmentosum patients who, because of their inability to repair damage induced by u.v. light, develop large numbers of skin lesions at an early age. This is a clear illustration of the importance of DNA repair in the prevention of cancer.

The major adducts which have been identified are those resulting from the dimerization of adjacent thymines and cytosines in DNA which appear to be responsible for the mutagenicity of u.v. radiation (Brash and Haseltine, 1982; Doetsch *et al.*, 1985). More recently purine-pyrimidine dimers have been detected (Rose *et al.*, 1983) although only as one of many photoproducts. Since these are formed by the reaction between normal components of DNA they will not be considered further here; they have been adequately reviewed elsewhere (Wang, 1976).

4.7.1. *Psoralens*

Many compounds produce phototoxic reactions and may be either natural or synthetic in origin. While such reactions are normally disadvantageous in some instances, e.g. the furocoumarins, advantage has been taken of these natural products, or analogous drugs developed (Isaacs *et al.*, 1977), for photochemical therapies of psoriasis (Parrish *et al.*, 1974). In addition, because the photochemical reaction is fairly specific to DNA, it has been used to treat T cell lymphomas by irradiating patients' blood extra-corporeally thereby avoiding damage to the individuals other nucleated cells (Edelson *et al.*, 1983). This approach reduces the risk of cancer induction by the psoralens which have been shown to be both mutagenic and carcinogenic (Ashwood-Smith *et al.*, 1980; Epstein, 1979; Giese, 1971; Stern *et al.*, 1979).

As with the polycyclic aromatic hydrocarbons there seems to be minimal toxicity from these compounds directly. Thus, although both classes of compounds intercalate strongly with DNA this produces little toxic effect. Unlike the hydrocarbons, however, the covalent binding to DNA of these compounds depends upon photochemical rather than metabolic activation. Light between 300 and 380 nm is the most effective yielding first mono adducts and then interstrand cross links. Early studies used the photochemical reaction with thymidine to investigate the chemistry of the adducts formed (for review see Song and Tapley, 1979). More recently adducts have been isolated from DNA irradiated in the presence of 8-methoxypsoralen or 4,5',8-trimethylpsoralen (Peckler *et al.*, 1982). The major products were two diastereomeric adducts formed by cycloaddition of the 4',5'-double bond of the furan ring to the 5,6-double-bond of thymidine. A corresponding adduct which involved the 3,4-double bond of the pyrone ring accounted for about 20% of the adducts in the case of the 8-methoxypsoralen but less than 3% in the case of the trimethyl derivative. All the adducts had *cis-syn* stereochemistry (Kanne *et al.*, 1982b). Thus of the sixty-four possible isomers only a limited number are found in DNA, presumably because of the constraints placed on the reactions by the conformation of the intercalated complex. The ratio of the two major diastereoisomers formed in DNA was 7:4 indicating preferential adduct formation when the psoralen is intercalated on one side of the T:A base pair with respect to its neighbors.

In general, the cross linked adducts are felt to be more important biologically and studies (Kanne *et al.*, 1982a) have shown that the furan but not the pyran side adducts can lead to cross linking. This again results in products having a *cis-syn* stereochemistry.

The selectivity of reaction is believed to result from the energy gained as the psoralen and nucleic acid residues maximize their potential overlap prior to the photo addition reactions. Models have been developed (Pearlman, 1985) which indicate that significant distortions in the DNA are induced by their crosslinking reactions, similar to but greater than those from thymine photodimerization.

4.7.2. *Other Photo Adducts*

Despite, as mentioned above, both benzo[a]pyrene and aflatoxin are believed to exert their carcinogenic effect through metabolic activation to products which bind to DNA, both have also been shown to undergo photochemical binding to DNA and to induce strand breaks (Blackburn *et al.*, 1972, 1977; Shieh and Sang, 1980; Strniste *et al.*, 1980). Since the latter is a hepatocarcinogen, it is unlikely that photochemistry can play an important role, although benzo[a]pyrene does produce photoallergic reactions and is a skin carcinogen. The use of coal tars containing particularly anthracene in the treatment of psoriasis has caused concern because of potential photoinduced carcinogenicity and binding of anthracene to DNA (Blackburn and Taussig, 1975).

4.8. CIS PLATINUM DNA ADDUCTS

Studies on the mechanism of action of platinum compounds as antineoplastic agents has led to the general conclusion that the activity of *cis*-[$PtCl_2(NH_3)_2$], as compared to the *trans* isomer or monofunctional analogs, results from the former being able to form intrastrand cross links. Detailed X-ray, NMR and CD studies with short oligodeoxynucleotides (Marcelis *et al.*, 1983; Sherman *et al.*, 1985), has confirmed the structure of these cross links. Two types were described in which cross links were formed between the N^7 positions of guanine residues in d(GpG) or d(GpXpG) sequences. 'X' may be any base in the case of DNA (Fichtinger-Schepman *et al.*, 1982) and, because of the intervening base in the latter example, the conformations of these two types of intrastrand cross links are different. One is a kinked structure (Den Hartog *et al.*, 1982) and one in which the intervening base is displaced (Fichtinger-Schepman *et al.*, 1982). It seems likely that the two types of adducts should be distinguishable by immunological methods which may be of interest in terms of monitoring patients on chemotherapy and their prognosis. Some antibodies are available showing the potential for these types of studies (Poirier *et al.*, 1982, 1985; Plooy *et al.*, 1985).

5. SUMMARY

The chemistry and molecular biology of DNA adducts is only one part of the carcinogenic process. Many other factors will determine whether a particular chemical will exert a carcinogenic effect. For example, the size of particles upon which a carcinogen may be adsorbed will influence whether or not, and if so where, deposition within the lung will occur. The simultaneous exposure to several different agents may enhance or inhibit the metabolism of a chemical to its ultimate carcinogenic form (Rice *et al.*, 1984; Smolarek and Baird, 1984). The ultimate carcinogenic metabolites may be influenced in their ability to react with DNA by a number of factors such as internal levels of detoxifying enzymes, the presence of other metabolic intermediates such as glutathione with which they could react either enzymatically or non-enzymatically, and the state of DNA which is probably most heavily influenced by whether or not the cell is undergoing replication or particular sequences being expressed. Replicating forks have been shown to be more extensively modified than other areas of DNA (Cordeiro-Stone et al., 1982). Another critical factor which can influence the final outcome of the DNA damage is whether or not the modifications can be repaired. If this occurs with high fidelity and the cell has not previously undergone replication then the effect of the damage by the carcinogen is likely to be minimal.

The major area in which progress is needed is an understanding of what this damage really does to the cell such that after an additional period of time, which may be as long as twenty or more years, these prior events are expressed and cell proliferation occurs. Clearly additional stimulatory factors, for example tumor promoting agents such as the phorbol esters or phenobarbital, are often needed. After such prolonged periods it seems likely that the DNA adducts would no longer be present. However, the way in which their earlier presence is remembered is not clear. Simple mutations do not explain all the characteristics of tumor progression and, when it occurs, regression. Even if a specific site mutation does occur then its expression must be under other types of control. Any explanation of the action of DNA modification at the molecular level also requires that account be taken of the diverse nature of the DNA adducts from simple modifications such as methylation to bulkier adducts such as benzo[a]pyrene, aflatoxin or aromatic amines. Currently attractive ideas include effects on normal DNA methylation patterns or that the DNA adducts can alter endogenous oncogenes such that they change from an obligatorily silent state to one in which, with appropriate additional stimulation, they can be expressed. Hopefully, research over the next few years will shed light on these critical aspects of our understanding of the carcinogenic process.

Acknowledgements—Part of the work described here was supported by NCI grant CA-02111. I thank Alycia Osborne for her assistance in preparing this manuscript.

REFERENCES

ABBOT, P. J. and COOMBS, M. (1981) DNA adducts of the carcinogen, 15,16-dihydro-11-methylcyclo-penta[a]phenanthrene-17-one, *in vivo* and *in vitro*: high pressure liquid chromatographic separation and partial characterization. *Carcinogenesis* 2: 629–636.
ADAMKIEWICZ, J., AHRENS, O. and RAJEWSKY, M. F. (1983) Quantitation of alkyldeoxyribonucleosides in DNA of individual cells by high-affinity monoclonal antibodies and electronically intensified, direct immunofluorescence. *J. Cancer Res. clin. Oncol.* 105: A15.
ALLABEN, W. T., WEIS, C. C., FULLERTON, N. F. and BELAND, F. A. (1983) Formation and persistence of DNA adducts from the carcinogen *N*-hydroxy 2-acetylaminofluorene in rat mammary gland *in vivo. Carcinogenesis* 4: 1067–1070.
AMES, B. N., LEE, F. D. and DURSTON, W. E. (1973) An improved bacterial test system for the detection and classification of mutagens and carcinogens. *Proc. natn. Acad. Sci U.S.A.* 70: 782–786.
AMES, B. N., MCCANN, J. and YAMASAKI, C. (1975) Methods for detecting carcinogens and mutagens with the Salmonella/mammalian-microsome mutagenicity test. *Mut. Res.* 31: 347–364.
ASHURST, S. W. and COHEN, G. M. (1981) The formation and persistence of benzo[a]pyrene metabolite-deoxyribonucleoside adducts in rat skin *in vivo. Int. J. Cancer* 28: 387–391.
ASHURST, S. W., COHEN, G. M., NESNOW, S., DIGIOVANNI, J. and SLAGA, T. J. (1983) Formation of benzo[a]pyrene/DNA adducts and their relationship to tumor initiation in mouse epidermis. *Cancer Res.* 43: 1025–1029.
ASHWOOD-SMITH, M. J., POULTON, G. A., BARKER, M. and MILDENBERGER, M. (1980) *Nature, Lond.* 285: 407–409.
AUGUSTO, O., FALJONI-ALARIO, A., LEITE, L. C. C. and NOBREGA, F. G. (1984) DNA strand scission by the carbon radical derived from 2-phenlethylhydrazine metabolism. *Carcinogenesis* 5: 781–784.
AUTRUP, H., BRADLEY, K. A., SHAMSUDDIN, A. K. M., WAKHISI, J. and WASUNNA, Q. (1983) Detection of putative adduct with fluorescence characteristics identical to 2,3-dihydro-2-(7′-guanyl)-3-hydroxyaflatoxin B_1 in human urine collected in Murang's district, Kenya. *Carcinogenesis* 4: 1193–1195.
AUTRUP, H., ESSIGMANN, J. M., CROY, R. G., TRUMP, B. F., WOGAN, G. N. and HARRIS, C. C. (1979) Metabolism of aflatoxin and B_1 identification of the major aflatoxin B_1-DNA adducts formed in cultured human bronchus and colon. *Cancer Res.* 39: 694–698.
BAER-DUBOWSKA, W., FRAYSSINET, C. and ALEXANDROV, K. (1981) Formation of covalent deoxyribonucleic acid benzo[a]pyrene 4,5-epoxide adduct in mouse and rat skin. *Cancer Lett.* 14: 125–129.
BAILLEUL, B., GALIEGUE, S. and LOUCHEUX-LEFEBVRE, M. H. (1981) Adducts from the reaction of *O, O′*-diacetyl or *O*-acetyl derivatives of the carcinogen 4-hydroxyaminoquinoline 1-oxide with purine nucleosides. *Cancer Res.* 41: 4559–4565.
BAIRD, W. M. and DIAMOND, L. (1977) The nature of benzo[a]pyrene-DNA adducts formed in hamster embryo

cells depends on the length of time of exposure to benzo[a]pyrene. *Biochem. biophys. Res. Commun.* **77**: 162–168.

BAIRD, W. M. and DIPPLE, A. (1977) Photosensitivity of DNA-bound 7,12-dimethylbenz[a]anthracene. *Int. J. Cancer* **20**: 427–431.

BAIRD, W. M., DUMASWALA, R. U. and DIAMOND, L. (1983) Time-dependent differences in the benzo[a]pyrene-DNA adducts present in cell cultures from different species. *Bas. Life Sci.* **24**: 565–586.

BAIRD, W. M., HARVEY, R. G. and BROOKES, P. (1975) Comparison of the cellular DNA-bound products of benzo[a]pyrene with the products formed by the reaction of benzo[a]pyrene 4,5-oxide with DNA. *Cancer Res.* **35**: 54–57.

BARRETT, J. C., TSUTSUI, T. and TS'O, P. O. P. (1978) Neoplastic transformation induced by a direct perturbation of DNA. *Nature, Lond.* **274**: 229–232.

BARROWS, L. R. and MAGEE, P. N. (1982) Nonenzymatic methylation of DNA by S-adenosylmethionine *in vitro*. *Carcinogenesis* **3**: 349–351.

BARROWS, L. R. and SHANK, R. C. (1981) Aberrant methylation of liver DNA in rats during hepatoxicity. *Toxicol. appl. Pharmac.* **60**: 334–345.

BASU, A. K. and MARNETT, L. J. (1984) Molecular requirements for the mutagenicity of malondialdehyde and related acroleins. *Cancer Res.* **44**: 2848–2854.

BECKER, R. A., BARROWS, L. R. and SHARK, R. C. (1981) Methylation of liver DNA guanine in hydrazine hepatoxicity: dose-response and kinetic characteristics of 7-methylguanine and O^6-methylguanine formation and persistence in rats. *Carcinogenesis* **2**: 1181–1188.

BELAND, F. A., DOOLEY, K. L. and CASCIANO, D. A. (1979) Rapid isolation of carcinogen-bound DNA and RNA by hydroxylapatite chromatography. *J. Chem.* **174**: 177–186.

BELAND, F. A., TULLIS, D. L., KADLUBAR, F. F., STRAUB, K. M. and EVANS, F. E. (1980) Characterization of DNA adducts of the carcinogen N-methyl-4-aminoazobenzene *in vitro* and *in vivo*. *Chem. Biol. Interact.* **31**: 1–17.

BELAND, F. A., TULLIS, D. L., KADLUBAR, F. F., STRAUB, K. M. and EVANS, F. E. (1981) Identification of the DNA adducts formed *in vitro* from N-benzoyloxy-N-methyl-4-aminoazobenzene and in rat liver *in vivo* after administration of N-methyl-4-aminoazobenzene. *J. natn. Cancer Inst.* **58**: 153–161.

BENNETT, R. A. and PEGG, A. E. (1981) Alkylation of DNA in rat tissues following administration of streptozotocin. *Cancer Res.* **41**: 2786–2790.

BENNETT, R. A., ESSIGMANN, J. M. and WOGAN, G. N. (1981) Excretion of an aflatoxin-guanine adduct in the urine of aflatoxin B_1-treated rats. *Cancer Res.* **41**: 650–654.

BERANEK, D. T., WEIS, C. C. and SWENSON, D. H. (1980) A comprehensive quantitative analysis of methylated and ethylated DNA using high pressure liquid chromatography. *Carcinogenesis* **1**: 595–606.

BERANEK, D. T., WEISS, C. C., EVANS, F. E., CHETSANGA, C. J. and KADLUBAR, F. F. (1983) Identification of N^5-methyl-N^5-2,5,6-triamino-4-hydroxypyrimidine as major adduct on rat liver DNA after treatment with the carcinogens, N,N-dimethylnitrosamine or 1,2-dimethylhydrazine. *Biochem. biophys. Res. Commun.* **110**: 625–631.

BERENBLUM, I. (1944) Irritation and carcinogenesis. *Archs Path.* **38**: 233–271.

BIGGER, C. A., SAWICKI, J. T., BLAKE, D. M., RAYMOND, L. G. and DIPPLE, A. (1983) Products of binding of 7,12-dimethylbenz[a]anthracene to DNA in mouse skin. *Cancer Res.* **43**: 5647–5651.

BLACKBURN, G. M. and TAUSSIG, P. E. (1975) The photocarcinogenicity of anthracene: photochemical binding to deoxyribonucleic acid in tissue culture. *Biochem. J.* **149**: 289–291.

BLACKBURN, G. M., FENWICK, R. G. and THOMPSON, M. H. (1972) The structure of thymidine: 3,4-benzopyrene photoproduct. *Tetrahedron Lett.* **7**: 589–592.

BLACKBURN, G. M., FENWICK, R. G., LOCKWOOD, G. and WILLIAMS, G. M. (1977) Photoproducts of DNA pyrimidine bases and polycyclic aromatic hydrocarbons. *Nucleic Acids Res.* **4**: 2467–2494.

BORGEN, A., DARVEY, H., CASTAGNOLI, N., CROCKER, T. T., RASMUSSEN, R. E. and WANG, I. Y. (1973) Metabolic conversion of benzo[a]pyrene by Syrian hamster liver microsomes and binding of metabolites to deoxyribonucleic acid. *J. med. Chem.* **16**: 502–506.

BOROUJERDI, M., KUNG, H., WILSON, A. G. and ANDERSON, M. W. (1981) Metabolism and DNA binding of benzo[a]pyrene *in vivo* in the rat. *Cancer Res.* **41**: 951–957.

BOSE, K., KARRAN, P. and STRAUSS, B. (1978) Repair of depurinated DNA *in vitro* by enzymes purified from human lymphoblasts. *Proc. natn. Acad. Sci. U.S.A.* **75**: 794–799.

BRASH, D. E. and HASELTINE, W. A. (1982) UV-induced mutation hotspots occur at DNA damage hotspots. *Nature, Lond.* **298**: 189–192.

BROOKES, P. (1979) The binding to mouse skin DNA of benzo[a]pyrene, its 7,8-diol and 7,8-diol-9,10-epoxides in relation to the tumorigenicity of these compounds. *Cancer Lett.* **6**: 285–289.

BROOKES, P. and LAWLEY, P. D. (1964) Evidence for the binding of polynuclear aromatic hydrocarbons to the nucleic acids of mouse skin: relationship between carcinogenic power of hydrocarbons and their binding to DNA. *Nature, Lond.* **202**: 781–784.

BROOKES, P. and OSBORNE, M. R. (1982) Mutation in mammalian cells by stereoisomers of anti-benzo[a]pyrene-diolepoxide in relation to the extent and nature of DNA reaction products. *Carcinogenesis* **3**: 1223–1226.

BROUDE, N .E. and BUDOWSKY, E. I. (1971) The reaction of glyoxal with nucleic acid components. III. Kinetics of the reaction with monomers. *Biochim. biophys. Acta* **254**: 380–388.

BROWN, H. S., JEFFREY, A. M. and WEINSTEIN, I. B. (1979) Formation of DNA adducts in 10T1/2 mouse embryo fibroblasts incubated with benzo[a]pyrene or dihydrodiol oxide derivatives. *Cancer Res.* **39**: 1673–1677.

BURFIELD, D. R., KHOO, T. K. and SMITHERS, R. (1981) External activation of epoxides by polarising groups borne by the nucleophile. *JCS Perkin I*: 8–11.

CARY, P. D., TURNER, C. H., COOPER, C. S., RIBEIRO, O., GROVER, P. L. and SIMS, P. (1980) Metabolic activation of benzo[a]anthracene in hamster embryo cells: the structure of a guanosine-anti-BA-8,9,-diol 10,11-oxide adduct. *Carcinogenesis* **1**: 505–512.

CASTLEMAN, H., HANAU, L. H. and ERLANGER, B. F. (1983). Stabilization of (dG-dC)n.(dG-dC)n in the Z conformation by a crosslinking reaction. *Nucl. Acids Res.* **11**: 8421–8429.

CAVALIERI, E., ROGAN, E. and ROTH, R. (1982) Multiple mechanisms of activation in aromatic hydrocarbon carcinogenesis. In: *Free Radicals and Cancer*, pp. 117–158, FLOYD, R. A. (ed.) Marcel Dekker, New York.

CHAW, Y. F. M., CRANE, I. E., LANGE, P. and SHAPIRO, R. (1980) Isolation and identification of cross-links from formaldehyde treated nucleic acids. *Biochemistry* **19**: 5525–5531.

CHEN, R., MIEYAL, J. J. and GODTHWAIT, D. A. (1981) The reaction of β-propiolacetone with derivatives of adenine and with DNA. *Carcinogenesis* **2**: 73–80.

CHETSANGA, C. J. and MAKAROFF, C. (1982) Alkaline opening of imidazole ring of 7-methylguanosine 2. Further studies on reaction mechanisms and products. *Chem. Biol. Interact.* **41**: 235–249.

CHETSANGA, C. J., BEARIE, B. and MAKAROFF, C. (1982) Alkaline opening of imidazole of 7-methylguanosine 1. Analysis of the resulting pyrimidine derivatives. *Chem. Biol. Interact.* **41**: 217–233.

CHUNG, F. L. and HECHT, S. S. (1983) Formation of cyclic 1,N^2-adducts by reaction of deoxyguanosine with alpha-acetoxy-N-nitrosopyrrolidine, 4-(carbethoxynitrosamino)butanal, or crotonaldehyde. *Cancer Res.* **43**: 1230–1235.

CHUNG, F. L., YOUNG, R. and HECHT, S. S. (1984) Formation of cyclic 1,N^2-propanodeoxyguanosine adducts in DNA upon reaction with acrolein or crotonaldehyde. *Cancer Res.* **44**: 990–995.

COHEN, G. M., MACLEOD, M. C., MOORE, C. and SELKIRK, J. K. (1980) Metabolism and macromolecular binding of carcinogenic and non carcinogenic metabolites of benzo[a]pyrene by hamster embryo cells. *Cancer Res.* **40**: 207–211.

COLBURN, N. H. and BOUTWELL, R. K. (1966) The binding of β-propiolactone to mouse skin DNA *in vivo*: its correlation with tumor-initiating activity. *Cancer Res.* **26**: 1701–1706.

COOPER, C. S., GERWIN, B. I. and SCHEINER, L. A. (1983) Sites of single-strand breaks in DNA treated with a diol-epoxide of benz[a]anthracene. *Carcinogenesis* **4**: 1645–1649.

COOPER, C. S., MACNICOLL, A. D., RIBEIRO, O., HEWER, A., WALSH, C., PAL, K., GROVER, P. L. and SIMS, P. (1980a) The involvement of a non-'bay-region' diol-epoxide in the metabolic activation of benz[a]anthracene in hamster embryo cells. *Cancer Lett.* **9**: 53–59.

COOPER, C. S., RIBEIRO, O., FARMER, P. B., HEWER, A., WALSH, C., PAL, K., GROVER, P. L. and SIMS, P. (1980b) The metabolic activation of benzo[a]anthracene in hamster embryo cells: evidence that diol-epoxides react with guanosine. *Chem. Biol. Interact.* **32**: 209–231.

COOPER, C. S., RIBEIRO, O., HEWER, A., WALSH, C., GROVER, P. L. and SIMS, P. (1980c) Additional evidence for the involvement of the 3,4-diol 1,2-oxides in the metabolic activation of 7,12-dimethylbenz[a]anthracene in mouse skin. *Chem. Biol. Interact.* **29**: 357–367.

COOPER, C. S., RIBEIRO, O., HEWER, A., WALSH, C., PAL, K., GROVER, P. L. and SIMS, P. (1980d) The involvement of a 'bay-region' and a non 'bay-region' diol epoxide in the metabolic activation of benz[a]anthracene in mouse skin and in hamster embryo cells. *Carcinogenesis* **1**: 233–243.

COOPER, C. S., VIGNY, P., KINDTS, M., GROVER, P. L. and SIMS, P. (1980e) Metabolic activation of 3-methylcholanthrene in mouse skin: fluorescence spectral evidence indicates the involvement of diol-epoxides formed in the 7,8,9,10-ring. *Carcinogenesis* **1**: 855–860.

COOPER, H. K., HAUENSTEIN, E., KOLAR, G. F. and KLEIHUES, P. (1978) DNA alkylation and neuro-oncogenesis by 3,3-dimethyl-1-phenyltrazene. *Acta Neuropath.* **43**: 105–109.

CORDEIRO-STONE, M., TOPAL, M. D. and KAUFMAN, D. G. (1982) DNA in proximity of the site of replication is preferentially alkylated in S phase 10T1/2 cells treated with N-methyl-N-nitrosourea. *Carcinogenesis* **3**: 1119–1127.

CROY, R. G. and WOGAN, G. N. (1981) Temporal patterns of covalent DNA adducts in rat liver after single and multiple doses of alfatoxin B_1. *Cancer Res.* **41**: 197–203.

CROY, R. G., ESSIGMANN, J. M., REINHOLD, V. N. and WOGAN, G. N. (1978) Identification of the principal aflatoxin B_1-DNA adduct formed *in vivo* in rat liver. *Proc. natn. Acad. Sci. U.S.A.* **75**: 1745–1749.

CZARNIK, A. W. and LEONARD, N. J. (1980) Unequivocal assignment of the skeletal structure of the guanine-glyoxal adduct. *J. org. Chem.* **45**: 3514.

DANIEL, F. B. and JOYCE, N. J. (1983) DNA adduct formation by 7,12-dimethylbenz[a]anthracene and its noncarcinogenic 2-fluoro analogue in female Sprague-Dawley rats. *J. natn. Cancer Inst.* **70**: 111–118.

DANIEL, F. B., SCHUT, H. A., SANDWITCH, D. W., SCHENCK, K. M., HOFFMANN, C. O., PATRICK, J. R. and STONER, G. D. (1983) Interspecies comparisons of benzo[a]pyrene metabolism and DNA-adduct formation in cultured human and animal bladder and tracheobronchial tissues. *Cancer Res.* **43**: 4723–4729.

DAUDEL, P., DUQUESNE, M., VIGNY, P., GROVER, P. L. and SIMS, P. (1975) Fluorescence spectral evidence that benzo[a]pyrene-DNA products in mouse skin arise from diol epoxides. *FEBS Lett.* **57**: 250–253.

DELCLOS, K. B., TARPLEY, W. G., MILLER, E. C. and MILLER, J. A. (1984). 4-Aminoazobenzene and N,N-dimethyl-4-aminoazobenzene as equipotent hepatic carcinogens in male C57BL/6 × C3H/HeF1 mice and characterisation of N-(deoxyguanosin-8-yl)-4-aminoazobenzene as the major persistent hepatic DNA-bound dye in these mice. *Cancer Res.* **44**: 2540–2550.

DEN HARTOG, J. H. J., ALTONA, C., CHOTTARD, J. C., GIRAULT, J. P., LALLEMAND, J. Y., DE LEEUW, F. A. A. M., MARCELIS, A. T. M. and REEDIJK, J. (1982) Conformational analysis of the adduct cis—[Pt(NH$_3$)$_2$(d)GpG)}]+ in aqueous solution. A high field (500–300 MHz) nuclear magnetic resonance investigation. *Nucleic Acids Res.* **10**: 4715–4728.

DIGIOVANNI, J., DIAMOND, L., HARVEY, R. G. and SLAGA, T. J. (1983a) Enhancement of the skin tumor-initiating activity of polycyclic aromatic hydrocarbons by methyl-substitution at non-benzo 'bay-region' positions. *Carcinogenesis* **4**: 403–407.

DIGIOVANNI, J., NEBZYDOSKI, A. P. and DECINA, P. C. (1983b) Formation of 7-hydroxymethyl-12-methyl-benz[a]anthracene-DNA adducts from 7,12-dimethylbenz[a]anthracene in mouse epidermis. *Cancer Res.* **43**: 4221–4226.

DIGIOVANNI, J., ROMSON, J. R., LINVILLE, D., JUCHAU, M. R. and SLAGA, T. J. (1979) Covalent binding of polycyclic aromatic hydrocarbons to adenine correlates with tumorigenesis in mouse skin. *Cancer Lett.* **7**: 39–43.

DIPPLE, A. and HAYES, M. E. (1979) Differential excision of carcinogenic hydrocarbon-DNA adducts in mouse embryo cell cultures. *Biochem. biophys. Res. Comm.* **91**: 1225–1231.

DIPPLE, A. and SCHULTZ, E. (1979) Excision of DNA damage arising from chemicals of different carcinogenic potencies. *Cancer Lett.* **72**: 103–108.

DIPPLE, A., MICHEJDA, C. J. and WEISBURGER, E. K. (1987) Metabolism of chemical carcinogens. This volume, p. 1

DIPPLE, A., PIGOTT, M. A. and ANDERSON, L. M. (1984a) 7,12-Dimethylbenz[a]anthracene-DNA adducts in cultures cells from mouse fetuses of different gestational stages. *Cancer Lett.* **21**: 285–292.

DIPPLE, A., PIGOTT, M. A., BIGGER, C. A. H. and BLAKE, D. M. (1984b) 7,12-Dimethylbenz[a]anthracene-DNA binding in mouse skin: response of different mouse strains and effects of various modifiers of carcinogenesis. *Carcinogenesis* **5**: 1087–1090.

DIPPLE, A., PIGOTT, M. A., MOSCHEL, R. C. and COSTANTINO, N. (1983) Evidence that binding of 7,12-dimethylbenz[a]anthracene to DNA in mouse embryo cell cultures results in extensive substitution of both adenine and guanine residues. *Cancer Res.* **43**: 4132–4135.

DIPPLE, A., TOMASZEWSKI, J. E., MOSCHEL, R. C., BIGGER, C. A., NEBZYDOSKI, J. A. and EGAN, M. (1979) Comparison of metabolism-mediated binding to DNA of 7-hydroxymethyl-12-methylbenz[a]anthracene and 7,12-dimethylbenz[a]anthracene. *Cancer Res.* **39**: 1154–1158.

DJURIC, Z., FIFER, E. K. and BELAND, F. A. (1985). Acetyl coenzyme A-dependent binding of carcinogenic and mutagenic dinitropyrenes to DNA. *Carcinogenesis* **6**: 941–944.

DOETSCH, P. W., CHAN, G. L. and HASELTINE, W. A. (1985). T4 DNA polymerase (3'–5') exonuclease, an enzyme for the detection and quantitation of stable DNA lesions: the ultraviolet light example. *Nucl. Acids Res.* **13**: 3285–3304.

DOOLEY, K. L., BELAND, F. A., BUCCI, T. J. and KADLUBAR, F. F. (1984) Local carcinogenicity, rates of absorption, extent and persistence of macromolecular binding, and acute histopathological effects of N-hydroxy-1-naphthylamine and N-hydroxy-2-naphthylamine. *Cancer Res.* **44**: 1172–1177.

DOVICHI, N. J., MARTIN, J. C. and KELLER, R. A. (1983) Attogram detection limit for aqueous dye samples by laser-induced fluorescence. *Science* **219**: 845–847.

DRINKWATER, N. R., J. A. MILLER, E. C. MILLER, and N. C. YANG (1978) Covalent intercalative binding to DNA in relation to the mutagenicity of hydrocarbon epoxides and N-acetoxy-2-acetylaminofl uorine. *Cancer Res* **38**: 3247–55.

DUNN, B. P. (1983) Wide-range linear dose-response curve for DNA binding of orally administered benzo[a]pyrene in mice. *Cancer Res.* **43**: 2654–2658.

EASTMAN, A. and BRESNICK, E. (1979) Metabolism and DNA binding of 3-methylcholanthrene. *Cancer Res.* **39**: 4316–4321.

EASTMAN, A., SWEETENHAM and BRESNICK, E. (1978) Comparison of *in vivo* and *in vitro* binding of polycyclic hydrocarbons to DNA. *Chem. Biol. Interact.* **23**: 345–353.

EDELSON, R. L., BERGER, C. L., GASPARRO. F. P., LEE, K. and TAYLOR, J. (1983) Treatment of leukemic T cell lymphoma with extra-corporeally-photoactivated 8-methoxy psoralen. *Clin. Res.* **31**: 467A.

EHLING, U. H., AVERBECK, D., CERUTTI, P., FRIEDMAN, J., GREIM, H., KOLBYE, A. C. and MENDELSOHN, M. L. (1983) Review for the evidence for the presence or absence of thresholds in the induction of genetic effect by genotoxic chemicals. *Mut. Res.* **123**: 281–341.

EHRENBERG, L. and HUSSAIN, S. (1981) Genetic toxicity of some important epoxides. *Mut. Res.* **86**: 1–113.

EISENSTADT, E., WARREN, A. J., PORTER J., ATKINS, D. and MILLER, J. H. (1982) Carcinogenic epoxides of benzo[a]pyrene and cyclopenta [cd] pyrene induce base substitutions via specific transversions. *Proc. natn. Acad. Sci. U.S.A.* **79**: 1945–1949.

EL-BAYOUMY, K. and HECHT, S. S. (1984) Identification of trans-1,2-dihydro-1,2-dihydroxy-6-nitrochrysene as major mutagenic metabolite of 6-nitrochrysene. *Cancer Res.* **44**: 3408–3413.

EL-BAYOUMY, K., HECHT, S. S., SACKL, T. and STONER, G. D. (1984) Tumorigenicity and metabolism of 1-nitropyrene in A/J mice. *Carcinogenesis* **5**: 1449–1454.

EPSTEIN, J. H. (1979) Risks and benefits of treatment of psoriasis. *New Engl. J. Med.* **300**: 852–853.

ESSIGMANN, J. M., BARKER, L. J., FOWLER, K. W., FRANCISCO, M. A., REINHOLD, V. N. and WOGAN, G. N. (1979) Sterigmatocystin-DNA interactions: identification of a major adduct formed after metabolic activation *in vitro*. *Proc. natn. Acad. Sci. U.S.A.* **76**: 179–183.

ESSIGMANN, J. M., CROY, R. G., BENNET, R. A. and WOGAN, G. N. (1982) Metabolic activation of aflatoxin B_1: patterns of DNA adduct formation, removal and excretion in relation to carcinogenesis. *Drug Metab. Rev.* **13**: 581–602.

ESSIGMANN, J. M., CROY, R. G., NADZAN, A. M., BUSBY, W. F., REINHOLD, V. N., BUCHI, G. and WOGAN, G. N. (1977) Structural identification of the major DNA adducts formed by alfatoxin B_1 *in vitro*. *Proc. natn. Acad. Sci. U.S.A.* **74**: 1870–1874.

ESSIGMANN, J. M., DONAHUE. P. R., STORY, D. L., WOGAN, G. N. and BRUNENGRABER, H. (1980) Use of isolated perfused rat liver to study carcinogen-DNA adduct formation from aflatoxin B_1 and sterigmatocystin. *Cancer Res.* **40**: 4085–4091.

EVERSON, R. B., RANDERATH, E., SANTELLA, R. M., CEFALO, R. C., AVITTS, T. A. and RANDERATH, K. (1986). Detection of smoking-related covalent DNA adducts in human placenta. *Science* **231**: 54–57.

FARMER, P. B., FOSTER, A. B., JARMAN, M. and TISDALE, M. J. (1973) The alkylation of 2'-deoxyguanosine and of thymidine with diazoalkanes. *Biochem. J.* **135**: 203–213.

FELDMAN, G., REMSEN, J., WANG, T. V. and CERUTTI, P. (1980) Formation and excision of covalent deoxyribonucleic acid adducts of benzo[a]pyrene 4,5-epoxide and benzo[a]pyrene diol-epoxide I in human lung cells A549. *Biochemistry* **19**: 1095–1101.

FENNELL, T. R., WISEMAN, R. W., MILLER, J. A. and MILLER, E. C. (1985). Major role of hepatic sulfotransferase activity in the metabolic activation, DNA formation, and carcinogenicity of 1'-hydroxy-2',3'-dehydroestragole in infant male C57BL/6J × C3H/HeJF1 mice. *Cancer Res.* **45**: 5310–5320.

FIGHTINGER-SCHEPMAN, A. J., LOHMAN, P. H. M. and REEDIJK, J. (1982) Detection and quantification of adducts formed upon interaction of diaminedichloroplatinum (II) with DNA, by anion-exchange chromatography after enzymatic degradation. *Nucleic Acids Res.* **10**: 5345–5356.

FIOLES, P. G., TRUSHIN, N. and CASTONGUAY, A. (1985). Measurement of O^6-methyldeoxyguanosine in DNA methylated by the tobacco-specific carcinogen 4-(methylnitroamino)-1-(3-pyridyl)-1-butanone using biotin-avidin enzyme-linked immunoabsorbent assay. *Carcinogenesis* **6**: 989–993.

FISCHER, S. M., OGLE, S., MARNETT, L. J., NESNOW, S. and SLAGA, T. J. (1983) The lack of initiating and/or promoting activity of sodium malondialdehyde on SENCAR mouse skin. *Cancer Lett.* **19**: 61–66.

FLAMMANG, T. J., WESTRA, J. G., KADLUBAR, F. F. and BELAND, F. A. (1985). DNA adducts formed from the probable proximate carcinogen, N-hydroxy-3,2'-dimethyl-4-aminobiphenyl, by acid catalysis or S-acetyl coenzyme A-dependent enzymatic esterification. *Carcinogenesis* **6**: 251–258.

FREDERICK, C. B., WEIS, C. C., FLAMMANG, T. J., MARTIN, C. N. and KADLUBAR, F. F. (1985). Hepatic N-oxidation, acetyl-transfer and DNA-binding of the acetylated metabolites of the carcinogen, benzidine. *Carcinogenesis* **6**: 959–965.

FREI, J. V. and LAWLEY, P. D. (1976) Tissue distribution and mode of DNA methylation in mice by methyl methanesulfonate and N-methyl-N'-nitro-N-nitrosoguanidine: lack of thymic lymphoma induction and low extent of methylation of target tissue DNA at 0–6 of guanine. *Chem. Biol. Interact.* **13**: 215–222.

FREI, J. V., SWENSON, D. H., WARREN, W. and LAWLEY, P. D. (1978) Alkylation of deoxyribonucleic acid *in vivo* in various organs of C57BL mice by the carcinogens N-methyl-N-nitrosourea, N-ethyl-N-nitrosourea and ethyl methanesulfonate in relation to induction of thymic lymphoma. Some applications of HPLC. *Biochem. J.* **174**: 1031–1044.

FU, P. P. and HARVEY, R. G. (1977) Synthesis of the diols and diol epoxides of carcinogenic hydrocarbons. *Tetrahedron Lett.* **12**: 2059.

GALIEGUE-ZOUITINA, S., BAILLEUL, B. and LOUCHEUX-LEFEBVRE, M. H. (1983) *In vitro* DNA reaction with an ultimate carcinogen model of 4-nitroquinoline-1-oxide: the 4-acetoxyaminoquinoline-1-oxide enzymatic degradation of the modified DNA. *Carcinogenesis* **4**: 249–254.

GAMPER, H. B., BARTHOLOMEW, J. C. and CALVIN, M. (1980) Mechanism of benzo[a]pyrene diol epoxide induced deoxyribonucleic acid strand scission. *Biochemistry* **19**: 3948–3956.

GAMPER, H. B., TUNG, A. S. C., STRAUB, K., BARTHOLOMEW, J. C. and CALVIN, M. (1977) DNA strand scission by benzo[a]pyrene diol epoxides. *Science* **197**: 671–674.

GARNER, R. C., MARTIN, C. N., SMITH, J. R. L., COLES, B. G. and TOLSON, M. R. (1979) Comparison of aflatoxin B_1 and aflatoxin G_1 binding to cellular macromolecules *in vitro*, *in vivo* and after peracid oxidation: characterization of the major nucleic acid adducts. *Chem. Biol. Interact.* **26**: 57–73.

GEACINTOV, N. E., HITSHOOSH, H., IBANEZ, V., BENJAMIN, M. J. and HARVEY, R. G. (1984) Mechanism of reaction of 7,8-dihydroxy-9,10-epoxybenzo[a]pyrene with DNA in aqueous solution. *Biophys. Chem.* **20**: 121–133.

GEACINTOV, N. E., IBANEZ, V., GAGLIANO, A. G., YOSHIDA, H. and HARVEY, R. G. (1980) Kinetics of hydrolysis to tetraols and binding of benzo[a]pyrene-7,8-dihydrodiol-9.10-oxide and its tetraol derivatives of DNA. Conformation of adducts. *Biochem. biophys. Res. Commun.* **92**: 1335–1342.

GEACINTOV, N. E., YOSHIDA, H., IBANEZ, V. and HARVEY, R. G. (1981) Non-covalent intercalative binding of 7,8-dihydroxy-9,10-epoxybenzo[a]pyrene to DNA. *Biochem. biophys. Res. Commun.* **100**: 1569–1577.

GEHLY, E. B. and HEIDELBERGER, C. (1982) Metabolic activation of benzo[a]pyrene by transformable and nontransformable C3H mouse fibroblasts in culture. *Cancer Res.* **42**: 2697–2704.

GIESE, A. C. (1971) Photosensitization of natural pigments. *Photophysiology* **6**: 77–129.

GLATT, H., COOPER, C. S., GROVER, P. L., SIMS, P., BENTLEY, P., MERDES, M., WAECHTER, F. and VOGEL, K. (1982) Inactivation of a diol epoxide by dihydrodiol dehydrogenase but not by two epoxide hydrolases. *Science* **215**: 1507–1509.

GLATT, H., FRIEDBERG, T., GROVER, P. L., SIMS, P. and OESCH, F. (1983) Inactivation of a diol-epoxide and K-region epoxide with high efficiency by glutathione transferase X. *Cancer Res.* **43**: 5713–5717.

GOLDSCHMIDT, B. M., BIAZEJ, T. P. and VANDUUREN, B. L. (1978) The reaction of guanosine and deoxyguanosine with glycidaldehyde. *Tetrahedron Lett.* **13**: 1583–1586.

GOMBAR, C. T., TONG, W. P. and LUDLUM, D. B. (1980) Mechanism of action of the nitrosoureas IV. Reactions of BCNU and CCNU with DNA. *Biochem. Pharmac.* **29**: 2639–2643.

GOSHMAN, L. M. and HEIDELBERGER, C. (1967) Binding of tritium labelled polycyclic hydrocarbons to DNA of mouse skin. *Cancer Res.* **27**: 1678–1688.

GRATTON, A. and WISE, R. A. (1985) Multiple organ carcinogenicity of 1,3-butadiene in B6C3F1 mice after 60 weeks of inhalation exposure. *Science* **227**: 548–549.

GRAVES, D. E., PATTARONI, C., KRISHNAN, B. S., OSTRANDER, J. M., HURLEY, L. H. and KRUGH, T. R. (1984). The reaction of anthramycin with DNA. Proton and carbon nuclear magnetic resonance studies on the structure of the anthramycin-DNA adduct. *J. Biol. Chem.* **259**: 8202–8209.

GRAVES, D. E., STONE, M. P. and KRUGH, T. R. (1985). NMR analysis of an oligonucleotide-drug adduct anthramycin-d(ATGCAT). *Prog. Clin. Biol. Res.* **172B**: 193–205.

GREEN, T. and HATHAWAY, D. E. (1978) Interactions of vinyl chloride with rat liver DNA *in vivo*. *Chem. Biol. Interact.* **22**: 211–224.

GROOPMAN, J. D., HAUGER, A., GOODRICH, G. R., WOGAN, G. N. and HARRIS, C. C. (1982) Quantitation of aflatoxin B_1-modified DNA using monoclonal antibodies. *Cancer Res.* **42**: 3120–3124.

GRUNBERGER, D., NELSON, J. H., CANTOR, C. R. and WEINSTEIN, I. B. (1970) Coding and conformational properties of oligonucleotides modified with the carcinogen N-2-acetylaminofluorene. *Proc. natn. Acad. Sci. U.S.A.* **66**: 488–494.

GUPTA, R. C., REDDY, M. V. and RANDERATH, K. (1982) 32P-Postlabeling analysis of non-radioactive aromatic carcinogen-DNA adducts. *Carcinogenesis* **3**: 1081–1092.

GUPTA, R. C. (1985) Enhanced sensitivity of 32P-postlabeling analysis of aromatic carcinogen-DNA adducts. *Cancer Res.* **45**: 5656–5662.

GURTOO, H. L., DAHMS, R. P. and PAIGEN, B. (1978) Metabolic activation of aflatoxins related to their mutagenicity. *Biochem. biophys. Res. Commun.* **81**: 965–972.

HADFIELD, S. T., BHATT, T. S. and COOMBS, M. M. (1984) The biological activity and activation of 15,16-dihydro-1,11-methanocyclopenta[a]phenanthren-17-one, a carcinogen with an obstructed bay region. *Carcinogenesis* **5**: 1485–1492.

HARRIS, C. C., FRANK, A. L., VAN HAAFTEN, C., KAUFMAN, D. G., JACKSON, F. and BARRETT, L. A. (1976) Binding of benzo[a]pyrene to DNA in cultured human bronchus. *Cancer Res.* **36**: 1011–1018.

HASELTINE, W. A., LO, K. M. and D'ANDREA, A. D. (1980) Preferred sites of strand scission in DNA modified by anti-diol epoxide of benzo(a)pyrene. *Science* **209**: 929–931.

HASHIMOTO, Y., SHUDO, K. and OKAMOTO, T. (1979) Structural identification of a modified base in DNA covalently bound with mutagenic 3-amino-1-methyl-5H-pyrido[4,3-b]indole. *Chem. pharm. Bull.* **27**: 1058–1060.

HASHIMOTO, Y., SHUDO, K. and OKAMOTO, T. (1980) Metabolic activation of a mutagen, 2-amino-6-methyldiprido-[1,2-a:3',2'-d]imidazole. Identification of 2-hydroxyamino-6-methyldipyrido[1,2-a:3',2'-d]imidazole and its reaction with DNA. *Biochem. biophys. Res. Commun.* **92**: 971–976.

HASHIMOTO, Y., SHUDO, K. and OKAMOTO, T. (1982) Structures of modified nucleotides isolated from calf thymus DNA alkylated with reductively activated mitomycin C. *Tetrahedron Lett.* **23**: 677–680.

HECHT, S. S., AMIN, S., RIVENSON, A. and HOFFMANN, D. (1979) Tumor initiating activity of 5,11-dimethylchrysene and the structural requirements favoring carcinogenicity of methylated polynuclear aromatic hydrocarbons. *Cancer Lett.* **8**: 65–70.

HECHT, S. S., BONDINELL, W. E. and HOFFMANN, D. (1974) Chrysene and methylchrysenes: presence in tobacco smoke and carcinogenicity. *J. natn. Cancer Inst.* **53**: 1121–1133.

HEISIG, V., JEFFREY, A. M., MCGLADE, M. J. and SMALL, G. J. (1984) Fluorescence line narrowed spectra of polycyclic aromatic carcinogen-DNA adducts. *Science* **223**: 289–291.

HEISIG, V., SANTELLA, R., CORTEZ, C., HARVEY, R. G. and JEFFREY, A. M. (1983) 9-Hydroxybenzo[a]pyrene 4,5-oxide: its role in the modification of DNA by benzo[a]pyrene. *Proc. Am. Ass. Cancer Res.* **24**: 68.

HEMMINKI, K. (1980) Identification of guanine adducts of carcinogens by their fluorescence. *Carcinogenesis* **1**: 311–316.

HEMMINKI, K. (1983) Sites of reaction of propane sultone with guanosine and DNA. *Carcinogenesis* **4**: 901–904.

HEMMINKI, K. (1984) Reactions of ethyleneimine with guanosine and deoxyguanosine. *Chem. Biol. Interact.* **48**: 249–260.

HEMMINKI, K. and HESSO, A. (1984) Reaction products of styrene oxide with guanosine in aqueous media. *Carcinogenesis* **5**: 601–607.

HEMMINKI, K. and VAINIO, H. (1984) Genotoxicity of epoxides and epoxy compounds. *Prog. clin. Biol. Res.* **141**: 373–384.

HEMMINKI, K., COOPER, C. S., RIBEIRO, O., GROVER, P. L. and SIMS, P. (1980a) Reactions of 'bay-region' and non-'bay-region' diol-epoxides of benz[a]anthracene with DNA: evidence indicating that the major products are hydrocarbon-N^2-guanine adducts. *Carcinogenesis* **3**: 277–286.

HEMMINKI, K., PAASIVIRTA, J., KURKIRINNE, T. and VIRKKI, L. (1980b) Alkylation products of DNA bases by simple epoxides. *Chem. Biol. Interact.* **30**: 259–270.

HERRON, D. C. and SHANK, R. C. (1979) Quantitative high-pressure liquid chromatography analysis of methylated purines in DNA of rats treated with chemical carcinogens. *Analyt. Biochem.* **100**: 58–63.

HERTZOG, P. J., SMITH, J. R. L. and GARNER, R. C. (1982) Production of monoclonal antibodies to guanine imidazole ring opened aflatoxin B_1 DNA, the persistent DNA adduct *in vivo*. *Carcinogenesis* **3**: 825–828.

HESSE, S., JERNSTROM, B., MARTINEZ, M., MOLDEUS, P., CHRISTODOULIDES, L. and KETTERER, B. (1982) Inactivation of DNA-binding metabolites of benzo[a]pyrene and benzo[a]pyrene-7,8-dihydrodiol by glutathione and glutathione S-transferases. *Carcinogenesis* **3**: 757–761.

HIROTA, F. and SUGIMURA, T. (1983) Potent new mutagens from the pyrolysis of proteins and naturally occurring potent tumor promoters. In: *Genes and Proteins in Oncogenesis*, pp. 111–123, WEINSTEIN, I. B. and VOGEL, A. J. (eds) Academic Press, New York.

HODGSON, R. M., CARY, P. D., GROVER, P. L. and SIMS, P. (1983a) Metabolic activation of chrysene by hamster embryo cells: evidence for the formation of a 'bay region' diol-epoxide-N^2-guanine adduct in RNA. *Carcinogenesis* **4**: 1153–1158.

HODGSON, R. M., PAL, K., GROVER, P. L. and SIMS, P. (1983b) The metabolic activation of chrysene in mouse skin: evidence for the involvement of a triol-epoxide. *Carcinogenesis* **4**: 1639–1643.

HODGSON, R. M., SEIDEL, A., BOCHNITSCHEK, W., GLATT, H. R., OESCH, F. and GROVER, P. L. (1985). The formation of 9-hydroxychrysene-1,2-diol as an intermediate in the metabolic activation of chrysene. *Carcinogenesis* **6**: 135–139.

HOGAN, M. E., DATTAGUPTA, N. and WHITLOCK, J. P. (1981). Carcinogen-induced alteration of DNA structure. *J. Biol. Chem.* **256**: 4504–4513.

HOWARD, P. C., GERRARD, J. A., MILO, G. E., FU, P. P., BELAND, F. A. and KADLUBAR, F. F. (1983a) Transformation of normal human skin fibroblasts by 1-nitropyrene and 6-nitrobenzo[a]pyrene. *Carcinogenesis* **4**: 353–355.

HOWARD, P. C., HEFLICH, R. H., EVANS, F. E. and BELAND, F. A. (1983b) Formation of DNA adducts *in vitro* and in *Salmonella typhimurium* upon metabolic reaction of the environmental mutagen 1-nitropyrene. *Cancer Res.* **43**: 2052–2058.

HSU, I. C., POIRIER, M. C., YUSPA, S. H., GRUNBERGER, D., WEINSTEIN, I. B., YOLKEN, R. H. and HARRIS, C. C. (1981) Measurement of benzo[a]pyrene-DNA adducts by enzyme immunoassays and radio-immunoassay. *Cancer Res.* **41**: 1090–1095.

HUGGINS, C., PATAKI, J. and HARVEY, R. G. (1967) Geometry of carcinogenic polycyclic aromatic hydrocarbons. *Proc. natn. Acad. Sci. U.S.A.* **58**: 2253–2260.

HULBERT, P. B. and GROVER, P. L. (1983) Chemical rearrangement of phenol-epoxide metabolites of polycyclic aromatic hydrocarbons to quinone-methides. *Biochem. biophys. Res. Commun.* **117**: 129–134.

HURLEY, L. H., REYNOLDS, V. L., SWENSON, D. H., PETZOLD, G. L. and SCAHILL, T. A. (1984). Reaction of the antitumor antibiotic CC-1065 with DNA: structure of a DNA adduct with DNA sequence specificity. *Science* **226**: 843–4.

IARC (1982) IARC Monographs on the evaluation of the carcinogenic risk of chemicals to humans. *Supplement* 4.
INSKEEP, P. B. and GENERICH, F. P. (1984) Glutathione-mediated binding of dibromoalkanes to DNA: specificity of rat glutathione-*S*-transferases and dibromoalkane structure. *Carcinogenesis* **5**: 805–808.
INSTITORIS, E. (1981) *In vivo* study of alkylation site in DNA by the bifunctional dianhydrogalactitol. *Chem. Biol. Interact.* **35**: 207–216.
INSTITORIS, E. and TAMAS, J. (1980) Alkylation by 1,2:5,6-dianhydrogalactitol of deoxyribonucleic acid and guanosine. *Biochem. J.* **185**: 659.
IRVING, C. C., VEAZEY, R. A. and RUSSELL, L. T. (1969) Possible role of the glucuronide conjugate in the biochemical mechanism of binding of the carcinogen *N*-hydroxy-2-acetylaminofluorene to rat liver deoxyribonucleic acid *in vivo*. *Chem. Biol. Interact.* **1**: 19–26.
ISSACS, S. T., SHEN C. J., HEARST, J. E. and RAPPORT, H. (1977) Synthesis and characterization of new psoralen derivatives with superior photoreactivity with DNA and RNA. *Biochemistry* **16**: 1058–1064.
IVANOVIC, V. and WEINSTEIN, I. B. (1980) Genetic factors in *Escherichia coli* that affect killing and mutagenicity induced by benzo[a]pyrene 7,8-dihydrodiol 9,10-oxide. *Cancer Res.* **40**: 3508–3511.
IVANOVIC, V., GEANCINTOV, N. and WEINSTEIN, I. B. (1976) Cellular binding of benzo[a]pyrene to DNA characterized by low temperature fluorescence. *Biochem. biophys. Res. Commun.* **70**: 1172–1179.
IVANOVIC, V., GEANCINTOV, N. E., JEFFREY, A. M., FU, P. P., HARVEY, R. G. and WEINSTEIN, I. B. (1978) Cell and microsome mediated binding of 7,12-dimethylbenz[a]anthracene to DNA studied by fluorescence spectroscopy. *Cancer Lett.* **4**: 131–140.
JEFFREY, A. M. and FU, P. P. (1977) Isotopic separations of tritium substituted compounds from protium analogs by high pressure liquid chromatography. *Analyt. Biochem.* **77**: 298–302.
JEFFREY, A. M., BLOBSTEIN, S. H., WEINSTEIN, I. B. and HARVEY, R. G. (1976a) High-pressure liquid chromatography of carcinogen-nucleoside conjugates: separation of 7,12-dimethylbenzanthracene derivatives. *Analyt. Biochem.* **73**: 378–385.
JEFFREY, A. M., BLOBSTEIN, S. H., WEINSTEIN, I. B., BELAND, F. A., HARVEY, R. G., KASAI, H. and NAKANISHI, K. (1976b) Structure of 7,12-dimethylbenz[a]anthracene-guanosine adducts. *Proc. natn. Acad. Sci. U.S.A.* **73**: 2311–2315.
JEFFREY, A. M., GRZESKOWIAK, K., WEINSTEIN, I. B., NAKANISHI, K., ROLLER, P. and HARVEY, R. G. (1979) Benzo[a]pyrene-7,8-dihydrodiol-9,10-oxide adenosine and deoxyadenosine adducts: structure and stereochemistry. *Science* **206**: 1309–1311.
JEFFREY, A. M., JENNETTE, K. W., BLOBSTEIN, S. H., WEINSTEIN, I. B., BELAND, F. A., HARVEY, R. G., KASAI, H. and NAKANISHI, K. (1976c) Benzo[a]pyrene-nucleic acid derivative found *in vivo*: structure of a benzo[a]pyrenetetrahydrodiol epoxide-guanosine adduct. *J. Am. chem. Soc.* **98**: 5714–5715.
JEFFREY, A. M., KINOSHITA, T., SANTELLA, R. M. and WEINSTEIN, I. B. (1980) The chemistry of polycyclic aromatic hydrocarbon-DNA adducts. In: *Carcinogenesis: Fundamental Mechanisms and Environmental Effects*, p. 565, Ts'o, P. P. and GELBOIN, H. (eds) Reidel, Boston.
JEFFREY, A. M., WEINSTEIN, I. B., JENNETTE, K. W., GRZESKOWIAK, K., NAKANISHI, K., HARVEY R. G., AUTRUP, H. and HARRIS, C. (1977) Structures of benzo[a]pyrene-nucleic acid adducts formed in human and bovine bronchial explants. *Nature, Lond.* **269**: 348–350.
JENNETTE, K. W., JEFFREY, A. M., BLOBSTEIN, S. H., BELAND, F. A., HARVEY, R. G. and WEINSTEIN, I. B. (1977) Nucleoside adducts from the *in vitro* reaction of benzo[a]pyrene-7,8-dihydrodiol 9,10-oxide or benzo[a]pyrene 4,5-oxide with nucleic acids. *Biochemistry* **16**: 932–938.
JERINA, D. M., SAYER, J. M., THAKKER, D. R. and YAGI, H. (1980) Carcinogenicity of polycyclic aromatic hydrocarbons: the bay-region theory. In: *Carcinogenesis: Fundamental Mechanisms and Environmental Effects*, pp. 1–12, PULLMAN, B., Ts'o, P. O. P. and GELBOIN, H. (eds) Reidel, Boston.
JOYCE, N. J. and DANIEL, F. B. (1982) 7,12-dimethylbenz[a]anthracene-deoxyribonucleoside adduct formation *in vivo*: evidence for the formation and binding of a mono-hydroxymethyl-DMBA metabolite to rat liver DNA. *Carcinogenesis* **3**: 297–301.
KADLUBAR, F. F., ANSON, J. F., DOOLEY, K. L. and BELAND, F. A. (1981a) Formation of urothelial and hepatic DNA adducts from carcinogen 2-naphthylamine. *Carcinogenesis* **2**: 467–470.
KADLUBAR, F. F., BERANEK, D. T., WEIS, C. C., EVANS, F. E., COX, R. and IRVING, C. C. (1984) Characterization of the purine ring-opened 7-methylguanine and its persistence in rat bladder epithelial DNA after treatment with the carcinogen *N*-methylnitrosourea. *Carcinogenesis* **5**: 587–592.
KADLUBAR, F. F., MELCHOIR, W. B., FLAMMANG, T. J., GAGLIANO, A. G., YOSHIDA, H. and GEACINTOV, N. E. (1981b) Structural consequences of modification of the oxygen atom of guanine in DNA by the carcinogen *N*-hydroxyl-1-naphthylamine. *Cancer Res.* **41**: 2168–2174.
KADLUBAR, F. F., UNRUH, L. E., BELAND, F. A., STRAUB, K. M. and EVANS, F. E. (1980) *In vitro* reaction of the carcinogen, *N*-hydroxy-2-naphthylamine, with DNA at the C-8 and N^2 atoms of guanine and at the N^6 atom of adenine. *Carcinogenesis* **1**: 139–150.
KADLUBAR, F. F., UNRUH, L. E., BELAND, F. A., STRAUB, K. M. and EVANS, F. E. (1981c) Formation of DNA adducts by the carcinogen *N*-hydroxy-2-naphthylamine. *Natn. Cancer Inst. Monogr.* **58**: 143–152.
KANG, D. S., HARVEY, S. C. and WELLS, R. D. (1985). Diepoxybutane forms a monoadduct with B-form (dG-dC)n.(dG-dC)n and a crosslinked diadduct with the left-handed Z-form. *Nucl. Acids Res.* **13**: 5645–5657.
KANNE, D., STRAUB, K. M., HEARST, J. E. and RAPOPORT, H. (1982a) Isolation and characterization of pyrimidine-psoralen-pyrimidine photodiadducts from DNA. *J. Am. chem. Soc.* **104**: 6754–6764.
KANNE, D., STRAUB, K. M., RAPOPORT, H. and HEARST, J. E. (1982b) Psoralendeoxyribonucleic acid photoreaction. Characterization of the monoaddition products of 8-methoxypsoralen and 4,5',8-trimethylpsoralen. *Biochemistry* **21**: 861–871.
KAPITULNIK, J., LEVIN, W., CONNEY, A. H., YAGI, H. and JERINA, D. M. (1977) Benzo[a]pyrene 7,8-dihydrodiol is more carcinogenic than benzo[a]pyrene in newborn mice. *Nature, Lond.* **266**: 378–380.
KAPITULNIK, J., LEVIN, W., YAGI, H., JERINA, D. M. and CONNEY, A. H. (1976) Lack of carcinogenicity of 4-, 5-, 6-, 7-, 8-, 9-, and 10-hydroxybenzo[a]pyrene on mouse skin. *Cancer Res.* **36**: 3625–3628.
KAWAZOE, Y., ARAKI, M., HUNAG, G. F., OKAMOTO, T. and TADA, M. (1975) Chemical structure of QA11, one of

the covalently bound adducts of carcinogenic 4-nitroquinoline 1-oxide with nucleic acid bases of cellular nucleic acids. *Chem. pharm. Bull., Tokyo* **23**: 3041–3043.

KENNELLY, J. C., BELAND, F. A., KADLUBAR, F. F. and MARTIN, C. N. (1984) Binding of N-acetylbenzidine and N,N'-diacetylbenzidine to hepatic DNA of rat and hamster *in vivo* and *in vitro*. *Carcinogenesis* **5**: 407–412.

KIKUCHI O. and HOPFINGER, A. J. (1980) Chemical reactivity of protonated aziridine with nucleophilic centers of DNA bases. *Biopolymers* **19**: 325–340.

KING, H. W. S. and BROOKES, P. (1984) On the nature of the mutations induced by the diolepoxide of benzo[a]pyrene in mammalian cells. *Carcinogenesis* **5**: 965–970.

KING, H. W. S., OSBORNE, M. R. and BROOKES, P. (1978) The identification of 3-methylcholanthrene-9,10-dihydrodiol as an intermediate in the binding of 3-methylcholanthrene to DNA in cells in culture. *Chem. Biol. Interact.* **20**: 367–371.

KING, H. W. S., OSBORNE, M. R. and BROOKES, P. (1979) The *in vitro* and *in vivo* reaction at the N^7 position of guanine of the ultimate carcinogen derived from benzo[a]pyrene. *Chem. Biol. Interact.* **24**: 345–353.

KING, H. W. S., THOMPSON, M. H. and BROOKES, P. (1976) The role of 9-hydroxybenzo[a]pyrene in the microsome mediated binding of benzo[a]pyrene to DNA. *Int. J. Cancer* **18**: 339–344.

KINOSHITA, T., LEE, H. M., HARVEY, R. G. and JEFFREY, A. M. (1982) Structures of covalent adducts derived from the reactions of the 9,10-epoxides of 7,8,9,10-tetrahydrobenzo[a]pyrene and 9,10,11,12-tetrahydrobenzo[a]-pyrene with DNA. *Carcinogenesis* **3**: 255–260.

KLEIHUES, P. and RAJEWSKY, M. F. (1984) Chemical neuro-oncogenesis: role of structural DNA modifications, DNA repair and neural target cell populations. *Prog. exp. Tumor Res.* **27**: 1–16.

KLOPMAN, G., GRINBERG, H. and HOPFINGER, A. J. (1979) MINDO/3 calculations of the conformation and carcinogenicity of epoxy-metabolites of aromatic hydrocarbons: 7,8-dihydroxy-9,10-oxy-7,8,9,10-tetrahydrobenzo[a]pyrene. *J. theor. Biol.* **79**: 355–365.

KOHN, K. W., ERICKSON, L. C., EWIG, R. A. G. and FRIEDMAN, C. A. (1976) Fractionation of DNA from mammalian cells by alkaline elution. *Biochemistry* **15**: 4629–4637.

KOREEDA, M., MOORE, P. D., WISLOCKI, P. G., LEVIN, W., CONNEY, A. H., YAGI, H. and JERINA, D. M. (1976) Alkylation of polyguanylic acid at the 2-amino group and phosphate by the potent mutagen 7 alpha, 8 beta-dihydroxy-9 alpha, 10 beta epoxy-7,8,9,10-tetrahydrobenzo[a]pyrene. *J. Am. chem. Soc.* **98**: 6720–6722.

KOREEDA, M., MOORE, P. D., WISLOCKI, P. G., LEVIN, W., CONNEY, A. H., YAGI, H. and JERINA, D. M. (1978) Binding of benzo[a]pyrene 7,8-diol-9,10-epoxide to DNA, RNA and protein of mouse skin occurs with high stereoselectivity. *Science* **199**: 778–781.

KRAUSS, R. S. and ELING, T. E. (1985). Formation of unique arylamine: DNA adducts from 2-aminofluorene activated by prostaglandin H synthase. *Cancer Res.* **45**: 1680–6.

KRIEK, E. and HENGEVELD, G. M. (1978). Reaction products of the carcinogen N-hydroxy-4-acetyl-4-fluoro with DNA in liver and kidney of the rat. *Chem. Biol. Interact.* **21**: 179–201.

KRIEK, E. and WESTRA, J. G. (1980) Structural identification of the pyrimidine derivatives formed from N-(deoxyguanosin-8-yl)-2-aminofluorene in aqueous solution at alkaline pH^+. *Carcinogenesis* **1**: 459–468.

KRIEK, E., MILLER, J. A., JUHL, U. and MILLER, E. C. (1967) 8-(N-2-Fluorenylacetamido)guanosine, an arylamidation reaction product of guanosine and the carcinogen N-acetoxy-N-2-fluorenylacetamide in neutral solution. *Biochemistry* **6**: 177–182.

KRUGER, F. W. and BERTRAM, B. (1973a) Metabolism of nitroamines *in vivo*. II. On the methylation of nucleic acids by aliphatic di-n-alkyl-nitroamines *in vivo*, caused by β-oxidation. *Z. Krebsforsch.* **79**: 90–97.

KRUGER, F. W. and BERTRAM, B. (1973b) Metabolism of nitrosamines *in vivo*. III. On the methylation of nucleic acids by aliphatic di-n-alkyl-nitrosamines *in vivo* resulting from β-oxidation. *Z. Krebsforch.* **80**: 189–196.

KRUGER, F. W., PREUSSMANN, R. and NEIPELT, N. (1971) Mechanism of carcinogenesis with 1-aryl-3,3-dialkyl-triazenes—III *in vivo* methylation of RNA and DNA with 1-phenyl-3,3-[14C]-dimethyltriazene. *Biochem. Pharmac.* **20**: 529–533.

KURTH, P. D. and BUSTIN, M. (1985). Site-specific carcinogen binding to DNA in polytene chromosomes. *Proc. Natl. Acad. Sci. U.S.A.* **82**: 7076–7080.

KUSMIEREK, J. T. and SINGER, B. (1982) Chloroacetaldehyde-treated ribo- and deoxyribonucleotides. 1. Reaction products. *Biochemistry* **21**: 5717–5722.

LAI, C. C., MILLER, J. A., MILLER, E. C. and LIEM, A. (1985). N-Sulfooxy-2-aminofluorene is the major ultimate electrophilic and carcinogenic metabolite of N-hydroxy-2-acetylaminofluorene in the livers of infant male C57BL/6J × C3H/HeJ F1 (B6C3F1) mice. *Carcinogenesis* **6**: 1037–1045.

LAIB, R. J., GWINNER, L. M. and BOLT, H. M. (1981) DNA alkylation by vinyl chloride metabolites: etheno derivatives or 7-alkylation of guanine? *Chem. Biol. Interact.* **37**: 219–231.

LAQUER, G. L. and SPATZ, M. (1975) Oncogenicity of cycasin and methylazoxymethanol. *GANN Monogr. Cancer Res.* **17**: 189–204.

LAWLEY, P. D. (1984) Carcinogenesis by alkylating agents. In: *Chemical Carcinogens: American Chem. Soc. Monograph*, pp. ud. 182, pp. 325–484, SEARLE, C. E. (ed) Am. Chem. Soc., Washington, DC.

LAWLEY, P. D. and BROOKES, P. (1963) Further studies on the alkylation of nucleic acids and their constituent nucleotides. *Biochem. J.* **89**: 127–138.

LAWLEY, P. D. and JARMAN, M. (1972) Alkylation by propylene oxide of deoxyribonucleic acid, adenine, guanosine and deoxyguanylic acid. *Biochem. J.* **126**: 893–900.

LAWLEY, P. D. and WARREN, W. (1976) Removal of minor methylation products 7-methyl adenine and 3-methylguanine from DNA of *Escherichia coli* treated with dimethyl sulphate. *Chem. Biol. Interact.* **12**: 211–220.

LENG, M., SAGE, E., FUCHS, R. P. and DUANE, M. P. (1978) Antibodies to DNA modified by the carcinogen N-acetoxy-N-2-acetylaminofluorene. *FEBS Lett.* **92**: 207–210.

LEOPOLD, W. R., MILLER, J. A. and MILLER, E. C. (1982) Comparison of some carcinogenic, mutagenic and biochemical properties of S-vinylhomocysteine and ethionine. *Cancer Res.* **42**: 4364–4374.

LEUNG, K. H. and ARCHER, M. C. (1984). Studies on the metabolic activation of beta-keto nitrosamines: mechanisms of DNA methylation by N-(2-oxopropyl)-N-nitrosourea and N-nitroso-N-acetoxymethyl-N-2-oxopropylamine. *Chem. Biol. Interact.* **48**: 169–179.

LEVIN, W., BUENING, M. K., WOOD, A. W., CHANG, R. L., KEDZIERSKI, B., THAKKER, D. R., BOYD, D. R. and GADADINAMATH, G. S. (1980) An enantiomeric interaction in the metabolism and tumorigenicity of (+)- and (−)-benzo[a]pyrene-7,8-oxide. *J. biol. Chem.* **255**: 9067–9074.

LEVIN, W., BUENING, M. K., WOOD, A. W., CHANG, R. L., THAKKER, D. R., JERINA, D. M. and CONNEY, A. H. (1979) Tumorigenic activity of 3-methylcholanthrene metabolites on mouse skin and newborn mice. *Cancer Res.* **39**: 3540–3553.

LEVIN, W., WOOD, A. W., CHANG, R. L., NEWMAN, M. S., THAKKER, D. R., CONNEY, A. H. and JERINA, D. M. (1983) The effect of steric strain in the bay-region of polycyclic aromatic hydrocarbons: tumorigenicity of alkyl-substituted benz[a]anthracenes. *Cancer Lett.* **20**: 139–146.

LIKHACHEV, A. J., MARGISON, G. P. and MONTESANO, R. (1977) Alkylated purines in the DNA in various rat tissues after administration of 1,2-dimethyl-hydrazine. *Chem. Biol. Interact.* **18**: 235–240.

LIN, J., MILLER, J. A. and MILLER, E. C. (1977) 2,3-Dihydro-2-(guan-7-yl)-3-hydroxy-aflatoxin B_1, a major acid hydrolysis product of aflatoxin B_1-DNA or ribosomal RNA adducts formed in hepatic microsome-mediated reactions and rat liver *in vivo*. *Cancer Res.*, **37**: 4430–4438.

LINDAMOOD, C., BEDELL, M. A., BILLINGS, K. C. and SWENBERG, J. A. (1982) Alkylation and *de novo* synthesis of liver cell DNA of C3H mice during continuous dimethylnitrosamine exposure. *Cancer Res.* **42**: 4153–4157.

LIPKOWITZ, K. B., CHEVALIER, T., WIDDIFIELD, M. and BELAND, F. A. (1982) Force field conformational analysis of aminofluorene and acetylaminofluorene substituted deoxyguanosine. *Chem. Biol. Interact.* **40**: 57–76.

LO, K. Y. and KAKUNAGA, T. (1982) Similarities in the formation and removal of covalent DNA adducts in benzo[a]pyrene-treated BALB/3T3 variant cells with different induced transformation frequencies. *Cancer Res.* **42**: 2644–2650.

LOFROTH, I., HEFNER, E., ALFHEIM, I. and MOLLER, M. (1980) Mutagenic activity in photocopies. *Science* **209**: 1037–1039.

LOUKAKOU, B., HEBERT, E., SAINT-RUF, G. and LENG, M. (1985). Reaction of DNA with a mutagenic 3-*N,N*-acetoxyacetylamino-4,6-dimethyldipyrido[1,2-a:3′,2′-d] imidazole (N-AcO-AGlu-P-3) related to glutamic acid pyrolysates. *Carcinogenesis* **6**: 377–383.

LUTZ, W. K., VIVIANI, A. and SCHLATTER, C. (1978) Nonlinear dose-response relationship for the binding of the carcinogen benzo[a]pyrene to rat liver DNA *in vivo*. *Cancer Res.* **38**: 575–578.

MACLEOD, M. C. and TANG, M-S. (1985) Interactions of benzo[a]pyrene diol-epoxides with linear and supercoiled DNA. *Cancer Res.* **45**: 51–56.

MACLEOD, M. C., COHEN, G. M. and SELKIRK, J. K. (1979) Metabolism and macromolecular binding of the carcinogen benzo[a]pyrene and its relatively inert isomer benzo[a]pyrene by hamster embryo cells. *Cancer Res.* **39**: 3463–3470.

MACNICOLL, A. D., COOPER, C. S., RIBEIRO, O., GERVASI, P. G., HEWER, A., WALSH, C., GROVER, P. L. and SIMS, P. (1979) The involvement of a non-'bay-region' diol-epoxide in the formation of benz[a]anthracene-DNA adducts in a rat-liver microsomal system. *Biochem. biophys. Res. Commun.* **91**: 490–497.

MACNICOLL, A. D., COOPER, C. S., RIBEIRO, O., PAL, K., HEWER, A., GROVER, P. L. and SIMS, P. (1981) The metabolic activation of benz[a]anthracene in three biological systems. *Cancer Lett.* **11**: 243–249.

MAHER, V. M. and MCCORMICK, J. J. (1987) Role of DNA lesions and repair in the transformation of human cells. This volume, p. 135

MALAVEILLE, C., TIERNEY, B., GROVER, P. L., SIMS, P. and BARTSCH, H. (1977) High microsome-mediated mutagenicity of the 3,4-dihydrodiol of 7-methylbenz[a]anthracene in *S. typhimurium* TA 98. *Biochem. biophys. Res. Commun.* **75**: 427–433.

MARCELIS, T. M., DEN HARTOG, H. J., VAN DER MAREL, G. A., WILLE, G. and REEDIJK, J. (1983) Interaction of platinum compounds with short oligodeoxynucleotides containing guanine and cytosine. *Eur. J. Biochem.* **136**: 343–349.

MARGISON, G. P., LIKHACHEV, A. J. and KOLAR, G. F. (1979) *In vivo* alkylation of foetal, maternal and normal rat tissue nucleic acids by 3-methyl-1-phenyltriazene. *Chem. Biol. Interact.* **25**: 345–353.

MARSHALL, C. J., VOUSDEN, K. H. and PHILLIPS, D. H. (1984) Activation of c-Ha-ras-a proto-oncogene by *in vitro* modification with a chemical carcinogen, benzo[a]pyrene diol epoxide. *Nature, Lond.* **310**: 586–589.

MARTIN, C. N. and GARNER, R. C. (1977) Aflatoxin B_1-oxide generated by chemical or enzymic oxidation of aflatoxin B_1 causes guanine substitution in nucleic acids. *Nature, Lond.* **267**: 863–865.

MARTIN, C. N., BELAND, F. A., KENNELLY, J. C. and KADLUBAR, F. F. (1983) Binding of benzidine, *N*-acetylbenzidine and *N,N*′-diacetylbenzidine and direct blue 6 to rat liver DNA. *Envir. Hlth Perspect.* **49**: 101–106.

MARTIN, C. N., BELAND, F. A., ROTH, R. W. and KADLUBAR, F. F. (1982) Covalent binding of benzidine and *N*-acetylbenzidine to DNA at the C-8 atom of deoxyguanosine *in vivo* and *in vitro*. *Cancer Res.* **42**: 2678–2686.

MATAGNE, R. (1969) Toxicity of and induction of mutations by isomers of diepoxybutane in *Arabidopsis thaliana*. *Bull. Soc. Bot. Belg.* **102**: 239–248.

MATE, U., SOLOMON, J. J. and SEGAL, A. (1977) *In vitro* binding of beta-propiolactone to calf thymus DNA and mouse liver DNA to form 1(2-carboxyethyl)adenine. *Chem. Biol. Interact.* **18**: 327–336.

MATSUMOTO, H. and HIGA, H. H. (1966) Studies on methylazoxymethanol, the aglycone of cycasin: methylation of nucleic acids *in vitro*. *Biochem. J.* **98**: 20C.

MATTOCKS A. R. (1986) *Chemistry and Toxicology of Pyrrolizidine Alkaloids* Academic Press, N.Y.

MAXAM, A. M. and GILBERT, W. (1977) A new method for sequencing DNA. *Proc. natn. Acad. Sci. U.S.A.* **74**: 650–654.

MCCANN, J., CHOI, E., YAMASAKI, E. and AMES, B. N. (1975a) Detection of carcinogens as mutagens in the Salmonella microsome test: assay of 300 chemicals. *Proc. natn. Acad. Sci. U.S.A.* **72**: 5135–5139.

MCCANN, J., SPRINGARN, N. E., JOBORI, J. and AMES, B. N. (1975b) Detection of carcinogens as mutagens: bacterial tester strains with R factor plasmids. *Proc. natn. Acad. Sci. U.S.A.* **72**: 979–983.

MEEHAN, T. and STRAUB, K. (1979) Double-stranded DNA stereoselectivity binds to benzo[a]pyrene diol epoxides. *Nature, Lond.* **277**: 410–412.

MEEHAN, T., STRAUB, K. and CALVIN, M. (1976a) Elucidation of hydrocarbon structure in an enzyme-catalyzed benzo[a]pyrene-poly (G) covalent complex. *Proc. natn. Acad. Sci. U.S.A.* **73**: 1437–1441.

MEEHAN, T., STRAUB, K. and CALVIN, M. (1977) Benzo[a]pyrene diol epoxide covalently binds to deoxyguanosine and deoxyadenosine in DNA. *Nature, Lond.* **269**: 725–727.

MEEHAN, T., WARSHAWSKY, D. and CALVIN, M. (1976b) Specific positions involved in enzyme catalyzed covalent binding of benzo[a]pyrene to poly (G). *Proc. natn. Acad. Sci. U.S.A.* **73**: 1117–1120.

MELIKIAN, A. A., AMIN, S., HECHT, S. S., HOFFMANN, D., PATAKI, J. and HARVEY, R. G. (1984) Identification of the major adducts formed by reaction of 5-methylchrysene anti-dihydrodiol-epoxides with DNA *in vitro*. *Cancer Res.* **44**: 2524–2529.

MELIKIAN, A. A., HECHT, S. S., HOFFMANN, D., PATAKI, J. and HARVEY, R. G. (1985). Analysis of syn- and anti-1,2-dihydroxy-3,4-epoxy-1,2,3,4-tetrahydro-5-methylchrysene- deoxyribonucleoside adducts by boronate chromatography. *Cancer Lett.* **27**: 91–97.

MELIKIAN, A. A., LAVOIE, E. J., HECHT, S. S. and HOFFMANN, D. (1982) Influence of bay-region methyl group on formation of 5-methylchrysene dihydrodiol epoxide: DNA adducts in mouse skin. *Cancer Res.* **42**: 1239–1242.

MELIKIAN, A. A., LAVOIE, E. J., HECHT, S. S. and HOFFMANN, D. (1983) 5-Methylchrysene metabolism in mouse epidermis *in vivo*, diol epoxide-DNA adduct persistence and epoxide reactivity with DNA as potential factors influencing the predominance of 5-methylchrysene-1,2-diol-3,4-epoxide-DNA adducts in mouse epidermis. *Carcinogenesis* **4**: 843–849.

MELIKIAN, A. A., LESZCZYNSKA, J. M., AMIN, S., HECHT, S. S., HOFFMANN, D., PATAKI, J. and HARVEY, R. G. (1985a). Rates of hydrolysis and extents of DNA binding of 5-methylchrysene dihydrodiol epoxides. *Cancer Res.* **45**: 1990–1996.

MESSIER, F., LU, C., ANDREWS, P., MCCARRY, B. E. and QUILLIAM, M. A. (1981) Metabolism of 1-nitropyrene and formation of DNA adducts in *Salmonella typhimurium*. *Carcinogenesis* **2**: 1007–1011.

MILLER, E. C. and MILLER, J. A. (1947) The presence and significance of bound aminoazo dyes in the livers of rats fed *p*-dimethylaminoazobenzene. *Cancer Res.* **7**: 468–480.

MILLER, E. C. and MILLER, J. A. (1974) Biochemical mechanisms of chemical carcinogenesis. In: *The Molecular Biology of Cancer*, pp. 377–402, BUSCH, H. (ed.) Academic Press, New York.

MILLER, E. C. and MILLER, J. A. (1981) Mechanisms of chemical carcinogenesis. *Cancer* **47**: 1055–1064.

MILLER, K. J., TAYLOR, E. R., DOMMEN, J. and BURBAUM, J. J. (1985). Stereoselectivity of benzo[a]pyrene diol epoxides by DNA for adducts formation with N2 on guanine. *Prog. Clin. Biol. Res.* **172A**: 187–197.

MOORE, P. and STRAUSS, B. S. (1979) Sites of inhibition of *in vitro* DNA synthesis in carcinogen and UV treated OX174 DNA. *Nature, Lond.* **278**: 664–666.

MOSCHEL, R. C. and LEONARD, N. J. (1976) Fluorescent modification of guanine. Reaction with substituted malondialdehydes. *J. org. Chem.* **41**: 294.

MOSCHEL, R. C., HUGGINS, W. R. and DIPPLE, A. (1979) Selectivity in nucleoside alkylation and aralkylation in relation to chemical carcinogenesis. *J. org. Chem.* **44**: 3324–3328.

MOSCHEL, R. C., HUGGINS, W. R. and DIPPLE, A. (1980) Aralkylation of guanosine by the carcinogen *N*-nitro-*N*-benzylurea. *J. org. Chem.* **45**: 533–535.

MOSCHEL, R. C., PIGOTT, M. A., COSTANTINO, N. and DIPPLE, A. (1983) Chromatographic and fluorescence spectroscopic studies of individual 7,12-dimethylbenz[a]anthracene-deoxyribonucleoside adducts. *Carcinogenesis* **4**: 1201–1204.

MULLER, R. J. and RAJEWSKY, M. F. (1978) Sensitive radioimmunoassay for detection of O^6-ethydeoxyguanosine in DNA exposed to the carcinogen ethylinitrosourea *in vivo* or *in vitro*. *Z. Naturfosch.* **33**: 897–901.

MULLER, R. J. and RAJEWSKY, M. F. (1980) Immunological qualification by high-affinity antibodies of O^6-ethydeoxyguanosine in DNA exposed to *N*-ethyl-*N*-nitrosourea. *Cancer Res.* **40**: 887–896.

MULLER, R. J., ADAMKIEWICZ, J. and RAJEWSKY, M. F. (1982) Immunological detection and quantitation of carcinogen-modified DNA components. *IARC Sci. Pub.* **39**: 463–479.

MUYSKEN-SCHOEN, M. A., BAAN, R. A. and LOHMAN, P. H. M. (1985). Detection of DNA adducts in *N*-acetoxy-2-acetylaminofluorene-treated human fibroblasts by means of immunofluorescence microscopy and quantitative immunoautoradiography. *Carcinogenesis* **6**: 999–1004.

NAKANISHI, K., KASAI, H., CHO, H., HARVEY, R. G., JEFFREY, A. M., JENNETTE, K. W. and WEINSTEIN, I. B. (1977) Absolute configuration of a ribonucleic acid adduct formed *in vivo* by metabolism of benzo[a]pyrene. *J. Am. chem. Soc.* **99**: 258–260.

NATIONAL ACADEMY OF SCIENCES (1983) *Identifying and Estimating the Genetic Impact of Chemical Mutagens*. National Academy Press, Washington, DC.

NEHLS, P., ADAMKIEWICZ, J. and RAJEWSKY, M. F. (1984). Immuno-slot-blot: a highly sensitive immunoassay for the quantitation of carcinogen-modified nucleosides in DNA. *J. Cancer Res. Clin. Oncol.* **108**: 23–29.

O'CONNOR, P. J., CAPPS, M. J. and CRAIG, A. W. (1973) Comparative studies of the hepatocarcinogen *N,N*-dimethylnitrosamine *in vivo*: reaction sites in rat liver DNA and the significance of their relative stabilities. *Br. J. Cancer.* **27**: 153–166.

OESCH, F. and DOERJER, G. (1982) Detection of N^2,3-ethanoguanine in DNA after treatment with chloroacetaldehyde *in vitro*. *Carcinogenesis* **3**: 663–665.

OESCH, F., SCHASSMANN, H. and BENTLEY, P. (1978) Specificity of human, rat and mouse skin epoxide hydratase towards K-region epoxides of polycyclic hydrocarbons. *Biochem. Pharmac.* **27**: 17–20.

OGHAKI, H., NEGISHI, C., WAKABAYASHI, K., KUSAMA, K., SATO, S. and SUGIMARA. T. (1984) Induction of sarcomas in rats by subcutaneous injection of dinitropyrenes. *Carcinogenesis* **5**: 587–592.

OSBORNE, M. R., HARVEY, R. G. and BROOKES, P. (1978) The reaction of trans-7,8-dihydroxy-anti-9,10-epoxy-7,8,9,10-tetrahydrobenzo[a]pyrene with DNA involves attack at the *N*-7 position of guanine moieties. *Chem. Biol. Interact.* **20**: 123–130.

OSBORNE, M. R., JACOBS, S., HARVEY, R. G. and BROOKES, P. (1981) Minor products from the reaction of (+) and (−) benzo[a]pyrene-anti-diol epoxide with DNA. *Carcinogenesis* **2**: 553–558.

OSTERMAN-GOLKAR, S., HULTMARK, D., SEGARBACK, D., CALLEMAN, C. J., GOTHE, R., EHRENBERG, L. and WACHTMAESTER, P. A. (1977) Alkylation of DNA and proteins in mice exposed to vinyl chloride. *Biochem. biophys. Res. Commun.* **76**: 259–266.

Parrish, J. A., Fitzpatrick, T. B., Tannenbaum, L. and Pathak, M. A. (1974) Photochemistry of psoriasis with oral methoxsalen and long wavelength ultraviolet light. *New Engl. J. Med.* **291**: 1207–1211.

Paules, R. S., Poirier, M. C., Mass, M. J., Yuspa, S. H. and Kaufman, D. G. (1985). Quantitation by electron microscopy of the binding of highly specific antibodies to benzo[a]pyrene-DNA adducts. *Carcinogenesis* **6**: 193–8.

Pawlak, J. W., Pawlak, K. and Konopa, J. (1983) The mode of action of cytotoxic and antitumor 1-nitroacridines. *Chem. Biol. Interact.* **43**: 151–173.

Pearlman, D. A., Holbrook, S. R., Pirkle, D. H. and Kim, S. H. (1985). Molecular models for DNA damaged by photoreaction. *Science* **227**: 1304–1308.

Pearson, R. G. and Songstad, J. (1967) Application of the principle of hard and soft acids and bases to organic chemistry. *J. Am. chem. Soc.* **89**: 1827–1836.

Peckler, S., Graves, D., Kanne, D., Rapoport, H., Hearst, J. E. and Kin, S. (1982) Structure of a psoralen-thymine monoadduct formed in photoreaction with DNA. *J. molec. Biol.* **162**: 157–172.

Pegg, A. E. and Balog, B. (1979) Formation and subsequent excision of O^6-ethylguanine from DNA of rat liver following administration of diethylnitosamine. *Cancer Res.* **39**: 5003–5009.

Perera, P. P., Poirier, M. C., Yuspa, S. H., Nakayama, J., Jaretzki, A., Curmen, M. M., Knowles, D. M. and Weinstein, I. B. (1982) A pilot project in molecular cancer epidemiology: determination of benzo[a]pyrene-DNA adducts in animal and human tissues by immunoassays. *Carcinogenesis* **3**: 1405–1410.

Petrusek, R. L., Anderson, G. L., Garner, T. F., Pannin, Q. L., Kaplan, D. J., Zimmer, S. G. and Hurley, L. H. (1981) Pyrrolo[1,4] benzodiazepine antibiotics. Proposed structures and characteristics of the *in vitro* deoxyribonucleic acid adducts of anthramycin, tomaymycin, sibiromycin and neothramycins A and B. *Biochemistry* **20**: 1111–1119.

Phillips, D. H., Grover, P. L. and Sims, P. (1979) A quantitative determination of the covalent binding of a series of polycyclic hydrocarbons to DNA in mouse skin. *Int. J. Cancer* **23**: 201–208.

Phillips, D. H., Miller, J. A., Miller, E. C. and Adams, B. (1981a) Structures of the DNA adducts formed in mouse liver after administration of the proximate hepatocarcinogen 1′-hydroxyestragole. *Cancer Res.* **41**: 176–186.

Phillips, D. H., Miller, J. A., Miller, E. C. and Adams, B. (1981b) N^2 atom of guanine and N^6 atom of adenine residues as sites for covalent binding of metabolically activated 1′-hyroxysafrole to mouse liver DNA *in vivo*. *Cancer Res.* **41**: 2664–2671.

Phillips, D. H., Reddy, M. V. and Randerath, K. (1984) 32P Postlabelling analysis of DNA adducts formed in the livers of animals treated with safrole, estragole and other naturally-occurring alkylbenzenes, II. Newborn male B6C3F1 mice. *Carcinogenesis* **5**: 1623–1628.

Plooy, A. C., Fichtinger-Schepman, A. M., Schutte, H. H., van Dijk, M. and Lohman, P. H. (1985). The quantitative detection of various Pt-DNA-adducts in Chinese hamster ovary cells treated with cisplatin: application of immunochemical techniques. *Carcinogenesis* **6**: 561–566.

Poirier, M. C. (1981) Antibodies to carcinogen-DNA adducts. *J. natn. Cancer Inst.* **67**: 515–519.

Poirier, M. C., Lippard, S., Zwelling, L. A., Ushay, M., Kerrigan, D., Thill, C. C., Santella, R. M. and Grunberger, D. (1982) Antibodies elicited against *cis*-diaminedichloroplatinum (II)-modified DNA are specific *cis*-diameneplatinum (II)-DNA adducts formed *in vivo* and *in vitro*. *Proc. natn. Acad. Sci. U.S.A.* **79**: 6443–6447.

Poirier, M. C., Reed, E., Zwelling, L. A., Ozols, R. and Yuspa, S. H. (1985) The use of polyclonal antibodies to quantitate *cis*-platinum drug DNA adducts in cancer patients. *Envir. Hlth Perspect.* **62**: 89–94.

Poulos, A. T., Kuzmin, V. and Geacintov, N. E. (1982) Probing the microenvironment of benzo[a]pyrene diol epoxide-DNA adducts by triplet excited state quenching methods. *J. biochem. Biophys. Meth.* **6**: 269–281.

Preussmann, R. and Hodenberg, A. V. (1969) Mechanism of carcinogenesis with 1-aryl-3,3-diakyltriazines—II *In vitro*-alkylation of guanosine, RNA and DNA with aryl-monoalkyltriazenes to form 7-alkylguanine. *Biochem. Pharmac.* **19**: 1505–1508.

Probst, G. S., McMahon, R. E., Hill, L. E., Thompson, C. Z., Epp, J. K. and Neal, S. B. (1981) Chemically-induced unscheduled DNA synthesis in primary rat hepatocytes cultures: a comparison with bacterial mutagenicity using 218 compounds. *Environ. Mut.* **33**: 11–32.

Rahn, R. O., Chang, S. S., Holland, J. M. and Shugart, I. R. (1982) A fluorometric-HPLC assay for quantitating the binding of benzo[a]pyrene metabolites to DNA. *Biochem. biophys. Res. Commun.* **109**: 262–268.

Rahn, R. O., Chang, S. S., Holland, J. M., Stephens, T. J. and Smith, L. H. (1980) Binding of benzo[a]pyrene to epidermal DNA and RNA as detected by synchronous luminescence spectrometry at 77K. *J. biochem. Biophys. Meth.* **3**: 285–291.

Rajewsky, M. F., Muller, R., Adamiewicz, J. and Drosdziok, W. (1980) Immunological detection and quantification of DNA components structurally modified by alkylating carcinogens ethylnitrosourea. In: *Carcinogenesis: Fundamental Mechanisms and Environmental Effects*, pp. 207–218, Pullman, P., Ts'o, P. O. P. and Gelboin, H. V. (eds) Reidel Press, Doudrecht, Holland.

Randerath, K., Haglund, R. E., Phillips, D. H. and Reddy, M. V. (1984) 32P Postlabelling analysis of DNA adducts formed in the livers of animals treated with safrole, estragole and other naturally-occurring alkenylbenzenes. I. Adult female CD-1 mice. *Carcinogenesis* **5**: 1613–1622.

Reddy, M. V., Gupta, R. C., Randerath, E. and Randerath, K. (1984) 32P-Postlabeling technique for covalent DNA binding of chemicals *in vivo*: application to a variety of aromatic carcinogens and methylating agents. *Carcinogenesis* **5**: 231–243.

Reddy, M. V. and Randerath, K. (1986) Nuclease P1-mediated enhancement of sensitivity of 32p-postlabeling test for structurally diverse DNA adducts. *Carcinogenesis* **7**: 1543–51.

Ribovich, M. L., Miller, J. A., Miller, E. C. and Timmins, L. G. (1982) Labeled 1, N^6-ethenocytidine in hepatic RNA of mice given (ethyl-1,2-3H or ethyl-1-14C) ethyl carbamate (urethan). *Carcinogenesis* **3**: 539–546.

Rice, J. E., Hosted, T. J. and La Voie, E. J. (1984) Fluoranthene and pyrene enhance benzo[a]pyrene-DNA adduct formation *in vivo* in mouse skin. *Cancer Lett.* **24**: 327–333.

ROBERTS, J. J. and WARWICK, G. P. (1963) The reactions of B-propiolactone with guanosine, deoxyguanylic acid and RNA. *Biochem. Pharmac.* **12**: 1441–1442.

ROBERTSON, J. A., NORDENSKJOLD, M. and JERNSTROM, B. (1984) The identification of bases in DNA involved in covalent binding of the reactive metabolite from 9-hydroxybenzo[a]pyrene. *Carcinogenesis* **5**: 821–826.

ROBERTSON, K. A. (1982) Alkylation of N^2 in deoxyguanosine by dehydroretronecine, a carcinogenic metabolite of the pyrrolizidine alkaloid monocrotaline. *Cancer Res.* **42**: 8–14.

ROBINS, D. J. (1982) The pyrrolizidine alkaloids. *Fortschr Chem. org. Natstoffe.* **41**: 115–202.

ROGAN, E., MAILANDER, P. and CAVALIERI, E. (1976) Metabolic activation of aromatic hydrocarbons in purified rat liver nuclei: induction of enzyme activities and binding to DNA with and without monooxygenase-catalyzed formation of active oxygen. *Proc. natn. Acad. Sci. U.S.A.* **73**: 457–461.

ROGAN E., ROTH, R., KATOMSKI, J., BENDERSON, J. and CAVALIERI, E. (1978) Binding of benzo[a]pyrene at the 1,3,6 positions to nucleic acids *in vivo* on mouse skin and *in vitro* with rat liver microsomes and nuclei. *Chem. Biol. Interact.* **22**: 35–51.

ROGERS, K. J. and PEGG, A. E. (1977) Formation of O^6-methylguanine by alkylation of rat liver, colon, and kidney DNA following administration of 1,2-dimethylhydrazine. *Cancer Res.* **37**: 4082–4087.

ROSE, S. N., DAVIES, R. J. H., SETHI, S. K. and MCCLOSKY, J. A. (1983) Formation of an adenine-thymidine photoproduct in the deoxyribonucleoside monophosphate d(TPA) and in DNA. *Science* **220**: 723–725.

ROSENKRANZ, H. S. (1984) Mutagenic and carcinogenic nitroarenes in diesel emissions: risk identification. *Mut. Res.* **140**: 1–6.

RUDIGER, H. W., NOWAK, D., HARTMANN, K. and CERUTTI, P. (1985) Enhanced formation of benzo[a]pyrene:DNA adducts in monocytes of patients with a presumed predisposition to lung cancer. *Cancer Res.* **45**: 5890–5894.

SAGE, E. and HASELTINE, W. A. (1984) High ratio of alkali-sensitive lesions to total DNA modification induced by benzo[a]pyrene diol epoxide. *J. biol. Chem.* **259**: 11098–11102.

Sanders, M. J., Cooper, R. S., Jankowiak, R., Small, G. J., Heisig, V. and Jeffrey, A. M. Identification of polycyclic aromatic hydrocarbon metabolites and DNA adducts in mixtures using fluorescence line narrowing spectrometry, *Anal. Chem.* **58**: 816–820, 1986.

SANTELLA, R. M., GRUNBERGER, D., BROYDE, S. and HINGERTY, B. E. (1981) Z-DNA conformation of N-2-acetylaminofluorene modified poly(dG-dC).poly(dG-dC) determined by reactivity with anti cytidine antibodies and minimized potential energy calculations. *Nucleic Acids Res.* **9**: 5459–5467.

SANTELLA, R. M., GRUNBERGER, D. and WEINSTEIN, I. B. (1979) DNA-Benzo[a]pyrene adducts formed in a *Salmonella typhimurium* mutagenesis assay system. *Mut. Res.* **61**: 181–189.

SANTELLA, R. M., LIN, C. D., CLEVELAND, W. L. and WEINSTEIN, I. B. (1984) Monoclonal antibodies to DNA modified by benzo[a]pyrene diol epoxide. *Carcinogenesis* **5**: 373–377.

SAWICKI, J. T., MOSCHEL, R. C. and DIPPLE, A. (1983) Involvement of both *syn*- and anti-dihydrodiol-epoxides in the binding of 7,12-dimethylbenz[a]anthracene to DNA in mouse embryo cell cultures. *Cancer Res.* **43**: 3212–3218.

SCHERER, E., STEWARD, A. P. and EMMELOT, P. (1977) Kinetics of the formation of O^6-ethylguanine in, and its removal from, liver DNA of rats receiving diethylnitrosamine. *Chem. Biol. Interact.* **19**: 1–11.

SCHERER, E., STEWARD, A. P. and EMMELOT, P. (1980) Formation of precancerous islands in rat liver and modification of DNA by ethyl carbamate: implications for its metabolism. In: *Mechanisms of Toxicity and Hazard Evaluation,* pp. 249–254, HOLMSTEDT, B., LAUWERYS, R., MERCIER, M. and ROBERFROID, M. (eds) Elsevier, North Holland.

SCHERER, E., VAN-DER LAKEN, C. J., GWINNER, L. M., LAIB, R. J. and EMMELOT, P. (1981) Modification of deoxyguanosine by chloroethylene oxide. *Carcinogenesis* **2**: 671–677.

SCHOENTAL, R. (1982) Health hazards of pyrrolizidine alkaloids: a short review. *Toxicol. Lett.* **10**: 323–326.

SCHUT, H. A. J. (1984) Metabolism of carcinogenic amino derivatives in various species and DNA alkylation by their metabolites. *Drug Met. Rev.* **15**: 753–839.

SCRIBNER, J. D. and NAIMY, N. K. (1975). Adducts between the carcinogen 2-acetamidophenanthrene and adenine and guanine in DNA. *Cancer Res.* **35**: 1414–1421.

SEGAL, A., SOLOMON, J. J., MIGNANO, J. and DINO, J. (1981) The isolation and characterization of 3-(2-carboxyethyl)cytosine following *in vitro* reaction of beta-propiolactone with calf thymus DNA. *Chem. Biol. Interact.* **35**: 349–361.

SEGERBACK, D. (1983) Alkylation of DNA and hemoglobin in the mouse following exposure to ethene and ethene oxide. *Chem. Biol. Interact.* **45**: 139–151.

SELKIRK, J. K., CROY, R. G. and GELBOIN, H. V. (1975) Isolation by high pressure liquid chromatography and characterization of benzo[a]pyrene 4,5-epoxide as a metabolite of benzo[a]pyrene. *Archs Biochem. Biophys.* **168**: 322–326.

SHAIKH, B., HUANG, S. S. and PONTZER, N. J. (1980) Urinary excretion of methylated purines and 1-methyl nicotinamide following administration of methylating carcinogens. *Chem. Biol. Interact.* **30**: 253–256.

SHAMAUDDIN, A. M. and HARRIS, C. C. (1983) Improved enzyme immunoassays using biotin-avidin-enzyme complex. *Archs Pathol. lab. Med.* **107**: 514–517.

SHAPIRO, R. and HACHMANN, J. (1966) The reaction of guanine derivatives with 1,2-dicarbonyl compounds. *Biochemistry* **5**: 2799–2807.

SHERMAN, S. E., GIBSON, D., WANG, A. H. J. and LIPPARD, S. J. (1985). X-ray structure of the major adduct of the anticancer drug cisplatin with DNA: *cis*-[Pt(NH3)2{d(pGpG)}]. *Science* **230**: 412–417.

SHIEH, J. C. and SONG, P. S. (1980) Photochemically induced binding of aflatoxins to DNA and its effect on template activity. *Cancer Res.* **40**: 689–695.

SHOYAB, M. (1979) Binding of polycyclic aromatic hydrocarbons to DNA in cells in culture: a rapid method for its analysis using hydroxyalpatite column chromatography. *Chem. Biol. Interact.* **25**: 71–85.

SHUGART, L., HOLLAND, J. M. and RHAN, R. O. (1983) Dosimetry of PAH skin carcinogenesis: covalent binding of benzo[a]pyrene to mouse epidermal DNA. *Carcinogenesis* **4**: 195–198.

SILVERMAN, B. D. (1981) Carcinogenicity of methylated benzo[a]pyrene: calculated ease of formation of the bay-region carbonium ion. *Cancer Biochem. biophys.* **5**: 201–212.

SIMS, P., GROVER, P. L., SWAISLAND, A., PAL, K. and HEWER, A. (1974) Metabolic activation of benzo[a]pyrene proceeds by a diol epoxide. *Nature, Lond.* **252**: 326–327.

SINGER, B. and GRUNBERGER, D. (1983) *Molecular Biology of Mutagens and Carcinogens.* Plenum Press, New York.

SLAGA, T. J., BRACKEN, W. M., DRESNER, S., LEVIN, W., YAGI, H., JERINA, D. M. and CONNEY, A. H. (1978) Skin tumor initiating activities of the twelve isomeric phenols of benzo[a]pyrene. *Cancer Res.* **38**: 678–681.

SLAGA, T. J., FISHER, S. M., NELSON, K. and GLEASON, G. L. (1980) Studies on the mechanism of skin tumor promotion. Evidence for several stages in promotion. *Proc. natn. Acad. Sci. U.S.A.* **77**: 3659–3663.

SLAGA, T. J., GLEASON, G. L., DIGIOVANNI, J., SUKUMARAN, K. B. and HARVEY, R. G. (1979a) Potent tumor-initiating activity of the 3,4-dihydrodiol of 7,12-dimethylbenz[a]anthracene in mouse skin. *Cancer Res.* **39**: 1934–1936.

SLAGA, T. J., HUBERMAN, E., DIGIOVANNI, J., GLEASON, G. and HARVEY, R. G. (1979b) The importance of the 'bay-region' diol-epoxide in 7,12-dimethylbenz[a]anthracene skin tumor initiation and mutagenesis. *Cancer Lett.* **6**: 213–220.

SLOR, H., MIZUSAWA, H., NEIHART, N., KAFEFUDA, T., DAY, R. S. and BUSTIN, M. (1981) Immunochemical visualization of binding of the chemical carcinogen benzo[a]pyrene diol-epoxide 1 to the genome. *Cancer Res.* **41**: 3111–3117.

SMOLAREK, T. A. and BAIRD, W. M. (1984) Benzo[e]pyrene-induced alterations in the binding of benzo[a]pyrene to DNA in hamster embryo cell cultures. *Carcinogenesis* **5**: 1065–1070.

SOLOMON, J. J., COTE, I. L., WORTMAN, M., DECKER, K. and SEGAL, A. (1984) In vitro alkylation of calf thymus DNA by acrylonitrile. Isolation of cyanoethyl-adducts of guanine and thymine and carboxyethyl-adducts of adenine and cytosine. *Chem. Biol. Interact.* **51**: 167–190.

SONG, P. S. and TAPLEY, K. J. (1979) Photochemistry and photobiology of psoralens. *Photochem. Photobiol.* **29**: 1177–1197.

STERN, R. S., THIBODEAU, L. A., KLEINERMAN, R. A., PARRISH, J. A. and FITZPATRICK, T. B. (1979) Risk of cutaneous carcinoma in patients treated with oral methoxsalen photochemotherapy for psoriasis. *New Engl. J. Med.* **300**: 809–813.

STOLLEY, P. D., SOPER, K. A., GALLOWAY, S. M., SMITH, J. G., NICHOLS, W. W. and WOLMAN, S. R. (1984) Sister-chromatid exchanges in association with occupational exposure to ethylene oxide. *Mut. Res.* **129**: 89–102.

STONER, G. D., DANIEL, F. B., SCHENCK, K. M., SCHUT, H. A., SANDWISCH, D. W. and GOHARA, F. (1982) DNA binding and adduct formation of alfatoxin B_1 cultured human and animal tracheobronchial and bladder tissues. *Carcinogenesis* **3**: 1345–1348.

STRAUB, K. M., MEEHAN, T., BURLINGAME, A. L. and CALVIN, M. (1977) Identification of the major adducts formed by reaction of benzo[a]pyrene diol epoxide with DNA *in vitro. Proc. natn. Acad. Sci. U.S.A.* **74**: 5285–5289.

STRICKLAND, P. T. and BOYLE, J. M. (1981) Characterization of two monoclonal antibodies specific for dimerised and non-dimerised adjacent thymidines in single stranded DNA. *Photochem. Photobiol.* **34**: 595–601.

STRNISTE, G. F., MARTINEZ, E., MARTINEZ, A. M. and BRAKE, R. J. (1980) Photo-induced reactions of benzo[a]pyrene with DNA *in vitro. Cancer Res.* **40**: 245–252.

STROUPE, R. C., TOKOUSBALIDES, P., DICKINSON, R. B., WEHRY, E. L. and MAMANTOV, G. (1977) Low-temperature fluorescence spectrometric determination of polycyclic aromatic hydrocarbons by matrix isolation. *Analyt. Chem.* **49**: 701–705.

SUBBIAH, A., ISLAM, S. A. and NEIDLE, S. (1983) Molecular modeling studies on the non-covalent intercalative interactions between DNA and the enantiomers of antibenzo[a]pyrene 7,8 diol-9,10-epoxide. *Carcinogenesis* **4**: 211–215.

SUGIMURA, T., SATO, S., NAGAGO, M., YAHAGI, T., MATSUSHIMA, T., SEINO, Y., TAKEUCHI, M. and KAWACHI, T. (1976) Overlapping of carcinogens and mutagens. In: *Fundamentals in Cancer Prevention*, pp. 191–215, MAGEE, P. N. (ed.) University of Tokyo Press, Tokyo.

SUKUMARAN, S., NOTANO, V., MARTIN-ZANCA, D. and BARBARID, M. (1983) Induction of mammary carcinomas in rats by nitrosomethyl urea involves malignant activation of H-ras-1 locus by a single point mutation. *Nature, Lond.* **306**: 658–661.

SUN, L. and SINGER, B. (1975) The specificity of different classes of ethylating agents toward various sites of HeLa cell DNA *in vitro* and *in vivo. Biochemistry* **14**: 1795–1802.

SVENSON, D. H., HURLEY, L. H., ROKEM, J. S., PETZOLD, G. L., DAYTON, B. D., WALLACE, T. L. and KRUEGER, W. C. (1982) Mechanism of interaction of CC-1065 with DNA. *Cancer Res.* **42**: 2821–2828.

SWAIN, C. G. and SCOTT, C. B. (1953) Quantitative correlation of relative rates. Comparison of hydroxide ion and other nucleophilic reagents toward alkyl halides, esters, epoxides and acyl halides. *J. Am. chem. Soc.* **75**: 141–147.

SWANN, P. F. and MAGEE, P. N. (1968) The alkylation of nucleic acids of the rat by *N*-methyl-*N*-nitrosourea, dimethylnitrosamine, dimethyl sulphate and methyl methanesulfonate. *Biochem. J.* **110**: 39–47.

SWANN, P. F., PEGG, A. E., HAWKS, A., FARBER, E. and MAGEE, P. N. (1971) Evidence for ethylation of rat liver deoxyribonucleic acid after administration of ethionine. *Biochem. J.* **123**: 175–181.

SWENBERG, J. A., DYROFF, M. C., BEDELL, M. A., POPP, J. A., HUH, N., KIRSTEIN, U. and RAJEWSKY, M. F. (1984) O^4-ethyldeoxythymidine, but not O^6-ethyldeoxyguanosine, accumulates in hepatocyte DNA of rats exposed continuously to diethylnitrosamine. *Proc. natn. Acad. Sci. U.S.A.* **81**: 1692–1695.

TADA, M. and TADA, M. (1971) Interaction of a carcinogen 4-nitroquinoline-1-oxide, with nucleic acid. Chemical degradation of the adduct. *Chem. Biol. Interact.* **3**: 225–229.

TADA, M. and TADA, M. (1976) Main binding sites of the carcinogen, 4-nitroquinoline 1-oxide in nucleic acids. *Biochim. biophys. Acta.* **454**: 558–566.

TARPLEY, W. G., MILLER, J. A. and MILLER, E. C. (1980) Adducts from the reaction of *N*-benzoyloxy-*N*-methyl-4-aminoazobenzene with deoxyguanosine or DNA *in vitro* and from hepatic DNA of mice treated with *N*-methyl- or *N,N*-dimethyl-4-aminoazobenzene. *Cancer Res.* **40**: 2493–2499.

TARPLEY, W. G., MILLER, J. A. and MILLER, E. C. (1981) DNA adducts formed from *N*-benzoloxy-*N*-methyl-4-

aminoazobenzene *in vitro* and from *N*,*N*-dimethyl-4-aminoazobenzene in mouse liver. *J. natn. Cancer Inst. Monogr.* **58**: 163–164.

TARPLEY, W. G., MILLER, J. A. and MILLER, E. C. (1982) Rapid release of carcinogen-guanine adducts from DNA after reaction with *N*-acetoxy-2-acetylaminofluorene or *N*-benzoyloxy-*N*-methyl-4-aminoazobenzene. *Carcinogenesis* **3**: 81–88.

TAYLOR, E. R., MILLER, K. J. and BLEYER, A. J. (1983) Interactions of molecules with nucleic acids X. Covalent intercalative binding of the carcinogenic BPDE I (+) to kinked DNA. *J. Biomolec. Struct. Dynam.* **1**: 883–904.

TEJWAN, R., JEFFREY, A. M. and MILO, G. E. (1982) Benzo[a]pyrene diol epoxide DNA adduct formation in transformable and non transformable human fibroblasts cells *in vitro*. *Carcinogenesis* **3**: 727–732.

TIERNY, B., HEWER, A., MACNICOLL, A. D., GERVASI, P. G., RATTLE, H., WALSH, C., GROVER, P. L. and SIMS, P. (1978) The formation of dihydrodiols by the chemical or enzymic oxidation of benz[a]anthracene and 7,12-dimethylbenz[a]anthracene. *Chem. Biol. Interact.* **23**: 243–257.

TILLIS, D. L., STRAUB, K. M. and KADLUBAR, F. F. (1981) A comparison of the carcinogen-DNA adducts formed in rat liver *in vivo* after administration of single or multiple doses of *N*-methyl-4-aminoazobenzene. *Chem. Biol. Interact.* **38**: 15–27.

TOMASZ, M., LIPMAN, R., SNYDER, J. K. and NAKANISHI, K. (1983) Full structure of a mitomycin C dinucleoside phosphate adduct. Use of differential FT-IR spectroscopy in microscale structural studies. *J. Am. chem. Soc.* **105**: 2059–2063.

TOMASZ, M., LIPMAN, R., VERDINE, G. L. and NAKANISHI, K. (1985). Nature of the destruction of deoxyguanosine residues by mitomycin C activated by mild acid pH. *J. Amer. Chem. Soc.* **107**: 6120–6121.

TONG, W. P. and LUDLUM, D. B. (1978) Mechanism of action of the nitrosoureas. I. Role of fluoroethyl cytidine in the reaction of BFNU with nucleic acids. *Biochem. Pharmac.* **27**: 77–81.

TONG, W. P. and LUDLUM, D. B. (1981) Formation of the cross-linked base, diguanylethane, in DNA treated with *N*,*N'*-bis(chloroethyl)-*N*-nitrosourea. *Cancer Res.* **41**: 380–382.

TONG, W. P., KIRK, M. C. and LUDLUM, D. B. (1982) Formation of the cross-linked 1-[N^3-deoxycytidyl],2-[N^1-deoxyguanosinyl]-ethane in DNA treated with *N*,*N'*-bis(2-chloroethyl)-*N*-nitrosourea. *Cancer Res.* **42**: 3102–3105.

TOTH, B. (1980) Actual new cancer-causing hydrazines, hydrazides and hydrazones. *J. Cancer Res. clin. Onc.* **97**: 97–108.

TSURUTA, Y., SUBRAHMANYAM, V. V., MARSHALL, W. and O'BRIEN, P. J. (1985). Peroxidase-mediated irreversible binding of arylamine carcinogens to DNA in intact polymorphonuclear leukocytes activated by a tumor promoter. *Chem. Biol. Interact.* **53**: 25–35.

VAHAKANGAS, K., HAUGEN, A. and HARRIS, C. C. (1985). An applied synchronous fluorescence spectrophotometric assay to study benzo[a]pyrene-diolepoxide-DNA adducts. *Carcinogenesis* **6**: 1109–1116.

VAN DER LAKEN. C. L., HAGENAARS, A. M., HERMSEN, G., KRIEK, E., KUIPERS, A. J., NAGEL, J., SCHERER, E. and WELING, M. (1982) Measurement of O^6-ethyl-deoxyguanosine and *N*-(deoxyguanosine-8-yl)-*N*-acetyl-2-aminofluorene in DNA by high-sensitive enzyme immunoassays. *Carcinogenesis* **5**: 569–572.

VERDINE, G. L. and NAKANISHI, K. (1985). Use of differential second-derivative UV and FTIR spectroscopy in structural studies of multichromophoric compounds. *J. Amer. Chem. Soc.* **107**: 6118–6120.

VERLY, W. G., BRAKIER, L. and FEIT, P. W. (1971) Inactivation of the T7 coliphage by the diepoxybutane stereoisomers. *Biochim. biophys. Acta* **228**: 400–406.

VIGNY, P., DUQUESNE, M., COULOMB, H., TIERNEY, B., GROVER, P. L. and SIMS, P. (1977) Fluorescence spectral studies on the metabolic activation of 3-methylcholanthrene and 7,12-dimethylbenz[a]anthracene. *FEBS Lett.* **82**: 278–282.

VIGNY, P., GINOT, Y. M., KINDTS, M., COOPER, C. S., GROVER, P. L. and SIMS, P. (1980a) Fluorescence spectral evidence that benzo[a]pyrene is activated by metabolism in mouse skin to a diol-epoxide and a phenol-epoxide. *Carcinogenesis* **1**: 945–954.

VIGNY, P., KINDTS, M., COOPER, C. S., GROVER, P. L. and SIMS, P. (1981) Fluorescence spectra of nucleoside-hydrocarbon adducts formed in mouse skin treated with 7,12-dimethylbenz[a]anthracene. *Carcinogenesis* **2**: 115–119.

VIGNY, P., KINDTS, M., DUQUESNE, M., COOPER, C. S., GROVER, P. L. and SIMS, P. (1980b) Metabolic activation of benz[a]anthracene: fluorescence spectral evidence indicates the involvement of a non bay region diol epoxide. *Carcinogenesis* **1**: 33–36.

VIGNY, P., SPIRO, M., HODGSON, R. M., GROVER, P. L. and SIMS, P. (1982) Fluorescence spectral studies on the metabolic activation of chrysene by hamster embryo cells. *Carcinogenesis* **3**: 1491–1493.

WALLIN, H., BORREBAECK, C. A. K., GLAD, C., MATTIASSON, B. and JERGIL, B. (1984) Enzyme immunoassay of benzo[a]pyrene conjugated to DNA, RNA and microsomal proteins using a monoclonal antibody. *Cancer Lett.* **22**: 163–170.

WANG, S. Y. (1976) *Photochemistry and Photobiology*, Vol. 2. Academic Press, New York.

WATABE, T., ISHIZUKA, T., ISOBE, M. and OZAWA, N. (1982) A 7-hydroxymethyl sulfate ester as an active metabolite of 7,12-dimethylbenz[a]anthracene. *Science* **215**: 403–404.

WEINSTEIN, I. B., JEFFREY, A. M., JENNETTE, K. W., BLOBSTEIN, S. H., HARVEY, R. G., HARRIS, C., AUTRUP, H. and NAKANISHI, K. (1976) Benzo[a]pyrene diol epoxides as intermediates in nucleic acid binding *in vitro* and *in vivo*. *Science* **193**: 592–595.

WESTRA, J. G., KRIEK, E. and HITTENHAUSEN, H. (1976) Identification of the persistently bound form of the carcinogen *N*-acetyl-2-aminofluorene to rat liver DNA *in vivo*. *Chem. Biol. Interact.* **15**: 149–164.

WIEBERS, J. L., ABBOTT, P. J., COOMBS, M. M. and LIVINGSTON, D. C. (1981) Mass spectral characterization of the major DNA-carcinogen adduct formed from the metabolically activated carcinogen 15,16-dihydro-11-methylcyclopenta[a]phenanthren-17-one. *Carcinogenesis* **2**: 637–643.

WIGLEY, C. B., NEWBOLD, R. F., AMOS, J. and BROOKES, P. (1979) Cell-mediated mutagenesis in cultured Chinese hamster cells by polycyclic hydrocarbons: mutagenicity and DNA reaction related to carcinogenicity in a series of compounds. *Int. J. Cancer* **23**: 661–669.

WILD, C. P., SMART, G., SAFFHILL, R. and BOYLE, J. M. (1983) Radioimmunoassay of O^6 methyl-deoxyguanosine in DNA of cells alkylated *in vitro* and *in vivo*. *Carcinogenesis* **4**: 1605–1609.

WILLIAMS, G. (1977) Detection of chemical carcinogens by unscheduled DNA synthesis in rat liver primary cell cultures. *Cancer Res.* **37**: 1845–1851.

WOOD, A. W., CHANG, R. L., HUANG, M., LEVIN, W., LEHR, R. E., KUMAR, S., THAKKER, D. R. and YAGI, H. (1980) Mutagenicity of benzo[e]pyrene and triphenylene tetrahydroepoxides and diol epoxides in bacterial and mammalian cells. *Cancer Res.* **40**: 1985–1989.

WOOD, A. W., CHANG, R. L., LEVIN, W., RYAN, D. E., THOMAS, P. E., MAH, H. D., KARLE, J. M. and YAGI, J. (1979a) Mutagenicity and tumorigenicity of phenanthrene and chrysene epoxides and diol epoxides. *Cancer Res.* **39**: 4069–4077.

WOOD, A. W., LEVIN, W., THAKKER, D. R., YAGI, H., CHANG, R. L., RYAN, D. E., THOMAS, P. E. and DANSETTE, P. M. (1979b) Biological activity of benzo[a]pyrene. An assessment based on mutagenic activities and metabolic profiles of the polycyclic hydrocarbon and its derivatives. *J. biol. Chem.* **254**: 4408–4415.

YAGI, H., THAKKER, D. R., HERNANDEZ, O., KOREEDA, M. and JERINA, D. M. (1977) Synthesis and reaction of the highly mutagenic 7,8-diol-9,10-epoxides of the carcinogen benzo[a]pyrene. *J. Am. chem. Soc.* **99**: 1604–1611.

YAGI, T., YAROSH, D. B. and DAY, R. S. (1984) Comparison of repair of O^6-methylguanine produced by N-methyl-N'-nitro-N-nitrosoguanidine in mouse and human cells. *Carcinogenesis* **5**: 593–600.

YAMASAKI, E. and AMES, B. N. (1977) Concentration of mutagens from urine by absorption with the nonpolar main XAD-2: cigarette smokers have mutagenic urine. *Proc. natn. Acad. Sci. U.S.A.* **74**: 3555–3559.

YAMASAKI, H., PULKRABEK, P., GRUNBERGER, D. and WEINSTEIN, I. B. (1977) Differential excision from DNA of the C-8 and N^2 guanosine adducts of N-acetyl-a-aminofluorene by single strand specific endonucleases. *Cancer Res.* **37**: 3756–3760.

YEH, C. Y., FU, P. P., BELAND, F. A. and HARVEY, R. G. (1978) CNDO/2 theoretical calculations to interpretation of the chemical reactivity and biological activity of the Syn and Anti diolepoxides of benzo[a]pyrene. *Bioorganic Chem.* **7**: 497–501.

ZURLO, J., CURPHEY, T. J., HILEY, R. and LONGNECKER, D. S. (1982) Identification of 7-carboxymethylguanine in DNA from pancreatic acinar cells exposed to azaserine. *Cancer Res.* **42**: 1286–1288.

CHAPTER 3

TUMOR PROMOTERS AND CELL TRANSFORMATION

Leila Diamond

The Wistar Institute of Anatomy and Biology, 36th Street at Spruce, Philadelphia, PA 19104, U.S.A.

PAH	Polycyclic aromatic hydrocarbon	PDBu	Phorbol-12,13-dibutyrate
B(a)P	Benzo(*a*)pyrene	CEF	Chick embryo fibroblasts
TPA	12-*O*-Tetradecanoylphorbol-13-acetate	EGF	Epidermal growth factor
DMBA	7,12-Dimethylbenz(*a*)anthracene	EBV	Epstein-Barr virus
TPCK	Tosylphenylalanine chloromethyl ketone	RSV	Rous sarcoma virus
RA	Retinoic acid	FELC	Friend erythroleukemia cells
ODC	Ornithine decarboxylase	PDGF	Platelet-derived growth factor
FA	Fluocinolone acetonide	IL-1	Interleukin-1
SOD	Superoxide dismutase	IL-2	Interleukin-2
PG	Prostaglandin	CSF	Colony-stimulating factors
DFMO	α-Difluoromethylornithine	SCE	Sister chromatid exchanges
BHA	2(3)-*tert*-butyl-4-hydroxyanisole	MTX	Methotrexate
TLCK	Tosyl lysine chloromethyl ketone	DHFR	Dihydrofolate reductase
PB	Phenobarbital	SV40	Simian virus 40
2-AAF	2-Acetylaminofluorene	DES	Diethylstilbestrol
DEN	Diethylnitrosamine	MNNG	*N*-Methyl-*N'*-nitro-*N*-nitrosoguanidine
TCDD	2,3,7,8-Tetrachlorodibenzo-*p*-dioxin	3-MCA	3-Methylcholanthrene
DHTB	Dihydroteleocidin B	SBTI	Soybean trypsin inhibitor
HEC	Hamster embryo cells	G_T	Trisialoganglioside

Abstract—Tumor promoters are a class of cocarcinogens that are not themselves mutagenic or carcinogenic but accelerate the rate of tumor development when applied repeatedly after an irreversible event has been initiated by a carcinogen. This article describes the attributes and limitations of models of two-stage transformation *in vitro* that are analogous to models of two-stage tumorigenesis by initiators and promoters *in vivo*. Also described are some models of initiation/promotion *in vivo*, the chemistry of tumor promoters, and the other biological and biochemical effects of tumor promoters on cells in culture that may or may not be related to the promotion of cell transformation or tumorigenesis.

1. INTRODUCTION

The concept that cancer is the end result of a multistage, multifactorial process has been reinforced by a wealth of epidemiological and experimental evidence that has appeared in the last decade or so. The factors involved may be chemical, physical or biological in nature and their sources may be endogenous or exogenous. Thus, one can consider that in humans one's entire lifestyle plays a part in the evolution of neoplastic disease, as Higginson (1983) has discussed. Perhaps the best example of the multistage, multifactorial nature of cancer in humans is the interaction of asbestos and cigarette smoking in the etiology of lung cancer, with the combined factors increasing the risk almost 900-fold compared to that in nonsmoking, nonasbestos workers (Selikoff and Hammond, 1975).

The virologist, Peyton Rous, was one of the first to question whether 'carcinogenic agents act in other ways besides bringing on neoplastic changes'. He asked whether some may act in two ways: first, changing normal cells into neoplastic cells and, second, encouraging the multiplication of the latter. In a 1944 publication that was the first to use the terms *initiation* and *promotion* in the context of tumor development, Friedewald and

Rous presented a superb analysis and discussion of the separate and different effects involved in carcinogenesis. They demonstrated that latent tumour cells initiated in rabbit skin by one treatment could be forced or *promoted* to reveal themselves by subsequent treatment of the skin with agents which did not themselves *initiate* neoplastic change.

Subsequently, the terms 'initiation' and 'promotion' have been defined by the protocol for two-stage carcinogenesis in mouse skin developed by Mottram (1944), Berenblum and Shubik (1947, 1949), and Berenblum (1954). In this now classic model system, a low subcarcinogenic dose of a carcinogen (the initiator) is applied to the skin and this is followed by frequent, repeated applications of a promoter. Eventually papillomas develop and, if the promoter treatment is continued, some carcinomas develop. The initiating action by the carcinogen is irreversible, whereas the effect of the promoter, which is not itself carcinogenic, is, at least in part, reversible. Thus, tumor-promoting compounds may be defined as a class of cocarcinogens that are not themselves mutagenic or carcinogenic but cause the formation of tumors, that is, accelerate the rate of tumor development when applied repeatedly *after* an irreversible event has been 'initiated' by a carcinogen (reviewed in Boutwell, 1964; Van Duuren, 1969; Stenbäck *et al.*, 1974; Scribner and Süss, 1978).

There are now a number of other *in vivo* model systems in which one can demonstrate at least two steps or stages in the carcinogenic process that can be separated in time and mediated or modulated by completely different factors or agents. Some of these experimental systems are still too new to be certain they fulfill the strict criteria for initiation and promotion defined by the mouse skin model. Hicks (1983) recently discussed the rather common problem of considering almost any factor that enhances tumor formation a promoter without first demonstrating that it meets at least the operational requirements defined in the classical model. There is a similar problem with some *in vitro* models of initiation/promotion. Several cell and organ culture systems have been described that are analogous with respect to timing and dosage to two-stage carcinogenesis *in vivo*. With others, the dose schedule does not allow one to distinguish what may simply be an enhancing effect of a particular compound on transformation rather than a promoting effect. These are important considerations if the goal is to have *in vivo* and *in vitro* models to study the mechanism(s) of tumor promotion and be able to screen for compounds with potential tumor-promoting activity in humans.

Recent reviews by Boutwell (1974), Diamond *et al.* (1980), Weinstein *et al.* (1980), Slaga (1980a), Slaga *et al.* (1982) and Shubik (1984), and the reports from two meetings (Slaga *et al.*, 1978; Hecker *et al.*, 1982) discuss the concept of tumor promotion and various theories on the biochemical and cellular mechanisms of promotion. These articles cover the literature prior to 1982 and should be consulted for original articles on specific aspects of the field. In addition, the new series 'Mechanisms of Tumor Promotion', edited by Slago (1983b, 1984a,b,c), provides reviews of the past, as well as more recent literature.

A short history of chemical carcinogenesis *in vitro*, and a discussion of this valuable tool in studying mechanisms of carcinogenesis, is presented by Heidelberger (1981) in Mishra *et al.* (1981). There are several recent reviews of transformation by chemicals in rodent cell cultures (International Agency for Research on Cancer Monograph, 1980; Dunkel *et al.*, 1981; Sivak and Tu, 1982; Meyer, 1983; Heidelberger *et al.*, 1983). These describe the major assays that have been used to screen for carcinogens in government and private laboratories, and discuss the advantages and disadvantages of each system. More details about transforming rodent and human cells by chemical and physical agents, and studying the mechanisms involved, can be found in Pienta (1980), Mishra *et al.* (1981), 'Oncology Overview' (1982) and Barrett and Tennant (1985). Transformation of human cells by chemical carcinogens has only recently been successful, and DiPaolo (1983) has discussed some of the possible reasons for the relative difficulties in transforming human and animal cells.

The effects of the phorbol diester tumor promoters on cells in culture have been reviewed most recently by Diamond *et al.* (1978a, 1980), Blumberg (1980, 1981) and Mastro (1982). Meyer (1983), Sivak (1982) and Bohrman (1983) have specifically reviewed the literature on two-stage cell transformation assays that are somewhat analogous to two-stage carcinogenesis *in vivo*. Screening for compounds with promoting activity in these assays is not done routinely as yet and the data base is too small to evaluate them.

The main focus of this article is models of two-stage transformation *in vitro* and their attributes and limitations. As background, there are sections on models of tumor promotion *in vivo*, the chemistry of tumor promoters, and the other biological and biochemical effects of tumor promoters on cells in culture that may or may not be related to the promotion of transformation or tumorigenesis. Because the mouse skin model is the most well-characterized model of tumor promotion and most studies with tumor promoters *in vitro* have been done with the phorbol diester series of mouse skin tumor promoters, the background material emphasizes this model and these compounds. This review is not meant to be all-inclusive, and a number of additional reviews are cited that can provide more in-depth discussion and literature references regarding specific aspects of cell transformation, tumor promotion, and current concepts concerning the mechanism of promotion.

2. TUMOR PROMOTION *IN VIVO*

2.1. Mouse Skin Model

2.1.1. *Concept of Two-Stage Carcinogenesis*

The protocol used in most two-stage carcinogenesis studies in mouse skin was first described by Mottram (1944); it is the basis of the operational criteria defining a promoter. Mottram elicited skin tumors by treating the backs of mice with a single, *subcarcinogenic* dose of the polycyclic aromatic hydrocarbon (PAH), benzo(*a*)pyrene [B(*a*)P], and following this with repeated applications of croton oil, obtained from the seeds of *Croton tiglium*, a member of the Euphorbiaceae family. Subsequently, the most active tumor-promoting component of croton oil was identified as the phorbol diester, 12-*O*-tetradecanoylphorbol-13-acetate (TPA) (Van Duuren and Orris, 1965; Van Duuren, 1969; Hecker, 1968, 1971), and today it is the most frequently used promoter for mechanistic studies.

As can be seen with this protocol (reviewed in Boutwell, 1964, 1974; Scribner and Süss, 1978; Diamond *et al.*, 1980), initiation is a relatively rapid process that produces no apparent morphological alterations in the epidermis. The initiation event is currently assumed to be the result of the interaction of an electrophilic form of the carcinogen with DNA. It can probably persist in the epidermal cells for the lifetime of the animal, even though the epidermis is a continually renewing tissue. In contrast, treatment with promoter produces profound biological and biochemical effects on the skin, and promotion itself is a slow process requiring frequent and repeated exposure to the promoter over a latent period of several weeks or months. This requirement implies that promotion is a reversible stage in carcinogenesis but it is not clear whether all or only some of the events in promotion are reversible (Fürstenberger *et al.*, 1983). The final tumor yield is related to the dose of initiator rather than promoter, provided that the dose of promoter used is not limiting for promotion of all initiated cells.

Two-stage carcinogenesis in which initiation and promotion are clearly separate events mediated by different agents, as with this experimental protocol, is distinct from protocols in which a single, high dose of a carcinogen is sufficient to elicit tumor formation without subsequent promoter treatment. In the latter case, the carcinogen is assumed, perhaps falsely, to act as a complete carcinogen which has both initiating and promoting activity (Scribner and Scribner, 1980; Verma *et al.*, 1982, 1983). It can also be shown that some carcinogens have weak promoting activity in skin initiated by other chemically related or unrelated carcinogens (Scribner and Scriber, 1980; Scribner *et al.*, 1983). However, it is not known whether the mechanism of promotion in these cases is the same as when a promoter such as TPA is used. By the same token, multiple, high doses of TPA, which is usually not considered a carcinogen, can evoke a low incidence of tumor formation in skin not initiated experimentally with a carcinogen (Iversen and Iversen, 1979; Astrup et al., 1980; Hennings et al., 1981), but whether this involves the same mechanisms of initiation and promotion as in a two-stage carcinogenesis protocol is not known.

Whereas protocols that use only carcinogen treatment to induce mouse skin tumors yield primarily carcinomas, the first tumors seen with the two-stage protocol are multiple, benign papillomas. Early discontinuation of promoter treatment can result in regression of many of these tumors; with continued promoter treatment nonregressing basal cell and squamous cell carcinomas usually appear on the skin. Promotion can be carried to a point at which formation of the malignant, nonregressing tumors is inevitable; if the promoter treatments are discontinued at that time, even before such tumors are visible, a significant number of carcinomas will develop, although fewer than if the treatments had been continued.

Hennings et al. (1983b) reported that when mice bearing papillomas induced by a single treatment with 7,12-dimethylbenz(a)anthracene (DMBA) and a relatively short period of TPA treatment were subsequently exposed to any of three initiators, the yield of carcinomas was increased and the rate of carcinoma development decreased, compared to control groups treated continuously with either TPA or acetone; TPA produced no significant increase in carcinoma formation compared with the acetone-treated group in these experiments. They concluded that malignant conversion of mouse skin papillomas is increased by tumor initiators and unaffected by TPA. This interpretation of the data is predicated on the assumption that, under the conditions of their experiments, most carcinomas develop from papillomas, which need not be the case (Burns et al., 1978; Scribner and Scribner, 1982). Interpretation of the findings is also complicated by the fact that the mice were treated with multiple doses of the second initiator, rather than with a single dose.

The important point, however, is that the results are consistent with the suggestions of others that malignant tumors do not arise as a direct consequence of promotion of papillomas (Scribner and Scribner, 1982) but are the result of at least two mutagenic or initiating events (Rous and Kidd, 1941; Potter, 1980, 1981). Potter (1980) has proposed that the role of promoters may be to enhance the proliferation of cells with one relevant mutation and thereby to increase the probability that a second exposure to an initiator will result in a promoter-independent cell with two relevant mutations, the two being required for malignancy. Thus, it is perhaps more appropriate to refer to Mottram's protocol as a model for two-stage *tumorigenesis* rather than carcinogenesis in mouse skin.

Tumor promotion in mouse skin can itself be divided into at least two steps that can each be mediated by TPA. Boutwell (1964) was the first to recognize an early step involving the 'conversion' of an initiated cell into a dormant or latent tumor cell and a second step in which the 'propagation' of that cell is promoted. These steps or stages were further defined by Slaga and coworkers (reviewed in Slaga et al., 1982; Slaga, 1983a). They found that mezerein, a diterpene ester similar in structure to TPA but a weak skin tumor promoter, acts as a potent second stage promoter if applied to initiated skin that first has been treated with TPA, even if with only a single application (Mufson et al., 1979; Slaga et al., 1980a, 1982). Similar experiments have defined the semi-synthetic compound phorbol-12-retinoate-13-acetate as a second stage promoter (Fürstenberger et al., 1981a).

Using TPA and mezerein in a two-stage promotion protocol, Slaga et al. (1980b, 1982) demonstrated that compounds known to inhibit phorbol ester promotion of mouse skin affect specific stages of promotion, presumably by inhibiting events that are important for those stages. For example, the protease inhibitor, tosylphenylalanine chloromethyl ketone (TPCK), is a potent inhibitor specifically of Stage I of promotion, whereas retinoic acid (RA), an inhibitor of promoter-induced ornithine decarboxylase (ODC) activity, specifically inhibits the second stage. Hennings and Yuspa (1985) have analyzed the published data from these and other two-stage promotion experiments and offer alternative interpretations. They propose, on the basis of tissue kinetics, that promotion in mouse skin does not require two mechanistically distinct stages and suggest experiments to resolve the question.

Many of the biological and biochemical effects induced in mouse skin by phorbol diester tumor promoters are also induced by other agents that cause inflammation and hyperplasia (reviewed in Boutwell, 1974; Diamond et al., 1980; Slaga et al., 1982). Thus, it has been extremely difficult to identify the specific effects of promoters in mouse skin which distinguish them from other growth-stimulating factors.

2.1.2. Biology of Tumor Promotion in Mouse Skin

Within a few hours after application of a single effective dose of a promoter such as TPA to mouse skin, localized edema and erythema characteristic of inflammation and irritation are evident, whether or not the skin has been initiated, and by 24 hr, there is leukocytic infiltration of the dermis (see Stenbäck et al., 1974; Scribner and Süss, 1978). At that time, there is also a 5- to 10-fold increase in the percentage of dark basal keratinocytes in the interfollicular epidermis (Raick, 1973a; Klein-Szanto et al., 1980; Klein-Szanto and Slaga, 1981). These 'dark cells' are characterized by their strong basophilia, dense chromatin and large numbers of free ribosomes. They increase in number in TPA-induced hyperplasia to a greater extent than in hyperplasia induced by mezerein or weakly promoting, hyperplastic agents (Raick and Burdzy, 1973; Klein-Szanto et al., 1980); the increase can be prevented by simultaneous treatment of the skin with TPCK or fluocinolone acetonide (FA) (Slaga et al., 1980b), inhibitors of Stage I of promotion. Some investigators (Slaga et al., 1980b, 1982) consider dark cells to be primitive stem cells and an increase in their number an important component of Stage I of promotion; others question these conclusions (Parsons et al., 1983).

Within one to two days after a single promoter treatment, there is a stimulation of mitotic activity in the basal cell layer of the epidermis which continues for several days and results in an increased number of nucleated cell layers (Raick, 1973a). This is followed by a phase of increased keratinization of the upper layers of the epidermis (Bach and Goerttler, 1971; Raick, 1973a; Balmain, 1976). Without additional promoter treatments, all these responses to the promoter gradually subside and the epidermis regains its normal appearance within two to three weeks of treatment (Raick, 1973b). Repeated promoter treatment, however, prevents this subsidence and the skin appears to be in a chronic state of irritation and regenerative hyperplasia (Dammert, 1961). With repeated TPA treatment of initiated skin benign tumors begin to appear in about six weeks (Boutwell, 1964); with some mouse strains such as SENCAR, there may be an average of 20–30 papillomas per mouse after 18 weeks of promoter treatment (DiGiovanni et al., 1980).

Argyris (1983) studied the nature of the epidermal hyperplasia produced by various promoting agents and found that during chronic administration of mezerein, the hyperplasia produced is significantly different from that produced by TPA and eventually becomes 'abnormal' and regresses. He suggests that mezerein may be a poor promoter because it cannot maintain the epidermal hyperplasia it initially produces.

2.1.3. Biochemical Effects of Tumor Promoters on Mouse Skin

Phorbol diesters with promoting activity produce an initial inhibition of tritiated thymidine incorporation into skin or epidermal DNA (Paul and Hecker, 1969; Baird et al., 1971; Raick, 1973a). This is soon followed by greatly increased rates of nucleic acid and protein synthesis (Paul and Hecker, 1969; Hennings and Boutwell, 1970; Baird et al., 1971). Promoter-treated skin shows an increase in phospholipid turnover (Süss et al., 1971; Rohrschneider et al., 1972; Balmain and Hecker, 1974) and prostaglandin accumulation (Verma et al., 1980; Fürstenberger and Marks, 1980); a decreased responsiveness to epidermal chalones (Marks, 1976; Marks et al., 1978) and β-adrenergic agonists (Marks and Grimm, 1972; Grimm and Marks, 1974; Verma and Murray, 1974; Mufson et al., 1977); and a decrease in the basal activities of epidermal superoxide dismutase (SOD) and catalase (Solanki et al., 1981a). Among the effects of promoter treatment on specific proteins are a decrease in epidermal histidase (Colburn et al., 1975); modification of epidermal keratins (Balmain, 1976; Schweizer and Winter, 1982; Nelson and Slaga, 1982); increased synthesis and phosphorylation of histones (Raineri et al., 1973, 1978; Link and Marks, 1981); and a large induction of ODC (O'Brien et al., 1975a,b), the rate-limiting enzyme in polyamine biosynthesis.

Boutwell and coworkers (see Boutwell, 1974, 1983 for review), together with many others, have done extensive studies which point to an essential role for ODC and/or its product, putrescine, in mouse skin tumor promotion, perhaps specifically in the second stage of promotion (Slaga et al., 1982). There have also been many studies in cell culture to define the role of ODC in promotion; these are discussed in Section 4.2. O'Brien et al. (1975a) first showed that a single application of TPA induces more than a 200-fold increase in ODC activity in the epidermis, with a peak at 5–6 hr and a return to the very low basal levels of untreated epidermis by 24 hr. The extent of induction correlates well with the promoting activities of different doses of TPA and different phorbol diesters (O'Brien et al., 1975b). Skin tumors produced by a two-stage tumorigenesis protocol have permanently elevated levels of ODC (O'Brien et al., 1975b, 1976). Second stage promoters such as mezerein also induce epidermal ODC, and inhibitors of Stage II but not Stage I of promotion (e.g. RA) inhibit the induction of ODC by TPA or mezerein (Slaga et al., 1982).

2.1.4. Modifiers of Tumor Promotion in Mouse Skin

High doses of indomethacin, a potent inhibitor of a key enzyme in prostaglandin (PG) synthesis (Robinson and Vane, 1974), inhibit the induction of ODC and skin tumor promotion by TPA (Verma et al., 1977, 1980; Viaje et al., 1977). These effects of indomethacin are prevented by treatment of the skin with prostaglandins of the E series (Verma et al., 1980; Nakadate et al., 1982b,c). Retinoids applied to the skin with TPA also inhibit ODC induction and promotion (Verma et al., 1978, 1979). The role of polyamine biosynthesis in mouse skin tumor promotion has been examined most recently with the use of a α-difluoromethylornithine (DFMO), an irreversible inhibitor of ODC and the accumulation of putrescine (Metcalf et al., 1978). When applied to the skin of initiated mice or administered via the drinking water, DFMO inhibits TPA-induced ODC activity and polyamine accumulation and TPA-promoted papilloma formation (Takigawa et al., 1982; Weeks et al., 1982). In two-stage promotion protocols, DFMO has more of an inhibitory effect on tumorigenesis when administered during the TPA portion (Stage I) of the protocol than when administered during the mezerein portion (Stage II) (Weeks et al., 1982).

Lipophilic phenolic antioxidants such as 2(3)-tert-butyl-4-hydroxyanisole (BHA) also inhibit the induction of ODC by TPA in mouse epidermis (Kozumbo et al., 1983) and a copper-coordination complex with SOD-mimetic activity is an inhibitor of both TPA-induced ODC and tumor promotion (Kensler et al., 1983). ODC induction by TPA is also blocked by inhibitors of phospholipase A_2 and lipoxygenase (Nakadate et al., 1982a,b,c; Kato et al., 1983); some of these compounds also inhibit skin tumor promotion by TPA (Kato et al., 1983; Nakadate et al., 1982a). Such studies have suggested that reactive oxygen and/or free radical species; cyclooxygenase and lipoxygenase products; and the stimulation of phospholipase A_2 may all play roles in the mechanism of tumor promotion through their effects on ODC induction by the promoters. However, as with other biological processes, the results of studies with inhibitors of the biological and biochemical effects of promoters are difficult to interpret. Nevertheless, experimental modification of tumor promotion has been very important for understanding the basic mechanisms involved.

Other chemical modifiers of mouse skin tumor promotion by phorbol esters include cortisone and the synthetic glucocorticoids dexamethasone and FA (Belman and Troll, 1972; Scribner and Slaga, 1973; Schwarz et al., 1977; Slaga et al., 1980b); prostaglandins and arachidonic acid (Fischer et al., 1980); synthetic [tosyl lysine chloromethyl ketone (TLCK) and TPCK] and natural (leupeptin) protease inhibitors (Troll et al., 1970; Hozumi et al., 1972); and 1α,25-dihydroxycholecalciferol (Wood et al., 1983), the hormonally active metabolite of vitamin D_3. Some modifiers can inhibit or enhance tumor formation depending on the conditions of the experiment. For example, low doses of indomethacin inhibit TPA-induced ODC and *enhance* tumor formation (Slaga *et al.*, 1982) whereas high doses inhibit both activities (Verma *et al.*, 1977, 1980). Promotion by TPA is inhibited by PGE_1, enhanced by PGF_{2a} and either enhanced or inhibited by PGE_2, depending on the time of its application with respect to TPA (Fischer *et al.*, 1980). RA inhibits promotion by TPA (Verma *et al.*, 1978; Verma *et al.*, 1979) but is itself a promoter of tumor formation in DMBA-initiated mouse skin (Hennings *et al.*, 1982). Other weak promoters such as chrysarobin, applied at nonpromoting doses, can potentiate the tumor-promoting activity of TPA (DiGiovanni and Boutwell, 1983).

The effects of these and other modifiers of promotion are reviewed and described in Slaga (1980b), Slaga *et al.* (1982), Slaga (1983a), Flavin and Kolbye, Jr. (1983), Kensler and Trush (1984), Cerutti (1985) and Troll and Wiesner (1985). Flavin and Kolbye, Jr. (1983) discuss modulation of promotion in terms of the nutritional factors which might affect the critical pathways that lead to tumor formation.

2.2. Other Models of Two-Stage Carcinogenesis

The first definitive evidence that the development of neoplasia in a nonepidermal system can be a multistage process was obtained with a rat liver model of hepatocarcinogenesis. Peraino *et al.* (1971, 1973) showed that dietary phenobarbital (PB) has a promoting effect on liver tumorigenesis in rats previously fed the carcinogen 2-acetylaminofluorene (2-AAF); that the enhancing effect of PB is most clearly demonstrated when the prior regimen of 2-AAF itself results in only a few tumors; that tumor formation is inhibited rather than enhanced when PB is administered at the same time as the carcinogen; and that the promoting action of PB can be demonstrated even when several weeks of control diet intervene between carcinogen and PB treatment. Thus, the model fulfilled all the criteria for an initiation-promotion protocol and has been extremely valuable in demonstrating that there are distinct stages of hepatic tumorigenesis.

Using modifications of the models of Peraino and others, Pitot *et al.* (1978) described a model in which relatively early hepatic changes leading to neoplasia can be analyzed; histochemical tests for a broad spectrum of markers are used to follow the sequential development of enzyme-altered foci. Rats are exposed by gastric intubation to a single nonnecrogenic dose of diethylnitrosamine (DEN) 24 hr after partial hepatectomy. This treatment, which produces foci of presumptive preneoplastic foci, is then followed by prolonged exposure to the promoter. Pitot *et al.* (1978) showed that dietary PB is also a promoter with this protocol, as is 2,3,7,8-tetrachlorodibenzo-*p*-dioxin (TCDD) injected s.c. at very low doses biweekly (Pitot *et al.*, 1980). Goldsworthy *et al.* (1984) recently analyzed the natural history and stability of enzyme-altered foci induced in rat liver by DEN and various doses of dietary PB.

The model of liver carcinogenesis developed by Farber, Solt and coworkers (Solt and Farber, 1976; Solt *et al.*, 1977) is based on creating conditions that enhance the selective growth of putative preneoplastic hepatocytes; earlier evidence has suggested that these are more resistant to the cytotoxic effects of carcinogens than normal hepatocytes. Rats are exposed to a necrogenic dose of a carcinogen such as DEN. Following recovery from any immediate toxicity, the 'selective' conditions are created by feeding the rats a low level of 2-AAF for seven days to inhibit the proliferation of normal hepatocytes and then subjecting them to a stimulus for liver cell proliferation such as partial hepatectomy or CCl_4. Within a short time, some hepatocytes begin rapidly to proliferate, forming foci and visible nodules; at the same time, regeneration of the bulk of the liver is blocked. The advantages of this model are the rapidity and intensity of production of carcinogen-altered hepatocytes, but it is a difficult model in which to demonstrate that there are specific initiation and promotion steps in tumorigenesis. Each of the components (high dose of DEN, 2-AAF, partial hepatectomy or CCl_4) has the potential for both initiating and promoting activity (see

Williams *et al.*, 1981) and it is almost impossible to dissect which activity of any single component may be operative at a particular time and which activities of the multiple components may be overlapping.

Kitagawa *et al.* (1980) developed an *in vivo–in vitro* assay for hepatocarcinogenesis. Cultures of hepatic cells are initiated from the livers of rats treated with carcinogen for various periods of time; the cultures are then observed for the development of foci of proliferating hepatocytes when maintained with and without the addition of PB to the medium. Kitagawa *et al.* (1980) obtained foci of epithelial-like cells when rats had been fed 2-AAF for at least nine weeks; after 4–12 months in culture, the foci had evolved into cell lines that were able to grow in soft agar and to produce hepatocellular carcinomas when injected into newborn syngeneic rats. The number and size of the foci were markedly enhanced by culturing the hepatocytes in PB.

Pitot and Sirica (1980), Farber (1980) and Peraino *et al.* (1983) have recently reviewed their models of hepatocarcinogenesis and discussed the advantages and disadvantages of each. Examples of other promoters of liver tumorigenesis can be found in Pitot and Sirica (1980) and Peraino *et al.* (1983).

There are now several other models of two-stage carcinogenesis *in vivo* and descriptions of these and examples of compounds with tumor-promoting activity in tissues other than skin and liver can be found in Diamond *et al.* (1980), Hecker *et al.* (1982) and Slaga (1983b).

3. CHEMISTRY OF PHORBOL DIESTERS AND OTHER SKIN TUMOR PROMOTERS

3.1. Phorbol and Other Diterpene Esters

The pure skin tumor-promoting compounds in croton oil, obtained from the seeds of *Croton tiglium L.*, were first isolated by the research groups of Hecker (reviewed in Hecker, 1968, 1971) and Van Duuren (reviewed in Van Duuren and Orris, 1965; Van Duuren, 1969). They were identified as 12,13-diesters of phorbol (Fig. 1), an alcohol with the carbon skeleton tigliane; phorbol itself is inactive in mouse skin (Hecker, 1971; Baird and Boutwell, 1971; Baird *et al.*, 1972; Slaga *et al.*, 1976). Two major groups of esters, with the long-chain ester either on C-12 (A group) or on C-13 (B group), were isolated. TPA (Hecker's Fraction A_1, Van Duuren's Fraction C) had the highest tumor-promoting activity of all the phorbol diesters isolated (Van Duuren and Orris, 1965; Van Duuren,

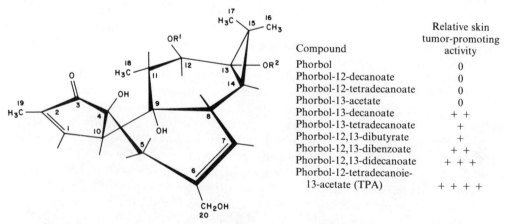

Compound	Relative skin tumor-promoting activity
Phorbol	0
Phorbol-12-decanoate	0
Phorbol-12-tetradecanoate	0
Phorbol-13-acetate	0
Phorbol-13-decanoate	+ +
Phorbol-13-tetradecanoate	+
Phorbol-12,13-dibutyrate	+
Phorbol-12,13-dibenzoate	+ +
Phorbol-12,13-didecanoate	+ + +
Phorbol-12-tetradecanoie-13-acetate (TPA)	+ + + +

Fig. 1. Structure of phorbol as described by Hecker (1978) and the relative skin tumor-promoting activity of some phorbol esters (reviewed in Hecker, 1971, 1978; Diamond *et al.*, 1980). The A series of esters has a long-chain fatty acid on C-12 (R^1) and a short-chain fatty acid on C-13 (R^2); the B series of esters has a long-chain fatty acid on C-13 (R^2) and a short-chain fatty acid on C-12 (R^1).

1969; Hecker, 1968, 1971). [Another numbering system designates the 12- and 13-positions as 9- and 9a-, respectively, and TPA as phorbol-9-myristate-9a-acetate, abbreviated in the literature as phorbol myristate acetate or PMA (see Van Duuren et al., 1978, 1979).]

Symmetrical 12,13-phorbol diesters and monoesters with a range of ester chain lengths have been synthesized, making it possible to study the role of ester size in the induction of biological activity (see Fig. 1) (reviewed in Hecker, 1978; Diamond et al., 1980). These studies showed that biological activity requires esterification with either a long-chain fatty acid on the 13-position or a combination of two fatty acids with a combined length of 14 to 20 carbons on the 12- and 13-positions.

Other studies have identified the portions of the phorbol nucleus required for tumor-promoting activity. The presence of the free allylic hydroxyl group at C-20 is critical, as is the steric configuration at the ring junction of C-4 and C-10; when the C-10-hydrogen and C-4-hydroxyl are *cis* rather than *trans*, as in phorbol (see Fig. 1), the esters are inactive. Loss of the 4-hydroxyl group does not affect biological activity but methylation of the 4-position results in an analog, 4-O-methyl-TPA, that may be a weak Stage I but not a complete tumor promoter (Slaga et al., 1980a; Fürstenberger et al., 1982).

Plants other than *Croton tiglium* contain different types of diterpene esters with tumor-promoting activity in mouse skin. Hecker (1978) has grouped these according to the nature of the carbon skeleton of the terpene molecule as follows: (1) esters of tigliane, including the phorbol diesters of *C. tiglium* described above and esters from other plants with different configurations of hydroxyl groups on tigliane; (2) esters of daphnane, including esters of resiniferonol or modified resiniferonol molecules such as mezerein, gnidia esters, and pimelea esters; and (3) esters of ingenane, including esters of ingenol from *Euphorbia lathyris* and similar esters from other *Euphorbia*. The structures of these esters and their irritant and tumor-promoting activities are reviewed in Hecker (1978). In general, a potent diterpene ester tumor promoter has both a highly lipophilic portion of the molecule and a hydrophilic portion. Major alterations in the hydrophilic or lipophilic nature of the molecule result in large or complete losses of biological activity (see Hecker, 1978; Diamond et al., 1980).

The availability of series of diterpene esters with a wide range of promoting activity has been extremely valuable for studying the mechanism of tumor promotion and identifying specific biological and biochemical effects associated with promotion. These compounds have also been tested for activity in many cell culture systems using a variety of endpoints. In general, good correlations have been obtained between the relative activity in an *in vitro* assay and the relative promoting activity in mouse skin. However, the broad spectrum of biological and biochemical effects induced has made it very difficult to pinpoint those which are specifically related to the mechanism of promotion by these compounds.

3.2. OTHER SKIN TUMOR PROMOTERS

Sugimura and coworkers (reviewed in Sugimura, 1982) have isolated an interesting series of skin tumor promoters which differ in structure from phorbol diesters but induce many of the same biological and biochemical effects (Fig. 2). 'Teleocidin', isolated from *Streptomyces mediocidicus*, is an indole alkaloid composed of teleocidin A, teleocidin B, and their isomers. Both teleocidin and dihydroteleocidin B (DHTB), a catalytically hydrogenated derivative of teleocidin B, have promoting activity comparable to that of TPA (Fujiki et al., 1981, 1982a; Suganuma et al., 1982). Lyngbyatoxin was first isolated from the Hawaiian seaweed *Lyngbya majuscula* by Cardellina II et al. (1979). It was subsequently shown to be structurally identical to teleocidin A (Fig. 2) (Sugimura, 1982).

Two polyacetates which differ in their chemical structure only by the presence or absence of a bromine residue in the hydrophilic region of the molecule (Fig. 2) were isolated from the seaweed *Lyngbya gracilis* (see Sugimura, 1982). Both induce many of the same effects as TPA but aplysiatoxin is a good promoter whereas debromoaplysiatoxin which lacks the bromine residue is a very weak promoter (Fujiki et al., 1982b; Sugimura, 1982; Shimomura

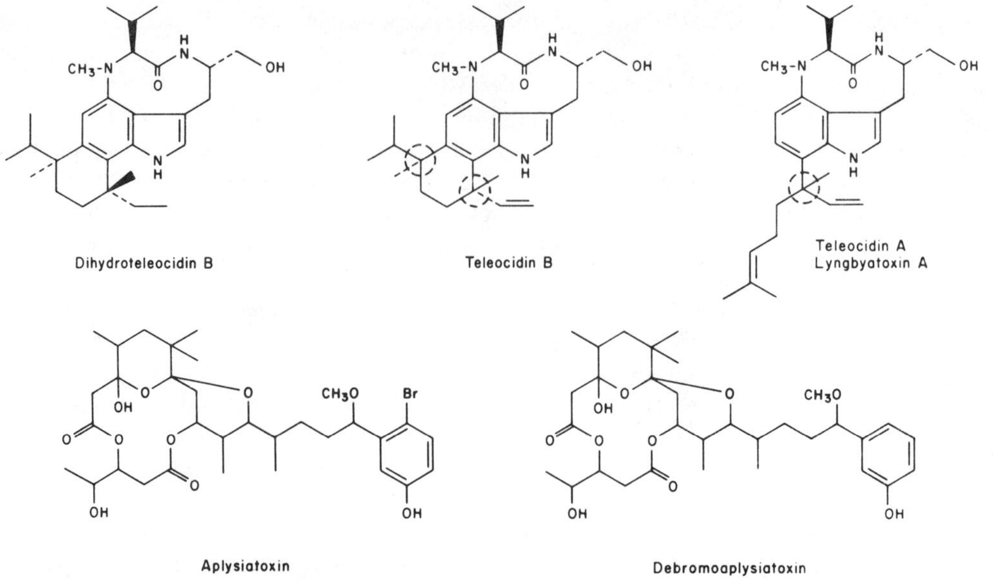

FIG. 2. Structures of various skin tumor promoters as described in Sugimura (1982). A circled carbon atom has R and S configurations so that teleocidin A and B consist of two and four isomers, respectively.

et al., 1983). As with diterpene ester analogs, this pair of compounds is useful for elucidating the mechanism of promotion and the function of the hydrophilic moiety.

Other examples of mouse skin tumor promoters are fatty acid methyl esters (Arffmann and Glavind, 1971); anthralin (Bock and Burns, 1963; Van Duuren *et al.*, 1978); chrysarobin (DiGiovanni and Boutwell, 1983); iodoacetic acid (Gwynn and Salaman, 1953); the weakly acidic fraction of cigarette smoke condensate (Wynder and Hoffman, 1961; Bock *et al.*, 1971; Hecht *et al.*, 1975, 1978); 7-bromomethylbenz(a)anthracene (Scribner and Scribner, 1980); benzoyl peroxide (Slaga *et al.*, 1981); retinoic acid (Hennings *et al.*, 1982); TCDD (Poland *et al.*, 1982); and wound healing (Argyris, 1982). Some (e.g. iodoacetic acid) are very weak and others (e.g. TCDD) are very strong promoters, compared to TPA. (See Diamond *et al.*, 1980, for additional compounds and references.)

3.3. Specific Binding Sites for Phorbol Diesters

Several investigators (Rohrschneider and Boutwell, 1973; Wilson and Huffman, 1976; Weinstein *et al.*, 1977) have suggested that promoters may act by usurping the action of some endogenous hormone or growth factor, perhaps by binding to its cellular receptor. However, the highly lipophilic nature of TPA made it difficult to demonstrate that it bound specifically to cellular targets. To overcome the problem of nonspecific partitioning of the ligand into membrane lipids, Driedger and Blumberg (1980a) developed a binding assay which used phorbol dibutyrate (PDBu) as the ligand. This phorbol derivative is much less lipophilic than TPA but still a good mouse skin tumor promoter (see Fig. 1).

Driedger and Blumberg (1980a,b) and Delclos *et al.* (1980) first characterized the specific binding of [^{3}H]PDBu to particulate preparations of chick embryo fibroblasts (CEF), whole mouse skin, and mouse epidermis. They demonstrated that (a) the structure-activity requirements for inhibiting binding of [^{3}H]PDBu correspond to biological activity in CEF *in vitro* and to promoting activity in mouse skin; (b) binding of [^{3}H]PDBu is only weakly inhibited by diterpene esters that are highly inflammatory but not promoting [the weakly inhibiting resiniferonol derivative, mezerein, has been shown to be a Stage II promoter (Mufson *et al.*, 1979)]; and (c) binding is not blocked by non-phorbol-related promoters (e.g. anthralin, iodoacetic acid), by compounds that inhibit promotion (e.g. dexamethasone, RA) or by compounds that mimic the effects of the phorbol esters *in vitro* or *in vivo* [e.g. epidermal growth factor (EGF), PGE$_2$]. In addition to the plasma membrane receptors, mouse epidermal nuclei contain saturable and specific TPA-binding components (Perrella *et al.*, 1982).

High-affinity and saturable membrane-associated receptor sites for phorbol diesters have also been identified in intact cells. Except for erythrocytes, specific binding has been demonstrated in all intact cells examined, including mouse fibroblasts and keratinocytes (Shoyab and Todaro, 1980; Solanki and Slaga, 1981); human fibroblasts, keratinocytes and melanocytes (Greenebaum *et al.*, 1983; Chida and Kuroki, 1983); and C3H-10T1/2 (Tran *et al.*, 1983); HL-60 (Solanki *et al.*, 1981b; Cooper *et al.*, 1982b) and Friend erythroleukemia (Yamasaki *et al.*, 1982) cells. Differences in binding affinities and/or the number of binding sites per cell do not seem to account for the cells' responsiveness or resistance to the effects of phorbol diesters or for the particular response induced (Sando *et al.*, 1982; Colburn *et al.*, 1982a,b, 1983a; Weinberg *et al.*, 1984). In some instances, down modulation of receptor binding has been associated with decreased cellular responsiveness to the phorbol diesters (Solanki *et al.*, 1981b) but this is not a general finding (Jaken *et al.*, 1981b).

Despite the differences in chemical structure (see Figs 1 and 2), teleocidin, aplysiatoxin and debromoaplysiatoxin are potent inhibitors of specific binding of PDBu (Umezawa *et al.*, 1981; Horowitz *et al.*, 1983; Shimomura *et al.*, 1983). Debromoaplysiatoxin, which has lower promoting and transforming activity than aplysiatoxin (see Sections 3.2 and 5.2), is equipotent with aplysiatoxin and TPA at inhibiting binding of PDBu to C3H/10T1/2 cells but is ~ 10-fold weaker at inhibiting binding to BALB/c 3T3 cells (Horowitz *et al.*, 1983; Shimomura *et al.*, 1983).

The fact that all nucleated animal cells tested bound PDBu or TPA implied that the function of the receptor was essential to life, but the identity of the receptor remained elusive until recently. Then Castagna *et al.* (1982) demonstrated that TPA can directly activate a Ca^{2+}- and phospholipid-dependent protein kinase from rat brain, termed protein kinase C. This is a widely distributed protein kinase in mammalian tissues, especially brain, which is cyclic nucleotide- and calmodulin-insensitive but Ca^{2+}- and phospholipid-dependent (see Nishizuka, 1983, 1984, for review). At physiologically low concentrations of Ca^{2+}, protein kinase C requires for activity, in addition to phospholipid, diacylglycerol which is transiently produced in membranes from phosphatidylinositol in response to a variety of extracellular messengers.

Castagna *et al.* (1982) showed that low concentrations of TPA can substitute for diacylglycerol in the activation *in vitro* of protein kinase C. Subsequently, it was found that saturable phorbol diester receptors in particulate or cytosolic preparations from several cells and tissues copurified with protein kinase C (Niedel *et al.*, 1983; Leach *et al.*, 1983; Ashendel *et al.*, 1983; Sando and Young, 1983; Vandenbark *et al.*, 1984). Sharkey *et al.* (1984) showed that diolein and other diacylglycerols competitively inhibit specific phorbol diester binding and suggested that diacylglycerol may be the 'endogenous analogs' of the phorbol diesters. However, still to be resolved are the exact cellular location of the phorbol diester receptor; the question of whether, after exposure to the phorbol diester *in vivo*, there is translocation of the receptor complex between plasma membrane, cytosol and nucleus or changes in the affinity of the receptor for phospholipid (Kraft *et al.*, 1982; Kraft and Anderson, 1983; Farrar *et al.*, 1985; Perrella *et al.*, 1985); and the functional significance of the receptor-ligand interaction.

Thus, phorbol diesters may exert their effects by initially binding to high-affinity sites on protein kinase C with a resultant increase in membrane-associated kinase activity and subsequent changes in cellular phosphoproteins (see Weinstein, 1983, and Nishizuka, 1984, for discussion). This in turn would imply a commonality, i.e. the enhanced phosphorylation of specific membrane proteins in the initial events induced by phorbol diester and related tumor promoters, oncogenic viruses, and various hormones and growth factors. Although there have been a number of recent reports on TPA-induced phosphorylation of specific protein residues (see Section 4.2), much remains to be done to identify the endogenous phosphate acceptor substrates for phorbol diester-activated protein kinase C and to determine how phosphorylation of those substrates results in specific biological and biochemical effects.

4. BIOLOGICAL AND BIOCHEMICAL EFFECTS OF PHORBOL DIESTERS ON CELLS IN CULTURE

Recent reviews of the biological and biochemical effects of phorbol diester tumor promoters *in vitro* have been published by Diamond *et al.* (1980), Blumberg (1980, 1981), and Mastro (1982, 1983). Examples of the effects of the most potent phorbol diester, TPA, are summarized, together with some appropriate references, in Tables 1–5 under somewhat arbitrary categories. An attempt has been made to point out those effects of TPA and other phorbol diesters which are particularly relevant to current concepts about the mechanism of tumor promotion and, consequently, perhaps also the mechanism of promoter-modulated cell transformation.

It is important to be aware that the phorbol diesters can have different effects on different cells, that even if only a single parameter is being examined, not all cells respond in the same way. A particular effect, or the magnitude of that effect, may depend on the cell type, its tissue and species of origin, and its state of differentiation and/or its stage in the cell cycle at the time of exposure to the promoter. In general, when a series of phorbol diesters has been tested in an *in vitro* cell system, there has been good correlation with tumor-promoting activity *in vivo*. On the other hand, compounds such as the Stage II promoter, mezerein,

TABLE 1. *Some Effects of Phorbol Diester Tumor Promoters on the Cell Surface and Plasma Membrane of Cells in Culture*

	References
Altered cell morphology	Diamond *et al.*, 1974; Driedger and Blumberg, 1977; Wilson and Reich, 1979; Boreiko *et al.*, 1980; Fey and Penman, 1984
Reduced cell volume	Diamond *et al.*, 1974; Driedger and Blumberg, 1977; O'Brien and Krzeminski, 1983
Altered cytoskeletal structures	Rifkin *et al.*, 1979; Toyama *et al.*, 1979; Phaire-Washington *et al.*, 1980a; Laszlo and Bissell, 1983
Induced cell aggregation and adherence	White *et al.*, 1974; Zucker *et al.*, 1974; Castagna *et al.*, 1979; Yamasaki *et al.*, 1979b; Hoshino *et al.*, 1980; Patarroyo *et al.*, 1982; Varani and Fantone, 1982; Badenoch-Jones, 1983
Induction of anchorage-independence	Kopelovich *et al.*, 1979; Colburn *et al.*, 1979, 1980; Jetten, 1983
Disruption of intercellular communication	Yotti *et al.*, 1979; Murray and Fitzgerald, 1979; Fitzgerald and Murray, 1980, 1982; Enomoto *et al.*, 1981; Yancey *et al.*, 1982; Friedman and Steinberg, 1982; Dorman *et al.*, 1983a; Yamasaki *et al.*, 1983
Transport of small molecules	Driedger and Blumberg, 1977; Moroney *et al.*, 1978; Lee and Weinstein, 1979; Dicker and Rozengurt, 1981; O'Brien, 1982; Wrighton and Mueller, 1982; O'Brien and Krzeminski, 1983; Amsler *et al.*, 1983; Nordenberg *et al.*, 1983
Inhibition of ligand binding	Lee and Weinstein, 1978; Shoyab *et al.*, 1979; Magun *et al.*, 1980; Lockyer *et al.*, 1981; King and Cuatrecasas, 1982; Grunberger and Gorden, 1982; Thomopoulos *et al.*, 1982; Osborne and Tashjian, Jr., 1982; Chen *et al.*, 1983
Stimulation of phospholipid metabolism	Süss *et al.*, 1972; Wertz and Mueller, 1978; Ohuchi and Levine, 1978a, 1978b; Hamilton, 1980; Crutchley *et al.*, 1980; Cabot *et al.*, 1980; Fürstenberger *et al.*, 1981b; Cassileth *et al.*, 1981; Beaudry *et al.*, 1982; Valone *et al.*, 1983; Wrighton *et al.*, 1983
Alteration of cell surface glycoproteins and glycolipids	Blumberg *et al.*, 1976; Huberman *et al.*, 1979; Delclos and Blumberg, 1979; Keski-Oja *et al.*, 1979; Ishimura *et al.*, 1980; Fukuda, 1981; Cossu *et al.*, 1982; Dion *et al.*, 1982; Srinivas and Colburn, 1982; Burczak *et al.*, 1983; Ullrich and Hawkes, 1983
Stimulation of protease secretion	Wigler and Weinstein, 1976; Loskutoff and Edgington, 1977; Vassalli *et al.*, 1977; Goldfarb and Quigley, 1978; Wilson and Reich, 1979; Brinckerhoff *et al.*, 1979; Moscatelli *et al.*, 1980; Jaken *et al.*, 1981a

TABLE 2. *Some Effects of Phorbol Diester Tumor Promoters on Intracellular Metabolism* In Vitro

	References
Increased glucose metabolism	Repine *et al.*, 1974; De Chatelet *et al.*, 1976; Zabros *et al.*, 1978; O'Brien *et al.*, 1979b
Inhibition of mitochondrial respiration	Backer *et al.*, 1982a
Increased production of superoxide anion radicals and related active species	De Chatelet *et al.*, 1976, 1982; Witz *et al.*, 1980; Lehrer and Cohen, 1981; Goldstein *et al.*, 1981; Birnboim, 1982a,b; Baxter *et al.*, 1983; Kinsella *et al.*, 1983
Phosphorylation of specific proteins	Feuerstein and Cooper, 1983; Kwong and Mueller, 1983; Gilmore and Martin, 1983; Bishop *et al.*, 1983; Rozengurt *et al.*, 1983; Naka *et al.*, 1983; Jacobs *et al.*, 1983; Iwashita and Fox, 1984; Cochet *et al.*, 1984; Werth and Pastan, 1984
Induction of ornithine decarboxylase	Yuspa *et al.*, 1976b; O'Brien and Diamond, 1977; Lichti *et al.*, 1978, 1981; O'Brien *et al.*, 1980; Wu *et al.*, 1981; O'Brien *et al.*, 1982; Landesman and Mossman, 1982; Sina *et al.*, 1983; Friedman and Cerutti, 1983
Interferon production	Adolf and Swetly, 1980; Yip *et al.*, 1981; Frankfort and Vilček, 1982
Production and release of lymphokines and hormones	Rosenstreich and Mizel, 1979; Farrar *et al.*, 1980a,b,c; Osborne and Tashjian, Jr., 1981; Mastro, 1982; Peterfreund and Vale, 1983; Sasaki *et al.*, 1983

and the indole alkaloid promoter, DHTB, have varied activity, depending on the cell system and/or effect being assayed. Some examples of this are cited in this Section and Section 5.

Also cited are examples of other compounds with potential promoting activity *in vivo* which have been tested for promotion of cell transformation and/or other effects on cells in culture. Testing for the ability to disrupt intercellular communication (see Section 4.1) or to activate latent Epstein-Barr virus (EBV) genomes (see Section 4.2) probably provides the largest data bases for such compounds at this time.

4.1. Cell Surface and Plasma Membrane

4.1.1. *Cell Morphology and Cytoskeleton*

TPA can have striking, early and reversible effects on cell morphology that are unrelated to specific effects on the cells' differentiation pathways (Diamond *et al.*, 1974; Driedger and Blumberg, 1977; Boreiko *et al.*, 1980). The mean cell volume of TPA-treated cells may be reduced 20–30% (Diamond *et al.*, 1974; Driedger and Blumberg, 1977; O'Brien and Krzeminski, 1983). Treated cells can assume a crisscrossed disoriented arrangement similar to but distinguishable from that induced by transformation with Rous sarcoma virus (RSV) (Driedger and Blumberg, 1977; Wilson and Reich, 1979) or chemicals (Popescu *et al.*, 1980), but subsequent treatments may induce less striking morphological alterations (Diamond *et al.*, 1977; Boreiko *et al.*, 1980).

CEF treated with TPA lose the ordered actin-containing cytoskeletal structures found in untreated normal cells (Rifkin *et al.*, 1979). There is also an alteration in the organization of the vimentin-containing intermediate filaments and an increase in the rate of vimentin synthesis (Laszlo and Bissell, 1983). In chick embryo multinucleated myotubes treated with TPA, there is a reversible disappearance of all striated myofibrils and an enhancement in the density of the 10-nm filaments (Toyama *et al.*, 1979).

In thioglycollate-elicited macrophages, TPA stimulates pinocytosis and membrane spreading and causes the extension and radial organization of microtubules and 10-nm filaments; it promotes the movement of secondary lysosomes from their perinuclear location to the peripheral cytoplasm (Phaire-Washington *et al.*, 1980a,b).

4.1.2. Cell Aggregation and Adherence

TPA triggers platelet aggregation and serotonin release from platelets (White *et al.*, 1974; Zucker *et al.*, 1974). TPA-induced activation of human platelets has recently been shown to be associated with protein kinase C-phosphorylation of myosin light chains (Naka *et al.*, 1983). Phorbol diesters rapidly induce aggregation and substrate-adhesion in normal and neoplastic human lymphoblastoid cells (Castagna *et al.*, 1979; Hoshino *et al.*, 1980; Patarroyo *et al.*, 1982); some clones of Friend erythroleukemia cells (FELC) (Yamasaki *et al.*, 1979b); Walker 256 carcinosarcoma cells (Varani and Fantone, 1982); and guinea pig macrophages (Badenoch-Jones, 1983). Patarroyo *et al.* (1983) analyzed the mechanisms, cell surface structures, and cell types involved in PDBu-induced intercellular binding between human lymphocytes; those authors concluded that this binding is mediated by cell surface proteins and requires Ca^{2+}, an intact membrane and functional microfilaments. Inhibitors of trypsin-like enzymes prevent TPA-induced adherence and agglutination of FELC (Fibach *et al.*, 1983).

4.1.3. Anchorage-Independence

Colburn *et al.* (1979, 1980) isolated mouse epidermal cell lines (JB6) that remain anchorage-dependent and nontumorigenic for many passages in culture but undergo an irreversible shift to anchorage-independence in response to exposure to tumor-promoting phorbol diesters. The mechanism of this effect induced by tumor promoters and other agents has been studied extensively by Colburn and coworkers (see Section 5.4). Kopelovich *et al.* (1979) reported that some strains of fibroblasts from individuals with hereditary adenomatosis of the colon and rectum, if treated chronically with TPA, form colonies with low frequency in TPA-supplemented soft agar medium, but cannot be serially maintained in agar beyond two passages. TPA reversibly induces colony formation in soft agar by normal rat kidney fibroblasts [NRK 536-3 (SA 6)]; the number of colonies and their average size are increased by simultaneous treatment with RA (Jetten, 1983).

4.1.4. Intercellular Communication

The first indication that phorbol diesters can disrupt functional intercellular communication was the finding in several cell systems that these agents can inhibit the metabolic cooperation which occurs via the cell–cell contact-dependent transfer of small molecules from donor to recipient cells. Yotti *et al.* (1979) found that TPA increases the recovery of 6-thioguanine-resistant Chinese hamster V79 cells cocultivated with wild type 6-thioguanine-sensitive cells. This observation became the basis of an *in vitro* assay in which a large number of compounds of varied structure have been screened for the ability to inhibit metabolic cooperation and, perhaps therefore, for potential tumor-promoting activity

TABLE 3. *Effects of Phorbol Diester Tumor Promoters on DNA, Chromosomes and Cell Replication* In Vitro

	References
DNA synthesis and cell division	Mueller and Kajiwara, 1965; Whitfield *et al.*, 1973; Diamond *et al.*, 1974; Mastro and Mueller, 1974, 1978; Yuspa *et al.*, 1976a,b; Sivak, 1977; Dicker and Rozengurt, 1978; Frantz *et al.*, 1979; Abb *et al.*, 1979; O'Brien *et al.*, 1979a; Tomei *et al.*, 1981; Eisinger and Marko, 1982
Sister chromatid exchanges	Kinsella and Radman, 1978; Loveday and Latt, 1979; Thompson *et al.*, 1980; Dzarlieva and Fusenig, 1982; Schwartz *et al.*, 1982; Nagasawa *et al.*, 1983
Chromosome aberrations and DNA damage	Emerit and Cerutti, 1981, 1982; Birnboim, 1982a,b; Dzarlieva and Fusenig, 1982
Gene amplification	Varshavsky, 1981b; Hayashi *et al.*, 1983; Bojan *et al.*, 1983

TABLE 4. *Examples of the Effects of Phorbol Diester Tumor Promoters on the Differentiation of Cells in Culture*

Cells	Effects	References
Chick embryo muscle	Inhibit myoblast differentiation and fusion of post-mitotic myoblasts	Cohen et al., 1977; Toyama et al., 1979; Croop et al., 1980; West and Holtzer, 1982
Human skeletal muscle	Inhibit myotube formation and creatine kinase isoenzyme transition	Fisher et al., 1983
Chick chondroblasts	Inhibit synthesis of chondroblast-specific glycosylated proteins	Pacifici and Holtzer, 1977; Lowe et al., 1978
Chick embryo ganglia	Inhibit nerve growth factor-stimulated neurite outgrowth	Ishii, 1978
Human neuroblastoma (SHSY5Y)	Enhance neurite outgrowth	Pahlman et al., 1981; Spinelli et al., 1982
Chick embryo melanocytes	Inhibit melanogenesis in presumptive melanoblasts and pigmented melanocytes	Payette et al., 1980
Mouse melanoma (B-16)	Delay melanogenesis	Fisher et al., 1981
Human melanoma (HO)	Stimulate melanogenesis and the formation of dendrite-like processes	Huberman et al., 1979
Mouse preadipocytes (BALB/c 3T3; ST13)	Inhibit adipose conversion and triglyceride accumulation	Diamond et al., 1977; Hiragun et al., 1981
Mouse epidermal keratinocytes	Inhibit or enhance terminal differentiation depending on basal cell's maturation state	Yuspa et al., 1982
Human epidermal keratinocytes	Stimulate terminal differentiation	Hawley-Nelson et al., 1982; Parkinson et al., 1983

(Umeda et al., 1980; Trosko et al., 1982, 1984). Decreased metabolic cooperation in this system may be due to the fact that, in TPA-treated cultures, the frequency of gap-junctional contacts between V79 cells is very much decreased and the area of membrane occupied by gap junctions is reduced more than 20-fold compared to untreated controls (Yancey et al., 1982).

Others have measured the transfer of radioactively labeled nucleotides from prelabeled to unlabeled cells to show that tumor promoters inhibit metabolic cooperation (Murray and Fitzgerald, 1979; Fitzgerald and Murray, 1980, 1982). Dorman et al. (1983a) used autoradiography to demonstrate that TPA induces a rapid, transient inhibition of

TABLE 5. *Effects of Phorbol Diester Tumor Promoters on the Differentiation of Malignant Hematopoietic Cells* In Vitro

Cells	Effects	References
Murine erythroid leukemia	Inhibit or stimulate erythroid differentiation, depending on cell line	Rovera et al., 1977; Yamasaki et al., 1977; Miao et al., 1978
Human promyelocytic leukemia (HL-60)	Induce monocyte/macrophage differentiation	Rovera et al., 1979a,b; Huberman and Callaham, 1979; Lotem and Sachs, 1979
Human myeloblastic leukemia	Decrease or increase macrophage-like differentiation, depending on dose	Chang and McCulloch, 1981; Takeda et al., 1982
Human thymic and lymphoblastic leukemia	Induce T cell differentiation	Nagasawa and Mak, 1980; Delia et al., 1982
Human B-lymphocyte lines	Stimulate secretion of IgM and IgG	Ralph and Kishimoto, 1981; Ralph et al., 1983

intercellular communication between untransformed and between untransformed and transformed C3H/10T1/2 cells. They found no correlation in this system between the dose responses of TPA for promotion of transformation and for reduction of cellular communication; they concluded that an inhibition of intercellular communication may not be sufficient for promotion of transformation of C3H/10T1/2 cells. Rivedal et al. (1985) reported that TPA also inhibits intercellular communication and enhances morphologic transformation of Syrian hamster embryo cells (HEC). Using a dye-transfer method of measuring gap-junctional communication, they found that HEC lines, whether sensitive or resistant to TPA-induced enhancement of morphologic transformation, showed communication. However, TPA inhibited communication only in the TPA-sensitive cells, again implying that blocked intercellular communication may be one mechanism among others involved in morphologic transformation.

Enomoto et al. (1981) and Yamasaki et al. (1983) showed that TPA can also cause a rapid, reversible inhibition of both the formation of electrical cell coupling and its maintenance. Ojakian (1981) found that TPA can induce changes in the structure and permeability of epithelial tight cell junctions. Friedman and Steinberg (1982) used microinjection of fluorescein to study cellular communication in epithelial cell cultures derived from benign and malignant tumors of the human colon. They observed that most premalignant cells within a colony were in communication and that TPA caused extensive uncoupling of cells in 'late-stage' but not 'early-stage' premalignant cultures. In contrast, malignant cells were extensively uncoupled in the absence of TPA and the promoter did not affect this.

The reader should consult Trosko et al. (1983) for a discussion of the evidence that intercellular communication has a role in tumor promotion.

4.1.5. Transport of Small Molecules

In Swiss 3T3 mouse fibroblasts, TPA stimulates uptake of ^{32}P and ^{86}Rb$^+$, a K$^+$ analog (Moroney et al., 1978). Both fluxes are ouabain-sensitive, suggesting that (Na$^+$/K$^+$)-ATPase activity is an early target of TPA. Dicker and Rozengurt (1981) proposed that the stimulation of pump activity by TPA is caused by an increased Na$^+$ influx which, coupled with H$^+$ exit, increases intracellular pH (Burns and Rozengurt, 1983). O'Brien and Krzeminski (1983) found that in BALB/c 3T3 preadipose mouse cells, TBA *decreases* the rate of ^{86}Rb$^+$ uptake; they suggest that a Na$^+$K$^+$/Cl$^-$ cotransport system is the target of TPA in these cells.

TPA can have marked effects on hexose transport and metabolism. For example, it stimulates transport of 2-deoxyglucose and 3-O-methylglucose in differentiated and undifferentiated preadipocytes by a mechanism which differs from insulin stimulation of hexose transport in these cells (O'Brien, 1982). Wrighton and Mueller (1982) suggest that stimulation of glucose transport by TPA in bovine lymphocytes is an oxygen-independent, early event resulting from a direct action on the cell membrane. TPA enhances 2-deoxyglucose uptake in mouse thymocytes, also apparently by direct interaction with membrane components, a mechanism different from that by which concanavalin A accelerates glucose uptake in thymocytes (Nordenberg et al., 1983).

4.1.6. Ligand Binding

TPA transiently inhibits binding of EGF to cellular receptors by interfering with binding to high but not low affinity sites (Lee and Weinstein, 1978; Shoyab et al., 1979; Magun et al., 1980; Lockyer et al., 1981). It does not bind directly to EGF receptors (Brown et al., 1979) but in at least one cell type can apparently delay the appearance at the plasma membrane of cryptic high affinity sites (King and Cuatrecasas, 1982). In some cell lines, TPA inhibits binding of insulin, also by altering the affinity rather than the number of insulin receptors (Grunberger and Gorden, 1982; Thomopoulos et al., 1982).

One question has been how TPA can inhibit binding of these ligands and yet act synergistically with them to induce synthesis of DNA in quiescent cells (Dicker and Rozengurt, 1978; Frantz et al., 1979). Magun et al. (1980) postulated that TPA only temporarily decreases affinity, binding and subsequent degradation of EGF, for example, so that when receptor affinity returns, more EGF is available in the medium of TPA-treated cultures than in control cultures. Other explanations may come from the recent evidence that protein kinase C can phosphorylate the receptors for ligands such as EGF and insulin; that TPA can stimulate this phosphorylation; and that C-kinase phosphorylation decreases auto-phosphorylation of the ligand receptor at tyrosine residues (Jacobs et al., 1983; Cochet et al., 1984; Iwashita and Fox, 1984; Decker, 1984) (see Section 4.2.2).

TPA also decreases the binding of EGF and thyrotropin- and somatostatin-releasing hormones to specific receptors on rat pituitary cells (Osborne and Tashjian, Jr., 1982) and decreases the binding of colony-stimulating factor-1 to mouse peritoneal macrophages (Chen et al., 1983). Inhibition of binding of these ligands is due to either the decreased affinity or number of receptor sites or to both.

4.1.7. Phospholipid Metabolism

TPA has been shown in a number of cell systems to produce rapid changes in phospholipid metabolism. In the dog kidney cell line MDCK, TPA stimulates deacylation of cellular phospholipids and biosynthesis of prostaglandins, measured as the release of arachidonic acid and its metabolites into the culture medium (Ohuchi and Levine, 1978a,b). The stimulation of release of radioactivity from arachidonate-labeled cells is inhibited by indomethacin (Ohuchi and Levine, 1978a,b). TPA-stimulated deacylation of phospholipids in MDCK cells is not specific for arachidonic acid, and membrane acyl composition controls the particular series of prostaglandin produced (Beaudry et al., 1982).

TPA also stimulates prostaglandin synthesis in other cells, including HeLa cells (Crutchley et al., 1980), macrophages (Hamilton, 1980) and mouse epidermal cells (Fürstenberger et al., 1981b). In addition to enhancing cyclooxygenation of arachidonic acid, TPA can enhance arachidonic acid lipoxygenation (Valone et al., 1983; Wrighton et al., 1983). In the myeloid leukemia cell line HL-60, TPA produces several alterations in lipid metabolism and stimulates phospholipid metabolism (Cabot et al., 1980). It inhibits phosphatidylcholine synthesis by methylation of phosphatidylethanolamine but enhances phosphatidylcholine synthesis from endogenous choline (Cassileth et al., 1981).

4.1.8. Surface Glycoproteins and Glycolipids

Human leukemic cell lines in which TPA induces differentiative changes show early (within several hours) alterations in the synthesis of high molecular weight cell surface glycoproteins (Ishimura et al., 1980; Fukuda, 1981; Cossu et al., 1982). On the other hand, primary hairy leukemia cells treated with TPA show no change in their characteristic cell surface glycoprotein pattern but do undergo alterations in cell morphology and cell-substratum adhesion (Lockney et al., 1982).

TPA can induce the release into the medium of the major surface glycoprotein, fibronectin (Blumberg et al., 1976; Keski-Oja et al., 1979). Production of another membrane glycoprotein, collagen, is decreased in normal and RSV-transformed CEF treated with TPA (Delclos and Blumberg, 1979), whereas synthesis of hyaluronic acid is increased (Ullrich and Hawkes, 1983). In the JB6 mouse epidermal cell line, TPA inhibits the synthesis of a 180,000 MW glycoprotein which has been identified as a specific procollagen (Dion et al., 1982).

In primary mouse epidermal cells, TPA induces the synthesis of a secreted 35,000 MW glycoprotein (Gottesman and Yuspa, 1981). Synthesis of this protein, which has been shown in other cell systems to be associated with transformation, is also increased in mouse skin exposed to TPA (Gottesman and Yuspa, 1981).

TPA can also affect metabolism of cell surface gangliosides (Huberman *et al.*, 1979; Srinivas and Colburn, 1982; Burczak *et al.*, 1983). It decreases the *de novo* synthesis of trisialoganglioside in 'promotable' JB6 mouse epidermal cells but not in promoter-resistant variants (Srinivas *et al.*, 1982).

4.1.9. *Protease Secretion*

TPA can stimulate synthesis of the serine protease plasminogen activator in many cell types including CEF (Wigler and Weinstein, 1976; Wilson and Reich, 1979), macrophages (Vassalli *et al.*, 1977), endothelial cells (Loskutoff and Edgington, 1977) and human embryonic lung cells (Jaken *et al.*, 1981a). Treatment of RSV-transformed CEF with TPA has a synergistic effect on plasminogen activator production; under serum-free conditions, such cells can secrete extremely high levels of plasminogen activator for 4–6 days (Goldfarb and Quigley, 1978). TPA stimulates the secretion of latent collagenase by rabbit synovial fibroblasts (Brinckerhoff *et al.*, 1979) and human endothelial cells (Moscatelli *et al.*, 1980).

4.2. Intracellular Metabolism

4.2.1. *Glucose and Oxygen Metabolism*

In normal human polymorphonuclear leukocytes, TPA mimics the effects of phagocytosis so that increases in oxygen utilization, hexose monophosphate shunt activity, nitroblue tetrazolium dye reductase, and superoxide anion release are observed (reviewed in Goldstein, 1978; Mastro, 1982; Baxter, 1984). Effects similar to those seen in normal phagocytic cells can be observed when cell lines derived from granulocyte-macrophage cancers are treated with TPA or when leukemia cell lines are induced by TPA to differentiate to macrophage-like cells (Greenberger *et al.*, 1978; Rovera *et al.*, 1979a,b; Huberman and Callaham, 1979).

Several workers (Goldstein *et al.*, 1979; Emerit and Cerutti, 1981; Birnboim, 1982a,b; Kensler and Trush, 1984; Cerutti, 1985; Troll and Wiesner, 1985) have suggested that increased production of leukocyte superoxide anion radicals in response to treatment with phorbol diesters plays a role in mouse skin tumor promotion. This view is supported by the fact that oxygen radical metabolism in TPA-activated leukocytes and TPA promotion of mouse skin tumorigenesis are inhibited by RA and antiproteases, inhibitors of superoxide anion production, and by a copper-coordinated complex that is an antioxidant (Troll *et al.*, 1970; Verma *et al.*, 1979; Witz *et al.*, 1980; Kensler and Trush, 1983; Kensler *et al.*, 1983). In addition, it has been reported that TPA-enhanced cell transformation is inhibited when retinoids or the free radical scavenger, SOD, are present in the medium (Miller *et al.*, 1981; Borek and Troll, 1983). However, Kinsella *et al.* (1983) failed to observe any gross chromosome aberrations associated with increased superoxide anion production in leukocytes treated with TPA, and found only a slight increase in chromosome aberrations after repeated and prolonged exposure of human fibroblasts to TPA. They concluded that chromosomes are not the primary site of action for the superoxide anions produced by TPA (see also Section 4.3.2).

4.2.2. *Protein Phosphorylation*

The recent findings that the phorbol diester binding site copurifies with, and activates, a Ca^{2+}-sensitive, phospholipid-dependent protein kinase (see Section 3.3.3) stimulated interest in determining the early effects of promoters on protein phosphorylation. Several groups (Bishop *et al.*, 1983; Gilmore and Martin, 1983) reported that treatment of normal CEF with TPA results in a very rapid ($\leqslant 5$ min) phosphorylation on tyrosine of a 40,000–43,000 MW polypeptide. Tyrosine-specific phosphorylation of a polypeptide of similar molecular weight is also stimulated in CEF by exposure to mitogenic hormones or by transformation with RSV (Cooper and Hunter, 1981; Cooper *et al.*, 1982a; Nakamura *et al.*, 1983). In contrast to the effects of these other agents, however, TPA also alters phosphorylation of serine and threonine residues (Gilmore and Martin, 1983).

Rozengurt *et al.* (1983) found that in quiescent 3T3 mouse cells, PDBu stimulates rapid phosphorylation of an 80,000 MW protein, an effect which could be mimicked by treating the cells with phospholipase C or platelet-derived growth factor (PDGF), an activator of endogenous phospholipase C activity. Kwong and Mueller (1983) examined the effects of TPA treatment of bovine lymphocytes on the ability of isolated membrane preparations to phosphorylate proteins. Those authors observed that TPA selectively stimulates the phosphorylation by *in situ* protein kinases of two proteins of MW 65,000 and 74,000 and suppresses the phosphorylation of a 130,000 MW protein. They suggested that in intact cells TPA may alter the exposure of some membrane proteins to the action of lipid-dependent protein kinases. In HL-60 leukemia cells, TPA induces rapid phosphorylation at serine residues of a 17,000 and a 27,000 MW cytosolic protein (Feuerstein and Cooper, 1983). TPA-induced activation of blood platelets is associated with the phosphorylation of the 20,000 MW light chain of myosin (Naka *et al.*, 1983), but in this reaction apparently mediated by protein kinase C, the site of phosphorylation is distinct from that phosphorylated by the myosin light chain kinase, a Ca^{2+}/calmodulin-dependent enzyme.

Jacobs *et al.* (1983) have suggested that the decreased affinity for insulin of intact TPA-treated cells may be due to an effect of the promoter on phosphorylation of insulin receptors. Indeed, they found that in intact cells of a human B-lymphocyte line (IM-9), TPA stimulates phosphorylation of the β-subunit of both the insulin and somatomedin C (insulin-like growth factor I) receptors. Insulin also enhanced phosphorylation of its own receptor with some of the phosphorylated residues being distinct from those phosphorylated by TPA. Those authors propose a role for protein kinase C in the regulation of insulin and somatomedin C receptors. Similar observations and suggestions have been made very recently concerning the regulation of the activity of EGF and PDGF receptors (Cochet *et al.*, 1984; Iwashita and Fox, 1984; Whiteley *et al.*, 1985).

4.2.3. *Ornithine Decarboxylase*

Early studies (Yuspa *et al.*, 1976b; Lichti *et al.*, 1978) had shown that treatment of fresh primary cultures of newborn mouse epidermal cells with TPA induces a 5–10-fold increase in ODC activity, but that after 48 hr in culture the cells are refractory to ODC induction. This loss of responsiveness was later explained by the finding that the primary cultures consist of two populations, proliferating cells and differentiating cells (reviewed in Yuspa, 1983, 1984). Medium with a reduced calcium concentration (0.02–0.1 mM) supports cell proliferation (Hennings *et al.*, 1980, 1983a), and cells growing in this medium remain sensitive to TPA induction of ODC for at least four weeks (Lichti *et al.*, 1981). Under such conditions, DNA synthesis is not stimulated, presumably because the cells are already proliferating maximally. If the cells are switched from low- to high (1.2 mM)-calcium medium, they undergo rapid differentiative changes and loss of ODC inducibility (Hennings *et al.*, 1980; Lichti *et al.*, 1981). Extensive studies with this system (reviewed in Yuspa, 1983, 1984) have led to the conclusion that there are subpopulations of epidermal cells which differ in their sensitivities to both the stimulation of proliferation and the induction or acceleration of terminal differentiation by TPA.

In cultures of transformed HEC, TPA potentiates induction of ODC by fresh serum-containing medium, whereas in normal HEC the effects of the two inducers are additive (O'Brien and Diamond, 1977; O'Brien *et al.*, 1980). Under the conditions of these experiments, the induction of ODC by TPA in the hamster cells is followed by increased cellular concentrations of polyamines but not by an increase in DNA synthesis. TPA also acts synergistically with serum factors on ODC induction in a hamster tracheal epithelial cell line (Landesman and Mossman, 1982) and in transformed, but not normal, hamster epidermal cells (Sina et al., 1983). O'Brien et al. (1982) and Sina et al. (1983) note that the altered regulation of polyamine biosynthesis detected by the cells' response to TPA-induction of ODC has been observed in all 'preneoplastic' hamster cell lines so far examined and may be an important factor in the progression of these cells from the normal to malignant phenotype.

The induction of ODC by TPA in a mouse mammary tumor cell line (Mm5mt/Cl) is inhibited by antioxidants (Friedman and Cerutti, 1983).

4.2.4. Interferon

TPA has been reported both to enhance and inhibit interferon production. In a Burkitt lymphoma cell line (Namalwa), TPA treatment increases interferon production in response to Sendai virus $\sim$20-fold (Adolf and Swetly, 1980). TPA enhances the production of γ-interferon in human lymphocyte cultures stimulated with phytohemagglutinin or other T cell mitogens (Yip et al., 1981). In human fibroblast cultures, the promoter inhibits both polyinosinate-polycytidylate- and Newcastle disease virus-induced interferon production (Frankfort and Vilček, 1982).

4.2.5. Lymphokines and Hormones

There is a large body of literature, recently reviewed by Mastro (1982, 1983) and Baxter (1984), on the effects of phorbol diester tumor promoters on lymphokine activity in vitro. For example TPA can stimulate release of the macrophage lymphocyte-activating factor, interleukin-1 (IL-1) (Rosenstreich and Mizel, 1979; Farrar et al., 1980c). TPA can replace IL-1 in the activation process, perhaps by triggering T cells to release interleukin-2 (IL-2) (Mizel et al., 1978; Rosenstreich and Mizel, 1979; Farrar et al., 1980a,b; Koretsky et al., 1982), the lymphokine required for clonal expansion of T cells. TPA can enhance IL-2 production in concanavalin A-induced mouse spleen cell cultures (Fuller-Farrar et al., 1981), enhance IL-2 production and mitogenesis induced by lectins in human T cell lymphoproliferative diseases (Vyth-Dreese et al., 1982), and stimulate IL-2 production but inhibit cell proliferation in a line of mouse thymoma cells (EL-4) (Farrar et al., 1980a; Sando et al., 1982). Pearlstein et al. (1983) have evidence that stimulation of IL-2 production by TPA may be related to the promoter's perturbation of the cell cycle. Because of the multiple effects of TPA on lymphokine activities, Mastro (1982) and Orosz et al. (1983) point out that caution should be exercised when using TPA to study lymphocyte functions and when interpreting the experimental results.

In a strain of rat pituitary cells (GH_4C_1) TPA stimulates the synthesis and release of prolactin but inhibits the production of growth hormone (Osborne and Tashjian, Jr., 1981). TPA stimulates somatostatin secretion from cultured brain cells (Peterfreund and Vale, 1983) and gonadotropin and progesterone secretion from a human choriocarcinoma cell line (T3M-3) (Sasaki et al., 1983).

4.3. DNA, Chromosomes and Cell Proliferation

4.3.1. DNA Synthesis and Cell Proliferation

TPA stimulates DNA synthesis in quiescent cultures of many but not all normal cell strains, sometimes after a period of transient inhibition of thymidine incorporation (reviewed in Diamond et al., 1980; Mastro, 1982). In mouse epidermal cell cultures, e.g. a transient decrease in thymidine incorporation is followed by a 5- to 10-fold stimulation of incorporation and an actual increase in the proliferative cell population (Yuspa et al., 1976a,b; Fusenig and Samsel, 1978). Under low calcium conditions in which the proliferative compartment is already large, TPA has little effect on thymidine incorporation (Lichti et al., 1981).

Dicker and Rozengurt (1978) and Frantz *et al.* (1979) showed that in quiescent cultures of either Swiss or BALB/c 3T3 mouse fibroblasts, TPA has little effect on DNA synthesis in the absence of serum. However, in combination with growth factors such as fibroblast growth factor, insulin, EGF, PDGF, or the defined serum growth fraction, platelet-poor plasma, TPA synergistically stimulates DNA synthesis. Thus, the mechanism of the mitogenic effect of TPA appears not to be identical to that of any of these growth factors. Dicker and Rozengurt (1980) suggested that the mechanism may be the same as that used by the neurohypophyseal hormone, vasopressin, the only mitogen of those they tested which did not act synergistically with TPA. Subsequent studies on ion transport and mitogenesis in TPA-treated 3T3 cells support this suggestion (O'Brien and Krzeminski, 1983; Collins and Rozengurt, 1984).

TPA has only slight effects on the growth of C3H/10T1/2 cells in medium supplemented with 10% serum, whereas in low serum medium it can decrease the population doubling time and increase the saturation density (Boreiko *et al.*, 1980; Tomei *et al.*, 1981). In transformed C3H/10T1/2 cells, TPA also has minimal effects in medium with 10% serum but inhibits cell growth in 1% serum. Several investigators have observed that in C3H/10T1/2 and other cells, TPA can have differential effects on DNA synthesis depending on the stage of the cell cycle (Magun and Bowden, 1979; Tomei *et al.*, 1981; Kinzel *et al.*, 1981). TPA can stimulate DNA synthesis in nonneoplastic cells blocked at the G_1/S boundary by Ca^{2+} deprivation (Boynton *et al.*, 1976; Boynton and Whitfield, 1980); it cannot replace Ca^{2+} but probably sensitizes the blocked cells to the ion.

Interesting differential effects of TPA on cell division have been observed in normal human cells and premalignant cells of the same histological origin. Friedman (1981) found that TPA enhanced the size of the proliferative fraction in colonies of early-stage premalignant colonic epithelial cells whereas in intermediate- and late-stage premalignant cells, promoter treatment disrupted the monolayer and induced the formation of multicellular clusters and the release of protease.

In primary human epidermal cell cultures, TPA selectively inhibits proliferation of keratinocytes and enhances outgrowth of melanocytes (Eisinger and Marko, 1982; Eisinger *et al.*, 1983). The melanocytes have a population doubling time of about four days and can be maintained in medium supplemented with TPA for at least 30 passages. In contrast, melanocytes derived from benign melanocytic nevi can be maintained in continuous culture in the absence of TPA (Herlyn *et al.*, 1983). This implies that nevic cells have undergone some change from normal so that they no longer require for *in vitro* proliferation an unidentified growth factor(s) whose production can be stimulated by TPA or, alternatively, can constitutively produce sufficient amounts of such a factor.

TPA stimulates DNA synthesis in lymphocyte populations from normal donors of several species including humans, as first shown by Mueller and Kajiwara (1965) and Whitfield *et al.* (1973). These effects of TPA, as well as its interactions with lectins and lymphokines in the induction of mitogenesis in lymphocytes, are reviewed in Mastro (1982, 1983) and Baxter (1984).

With some lymphocyte populations, TPA is only a weak mitogen but can be a potent comitogen with lectins and lipopolysaccharide (see, e.g. Mastro and Mueller, 1974; Fish *et al.*, 1981; Kabelitz *et al.*, 1982; Mastro and Pepin, 1982). However, TPA can also be a potent inhibitor of DNA synthesis in lymphocyte cultures undergoing the mixed lymphocyte response (Mastro and Mueller, 1978; Yamasaki and Martel, 1981; Baxter *et al.*, 1981).

TPA is also directly mitogenic for purified human B lymphocytes and enhances the effects of other B cell stimuli (Abb *et al.*, 1979; Bertoglio, 1983). It stimulates colony formation *in vitro* by murine granulocyte-macrophage progenitors without added colony-stimulating factors (CSF) (Stuart *et al.*, 1981, 1983). Stuart *et al.* (1983) suggest that this effect of TPA involves direct stimulation of the progenitors and/or enhancement of their response to CSF, rather than stimulation of the production of CSF by TPA. With normal human myeloid cells, TPA stimulates cluster formation in agar in the absence of exogenous CSF, whereas it inhibits myeloid colony and cluster formation in the presence of CSF (Abrahm and Smiley, 1981; Griffin *et al.*, 1983; Ozawa *et al.*, 1983).

4.3.2. Sister Chromatid Exchanges and Chromosome Aberrations

In 1978, Kinsella and Radman reported that TPA induced sister chromatid exchanges (SCE) in V79 Chinese hamster lung fibroblasts and proposed that the irreversible step in tumor promotion might be the result of an aberrant mitotic segregation event leading to the expression of carcinogen/mutagen-induced recessive genetic or epigenetic chromosomal changes. Since then, a number of investigators have studied the effects of tumor promoters on SCE with conflicting results. Even when using the same target cells, some investigators have found no effect of TPA on SCE (Thompson et al., 1980; Loveday and Latt, 1979; Gainer and Kinsella, 1983) and others have found slight to moderate effects (Nagasawa and Little, 1981; Schwartz et al., 1982; Ray-Chaudhuri et al., 1982; Dzarlieva and Fusenig, 1982; Nagasawa et al., 1983). Nagasawa and Little (1981) analyzed the factors influencing the induction of SCE by TPA in different cells. They concluded that the lot and concentration of TPA and heat-inactivation of the serum were critical for demonstrating a positive effect of TPA; they implicated oxygen radicals as the actual inducers of the SCE.

Emerit and Cerutti (1981, 1982, 1983) observed that TPA induced chromosomal aberrations and polyploidy in phytohemagglutinin-stimulated human lymphocyte cultures. They found that the promoter induced the production of a low molecular weight, diffusible clastogenic factor which caused aberrations in fresh blood cultures never exposed to TPA. The presence of monocytes, polymorphonuclear leukocytes, or platelets was required for formation of the clastogenic factor by TPA. Since Cu-Zn SOD and inhibitors of arachidonic acid metabolism inhibited both the clastogenicity of TPA and the clastogenic activity of any previously formed factor, the authors suggested that superoxide radicals and stimulation of the arachidonic acid cascade play a role in the clastogenic action of TPA. They propose that this effect of TPA could be involved in mouse skin tumor promotion through the leukocytic infiltration of the dermis which occurs in response to promoter treatment (see also Section 4.2.1).

Birnboim (1982a,b) found that TPA induced extensive DNA strand breakage in human leukocyte preparations rich in polymorphonuclear leukocytes. The effect correlated with a TPA-stimulated 'respiratory burst' and studies with inhibitors are consistent with a mechanism involving superoxide anion and hydrogen peroxide. Weitberg et al. (1983) reported that SCE were increased in Chinese hamster ovary (CHO) target cells cocultivated for 30 min with human leukocytes and TPA; they presented evidence in support of the hypothesis that the effect was due to TPA-stimulated oxygen metabolism in the leukocytes.

Kinsella et al. (1983) and Gainer and Kinsella (1983) observed that after chronic exposure to TPA (nine passages), there was a slight increase in chromosomal aberrations and polyploidy in skin fibroblasts from both normal individuals and patients with hereditary retinoblastoma. Interestingly, several quadriradial figures which were never seen in control cultures were found in TPA-treated cultures.

There is as yet no direct evidence that the induction of SCE or chromosomal aberrations by TPA is a factor in the enhancement of mutation or transformation by TPA (Thompson et al., 1980; Popescu et al., 1980, 1982; Miller et al., 1981; Gensler and Bowden, 1983). The only direct evidence that effects of TPA on the genetic material of the target cells may play a role in the mechanism of skin tumor promotion in vivo is that of Dzarlieva and Fusenig (1982), who found that TPA significantly enhanced SCE and numerical and structural chromosomal aberrations in mouse epidermal cell cultures.

4.3.3. Gene Amplification

Varshavsky (1981a) has proposed that tumor promotion and carcinogenesis may involve a generalized increase in the frequency of disproportionate DNA replication within a single cell cycle, resulting in accelerated gene amplification. In support of this hypothesis, he (Varshavsky, 1981b) presented evidence that TPA can facilitate gene amplification. During a single-step selection for resistance to methotrexate (MTX), a specific inhibitor of dihydrofolate reductase (DHFR), the incidence of MTX-resistant colony-forming mouse 3T6 cells was increased up to 100-fold when selection was carried out in the presence of TPA.

MTX-resistance persisted in the absence of TPA, and DHFR gene copy numbers per cell in MTX-resistant clones were higher than in parental 3T6 cells and independent of the presence or absence of TPA during MTX selection. Other phorbol esters with tumor-promoting activity *in vivo* increased the incidence of MTX-resistant 3T6 colonies, as did some nonphorbol promoters and mitogenic hormones such as EGF, insulin, and vasopressin (Barsoum and Varshavsky, 1983). Mezerein was extremely effective, suggesting that if there is any role of facilitated gene amplification in tumor promotion, it is at a late, probably irreversible stage of promotion.

Hayashi *et al.* (1983) obtained similar data using another cell system. They found that tumor promoters of various classes enhanced the frequency of cadmium-resistant colonies of Chinese hamster lung cells when the parental cells were exposed to cytotoxic levels of cadmium chloride. Cadmium-resistance was associated with overproduction of metallothionein mRNA and amplified metallothionein I genes.

Bojan *et al.* (1983) compared the effects of TPA on the resistance of mouse (3T3 and 3T6) and Chinese hamster (V79 and CHO) cells to three drugs [MTX, cadmium, and *N*-(phosphonacetyl)-L-aspartate]. TPA treatment resulted in greater recovery of drug-resistant mouse cells than hamster cells, and resistance was much more stable in mouse clones surviving MTX treatment than in hamster clones. The authors discuss these findings in relation to the species specificity of skin tumor promotion by TPA.

4.4. Cell Differentiation

Phorbol diester tumor promoters can modulate the differentiation *in vitro* of normal and transformed cells of a wide spectrum of phenotypes. This finding suggested several possible mechanisms for skin tumor promotion based on promoter-modulated epidermal differentiation (see Diamond *et al.*, 1980; Yuspa, 1983). It also showed that this class of compounds was not only important for the study of promotion but also a valuable tool for other fields of investigation such as differentiation and immunology. Some examples of the inhibition or stimulation of differentiation by phorbol esters in hematopoietic and non-hematopoietic cells are discussed herein and cited in Tables 4 and 5. Interferon, another pleiotropic agent that can modulate differentiation, acts synergistically with phorbol diesters in some cell systems and, in others, has an opposite effect; a few examples of each situation are cited.

4.4.1. *Nonhematopoietic Cells*

Cohen *et al.* (1977) were the first to show that TPA can inhibit the differentiation program of normal cells in culture. They found that TPA is mitogenic for presumptive myoblasts and inhibits their conversion to definitive myoblasts synthesizing muscle-specific myosin heavy and light chains. TPA also blocks the fusion of postmitotic myoblasts to form myotubes, presumably by acting directly on the cell surface (Cohen *et al.*, 1977). When added to cultures in which postmitotic multinucleated myotubes have already formed, TPA reversibly enhances the degradation of striated myofibrils and the accumulation of muscle-specific 10 nm filaments (Toyama *et al.*, 1979; Croop *et al.*, 1980; West and Holtzer, 1982). These latter experiments demonstrated that TPA can alter the differentiation process in cells not traversing the cell cycle.

In normal human myoblast cultures derived from mature skeletal muscle, TPA inhibits myotube formation and creatine kinase isoenzyme transition from immature CK-BB to more mature CK-MM (Fisher *et al.*, 1983). In contrast, interferon accelerates myotube formation and the isoenzyme transition in these cells (Fisher *et al.*, 1983).

In normal chick chondroblasts, TPA inhibits synthesis of two unique proteins, the type IV sulfated proteoglycan and a glycosylated protein of 180,000 MW (Pacifici and Holtzer,

1977; Lower *et al.*, 1978). In CEF, TPA inhibits collagen synthesis (Delclos and Blumberg, 1979) but does not affect synthesis of the unique type III sulfated proteoglycan (Lower *et al.*, 1978). The differential effects of TPA on proteoglycan synthesis in these two cell types that have a close lineal relationship demonstrate how its effects may depend to a large extent on the differentiative state and phenotypic expression of the responding cell.

In cultured embryonic chick and neonatal rat ganglia, TPA reversibly inhibits nerve growth factor-stimulated neurite outgrowth (Ishii, 1978; Spinelli *et al.*, 1982). It also reversibly inhibits the spontaneous neurite formation that occurs in response to serum deprivation in the mouse C1300 neuroblastoma cell line (Ishii *et al.*, 1978). In contrast, TPA reversibly enhances neurite outgrowth and induces other differentiation-specific biochemical and morphological changes in the human neuroblastoma cell line SH-SY5Y; the combined effects of TPA and nerve growth factor on these cells are greater than those of either compound alone (Pählman *et al.*, 1981; Spinelli *et al.*, 1982).

TPA can also have inhibitory or stimulatory effects on the differentiation of melanocytic cells of diverse origins. It reversibly blocks melanogenesis in normal unpigmented presumptive chick melanoblasts and in replicating pigmented chick melanocytes from three different tissues (Payette *et al.*, 1980). It delays both spontaneous melanogenesis and melanogenesis stimulated by melanocyte-stimulating hormone in a mouse melanoma cell line (B-16) (Fisher *et al.*, 1981); a similar inhibitory effect of interferon in these cells is synergistic with that of TPA (Fisher *et al.*, 1981).

With primary cultures of human skin, TPA inhibits growth of keratinocytes and favors outgrowth of melanocytes (Eisinger and Marko, 1982; Parkinson and Emmerson, 1982; Eisinger *et al.*, 1983; Fischer *et al.*, 1984) (see below). In contrast to its proliferative effects on normal human melanocytes, TPA induces terminal differentiation in a human melanoma cell line (HO), with the stimulation of melanogenesis and formation of dendrite-like processes (Huberman *et al.*, 1979).

Another cell system in which TPA and interferon have a synergistic effect on the inhibition of differentiation is a clone (A31T) of BALB/c 3T3 cells which can undergo spontaneous or insulin-stimulated adipocyte differentiation and triglyceride accumulation (Diamond *et al.*, 1977; Cioé *et al.*, 1980). The biological and biochemical mechanisms of the inhibition of differentiation and stimulation of DNA synthesis by TPA in these cells have been studied by Yun and Scott (1983) and O'Brien and coworkers (O'Brien *et al.*, 1979b; O'Brien and Saladik, 1980, 1982; O'Brien, 1982). The latter group found that both the differentiation program and its inhibition by TPA are dependent on the pH of the medium, that TPA stimulates lactate production by the cells, and that TPA's inhibitory effect on triglyceride accumulation is related to its interference with the cells' normal pattern of glucose metabolism. Scott and Wille, Jr. (1984) have developed a hypothesis for the mechanisms of initiation and promotion which is based, in part, on their studies in this cell system.

With the observation that extracellular Ca^{2+} was an important regulator of the growth and differentiation of mouse epidermal cells in culture (see Section 4.2.3), it became possible to dissect more thoroughly the effects of TPA on these cells. When low-calcium medium which favors cell proliferation is used, there is a heterogeneous response to TPA in which some cells are stimulated to proliferate and others induced to differentiate with increased epidermal transglutaminase and cornification (Yuspa *et al.*, 1980, 1982). Those authors (Yuspa *et al.*, 1982; Yuspa, 1984) have proposed that the maturation state of the cells at the time of TPA exposure is one of the determinants of the response. They suggest that TPA accelerates differentiation of cells in a more advanced maturation state, whereas it prevents differentiation and stimulates proliferation of less mature cells. They (Yuspa, 1983, 1984) have also proposed that such divergent responses *in vivo* could cause cell selection and serve as the basis for clonal expansion of initiated cells.

When normal epidermal cell cultures are switched from low- to high-calcium medium, the cells cease proliferation and undergo differentiation (Hennings *et al.*, 1980, 1983a). TPA, added up to 5 hr after the switch, markedly accelerates the differentiation program (Yuspa *et al.*, 1983). In cell cultures treated with carcinogens or derived from mouse skin treated with carcinogen *in vivo*, there are subpopulations of cells resistant to the inhibition of proliferation and induction of differentiation by high Ca^{2+} (Kulesz-Martin *et al.*, 1980; Yuspa and Morgan, 1981; Yuspa, 1983; Kawamura *et al.*, 1985). The resistant cells derived from skin initiated with carcinogen *in vivo* are presumably the progeny of initiated cells that, if maintained *in vivo* with continuous promoter treatment, would develop into papillomas.

TPA stimulates terminal differentiation of human epidermal keratinocytes and inhibits cell multiplication (Eisinger and Marko, 1982; Parkinson and Emmerson, 1982; Parkinson *et al.*, 1983; Fischer *et al.*, 1984). After TPA treatment of human keratinocytes, Hawley-Nelson *et al.* (1982) recovered a morphologically distinct, unidentified cell type, the growth of which was enhanced by TPA. Parkinson *et al.* (1983) found that more than 90% of colony-forming human keratinocytes lose colony-forming ability and form cornified envelopes after TPA treatment and that the resistant cells are the precursors of the sensitive cells. As in the mouse system, after 8 days in the absence of TPA, the population was once again heterogeneous in its response to TPA. Parkinson *et al.* (1983) also found that cultures of human squamous cell carcinoma lines and a simian virus 40 (SV40)-transformed human keratinocyte line contained 3- to 17-times more TPA-resistant keratinocytes than normal keratinocyte strains and showed a 3- to 25-fold smaller increase in cornified envelope formation after TPA exposure.

More recently, Stanley *et al.* (1985) compared the responses of human cervical keratinocytes to diethylstilbestrol (DES) and the promoters of TPA and mezerein, using the loss of colony-forming efficiency as a measure of terminal differentiation *in vitro*. All 3 compounds inhibited growth of normal keratinocytes ($> 2.5 \times 10^{-5}$ M DES, $> 10^{-8}$ M TPA, $> 10^{-9}$ M mezerein) and increased cornified envelope formation. About 10% of the normal cell population was resistant to 10^{-6} M TPA but not to 10^{-6} M mezerein or 5×10^{-5} M DES. In contrast, cornified envelope formation was negligible when a malignant cervical keratinocyte cell line was grown in the presence of DES or the promoters. About 85% of these cells retained colony-forming ability after exposure to 10^{-6} M TPA or mezerein but not DES. The authors concluded that the mechanisms by which DES inhibits growth and induces cornified envelope formation in human cervical keratinocytes appear to be distinct from those activated by TPA and mezerein. Their observations on the differential effects of promoters on normal and malignant cervical keratinocytes and the heterogeneity of the normal population in its response to TPA are similar to the findings with mouse keratinocytes (*supra vide*). Stanley *et al.* (1985) discuss the relevance of their findings for understanding the mechanism of mouse skin tumor promotion and the fact that mezerein is a second-stage promoter but not a complete promoter in mouse skin (Mufson *et al.*, 1979; Slaga *et al.*, 1980a, 1982).

In an analogous system, Yoakum *et al.* (1985) took advantage of the differential sensitivity of normal and transformed human bronchial epithelial cells to TPA-induced differentiation to select transformants following transfection of normal cells with a plasmid carrying the *ras* oncogene of Harvey murine sarcoma virus.

Hamster epidermal cells grown in the presence of TPA show a marked reversible inhibition of terminal differentiation; the number of cells with cornified envelopes is reduced by as much as 70% (Sisskin and Barrett, 1981).

McIlhinney *et al.* (1983) recently examined the effects of TPA on a clonal teratoma-derived undifferentiated embryonal carcinoma cell line (LICR LON HT 39/7), in which other chemical inducers of differentiation produce no morphological changes indicative of cellular differentiation. They found that TPA induced irreversible morphological, biochemical and ultrastructural alterations and have characterized these in detail.

4.4.2. *Hematopoietic Cells*

One of the earliest demonstrations that TPA could modulate cell differentiation was the observation that it dramatically inhibits terminal differentiation of FELC (Friend *et al.*,

1971; Rovera *et al.*, 1977; Yamasaki *et al.*, 1977). There have since been many reports describing the effects of TPA on the maturation of other leukemic cell lines and leukemic cells from peripheral blood. They show that TPA can affect different types of leukemic cells in different ways and, in fact, it has been proposed to use TPA as a diagnostic tool to classify blasts in human leukemias (Pegoraro *et al.*, 1980; Polliack *et al.*, 1982). The reader is referred to the reviews of Mastro (1982, 1983) and Baxter (1984) for in-depth discussions of the effects of tumor promoters on the differentiation of leukemic, as well as normal, hematopoietic cells.

TPA is effective at inhibiting terminal differentiation both in clones of FELC with a high percentage of cells undergoing spontaneous erythroid differentiation (Rovera *et al.*, 1977) and when it clones with a low percentage of spontaneously differentiating cells which are induced to differentiate with agents such as hexamethylene bisacetamide (Yamasaki *et al.*, 1977; Diamond *et al.*, 1978b; Fibach *et al.*, 1978). In both cases, TPA-treated cells maintained in continuous culture for many months retain the potential for differentiation and differentiate when released from the TPA block (Rovera *et al.*, 1977; Yamasaki *et al.*, 1979a; Fibach *et al.*, 1979). In contrast to the inhibitory effect on FELC, high concentrations of TPA *stimulate* differentiation of Rauscher virus-transformed murine erythroid cells (Miao *et al.*, 1978).

Human leukemic cells also can be induced to differentiate by TPA. The promyelocytic leukemia cell line, HL-60 (Collins *et al.*, 1977; Gallagher *et al.*, 1979), which normally grows in suspension, is induced to differentiate along the myeloid pathway by a variety of inducing agents including dimethyl sulfoxide, butric acid, and RA (reviewed in Gallo *et al.*, 1982). TPA, however, induces cell-substratum adherence and differentiation along the monocytic pathway; the adherent cells have many, but not all, the morphological, biological and functional characteristics of macrophages (Rovera *et al.*, 1979a,b; Huberman and Callaham, 1979; Lotem and Sachs, 1979; Newburger *et al.*, 1981). Pretreatment with interferon or dimethyl sulfoxide sensitizes HL-60 cells to TPA and enhances differentiation to macrophage-like cells (Tomida *et al.*, 1982; Fibach *et al.*, 1982). The HL-60 cell system is being used extensively to study the interaction of phorbol esters with cell receptors and how such interactions trigger biological effects (e.g. Solanki *et al.*, 1981b; Cooper *et al.*, 1982b; Vandenbark *et al.*, 1984; Vandenbark and Niedel, 1984) (see Section 3.3).

Kasukabe *et al.* (1981) reported that with a line of mouse myeloid leukemia cells (M1), TPA either inhibits or enhances induction of morphological and functional differentiation by dexamethasone and other agents depending on the type of serum in which the cells are grown. With medium containing 10% calf serum, TPA inhibits differentiation and with 10% fetal calf serum, it enhances differentiation.

In human myeloblast leukemia cells, TPA at low concentrations decreases differentiation and increases the frequency of cells capable of forming blast cell colonies in culture (Chang and McCullock, 1981). At higher doses of TPA, blast progenitors lose proliferative capacity and undergo differentiation to macrophage-like cells (Chang and McCulloch, 1981; Takeda *et al.*, 1982).

TPA induces T cell differentiation of human thymic and lymphoblastic leukemia cells (Nagasawa and Mak, 1980; Delia *et al.*, 1982). It stimulates immunoglobulin production by EBV-positive B cell lines (Ralph and Kishimoto, 1981; Ralph *et al.*, 1983).

5. MODELS OF TWO-STAGE TRANSFORMATION *IN VITRO*

5.1. Mouse C3H-10T1/2 Cells

Probably the clearest demonstration of two-stage transformation *in vitro* analogous with respect to timing and dosage to two-stage initiation/promotion protocols *in vivo* has been obtained with the C3H/10T1/2 mouse embryo cell line developed by Reznikoff *et al.* (1973b). This is one of the few cell culture systems in which it has been shown that a promoter such as TPA can be effective when treatment with it begins a long time (48–96 hr) after exposure to the initiator (if the cells have not yet reached confluence) and that the enhancement of transformation is not due to a simple stimulation of cell division (Mondal *et al.*, 1976; Kennedy *et al.*, 1980b).

C3H/10T1/2 is an aneuploid cell line, similar to mouse 3T3 cell lines, which shows strong density-dependent inhibition of cell division (Reznikoff *et al.*, 1973b). In the basic transformation assay with these cells (Reznikoff *et al.*, 1973a), the cells are seeded at low density, treated with carcinogen 24 hr later, and maintained with frequent refeeding for 5–6 weeks, at which time foci of piled up, transformed cells can be seen and scored. Three types (designated types I, II and III) of morphologically distinct foci are obtained; only cells from type III foci are consistently tumorigenic when inoculated into irradiated, syngeneic mice (Mondal, 1980). A number of workers have used this assay for studying transformation induced by chemical and physical agents and analyzed the basic requirements for success (see, e.g. Mondal, 1980). However, Thomassen *et al.* (1983) point out that this cell system may be detecting the induction of preneoplastic cells to neoplastic cells rather than the induction of normal cells to preneoplastic cells; the latter may require the use of freshly isolated, early passage cells.

Mondal and Heidelberger (1976) and Mondal *et al.* (1976) showed that when C3H/10T1/2 cells are treated with non-transforming (initiating) doses of carcinogenic PAHs or u.v.-irradiation, transformed foci develop if the cells are treated subsequently and continuously with phorbol ester tumor promoters. It has since been found that TPA also enhances transformation of these cells when they are initiated by such agents as X-irradiation (Kennedy *et al.*, 1978; Miller *et al.*, 1981; Han and Elkind, 1982), fission-spectrum neutrons (Han and Elkind, 1982), aflatoxin B_1 (Boreiko *et al.*, 1982), N-methyl-N'-nitro-N-nitrosoguanidine (MNNG), and many others (Frazelle *et al.*, 1983b, 1984; Boreiko, 1985). It has also been found that saccharin and diethylstilbestrol promote transformation of C3H/10T1/2 cells initiated by 3-methylcholanthrene (3-MCA) (Mondal *et al.*, 1978; Lillehaug and Djurhuus, 1982). Formaldehyde acts as both a weak initiator and weak promoter in these cells (Ragan and Boreiko, 1981; Frazelle *et al.*, 1983a). However, one should be aware that the treatment schedules used in some experiments (e.g. Miller *et al.*, 1981; Frazelle *et al.*, 1983a) may be such that they detect compounds that are acting as cocarcinogens or growth enhancers rather than promoters, as defined by the strict definition in Hicks' review (1983).

Frazelle *et al.* (1983b; 1984) recently analzyed some of the factors which influence the promotion of transformation of C3H/10T1/2 cells. The lot of fetal calf serum used was a major determinant of focus formation and each lot had to be carefully screened for the ability to support: synergistic increases in focus formation in cultures treated with MNNG and 0.25 μg/ml TPA; the production of large foci with distinctive morphology; and a low incidence of foci in control dishes. Of nine serum lots tested, only two were suitable for initiation/promotion experiments, although seven were suitable for transformation studies with 3-MCA alone. The concentration of TPA that was found to be optimal is somewhat higher than the 0.1 μg/ml concentration usually used. As has been reported for transformation of C3H/10T1/2 cells by 3-MCA alone, the use of high-passage stock cultures and the use of fungizone (Amphotericin B) in the medium adversely affected focus production. The presence of TPA was required both during the period of logarithmic growth and throughout the period of maintenance at confluence, providing further evidence that the mechanism of promotion in these cells involves more than simply a selective mechanism in which the ability of transformed cells to overcome suppression of growth by confluent monolayers of normal cells is enhanced.

Thus, in a representative assay of Frazelle *et al.* (1983b), 10^3 cells are seeded in 60-mm plates and treated with MNNG (3.4 μM) on day 1 and TPA (0.25 μg/ml) beginning on day 6. The cells reach confluence on about day 14, macroscopic foci develop between days 33 and 36, and the assays are terminated on day 43. By closely controlling what they found to be the important variables for successful initiation/promotion experiments, those authors (Boreiko *et al.*, 1982; Frazelle *et al.*, 1983b, 1984) determined that many alkylating and chemotherapeutic agents can act as initiators in C3H/10T1/2 cells, even though the same compounds may not produce transformed foci in the absence of TPA treatment.

Kennedy and Little (1981) studied the effects of protease inhibitors on different phases of X-ray-induced transformation of C3H/10T1/2 cells. Antipain suppressed transformation when present for only one day following irradiation, i.e. during the fixation phase when DNA repair processes are presumably active. The expression phase of transformation, the 6-week, post-irradiation period required for focus development, was affected by antipain and, to a lesser extent, by leupeptin but not by soybean trypsin inhibitor (SBTI). Antipain and SBTI prevented the enhancement of transformation induced by exposure to TPA during the expression phase.

The most effective of the inhibitors in all three phases, antipain, was also an inhibitor of TPA-induced plasminogen activator activity in normal C3H/10T1/2 cells and of the normally high activity of this serine protease in radiation-transformed C3H/10T1/2 cells; leupeptin and SBTI did not inhibit these activities (Long et al., 1981). These and other findings led Long et al. (1981) to suggest a role for proteases in the transformation event itself or in the maintenance of the transformed state.

Miller et al. (1981) reported that the transformation of C3H/10T1/2 cells induced by 400 rads of X-rays was enhanced about two-fold by exposure of the cells to TPA throughout the course of the experiment. Both radiation-induced transformation and its enhancement by TPA were inhibited by exposure to a retinoid for four days.

The glucocorticoid hormone, dexamethasone, an inhibitor of promotion in mouse skin *in vivo* (Scribner and Slaga, 1973), does not suppress X-ray-induced transformation in C3H/10T1/2 cells (Kennedy and Weichselbaum, 1981). Another glucocorticoid, cortisone, itself induces transformation of C3H/10T1/2 cells and increases synergistically the yield of radiation-induced transformants (Kennedy and Weichselbaum, 1981).

In experiments designed to investigate the mechanism of enhancement of radiation transformation by TPA, Kennedy and Little (1980) either irradiated cultures at differing initial cell densities or irradiated them at the same cell densities and, after they reached confluence, reseeded at different densities; in both cases, TPA treatment was started after exposure to 100 rads of X-rays. They found that the number of transformed foci which ultimately developed per dish was approximately the same even though the cell densities (initial or reseeded) varied over several orders of magnitude. This number was also similar to that obtained in cultures exposed to higher doses (400–600 rads) of X-rays alone, again under conditions which demonstrated that the yield was independent of the initial number of irradiated cells (Kennedy et al., 1980a). On the basis of these results, they proposed that transformation of C3H/10T1/2 cells by radiation (600 rads) alone occurs via a two-step mechanism in which the first step involves a heritable alteration occurring in a large proportion (perhaps 100%) of the cells and the second step occurs with low probability and usually after confluence is reached. They speculate that TPA treatment following exposure to low doses of X-ray, which interestingly yields primarily type II foci, modifies the first step in this process, perhaps by enhancing the degree of change induced by radiation or by giving cells altered by radiation a selective growth advantage (Kennedy and Little, 1980).

From similar experiments in which C3H/10T1/2 cells at low cell densities were treated with chemical carcinogens, Fernandez et al. (1980) proposed a similar model and estimated that the second step occurs after many cell generations subsequent to the first and at a probability of about 10^{-6} per generation. However, Backer et al. (1982b) used a different protocol from Kennedy and Little (1980) and Fernandez et al. (1980) and obtained very different data. They treated C3H/10T1/2 cells at high cell density (5×10^4 cells per 60-mm dish) with the active metabolite of B(a)P, ($\pm$)7β,8α-dihydroxy-9α,10α-epoxy-7,8,9,10-tetrahydro-B(a)P, and at various times thereafter reseeded the cells without diluting them. The number of transformed foci on the secondary plates began to increase by 48 hr and reached a plateau at 8–12 days; the foci were not seen in the primary cultures that had not been reseeded. They did not determine the effect of TPA or other promoters in the high cell density treatment protocol.

One interpretation of the data of Backer et al. (1982b) is that transformation of C3H/10T1/2 cells involves only a single (mutational?) event that needs 48 hr for its fixation and occurs at a low probability requiring that at least 10^4 cells be treated initially. Transformants occurring via this mechanism would not be detected under the experimental conditions used by Kennedy and Little (1980) and Fernandez et al. (1980). Backer et al. (1982b) suggest that their data would fit the two-step model of the other investigators if a high cell density at the time of treatment increases the probability that a second step occurs more rapidly. As yet, the reasons for the striking differences in the transformation data of these investigators have not been resolved.

Mordan et al. (1982) and Mordan and Bertram (1983) isolated a 3-MCA-treated clone of C3H/10T1/2 cells which, in the presence of retinyl acetate, is contact-inhibited and morphologically indistinguishable from the parental cells. Cells of this clone will undergo neoplastic transformation about three weeks after removal of the retinyl acetate. The authors consider that the clone has many of the characteristics of 'initiated' cells and have designated it INIT/10T1/2.

In reconstruction experiments in which normal C3H/10T1/2 and INIT/10T1/2 cells were cocultured in the absence of retinyl acetate, almost 100% of the 'initiated' cells formed type I, II or III foci after a latent period of 3-4 weeks. The formation of transformed foci was accelerated by treating the cells with TPA. Inclusion of retinoid in the medium negated the shorter latent period that TPA produced and reduced the total number of transformed foci that appeared. Using this model, Mordan et al. (1983) find that the expression of transformation in 'initiated' C3H/10T1/2 cells is directly related to the size of the colony of initiated cells at the time the monolayer reaches confluence, i.e. to the number of cell generations to confluence. They found that INIT/10T1/2 cells must form a colony at confluence in excess of approximately 40 cells in order to attain the competence to transform and that, thereafter, there is a direct relationship beween colony size at confluence and the probability of transformation. Their findings support similar conclusions made earlier by Haber et al. (1977) but argue against the conclusion of Kennedy et al. (1980a) and Fernandez et al. (1980) that carcinogen treatment may induce an initiating event(s) in *all* exposed cells. Mordan et al. (1983) point out that the data of the other two groups fit their model with regard to the colony size dependency of transformation.

Constantinides et al. (1978) and Taylor and Jones (1979) showed that normal C3H/10T1/2 cells can be induced to differentiate into muscle cells, adipocytes and chondrocytes by treatment with 5-azacytidine. Mondal and Heidelberger (1980) found that TPA inhibited at least one of these differentiation programs, muscle formation, when added to the cultures 48 hr after removing the 5-azacytidine. They tested a number of other compounds of varied chemical structures and found a rather good correlation between the ability to inhibit differentiation in this system and promoting activity in C3H/10T1/2 cells and/or tumor-promoting activity in rodents *in vivo*, e.g., anthralin, mezerein and saccharin were strong inhibitors, but phenobarbital was not. However, several *inhibitors* of transformation and promotion such as antipain, dexamethasone, and retinyl acetate were also strong inhibitors of muscle cell formation and did not affect the inhibitory action of TPA.

5.2. MOUSE 3T3 CELL LINES

Sivak and Van Duuren (1967) were the first to show that a tumor-promoting compound can enhance phenotypic expression of transformation. They cocultivated contact-inhibited Swiss mouse 3T3 fibroblasts with SV40-transformed cells and found that outgrowth of the transformed cells was enhanced by the presence of TPA in the medium. Sivak and Tu (1980) subsequently described an assay which, with respect to timing, is analogous to initiation/promotion assays in mouse skin and in C3H/10T1/2 cells. It uses the BALB/c 3T3 transformation system developed by Kakunaga (1973) (see also Sivak et al., 1980). The cells are treated with 3-MCA (0.1 µg/ml) for three days and treatment with promoter is started three days after removing the 3-MCA. Sivak and Tu (1980) found that under these conditions TPA was very active in enhancing formation of type III transformed foci but saccharin was inactive, in contrast to what had been observed in C3H/10T1/2 cells (Mondal et al., 1978).

So far, only a few other compounds have been tested for promoting activity in the BALB/c 3T3 system. One is the indole alkaloid, DHTB, which Hirakawa et al. (1982) found to be at least 100 times more active than TPA in promoting transformation initiated with 3-MCA (1 μg/ml). The polyacetate, aplysiatoxin, is as active as phorbol-12,13-didecanoate whereas debromoaplysiatoxin has no effect on transformation frequency in BALB/c 3T3 cells (Shimomura et al., 1983). 3-MCA-induced transformation of these cells is markedly enhanced by 1α,25-dihydrovitamin D_3 (Kuroki et al., 1983). Interestingly, this compound, which also induces monocyte/macrophage differentiation of HL-60 cells *in vitro* (Murao et al., 1983; Tanaka et al., 1983), is an *inhibitor* of TPA promotion of mouse skin tumor formation *in vivo* (Wood et al., 1983).

5.3. Hamster Embryo Cells

In the quantitative colony transformation assay first described by Berwald and Sachs (1965), primary or secondary Syrian hamster embryo cells (HEC) are seeded on irradiated feeder cells at densities that allow outgrowth of discrete colonies. Recently, Evans and DiPaolo (1982) showed that endothelial cell growth factor can substitute for feeder cells in supporting growth of the HEC. One day after seeding, the cultures are treated with carcinogen for times ranging from less than one hour to the entire period of the assay, depending on the type of carcinogen used. After a total incubation time of 7–9 days, the cultures are fixed and stained and the colonies counted and evaluated morphologically. The normal cells form flat colonies of fibroblastic or epithelioid cells growing in an orderly arrangement; the transformed cells are fibroblastic and grow with random orientation to form colonies of piled-up, crisscrossed cells. DiPaolo et al. (1969, 1971) showed that cells derived from the morphologically transformed colonies are tumorigenic when inoculated into appropriate hosts. However, it may require several weeks or months of subcultivation before morphologically transformed cells progress to the neoplastic or malignantly transformed phenotype (Barrett et al., 1977; Barrett and Ts'o, 1978; O'Brien et al., 1982).

Popescu et al. (1980) showed that treatment of HEC with TPA 24–72 hr after exposure to low doses of MNNG increases the absolute number of transformed colonies/dish and the average transformation frequency (transformed colonies/total colonies). They noted that the cells of some TPA-treated colonies formed a swirling pattern throughout the entire colony or only in the center, but transformed colonies obtained after treatment with carcinogen and TPA were indistinguishable from those transformed by carcinogen alone. Three weak or noncarcinogenic PAHs were also tested in this system by Popescu et al. (1980). Alone, these hydrocarbons produced no or few transformed colonies but with TPA, transformation occurred. The authors conclude that, because the number of cells in which transformation is initiated by low doses of carcinogen is much larger than the number of cells giving rise to transformed colonies in the absence of TPA, the frequency of the initial event is greater than can be expected from point mutations.

DiPaolo et al. (1981) also proposed, from the results of experiments in which transformation was induced in HEC with X-irradiation and TPA, that there might be more than one type of initiation of radiation carcinogenesis. One type of initiation would be TPA-independent and lead to transformation; the second type would be TPA-dependent with transformation being recognizable only after addition of, and maintenance in, TPA. The hypothesis is based on their observation that there was a 15- to 20-fold enhancement of morphologic transformation of HEC treated with TPA after irradiation but that some colonies reverted to normal morphology when TPA was removed. Isolation of transformed colonies from X-irradiated, TPA-treated cultures resulted in the continuance of the transformed morphology for 10 population doublings as long as TPA treatment was maintained. Unfortunately, morphology was the only criterion used to evaluate transformation and its reversal in these experiments, making it difficult to interpret the meaning of the observations.

Popescu *et al.* (1980, 1982) found that under conditions in which TPA enhanced transformation of HEC by low doses of MNNG, the promoter did not affect the frequency of SCE and chromosome aberrations produced by MNNG alone. TPA also had no effect on the frequency of SCE induced in HEC by u.v.- or X-irradiation.

Rivedal and Sanner (1980, 1981, 1982, 1985) have tested a number of compounds for cocarcinogenic (compound added together with initiator) and promoting (compound added immediately after treatment with initiator) activity in the HEC colony transformation system. In cells treated sequentially with B(a)P and phorbol esters, those phorbol esters with tumor-promoting activity in mouse skin enhanced the frequency of morphologic transformation, whereas analogs which are not promoters *in vivo* did not (Rivedal and Sanner, 1982). A low dose of B(a)P also enhanced transformation when applied subsequent to a higher dose, suggesting that B(a)P may act as a promoter as well as an initiator in this system (Rivedal and Sanner, 1982). Retinoids were found to induce morphologic transformation of HEC and to increase the transformation frequency when applied in combination with or subsequent to B(a)P. However, the retinoids decreased anchorage-independent growth of a B(a)P-transformed HEC line (Rivedal and Sanner, 1985).

Cigarette smoke extract potentiated morphologic transformation either when combined with low concentrations of B(a)P or added subsequently to B(a)P (Rivedal and Sanner, 1980). Combined treatment of HEC with organic carcinogens [(B(a)P, N-hydroxy-2-AAF, 4-nitroquinoline 1-oxide] and several metal salts (nickel sulfate, cadmium acetate, potassium chromate) enhanced transformation synergistically (Rivedall and Sanner, 1982). Nickel sulfate and cadmium acetate enhanced transformation of cells pretreated with B(a)P and also initiated cells in which promotion was mediated by TPA or B(a)P. The two metal salts appeared to be more potent as promoters than initiators.

A lymphotoxin derived from Syrian hamster peritoneal leukocytes reduces the level of TPA-enhanced morphologic transformation of HEC initiated by either u.v.- or X-irradiation (Evans and DiPaolo, 1981). It is effective when added prior to or with TPA and appears permanently to alter the cellular response to TPA.

SOD prevents a two-fold enhancement of X-ray-induced morphologic transformation of HEC by TPA when added to the medium together with TPA 24 hr after irradiation and left throughout the experiment (Borek and Troll, 1983). Retinoids also block the enhancement of TPA of radiation-induced transformation of HEC (Miller *et al.*, 1981).

Dexamethasone and hydrocortisone, which inhibit morphologic transformation of HEC induced by B(a)P and u.v.-irradiation (Evans and Greiner, 1981), also inhibit TPA-enhanced morphologic transformation of HEC pretreated with a low dose of B(a)P (Rivedal, 1982). In the latter case, the glucocorticoids are effective only when added with TPA after the cells have been exposed to B(a)P; they are ineffective when present only during the period of pretreatment with B(a)P. Phytohemagglutinin does not affect transformation of HEC induced by u.v.-irradiation alone, but does inhibit TPA-enhanced morphologic transformation of HEC initiated by u.v.-irradiation (DiPaolo *et al.*, 1982). Both dexamethasone and phytohemagglutinin will reduce the frequency of transformation promoted by TPA even if present for only 6 hr prior to staining the cells and terminating an experiment (DiPaolo *et al.*, 1982). Since the only criterion of transformation used in all these experiments was the random orientation of the cells that make up the transformed colonies, it is difficult to interpret the significance of either the enhancement of morphologic transformation of HEC by TPA or the inhibition of this enhancement by other compounds.

Phorbol ester tumor promoters can also increase the *in vitro* lifespan of HEC. If continuously present in the medium, they increase the total number of population doublings attained by serially cultivated cells but do not prevent eventual senescence (Umeda and Enaka, 1981; Bruce *et al.*, 1983). Continuous exposure of *carcinogen-treated* cells to tumor promoter increases the probability that a culture will not senesce but will instead develop into an 'immortal' cell line (O'Brien *et al.*, 1982). Little is known as yet about the effects of the phorbol esters on the progression of 'immortal' cell lines to neoplastic transformants able to grow in soft agar and/or produce tumors in appropriate hosts.

As Thomassen *et al.* (1985b) point out, the HEC system is a good one in which to compare various events in carcinogen- and oncogene-induced transformation. With this in mind, they tested the ability of the viral Harvey-*ras* oncogene, alone in combination with the viral *myc* oncogene, to transform normal and carcinogen-induced, preneoplastic HEC to the anchorage-independent and tumorigenic phenotype. The *ras* oncogene alone was insufficient to cause neoplastic transformation of the normal or preneoplastic cells. The data provide evidence that, as with carcinogen-induced neoplastic transformation, multiple steps are required for neoplastic transformation of normal or preneoplastic HEC transfected with and expressing the Harvey-*ras* oncogene.

5.4. Mouse Epidermal Cells

5.4.1. *Primary Cells*

Studies on the selective growth of mouse basal epidermal cells in low-calcium medium; the triggering of differentiative changes when the cells are switched from low- to high-calcium medium; the heterogeneous sensitivities of subpopulations of epidermal cells to both the stimulation of proliferation and the induction of terminal differentiation by TPA; and the resistance of carcinogen-treated epidermal cells to the inhibition of proliferation and induction of differentiation by high calcium concentrations are discussed in Sections 4.2.3 and 4.4.1. There is as yet no *in vitro* assay for two-stage transformation of primary mouse epidermal cells. In particular, there is no assay in which the recovery of transformed foci is based on the heterogeneous responses to TPA of normal and initiated basal cells, the basis of the mechanism of skin tumor promotion *in vivo* proposed by Yuspa (1983, 1984) (see Section 4.4.1).

5.4.2. *Cell Lines*

Colburn and coworkers have developed an *in vitro* mouse epidermal cell model system that may be analogous to late-stage tumor promotion in mouse skin *in vivo* (see Colburn *et al.*, 1982b). They isolated anchorage-dependent clonal cell lines that, in response to brief exposure to phorbol ester tumor promoters, undergo a shift in phenotype to anchorage-independence and tumorigenicity. The rapidity with which the shift occurs and the irreversibility of the shift suggest that, prior to exposure to promoter, the cells had already undergone the events necessary for initiation and the first stage of promotion.

The JB6 line, the one which has been studied most extensively, is derived from primary epidermal cell cultures prepared from newborn BALB/c mice and exposed initially only to solvent (Colburn *et al.*, 1979). After about 30 passages in culture, the cells were still anchorage-dependent and nontumorigenic in nude mice but had become responsive to the induction of anchorage-independent growth by TPA, as determined by colony-formation in TPA-containing soft agar medium. Cells isolated from colonies growing in agar retained the capacity for anchorage-independent growth in the absence of TPA. The shift to anchorage-independence in response to TPA, which the authors refer to as promotion, is thought to occur by an inductive rather than a selective mechanism, for when exposed to promoter in agar, single cells of the JB6 line or its clonal derivatives form colonies with about the same efficiency (10–35%) as mass cultures (10^4 cells/60 mm plate) seeded in agar (Colburn, 1980). Anchorage-independence is induced in JB6 cells by EGF, human transforming growth factor and mezerein, but not by the first-stage mouse skin tumor promoter, A23187, a calcium ionophore; it is inhibited by simultaneous exposure to retinoids (Colburn *et al.*, 1982b).

The fact that there are TPA-'promotable' (P^+) JB-6 variants which are resistant to TPA-induced mitogenesis at plateau density (Colburn, 1980; Colburn *et al.*, 1981) seems to rule out a requirement for TPA-induced mitogenesis in the induction of the anchorage-independent phenotype. Variants that are resistant to TPA promotion (P^-) bind phorbol esters and undergo down modulation of the receptor following ligand binding, and TPA induces a decrease in the binding of EGF in both TPA-sensitive and -resistant variants

(Colburn et al., 1982a). These and other studies with the variants imply that TPA conversion of JB6 cells to anchorage-independence requires phorbol ester but not EGF receptors, whereas TPA-induced mitogenesis requires receptors for both ligands.

TPA reduces synthesis of procollagen in P^+ JB6 clones; with continuous exposure to TPA, the decrease in collagen synthesis persists for at least eight days (Dion et al., 1982). Since TPA has a similar effect on P^- JB6 clones, the collagen decrease may be necessary but not sufficient to produce the promotion response. TPA does produce differential effects on cell surface gangliosides in P^+ and nonpromotable variants; it decreases the de novo synthesis of trisialoganglioside (G_T) only in promotable variants (Srinivas and Colburn, 1982; Srinivas et al., 1982). When inserted in liposome form into cell membranes, G_T, but not other sialoglycoconjugates, blocks TPA-mediated conversion of JB6 cells to anchorage-independence; G_T does not inhibit specific binding of phorbol esters. Since it does not block growth in agar of transformed cell lines, it appears to act at the level of induction rather than expression of the transformed phenotype.

The results of recent experiments indicate that the DNA of promotable JB6 cells has sequences that confer promotion-sensitivity on promotion-insensitive JB6 variants. The P^+ trait was transferred to P^- cells by transfection of DNA from P^+ cell lines and genes conferring the P^+ trait were cloned (Colburn et al., 1983, 1984; Gindhart et al., 1985). These genes, which determine sensitivity to transformation by tumor promoters but do not alone produce transformation, have been termed pro genes for promotogenes, in analogy with oncogenes. They appear to be unrelated to 11 common viral oncogenes. The possible mechanisms by which pro genes might determine sensitivity to transformation by tumor promoters are discussed in Gindhart et al. (1985).

5.5. RAT TRACHEAL EPITHELIAL CELLS

Nettesheim and Marchok (1983) recently published an excellent review on 'Neoplastic Development in Airway Epithelium', in which they consider and discuss observations made in both humans and experimental models. Steele and Nettesheim (1983) also have a brief review concerned specifically with models of two-stage carcinogenesis in the respiratory tract.

Nettesheim and coworkers (see Steele and Nettesheim, 1983) developed a heterotopic transplant system in which tracheas transplanted subcutaneously onto the backs of syngeneic rats can be exposed sequentially to carcinogen and promoter. When tracheas were exposed first to DMBA and then to asbestos, those authors found a definite enhancement of carcinoma formation compared to that in tracheas exposed to only DMBA or asbestos. They concluded that the tumor-enhancing effect of asbestos fibers in these experiments and those of others (e.g. Mossman and Craighead, 1979), as well as in asbestos workers who smoke (Selikoff and Hammond, 1975), may have two components: promotional activity and a carcinogen (e.g. PAH)-carrier function.

Exposure of heterotopic tracheal transplants to DMBA and subsequently to TPA released from intraluminal pellets also results in a greatly increased tumor yield with a shorter latent period compared to results with DMBA alone (Topping and Nettesheim, 1980; Steele and Nettesheim, 1983). Terzaghi et al. (1983) have used this model to analyze the effects of TPA on the evolution of carcinogen-initiated cells. They had previously shown (Terzaghi and Nettesheim, 1979) that soon after exposure of tracheas to DMBA in vivo, carcinogen-altered cells appear which can rapidly proliferate in vitro under culture conditions that are 'nonpermissive' for normal tracheal epithelial cells. Cells from some epithelial cell colonies escape senescence and can be propagated in vitro indefinitely. With time, some cell lines show anchorage-independent growth and become neoplastically transformed, producing invasive carcinomas when inoculated into appropriate hosts.

With explants derived from tracheas that had been exposed to TPA for up to 18 months beginning two weeks after exposure to DMBA, it was found (Terzaghi et al., 1983) that exposure to TPA in vivo had no effect on the yield of either epithelial foci or immortal cell lines from DMBA-treated tracheas. However, with cultures derived from tracheas exposed to TPA for 12 months following exposure to DMBA, the percentage of epithelial foci which gave rise to anchorage-independent progeny was much higher than with tracheas exposed to DMBA alone. Since the percentage of epithelial foci that were anchorage-independent was the same in DMBA and DMBA-TPA exposure groups in cultures established three months after exposure, Terzaghi et al. (1983) concluded that some phenotypically altered cells may revert to a more normal state and/or fail to replicate and that exposure to TPA may enhance the persistence of these altered cell populations in intact tissue. The authors suggest that they were able to detect reversibility of initiation, which is usually thought to involve an irreversible process, because of the low dose of initiator used. Their experiments cannot determine whether the effect of TPA is to inhibit reversal of the potentially neoplastic state or to enhance those processes (progression and/or expansion) which increase the conversion rate of subculturable epithelial foci to anchorage-independent foci or increase the growth of anchorage-independent epithelial foci *in vivo*.

Steele et al. (1980) found that TPA enhances malignant transformation of rat tracheal epithelium in an organ culture-cell culture system. Tracheal explants were exposed in organ culture to MNNG followed by multiple exposures to TPA, and primary cultures were established from epithelial cell outgrowths. Morphologically altered cells that could be repeatedly subcultured and that formed carcinomas when inoculated into immunosuppressed isogenic hosts were obtained from explants exposed to MNNG or MNNG and TPA; the number of explants that yielded tumorigenic cell lines was greater and the time to tumorigenic potential was shorter in the MNNG-TPA group. Explants exposed only to TPA gave rise to nontumorigenic permanent cell lines, confirming the earlier report of Steele et al. (1978) that multiple exposures of rat tracheal epithelium in organ culture to TPA can result in escape from senescence. They had shown the epithelial nature of TPA-derived cell lines by demonstrating that the cells could repopulate heterotopically transplanted tracheas stripped of their epithelium. The repopulated tracheas displayed typical epithelial morphology; ciliated and secretory cells were found in large numbers, although they were not evident during culture *in vitro*.

Following their experiences with the *in vivo* and *in vivo/in vitro* tracheal epithelium systems, Nettesheim's group (Pai et al., 1982, 1983; Steele et al., 1983) developed a quantitative assay for transformation of primary tracheal epithelial cells in monolayer culture. The cells are plated on collagen-coated dishes, treated with carcinogen (MNNG), and monitored for the formation of densely packed, rapidly proliferating epithelial foci. Increased DNA content is also an early marker of transformation of these cells (Vanderlaan et al., 1983). Cultures containing carcinogen-altered foci continue to grow after control cultures cease proliferation and senesce; they can be subcultured, presumably indefinitely, on dishes not coated with collagen; and eventually after 10–20 passages develop the capacity for anchorage-independent growth and tumorigenicity.

Another quantitative assay for transformation of rat tracheal epithelial cells is based on growing the normal cells at clonal density on plastic with 3T3 feeder layers rather than on collagen (reviewed in Nettesheim and Barrett, 1984). After treatment with carcinogen and an appropriate interval for fixation and expression of carcinogen-induced events, the feeder cells are removed by EDTA at low concentration. This results in enlargement and sloughing of most of the normal cells and selection in 3–6 weeks of variants that form large colonies of small, hyperchromatic cells. In addition to their increased growth potential and acquired ability to grow directly on plastic, these variants have a reduced propensity to senesce. Changes in stem cell populations during this early preneoplastic stage have been described (Thomassen et al., 1985a) and the rate of spontaneous change of preneoplastic variant cell lines from anchorage-dependence to anchorage-independence has been determined (Thomassen et al., 1985c). At concentrations that do not inhibit colony-forming efficiency, RA decreases the frequency of transformation induced by exposure to MNNG (Mass et al., 1984).

Preliminary reports (Steele et al., 1983; Steele 1985; Vanderlaan et al., 1983) indicate that primary tracheal epithelial cell cultures exposed only to TPA can develop into anchorage-dependent, non-tumorigenic cell lines. Exposure of primary cell cultures to low doses of

MNNG followed by multiple exposures to TPA increases the number of morphologically altered foci that develop compared to cultures exposed only to MNNG. TPA also shortens the time required for MNNG-treated cultures to express anchorage-independence and tumorigenicity. Interestingly, TPA has completely opposite effects on normal and transformed rat tracheal epithelial cells. It markedly stimulates the colony-forming efficiency of normal cells but triggers a rapid cytotoxic response in transformed cells (Nettesheim *et al.*, 1985). The possible significance of this differential effect for promotion *in vivo* is discussed by Nettesheim *et al.* (1985).

Braslawsky *et al.* (1984) recently described the characterization of monoclonal antibodies which detect antigenic determinants expressed on carcinogen-altered preneoplastic and neoplastic rat tracheal cells in culture but do not detect them on normal, primary tracheal cells. It will be interesting to see how TPA affects expression of these determinants on carcinogen-treated and -untreated cells.

5.6. Mouse Submandibular Gland

Another *in vitro* epithelial cell system in which the effects of TPA on transformation have been assessed is the mouse submandibular gland. Exposure of gland explants to PAHs results in transformation that can be quantitated by the number of foci of altered epithelial cells which arise (Knowles and Franks, 1977; Knowles, 1979; Wigley, 1983). Many of these foci can be subcultured and some evolve into relatively stable preneoplastic cell lines from which tumorigenic variants eventually emerge. TPA is a potent mitogen for epithelial cell outgrowths from gland explants (Knowles, 1979; Wigley, 1983). It has no effect on the induction of foci of preneoplastic cells when present during the transformation which occurs in response to treatment with a dose of B(a)P that itself is sufficient for transformation. However, multiple treatments with TPA do increase the frequency of foci arising from explants pretreated with only a subthreshold, initiating dose of B(a)P. Foci from all TPA-treated cultures appear earlier and enlarge more rapidly than those induced by carcinogen alone but there is no evidence that foci from TPA-treated cultures are more likely to give rise to permanent cell lines or that these cell lines become tumorigenic earlier. TPA alone does not induce high frequencies of foci, implying that, as in mouse skin tumorigenesis, TPA can act only on already initiated cells, spontaneous or carcinogen-induced, to promote expression of their altered state.

5.7. Human Cells

In monolayer cultures of human endometrial stromal cells pretreated with MNNG, TPA enhances the expression of several phenotypes usually associated with transformation; it has little effect on the same phenotypes in control cultures (Siegfried and Kaufman, 1983). The phenotypic markers analyzed in these studies were increased saturation density, morphologic atypia, γ-glutamyltranspeptidase expression, and colony-forming ability in restrictive media. Chronic exposure to TPA decreased the amount of pretreatment with MNNG required for expression of these cell characteristics. Siegfried and Kaufman (1983) suggest that this enhancement by TPA may depend on the increased ability of carcinogen-treated cells to survive and grow (Dorman *et al.*, 1983b) and may result, at least in part, from selection of carcinogen-altered cells. The latter suggestion is based on the fact that (1) MNNG-treated cultures were more sensitive to the immediate, reversible morphologic changes induced by TPA and (2) the colony-forming ability of carcinogen-treated cells was increased by prolonged exposure to TPA. The authors did not demonstrate that either the MNNG- or MNNG + TPA-treated cultures with altered phenotypes have the capacity for anchorage-independent growth or tumorigenicity in nude mice. Chronic exposure of MNNG-treated human endometrial stromal cells to DES increases expression of phenotypic alterations similar to those seen with chronic exposure to TPA (Siegfried *et al.*, 1984).

Several other new human cell systems have been described that will be valuable for studying the effects of tumor promoters on progression of human cells from the preneoplastic to the neoplastic state and the multistep process of human carcinogenesis. For example, Parsa et al. (1981, 1984) have developed chemically defined media for maintaining explants of human pancreas in organ culture for long periods of time. They have described the proliferation and differentiation of fetal and adult human pancreas under these conditions, the development of foci of enhanced proliferation and carcinoma in explants treated with methylnitrosourea, and the modulation of cytotypic cell surface markers in the fetal pancreas model.

Hammond et al. (1984) developed a serum-free medium with bovine pituitary extract as the only undefined supplement which supports serial subcultivation of normal human mammary epithelial cells for 10–20 passages (1:10 split ratios) and rapid clonal growth of the cells with plating efficiencies up to 35%. Treatment of primary cultures with B(a)P results in a population of cells with an extended lifespan in culture (Stampfer and Bartley, 1985). These 'extended life' cultures show widespread heterogeneity in cellular morphology and growth patterns both initially and during subsequent subcultivation. Two apparently immortal cell lines have so far evolved from the 'extended life' cultures: the cells of these continuous lines show little or no anchorage-independent growth and do not form tumors in nude mice. Paraskeva et al. (1984) have determined the conditions for routine primary culture and serial transfer of colorectal adenomatous epithelium derived from familial polyposis coli patients. They isolated 4 colorectal epithelial cell lines at different stages in malignant transformation and compared their morphological characteristics, growth potential, tumorigenicity and karotypes.

Normal human fibroblasts which have a limited lifespan *in vitro* can now be transformed to anchorage-independence by treatment with chemical and physical carcinogens or by transfection of genomic DNA from malignant human cell lines (Milo et al., 1981; Maher et al., 1982; Sutherland and Bennett, 1984; Sutherland et al., 1985). However, in all cases the transformed cells are still programmed to senesce and have limited tumorigenic potential in nude mice. It is possible that protocols may eventually be devised whereby treatment with tumor promoters, either alone or in combination with other agents, will immortalize these transformed cells and/or increase their malignant potential.

6. INTERACTIONS BETWEEN TUMOR PROMOTERS AND TUMOR VIRUSES

One can suggest several mechanisms by which tumor promoters might affect cell transformation induced by tumor viruses. On the one hand, promoters could enhance expression of the transformed phenotype, perhaps as they 'promote' expression of the chemically transformed phenotype. On the other hand, they could act directly on the virus and enhance virus synthesis in productively infected cells or virus expression in cells with latent virus genomes, with a resultant increase in transformation frequency. There are now many reports (reviewed in Yamamoto, 1984) in which these and other effects of tumor promoters on virus-cell interactions have been described (Table 6).

6.1. Enhancement of Virus-Induced Transformation

There are several examples of tumor promoters enhancing either virus-induced transformation or phenotypic expression of the transformed state (reviewed in Fisher, 1983). Sivak and Van Duuren reported in 1967 that TPA enhanced outgrowth of the transformed cells in cocultures of contact-inhibited Swiss mouse 3T3 fibroblasts and spontaneously or SV40-transformed cells. Subsequently, Fisher et al. (1978) observed that TPA added to rat embryo cells infected with adenovirus type 5 enhances the number of transformed foci 2- to 3-fold. They showed that TPA did not affect virus uptake or the integration pattern of viral DNA sequences but did enhance the cloning efficiency of adenovirus-transformed cells grown either alone or with an excess of normal rat embryo cells (Fisher et al., 1978, 1979).

TABLE 6. *Effects of Phorbol Diester Tumor Promoters on the Synthesis and Biological Activity of Tumor Viruses* In Vitro

	References
Enhancement of Virus-Induced Transformation	
SV40	Sivak and Van Duuren, 1967; Martin *et al.*, 1979a,b; Daya-Grosjean *et al.*, 1982
Adenovirus	Fisher *et al.*, 1978, 1979
Epstein-Barr virus	Yamamoto and zur Hausen, 1979; Yamamoto *et al.*, 1981; Harada *et al.*, 1981
Murine sarcoma virus	Lipp *et al.*, 1982
Activation of latent viral genomes	
Herpesviruses	zur Hausen *et al.*, 1978, 1979; Lin *et al.*, 1979, 1983; Hudewentz *et al.*, 1980; Datta *et al.*, 1980; Ablashi *et al.*, 1980; Ito *et al.*, 1981a
Papovaviruses	Krieg *et al.*, 1981; Amtmann and Sauer, 1982; Nomura *et al.*, 1983
Enhancement of virus expression	
Murine retroviruses	
Friend leukemia virus	Colletta *et al.*, 1908; Lipp *et al.*, 1982
Mammary tumor virus	Arya, 1980; Chan and Buehring, 1983
Endogenous type C virus	Hellman and Hellman, 1981; Lipp *et al.*, 1982
Primate retroviruses	
Simian	Wunderlich *et al.*, 1981
Human (HTLV, human T-cell leukemia/lymphoma virus)	Vyth-Dreese and de Vries, 1983
Modulation of expression of cellular oncogenes	
c-myc	Kelly *et al.*, 1983; Grosso and Pitot, 1984
c-myb	Craig and Bloch, 1984
c-fos	Müller *et al.*, 1985

They concluded that the enhancement of adenovirus transformation by TPA was probably due to its ability to facilitate *expression* of the transformed state.

Transformation of Chinese hamster lung cells by SV40 mutants defective in the synthesis of 20,000 MW t-antigen is slightly enhanced by TPA (Martin *et al.*, 1979a,b). The promoter has a dramatic effect on expression of transformation in Swiss mouse 3T3 cells infected with SV40 temperature-sensitive (tsA) mutants in which the large T protein is inactivated (Daya-Grosjean *et al.*, 1982). At the non-permissive temperature for transformation (39°C), infected cells undergo 'abortive' transformation in soft agar medium and do not form large colonies but, if TPA is incorporated into the agar, macroscopic colonies form at a frequency of $2-5 \times 10^{-3}$, as compared to a frequency of 10^{-5} without TPA. However, cells isolated from such colonies are not permanently transformed; cell lines derived from them do not contain integrated SV40 DNA sequences and are not anchorage-independent. At the permissive temperature for transformation (33°C), tsA-infected cells form colonies in agar in the absence of TPA but, as with adenovirus-transformed rat cells, the transformation efficiency is increased 3- to 4-fold if TPA is included in the agar medium.

TPA enhances the transformation of human leukocytes by the oncogenic herpesvirus, EBV, as determined by the increased ability of infected cells to form colonies in soft agar (Yamamoto and zur Hausen, 1979). By a mechanism involving different effects on different cell populations, TPA also increases the incidence of spontaneous establishment of lymphoblastoid cell lines from the peripheral blood of EBV-seropositive donors (Yamamoto *et al.*, 1981). The promoter both stimulates the proliferation of EBV-nuclear antigen-positive and/or -DNA synthesizing B cells and suppresses the usual regression of transformation due to the EBV-specific and -nonspecific cytotoxicity of T cells (Harada *et al.*, 1981).

Hsiao *et al.* (1984) reported that TPA and teleocidin enhanced the number of transformed foci obtained when C3H/10T1/2 cells were transfected with the cloned human bladder c-Harvey-*ras* oncogene, pT24. On the other hand, TPA reduced the number of foci seen when NIH/3T3 cells were transfected with the same plasmid. The different effects of

TPA in the two cell types did not appear to be due to differences at the level of cellular uptake or integration of the transfected oncogene. Connan et al. (1985) found that focus formation could be induced in primary rat embryo fibroblast cultures by transfer of the gene encoding the large-T protein of polyoma virus and subsequent treatment of the cells with TPA; transfer of this or the *myc* oncogene to the primary cells without TPA treatment induced 'immortality' but not transformed foci.

6.2. Activation and Enhancement of Virus Expression

The most striking example of TPA's ability to induce expression of latent viral genomes is seen in lymphoblastoid cell lines that have persisting EBV genomes, but neither physical nor chemical carcinogens induce significant amounts of virus replication (zur Hausen et al., 1978, 1979). In EBV-transformed lymphoblastoid cell lines that are not producing any virus, TPA causes abortive induction of virus synthesis with increased synthesis of only the early antigen complex (zur Hausen et al., 1979; Lin et al., 1979; Hudewentz et al., 1980). However, with cell lines in which a small proportion of cells are spontaneously producing virus, treatment with TPA can increase the proportion of cells producing virus capsid antigen to 80–90%, increase the amount of viral DNA per cell more than 20-fold and result in rapid induction of virus-associated DNA polymerase activity (zur Hausen et al., 1978, 1979; Lin et al., 1979; Hudewentz et al., 1980; Datta et al., 1980).

Recently, Lin et al. (1983) observed a close association between activation of latent EBV genomes by TPA and selective stimulation of synthesis of specific chromosomal proteins, despite an overall reduction in total cellular protein synthesis. The viral DNA isolated from TPA-induced cells has the physicochemical properties and restriction endonuclease cleavage pattern of viral DNA from uninduced cells and that produced as the result of superinfection (zur Hausen et al., 1979; Lin et al., 1979). Virus induction by TPA, but not by superinfection, is inhibited by RA (Yamamoto et al., 1979). TPA can also increase production of infectious EBV and early and late virus antigens in human and simian cell cultures infected with other oncogenic primate herpesviruses (Ablashi et al., 1980).

Ito and coworkers (Ito et al., 1981a,b, 1983) have designed a short-term *in vitro* assay to detect substances with potential promoting activity which is based on the ability of the compounds to induce expression of early antigen in a nonproducer human lymphoblastoid cell line carrying the EBV genome. A wide range of compounds, structurally related and unrelated to the phorbol diesters, is active in this assay.

TPA also has striking effects on the expression of another class of DNA tumor viruses, the papovaviruses. In mouse embryo fibroblast cultures in which genomes of bovine papilloma virus-1 reside in a nonexpressed episomal state after infection, treatment with TPA results in transcription of viral mRNA and viral DNA replication (Krieg et al., 1981; Amtmann and Sauer, 1982). In addition, the cells acquire characteristics of transformation not evident in control cultures such as high saturation density and the ability to grow in low serum concentrations (Amtmann and Sauer, 1982). TPA also enhances the viral DNA content of monkey cell lines in which the papovavirus, stump-tailed macaque virus, exists in a latent episomal state (Krieg et al., 1981). It has no effect on either the viral DNA copy number in cells productively infected with this virus or the state of the integrated virus in transformed cells.

TPA is also reported to enhance the synthesis of infectious virus in cultures of SV40-transformed hamster kidney cells induced to synthesize virus by treatment with DNA-damaging agents and carcinogens (Nomura et al., 1983).

TPA can also enhance the expression of several RNA tumor viruses. In mouse mammary tumor cells persistently infected with mouse mammary tumor virus, a type B retrovirus, TPA stimulates a 10- to 20-fold increase in virus production (Arya, 1980). The glucocorticoid, dexamethasone, also stimulates virus production, and the result of the combined effects of the two inducers is a 100- to 200-fold enhancement of virus production (Arya, 1980). Antiglucocorticoids that suppress enhancement of virus production by dexamethasone also suppress enhancement by TPA (Chan and Buehring, 1983). Chan and Buehring

(1983) also found that TPA competes with dexamethasone for binding to the cellular glucocorticoid receptor and suggest that it may stimulate virus expression by mimicking a glucocorticoid hormone.

Murine FELC treated with TPA show about a two-fold increase in the amount of type C virus-specific extracellular reverse transcriptase activity, whether or not the cells are simultaneously induced by dimethyl sulfoxide to undergo terminal differentiation (Colletta et al., 1980; Lipp et al., 1982). TPA also induces release of endogenous type C retroviruses from BALB/c mouse cells (Hellman and Hellman, 1981; Lipp et al., 1982). In one system, it was shown that TPA synergistically enhanced the level of endogenous virus induction by 5-iodo-2-deoxyuridine and that virus induction by both agents was inhibited by the protease inhibitors, antipain and leupeptin (Hellman and Hellman, 1981). The promoter also stimulates virus production in primate cell cultures persistently infected with types C and D simian retroviruses (Wunderlich et al., 1981). Recently, Vyth-Dreese and deVries (1983) reported that the expression of the p19 structural core protein of the human T cell leukemia/lymphoma retrovirus (HTLV) was greatly enhanced in neoplastic (T-cell chronic lymphoma leukemia) T cells induced to proliferate by TPA's generation of IL-2 activity. This is another example of the dual effects TPA may have on complex virus-cell interactions.

6.3. MODULATION OF ONCOGENE EXPRESSION

There are a number of excellent, recent reviews on viral oncogenes, their products and functions (Hunter, 1984; Heldin and Westermark, 1984; Bishop, 1984). As yet, there have been only a few studies of the effects of tumor promoters on oncogene expression and their biological significance, but it can be anticipated that there will soon be a dramatic increase in such studies. Some of the reports on the interaction of phorbol esters with growth factors and growth factor receptors encoded by retroviral oncogenes and their normal cellular homologues (e.g. the gene product of *sis*, PDGF, and the receptor for EGF encoded by *erb*B) are cited in Section 4. Bissell et al. (1979) had originally reported that chick cells infected with a temperature-sensitive mutant of RSV defective in the *src* gene had an increased sensitivity to TPA at the non-permissive temperature; the cells exhibited a 'more complete transformed phenocopy' than normal, uninfected cells treated with TPA. Subsequently, this group (Laszlo et al., 1981) found no evidence that TPA's effects on either the normal or uninfected cells were mediated by activation of the phosphorylating activity of the *proto-src* gene, the cellular homologue of the viral *src* gene, or by the activation of other cellular phosphotyrosine-specific kinases. Pietropaolo et al. (1981) also concluded that the effects of TPA on the morphology and increased plasminogen activator activity of avian sarcoma virus-transformed and normal CEF could not be ascribed to an effect on *src* or *proto-src* gene-associated protein kinases. Recently, however, with the use of more sensitive techniques, phosphorylation at tyrosine residues in chick cells treated with TPA has been reported (Gilmore and Martin, 1983; Bishop et al., 1983).

It has been shown recently that TPA can regulate expression of the *c-myc* gene (Kelly et al., 1983). Within 1-3 hr after addition of TPA to quiescent cultures of BALB/c 3T3 fibroblasts, *c-myc* mRNA is increased about eight-fold. This effect of TPA is similar to the effects on *c-myc* expression of other mitogens that can initiate the first phase of a proliferative response in quiescent cultures, i.e. stimulation of lymphocytes by lipopolysaccharide or concanavalin A and PDGF-stimulation of fibroblasts (see Pledger et al., 1977; Baserga, 1984). HL-60 cells induced to differentiate with TPA show a very early increase in expression of *c-fos* (Müller et al., 1985) and a subsequent decrease in *c-myc* expression (Grosso and Pitot, 1984). With a human myeloblastic leukemia cell line (ML-1) which expresses *c-myb*, proliferating cells induced by TPA to differentiate to non-dividing monocyte-macrophages undergo a very rapid decline in the expression of *c-myb*, with a 50% decline in mRNA in 3 hr and a subsequent decrease on DNA synthesis (Craig and Bloch, 1984). Whether effects of TPA on expression of these proto-oncogenes are in any way related to either the conversion or proliferative phases of promotion, or to neither, remains to be determined.

Balmain et al. (1984a,b) transfected NIH/3T3 cells with DNA for normal mouse epidermis and from benign papillomas and malignant carcinomas induced by DMBA and TPA using the two-stage initiation/promotion protocol. The transfected tumor DNA induced transformants; analysis of the DNA showed that the tumors, even at the papilloma stage contained an activated cellular Harvey-*ras* gene. Balmain et al. (1984b) also found that the steady-state levels of c-Harvey *ras* transcripts were elevated to varying degrees in some primary tumors. On the other hand, Toftgard et al. (1985) observed no increase in expression of the oncogenes Harvey-*ras*, Kirsten-*ras, fos, myc, abl* and *raf* in chemically induced mouse skin tumors, or in mouse epidermis after treatment with TPA; the expression of *abl* was reduced in tumors and after repeated applications of TPA. Treatment of primary mouse epidermal cells with TPA did not alter expression of the cellular homologues of these oncogenes and cell lines derived by treatment of epidermal cells with chemical carcinogens showed expression levels similar to those of untreated primary epidermal basal cells. Yuspa et al. (1985) have found that infection with Kirsten sarcoma virus and expression of the activated *ras* gene blocks keratinocyte maturation at an early step prior to irreversible commitment to terminal differentiation. Virus-infected cells respond to TPA with an increase in thymidine incorporation into DNA, suggesting that the promoter induces reversion to an even less mature basal cell phenotype. The authors discuss how these observations are consistent with the possibility that *ras* activation is involved in epidermal tumorigenesis, in particular, in the formation of promoter-dependent papillomas. They suggest that, in contrast, the characteristics of differentiation-resistant *chemically* altered cells are consistent with changes in autonomous papillomas that do not require further promoter treatment for maintenance.

7. CONCLUSIONS

In a previous review of tumor promotion (Diamond et al., 1980), we discussed several hypotheses for the mechanism of mouse skin tumor promotion. These hypotheses, which were based on the biological and biochemical changes induced in cells by the phorbol diesters and structurally related tumor promoters, have been considered, where appropriate, throughout this chapter. All involve mechanisms by which tumor promoters might endow initiated cells with the selective advantage required for tumor formation to occur. The theories considered include altered gene expression such as enhanced expression of specific regulatory genes; altered response of initiated cells to differentiation signals; and enhanced production of growth-promoting factors. Clearly, such hypotheses are still viable possibilities for the mechanism of phorbol diester-mediated promotion of skin tumor formation *in vivo* and transformation *in vitro*. Except perhaps for compounds such as dihydroteleocidin B, there is still too little data with other compounds that promote transformation even to speculate about whether they act by the same mechanism as the phorbol diesters or by different mechanisms.

It is probably fair to say, however, that we do not yet have cell culture models in which adequately to test theories of the mechanism of promotion of transformation. This is particularly true if promotion requires that there be interaction between different cell types (see, e.g. Section 4.3) or that within the same cell population, there be heterogeneous responses to promoter-induced effects (see Section 5.4.1.). If either of these possibilities be the case, then the rat tracheal epithelial cell models (Section 5.5) and the mouse epidermal cell model (Section 5.4.1) fulfill many, but still not all, of the criteria for a cell culture model of promotion of transformation.

There are, on the other hand, many cell culture systems in which to analyze the mechanism(s) by which the phorbol diesters exert their myriad effects on cells, some of which may be involved in the mechanism of promotion *per se*. The major findings in the last few years regarding the mechanism of action of the phorbol diesters were that they have specific cellular binding sites and that at least some of these receptors sites copurify with a specific protein kinase (see Section 3.3). These findings suggested that phosphorylation of specific protein residues may be a common factor in the mechanism of action of tumor promoters, tumor viruses, growth factors and hormones. However, the similarities and differences in the protein kinases activated by each of these factors and their possible interactions with one another must still be determined. In addition, the critical steps that follow the initial receptor-ligand interaction and lead to a specific effect induced by a specific factor at any one point in time must still be determined. With regard to the phorbol diesters, it may be that the mediation of each stage of promotion involves a different reaction subsequent to binding, that is, the reaction mediating the conversion stage (Stage I) of promotion may very well differ from the reaction mediating proliferation of initiated, preneoplastic cells (Stage II). The concept of multistage promotion has been delineated quite clearly in the mouse skin model (see, however, Hennings and Yuspa, 1985) but the equivalent steps have not been as well defined *in vitro*.

There are primary cell culture models such as the hamster embryo cell system (Section 5.3) in which tumor promoters enhance the frequency of carcinogen-induced morphologic transformation. In this and the rat tracheal epithelial cell systems, it has been shown that TPA also can increase the proliferative lifespan *in vitro* and the probability of immortalization. It remains to be determined whether these observations are in any way related to the altered expression of an oncogene in promoter-treated cells. Recent reports indicate that neoplastic transformation by oncogenes requires that cells undergo not only a transforming but also an immortalization event, which may occur 'spontaneously' or be induced by carcinogen treatment or by transfection with specific oncogenes (reviewed in Land *et al.*, 1983). One can also ask whether oncogenes play a role in the morphologic transformation induced by carcinogens in the immortal, perhaps preneoplastic (Thomassen *et al.*, 1983) mouse cell lines, C3H/10T1/2 and BALB/c 3T3, and/or in the enhancement of the frequency of that transformation by tumor promoters (see Sections 5.1 and 5.2).

The increase of morphologic transformation induced by promoters in these cell systems appears to be reversible and may be equivalent to Stage I of promotion in mouse skin. However, it cannot be assumed that the action of the promoter is the same in primary cells as it is in immortal cell lines, although the investigator may be scoring the same endpoints. Promoter-induced anchorage-independence in the mouse epidermal cell line, JB6 (Section 5.4.2), is irreversible, clearly different from promoter-enhanced morphologic transformation, and probably related to the late stages of promotion *in vivo*. As yet, the effects of promoter-treatment on the progression of cells from morphologic to neoplastic transformants capable of anchorage-independence and tumorigenicity have not been thoroughly analyzed in any single cell system. It is still unclear whether promoter treatment increases the frequency and decreases the latency of this progression, or whether such enhancement may sometimes be promoter-independent and perhaps require a second exposure to a carcinogen (see Section 2.1.1). In the case of the transformation of human cells by chemical carcinogens *in vitro*, where anchorage-independence is an early event and the acquisition of immortality very rare (reviewed in DiPaolo, 1983), even less is known about the effects of promoter treatment on the various stages of transformation.

Although perhaps none of the cell culture models reviewed herein is ideal for studying the effects of tumor promoters on all the steps of transformation and progression, each has been able, and will continue, to tell us something about the individual steps. Recent advances in cell culture technology, such as the formulation of media for growing human epithelial cells (e.g. Reznikoff *et al.*, 1983; Hammond *et al.*, 1984), should lead to the development of other model systems for studying transformation and promotion (Stampfer and Bartley, 1985; Yoakum *et al.*, 1985). These models are invaluable for elucidating the mechanism(s) of promotion by chemicals of varied structures, both *in vivo* and *in vitro*, and learning how to prevent their harmful effects. Two-stage transformation assays, and other short-term assays that detect compounds with the potential for promoting activity *in vivo*, are also extremely important for identifying promoters in the environment and controlling exposure to them.

Acknowledgements—The author's research on tumor promoters is supported by grants CA 23413 and CA 10815 from the National Cancer Institute, Department of Health and Human Services.

REFERENCES

ABB, J., BAYLISS, G. J. and DEINHARDT, F. (1979) Lymphocyte activation by the tumor-promoting agent 12-O-tetradecanoylphorbol-13-acetate (TPA). *J. Immunol.* **122**: 1639–1642.

ABLASHI, D. V., BENGALI, Z. H., EICHELBERGER, M. A., SUNDAR, K. S., ARMSTRONG, G. R., DANIEL, M. and LEVINE, P. H. (1980) Increased infectivity of oncogenic herpes viruses of primates with tumor promoter 12-O-tetradecanoylphorbol-13-acetate. *Proc. Soc. exp. Biol. Med.* **164**: 485–490.

ABRAHM, J. L. and SMILEY, R. (1981) Modification of normal human myelopoiesis by12-O-tetradecanolyphorbol-13-acetate (TPA). *Blood* **58**: 1119–1126.

ADOLF, G. R. and SWETLY, P. (1980) Tumor-promoting phorbol esters inhibit DNA synthesis and enhance virus-induced interferon production in a human lymphoma cell line. *J. gen. Virol.* **51**: 61–67.

AMSLER, K., SHAFFER, C. and COOK, J. S. (1983) Growth-dependent AIB and meAIB uptake in LLC-PK$_1$ cells: effects of differentiation inducers and of TPA. *J. cell. Physiol.* **114**: 184–190.

AMTMANN, E. and SAUER, G. (1982) Activation of non-expressed bovine papilloma virus genomes by tumour promoters. *Nature* **296**: 675–677.

ARFFMANN, E. and GLAVIND, J. (1971) Tumour-promoting activity of fatty acid methyl esters in mice. *Experinetia* **27**: 1465–1466.

ARGYRIS, T. S. (1982) Epidermis tumor promotion by regeneration. In: *Carcinogenesis—A Comprehensive Survey, Volume 7: Cocarcinogenesis and Biological Effects of Tumor Promoters*, pp. 43–48, HECKER, E., KUNZ, W., FUSENIG, N. E., MARKS, F. and THIELMANN, H. W. (eds). Raven Press, New York.

ARGYRIS, T. S. (1983) Nature of the epidermal hyperplasia produced by mezerein, a weak tumor promoter, in initiated skin of mice. *Cancer Res.* **43**: 1768–1773.

ARYA, S. K. (1980) Phorbol ester-mediated stimulation of the synthesis of mouse mammary tumour virus. *Nature* **284**: 71–72.

ASHENDEL, C. L., STALLER, J. M. and BOUTWELL, R. K. (1983) Protein kinase activity associated with a phorbol ester receptor purified from mouse brain. *Cancer Res.* **43**: 4333–4337.

ASTRUP, E. G., IVERSEN, O. H. and ELGJO, K. (1980) The tumorigenic and carcinogenic effect of TPA (12-O-tetradecanoylphorbol-13-acetate) when applied to the skin of BALB/cA mice. *Virchows Arch. B Cell Path.* **33**: 303–304.

BACH, H. and GOERTTLER, K. (1971) Morphologische Untersuchungen zur hyperplasiogenen Wirkung des biologisch aktiven Phorbol esters A$_1$. *Virchows Arch.* [*Zellpathol.*] **8**: 196–205.

BACKER, J. M., BOERSIG, M. R. and WEINSTEIN, I. B. (1982a) Inhibition of respiration by a phorbol ester tumor promoter in murine cultured cells. *Biochem. biophys. Res. Commun.* **105**: 855–860.

BACKER, J. M., BOERZIG, M. and WEINSTEIN, I. B. (1982b) When do carcinogen-treated 10T1/2 cells acquire the commitment to form transformed foci? *Nature* **299**: 458–460.

BADENOCH-JONES, P. (1983) Phorbol myristate acetate-induced macrophage aggregation. *Exp. Cell Biol.* **51**: 38–43.

BAIRD, W. M. and BOUTWELL, R. K. (1971) Tumor-promoting activity of phorbol and four diesters of phorbol in mouse skin.*Cancer Res.* **31**: 1074–1079.

BAIRD, W. M., MELERA, P. W. and BOUTWELL, R. K. (1972) Acrylamide gel electrophoresis studies of the incorporation of cytidine-^{3}H into mouse skin RNA at early times after treatment with phorbol esters. *Cancer Res.* **32**: 781–788.

BAIRD, W. M., SEDGWICK, J. A. and BOUTWELL, R. K. (1971) Effects of phorbol and four diesters of phorbol on the incorporation of tritiated precursors into DNA, RNA, and protein in mouse epidermis. *Cancer Res.* **31**: 1434–1439.

BALMAIN, A. (1976) The synthesis of specific proteins in adult mouse epidermis during phases of proliferation and differentiation induced by the tumor promoter TPA, and in basal and differentiating layers of neonatal mouse epidermis. *J. invest. Dermatol.* **67**: 246–253.

BALMAIN, A. and HECKER, E. (1974) On the biochemical mechanism of tumorigenesis in mouse skin. VI. Early effects of growth-stimulating phorbol esters on phosphate transport and phospholipid synthesis in mouse epidermis. *Biochim. biophys. Acta* **362**: 457–468.

BALMAIN, A., RAMSDEN, M., BOWDEN, G. T. and SMITH, J. (1984a) Activation of the mouse cellular Harvey-*ras* gene in chemically induced benign skin papillomas. *Nature* **307**: 658–660.

BALMAIN, A., SAUERBORN, R., RAMSDEN, M., PRAGNELL, I. B., BOWDEN, G. T., SMITH, J. and COLE, G. (1984b) Oncogene activation at different stages of chemical carcinogenesis in mouse skin. In: *Banbury Report 16: Genetic Variability in Reponses to Chemical Exposure*, pp. 243–255, Cold Spring Harbor Laboratory, New York.

BARRETT, J. C. and TENNANT, R. W. (eds). (1985) *Carcinogenesis: A Comprehensive Survey, Vol. 9*, Mammalian Cell Transformation: Mechanisms of Carcinogenesis and Assays for Carcinogens. Raven Press, New York.

BARRETT, J. C. and Ts'O, P. O. P. (1978) Evidence for the progressive nature of neoplastic transformation *in vitro*. *Proc. natn. Acad. Sci. U.S.A.* **75**: 3761–3765.

BARRETT, J. C., CRAWFORD, D. B., GRADY, D. L., HESTER, L. D., JONES, P. A., BENEDICT, W. F. and Ts'O, P. O. P. (1977) Temporal acquisition of enhanced fibrinolytic activity by Syrian hamster embryo cells following treatment with benzo(a)pyrene. *Cancer Res.* **37**: 3815–3823.

BARSOUM, J. and VARSHAVSKY, A. (1983) Mitogenic hormones and tumor promoters greatly increase the incidence of colony-forming cell bearing amplified dihydrofolate reductase genes. *Proc. natn. Acad. Sci. U.S.A.* **80**: 5330–5334.

BASERGA, R. (1983) Growth in size and cell DNA replication. *Exp. Cell Res.* **151**: 1–5.

BAXTER, C. S. (1984) Interaction of phorbol diesters and other tumor-promoting agents with immunofunctional cells *in vitro*. In: *Mechanisms of Tumor Promotion*, SLAGA, T. J. (ed.). CRC Press, Boca Raton, Florida, in press.

BAXTER, C. S., FISH, L. A. and BASH, J. A. (1981) Parallel orders of reactivity in murine cells of mezerein and phorbol esters towards the mixed-lymphocyte response and promotion of tumorigenesis. *Tox. appl. Pharmac.* **59**: 173–176.

BAXTER, M. A., LESLIE, R. G. Q. and REEVES, W. G. (1983) The stimulation of superoxide anion production in guinea-pig peritoneal macrophages and neutrophils by phorbol myristate acetate, opsonized zymosan and IgG2-containing soluble immune complexes. *Immunology* **48**: 657–665.

BEAUDRY, G. A., KING, L., DANIEL, L. W. and WAITE, M. (1982) Stimulation of deacylation in Madin-Darby canine kidney cells. Specificity of deacylation and prostaglandin production in 12-O-tetradecanoylphorbol-13-acetate-treated cells. *J. biol. Chem.* **257**: 10973–10977.

BELMAN, S. and TROLL, W. (1972) The inhibition of croton oil-promoted mouse skin tumorigenesis by steroid hormones. *Cancer Res.* **32**: 450–454.

BERENBLUM, I. (1954) Carcinogenesis and tumor pathogenesis. *Adv. Cancer Res.* **2**: 129–175.

BERENBLUM, I, and SHUBIK, P. (1947) The role of croton oil applications, associated with a single painting of a carcinogen, in tumour induction of the mouse's skin. *Br. J. Cancer* **1**: 379–382.

BERENBLUM, I, and SHUBIK, P. (1949) The persistence of latent tumor cells induced in the mouse's skin by a single application of 9,10-dimethyl-1,2-benzanthracene. *Br. J. Cancer* **3**: 384–386.

BERTOGLIO, J. H. (1983) Monocyte-independent stimulation of human B lymphocytes by phorbol myristate acetate. *J. Immunol.* **131**: 2279–2281.

BERWALD, Y. and SACHS, L. (1965) *In vitro* transformation of normal cells to tumor cells by carcinogenic hydrocarbons. *J. natn. Cancer inst.* **35**: 641–661.

BIRNBOIM, H. C. (1982a) DNA strand breakage in human leukocytes exposed to a tumor promoter, phorbol myristate acetate. *Science* **215**: 1247–1249.

BIRNBOIM, H. C. (1982b) Factors which affect DNA strand breakage in human leukocytes exposed to a tumor promoter, phorbol myristate acetate. *Can. J. Physiol. Pharmac.* **60**: 1359–1366.

BISHOP, J. M. (1985) Viral oncogenes. *Cell 42*i: 23–38.

BISHOP, R., MARTINEZ, R., NAKAMURA, K. D. and WEBER, M. J. (1983) A tumor promoter stimulates phosphorylation on tyrosine. *Biochem. biophys. Res. Commun.* **115**: 536–543.

BISSELL, M. J., HATIÉ, C. and CALVIN, M. (1979) Is the product of the src gene a promoter? *Proc. natn. Acad. Sci. U.S.A.* **76**: 348–352.

BLUMBERG, P. M. (1980) *In vitro* studies on the mode of action of the phorbol esters, potent tumor promoters: Part 1. *CRC Crit. Rev. Tox.* **8**: 153–197.

BLUMBERG, P. M. (1981) *In vitro* studies on the mode of action of the phorbol esters, potent tumor promoters. Part 2. *CRC Crit. Rev. Tox.* **8**: 199–234.

BLUMBERG, P. M., DRIEDGER, P. E. and ROSSOW, P. W. (1976) Effect of a phorbol ester on a transformation-sensitive surface protein of chick fibroblasts. *Nature* **264**: 446–447.

BOCK, F. G. and BURNS, R. (1963) Tumor-promoting properties of anthralin (1,8,9-Anthratriol). *J. natn. Cancer Inst.* **30**, 393–398.

BOCK, F. G., SWAIN, A. P. and STEDMAN, R. L. (1971) Composition studies on tobacco. XLIV. Tumor-promoting activity of subfractions of the weak acid fraction of cigarette smoke condensate. *J. natn. Cancer. Inst.* **47**: 429–436.

BOHRMAN, J. S. (1983) Identification and assessment of tumor-promoting and cocarcinogenic agents: state-of-the-art *in vitro* methods. *CRC Crit. Rev. Tox.* **11**: 121–167.

BOJAN, F., KINSELLA, A. R. and FOX, M. (1983) Effect of tumor promoter 12-O-tetradecanoylphorbol-13-acetate on recovery of methotrexate-, N-(phosphonacetyl)-L-aspartate-, and cadmium-resistant colony-forming mouse and hamster cells. *Cancer Res.* **43**: 5217–5221.

BOREIKO, C. J. (1985) Initiation and promotion in cultures of C3H10T1/2 mouse embryo fibroblasts. In: *Carcinogenesis—A Comprehensive Survey, Vol. 8, Cancer of the Respiratory Tract: Predisposing Factors*, pp. 329–340, MASS, M. J., KAUFMAN, D. G., SIEGFRIED, J. M., STEELE, V. E. and NESNOW, S. (eds), Raven Press, New York.

BOREIKO, C., MONDAL, S., NARAYAN, K. S. and HEIDELBERGER, C. (1980) Effect of 12-O-tetradecanoylphorbol-13-acetate on the morphology and growth of C3H/10T1/2 mouse embryo cells. *Cancer Res.* **40**: 4709–4716.

BOREIKO, C. J., RAGAN, D. L., ABERNETHY, D. J, and FRAZELLE, J. H. (1982) Initiation of C3H/10T1/2 cell transformation by N-methyl-N'-nitro-N-nitrosoguanidine and aflatoxin B$_1$. *Carcinogenesis* **3**: 391–395.

BOREK, C. and TROLL, W. (1983) Modifiers of free radicals inhibit *in vitro* the oncogenic actions of X-rays, bleomycin, and the tumor promoter 12-O-tetradecanoylphorbol-13-acetate. *Proc. natn. Acad. Sci. U.S.A.* **80**: 1304–1307.

BOUTWELL, R. K. (1964) Some biological aspects of skin carcinogenesis. *Prog. exp. Tumor Res.* **4**: 207–250.

BOUTWELL, R. K. (1974) The function and mechanism of promoters of carcinogenesis. *CRC Crit. Rev. Tox.* **2**: 419–443.

BOUTWELL, R. K. (1983) Biology and biochemistry of the two-step model of carcinogenesis. In: *Modulation and Mediation of Cancer by Vitamins*, pp. 2–9, MEYSKENS, F. L. (ed.), S. Karger AG, Basel.

BOYNTON, A. L. and WHITFIELD, J. F. (1980) Stimulation of DNA synthesis in calcium-deprived T51B liver cells by the tumor promoters phenobarbital, saccharin, and 12-O-tetradecanoylphorbol-13-acetate. *Cancer Res.* **40**: 4541–4545.

BOYNTON, A. L., WHITFIELD, J. F. and ISAACS, R. J. (1976) Calcium-dependent stimulation of BALB/c 3T3 mouse cell DNA synthesis by a tumor-promoting phorbol ester (PMA). *J. cell. Physiol.* **87**: 25–32.

BRASLAWSKY, G. R., KENNEL, S. J., HAND, R. E. and NETTESHEIM, P. (1984) Monoclonal antibodies directed against rat tracheal epithelial cells transformed *in vitro*. *Int. J. Cancer.* **33**: 131–138.

BRINCKERHOFF, C. E., MCMILLAN, R. M., FAHEY, J. V. and HARRIS, E. D. JR. (1979) Collagenase production by synovial fibroblasts treated with phorbol myristate acetate. *Arthrit. Rheumat.* **22**: 1109–1116.

BROWN, K. D., DICKER, P. and ROZENGURT, E. (1979) Inhibition of epidermal growth factor binding to surface receptors by tumor promoters. *Biochem. biophys. Res. Commun.* **86**: 1037–1043.

BRUCE, S., DEAMOND, S., UEO, H. and TS'O, P. O. P. (1983) Age-related differences in promoter-induced extension of *in vitro* life span of Syrian hamster (SH) cells. *J. Cell Biol.* **97**: 346a.

BURCZAK, J. D., MOSKAL, J. R., TROSKO, J. E., FAIRLEY, J. L. and SWEELEY, C. C. (1983) Phorbol ester-associated changes in ganglioside metabolism. *Exp. cell. Res.* **147**: 281–286.

Burns, C. P. and Rozengurt, E. (1983) Serum, platelet-derived growth factor vasopressin and phorbol esters increase intracellular pH in Swiss 3T3 cells. *Biochem. biophys. Res. Commun.* **116**: 931–938.

Burns, F. J., Vanderlaan, M., Snyder, E. and Albert, R. E. (1978) Induction and progression kinetics of mouse skin papillomas. In: *Carcinogenesis, Volume 2. Mechanisms of Tumor Promotion and Cocarcinogenesis* pp. 91–96, Slaga, T. J., Sivak, A. and Boutwell, R. K. (eds). Raven Press, New York.

Cabot, M. C., Welsh, C. J., Callaham, M. F. and Huberman, E. (1980) Alterations in lipid metabolism induced by 12-*O*-tetradecanoylphorbol-13-acetate in differentiating human myeloid leukemia cells. *Cancer Res.* **40**: 3674–3679.

Cardellina II, J. H., Marner, F-J. and Moore, R. E. (1979) Seaweed dermatitis: structure of lyngbyatoxin A. *Science* **204**: 193–195.

Cassileth, P. A., Suholet, D. and Cooper, R. A. (1981) Early changes in phosphatidylcholine metabolism in human acute promyelocytic leukemia cells stimulated to differentiate by phorbol ester. *Blood* **58**: 237–243.

Castagna, M., Rochette-Egly, C. and Rosenfeld, C. (1979) Tumor-producing phorbol diester induces substrate-adhesion and growth inhibition in lymphoblastoid cells. *Cancer Lett.* **6**: 227–234.

Castagna, M., Takai, Y., Kaibuchi, K., Sano, K., Kikkawa, U. and Nishizuka, Y. (1982) Direct activation of calcium-activated, phospholipid-dependent protein kinase by tumor-promoting phorbol esters. *J. biol. Chem.* **257**: 7847–7851.

Cerutti, P. A. (1985) Prooxidant states and tumor promotion. *Science* **227**: 375–381.

Chan, R. and Buehring, G. C. (1983) Anti-glucocorticoids block the enhancement of mouse mammary tumor virus production by 12-*O*-tetradecanoylphorbol-13-acetate. *Carcinogenesis* **4**: 1611–1614.

Chang, L. J-A. and McCulloch, E. A. (1981) Dose-dependent effects of a tumor promoter on blast cell progenitors in human myeloblastic leukemia. *Blood* **57**: 361–367.

Chen, B. D-M., Lin, H-S. and Hsu, S. (1983) Tumor-promoting phorbol esters inhibit the binding of colony-stimulating factor (CSF-1) to murine peritoneal exudate macrophages. *J. cell. Physiol.* **116**: 207–212.

Chida, K. and Kuroki, T. (1983) Presence of specific binding sites for phorbol ester tumor promoters in human epidermal and dermal cells in culture but lack of down regulation in epidermal cells. *Cancer Res.* **43**: 3638–3642.

Cioe, L., O'Brien, T. G. and Diamond, L. (1980) Inhibition of adipose conversion of BALB/c 3T3 cells by interferon and 12-*O*-tetradecanoylphorbol-13-acetate. *Cell. Biol. int. Rep.* **4**: 255–264.

Cochet, C., Gill, G. N., Meisenhelder, J., Cooper, J. A. and Hunter, T. (1984) C-kinase phosphorylates the epidermal growth factor receptor and reduces its epidermal growth factor-stimulated tyrosine protein kinase activity. *J. biol. Chem.* **259**: 2553–2558.

Cohen, R., Pacifici, M., Rubinstein, N., Biehl, J. and Holtzer, H. (1977) Effect of a tumor promoter on myogenesis. *Nature* **266**: 538–540.

Colburn, N. H. (1980) Tumor promoter produces anchorage independence in mouse epidermal cells by an induction mechanism. *Carcinogenesis* **1**: 951–954.

Colburn, N. H., Former, B. F., Nelson, K. A. and Yuspa, S. H. (1979) Tumour promoter induces anchorage independence irreversibly. *Nature* **281**: 589–591.

Colburn, N. H., Gindhart, T. D., Dalal, B. and Hegamyer, G. A. (1983a) The role of phorbol ester receptor binding in responses to promoters by mouse and human cells. In: *Organ and Species Specificity in Chemical Carcinogenesis*, pp. 189–200, Langenbach, R., Nesnow, S. and Rice, J. M. (eds), Plenum Press, New York.

Colburn, N. H., Gindhart, T. D., Hegamyer, G. A., Blumberg, P. M., Delclos, K. B., Magun, B. E. and Lockyer, J. (1982a) Phorbol diester and epidermal growth factor receptors in 12-*O*-tetradecanoylphorbol-13-acetate-resistant and -sensitive mouse epidermal cells. *Cancer Res.* **42**: 3093–3097.

Colburn, N. H., Koehler, B. A. and Nelson, K. J. (1980) A cell culture assay for tumor-promoter-dependent progression toward neoplastic phenotype: detection of tumor promoters and promotion inhibitors. *Terat. Carc. Mutag.* **1**: 87–96.

Colburn, N. H., Lau, S. and Head, R. (1975) Decrease of epidermal histidase activity by tumor-promoting phorbol esters. *Cancer Res.* **35**: 3154–3159.

Colburn, N. H., Lerman, M. I., Hegamyer, G. A., Wendel, E. and Gindhart, T. D. (1984) Genetic determinants of tumor promotion—studies with promoter resistant variants of JB6. In: *Genes and Cancer, UCLA Symposia on Molecular and Cellular Biology—New Series. Vol. 17*, pp. 137–155, Bishop, J. M., Rowley, J. D. and Greaves, M. (eds), Alan R. Liss, New York.

Colburn, N. H., Talmadge, C. B. and Gindhart, T. D. (1983b) Transfer of sensitivity to tumor promoters by transfection of DNA from sensitive into insensitive mouse JB6 epidermal cells. *Mol. cell. Biol.* **3**: 1182–1186.

Colburn, N. H., Wendel, E. J. and Abruzzo, G. (1981) Dissociation of mitogenesis and late-stage promotion of tumor cell phenotype by phorbol esters: mitogen-resistant variants are sensitive to promotion. *Proc. natn. Acad. Sci. U.S.A.* **78**: 6912–6916.

Colburn, N. H., Wendel, E. and Srinivas, L. (1982b) Responses of preneoplastic epidermal cells to tumor promoters and growth factors: use of promoter-resistant variants for mechanism studies. *J. cell. Biochem.* **18**: 261–270.

Colletta, G., Di Fiore, P. P., Ferrentino, M., Pietropaolo, C., Turco, M. C. and Vecchio, G. (1980) Enhancement of viral gene expression in Friend erythroleukemic cells by 12-*O*-tetradecanoylphorbol-13-acetate. *Cancer Res.* **40**: 3369–3373.

Collins, M. K. L. and Rozengurt, E. (1984) Homologous and heterologous mitogenic desensitization of Swiss 3T3 cells to phorbol esters and vasopressin: role of receptor and postreceptor steps. *J. cell. Physiol.* **118**: 133–142.

Collins, S. J., Gallo, R. C. and Gallagher, R. E. (1977) Continuous growth and differentiation of human myeloid leukaemic cells in suspension culture. *Nature* **270**: 347–349.

Connan, G., Rassoulzadegan, M. and Cuzin, F. (1985) Focus formation in rat fibroblasts exposed to a tumor promoter after transfer of polyoma *plt* and *myc* oncogenes. *Nature* **314**: 277–279.

Constantinides, P. G., Taylor, S. M. and Jones, P. A. (1978) Phenotypic conversion of cultured mouse embryo cells by aza pyrimidine nucleosides. *Dev. Biol.* **66**: 57–71.

COOPER, J. A. and HUNTER, T. (1981) Changes in protein phosphorylation in Rous sarcoma virus-transformed chicken embryo cells. *Molec. cell. Biol.* **1**: 165–178.

COOPER, J. A., BOWEN-POPE, D. F., RAINES, E., ROSS, R. and HUNTER, T. (1982a) Similar effects of platelet-derived growth factor and epidermal growth factor on the phosphorylation of tyrosine in cellular proteins. *Cell* **31**: 263–273.

COOPER, R. A., BRAUNWALD, A. D. and KUO, A. L. (1982b) Phorbol ester induction of leukemic cell differentiation is a membrane-mediated process. *Proc. natn. Acad. Sci. U.S.A* **79**: 2865–2869.

COSSU, G., KUO, A. L., PESSANO, S., WARREN, L. and COOPER, R. A. (1982) Decreased synthesis of high-molecular-weight glycopeptides in human promyelocytic leukemic cells (HL-60) during phorbol ester-induced macrophage differentiation. *Cancer Res.* **42**: 484–489.

CRAIG, R. W. and BLOCH, A. (1984) Early decline in c-myb oncogene expression in the differentiation of human myeloblastic leukemia (ML-1) cells induced with 12-O-tetradecanoylphorbol-13-acetate. *Cancer Res.* **44**: 442–446.

CROOP, J., TOYAMA, Y., DLUGOSZ, A. A. and HOLTZER, H. (1980) Selective effects of phorbol 12-myristate 13-acetate on myofibrils and 10-nm filaments. *Proc. natn. Acad. Sci. U.S.A.* **77**: 5273–5277.

CRUTCHLEY, D. J., CONANAN, L. B. and MAYNARD, J. R. (1980) Induction of plasminogen activator and prostaglandin biosynthesis in HeLa cells by 12-O-tetradecanoylphorbol-13-acetate. *Cancer Res.* **40**: 849–852.

DAMMERT, K. (1961) A histological and cytological study of different methods of skin tumorigenesis in mice. *Acta path. microbiol. Scand.* **53**: 33–49.

DATTA, A. K., FEIGHNY, R. J. and PAGANO, J. S. (1980) Induction of Epstein-Barr virus-associated DNA polymerase by 12-O-tetradecanoylphorbol-13-acetate. *J. biol. Chem.* **255**: 5120–5125.

DAYA-GROSJEAN, L., SARASIN, A. and MONIER, R. (1982) Effect of tumor promoters on soft-agar growth of Swiss 3T3 cells infected with SV40 tsA mutants. *Carcinogenesis* **3**: 833–835.

DECHATELET, L. R., LEES, C. J., WALSH, C. E., LONDON, G. D. and SHIRLEY, P. S. (1982) Comparison of the calcium ionophore and phorbol myristate acetate on the initiation of the respiratory burst in human neutrophils. *Infect. Immun.* **38**: 969–974.

DECHATELET, L. R., SHIRLEY, P. S. and JOHNSTON, R. B. JR. (1976) Effect of phorbol myristate acetate on the oxidative metabolism of human polymorphonuclear leukocytes. *Blood* **47**: 545–554.

DECKER, S. J. (1984) Effects of epidermal growth factor and 12-O-tetradecanoylphorbol-13-acetate on metabolism of the epidermal growth factor receptor in normal human fibroblasts. *Mol. Cell. Biol.* **4**: 1718–1724.

DELCLOS, K. B. and BLUMBERG, P. M. (1979) Decrease in collagen production in normal and Rous sarcoma virus-transformed chick embryo fibroblasts induced by phorbol myristate acetate. *Cancer Res.* **39**: 1667–1672.

DELCLOS, K. B., NAGLE, D. S. and BLUMBERG, P. M. (1980) Specific binding of phorbol ester tumor promoters to mouse skin. *Cell* **19**: 1025–1032.

DELIA, D., GREAVES, M. F., NEWMAN, R. A., SUTHERLAND, D. R., MINOWADA, J., KUNG, P. and GOLDSTEIN, G. (1982) Modulation of T leukaemic cell phenotype with phorbol ester. *Int. J. Cancer* **29**: 23–31.

DIAMOND, L., O'BRIEN, S., DONALDSON, C. and SHIMIZU, Y. (1974) Growth stimulation of human diploid fibroblasts by the tumor promoter, 12-O-tetradecanoylphorbol-13-acetate. *Int. J. Cancer* **13**: 721–730.

DIAMOND, L., O'BRIEN, T. G. and BAIRD, W. M. (1980) Tumor promoters and the mechanism of tumor promotion. *Adv. Cancer Res.* **32**: 1–74.

DIAMOND, L., O'BRIEN, T. G. and ROVERA, G. (1977) Inhibition of adipose conversion of 3T3 fibroblasts by tumour promoters. *Nature* **269**: 247–249.

DIAMOND, L., O'BRIEN, T. G. and ROVERA, G. (1978a) Tumor promoters: effects of proliferation and differentiation of cells in culture. *Life Sci.* 1979–1988.

DIAMOND, L., O'BRIEN, T. and ROVERA, G. (1978b) Tumor promoters inhibit terminal cell differentiation in culture. In: *Carcinogenesis, Volume 2. Mechanisms of Tumor Promotion and Cocarcinogenesis*, pp. 335–341, SLAGA, T. J., SIVAK, A. and BOUTWELL, R. K. (eds). Raven Press, New York.

DICKER, P. and ROZENGURT, E. (1978) Stimulation of DNA synthesis by tumor promoter and pure mitogenic factors. *Nature* **276**: 723–726.

DICKER, P. and ROZENGURT, E. (1980) Phorbol esters and vasopressin stimulate DNA synthesis by a common mechanism. *Nature* **287**: 607–612.

DICKER, P. and ROZENGURT, E. (1981) Phorbol ester stimulation of Na influx and Na–K pump activity in Swiss 3T3 cells. *Biochem. biophys. Res. Commun.* **100**, 433–441.

DIGIOVANNI, J. and BOUTWELL, R. K. (1983) Tumor promoting activity of 1,8-dihydroxy-3-methyl-9-anthrone (chrysarobin) in female SENCAR mice. *Carcinogenesis* **4**: 281–284.

DIGIOVANNI, J., SLAGA, T. J. and BOUTWELL, R. K. (1980) Comparison of the tumor-initiating activity of 7,12-dimethylbenz[a]anthracene and benzo[a]pyrene in female SENCAR and CD-1 mice. *Carcinogenesis* **1**: 381–389.

DION, L. D., BEAR, J., BATEMAN, J., DE LUCA, L. M. and COLBURN, N. H. (1982) Inhibition by tumor-promoting phorbol esters of procollagen synthesis in promotable JB6 mouse epidermal cells. *J. natn. Cancer Inst.* **69**: 1147–1154.

DIPAOLO, J. A. (1983) Relative difficulties in transforming human and animal cells in vitro. *J. natn. Cancer Inst.* **70**: 3–8.

DIPAOLO, J. A., DEMARINIS, A. J., EVANS, C. H. and DONIGER, J. (1981) Expression of initiated and promoted stages of irradiation carcinogenesis in vitro. *Cancer Lett.* **14**: 243–249.

DIPAOLO, J. A., DONOVAN, P. and NELSON, R. (1969) Quantitative studies of in vitro transformation by chemical carcinogens. *J. natn. Cancer Inst.* **42**: 867–876.

DIPAOLO, J. A., EVANS, C. H., DEMARINIS, A. J. and DONIGER, J. (1982) Phytohemagglutinin inhibits phorbol diester promotion of UV-irradiation initiated transformation in Syrian hamster embryo cells. *Int. J. Cancer* **30**: 781–785.

DIPAOLO, J. A., NELSON, R. L. and DONOVAN, P. J. (1971) Morphological, oncogenic, and karyological characteristics of Syrian hamster embryo cells transformed in vitro by carcinogenic polycyclic hydrocarbons. *Cancer Res.* **31**: 1118–1127.

DORMAN, B. H., BUTTERWORTH, B. E. and BOREIKO, C. J. (1983a) Role of intercellular communication in the promotion of C3H/10T1/2 cell transformation. *Carcinogenesis* **4**: 1109–1115.

DORMAN, B. H., SIEGFRIED, J. M. and KAUFMAN, D. G. (1983b) Alterations of human endometrial stromal cells produced by N-methyl-N'-nitro-N-nitrosoguanidine. *Cancer Res* **43**: 3348–3357.

DRIEDGER, P. E. and BLUMBERG, P. M. (1977) The effect of phorbol diesters on chicken embryo fibroblasts. *Cancer Res.* **37**: 3257–3265.

DRIEDGER, P. E. and BLUMBERG, P. M. (1980a) Specific binding of phorbol ester tumor promoters. *Proc. natn. Acad. Sci. U.S.A.* **77**: 567–571.

DRIEDGER, P. E. and BLUMBERG, P. M. (1980b) Structure-activity relationships in chick embryo fibroblasts for phorbol-related diterpene esters showing anomalous activities *in vitro*. *Cancer Res.* **40**: 339–346.

DUNKEL, V. C., PIENTA, R. J., SIVAK, A. and TRAUL, K. A. (1981) Comparative neoplastic transformation responses of Balb/3T3 cells, Syrian hamster embryo cells, and Rauscher murine leukemia virus-infected Fischer 344 rat embryo cells to chemical carcinogens. *J. natn. Cancer Inst.* **67**: 1303–1315.

DZARLIEVA, R. T. and FUSENIG, N. E. (1982) Tumor promoter 12-O-tetradecanoyl-phorbol-13-acetate enhances sister chromatid exchanges and numerical and structural chromosome aberrations in primary mouse epidermal cell cultures. *Cancer Lett.* **16**: 7–17.

EISINGER, M. and MARKO, O. (1982) Selective proliferation of normal human melanocytes *in vitro* in the presence of phorbol ester and cholera toxin. *Proc. natn. Acad. Sci. U.S.A.* **79**: 2018–2022.

EISINGER, M., MARKO, O. and WEINSTEIN, I. B. (1983) Stimulation of growth of human melanocytes by tumor promoters. *Carcinogenesis* **4**: 779–781.

EMERIT, I. and CERUTTI, P. A. (1981) Tumour promoter phorbol-12-myristate-13-acetate induces chromosomal damage via indirect action. *Nature* **293**: 144–146.

EMERIT, I. and CERUTTI, P. A. (1982) Tumor promoters phorbol 12-myristate 13-acetate induces a clastogenic factor in human lymphocytes. *Proc. natn. Acad. Sci. U.S.A.* **79**: 7509–7513.

EMERIT, I. and CERUTTI, P. (1983) Clastogenic action of tumor promoter phorbol-12-myristate-13 acetate in mixed human leukocyte cultures. *Carcinogenesis* **4**: 1313–1316.

ENOMOTO, T., SASAKI, Y., SHIBA, Y., KANNO, Y. and YAMASAKI, H. (1981) Tumor promoters cause a rapid and reversible inhibition of the formation and maintenance of electrical cell coupling in culture. *Proc. natn. Acad. Sci. U.S.A.* **78**: 5628–5632.

EVANS, C. H. and DIPAOLO, J. A. (1981) Lymphotoxin: an anticarcinogenic lymphokine as measured by inhibition of chemical carcinogen or ultraviolet-irradiation induced transformation of Syrian hamster cells. *Int. J. Cancer* **27**: 45–49.

EVANS, C. H. and DIPAOLO, J. A. (1982) Equivalency of endothelial cell growth supplement to irradiated feeder cells in carcinogen-induced morphologic transformation of Syrian hamster embryo cells. *J. natn. Cancer Inst.* **68**: 127–131.

EVANS, C. H. and GREINER, J. W. (1981) Corticosteroid prevention of carcinogenesis in benzo[a]pyrene or ultraviolet irradiation treated Syrian hamster cells in vitro. *Cancer Lett.* **12**: 23–27.

FARBER, E. (1980) The sequential analysis of liver cancer induction. *Biochim. biophys. Acta* **605**: 149–166.

FARRAR, J. J., FULLER-FARRAR, J., SIMON, P. L., HIKFIKER, M. L., STADLER, B. M. and FARRAR, W. L. (1980a) Thymoma production of T cell growth factor (Interleukin 2). *J. Immunol.* **125**: 2555–2558.

FARRAR, J. J., MIZEL, S. B., FULLER-FARRAR, J., FARRAR, W. L. and HILFIKER, M. L. (1980b) Macrophage-independent activation of helper T cells. I. Production of Interleukin 2. *J. Immunol.* **125**: 793–798.

FARRAR, W. L., MIZEL, S. B. and FARRAR, J. J. (1980c) Participation of lymphocyte activating factor (Interleukin 1) in the induction of cytotoxic T cell responses. *J. Immunol.* **124**: 1371–1377.

FARRAR, W. L., THOMAS, T. P. and ANDERSON, W. B. (1985) Altered cytosol/membrane enzyme redistribution on interleukin-3 activation of protein kinase C. *Nature* **315**: 235–237.

FERNANDEZ, A., MONDAL, S. and HEIDELBERGER, C. (1980) Probabilistic view of the transformation cultured C3H/10T1/2 mouse embryo fibroblasts by 3-methylcholanthrene. *Proc. natn. Acad. Sci. U.S.A.* **77**: 7272–7276.

FEUERSTEIN, N. and COOPER, H. L. (1983) Rapid protein phosphorylation induced by phorbol ester in HL-60 cells. Unique alkali-stable phosphorylation of a 17,000-dalton protein detected by two-dimensional gel electrophoresis. *J. Biol. Chem.* **258**: 10786–10793.

FEY, E. G. and PENMAN, S. (1984) Tumor promoters induce a specific morphological signature in the nuclear matrix-intermediate filament scaffold of Madin-Darby canine kidney (MDCK) cell colonies. *Proc. natn. Acad. Sci. USA* **81**: 4409–4413.

FIBACH, E., GAMBARI, R., SHAW, P. A., MANIATIS, G., REUBEN, R. C., SASSA, S., RIFKIND, R. A. and MARKS, P. A. (1979) Tumor promoter-mediated inhibition of cell differentiation: suppression of the expression of erythroid functions in murine erythroleukemia cells. *Proc. natn. Acad. Sci. U.S.A.* **76**: 1906–1910.

FIBACH, E., KIDRON, M., NACHSHON, I. and MAYER, M. (1983) Phorbol ester-induced adhesion of murine erythroleukemia cells: possible involvement of cellular proteases. *Carcinogenesis* **4**: 1395–1399.

FIBACH, E., PELED, T., TREVES, A., KORNBERG, A. and RACHMILEWITZ, E. A. (1982) Modulation of the maturation of human leukemic promyelocytes (HL-60) to granulocytes or macrophages. *Leukemia Res.* **6**: 781–790.

FIBACH, E., YAMASAKI, H., WEINSTEIN, I. B., MARKS, P. A. and RIFKIND, R. A. (1978) Heterogeneity of murine erythroleukemia cells with respect to tumor promoter-mediated inhibition of cell differentiation. *Cancer Res.* **38**: 3685–3688.

FISCHER, S. M., GLEASON, G. L., HARDIN, L. G., BOHRMAN, J. S. and SLAGA. T. J. (1980) Prostaglandin modulation of phorbol ester skin tumor promotion. *Carcinogenesis* **1**: 245–248.

FISCHER, S. M., VIAJE, A., MILLS, G. D., WONG, E. W., WEEKS, C. E. and SLAGA, T. J. (1984) The growth of cultured human foreskin keratinocytes is not stimulated by a tumor promoter. *Carcinogenesis* **5**: 109–112.

FISH, L. A., BAXTER, C. S. and BASH, J. A. (1981) Murine lymphocyte comitogenesis by phorbol esters and its inhibition by retinoic acid and inhibitors of polyamine biosynthesis. *Tox. appl. Pharmac.* **58**: 39–47.

FISHER, P. B. (1983) Chemical-viral interactions in cell transformation. *Cancer Invest.* **1**: 495–509.

FISHER, P. B., DORSCH-HÄSLER, K., WEINSTEIN, I. B. and GINSBERG, H. S. (1979) Tumour promoters enhance anchorage-independent growth of adenovirus-transformed cells without altering the integration pattern of viral sequences. *Nature* **281**: 591–594.

FISHER, P. B., MIRANDA, A. F., BABISS, L. E., PESTKA, S. and WEINSTEIN, I. B. (1983) Opposing effects of interferon produced in bacteria and of tumor promoters on myogenesis in human myoblast cultures. *Proc. natn. Acad. Sci. U.S.A.* **80**: 2961–2965.

FISHER, P. B., MUFSON, R. A. and WEINSTEIN, I. B. (1981) Interferon inhibits melanogenesis in B-16 mouse melanoma cells. *Biochem. biophys. Res. Commun.* **100**: 823–830.

FISHER, P. B., WEINSTEIN, I. B., EISENBERG, D. and GINSBERG, H. S. (1978) Interactions between adenovirus, a tumor promoter, and chemical carcinogens in transformation of rat embryo cell cultures. *Proc. natn. Acad. Sci. U.S.A.* **75**: 2311–2314.

FITZGERALD, D. J. and MURRAY, A. W. (1980) Inhibition of intercellular communication by tumor-promoting phorbol esters. *Cancer Res.* **40**: 2935–2937.

FITZGERALD, D. J. and MURRAY, A. W. (1982) A new intercellular communication assay: its use in studies on the mechanism of tumour promotion. *Cell Biol. int. Rep.* **6**: 235–242.

FLAVIN, D. F. and KOLBYE, A. C. JR. (1983) Nutritional factors with the potential to inhibit critical pathways of tumour promotion. In: *Modulation and Mediation of Cancer by Vitamins*, pp. 24–38, MEYSKENS, F. L. and PRASAD, K. N. (eds). S. Karger AG, Basel.

FRANKFORT, H. M. and VILČEK, J. (1982) Inhibition of interferon production in human fibroblasts by a tumor promoting phorbol ester. *Arch. Virol.* **73**: 295–309.

FRANTZ, C. N., STILES, C. D. and SCHER, C. D. (1979) The tumor promoter 12-O-tetradecanoyl-phorbol-13-acetate enhance the proliferative response of Balb/c-3T3 cells to hormonal growth factors. *J. cell. Physiol.* **100**: 413–424.

FRAZELLE, J. H., ABERNETHY, D. J. and BOREIKO, C. J. (1983a) Weak promotion of C3H/10T1/2 cell transformation by repeated treatments with formaldehyde. *Cancer Res.* **43**: 3236–3239.

FRAZELLE, J. H., ABERNETHY, D. J. and BOREIKO, C. J. (1983b) Factors influencing the promotion of transformation in chemically-initiated C3H/10T1/2 Cl 8 mouse embryo fibroblasts. *Carcinogenesis* **4**: 709–715.

FRAZELLE, J. H., ABERNETHY, D. J. and BOREIKO, C. J. (1984) Enhanced sensitivity of the C3H/10T1/2 cell transformation system to alkylating and chemotherapeutic agents by treatment with 12-O-tetradecanoyl-phorbol-13-acetate. *Environ. Mutag.* **6**: 81–89.

FRIEDEWALD, W. F. and ROUS, P. (1944) The initiating and promoting elements in tumor production. An analysis of the effects of tar, benzpyrene, and methylcholanthrene on rabbit skin. *J. exp. Med.* **8**: 101–126.

FRIEDMAN, E. A. (1981) Differential response of premalignant epithelial cell classes to phorbol ester tumor promoters and to deoxycholic acid. *Cancer Res.* **41**: 4588–4599.

FRIEDMAN, E. A. and STEINBERG, M. (1982) Disrupted communication between late-stage premalignant human colon epithelial cells by 12-O-tetradecanoylphorbol-13-acetate. *Cancer Res.* **42**: 5096–5105.

FRIEDMAN, J. and CERUTTI, P. (1983) The induction of ornithine decarboxylase by phorbol 12-myristate 13-acetate or by serum is inhibited by antioxidants. *Carcinogenesis* **4**: 1425–1427.

FRIEND, C., SCHER, W., HOLLAND, J. G. and SATO, T. (1971) Hemoglobin synthesis in murine virus-induced leukemic cells *in vitro*: stimulation of erythroid differentiation by dimethyl sulfoxide. *Proc. natn. Acad. Sci. U.S.A.* **68**: 378–382.

FUJIKI, H., MORI, M., NAKAYASU, M., TERADA, M., SUGIMURA, T. and MOORE, R. E. (1981) Indole alkaloids: Dihydroteleocidin B, teleocidin, and lyngbyatoxin A as members of a new class of tumor promoters. *Proc. natn. Acad. Sci. U.S.A.* **78**: 3872–3876.

FUJIKI, H., SUGANUMA, M., MATSUKURA, N., SUGIMURA, T. and TAKAYAMA, S. (1982a) Teleocidin from *Streptomyces* is a potent promoter of mouse skin carcinogenesis. *Carcinogenesis* **3**: 895–898.

FUJIKI, H., SUGANUMA, M., NAKAYASU, M., HOSHINO, H., MOORE, R. E. and SUGIMURA, T. (1982b) The third class of new tumor promoters, polyacetates (debromoaplysiatoxin and aplysiatoxin), can differentiate biological actions relevant to tumor promoters. *Gann* **73**: 495–497.

FUKUDA, M. (1981) Tumor-promoting phorbol diester-induced specific change in cell surface glycoprotein profile of K562 human leukemic cells. *Cancer Res.* **41**: 4621–4628.

FULLER-FARRAR, J., HILFIKER, M. L., FARRAR, W. L. and FARRAR, J. J. (1981) Phorbol myristic acetate enhances the production of interleukin 2. *Cell Immunol.* **58**: 156–164.

FÜRSTENBERGER, G. and MARKS, F. (1980) Early prostaglandin E synthesis is an obligatory event in the induction of cell proliferation in mouse epidermis *in vivo* by the phorbol ester TPA. *Biochem. biophys. Res. Commun.* **92**: 749–756.

FÜRSTENBERGER, G., BERRY, D. L., SORG, B. and MARKS, F. (1981a) Skin tumor promotion by phorbol esters is a two-stage process. *Proc. natn. Acad. Sci. U.S.A.* **78**: 7722–7726.

FÜRSTENBERGER, G., RICHTER, H., ARGYRIS, T. S. and MARKS, F. (1982) Effects of the phorbol ester 4-O-methyl-12-O-tetradecanoylphorbol-13-acetate on mouse skin *in vivo*: evidence for its uselessness as a negative control compound in studies on the biological effects of phorbol ester tumor promoters. *Cancer Res.* **42**: 342–348.

FÜRSTENBERGER, G., RICHTER, H., FUSENIG, N. E. and MARKS, F. (1981b) Arachidonic acid and prostaglandin E_2 release and enhanced cell proliferation induced by the phorbol ester TPA in a murine epidermal cell line. *Cancer Lett.* **11**: 191–198.

FÜRSTENBERGER, G., SORG, B. and MARKS, F. (1983) Tumor promotion by phorbol esters in skin: evidence for a memory effect. *Science* **220**: 89–91.

FUSENIG, N. E. and SAMSEL, W. (1978) Growth-promoting activity of phorbol ester TPA on cultured mouse skin keratinocytes, fibroblasts, and carcinoma cells. In: *Carcinogenesis, Volume 2. Mechanisms of Tumor Promotion and Cocarcinogenesis*, pp. 203–220, SLAGA, T. J., SIVAK, A., and BOUTWELL, R. K. (eds). Raven Press, New York.

GAINER, H. S. and KINSELLA, A. R. (1983) Analysis of spontaneous, carcinogen-induced and promoter-induced chromosomal instability in patients with hereditary retinoblastoma. *Int. J. Cancer* **32**: 449–453.

GALLAGHER, R., COLLINS, S., TRUJILLO, J., MCCREDIE, K., AHEARN, M., TSAI, S., METZGAR, R., AULAKH, G., TING, R., RUSCETTI, F. and GALLO, R. (1979) Characterization of the continuous, differentiating myeloid cell line (HL-60) from a patient with acute promyelocytic leukemia. *Blood* **54**: 713–733.

GALLO, R. C., BREITMAN, T. R. and RUSCETTI, F. W. (1982) Proliferation and differentiation of human myeloid leukemia cell lines *in vitro*. In: *Maturation Factors and Cancer*, pp. 255–271, MOORE, M. A. S. (ed.). Raven Press, New York.

GENSLER, H. L. and BOWDEN, G. T. (1983) Evidence suggesting a dissociation of DNA strand scissions and late-stage promotion of tumor cell phenotype. *Carcinogenesis* **4**: 1507–1511.

GILMORE, T. and MARTIN, G. S. (1983) Phorbol ester and diacylglycerol induce protein phosphorylation at tyrosine. *Nature* **306**: 487–490.

GINDHART, T. D., NAKAMURA, Y., STEVENS, L. A., HEGAMEYER, G. A., WEST, M. W., SMITH, B. M. and COLBURN, N. H. (1985) Genes and signal transduction in tumor promotion: conclusions from studies with promoter resistant variants of JB-6 mouse epidermal cells. In: *Carcinogenesis—A Comprehensive Survey, Vol. 8, Cancer of the Respiratory Tract: Predisposing Factors*, pp. 341–367, MASS, M. J., KAUFMAN, D. G., SIEGFRIED, J. M., STEELE, V. E. and NESNOW, S. (eds), Raven Press, New York.

GOLDFARB, R. H. and QUIGLEY, J. P. (1978) Synergistic effect of tumor virus transformation and tumor promoter treatment on the production of plasminogen activator by chick embryo fibroblasts. *Cancer Res.* **38**: 4601–4609.

GOLDSTEIN, B. D., WITZ, G., AMORUSO, M. and TROLL, W. (1979) Protease inhibitors antagonize the activation of polymorphonuclear leukocyte oxygen consumption. *Biochem. biophys. Res. Commun.* **88**: 854–860.

GOLDSTEIN, B. D., WITZ, G., AMORUSO, M., STONE, D. S. and TROLL, W. (1981) Stimulation of human polymorphonuclear leukocyte superoxide anion radical production by tumor promoters. *Cancer Lett.* **11**: 257–262.

GOLDSTEIN, I. M. (1978) Effects of phorbol esters on polymorphonuclear leukocyte functions *in vitro*. In: *Carcinogenesis, Volume 2. Mechanisms of Tumor Promotion and Cocarcinogenesis*, pp. 389–400. SLAGA, T. J., SIVAK, A. and BOUTWELL, R. K. (eds). Raven Press, New York.

GOLDSWORTHY, T., CAMPBELL, H. A. and PITOT, H. C. (1984) The natural history and dose-response characteristics of enzyme-altered foci in rat liver following phenobarbital and diethylnitrosamine administration. *Carcinogenesis* **5**: 67–71.

GOTTESMAN, M. M. and YUSPA, S. H. (1981) Tumor promoters induce the synthesis of a secreted glycoprotein in mouse skin and cultured primary mouse epidermal cells. *Carcinogenesis* **2**: 971–976.

GREENBERGER, J. S., NEWBURGER, P. E., KARPAS, A. and MOLONEY, W. C. (1978) Constitutive and inducible granulocyte-macrophage functions in mouse, rat, and human myeloid leukemia-derived continuous tissue culture lines. *Cancer Res.* **33**: 3340–3348.

GREENEBAUM, E., NICOLAIDES, M., EISINGER, M., VOGEL, R. H. and WEINSTEIN, I. B. (1983) Binding of phorbol dibutyrate and epidermal growth factor to cultured human epidermal cells. *J. natn. Cancer Inst.* **70**: 435–441.

GRIFFIN, J. D., BEVERIDGE, R. P. and SCHLOSSMAN, S. F. (1983) Effect of phorbol ester on differentiation of human myeloid colony forming cells (CFU-C). *Leukemia Res.* **7**: 43–49.

GRIMM, W. and MARKS, F. (1974) Effect of tumor-promoting phorbol esters on the normal and the isoproterenol-elevated level of adenosine 3′,5′-cyclic monophosphate in mouse epidermis *in vivo*. *Cancer Res.* **34**: 3128–3134.

GROSSO, L. E. and PITOT, H. E. (1984) Modulation of c-*myc* expression in the HL-60 cell line. *Biochem. Biophys. Res. Commun.* **119**: 473–480.

GRUNBERGER, G. and GORDON, P. (1982) Affinity alteration of insulin receptor induced by a phorbol ester. *Am. J. Physiol.* **243**: E319–E324.

GWYNN, R. H. and SALAMAN, M. H. (1953) Studies on co-carcinogenesis SH-reactors and other substances tested for co-carcinogenic action in mouse skin. *Br. J. Cancer* **7**: 482–489.

HABER, D. A., FOX, D. A., DYNAN, W. S. and THILLY, W. G. (1977) Cell density dependence of focus formation in the C3H/10T1/2 transformation assay. *Cancer Res.* **37**: 1644–1648.

HAMILTON, J. A. (1980) Stimulation of macrophage prostaglandin and neutral protease production by phorbol esters as a model for the induction of vascular changes associated with tumor promotion. *Cancer Res.* **40**: 2273–2280.

HAMMOND, S. L., HAM, R. G. and STAMPFER, M. R. (1984) Serum-free growth of human mammary epithelial cells: rapid clonal growth in defined medium and extended serial passage with pituitary extract. *Proc. natn. Acad. Sci. USA* **81**: 5435–5439.

HAN, A. and ELKIND, M. M. (1982) Enhanced transformation of mouse 10T1/2 cells by 12-O-tetradecanoylphorbol-13-acetate following exposure to X-rays or to fission-spectrum neutrons. *Cancer Res.* **42**: 477–483.

HARADA, S., KATSUKI, T., YAMAMOTO, H. and HINUMA, Y. (1981) Mechanism of enhancement of Epstein-Barr virus-induced transformation of peripheral blood lymphocytes by a tumor promoter, TPA, with special reference to lowering of cytotoxicity of T cells. *Int. J. Cancer* **27**: 617–623.

HAWLEY-NELSON, P., STANLEY, J. R., SCHMIDT, J., GULLINO, M. and YUSPA, S. H. (1982) The tumor promoter, 12-O-tetradecanoylphorol-13-acetate, accelerates keratinocyte differentiation and stimulates growth of an unidentified cell type in cultured human epidermis. *Exp. Cell Res.* **137**: 155–167.

HAYASHI, K., FUJIKI, H. and SUGIMURA, T. (1983) Effects of tumor promoters on the frequency of metallothionein I gene amplification in cells exposed to cadmium. *Cancer Res.* **43**: 5433–5436.

HECHT, S. S., CARMELLA, S. and HOFFMAN, D. (1978) Chemical studies on tobacco smoke—LIV Determination of hydroxybenzyl alcohols and hydroxyphenyl ethanols in tobacco and tobacco smoke. *J. analyt. Tox.* **2**: 56–59.

HECHT S. S., THORNE, R. L., MARONPOT, R. R. and HOFFMAN, D. (1975) A study of tobacco carcinogenesis. XIII. Tumor-promoting subfractions of the weakly acidic fraction. *J. natn. Cancer inst.* **55**: 1329–1336.

HECKER, E. (1968) Cocarcinogenic principles from the seed oil of *Croton tiglium* and from other Euphorbiaceae. *Cancer Res.* **28**: 2338–2349.

HECKER, E. (1971) Isolation and characterization of the cocarcinogenic principles from croton oil. In: *Methods in Cancer Research, Vol. 6*, pp. 439–484, BUSCH, H. (ed.). Academic Press, New York.

HECKER, E. (1978) Structure-activity relationships in diterpene esters irritant and cocarcinogenic to mouse skin. In: *Carcinogenesis, Volume 2, Mechanisms of Tumor Promotion and Cocarcinogenesis*, pp. 11–48. SLAGA, T. J., SIVAK, A. and BOUTWELL, R. K. (eds). Raven Press, New York.

HECKER, E., KUNZ, W., FUSENIG, N. E., MARKS, F. and THIELMANN, H. W. (eds) (1982) *Carcinogenesis—A Comprehensive Survey, Volume 7. Cocarcinogenesis and Biological Effects of Tumor Promoters.* Raven Press, New York.
HEIDELBERGER, C. (1981) Cellular transformation as a basic tool for chemical carcinogenesis. In: *Adv. Modern Environ. Toxicol., Volume 1. Mammalian Cell Transformation by Chemical Carcinogens*, pp. 1–28, MISHRA, N., DUNKEL, V. and MEHLMAN, M. (eds). Senate Press, Inc., Princeton Junction, NJ.
HEIDELBERGER, C., FREEMAN, A. E., PIENTA, R. J., SIVAK, A., BERTRAM, J. S., CASTO, B. C., DUNKEL, V. C., FRANCIS, M. W., KAKUNAGA, T., LITTLE, J.B. and SCHECHTMAN, L. M. (1983) Cell transformation by chemical agents—a review and analysis of the literature. *Mutat. Res.* **114**: 283–385.
HELDIN, C.-H. and WESTERMARK, B. (1984) Growth factors: mechanism of action and relation to oncogenes. *Cell* **31**: 9–20.
HELLMAN, K. B. and HELLMAN, A. (1981) Induction of type-C retrovirus by the tumor promoter TPA. *Int. J. Cancer* **27**: 95–99.
HENNINGS, H. and BOUTWELL, R. K. (1970) Studies on the mechanism of skin tumor promotion. *Cancer Res.* **30**: 312–320.
HENNINGS, H. and YUSPA, S. H. (1985) Two-stage tumor promotion in mouse skin: an alternative interpretation. *J. natn. Cancer Inst.* **74**: 735–740.
HENNINGS, H., DEVOR, D., WENK, M. L., SLAGA, T. J., FORMER, B., COLBURN, N. H., BOWDEN, G. T., ELGJO, K. and YUSPA, S. H. (1981) Comparison of two-stage epidermal carcinogenesis initiated by 7,12-dimethylbenz-(a)anthracene or N-methyl-N'-nitro-N-nitrosoguanidine in newborn and adult SENCAR and BALB/c mice. *Cancer Res.* **41**: 773–779.
HENNINGS, H., HOLBROOK, K. A. and YUSPA, S. H. (1983a) Factors influencing calcium-induced terminal differentiation in cultured mouse epidermal cells. *J. cell. Physiol.* **116**: 265–281.
HENNINGS, H., MICHAEL, D., CHENG, C., STEINERT, P., HOLBROOK, K. and YUSPA, S. H. (1980) Calcium regulation of growth and differentiation of mouse epidermal cells in culture. *Cell* **19**: 245–254.
HENNINGS, H., SHORES, R., WENK, M. L., SPANGLER, E. F., TARONE, R. and YUSPA, S. H. (1983b) Malignant conversion of mouse skin tumours is increased by tumour initiators and unaffected by tumour promoters. *Nature* **304**: 67–69.
HENNINGS, H., WENK, M. L. and DONAHOE, R. (1982) Retinoic acid promotion of papilloma formation in mouse skin. *Cancer Lett.* **16**: 1–5.
HERLYN, M., HERLYN, D., ELDER, D. E., BONDI, E., LAROSSA, D., HAMILTON, R., SEARS, H. F., BALABAN, G., GUERRY, D., IV, CLARK, W. H. and KOPROWSKI, H. (1983) Phenotypic characteristics of cells derived from precursors of human melanoma. *Cancer Res.* **43**: 5502–5508.
HICKS, R. M. (1983) Pathological and biochemical aspects of tumour promotion. *Carcinogenesis* **4**: 1209–1214.
HIGGINSON, J. (1983) Developing concepts on environmental cancer: the role of geographical pathology. *Environ. Mutag.* **5**: 929–940.
HIRAGUN, A., SATO, M. and MITSUI, H. (1981) Prevention of tumor promoter-mediated inhibition of preadipocyte differentiation by dexamethasone. *Gann* **72**: 891–897.
HIRAKAWA, T., KAKUNAGA, T., FUJIKI, H. and SUGIMURA, T. (1982) A new tumor-promoting agent, dihydroteleocidin B, markedly enhances chemically induced malignant cell transformation. *Science* **216**: 527–529.
HOROWITZ, A. D., FUJIKI, H., WEINSTEIN, I. B., JEFFREY, A., OKIN, E., MOORE, R. E. and SUGIMURA, T. (1983) Comparative effects of aplysiatoxin, debromoaplysiatoxin, and teleocidin on receptor binding and phospholipid metabolism. *Cancer Res.* **43**: 1529–1535.
HOSHINO, H., MIWA, W., FUJIKI, H. and SUGIMURA, T. (1980) Aggregation of human lymphoblastoid cells by tumor-promoting phorbol esters and dihydroteleocidin B. *Biochem. biophys. Res. Commun.* **95**: 842–848.
HOZUMI, M., OGAWA, M., SUGIMURA, T., TAKEUCHI, T. and UMEZAWA, H. (1972) Inhibition of tumorigenesis in mouse skin by leupeptin, a protease inhibitor from *Actinomycetes*. *Cancer Res.* **32**: 1725–1728.
HSIAO, W.-L.W., GATTONI-CELLI, S. and WEINSTEIN, I. B. (1984) Oncogene-induced transformation of C3H 10T1/2 cells is enhanced by tumor promoters. *Science* **226**: 552–555.
HUBERMAN, E. and CALLAHAM, M. F. (1979) Induction of terminal differentiation in human promyelocytic leukemia cells by tumor-promoting agents. *Proc. natn. Acad. Sci. U.S.A.* **76**: 1293–1297.
HUBERMAN, E., HECKMAN, C. and LANGENBACH, R. (1979) Stimulation of differentiated functions in human melanoma cells by tumor-promoting agents and dimethyl sulfoxide. *Cancer Res.* **39**: 2618–2624.
HUDEWENTZ, J., BORNKAMM, G. W. and ZUR HAUSEN, H. (1980) Effect of the diterpene ester TPA on Epstein-Barr virus antigen- and DNA synthesis in producer and nonproducer cell lines. *J. Virol.* **100**: 175–178.
HUNTER, T. (1984) Oncogenes and proto-oncogenes: how do they differ? *J. natn. Cancer inst.* **73**: 773–786.
INTERNATIONAL AGENCY FOR RESEARCH ON CANCER (1980) Long-term and short-term screening assays for carcinogens: a critical appraisal. In: *IARC Monogr., Suppl. 2.*, Lyon.
ISHII, D. N. (1978) Effect of tumor promoters on the response of cultured embryonic chick ganglia to nerve growth factor. *Cancer Res.* **38**: 3886–3893.
ISHI, D. N., FIBACH, E., YAMASAKI, H. and WEINSTEIN, I. B. (1978) Tumor promoters inhibit morphological differentiation in cultured mouse neuroblastoma cells. *Science* **200**: 556–559.
ISHIMURA, K., HIRAGUN, A. and MITSUI, H. (1980) Specific changes in the surface glycoprotein pattern of a human leukemic null cell line NALL-1 associated with morphologic and biological alterations induced by phorbol-ester. *Biochem. biophys. Res. Commun.* **93**: 293–300.
ITO, Y., KAWANISHI, M., HARAYAMA, T. and TAKABAYASHI, S. (1981a) Combined effect of the extracts from *Croton tiglium, Euphorbia lathyris* or *Euphorbia tirucalli* and n-butyrate on Epstein-Barr virus expression in human lymphoblastoid P3HR-1 and Raji cells. *Cancer Lett.* **12**: 175–180.
ITO, Y., OHIGASHI, H., KOSHIMIZU, K. and YI, Z. (1983) Epstein-Barr virus-activating principle in the ether extracts of soils collected from under plants which contain active diterpene esters. *Cancer Lett.* **19**: 113–117.
ITO, Y., YANASE, S., FUJITA, J., HARAYAMA, T., TAKASHIMA, M. and IMANAKA, H. (1981b) A short-term *in vitro* assay for promoter substances using human lymphoblastoid cells latently infected with Epstein-Barr virus. *Cancer Lett.* **13**: 29–37.

IVERSEN, U. M. and IVERSEN, O. H. (1979) The carcinogenic effect of TPA (12-O-tetradecanoylphorbol-13-acetate) when applied to the skin of hairless mice. *Virchows Arch. B. Cell Path.* **30**: 33–42.

IWASHITA, S. and FOX, C. F. (1984) Epidermal growth factor and potent phorbol tumor promoters induce epidermal growth factor receptor phosphorylation in a similar but distinctively different manner in human epidermoid carcinoma A431 cells. *J. biol. Chem.* **259**: 2559–2567.

JACOBS, S., SAHYOUN, N. E., SALTIEL, A. R. and CUATRECASAS, P. (1983) Phorbol esters stimulate the phosphorylation of receptors for insulin and somatomedin C. *Proc. natn. Acad. Sci. U.S.A.* **80**: 6211–6213.

JAKEN, S., GEFFEN, C. and BLACK, P. H. (1981a) Dexamethasone inhibition and phorbol myristate acetate stimulation of plasminogen activator in human embryonic lung cells. *Biochem. biophys. Res. Commun.* **99**: 379–384.

JAKEN, S., TASHJIAN, A. H. JR. and BLUMBERG, P. M. (1981b) Relationship between biological responsiveness to phorbol esters and receptor levels in GH_4C_1 rat pituitary cells. *Cancer Res.* **41**: 4956–4960.

JETTEN, A. M. (1983) Action of retinoids on the anchorage-independent growth of normal rat kidney fibroblasts induced by 12-O-tetradecanoylphorbol-13-acetate or sarcoma growth factor. *Cancer Res.* **43**: 68–72.

KABELITZ, D., TÖTTERMAN, T. H., GIDLUND, M., NILSSON, K. and WIGZELL, H. (1982) Activation of human T lymphocytes by 12-O-tetradecanoylphorbol-13-acetate: role of accessory cells and interaction with lectins and allogeneic cells. *Cell. Immunol.* **70**: 277–286.

KAKUNA, A. T. (1973) A quantitative system for assay of malignant transformation by chemical carcinogens using a clone derived from BALB/3T3. *Int. J. Cancer* **12**: 463–473.

KASUKABE, T., HONMA, Y. and HOZUMI, M. (1981) The tumor promoter 12-O-tetradecanoyl-phorbol-13-acetate inhibits or enhances induction of differentiation of mouse myeloid leukemia cells depending on the type of serum in the medium. *Gann* **72**: 310–314.

KATO, R., NAKADATE, T., YAMAMOTO, S. and SUGIMURA, T. (1983) Inhibition of 12-O-tetradecanoylphorbol-13-acetate-induced tumor promotion and ornithine decarboxylase activity by quercetin: possible involvement of lipoxygenase inhibition. *Carcinogenesis* **4**: 1301–1305.

KAWAMURA, H., STRICKLAND, J. E. and YUSPA, S. H. (1985) Association of resistance to terminal differentiation with initiation of carcinogenesis in adult mouse epidermal cells. *Cancer Res.* **45**: 2748–2752.

KELLY, K., COCHRAN, B. H., STILES, C. D. and LEDER, P. (1983) Cell cycle-specific regulation of the *c-myc* gene by lymphocyte mitogens and platelet-derived growth factor. *Cell* **35**: 603–610.

KENNEDY, A. R. and LITTLE, J. B. (1980) Investigation of the mechanism for enhancement of radiation transformation *in vitro* by 12-O-tetradecanoylphorbol-13-acetate. *Carcinogenesis* **1**: 1039–1047.

KENNEDY, A. R. and LITTLE. J. B. (1981) Effects of protease inhibitors on radiation transformation *in vitro*. *Cancer Res.* **41**: 2103–2108.

KENNEDY, A. R. and WEICHSELBAUM, R. R. (1981) Effects of dexamethasone and cortisone with X-ray irradiation on transformation of C3H 10T1/2 cells. *Nature* **294**: 97–98.

KENNEDY, A. R., FOX, M., MURPHY, G. and LITTLE, J. B. (1980a) Relationship between X-ray exposure and malignant transformation in C3H 10T1/2 cells. *Proc. natn. Acad. Sci. U.S.A.* **77**: 7262–7266.

KENNEDY, A. R., MONDAL, S., HEIDELBERGER, C. and LITTLE, J. B. (1978) Enhancement of X-ray transformation by 12-O-tetradecanoyl-phorbol-13-acetate in a cloned line of C3H mouse embryo cells. *Cancer Res.* **38**: 439–443.

KENNEDY, A. R., MURPHY, G. and LITTLE, J. B. (1980b) Effect of time and duration of exposure to 12-O-tetradecanoylphorbol-13-acetate on X-ray transformation of C3H 10T1/1 cells. *Cancer Res.* **40**: 1915–1920.

KENSLER, T. W. and TRUSH, M. A. (1983) Inhibition of oxygen radical metabolism in phorbol ester-activated polymorphonuclear leukocytes by an antitumor promoting copper complex with superoxide dismutase-mimetic activity. *Biochem. Pharm.* **32**: 3485–3487.

KENSLER, T. W. and TRUSH, M. A. (1984) Role of oxygen radicals in tumor promotion. *Environmental Mutag.* **6**: 593–616.

KENSLER, T. W., BUSH, D. M. and KOZUMBO, W. J. (1983) Inhibition of tumor promotion by a biomimetic superoxide dismutase. *Science* **221**: 75–77.

KESKI-OJA, J., SHOYAB, M., DE LARCO, J. E. and TODARO, G. J. (1979) Rapid release of fibronectin from human lung fibroblasts by biologically active phorbol esters. *Int. J. Cancer* **24**: 218–224.

KING, A. C. and CUATRECASAS, P. (1982) Resolution of high and low affinity epidermal growth factor receptors. Inhibition of high affinity component by low temperature, cycloheximide, and phorbol esters. *J. biol. Chem.* **257**: 3053–3060.

KINSELLA, A. R. and RADMAN, M. (1978) Tumor promoter induces sister chromatid exchanges: relevance to mechanisms of carcinogenesis. *Proc. natn. Acad. Sci. U.S.A.* **75**: 6149–6153.

KINSELLA, A. R., GAINER, H. S. and BUTLER, J. (1983) Investigation of a possible role for superoxide anion production in tumor production. *Carcinogenesis* **4**: 717–719.

KINZEL, V., RICHARDS, J. and STÖHR, M. (1981) Early effects of the tumor-promoting phorbol ester 12-O-tetradecanoylphorbol-13-acetate on the cell cycle traverse of asynchronous HeLa cells. *Cancer Res.* **41**: 300–305.

KITAGAWA, T., WATANABE, R., KAYANO, T. and SUGANO, H. (1980) *In vitro* carcinogenesis of hepatocytes obtained from acetylaminofluorene-treated rat liver and promotion of their growth by phenobarbital. *Gann* **71**: 747–754.

KLEIN-SZANTO, A. J. P. and SLAGA, T. J. (1981) Numerical variation of dark cells in normal and chemically induced hyperplastic epidermis with age of animal and efficiency of tumor promoter. *Cancer Res.* **41**: 4437–4440.

KLEIN-SZANTO, A. J. P., MAJOR, S. K. and SLAGA, T. J. (1980) Induction of dark keratinocytes by 12-O-tetradecanoylphorbol-13-acetate and mezerein as an indicator of tumor-promoting efficiency. *Carcinogenesis* **1**: 399–406.

KNOWLES, M. A. (1979) Effects of the tumor-promoting agent 12-O-tetradecanoylphorbol-13-acetate on normal and 'preneoplastic' mouse submandibular gland epithelial cells *in vitro*. *J. natn. Cancer inst.* **62**: 349–352.

KNOWLES, M. A. and FRANKS, L. M. (1977) Stages in neoplastic transformation of adult epithelial cells by 7,12-dimethylbenz(a)anthracene *in vitro*. *Cancer Res.* **37**: 3917–3924.

KOPELOVICH, L., BIAS, N. E. and HELSON, L. (1979) Tumour promoter alone induces neoplastic transformation of fibroblasts from humans genetically predisposed to cancer. *Nature* **282**: 619–621.

KORETZKY, G. A., DANIELE, R. P. and NOWELL, P. C. (1982) A phorbol ester (TPA) can replace macrophages in human lymphocyte cultures stimulated with a mitogen but not with an antigen. *J. Immunol.* **128**: 1776–1780.

KOZUMBO, W. J., SEED, J. L. and KENSLER, T. W. (1983) Inhibition by 2(3)-*tert*-butyl-4-hydroxyanisole and other antioxidants of epidermal ornithine decarboxylase activity induced by 12-*O*-tetradecanoylphorbol-13-acetate. *Cancer Res.* **43**: 2555–2559.

KRAFT, A. S. and ANDERSON, W. B. (1983) Phorbol esters increase the amount of Ca^{2+}, phospholipid-dependent protein kinase associated with plasma membrane. *Nature* **301**: 621–623.

KRAFT, A. S., ANDERSON, W. B., COOPER, H. L. and SANDO, J. J. (1982) Decrease in cytosolic calcium/phospholipid-dependent protein kinase activity following phorbol ester treatment of EL4 thymoma cells. *J. Biol. Chem.* **257**: 13193–13196.

KRIEG, P., AMTMANN, E. and SAUER, G. (1981) Interaction between latent papovavirus genomes and the tumor promoter TPA. *FEBS Lett.* **128**: 191–194.

KULESZ-MARTIN, M., KOEHLER, B., HENNINGS, H. and YUSPA, S. H. (1980) Quantitative assay for carcinogen altered differentiation in mouse epidermal cells. *Carcinogenesis* **1**: 995–1006.

KUROKI, T., SASAKI, K., CHIDA, K., ABE, E. and SUDA, T. (1983) 1α,25-Dihydroxyvitamin D_3 markedly enhances chemically-induced transformation in BALB 3T3 cells. *Gann* **74**: 611–614.

KWONG, C. H. and MUELLER, G. C. (1983) Influence of tumor-promoting phorbol esters on the phosphorylation of membrane proteins in lymphocytes. *Carcinogenesis* **4**: 663–670.

LAND, H., PARADA, L. F. and WEINBERG, R. A. (1983) Cellular oncogenes and multistep carcinogenesis. *Science* **222**: 771–778.

LANDESMAN, J. M. and MOSSMAN, B. T. (1982) Induction of ornithine decarboxylase in hamster tracheal epithelial cells exposed to asbestos and 12-*O*-tetradecanoylphorbol-13-acetate. *Cancer Res.* **42**: 3669–3675.

LASZLO, A. and BISSELL, M. J. (1983) TPA induces simultaneous alterations in the synthesis and organization of vimentin. *Exp. Cell Res.* **148**: 221–234.

LASZLO, A., RADKE, K., CHIN, S. and BISSELL, M. J. (1981) Tumor promoters alter gene expression and protein phosphorylation in avian cells in culture. *Proc. natn. Acad. Sci. U.S.A.* **78**: 6241–6245.

LEACH, K. L., JAMES, M. L. and BLUMBERG, P. M. (1983) Characterization of a specific phorbol ester aporeceptor in mouse brain cytosol. *Proc. natn. Acad. Sci. U.S.A.* **80**: 4208–4212.

LEE, L-S. and WEINSTEIN, I. B. (1978) Tumor-promoting phorbol esters inhibit binding of epidermal growth factor to cellular receptors. *Science* **202**: 313–315.

LEE, L-S. and WEINSTEIN, I. B. (1979) Membrane effects of tumor promoters: stimulation of sugar uptake in mammalian cell cultures. *J. cell. Physiol.* **99**: 451–460.

LEHRER, R. I. and COHEN, L. (1981) Receptor-mediated regulation of superoxide production in human neutrophils stimulated by phorbol myristate acetate. *J. clin. Invest.* **68**: 1314–1320.

LICHTI, U., PATTERSON, E., HENNINGS, H. and YUSPA, S. H. (1981) The tumor promoter 12-*O*-tetradecanoylphorbol-13-acetate induces ornithine decarboxylase in proliferating basal cells but not in differentiating cells from mouse epidermis. *J. cell. Physiol.* **107**: 261–270.

LICHTI, U., YUSPA, S. H. and HENNINGS, H. (1978) Ornithine and S-adenosylmethionine decarboxylases in mouse epidermal cell cultures treated with tumor promoters. In: *Carcinogenesis. Volume 2. Mechanisms of Tumor Promotion and Cocarcinogenesis*, pp. 221–232, SLAGA, T. J., SIVAK, A. and BOUTWELL, R. K. (eds). Raven Press, New York.

LILLEHAUG, J. R. and DJURHUUS, R. (1982) Effect of diethylstilbestrol on the transformable mouse embryo fibroblast C3H/10T1/2C18 cells. Tumor promotion, cell growth. DNA synthesis, and ornithine decarboxylase. *Carcinogenesis* **3**: 797–799.

LIN, J-C., SHAW, J. E., SMITH, M. C. and PAGANO, J. S. (1979) Effect of 12-*O*-tetradecanoyl-phorbol-13-acetate on the replication of Epstein-Barr virus. I. Characterization of viral DNA. *J. Virol.* **99**: 183–187.

LIN, J-C., SMITH, M. C. and PAGANO, J. S. (1983) Activation of latent Epstein-Barr virus genomes: selective stimulation of synthesis of chromosomal proteins by a tumor promoter. *J. Virol.* **45**: 985–991.

LINK, R. and MARKS, F. (1981) Histone phosphorylation in phorbol ester stimulated and β-adrenergically stimulated mouse epidermis *in vivo* and characterization of an epidermal protein phosphorylation system. *Biochim. Biophys. Acta* **675**: 265–275.

LIPP, M., SCHERER, B., LIPS, G., BRANDNER, G. and HUNSMANN, G. (1982) Diverse effects: augmentation, inhibition, and non-efficacy of 12-*O*-tetradecanoylphorbol-13-acetate (TPA) on retrovirus genome expression *in vivo* and *in vitro*. *Carcinogenesis* **3**: 261–265.

LOCKNEY, M. W., GOLOMB, H. M. and DAWSON, G. (1982) Unique cell surface glycoprotein expression in hairy cell leukemia: effect of phorbol ester tumor promoters on surface glycoproteins, cell morphology and adherence. *Cancer Res.* **42**: 3724–3728.

LOCKYER, J. M., BOWDEN, G. T., MATRISIAN, L. M. and MAGUN, B. E. (1981) Tumor promoter-induced inhibition of epidermal growth factor binding to cultured mouse primary epidermal cells. *Cancer Res.* **41**: 2308–2314.

LONG, S. D., QUIGLEY, J. P., TROLL, W. and KENNEDY, A. R. (1981) Protease inhibitor antipain suppresses 12-*O*-tetradecanoylphorbol-13-acetate induction of plasminogen activator in transformable mouse embryo fibroblasts. *Carcinogenesis* **2**: 933–936.

LOSKUTOFF, D. J. and EDGINGTON, T. S. (1977) Synthesis of a fibrinolytic activator and inhibitor by endothelial cells. *Proc. natn. Acad. Sci. U.S.A.* **74**: 3903–3907.

LOTEM, J. and SACHS, L. (1979) Regulation of normal differentiation in mouse and human myeloid leukemic cells by phorbol esters and the mechanism of tumor promotion. *Proc. natn. Acad. Sci. U.S.A.* **76**: 5158–5162.

LOVEDAY, K. S. and LATT, S. A. (1979) The effect of a tumor promoter, 12-*O*-tetradecanoylphorbol-13-acetate (TPA), on sister-chromatid exchange formation in cultured Chinese hamster cells. *Mutat. Res.* **67**: 343–348.

LOWE, M. E., PACIFICI, M. and HOLTZER, H. (1978) Effects of phorbol-12-myristate-13-acetate on the phenotypic program of cultured chondroblasts and fibroblasts. *Cancer Res.* **38**: 2350–2356.

MAGUN, B. E. and BOWDEN, G. T. (1979) Effects of the tumor promoter TPA on the induction of DNA synthesis in normal and RSV-transformed rat fibroblasts. *J. supramol. Struct.* **12**: 63–72.

Magun, B. E., Matrisian, L. M. and Bowden, G. T. (1980) Epidermal growth factor: ability of tumor promoter to alter its degradation, receptor affinity and receptor number. *J. biol. Chem.* **255**: 6373–6381.

Maher, V. M., Rowan, L. A., Silinskas, K. C., Kateley, S. A. and McCormick, J. J. (1982) Frequency of UV-induced neoplastic transformation of diploid human fibroblasts is higher in xeroderma pigmentosum cells than in normal cells. *Proc. natn. Acad. Sci USA* **79**: 2613–2617.

Marks, F. (1976) Epidermal growth control mechanisms, hyperplasia, and tumor promotion in the skin. *Cancer Res.* **36**: 2636–2643.

Marks, F. and Grimm, W. (1972) Diurnal fluctuation and β-adrenergic elevation of cyclic AMP in mouse epidermis *in vivo*. *Nature New. Biol.* **240**: 178–179.

Marks, F., Bertsch, S., Grimm, W. and Schweizer, J. (1978) Hyperplastic transformation and tumor promotion in mouse epidermis: possible consequences of disturbances of endogenous mechanisms controlling proliferation and differentiation. In: *Carcinogenesis, Volume 2. Mechanisms of Tumor Promotion and Cocarcinogenesis*, pp. 97–116. Slaga, T. J., Sivak, A. and Boutwell, R. J. (eds). Raven Press, New York.

Martin, R. G., Setlow, V. P. and Edwards, C. A. F. (1979a) Roles of the Simian virus 40 tumor antigens in transformation of Chinese hamster lung cells: Studies with Simian virus 40 double mutants. *J. Virol.* **31**: 596–607.

Martin, R. G., Setlow, V. P., Edwards, C. A. F. and Vembu, D. (1979b) The roles of the Simian virus 40 tumor antigens in transformation of Chinese hamster lung cells. *Cell* **17**: 635–643.

Mass, M. J., Nettesheim, P., Beeman, D. K. and Barrett, J. C. (1984) Inhibition of transformation of primary rat tracheal epithelial cells by retinoic acid. *Cancer Res.* **44**: 5688–5691.

Mastro, A. M. (1982) Phorbol esters: Tumor promotion, cell regulation, and the immune response. *Lymphokines* **6**: 263–313.

Mastro, A. M. (1983) Phorbol ester tumor promoters and lymphocyte proliferation. *Cell. Biol. int. Rep.* **7**: 881–893.

Mastro, A. M. and Mueller, G. C. (1974) Synergistic action of phorbol esters in mitogen-activated bovine lymphocytes. *Exp. Cell Res.* **88**: 40–46.

Mastro, A. M. and Mueller, G. C. (1978) Inhibition of the mixed lymphocyte proliferative response by phorbol esters. *Biochim. biophys. Acta* **517**: 246–254.

Mastro, A. M. and Pepin, K. G. (1982) Effect of macrophages on phorbol ester-stimulated comitogenesis in bovine lymphocytes. *Cancer Res.* **42**: 1630–1635.

McIlhinney, R. A. J., Patel, S. and Monaghan, P. (1983) Effects of 12-O-tetradecanoylphorbol 13-acetate (TPA) on a clonal human teratoma-derived embryonal carcinoma cell line. *Exp. Cell Res.* **144**: 297–311.

Metcalf, B. W., Bey, P., Danzin, C., Jung, M. J., Casara, P. and Vevert, J. P. (1978) Catalytic irreversible inhibition of mammalian ornithine decarboxylase (EC 4.1.1.17) by substrate and product analogues. *J. Amer. chem. Soc.* **100**: 2551–2553.

Meyer, A. L. (1983) *In vitro* transformation assays for chemical carcinogens. *Mutat. Res.* **115**: 323–338.

Miao, R. M., Fieldsteel, A. H. and Fodge, D. W. (1978) Opposing effects of tumour promoters on erythroid differentiation. *Nature* **274**: 271–272.

Miller, R. C., Geard, C. R., Osmak, R. S., Rutledge-Freeman, M., Ong, A., Mason, H., Napholz, A., Perez, N., Harisiadis, L. and Borek, C. (1981) Modification of sister chromatid exchanges and radiation-induced transformation in rodent cells by the tumor promoter 12-O-tetradecanoylphorbol-13-acetate and two retinoids. *Cancer Res.* **41**: 655–659.

Milo, G. E., Oldham, J. W., Zimmerman, R., Hatch, G. G. and Weisbrode, S. A. (1981) Characterization of human cells transformed by chemical and physical carcinogens *in vitro*. *In Vitro* **17**: 719–729.

Mishra, N., Dunkel, V. and Mehlman, M. (eds) (1981) *Adv. Mod. Environ. Toxicol. Vol. 1: Mammalian Cell Transformation by Chemical Carcinogens*. Senate Press, Inc., Princeton Junction. NJ.

Mizel, S. B., Rosenstreich, D. L. and Oppenheim, J. J. (1978) Phorbol myristic acetate stimulates LAF production by the macrophage cell line, $P388D_1$. *Cell. Immunol.* **40**: 230–235.

Mondal, S. (1980) C3H/10T1/2Cl mouse embryo cell line: its use for the study of carcinogenesis and tumor promotion in cell culture. In: *Adv. Mod. Environ. Toxicol., Vol. 1: Mammalian Cell Transformation by Chemical Carcinogens*, pp. 181–211, Mishra, N., Dunkel, V. and Mehlman, M. (eds). Senate Press. Inc., Princeton Junction, NJ.

Mondal, S. and Heidelberger, C. (1976) Transformation of C3H/10T1/2 CL8 mouse embryo fibroblasts by ultraviolet irradiation and a phorbol ester. *Nature* **260**: 710–711.

Mondal, S. and Heidelberger, C. (1980) Inhibition of induced differentiation of C3H/10T1/2 clone 8 mouse embryo cells by tumor promoters. *Cancer Res.* **40**: 334–338.

Mondal, S., Brankow, D. W. and Heidelberger, C. (1976) Two-stage oncogenesis in cultures of C3H/10T1/2 cells. *Cancer Res.* **36**: 2254–2260.

Mondal, S., Brankow, D. W. and Heidelberger, C. (1978) Enhancement of oncogenesis in C3H/10T1/2 mouse embryo cell cultures by saccharin. *Science* **201**: 1141–1142.

Mordan, L. J. and Bertram, J. S. (1983) Retinoid effects on cell–cell interactions and growth characteristics of normal and carcinogen-treated C3H/10T1/2 cells. *Cancer Res.* **43**: 567–571.

Mordan, L. J., Bergin, L. M., Budnick, J. L., Meegan, R. R. and Bertram, J. S. (1982) Isolation of methylcholanthrene-'initiated' C3H/10T1/2 cells by inhibiting neoplastic progression with retinyl acetate. *Carcinogenesis* **3**: 279–285.

Mordan, L. J., Martner, J. E. and Bertram, J. S. (1983) Quantitative neoplastic transformation of C3H/10T1/2 fibroblasts: dependence upon the size of the initiated cell colony at confluence. *Cancer Res.* **43**: 4062–4067.

Moroney, J., Smith, A., Tomei, L. D. and Wenner, C. E. (1978) Stimulation of $^{86}Rb^-$ and $^{32}P_i$ movements in 3T3 cells by prostaglandins and phorbol esters. *J. cell. Physiol.* **95**: 287–294.

Moscatelli, D., Jaffe, E. and Rifkin, D. B. (1980) Tetradecanoyl phorbol acetate stimulates latent collagenase production by cultured human endothelial cells. *Cell* **20**: 343–351.

Mossman, B. T. and Craighead, J. E. (1979) Use of hamster tracheal organ cultures for assessing the cocarcinogenic effects of inorganic particulates on the respiratory epithelium. In: *Progr. Exp. Tumor Res.*,

Vol. 24: The Syrian Hamster in Toxicology and Carcinogenesis Research, pp. 37–47, HOMBURGER, F. (ed.). S. Karger, Basel.

MOTTRAM, J. C. (1944) A developing factor in experimental blastogenesis. *J. Pathol. Bacteriol.* **56**: 181–187.

MUELLER, G. C. and KAJIWARA, K. (1965) Regulatory steps in the replication of mammalian cell nuclei. In: *Developmental and Metabolic Control Mechanisms and Neoplasia*, pp. 452–474. Williams and Wilkens. Baltimore.

MUFSON, R. A., FISCHER, S. M., VERMA, A. K., GLEASON, G. L., SLAGA, T. J. and BOUTWELL, R. K. (1979) Effects of 12-O-tetradecanoylphorbol-13-acetate and mezerein on epidermal ornithine decarboxylase activity, isoproterenol-stimulated levels of cyclic adenosine 3':5'-monophosphate, and induction of mouse skin tumors in vivo. *Cancer Res.* **39**: 4791–4795.

MUFSON, R. A., SIMSIMAN, R. C. and BOUTWELL, R. K. (1977) The effect of the phorbol ester tumor promoters on the basal and catecholamine-stimulated levels of cyclic adenosine 3':5'-monophosphate in mouse skin and epidermis in vivo. *Cancer Res.* **37**: 665–669.

MÜLLER, R., CURRAN, T., MÜLLER, D. and GUILBERT, L. (1985) Induction of c-*fos* during myelomonocytic differentiation and macrophage proliferation. *Nature* **314**: 546–548.

MURAO. S., GEMMELL, M. A., CALLAHAM, M. F., ANDERSON, N. L. and HUBERMAN, E. (1983) Control of macrophage cell differentiation in human promyelocytic HL-60 leukemia cells by 1,25-dihydroxyvitamin D_3 and phorbol-12-myristate-13-acetate. *Cancer Res.* **43**: 4989–4996.

MURRAY, A. W. and FITZGERALD, D. J. (1979) Tumor promoters inhibit metabolic cooperation in cocultures of epidermal and 3T3 cells. *Biochem. biophys. Res. Commun.* **91**: 395–401.

NAGASAWA, H. and LITTLE, J. B. (1981) Factors influencing the induction of sister chromatid exchanges in mammalian cells by 12-O-tetradecanoylphorbol-13-acetate. *Carcinogenesis* **2**: 601–607.

NAGASAWA, H., FORNACE, D. and LITTLE, J. B. (1983) Induction of sister-chromatid exchanges by DNA-damaging agents and 12-O-tetradecanoylphorbol-13-acetate (TPA) in synchronous Chinese hamster ovary (CHO) cells. *Mutat. Res.* **107**: 315–327.

NAGASAWA, K. and MAK, T. W. (1980) Phorbol esters induce differentiation in human malignant T lymphoblasts. *Proc. natn Acad. Sci. U.S.A.* **77**: 2964–2968.

NAKA, M., NISHIKAWA, M., ADELSTEIN, R. S. and HIDAKA, H. (1983) Phorbol ester-induced activation of human platelets is associated with protein kinase C phosphorylation of myosin light chains. *Nature* **306**: 490–492.

NAKADATE, T., YAMAMOTO, S., ISEKI, H., SONODA, S., TAKEMURA, S., URA, A., HOSODA, Y. and KATO, R. (1982a) Inhibition of 12-O-tetradecanoylphorbol-13-acetate-induced tumor promotion by nordihydroguaiaretic acid, a lipoxygenase inhibitor, and p-bromophenacyl bromide, a phospholipase A_2 inhibitor. *Gann* **73**: 841–843.

NAKADATE, T., YAMAMOTO, S., ISHII, M. and KATO, R. (1982b) Inhibition of 12-O-tetradecanoylphorbol-13-acetate-induced epidermal ornithine decarboxylase activity by phospholipase A_2 inhibitors and lipoxygenase inhibitor. *Cancer Res.* **42**: 2841–2845.

NAKADATE, T., YAMAMOTO, S., ISHII, M. and KATO, R. (1982c) Inhibition of 12-O-tetradecanoylphorbol-13-acetate-induced epidermal ornithine decarboxylase activity by lipoxygenase inhibitors: possible role of product(s) of lipoxygenase pathway. *Carcinogenesis* **3**: 1411–1414.

NAKAMURA, K. D., MARTINEZ, R. and WEBER, M. J. (1983) Tyrosine phosphorylation of specific proteins after mitogen stimulation of chicken embryo fibroblasts. *Molec. cell. Biol.* **3**: 380–390.

NELSON, K. G. and SLAGA, T. J. (1982) Effects of inhibitors of tumor promotion on 12-O-tetradecanoylphorbol-13-acetate-induced keratin modification in mouse epidermis. *Carcinogenesis* **3**: 1311–1315.

NETTESHEIM, P. and BARRETT, J. C. (1984) Tracheal epithelial cell transformation: a model system for studies on neoplastic progression. *CRC Crit. Rev. Tox.* **12**: 215–239.

NETTESHEIM, P. and MARCHOK, A. (1983) Neoplastic development in airway epithelium. *Adv. Cancer Res.* **39**: 1–70.

NETTESHEIM, P., GRAY, T. E. and BARRETT, J. C. (1985) Contrasting responses of normal and transformed rat tracheal epithelial (RTE) cells to tumor promoter 12-O-tetradecanoylphorbol-13-acetate. *Carcinogenesis—A Comprehensive Survey, Vol. 8, Cancer of the Respiratory Tract: Predisposing Factors*, pp. 329–340, MASS, M. J., KAUFMAN, D. G., SIEGFRIED, J. M., STEELE, V. E. and NESNOW, S. (eds), Raven Press, New York.

NEWBURGER, P. E., BAKER, R. D., HANSEN, S. L., DUNCAN, R. A. and GREENBERGER, J. S. (1981) Functionally deficient differentiation of HL-60 promyelocytic leukemia cells induced by phorbol myristate acetate. *Cancer Res.* **41**: 1861–1865.

NIEDEL, J. E., KUHN, L. J. and VANDENBARK, G. R. (1983) Phorbol diester receptor copurifies with protein kinase C. *Proc. natn. Acad. Sci. U.S.A.* **80**: 36–40.

NISHIZUKA, Y. (1983) Phospholipid degradation and signal translation for protein phosphorylation. *Trends biochem. Sci.* **8**: 13–16.

NISHIZUKA, Y. (1984) The role of protein kinase C in cell surface signal transduction and tumour promotion. *Nature* **308**: 693–697.

NOMURA, S., SHOBU, N. and OISHI, M. (1983) Tumor promoter 12-O-tetradecanoylphorbol 13-acetate stimulates simian virus 40 induction by DNA-damaging agents and tumor initiators. *Molec. cell. Biol.* **3**: 757–760.

NORDENBERG, J., STENZEL, K. H. and NOVOGRODSKY, A. (1983) 12-O-Tetradecanoylphorbol-13-acetate and concanavalin A enhance glucose uptake in thymocytes by different mechanisms. *J. cell. Physiol.* **117**: 183–188.

O'BRIEN, T. G. (1982) Hexose transport in undifferentiated and differentiated BALB/c 3T3 preadipose cells: effects of 12-O-tetradecanoylphorbol-13-acetate and insulin. *J. cell. Physiol.* **110**: 63–71.

O'BRIEN, T. G. and DIAMOND, L. (1977) Ornithine decarboxylase induction and DNA synthesis in hamster embryo cell cultures treated with tumor-promoting phorbol diesters. *Cancer Res.* **37**: 3895–3900.

O'BRIEN, T. G. and KRZEMINSKI, K. (1983) Phorbol ester inhibits furosemide-sensitive potassium transport in BALB/c 3T3 preadipose cells. *Proc. natn. Acad. Sci. U.S.A.* **80**: 4334–4338.

O'BRIEN, T. G. and SALADIK, D. (1980) Inhibition of the adipose conversion of BALB/c 3T3 cells by 12-O-tetradecanoylphorbol-13-acetate: dependence on pH reduction via lactic acid production. *J. cell. Physiol.* **104**: 35–40.

O'BRIEN, T. G. and SALADIK, D. (1982) Regulation of hexose transport in BALB/c 3T3 preadipose cells: effects of glucose concentration and 12-O-tetradecanoylphorbol-13-acetate. *J. cell. Physiol.* **112**: 376–384.

O'BRIEN, T. G., LEWIS, M. A. and DIAMOND, L. (1979a) Ornithine decarboxylase activity and DNA synthesis after treatment of cells in culture with 12-O-tetradecanoylphorbol-13-acetate. *Cancer Res.* **39**: 4477–4480.

O'BRIEN, T. G., SALADIK, D. and DIAMOND, L. (1979b) The tumor promoter 12-O-tetradecanoylphorbol-13-acetate stimulates lactate production in BABL/c 3T3 preadipose cells. *Biochem. biophys. Res. Commun.* **88**: 103–110.

O'BRIEN, T. G., SALADIK, D. and DIAMOND, L. (1980) Regulation of polyamine biosynthesis in normal and transformed hamster cells in culture. *Biochim. biophys. Acta.* **632**: 270–283.

O'BRIEN, T. G., SALADIK, D. and DIAMOND, L. (1982) Effects of tumor-promoting phorbol diesters on neoplastic progression of Syrian hamster embryo cells. *Cancer Res.* **42**: 1233–1238.

O'BRIEN, T. G., SIMSIMAN, R. C. and BOUTWELL, R. K. (1975a) Induction of the polyamine-biosynthetic enzymes in mouse epidermis by tumor-promoting agents. *Cancer Res.* **35**: 1662–1670.

O'BRIEN, T. G., SIMSIMAN, R. C. and BOUTWELL, R. K. (1975b) Induction of the polyamine-biosynthetic enzymes in mouse epidermis and their specificity for tumor promotion. *Cancer Res.* **35**: 2426–2433.

O'BRIEN, T. G., SIMSIMAN, R. C. and BOUTWELL, R. K. (1976) The effect of colchicine on the induction of ornithine decarboxylase by 12-O-tetradecanoyl-phorbol-13-acetate. *Cancer Res.* **36**: 3766–3770.

OHUCHI, K. and LEVINE, L. (1978a) Stimulation of prostaglandin synthesis by tumor-promoting phorbol-12,13-diesters in canine kidney (MDCK) cells. Cycloheximide inhibits the stimulated prostaglandin synthesis, deacylation of lipids, and morphological changes. *J. biol. Chem.* **253**: 4783–4790.

OHUCHI, K. and LEVINE, L. (1978b) Tumor promoting phorbol diesters stimulate release of radioactivity from [^{3}H]arachidonic acid labeled—but not [^{14}C]linoleic acid labeled—cells. Indomethacin inhibits the stimulated release from [^{3}H]arachidonate labeled cells. *Prostagl. Med.* **1**: 421–431.

OJAKIAN, G. K. (1981) Tumor promoter-induced changes in the permeability of epithelial cell tight junctions. *Cell* **23**: 95–103.

Oncology Overview: Short-Term Test Systems for Potential Mutagens and Carcinogens: III. In Vitro Transformation (1982) U.S. Department of Health and Human Services.

OROSZ, C. G., ROOPERNIAN, D. C. and BACH, F. H. (1983) Phorbol myristate acetate and *in vitro* T lymphocyte function. 1. PMA may contaminate lymphokine preparations and can interfere with interleukin bioassays. *J. Immunol.* **130**: 1764–1769.

OSBORNE, R. and TASHJIAN, A. H., JR. (1981) Tumor-promoting phorbol esters affect production of prolactin and growth hormone by rat pituitary cells. *Endocrinology* **108**: 1164–1170.

OSBORNE, R. and TASHJIAN, A. H., JR. (1982) Modulation of peptide binding to specific receptors on rat pituitary cells by tumor-promoting phorbol esters: decreased binding of thyrotropin-releasing hormone and somatostatin as well as epidermal growth factor. *Cancer Res.* **42**: 4375–4381.

OZAWA, K., MIURA, Y., HASHIMOTO, Y., KIMURA, Y., URABE, A. and TAKAKU, F. (1983) Effects of 12-O-tetradecanoylphorbol-13-acetate on the proliferation and differentiation of normal and leukemic myeloid progenitor cells. *Cancer Res.* **43**: 2306–2310.

PACIFICI, M. and HOLTZER, H. (1977) Effects of a tumor-promoting agent on chondrogenesis. *Am. J. Anat.* **150**: 207–212.

PÄHLMAN, S., ODELSTAD, L., LARSSON, E., GROTTE, G. and NILSSON, K. (1981) Phenotypic changes of human neuroblastoma cells in culture induced by 12-O-tetradecanoylphorbol-13-acetate. *Int. J. Cancer* **28**: 583–589.

PAI, S. B., STEELE, V. E. and NETTESHEIM, P. (1982) Identification of early carcinogen-induced changes in nutritional and substrate requirements in cultured tracheal epithelial cells. *Carcinogenesis* **3**: 1201–1206.

PAI, S. B., STEELE, V. E. and NETTESHEIM, P. (1983) Neoplastic transformation of primary tracheal epithelial cell cultures. *Carcinogenesis* **4**: 369–374.

PARASKEVA, C., BUCKLE, B. G., SHEER, D., and WIGLEY, C. B. (1984) The isolation and characterization of colorectal epithelial cell lines at different stages in malignant transformation from familial polyposis coli patients. *Int. J. Cancer* **34**: 49–56.

PARKINSON, E. K. and EMMERSON, A. (1982) The effects of tumour promoters on the multiplication and morphology of cultured human epidermal keratinocytes. *Carcinogenesis* **3**: 525–531.

PARKINSON, E. K., GRABHAM, P. and EMMERSON, A. (1983) A subpopulation of cultured human keratinocytes which is resistant to the induction of terminal differentiation-related changes by phorbol, 12-myristate, 13-acetate: evidence for an increase in the resistant population following transformation. *Carcinogenesis* **4**: 857–861.

PARSA, I., BLOOMFIELD, R. D., FOYE, C. A. and SUTTON, A. L. (1984) Methylnitrosourea-induced carcinoma in organ-cultured fetal human pancreas. *Cancer Res.* **44**: 3530–3538.

PARSA, I., MARSH, W. H. and SUTTON, A. L. (1981) An *in vitro* model of human pancreas carcinogenesis: effects of nitroso compounds. *Cancer* **47**: 1543–1551.

PARSONS, D. F., MARKO, M., BRAUN, S. J. and WANSOR, K. J. (1983) 'Dark cells' in normal, hyperplastic, and promoter-treated mouse epidermis studied by conventional and high-voltage electron microscopy. *J. Invest. Dermatol.* **81**: 62–67.

PATARROYO, M., JONDAL, M., GORDON, J. and KLEIN, E. (1983) Characterization of the phorbol 12,13-dibutyrate (P(Bu)$_2$) induced binding between human lymphocytes. *Cell. Immunol.* **81**: 373–383.

PATARROYO, M., YOGEESWARAN, G., BIBERFELD, P., KLEIN, E. and KLEIN, G. (1982) Morphological changes, cell aggregation and cell membrane alterations caused by phorbol 12,13-dibutyrate in human blood lymphocytes. *Int. J. Cancer* **30**: 707–717.

PAUL, D. and HECKER, E. (1969) On the biochemical mechanism of tumorigenesis in mouse skin. II. Early effects on the biosynthesis of nucleic acids induced by initiating doses of DMBA and by promoting doses of phorbol-12,13-diester TPA. *Z. Krebsforsch.* **73**: 149–163.

PAYETTE, R., BIEHL, J., TOYAMA, Y., HOLTZER, S. and HOLTZER, H. (1980) Effects of 12-O-tetradecanoyl-phorbol-13-acetate on the differentiation of avian melanocytes. *Cancer Res.* **40**: 2465–2474.

PEARLSTEIN, K. T., STAIANO-COICO, L., MILLER, R. A., PELUS, L. M., KIRCH, M. E., STUTMAN, P. and PALLADINO,

M. A. (1983) Multiple lymphokine production by a phorbol ester-stimulated mouse thymoma: relationship to cell cycle events. *J. natn. Cancer. Inst.* **71**: 583–590.

PEGORARO, L., ABRAHM, J., COOPER, R. A., LEVIS, A., LANGE, B., MEO, P. and ROVERA, G. (1980) Differentiation of human leukemias in response to 12-O-tetradecanoylphorbol-13-acetate *in vitro. Blood* **55**: 859–862.

PERAINO, C., FRY, R. J. M. and STAFFELDT, E. (1971) Reduction and enhancement by phenobarbital of hepatocarcinogenesis induced in the rat by 2-acetylaminofluorene. *Cancer Res.* **31**: 1506–1512.

PERAINO, C., FRY, R. J. M., STAFFELDT, E. and KISIELESKI, W. E. (1973) Effects of varying the exposure to phenobarbital in its enhancement of 2-acetylaminofluorene-induced hepatic tumorigenesis in the rat. *Cancer Res.* **33**: 2701–2705.

PERAINO, C., RICHARDS, W. L. and STEVENS, F. J. (1983) Multistage hepatocarcinogenesis. In: *Mechanisms of Tumor Promotion, Volume 1. Tumor Promotion in Internal Organs*, pp. 1–53, SLAGA, T. J. (ed.). CRC Press, Inc., Boca Raton, FL.

PERRELLA, F. W., ASHENDEL, C. L. and BOUTWELL, R. K. (1982) Specific high-affinity binding of the phorbol ester tumor promoter 12-O-tetradecanoylphorbol-13-acetate to isolated nuclei and nuclear macromolecules in mouse epidermis. *Cancer Res.* **42**: 3496–3501.

PERRELLA, F. W., HELLMIG, B. D. and DIAMOND, L. (1986) Up-regulation of the phorbol ester receptor/protein kinase C in HL-60 variant cells. *Cancer Res.* **46**: 567–572.

PETERFREUND, R. A. and VALE, W. W. (1983) Phorbol diesters stimulate somatostatin secretion from cultured brain cells. *Endocrinology* **113**: 200–208.

PHAIRE-WASHINGTON, L., SILVERSTEIN, S. C. and WANG, E. (1980a) Phorbol myristate acetate stimulates microtubule and 10-nm filament extension and lysosome redistribution in mouse macrophages. *J. Cell Biol.* **86**: 641–655.

PHAIRE-WASHINGTON, L., WANG, E. and SILVERSTEIN, S. C. (1980b) Phorbol myristate acetate stimulates pinocytosis and membrane spreading in mouse peritoneal macrophages. *J. Cell Biol.* **86**: 634–640.

PIENTA, R. J. (1980) Transformation of Syrian hamster embryo cells by diverse chemicals and correlation with their reported carcinogenic and mutagenic activities. In: *Chemical Mutagens: Principles and Methods for Their Detection, Volume 6*, pp. 175–202, DE SERRES, F. J. and HOLLAENDER, A. (eds). Plenum Press, New York.

PIETROPAOLO, C., LASKIN, J. D. and WEINSTEIN, I. B. (1981) Effect of tumor promoters on *sarc* gene expression in normal and transformed chick embryo fibroblasts. *Cancer Res.* **41**: 1565–1571.

PITOT, H. C. and SIRICA, A. E. (1980) The stages of initiation and promotion in hepatocarcinogenesis. *Biochim. biophys. Acta.* **605**: 191–215.

PITOT, H. C., BARSNESS, L., GOLDSWORTHY, T. and KITAGAWA, T. (1978) Biochemical characterisation of stages of hepatocarcinogenesis after a single dose of diethylnitrosamine. *Nature* **271**: 456–458.

PITOT, H. C., GOLDSWORTHY, T., CAMPBELL, H. A. and POLAND, A. (1980) Quantitative evaluation of the promotion by 2,3,7,8-tetrachlorodibenzo-*p*-dioxin of hepatocarcinogenesis from diethylnitrosamine. *Cancer Res.* **40**: 3616–3620.

PLEDGER, W. J., STILES, C. D., ANTONIADES, H. N. and SCHER, C. D. (1977) Induction of DNA synthesis in BALB/c 3T3 cells by serum components: reevaluation of the commitment process. *Proc. natn. Acad. Sci. U.S.A.* **74**: 4481–4485.

POLAND, A., PALEN, D. and GLOVER, E. (1982) Tumour promotion by TCDD in skin of HRS/J hairless mice. *Nature* **300**: 271–273.

POLLIACK, A., LEIZEROWITZ, R., KORKESH, A., GURFEL, D., GAMLIEL, H. and GALILI, U. (1982) Exposure to phorbol diester (TPA) *in vitro* as an aid in the classification of blasts in human myelogenous and lymphoid leukemias: *in vitro* differentiation, growth patterns, and ultrastructural observations. *Am. J. Hematol.* **13**: 199–211.

POPESCU, N. C., AMSBAUGH, S. C. and DiPAOLO, J. A. (1980) Enhancement of *N*-methyl-*N*'-nitro-*N*-nitrosoguanidine transformation of Syrian hamster cells by a phorbol diester is independent of sister chromatid exchanges and chromosome aberrations. *Proc. natn. Acad. Sci. U.S.A.* **77**: 7282–7286.

POPESCU, N. C., AMSBAUGH, S. C., LARRAMENDY, M. L. and DiPAOLO, J. A. (1982) 12-O-Tetradecanoylphorbol-13-acetate and its relationship to SCE induction in Syrian and Chinese hamster cells. *Environ. Mutag.* **4**: 73–81.

POTTER, V. (1980) Initiation and promotion in cancer formation: the importance of studies on intercellular communication. *Yale J. Biol. Med.* **53**: 367–384.

POTTER, V. (1981) A new protocol and its rationale for the study of initiation and promotion of carcinogenesis in rat liver. *Carcinogenesis* **2**: 1375–1379.

RAGAN, D. L. and BOREIKO, C. J. (1981) Initiation of C3H/10T1/2 cell transformation by formaldehyde. *Cancer Lett.* **13**: 325–331.

RAICK, A. N. (1973a) Ultrastructural, histological, and biochemical alterations produced by 12-O-tetradecanoylphorbol-13-acetate on mouse epidermis and their relevance to skin tumor promotion. *Cancer Res.* **33**: 269–286.

RAICK, A. N. (1973b) Late ultrastructural changes induced by 12-O-tetradecanoyl-phorbol-13-acetate in mouse epidermis and their reversal. *Cancer Res.* **33**: 1096–1103.

RAICK, A. N. and BURDZY, K. (1973) Ultrastructural and biochemical changes induced in mouse epidermis by a hyperplastic agent, ethylphenylpropiolate. *Cancer Res.* **33**: 2221–2230.

RAINERI, R., SIMSIMAN, R. C. and BOUTWELL, R. K. (1973) Stimulation of the phosphorylation of mouse epidermal histones by tumor-promoting agents. *Cancer Res.* **33**: 134–139.

RAINERI, R., SIMSIMAN, R. C. and BOUTWELL, R. K. (1978) Stimulation of the synthesis of the H1 and H3 histone fractions of mouse epidermis by 12-O-tetradecanoylphorbol-13-acetate. *Cancer Lett.* **5**: 277–284.

RALPH, P. and KISHIMOTO, T. (1981) Tumor promoter phorbol myristic acetate stimulates immunoglobulin secretion correlated with growth cessation in human B lymphocyte cell lines. *J. clin. Invest.* **68**: 1093–1096.

RALPH, P., SAIKI, O., MAURER, D. H. and WELTE, K. (1983) IgM and IgG secretion in human B-cell lines regulated by B-cell-inducing factors (BIF) and phorbol ester. *Immunol. Lett.* **7**: 17–23.

RAY-CHAUDHURI, R., CURRENS, M. and IYPE, P. T. (1982) Enhancement of sister-chromatid exchanges by tumour promoters. *Br. J. Cancer* **45**: 769–777.

REPINE, J. E., WHITE, J. G., CLAWSON, C. C. and HOLMES, B. M. (1974) The influence of phorbol myristate acetate on oxygen consumption by polymorphonuclear leukocytes. *J. lab. clin. Med.* **83**: 911–920.

REZNIKOFF, C. A., BERTRAM, J. S., BRANKOW, D. W. and HEIDELBERGER, C. (1973a) Quantitative and qualitative studies of chemical transformation of cloned C3H mouse embryo cells sensitive to postconfluence inhibition of cell division. *Cancer Res.* **33**: 3239–3249.

REZNIKOFF, C. A., BRANKOW, D. W. and HEIDELBERGER, C. (1973b) Establishment and characterization of a cloned line of C3H mouse embryo cells sensitive to postconfluence inhibition of division. *Cancer Res.* **33**: 3231–3238.

REZNIKOFF, C. A., JOHNSON, M. D., NORBACK, D. H. and BRYAN, G. T. (1983) Growth and characterization of normal human urothelium *in vitro*. *In Vitro* **19**: 326–343.

RIFKIN, D. B., CROWE, R. M. and POLLACK, R. (1979) Tumor promoters induce changes in the chick embryo fibroblast cytoskeleton. *Cell* **18**: 361–368.

RIVEDAL, E. (1982) Reversal of the promotional effect of 12-O-tetradecanoylphorbol-13-acetate on morphological transformation of hamster embryo cells by glucocorticoids. *Cancer Lett.* **15**: 105–113.

RIVEDAL, E. and SANNER, T. (1980) Potentiating effect of cigarette smoke extract on morphological transformation of hamster embryo cells by benzo[a]pyrene. *Cancer Lett.* **10**: 193–198.

RIVEDAL, E. and SANNER, T. (1981) Metal salts as promoters of *in vitro* morphological transformation of hamster embryo cells initiated by benzo(a)pyrene. *Cancer Res.* **41**: 2950–2953.

RIVEDAL, E. and SANNER, T. (1982) Promotional effect of different phorbol esters on morphological transformation of hamster embryo cells. *Cancer Lett.* **17**: 1–8.

RIVEDAL, E. and SANNER, T. (1985) Retinoids have different effects on morphological transformation and anchorage independent growth of Syrian hamster embryo cells. *Carcinogenesis* **6**: 955–958.

RIVEDAL, E., SANNER, T., ENOMOTO, T. and YAMASAKI, H. (1985) Inhibition of intercellular communication and enhancement of morphological transformation of Syrian hamster embryo cells by TPA. Use of TPA-sensitive and TPA-resistant cell lines. *Carcinogenesis* **6**: 899–902.

ROBINSON, J. and VANE, J. R. (1974) *Prostaglandin Synthetase Inhibitors—Their Effects on Physiological Functions and Pathological States.* Raven Press, New York.

ROHRSCHNEIDER, L. R. and BOUTWELL, R. K. (1973) Phorbol esters, fatty acids and tumour promotion. *Nature New Biol.* **243**: 212–213.

ROHRSCHNEIDER, L. R., O'BRIEN, D. H. and BOUTWELL, R. K. (1972) The stimulation of phospholipid metabolism in mouse skin following phorbol ester treatment. *Biochim. biophys. Acta* **280**: 57–70.

ROSENSTREICH, D. L. and MIZEL, S. B. (1979) Signal requirements for T lymphocyte activation: replacement of macrophage function with phorbol myristic acetate. *J. Immunol.* **123**: 1749–1753.

ROUS, P. and KIDD, J. G. (1941) Conditional neoplasms and subthreshold neoplastic states. A study of the tar tumors of rabbits. *J. exp. Med.* **73**: 365–389.

ROVERA, G., O'BRIEN, T. G. and DIAMOND, L. (1977) Tumor promoters inhibit spontaneous differentiation of Friend erythroleukemia cells in culture. *Proc. natn. Acad. Sci. U.S.A.* **74**: 2894–2898.

ROVERA, G., O'BRIEN, T. G. and DIAMOND, L. (1979a) Induction of differentiation in human promyelocytic leukemia cells by tumor promoters. *Science* **204**: 868–870.

ROVERA, G., SANTOLI, D. and DAMSKY, C. (1979b) Human promyelocytic leukemia cells in culture differentiate into macrophage-like cells when treated with a phorbol diester. *Proc. natn. Acad. Sci. U.S.A.* **76**: 2779–2783.

ROZENGURT, E., RODRIGUEZ-PENA, M. and SMITH, K. A. (1983) Phorbol esters, phospholipase C. and growth factors rapidly stimulate the phosphorylation of a M_r 80,000 protein in intact quiescent 3T3 cells. *Proc. natn. Acad. Sci. U.S.A.* **80**: 7244–7248.

SANDO, J. J. and YOUNG, M. C. (1983) Identification of highly-affinity phorbol ester receptor in cytosol of EL4 thymoma cells: requirement for calcium, magnesium, and phospholipids. *Proc. natn. Acad. Sci. U.S.A.* **80**: 2642–2646.

SANDO, J. J., HILFIKER, M. L., PIACENTINI, M. J. and LAUFER, T. M. (1982) Identification of phorbol ester receptors in T-cell growth factor-producing and -nonproducing EL4 mouse thymoma cells. *Cancer Res.* **42**: 1676–1680.

SASAKI, N., MATSUZAKI, M., OKABE, T., IMAI, Y., KANEKO, Y. and MATSUZAKI, F. (1983) Stimulation of human chorionic gonadotropin and progesterone secretion by tumor promoters, phorbol ester and teleocidin B, in cultured choriocarcinoma cells. *Experientia* **39**: 330–331.

SCHWARTZ, J. L., BANDA, M. J. and WOLFF, S. (1982) 12-O-Tetradecanoylphorbol-13-acetate (TPA) induces sister-chromatid exchanges and delays in cell progression in Chinese hamster ovary and human cell lines. *Mutat. Res.* **92**: 393–409.

SCHWARZ, J. A., VIAJE, A., SLAGA, T. J., YUSPA, S. H., HENNINGS, H. and LICHTI, U. (1977) Fluocinolone acetonide: a potent inhibitor of mouse skin tumor promotion and epidermal DNA synthesis. *Chem.-Biol. Interactions* **17**: 331–347.

SCHWEIZER, J. and WINTER, H. (1982) Changes in regional keratin polypeptide patterns during phorbol ester-mediated reversible and permanently sustained hyperplasia of mouse epidermis. *Cancer Res.* **42**: 1517–1529.

SCOTT, R. E. and WILLE, J. J. JR. (1984) Mechanisms for the initiation and promotion of carcinogenesis: a review and a new concept. *Mayo Clin. Proc.* **59**: 107–117.

SCRIBNER, J. D. and SCRIBNER, N. K. (1982) Is the initiation-promotion regimen in mouse skin relevant to complete carcinogenesis? In: *Carcinogenesis*, Vol. 7, pp. 13–18, HECKER, E., KUNZ, W., FUSENIG, N. E., MARKS, F. and THEILMANN, H. W. (eds). Raven Press, New York.

SCRIBNER, J. D. and SLAGA, T. J. (1973) Multiple effects of dexamethasone on protein synthesis and hyperplasia caused by a tumor promoter. *Cancer Res.* **33**: 542–546.

SCRIBNER, J. D. and SÜSS, R. (1978) Tumor initiation and promotion. In: *International Review of Experimental Pathology*, pp. 137–198, RICHTER, G. W. and EPSTEIN, M. A. (eds), Vol. 18. Academic Press, New York.

SCRIBNER, J. D., SCRIBNER, N. K., MCKNIGHT, B. and MOTTET, N. K. (1983) Evidence for a new model of tumor progression from carcinogenesis and tumor promotion studies with 7-bromomethylbenz[a]anthracene. *Cancer Res* **43**: 2034–2041.

SCRIBNER, N. K. and SCRIBNER, J. D. (1980) Separation of initiating and promoting effects of the skin carcinogen 7-bromomethylbenz[a]anthracene. *Carcinogenesis* **1**: 97–100.

SELIKOFF, I. J. and HAMMOND, E. C. (1975) Multiple risk factors in environmental cancer. In: *Persons at High Risk of Cancer. An Approach to Cancer Etiology and Control*, pp. 467–483. FRAUMENI, J. F., JR. (ed.). Academic Press, New York.

SHARKEY, N. A., LEACH, K. L. and BLUMBERG, P. M. (1984) Competitive inhibition by diacylglycerol of specific phorbol ester binding, *Proc. natn. Acad. Sci. U.S.A.* **81**: 607–610.

SHIMOMURA, K., MULLINIX, M. G., KAKUNAGA, T., FUJIKI, H. and SUGIMURA, T. (1983) Bromine residue at hydrophilic region influences biological activity of aplysiatoxin, a tumor promoter. *Science* **222**: 1242–1244.

SHOYAB, M. and TODARO, G. J. (1980) Specific high affinity cell membrane receptors for biologically active phorbol and ingenol esters. *Nature* **288**: 451–455.

SHOYAB, M., DE LARCO, J. E. and TODARO, G. J. (1979) Biologically active phorbol esters specifically alter affinity of epidermal growth factor membrane receptors. *Nature* **279**: 387–391.

SHUBIK, P. (1984) Progression and promotion. *J. natn. Cancer Inst.* **73**: 1005–1011.

SIEGFRIED, J. M. and KAUFMAN, D. G. (1983) Enhancement by TPA of phenotypes associated with transformation in carcinogen-treated human cells: evidence for a selective mechanism. *Int. J. Cancer.* **32**: 423–429.

SIEGFRIED, J. M., NELSON, K. G., MARTIN, J. L. and KAUFMAN, D. G. (1984) Promotional effect of diethylstilbestrol on human endometrial stromal cells pretreated with a direct-acting carcinogen. *Carcinogenesis* **5**: 641–646.

SINA, J. F. BRADLEY, M. O., DIAMOND, L. and O'BRIEN, T. G. (1983) Ornithine decarboxylase activity and DNA synthesis in primary and transformed hamster epidermal cells exposed to tumor promoter. *Cancer Res.* **43**: 4108–4113.

SISSKIN, E. E. and BARRETT, J. C. (1981) Inhibition of terminal differentiation of hamster epidermal cells in culture by the phorbol ester 12-O-tetradecanoylphorbol-13-acetate. *Cancer Res.* **41**: 593–603.

SIVAK, A. (1977) Induction of cell division in BALB/c-3T3 cells by phorbol myristate acetate or bovine serum: effects of inhibitors of cyclic AMP phosphodiesterase and NA^+–K^+-ATPase. *In Vitro* **13**: 337–343.

SIVAK, A. (1982) An evaluation of assay procedures for detection of tumor promoters. *Mutat. Res.* **98**: 377–387.

SIVAK, A. and TU, A. S. (1980) Cell culture tumor promotion experiments with saccharin, phorbol myristate acetate and several common food materials. *Cancer Lett.* **10**: 27–32.

SIVAK, A. and TU, A. S. (1982) Transformation of somatic cells in culture. In: *Mutagenicity: New Horizons in Genetic Toxicology*, pp. 143–169, HEDDLE, J. A. (ed.). Academic Press, New York.

SIVAK, A. and VAN DUUREN, B. L. (1967) Phenotypic expression of transformation: induction in cell culture by a phorbol ester. *Science* **157**: 1443–1444.

SIVAK, A., CHAREST, M. C., RUDENKO, L., SILVEIRA, D. M., SIMONS, I. and WOOD, A. M. (1980) BALB/c-3T3 cells as target cells for chemically induced neoplastic transformation. In: *Adv. Mod. Environ. Toxicol., Vol. 1, Mammalian Cell Transformation by Chemical Carcinogens*, pp. 133–180, MISHRA, N., DUNKEL, V. and MEHLMAN, M. (eds). Senate Press, Princeton Junction, NJ.

SLAGA, T. J. (ed.) (1980a) *Carcinogenesis—A Comprehensive Survey, Vol. , Modifiers of Chemical Carcinogenesis: An Approach to the Biochemical Mechanism and Cancer Prevention.* Raven Press, New York.

SLAGA, T. J. (1980b) Antiinflammatory steroids: potent inhibitors of tumor promotion. In: *Carcinogenesis—A Comprehensive Survey, Vol. 5, Modifiers of Chemical Carcinogenesis*, pp. 111–126, SLAGA, T. J. (ed.). Raven Press, New York.

SLAGA, T. J. (1983a) Multistage skin carcinogenesis and specificity of inhibitors. In: *Modulation and Mediation of Cancer by Vitamins*, pp. 10–23, MEYSKENS, F. L. and PRASAD, K. N. (eds). S. Karger, Basel.

SLAGA, T. J. (1983b) *Mechanisms of Tumor Promotion, Volume 1: Tumor Promotion in Internal Organs.* CRC Press, Inc., Boca Raton, FL.

SLAGA, T. J. (1984a) *Mechanisms of Tumor Promotion, Volume 2: Tumor Promotion in Internal Organs.* CRC Press, Inc., Boca Raton, FL.

SLAGA, T. J. (1984b) *Mechanisms of Tumor Promotion, Volume 3: Tumor Promotion and Carcinogenesis In vitro.* CRC Press, Inc., Boca Raton, FL.

SLAGA, T. J. (1984c) *Mechanisms of Tumor Promotion, Volume 4: Cellular Responses to Tumor Promoters.* CRC Press, Inc., Boca Raton, FL.

SLAGA, T. J., FISCHER, S. M., NELSON, K. and GLEASON, G. L. (1980a) Studies on the mechanism of skin tumor promotion: evidence for several stages in promotion. *Proc. natn. Acad. Sci. U.S.A.* **77**: 3659–3663.

SLAGA, T. J., FISCHER, S. M., WEEKS, C. E., NELSON, K., MAMRACK, M. and KLEIN-SZANTO, A. J. P. (1982) Specificity and mechanism(s) of promoter inhibitors and multistage promotion. In: *Carcinogenesis—A Comprehensive Survey, Volume 7: Cocarcinogenesis and Biological Effects of Tumor Promoters*, pp. 19–34, HECKER, E., KUNZ, W., FUSENIG, N. E., MARKS, F. and THEILMANN, H. W. (eds). Raven Press, New York.

SLAGA, T. J., KLEIN-SZANTO, A. J. P., FISCHER, S. M., WEEKS, C. E., NELSON, K. and MAJOR, S. (1980b) Studies on mechanism of action of anti-tumor-promoting agents: their specificity in two-stage promotion. *Proc. natn. Acad. Sci. U.S.A.* **77**: 2251–2254.

SLAGA, T. J., KLEIN-SZANTO, A. J. P., TRIPLETT, L. L., YOTTI, L. P. and TROSKO, J. E. (1981) Skin tumor-promoting activity of benzoyl peroxide, a widely used free radical-generating compound. *Science* **213**: 1023–1025.

SLAGA, T. J., SCRIBNER, J. D., THOMPSON, S. and VIAJE, A. (1976) Epidermal cell proliferation and promoting ability of phorbol esters. *J. natn. Cancer Inst.* **57**: 1145–1149.

SLAGA, T. J., SIVAK, A. and BOUTWELL, R. K. (eds) (1978) *Carcinogenesis—A Comprehensive Survey. Volume 2; Mechanisms of Tumor Promotion and Cocarcinogenesis.* Raven Press, New York.

SOLANKI, V. and SLAGA, T. J. (1981) Specific binding of phorbol ester tumor promoters to intact primary epidermal cells from Sencar mice. *Proc. natn. Acad. Sci. U.S.A.* **78**: 2549–2553.

SOLANKI, V., RANA, R. S. and SLAGA, T. J. (1981a) Diminution of mouse epidermal superoxide dismutase and catalase activities by tumor promoters. *Carcinogenesis* **2**: 1141–1146.

SOLANKI, V., SLAGA, T. J., CALLAHAM, M. and HUBERMAN, E. (1981b) Down regulation of specific binding of [20-^{3}H]phorbol 12,13-dibutyrate and phorbol ester-induced differentiation of human promyelocytic leukemia cells. *Proc. natn. Acad. Sci. U.S.A.* **78**: 1722–1725.

SOLT, D. and FARBER, E. (1976) New principle for the analysis of chemical carcinogenesis. *Nature* **263**: 701–703.

SOLT, D. B., MEDLINE, A. and FARBER, E. (1977) Rapid emergence of carcinogen-induced hyperplastic lesions in a new model for the sequential analysis of liver carcinogenesis. *Am. J. Pathol.* **88**: 595–618.

SPINELLI, W., SONNENFELD, K. H. and ISHII, D. N. (1982) Effects of phorbol ester tumor promoters and nerve growth factor on neurite outgrowth in cultured human neuroblastoma cells. *Cancer Res.* **42**: 5067–5073.

SRINIVAS, L. and COLBURN, N. H. (1982) Ganglioside changes induced by tumor promoters in promotable JB6 mouse epidermal cells: antagonism by an antipromoter. *J. natn. Cancer Inst.* **68**: 469–473.

SRINIVAS, L., GINDHART, T. D. and COLBURN, N. H. (1982) Tumor-promoter-resistant cells lack trisialoganglioside response. *Proc. natn. Acad. Sci. U.S.A.* **79**: 4988–4991.

STAMPFER, M. R. and BARTLEY, J. C. (1985) Induction of transformation and continuous cell lines from normal human mammary epithelial cells after exposure to benzo[a]pyrene. *Proc. natn. Acad. Sci. USA* **82**: 2394–2398.

STANLEY, M. A., CROWCROFT, N. S., QUIGLEY, J. P. and PARKINSON, E. K. (1985) Responses of human cervical keratinocytes *in vitro* to tumour promoters and diethylstilboestrol. *Carcinogenesis* **6**: 1011–1015.

STEELE, V. E. (1985) Multistage-promotion and carcinogenesis studies in rat tracheal epithelial cells in culture. In: *Carcinogenesis—A Comprehensive Survey, Vol. 8, Cancer of the Respiratory Tract: Predisposing Factors*, pp. 191–205, MASS, M. J., KAUFMAN, D. G., SIEGFRIED, J. M., STEELE, V. E., and NESNOW, S. (eds.), Raven Press, New York.

STEELE, V. E. and NETTESHEIM, P. (1983) Tumor promotion in respiratory tract carcinogenesis. In: *Mechanisms of Tumor Promotion, Volume 1: Tumor Promotion in Internal Organs*, pp. 92–105, SLAGA, T. J. (ed.). CRC Press, Inc., Boca Raton, FL.

STEELE, V. E., MARCHOK, A. C. and NETTESHEIM, P. (1978) Establishment of epithelial cell lines following exposure of cultured tracheal epithelium to 12-*O*-tetradecanoyl-phorbol-13-acetate. *Cancer Res.* **38**: 3563–3565.

STEELE, V. E., MARCHOK, A. C. and NETTESHEIM, P. (1980) Enhancement of carcinogenesis in cultured respiratory tract epithelium by 12-*O*-tetradecanoylphorbol-13-acetate. *Int. J. Cancer* **26**: 343–348.

STEELE, V. E., TOPPING, D. C. and PAI, S. B. (1983) Tumor promotion studies in rat tracheal epithelium. *Environ. Health Perspect.* **50**: 259–266.

STENBACK, F., GARCIA, H. and SHUBIK, P. (1974) Present status of the concept of promoting action of cocarcinogenesis in skin. In: *The Physiopathology of Cancer, Vol. 1. Biology and Biochemistry*, pp. 155–225. Karger, Basel.

STUART, R. K., HAMILTON, J. A., SENSENBRENNER, L. L. and MOORE, M. A. S. (1981) Regulation of myelopoiesis *in vitro*: partial replacement of colony-stimulating factors by tumor-promoting phorbol esters. *Blood* **57**: 1032–1042.

STUART, R. K., SENSENBRENNER, L. L., SHADDUCK, R. K., WAHEED, A. and CARAMATTI, C. (1983) Phorbol ester-stimulated murine myelopoiesis: role of colony-stimulating factors. *J. cell. Physiol.* **117**: 30–38.

SUGANUMA, M., FUJIKI, H. and SUGIMURA, T. (1982) Existence of an optimal dose of dihydroteleocidin B for skin tumor promotion. *Gann* **73**: 531–533.

SUGIMURA, T. (1982) Potent tumor promoters other than phorbol ester and their significance. *Gann* **73**: 499–507.

SÜSS, R., KINZEL, V. and KREIBICH, G. (1971) Cocarcinogenic croton oil factor A_1 stimulates lipid synthesis in cell cultures. *Experientia* **27**: 46–47.

SÜSS, R., KREIBICH, G. and KINZEL, V. (1972) Phorbol esters as a tool in cell research? *Eur. J. Cancer* **8**: 299–304.

SUTHERLAND, B. M. and BENNETT, P. V. (1984) Transformation of human cells by DNA transfection. *Cancer Res.* **44**: 2769–2772.

SUTHERLAND, B. M., BENNETT, P. V., FREEMAN, A. G., MOORE, S. P. and STRICKLAND, P. T. (1985) Transformation of human cells by DNAs ineffective in transformation of NIH 3T3 cells. *Proc. natn. Acad. Sci. USA* **82**: 2399–2403.

TAKEDA, K., MINOWADA, J. and BLOCH, A. (1982) Kinetics of appearance of differentiation-associated characteristics in ML-1, a line of human myeloblastic leukemia cells, after treatment with 12-*O*-tetradecanoylphorbol-13-acetate, dimethyl sulfoxide, or 1-β-D-arabinofuranosylcytosine. *Cancer Res.* **42**: 5152–5158.

TAKIGAWA, M., VERMA, A. K., SIMSIMAN, R. C. and BOUTWELL, R. K. (1982) Polyamine biosynthesis and skin tumor promotion: inhibition of 12-*O*-tetradecanoylphorbol-13-acetate-promoted mouse skin tumor formation by the irreversible inhibitor of ornithine decarboxylase α-difluoromethylornithine. *Biochem. biophys. Res. Commun.* **105**: 969–976.

TANAKA, H., ABE, E., MIYAURA, C., SHIINA, Y. and SUDA. T. (1983) 1α,25-Dihydroxyvitamin D_3 induces differentiation of human promyelocytic leukemia cells (HL-60) into monocyte-macrophages, but not into granulocytes. *Biochem. biophys. Res. Commun.* **117**: 86–92.

TAYLOR, S. M. and JONES, P. A. (1979) Multiple new phenotypes induced in 10T1/2 and 3T3 cells treated with 5-azacytidine. *Cell* **17**: 771–779.

TERZAGHI, M. and NETTESHEIM, P. (1979) Dynamics of neoplastic development in carcinogen-exposed tracheal mucosa. *Cancer Res.* **39**: 4003–4010.

TERZAGHI, M., KLEIN-SZANTO, A. and NETTESHEIM, P. (1983) Effect of the promoter 12-*O*-tetradecanoylphorbol-13-acetate on the evolution of carcinogen-altered cell populations in tracheas initiated with 7,12-dimethylbenz(a)anthracene. *Cancer Res.* **43**: 1461–1466.

THOMASSEN, D., BARRETT, J. C., BEEMAN, D. K. and NETTESHEIM, P. (1985a) Changes in stem cell populations of rat tracheal epithelial cell cultures at an early stage in neoplastic progression. *Cancer Res.* **45**: 3322–3331.

THOMASSEN, D. G., GILMER, T. M., ANNAB, L. A. and BARRETT, J. C. (1985b) Evidence for multiple steps in neoplastic transformation of normal and preneoplastic Syrian hamster embryo cells following transfection with Harvey murine sarcoma virus oncogene (*v-Ha-ras*). *Cancer Res.* **45**: 726–732.

THOMASSEN, D. G., GRAY, T. E., MASS, M. J. and BARRETT, J. C. (1983) High frequency of carcinogen-induced early, preneoplastic changes in rat tracheal epithelial cells in culture. *Cancer Res.* **43**: 5956–5963.

THOMASSEN, D. G., NETTESHEIM, P., GRAY, T. E. and BARRETT, J. C. (1985c) Quantitation of the rate of spontaneous generation and carcinogen-induced frequency of anchorage-independent variants of rat tracheal epithelial cells in culture. *Cancer Res.* **45**: 1516–1524.

THOMOPOULOS, P., TESTA, U., GOURDIN, M.-F., HERVY, C., TITEUX, M. and VAINCHENKER, W. (1982) Inhibition of insulin receptor binding by phorbol esters. *Eur. J. Biochem.* **129**: 389–393.

THOMPSON, L. H., BAKER, R. M., CARRANO, A. V. and BROOKMAN, K. W. (1980) Failure of the phorbol ester 12-O-tetradecanoylphorbol-13-acetate to enhance sister chromatid exchange, mitotic segregation, or expression of mutations in Chinese hamster cells. *Cancer Res.* **40**: 3245–3251.

TOFTGARD, R., ROOP, D. R. and YUSPA, S. H. (1985) Proto-oncogene expression during two-stage carcinogenesis in mouse skin. *Carcinogenesis* **6**: 655–657.

TOMEI, L. D., CHENEY, J. C. and WENNER, C. E. (1981) The effect of phorbol esters on the proliferation of C3H-10T1/2 mouse fibroblasts: consideration of both stimulatory and inhibitory effects. *J. cell. Physiol.* **107**: 385–389.

TOMIDA, M., YAMAMOTO, Y. and HOZUMI, M. (1982) Stimulation by interferon of induction of differentiation of human promyelocytic leukemia cells. *Biochem. biophys. Res. Commun.* **104**: 30–37.

TOPPING, D. C. and NETTESHEIM, P. (1980) Promotion-like enhancement of tracheal carcinogenesis in rats by 12-O-tetradecanoylphorbol-13-acetate. *Cancer Res.* **40**: 4352–4355.

TOYAMA, Y., WEST, C. M. and HOLTZER, H. (1979) Differential response of myofibrils and 10-nm filaments to a cocarcinogen. *Am. J. Anat.* **156**: 131–137.

TRAN, P. L., CASTAGNA, M., SALA, M., VASSENT, G., HOROWITZ, A. D., SCHACHTER, D. and WEINSTEIN, I. B. (1983) Differential effect of tumor promoters on phorbol-ester-receptor binding and membrane fluorescence anisotropy in C3H 10T1/2 cells. *Eur. J. Biochem.* **130**: 155–160.

TROLL, W. and WIESNER, R. (1985) The role of oxygen radicals as a possible mechanism of tumor promotion. *Ann. Rev. Pharmacol. Toxicol.* **25**: 509–528.

TROLL, W., KLASSEN, A. and JANOFF, A. (1970) Tumorigenesis in mouse skin: inhibition by synthetic inhibitors of proteases. *Science* **169**: 1211–1213.

TROSKO, J. E., CHANG, C-C. and MEDCALF, A. (1983) Mechanisms of tumor promotion: potential role of intercellular communication. *Cancer Invest.* **1**: 511–526.

TROSKO, J. E., JONE, C. and CHANG, C-C. (1984) Detection of nonmutagenic and noncytotoxic hazardous chemicals. *In Vitro* **20**: 262.

TROSKO, J. E., YOTTI, L. P., WARREN, S. T., TSUSHIMOTO, G. and CHANG, C-C. (1982) Inhibition of cell–cell communication by tumor promoters. In: *Carcinogenesis—A Comprehensive Survey. Volume 7: Co-carcinogenesis and Biological Effects of Tumor Promoters*, pp. 565–585, HECKER, E., KUNZ, W., FUSENIG, N. E., MARKS, F. and THIELMANN, H. W. (eds). Raven Press, New York.

ULLRICH, S. J. and HAWKES, S. P. (1983) The effect of the tumor promoter, phorbol myristate acetate (PMA), on hyaluronic acid (HA) synthesis by chicken embryo fibroblasts. *Exp. Cell Res.* **148**: 377–386.

UMEDA, M. and ENAKA, K. (1981) Some aspects of *in vitro* carcinogenesis using Syrian hamster cells. *Gann Monogr. Cancer Res.* **7**: 183–194.

UMEDA, M., NODA, K. and ONO, T. (1980) Inhibition of metabolic cooperation in Chinese hamster cells by various chemicals including tumor promoters. *Gann* **71**: 614–620.

UMEZAWA, K., WEINSTEIN, I. B., HOROWITZ, A., FUJIKI, H., MATSUSHIMA, T. and SUGIMURA, T. (1981) Similarity of teleocidin B and phorbol ester tumour promoters in effects on membrane receptors. *Nature* **290**: 411–412.

VALONE, F. H., OBRIST, R., TARLIN, N. and BAST, R. C., JR. (1983) Enhanced arachidonic acid lipoxygenation by K562 cells stimulated with 12-O-tetradecanoylphorbol-13-acetate. *Cancer Res.* **43**: 197–201.

VANDENBARK, G. R. and NIEDEL, J. E. (1984) Phorbol diesters and cellular differentiation. *J. natn. Cancer Inst.* **73**: 1013–1019.

VANDENBARK, G. R., KUHN, L. J. and NIEDEL, J. E. (1984) Possible mechanism of phorbol diester-induced maturation of human promyelocytic leukemia cells. Activation of protein kinase C. *J. clin. Invest.* **73**: 448–457.

VANDERLAAN, M., STEELE, V. and NETTESHEIM, P. (1983) Increased DNA content as an early marker of transformation in carcinogen-exposed rat tracheal cell cultures. *Carcinogenesis* **4**: 721–727.

VAN DUUREN, B. L. (1969) Tumor-promoting agents in two-stage carcinogenesis. *Progr. exp. Tumor Res.* **11**: 31–68.

VAN DUUREN, B. L. and ORRIS, L. (1965) The tumor-enhancing principles of *Croton tiglium* L. *Cancer Res.* **25**: 1871–1875.

VAN DUUREN, B. L., TSENG, S. S., SEGAL, A., SMITH, A. C., MELCHIONNE, S. and SEIDMAN, I. (1979) Effects of structural changes on the tumor-promoting activity of phorbol myristate acetate on mouse skin. *Cancer Res.* **39**: 2644–2646.

VAN DUUREN, B. L., WITZ, G. and GOLDSCHMIDT, B. M. (1978) Structure–activity relationships of tumor promoters and cocarcinogens and interaction of phorbol myristate acetate and related esters with plasma membranes. In: *Carcinogenesis, Vol. 2: Mechanisms of Tumor Promotion and Cocarcinogenesis*, pp. 491–507, SLAGA, T. J., SIVAK, A. and BOUTWELL, R. K. (eds). Raven Press, New York.

VARANI, J. and FANTONE, J. C. (1982) Phorbol myristate acetate-induced adherence of Walker 256 carcinosarcoma cells. *Cancer Res.* **42**: 190–197.

VARSHAVSKY, A. (1981a) On the possibility of metabolic control of replicon 'misfiring'. Relationship to emergence of malignant phenotypes in mammalian cell lineages. *Proc. natn. Acad. Sci. U.S.A.* **78**: 3673–3677.

VARSHAVSKY, A. (1981b) Phorbol ester dramatically increases incidence of methotrexate-resistant mouse cells: possible mechanisms and relevance to tumor promotion. *Cell* **25**: 561–572.

VASSALLI, J.-D., HAMILTON, J. and REICH, E. (1977) Macrophage plasminogen activator: induction by concanavalin A and phorbol myristate acetate. *Cell* **11**: 695–705.

VERMA, A. K. and MURRAY, A. W. (1974) The effect of benzo(a)pyrene on the basal and isoproterenol-stimulated levels of cyclic adenosine 3',5'-monophosphate in mouse epidermis. *Cancer Res.* **43**: 3408–3413.

VERMA, A. K., ASHENDEL, C. L. and BOUTWELL, R. K. (1980) Inhibition by prostaglandin synthesis inhibitors of the induction of epidermal ornithine decarboxylase activity, the accumulation of prostaglandins, and tumor promotion caused by 12-O-tetradecanoylphorbol-13-acetate. *Cancer Res.* **40**: 308–315.

VERMA, A. K., CONRAD, E. A. and BOUTWELL, R. K. (1982) Differential effects of retinoic acid and 7,8-benzoflavone on the induction of mouse skin tumors by the complete carcinogenesis process and by the initiation-promotion regimen. *Cancer Res.* **42**: 3519–3525.

VERMA, A. K., GARCIA, C. T., ASHENDEL, C. L. and BOUTWELL, R. K. (1983) Inhibition of 7-bromoniethylbenz-[a]anthracene-promoted mouse skin tumor formation by retinoic acid and dexamethasone. *Cancer Res.* **43**: 3045–3049.

VERMA, A. K., RICE, H. M. and BOUTWELL, R. K. (1977) Prostaglandins and skin tumor promotion: inhibition of tumor promoter-induced ornithine decarboxylase activity in epidermis by inhibitors of prostaglandin synthesis. *Biochem. biophys. Res. Commun.* **79**: 1160–1166.

VERMA, A. K., RICE, H. M., SHAPAS, B. G. and BOUTWELL, R. K. (1978) Inhibition of 12-O-tetradecanoylphorbol-13-acetate-induced ornithine decarboxylase activity in mouse epidermis by vitamin A analogs (retinoids). *Cancer Res.* **38**: 793–801.

VERMA, A. K., SHAPAS, B. G., RICE, H. M. and BOUTWELL, R. K. (1979) Correlation of the inhibition by retinoids of tumor promoter-induced mouse epidermal ornithine decarboxylase activity and of skin tumor promotion. *Cancer Res.* **39**: 419–425.

VIAJE, A., SLAGA, T. J., WIGLER, M. and WEINSTEIN, I. B. (1977) Effects of antiinflammatory agents on mouse skin tumor promotion, epidermal DNA synthesis, phorbol ester-induced cellular proliferation, and production of plasminogen activator. *Cancer Res.* **37**: 1530–1536.

VYTH-DREESE, F. A. and DE VRIES, J. E. (1983) Enhanced expression of human T-cell leukemia/lymphoma virus in neoplastic T cells induced to proliferate by phorbol ester and interleukin-2. *Int. J. Cancer* **32**: 53–59.

VYTH-DREESE, F. A., VAN DER REIJDEN, H. J. and DE VRIES, J. E. (1982) Phorbol-ester-mediated induction and augmentation of mitogenesis and interleukin-2 production in human T-cell lymphoproliferative disease. *Blood* **60**: 1437–1446.

WEEKS, C. E., HERRMANN, A. L., NELSON, F. R. and SLAGA, T. J. (1982) α-Difluoromethylornithine, an irreversible inhibitor of ornithine decarboxylase, inhibits tumor promoter-induced polyamine accumulation and carcinogenesis in mouse skin. *Proc. natn. Acad. Sci. U.S.A.* **79**: 6028–6032.

WEINBERG, J. B., MISUKONIS, M. A. and GOODWIN, B. J. (1984) Human leukemia cell lines with comparable receptor binding characteristics but different phenotypic responses to phorbol diesters. *Cancer Res.* **44**: 976–980.

WEINSTEIN, I. B. (1983) Tumour promoters. Protein kinase, phospholipid and control of growth, *Nature* **302**: 750.

WEINSTEIN, I. B., MUFSON, R. A., LEE, L. S., FISHER, P. B., LASKIN, J., HOROWITZ, A. D. and IVANOVIC, V. (1980) Membrane and other biochemical effects of the phorbol esters and their relevance to tumor promotion. In: *Carcinogenesis: Fundamental Mechanisms and Environmental Effects*, pp. 543–563, PULLMAN, B., Ts'o, P. O. P. and GELBOIN, H. (eds). D. Reidel Publishing Company, Dordrecht, Holland.

WEINSTEIN, I. B., WIGLER, M. and PIETROPAOLO, C. (1977) The action of tumor-promoting agents in cell culture. In: *The Origins of Human Cancer*, pp. 751–772, HIATT, H. H., WATSON, J. D. and WINSTEIN, J. A. (eds). Cold Spring Harbor Conferences on Cell Proliferation, Vol. 4, Cold Spring Harbor, New York.

WEITBERG, A. B., WEITZMAN, S. A., DESTREMPES, M., LATT, S. A. and STOSSEL, T. P. (1983) Stimulated human phagocytes produce cytogenetic changes in cultured mammalian cells. *New Engl. J. Med.* **308**: 26–30.

WERTH, D. K. and PASTAN, I. (1984) Vinculin phosphorylation in response to calcium and phorbol esters in intact cells. *J. Biol. Chem.* **259**: 5264–5270.

WERTZ, P. W. and MUELLER, G. C. (1978) Rapid stimulation of phospholipid metabolism in bovine lymphocytes by tumor-promoting phorbol esters. *Cancer Res.* **38**: 2900–2904.

WEST, C. M. and HOLTZER, H. (1982) Protein synthesis and degradation in cultured muscle is altered by a phorbol diester tumor promoter. *Arch. Biochem. Biophys.* **219**: 335–350.

WHITE, J. G., RAO, G. H. R. and ESTENSEN, R. D. (1974) Investigation of the release reaction in platelets exposed to phorbol myristate acetate. *Am. J. Pathol.* **75**: 301–314.

WHITELEY, B., DEUEL, T. and GLASER, L. (1985) Modulation of the activity of the platelet-derived growth factor receptor by phorbol myristate acetate. *Biochem. Biophys. Res. Commun.* **129**: 854–861.

WHITFIELD, J. F., MACMANUS, J. P. and GILLAN, D. J. (1973) Calcium-dependent stimulation by a phorbol ester (PMA) of thymic lymphoblast DNA synthesis and proliferation. *J. cell. Physiol.* **82**: 151–156.

WIGLER, M. and WEINSTEIN, I. B. (1976) Tumour promoter induces plasminogen activator. *Nature* **264**: 446–447.

WIGLEY, C. B. (1983) TPA affects early and late stages of chemically-induced transformation in mouse submandibular salivary epithelial cells *in vitro*. *Carcinogenesis* **4**: 101–106.

WILLIAMS, G. M., KATAYAMA, S. and OHMORI, T. (1981) Enhancement of hepatocarcinogenesis by sequential administration of chemicals: summation versus promotion effects. *Carcinogenesis* **2**: 1111–1117.

WILSON, E. L. and REICH, E. (1979) Modulation of plasminogen activator synthesis in chick embryo fibroblasts by cyclic nucleotides and phorbol myristate acetate. *Cancer Res.* **39**: 1579–1586.

WILSON, S. R. and HUFFMAN, J. C. (1976) The structural relationship of phorbol and cortisol: a possible mechanism for the tumor promoting activity of phorbol. *Experientia* **32**: 1489–1490.

WITZ, G., GOLDSTEIN, B. D., AMORUSO, M., STONE, D. S. and TROLL, W. (1980) Retinoid inhibition of superoxide anion radical production by human polymorphonuclear leukocytes stimulated with tumor promoters. *Biochem. biophys. Res. Commun.* **97**: 883–888.

WOOD, A. W., CHANG, R. L., HUANG, M.-T., USKOKOVIC, M. and CONNEY, A. H. (1983) 1χ,25-Dihydroxyvitamin D_3 inhibits phorbol ester-dependent chemical carcinogenesis in mouse skin. *Biochem. biophys. Res. Commun.* **116**: 605–611.

WRIGHTON, S. A. and MUELLER, G. C. (1982) Rapid acceleration of deoxyglucose transport by phorbol esters in bovine lymphocytes. *Carcinogenesis* **3**: 1415–1418.

WRIGHTON, S. A., PAI, J.-K. and MUELLER, G. C. (1983) Demonstration of two unique metabolites of arachidonic acid from phorbol ester-stimulated bovine lymphocytes. *Carcinogenesis* **4**: 1247–1251.

WU, V. S., DONATO, N. J. and BYUS, C. V. (1981) Growth state-dependent alterations in the ability of 12-O-

tetradecanoylphorbol-13-acetate to increase ornithine decarboxylase activity in Reuber H35 hepatoma cells. *Cancer Res.* **41**: 3384–3391.
WUNDERLICH, V., BAUMBACH, L. and SYDOW, G. (1981) Phorbolester-induzierte expression von Primatenretroviren. *Arch. Geschwulstforsch.* **51**: S. 615–622.
WYNDER, E. L. and HOFFMAN, D. (1961) A study of tobacco carcinogenesis. VIII. The role of the acidic fractions as promoters. *Cancer* **14**: 1306–1315.
YAMAMOTO, N. (1984) Interaction of viruses with tumor promoters. *Rev. Physiol. Biochem. Pharmacol.* **101**: 111–159.
YAMAMOTO, N. and ZUR HAUSEN, H. (1979) Tumour promoter TPA enhances transformation of human leukocytes by Epstein-Barr virus. *Nature* **280**: 244–245.
YAMAMOTO, H., KATSUKI, T., HARADA, S. and HINUMA, Y. (1981) Enhancement of outgrowth of EB virus-transformed cells from normal human peripheral blood by a tumor promoter, TPA. *Int. J. Cancer* **27**: 161–166.
YAMAMOTO, N., BISTER, K. and ZUR HAUSEN, H. (1979) Retinoic acid inhibition of Epstein-Barr virus induction. *Nature* **278**: 553–554.
YAMASAKI, H. and MARTEL, N. (1981) Inhibitory effect of tumor promoting phorbol esters and mezerein on human mixed lymphocyte reaction. *Cancer Lett.* **12**: 43–52.
YAMASAKI, H., DREVON, C. and MARTEL, N. (1982) Specific binding of phorbol esters to Friend erythroleukemia cells—general properties, down regulation and relationship to cell differentiation. *Carcinogenesis* **3**: 905–910.
YAMASAKI, H., ENOMOTO, T., MARTEL, N., SHIBA, Y. and KANNO, Y. (1983) Tumour promoter-mediated reversible inhibition of cell–cell communication (electrical coupling). *Exp. Cell Res.* **146**: 297–308.
YAMASAKI, H., FIBACH, E., NUDEL, U., WEINSTEIN, I. B., RIFKIND, R. A. and MARKS, P. A. (1977) Tumor promoters inhibit spontaneous and induced differentiation of murine erythroleukemia cells in culture. *Proc. natn. Acad. Sci. U.S.A.* **74**: 3451–3455.
YAMASAKI, H., FIBACH, E., WEINSTEIN, I. B., NUDEL, U., RIFKIND, R. A. and MARKS, P. A. (1979a) Inhibition of Friend leukemia cell differentiation by tumor promoters. In: *Oncogenic Viruses and Host Cell Genes*, pp. 365–376, IKAWA, Y. and OKADA, T. (eds) Academic Press, New York.
YAMASAKI, H., WEINSTEIN, I. B., FIBACH, E., RIFKIND, R. A. and MARKS, P. A. (1979b) Tumor promoter-induced adhesion of the DS19 clone of murine erythroleukemia cells. *Cancer Res.* **39**: 1989–1994.
YANCEY, S. B., EDENS, J. E., TROSKO, J. E., CHANG, C.-C. and REVEL, J.-P. (1982) Decreased incidence of gap junctions between Chinese hamster V-79 cells upon exposure to the tumor promoter 12-O-tetradecanoylphorbol-13-acetate. *Exp. Cell Res.* **139**: 329–340.
YIP, Y. K., PANG, R. H. L., OPPENHEIM, J. D., NACHBAR, M. S., HENRIKSEN, D., ZEREBECKYJ-ECKHARDT. I. and VILČEK, J. (1981) Stimulation of human gamma interferon-production by diterpene esters. *Infect. Immunol.* **34**: 131–139.
YOAKUM, G. H., LECHNER, J. F., GABRIELSON, E. W., KORBA, B. E., MALAN-SHIBLEY, L., WILLEY, J. C., VALERIO, M. G., SHAMSUDDIN, A. M., TRUMP, B. F. and HARRIS, C. C. (1985) Transformation of human bronchial epithelial cells transfected by Harvey *ras* oncogene. *Science* **227**: 1174–1179.
YOTTI, L. P., CHANG, C. C. and TROSKO, J. E. (1979) Elimination of metabolic cooperation in Chinese hamster cells by a tumor promoter. *Science* **206**: 1089–1091.
YUN, K. and SCOTT, R. E. (1983) Biological mechanisms of phorbol myristate acetate-induced inhibition of proadipocyte differentiation. *Cancer Res.* **43**: 88–96.
YUSPA, S. H. (1983) Alterations in epidermal functions resulting from exposure to initiators and promoters of carcinogenesis. In: *Normal and Abnormal Epidermal Differentiation*, pp. 227–241. SEIJI, M. and BERNSTEIN, I. A. (eds). Proc. Japan-U.S. Seminar on Normal and Abnormal Epidermal Differentiation. University of Tokyo Press.
YUSPA, S. H. (1984) Tumor promotion and carcinogenesis in vitro. In: *Mechanisms of Tumor Promotion*, Vol. 3, pp. 1–11. SLAGA, T. J. (ed.). CRC Press, Inc., Boca Raton. FL.
YUSPA, S. H. and MORGAN, D. L. (1981) Mouse skin cells resistant to terminal differentiation associated with initiation of carcinogenesis. *Nature* **293**: 72–74.
YUSPA, S. H., BEN, T. and HENNINGS, H. (1983) The induction of epidermal transglutaminase and terminal differentiation by tumor promoters in cultured epidermal cells. *Carcinogenesis* **4**: 1413–1418.
YUSPA, S. H., BEN, T., HENNINGS, H. and LICHTI, U. (1980) Phorbol ester tumor promoters induce epidermal transglutaminase activity. *Biochem. biophys. Res. Commun.* **97**: 700–708.
YUSPA, S. H. BEN, T., HENNINGS, H. and LICHTI, U. (1982) Divergent responses in epidermal basal cells exposed to the tumor promoter 12-O-tetradecanoylphorbol-13-acetate. *Cancer Res.* **42**: 2344–2349.
YUSPA, S. H., BEN, T., PATTERSON, E., MICHAEL, D., ELGJO, K. and HENNINGS, H. (1976a) Stimulated DNA synthesis in mouse epidermal cell cultures treated with 12-O-tetradecanoyl-phorbol-13-acetate. *Cancer Res.* **36**: 4062–4068.
YUSPA, S. H., KILKENNAY, A. E., STANLEY, J. and LICHTI, U. (1985) Keratinocytes blocked in phorbol ester-responsive early stage of terminal differentiation by sarcoma related viruses. *Nature* **314**: 459–462.
YUSPA, S. H., LICHTI, U., BEN, T., PATTERSON, E., HENNINGS, H., SLAGA, T. J., COLBURN, N. and KELSEY, W. (1976b) Phorbol esters stimulate DNA synthesis and ornithine decarboxylase activity in mouse epidermal cell cultures. *Nature* **262**: 402–404.
ZABOS, P., KYNER, D., MENDELSOHN, N., SCHREIBER, C., WAXMAN, S., CHRISTMAN, J. and ACS, G. (1978) Catabolism of 2-deoxyglucose by phagocytic leukocytes in the presence of 12-O-tetradecanoyl phorbol-13-acetate. *Proc. natn. Acad. Sci. U.S.A.* **75**: 5422–5426.
ZUCKER, M. B., TROLL, W. and BELMAN, S. (1974) The tumor-promoter phorbol ester (12-O-tetradecanoylphorbol-13-acetate), a potent aggregating agent for blood platelets. *J. Cell Biol.* **60**: 325–336.
ZUR HAUSEN, H., BORNKAMM, G. W., SCHMIDT, R. and HECKER, E. (1979) Tumor initiators and promoters in the induction of Epstein-Barr virus. *Proc. natn. Acad. Sci. U.S.A.* **76**: 782–785.
ZUR HAUSEN, H., O'NEILL, F. J. and FREESE, U. K. (1978) Persisting oncogenic herpesvirus induced by the tumour promoter TPA. *Nature* **272**: 373–375.

CHAPTER 4

ROLE OF DNA LESIONS AND REPAIR IN THE TRANSFORMATION OF HUMAN CELLS

Veronica M. Maher and J. Justin McCormick

Carcinogenesis Laboratory—Fee Hall, Department of Microbiology and Department of Biochemistry, Michigan State University, East Lansing, MI 48824-1316, USA

Abstract—Results of studies on the *transformation* of diploid human *fibroblasts* in culture into tumor-forming cells by exposure to chemical carcinogens or radiation indicate that such transformation is a multi-stepped process and that at least one step, acquisition of *anchorage independence*, occurs as a mutagenic event. Studies comparing normal-repairing human cells with DNA *repair-deficient* cells, such as those derived from cancer-prone *xeroderma pigmentosum* patients, indicate that excision repair in human fibroblasts is essentially an error-free process and that the ability to excise potentially cytotoxic, mutagenic, or transforming lesions induced in DNA by carcinogens determines their ultimate biological consequences. Cells deficient in excision repair are abnormally sensitive to these agents. Studies with cells treated at various times in the *cell cycle* show that there is a certain limited amount of time available for DNA repair between the initial exposure and the onset of the cellular event responsible for mutation induction and transformation to anchorage independence. The data suggest that DNA replication on a template containing unexcised lesions (photoproducts, adducts) is the critical event.

ABBREVIATIONS

BPDE	($\pm$)-7β, 8a-dihydroxy-9a,10a-epoxy-7,8,9,10-tetrahydrobenzo(a)pyrene
Asynch.	asynchronously growing population
G_1	period in the cell cycle between mitosis and DNA synthesis phase
G_0	resting, non-cycling state
HPRT	hypoxanthine(guanine)phosphoribosyltransferase
N-AcO-AABP	N-acetoxy-4-acetylaminobiphenyl
N-AcO-AAF	N-acetoxy-2-acetylaminofluorene
N-AcO-AAP	N-acetoxy-2-acetylaminophenanthrene
N-AcO-AASh	N-acetoxy-4-acetylaminostilbene
NF	normal fibroblasts
S-phase	semi-conservative DNA synthesis phase of the cell cycle
TG	6-thioguanine
UV	ultraviolet radiation (254 nm)
XP	xeroderma pigmentosum

1. INTRODUCTION

It is clear from numerous studies on the induction of cancer in animals by chemical carcinogens that carcinogenesis is a multistepped process. Similarly, epidemiologists studying the induction of human cancer by radiation, chemicals, or unknown causes find that only a process with a number of steps fits the available evidence. The word 'multi-stage' is usually used by epidemiologists to describe this phenomenon. The number of stages involved will depend on whether one or more of the intermediate changes (stages) confers on the cell a selective advantage over the others (Peto, 1977). It has been difficult to determine the nature and number of the steps involved in carcinogenesis because of our inability to identify cells in the intact animal or human that have undergone only one or two of the discrete changes involved in the process. Even a fully malignant cell, i.e., one capable of forming an invasive tumor, usually cannot be recognized until it has given rise to a tumor. By that time numerous other cellular alterations may have occurred, making it hard to determine which cellular characteristics are essential for tumor formation.

The complexity of the problem has prompted the development of an alternative approach, viz., the use of cells in culture, an approach that we have taken. Studies with cells in culture permit a more direct experimental manipulation of the changes and facilitate the isolation and identification of cells that have undergone the particular changes of interest (see, for example, Perez-Rodrigues et al., 1981, 1982). An added advantage in using cells in culture instead of whole animals is that it allows one to manipulate experimentally human cells in ways that cannot ethically be done with human beings and permits direct comparisons between animal cells and human cells.

When we began our studies with diploid human skin fibroblasts in 1972, there were no quantitative assays or systems for measuring the frequency of transformation of human cells in culture by chemical carcinogens or radiation. In fact, there were only a few assays for use with non-human mammalian cells (see Heidelberger, 1973 for a review). One of the latter assays used primary or early passage Syrian (golden) hamster embryo cells (Berwald and Sachs, 1965; Borek and Sachs, 1967; Huberman and Sachs, 1966); a second used a fibroblast cell line with an unlimited lifespan derived from adult mouse prostate (Heidelberger and Iype, 1967); and the third used permanent cell lines derived from mouse embryos, e.g., Balb 3T3 cells (DiPaolo et al., 1972; Kakunaga, 1973), C3H 10T$\frac{1}{2}$ cells (Reznikoff et al., 1973). The end point for transformation of the permanent cell lines was formation of piled up cells on the top of a confluent monolayer (foci formation); that of the primary hamster embryo cells was formation of morphologically-altered colonies of cells growing in a criss-crossed pattern.

Two important reasons for developing a quantitative transformation assay for normal *human* cells (i.e., diploid fibroblasts with a limited life span in culture), rather than for animal cell lines were: one, because of their particular relevance to human cancer; and two, because at the time human tissue was virtually the only source of DNA repair-deficient mammalian cells for use in comparative studies aimed at determining the role of mutations and DNA repair in the carcinogenesis process (Maher et al., 1968, 1970, 1971). The 1968 finding by Cleaver that skin fibroblasts derived from xeroderma pigmentosum (XP) patients, who are characterized by an inherited predisposition to sunlight-induced skin cancer (see Robbins et al., 1974 for a review), are deficient in excision repair of DNA damage caused by ultraviolet (254 nm) light offered the opportunity to study the role of DNA lesions and DNA repair in the carcinogenesis process. If a quantitative human cell transformation system could be developed, one would be able to compare the frequency of transformation induced in excision repair-deficient XP cells with that of cells from normal persons. Significant differences could shed light on the role of repair. No such repair-deficient cells were available among the non-human animal cells lines.

Our working hypothesis at the time was that carcinogenic agents, such as chemicals and radiation, cause cancer by acting as mutagens and that an essential event leading ultimately to the transformation of a normal cell into a tumor-forming cell results from damage to DNA that gets converted into a mutation. Excision repair-deficient cells from XP patients were predicted to be abnormally sensitive to the mutagenic action of UV radiation and of other agents causing DNA damage that XP cells could not excise as rapidly as normal human cells. If genetic changes resulting from the presence of unrepaired lesions were causally involved in the neoplastic transformation process, XP cells should also be more sensitive than normal cells to transformation by UV radiation and by these other agents. This is because the XP cells would have more of the initial damage remaining in their DNA at the time such lesions were translated into permanent changes in DNA (mutations). To test these hypotheses, quantitative assays for induction of mutations in normal and XP cells were developed (Maher and McCormick, 1976; Maher et al., 1976a, b, 1977, 1979, 1980) and more recently assays for comparing the induction of transformation in such cells (Maher et al., 1982). Results indicate that DNA excision repair by human cells can eliminate potentially cytotoxic, mutagenic, and transforming damage before the lesions are transformed into permanent cellular effects. The transformation data further suggest that at least one step on the path to full neoplastic transformation, viz., loss of anchorage independence, results from a mutagenic event (Maher et al., 1982; Silinskas et al., 1981). In all these studies, the target cells were non-malignant, diploid fibroblasts derived from foreskin of normal neonates or from skin biopsies of apparently normal individuals or XP patients. All skin biopsies were taken from non-sunlight exposed areas and the cells are non-tumorigenic when assayed in athymic mice (see below).

2. EFFECT OF REPAIR ON THE CYTOTOXICITY OF CARCINOGENS IN HUMAN CELLS

Studies have shown that one of the best ways to determine the cytotoxic effect of an agent on cells in culture is to compare treated and untreated populations for their ability to form colonies when plated at low density (survival of reproductive capacity) (Roper and Drewinko, 1976). For such studies to be reliable, care must be taken to insure that the cloning efficiency of the untreated cells is high and that the cell densities used for the treated populations with expected low survivals do not affect the cloning efficiency. For example, if one plates more than 600 cells/cm^2 the cloning efficiency of the cells in those dishes can be increased by a 'feeder layer' effect. If that occurs, it is not valid to calculate the survival as percentage of the cloning efficiency of the control cells plated at a much lower density. The methods by which cloning efficiencies of greater than 50% with fibroblasts derived from normal persons and 20–70% with cells derived from persons with various genetic predispositions to cancer are achieved have been described (McCormick and Maher, 1981).

Two approaches were used to determine the effect of DNA excision repair on the cytotoxic action of carcinogens in human fibroblasts. One varied the rate of repair by comparing the survival curves of a series of cell lines, each with a different rate of excision repair of DNA lesions. The second varied the length of time available for excision repair following carcinogen treatment before resting cells are allowed to cycle. For the latter type of experiment, a series of cells were grown to confluence and starved for mitogens to prevent them from replicating (density inhibition) and then irradiated or treated with carcinogens. One set of treated cells was harvested immediately and assayed for the initial number of DNA lesions (or covalently bound DNA adducts) and a portion of the cells were plated at cloning density to measure the survival of their cloning ability as percentage of untreated controls. The cells in the other sets of cultures were maintained in the density-inhibited state for various periods to allow time for excision repair in the non-cycling state. They were subsequently assayed for the number of DNA adducts remaining unexcised and for survival of colony-forming ability.

Examples of the results obtained using the first approach are shown in Figs 1 and 2. There was a direct correlation between the cells' ability to repair DNA lesions and their ability to exhibit high survival after exposure to a potentially cytotoxic dose of an agent. For example, the most sensitive cells in Fig. 1, XP12BE cells, have been shown to be virtually incapable of excising UV-induced DNA damage (Petinga et al., 1977). A dose as low as 1 J/m^2 reduced their survival to less than 1%. The intermediate cells, XP2BE cells, have been shown to excise UV-induced damage slowly (Robbins et al., 1974). In these cells a dose of 1 J/m^2 resulted in 15% survival. Normal cells (NF) excise UV-induced damage very rapidly and exhibited 100% survival at that dose and even after doses that reduce the survival of the XP12BE cells far below 1% (Maher et al., 1979).

Examples of the second approach, i.e., treating cells in confluence and allowing time for repair before cycling, are shown in Figs 3 and 4A and B. The XP12BE cells proved unable to excise the large bulky multi-ringed DNA adducts formed by a series of reactive

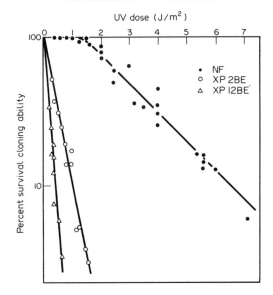

FIG. 1. Comparison of the cytotoxic effect of low doses of UV 254 nm radiation in diploid human fibroblasts with different rates of DNA excision repair of pyrimidine dimers. Cells from exponentially-growing cultures were plated into 60-mm diameter dishes at appropriate densities to yield a countable number of colonies (cloning densities), allowed 12 hr for attachment, irradiated *in situ* in a film of buffer, fed with Ham's F10 medium supplemented with fetal bovine serum, and allowed to develop colonies. (●) normal cells; (○) XP2BE cells from complementation group C; (△) XP12BE from group A. (Figure adapted from similar figure in Maher *et al.*, 1979. Recent data using an improved dosimeter suggest that the doses shown are 80% of the actual doses used for the study in 1979.)

derivatives of polycyclic aromatic amides (Fig. 3) (Heflich *et al.*, 1980; Maher *et al.*, 1981). They were also unable to remove the potentially cytotoxic lesions caused by these four carcinogens. In contrast, normally repairing cells treated in confluence gradually removed these DNA adducts and at each time point showed the correspondingly higher survival levels expected for cells with correspondingly lower numbers of adducts remaining in their DNA at the time of assay. A similar result was found with the anti-isomeric 7,8-diol-9,10-epoxide of benzo(a)pyrene (BPDE) (Fig. 4A and B) (Yang *et al.*, 1980).

Taken all together, the data of these comparative studies indicate that excision repair process in human fibroblasts protect cells from the potentially lethal effects of UV radiation and of various chemical carcinogens. More recently, Domoradzki *et al.* (1984) investigated the biological significance of another type of DNA repair system, i.e., O^6-alkylguanine DNA alkyltransferase, an acceptor protein that specifically removes methyl or ethyl groups from the O^6 position of guanine in alkylated DNA (Mehta *et al.*, 1981). Two human cell lines that are extremely deficient in this methyltransferase activity were identified. One is an XP cell line (XP12RO) transformed to an infinite life span by Simian virus 40 (SV40); the other, GM3314 from the Institute of Medical Research (Camden, NJ), is a fibroblast cell line from a skin biopsy of a patient with an inherited predisposition to colon cancer (Gardner's syndrome). A third cell line obtained by SV40 virus transformation of normal fibroblasts (GM637), exhibited an intermediate level of methyltransferase activity. These repair-deficient cells and a series of cell lines with normal levels of methyltransferase were compared for sensitivity to the killing effect of methylating or ethylating agents, i.e., N-methyl-N'-nitro-N-nitrosoguanidine (MNNG) and N-ethyl-N-nitrosourea (ENU). Figure 5A shows that the two repair-deficient cell lines were extremely sensitive to the cytotoxic action of MNNG compared to cells with a normal ability to remove this lesion and that the response of the GM637 cells was intermediate. These data suggest that O^6-methylguanine is a potentially cytotoxic lesion in human fibroblasts and that this repair system protects the cells.

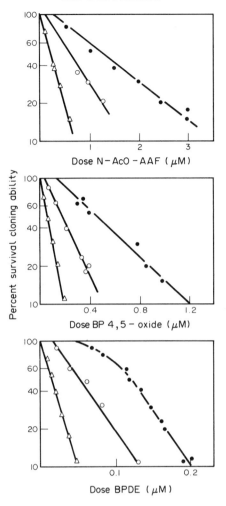

FIG. 2. Comparison of the cytotoxic effect of reactive derivates of chemical carcinogens in human fibroblasts with different rates of excision repair. Cells were plated as described in Fig. 1. The growth medium was exchanged for serum-free medium and the carcinogen, freshly-dissolved in anhydrous ethanol, was delivered by micropipette. After 2 hr this medium was replaced with fresh medium. Symbols used are as in Fig. 1.

3. EFFECT OF EXCISION REPAIR ON THE FREQUENCY OF MUTATIONS INDUCED IN HUMAN CELLS BY CARCINOGENS

These same approaches were used to determine the role of excision repair or methyltransferase repair on the mutagenic action of carcinogens. Resistance to 6-thioguanine (TG) was the genetic marker for the majority of these studies and the procedures used have been described (Konze-Thomas et al., 1982; Maher et al., 1979; McCormick and Maher, 1981). Cell lines that differed in excision repair capacity were compared for the frequency of TG-resistant cells induced by UV radiation (Konze-Thomas et al., 1982; Maher et al., 1979, 1982), by an aflatoxin derivative (Mahoney et al., 1984), by reactive metabolites of polycyclic aromatic amides (Maher et al., 1980) or hydrocarbons (Maher et al., 1977; Yang et al., 1980, 1982) or by the simple alkylating agents ENU (Simon et al., 1981) and MNNG (Domoradzki et al., 1984). Alternatively, cells were treated in the non-replicating (confluent) state and assayed immediately or after various times post-treatment for the number of lesions or adducts in DNA and, after a suitable expression period, for the frequency of TG-resistant cells induced (Heflich et al., 1980; Konze-Thomas et al., 1979; Maher et al., 1979, 1980; Yang et al., 1980).

Examples of results obtained with the first approach, i.e., comparing cell lines that differ

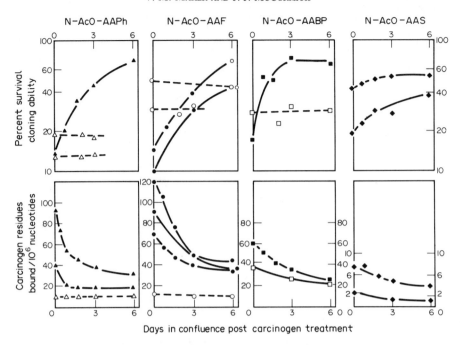

FIG. 3. Comparison of the rates of recovery from the potentially lethal effects of four aromatic amine derivatives with the rate of removal of radioactive labeled residues from the DNA of normal (closed symbols) or XP12BE cells (open symbols). Cells were grown to confluence and entered the G_0 state as described. They were treated in the density-inhibited state with radiolabeled carcinogens, i.e., N-acetoxy-4-acetylaminobiphenyl (N-AcO-AABP); N-acetoxy-2-acetylaminofluorene (N-AcO-AAF); N-acetoxy-2-acetylaminophenanthrene (N-AcO-AAPh); and N-acetoxy-4-acetylaminostilbene (N-AcO-AAS). The cells were harvested and assayed for survival and number of carcinogen residues (adducts) remaining covalently bound to DNA after the designated period of time in the G_0 state. (Taken from Maher et al., 1981.)

in rate of repair, are shown in Figs 5 and 6. The data in Fig. 5B show that mutations were induced by low doses of MNNG in the two cell lines that lack the ability to remove methyl groups from the O^6 position of guanine, but in repair-proficient cells there was no significant increase in the frequency of mutants at these low doses. Only at much higher concentrations did mutation induction occur. The GM637 cells with an intermediate level of methyltransferase activity showed an intermediate response to mutation induction (Domoradzki et al., 1984).

Figure 6B shows that a similar pattern of response was found in human cells that differ in rate of excision of UV-induced DNA lesions. For a given exposure to UV 254 nm radiation, XP7BE cells, which excise at a rate ~16% that of normal cells (Robbins et al., 1974) or XP12BE cells which have little or no ability to excise, exhibited a significantly higher frequency of mutants than did normal cells. Similar results were reported by Arlett and Harcourt (1983), Grosovsky and Little (1983), Maher et al. (1979, 1982) and Patton et al. (1984). Such increased sensitivity of XP cells compared to normal cells was also obtained with chemical carcinogens that produce bulky DNA adducts which normal cells can repair but which the XP cells are unable to excise or excise only very slowly (Maher et al., 1977, Yang et al., 1980). Taken all together, these results suggest that when cells receive DNA damage there is a finite amount of time for them to remove or excise the potentially mutagenic lesions before these are permanently converted into mutations.

Figure 4C gives an example of results obtained using the second approach, i.e., treating cells in confluence and allowing various lengths of time for excision repair before assaying the effect of the treatment. Yang et al. (1980) found a direct correlation between the cells' ability to excise BPDE-induced adducts from DNA and their ability to decrease the frequency of potentially mutagenic effects of this carcinogen. The excision repair-deficient XP12BE cells did not remove the damage (Fig. 4B) and the frequency of mutants did not change with time (Fig. 4C).

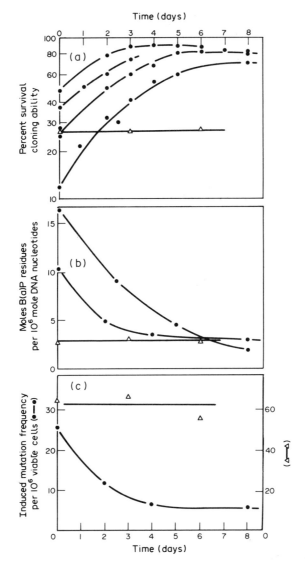

FIG. 4. Kinetics of removal of covalently bound adducts (B) and recovery of normal (circles) or XP12BE cells (triangles) from the potentially cytotoxic (A) or mutagenic (C) effects of anti-BPDE. The cells were treated in the G_0 state for 1 hr with radiolabeled BPDE, released immediately (Day 0) or after the designated time post carcinogen treatment and assayed for survival of colony-forming ability, for the number of residues bound to DNA, and after a suitable expression period, for the frequency of induced mutations to TG resistance. See text for details. (Taken from Yang et al. (1980) with permission.)

4. EFFECT OF ALLOWING TIME FOR EXCISION BEFORE DNA SYNTHESIS

These data suggested that the critical event for mutation induction is the replication of cellular DNA containing unexcised lesions. To test this hypothesis, normal cells and excision-repair deficient XP12BE cells were synchronized by growing them to confluence and starving them for mitogens for several days to cause cells to enter the G_0 resting stage. They were then released from density-inhibition and plated at lower densities. In such cells DNA synthesis begins approximately 22 hr later (Konze-Thomas et al., 1982). Normal and XP cells were treated just prior to the onset of DNA synthesis (S-phase); or in early G_1 (~18 hr prior to S-phase); or in the density-inhibited G_0 state with immediate release from confluence so that there would be ~24 hr before onset of DNA synthesis. Figure 7 shows

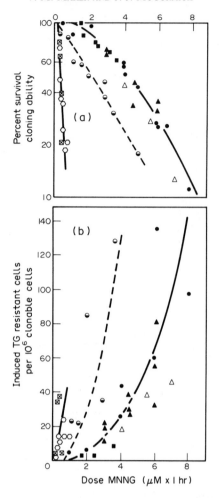

FIG. 5. Comparison of the cyctoxic (A) and mutagenic (B) effects of MNNG in human cells with different levels of O^6-alkylguanine-DNA alkyltransferase activity. (A) Cells were plated and treated with MNNG, dissolved in dimethylsulfoxide as described in Fig. 2 and allowed to form colonies *in situ*. (B) Cells were treated at densities from 3 to 7×10^3 cells/cm² and allowed to replicate and express the mutations before being pooled and assayed for TG-resistant cells. Cells lacking the repair protein: GM3314 (⊠) and SV-40 virus-transformed XP12RO cells (○); cells with intermediate levels of repair activity: GM637 (◐); cells with normal levels of activity: fibroblasts from a normal person (●), GM2355 (▲); GM3948 (■) and XP12BE cells (△). The data are taken from Domoradzki et al., 1984. The figure is adapted from Fig. 2 in that publication.

the results of such a study with UV radiation as the carcinogen (Konze-Thomas et al., 1982). As expected for cells that are unable to excise the DNA damage induced by UV, the frequency of mutations induced in XP12BE cells was the same whether they were irradiated just before S-phase, 18 hr prior to S-phase, or 24 hr before DNA synthesis began. But the slope of the mutation response for normal cells irradiated just prior to S-phase was 6-fold steeper than that of cells irradiated in early G_1 or of cells treated in confluence and immediately following irradiation plated at lower density and allowed to proceed to S-phase. These results support the idea that premutagenic lesions in DNA (i.e., adducts or photoproducts) are converted to mutations by the process of DNA replication and that the frequency of mutations is determined by the number of unexcised lesions remaining in the DNA at the time the gene of interest, i.e., hypoxanthine(guanine)phosphoribosyltransferase (HPRT), is replicated. Recently, Grossmann et al. (1985) used a population of human cells highly synchronized by exposure to the alpha polymerase inhibitor, aphidocolin, and showed that the HPRT gene may be replicated during the very early part of S-phase.

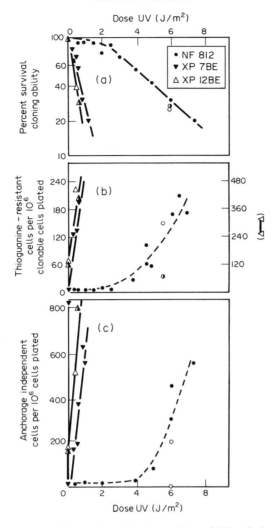

FIG. 6. Cytotoxicity, mutagenicity, and transforming activity of UV radiation in normal and XP cells. The frequency of thioguanine-resistant cells was assayed after six doublings; that of anchorage-independent cells after nine doublings. The former were corrected for cloning efficiency on plastic. Data for normal cells are symbolized by circles; for XP7BE by inverted triangles; for XP12BE by triangles. Solid symbols, populations irradiated in exponential growth; open symbols, cells synchronized by release from confluence and irradiated shortly before onset of S-phase; half-solid symbols, cells irradiated 18–20 hr prior to S. See text for details. The background frequencies have not been subtracted from the induced values. The open and half-solid circles in the middle panel represent data from a separate mutagenesis experiment, not identical to the one represented by the same symbols in the bottom panel. Otherwise, all data are from experiments in which cells were assayed for all three parameters. (Taken from McCormick and Maher (1983) with permission.)

5. STUDIES ON NEOPLASTIC TRANSFORMATION OF DIPLOID HUMAN FIBROBLASTS

These mutagenesis studies were part of a broader goal, i.e., to determine the nature of the steps involved in transformation. The genetic marker, resistance to thioguanine, served only as a model for detecting the kinds of changes by which radiation and chemicals were considered to transform normal cells into tumorigenic cells. Therefore, during this time efforts were made to develop methods for detecting transformation of human cells.

In 1977 Kakunaga reported the successful transformation of human fibroblasts into focus-forming cells using 4 nitroquinoline-1-oxide or MNNG (Kakunaga 1977, 1978). The

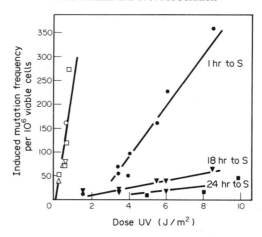

FIG. 7. Comparison of the mutagenicity of UV in normal (closed symbols) or XP12BE cells (open symbols) irradiated under conditions designed to allow various lengths of time for excision repair to take place prior to the onset of S-phase. Cells irradiated in confluence (G_0) and then released and plated at lower densities (■,□); cells released from G_0 and irradiated 6 hr later (▼,▽); cells released and irradiated 24 hr later (●,○); cells replated from asynchronously growing cultures, and irradiated ∼16 hr later (△). (Adapted from a figure in Konze-Thomas et al., 1982.)

progeny of one of these cells was tumorigenic when injected subcutaneously into athymic mice. However, no other workers have reported successfully producing malignant tumors by this technique. In 1978 Milo and Dipaolo (1978) reported the successful induction of anchorage independent growth (growth in semi-solid medium) of human fibroblasts. The progeny of these anchorage independent cells sometimes produced nodular growths in athymic mice, but progressively growing tumors were not observed. McCormick and coworkers (McCormick et al, 1980; Silinskas et al. 1981) confirmed that foci and anchorage independent growth could be included and decided that an assay in which the transformed cells were selected by their ability to form colonies in soft agar was the better assay for quantitative comparative studies with normal and XP cells.

In 1979, Bellett and Younghusband showed that in primary cultures of mouse embryo cells, anchorage independence arose as the result of a mutagenic event. McCormick and Maher, therefore, modeled their human cell transformation assay after their human cell mutagenesis assays, expecting anchorage independence to have in common with mutagenesis: (a) an expression period, i.e., a period of time between carcinogen damage of cells and their ability to express their transformed phenotype; (b) a dose response; (c) a concentration dependence for the carcinogenic agent which resembled that required for induction of mutations, so that strong mutagens would usually be strong transforming agents; (d) a low, but measurable frequency of transformed cells in non-carcinogen-treated cell populations just as one finds a low, but measurable, frequency of mutant cells in such populations; (e) a higher frequency of transformation per dose of UV in DNA repair-deficient XP cells than in normal cells, as they had shown for the induction of mutants; (f) a cell cycle dependence similar to that for mutation induction, so that populations of cells treated just before S-phase would show a higher frequency of transformation than cell populations treated with the same dose of the agent far from S-phase.

Conditions were, therefore, worked out to test each of these predictions. The assay as they are presently using it is as follows: diploid human fibroblasts (non-malignant cells) obtained from normal persons of various ages or from persons with a genetic predisposition to cancer, are treated in exponential growth with low doses of carcinogens or at various times during the cell cycle. The doses of radiation or chemicals are those previously determined to lower the cell survival to between 80 and 10% of the control population. The number of target cells is adjusted to insure at least 10^6 survivors. The surviving cells are allowed to undergo ∼5 population doublings in the original dishes before being pooled. A portion of cells (10^6) is assayed for ability to form colonies in 0.33% agar. The rest are subcultured and at a lower density allowed to undergo an additional 4–5 population doublings. At that time the cells are pooled and $1-2 \times 10^6$ cells are assayed for TG resistance and 10^6 cells are assayed for anchorage independence. Within 21–35 days, colonies of anchorage independent cells develop. Those with a diameter $\geq 60\,\mu m$ are scored using an inverted microscope equipped with an ocular micrometer (McCormick et al., 1985, 1986).

These soft agar colonies have repeatedly been tested for tumorigenicity by pooling colonies, propagating them into large populations and injecting 5×10^6 cells s.c. into immunologically depressed (sub-lethally X-irradiated) athymic mice. Occasionally tumors (nodules) of ~ 1 cm diameter develop at the site of injection. No tumors develop in animals injected with non-transformed control populations ($> 2 \times 10^7$/injection and > 50 mice so far). Pathology examination has identified representative tumors as fibrosarcomas. However, after attaining a diameter of ~ 1 cm, the nodules stop growing and if left in the animal, they regress. If these tumors are excised and their cells are returned to culture, the cells exhibit anchorage independence but retain a diploid number of chromosomes and a limited life span in culture. Therefore, they are considered to be 'partial transformants' i.e., cells that have undergone only one step in the multi-stepped pathway to neoplastic transformation.

Pouyssegur and his associates (Perez-Rodrigues et al., 1981, 1982) using sequential mutagenesis and selection techniques on a Chinese hamster cell line with an unlimited lifespan obtained a series of partially transformed cells, each more transformed than its antecedent. Such cells formed regressing tumors but these investigators were able to select fully-transformed aneuploid mutant cells capable of forming non-regressing tumors. These elegant studies shed light on the number and kinds of steps to transformation and indicate a role for somatic cell mutations. However, because of the lack of DNA repair-deficient mutants they do not permit investigation into the role of DNA repair in the process.

6. ROLE OF DNA EXCISION REPAIR IN THE TRANSFORMATION PROCESS

To examine the effect of DNA excision repair on the frequency of transformation, the frequency of anchorage independence induced by UV was compared in normal human cells and in two strains of excision repair-deficient XP cells, i.e., XP7BE and XP12BE (Maher et al., 1982). (XP7BE cells excise UV damage at $\sim 16\%$ the normal rate; XP12BE cells at $< 0.5\%$ the normal rate.) XP7BE cells and normal cells were irradiated in exponential growth and carried through the protocol outlined above. The results are shown in Fig. 6. The XP7BE population was assayed for TG resistance and for anchorage independence; the normal cells were assayed only for survival and for growth in soft agar since the frequency of UV-induced TG-resistant cells had been determined previously (Maher et al., 1979).

The results indicate that to achieve a particular degree of cell killing, mutagenesis, and transformation, the repair-proficient normal cells have to be exposed to 8- to 10-fold higher doses of UV radiation than XP cells (Maher et al., 1982). This is the result expected if induction of anchorage independence, as well as thioguanine resistance, results ultimately from DNA damage remaining unexcised in the cell at some critical time after irradiation and if, because of the difference in their respective rates of excision repair, the average number of lesions remaining at this critical time is approximately equal in the three populations. McCormick et al. (1986) report additional data suggesting that the event responsible for induction of anchorage independence is related to mutation induction. They showed that XP variant cells are equally sensitive to UV induction of either event.

7. EFFECT ON TRANSFORMATION OF ALLOWING TIME FOR EXCISION REPAIR BEFORE ONSET OF DNA SYNTHESIS

If anchorage independence results from an event 'fixed' during semi-conservative DNA synthesis on a template which still contains unexcised lesions, the frequency of such cells should be much higher in populations of normal human cells UV-irradiated just before the onset of S-phase than in cells irradiated 18–20 hr prior to S-phase. XP12BE cells should not show such a cell-cycle dependence. This prediction was tested by synchronizing the cells as described above and irradiating them just prior to S-phase or in early G_1 and assaying them for anchorage independence. The results are included in Fig. 6C as open symbols (1–3 hr prior to S-phase) or half-solid symbols (18–24 hr prior to S) (Maher et al., 1982). Irradiation of normal cells with $6 J/m^2$ ~3 hr prior to onset of S-phase yielded 200 anchorage-independent cells per 10^6 cells plated. The population irradiated ~18 hr prior to S-phase showed no colonies out of 10^6 cells plated into semi-solid medium. The control cells in this experiment also gave no colonies out of 2×10^6. In contrast, the frequency of anchorage-independent cells in the XP12BE population irradiated in early G_1 did not decrease; in fact it was somewhat higher. In the corresponding mutation experiment (Fig. 6B), the frequencies of TG-resistant cells in the XP12BE population irradiated at the two times were equal.

The mutagenesis data for the normal cells treated 18 hr prior to S-phase (half-solid circle in Fig. 6B) were taken from one of the experiments shown in Fig. 7 in which the frequency of mutant cells did not decrease completely the background level. However, in this mutagenesis experiment, the cells irradiated in G_1 had somewhat less time for excision repair before onset of S than was available in these transformation experiments. The fact that allowing substantial time for excision before DNA synthesis eliminated the potentially mutagenic and transforming effect of UV radiation in normal cells, but not in XP12BE cells, suggests that DNA synthesis on a template still containing unexcised lesions is the cellular event responsible for 'fixing' the mutations and transformation.

Earlier experiments reported by Kakunaga in 1974 and 1975 are consistent with this interpretation. He showed that when confluent cultures of a mouse cell line were exposed to 4-nitroquinoline-1-oxide or 3 methylcholanthrene and then allowed to carry out excision repair but not to replicate, the potential for foci formation was gradually eliminated. However, when the cells were allowed to undergo a single population doubling after treatment before attaining confluence, additional time in confluence did not decrease the transformation frequency.

McCormick and co-workers are currently extending these kinds of studies to include the effect of time for repair of damage caused by low doses of ionizing radiation in human cells. Presumably, the replication converted the potentially transforming lesion into a permanent change. A number of other investigators are also currently studying carcinogen-induced transformation of human cells (Borek, 1980; Milo et al., 1981; Sutherland et al., 1980, 1981; Zimmerman and Little, 1983). However, they have not examined the effect of DNA repair on these processes.

8. CONCLUSION

In summary, our results suggest that the transformation of diploid human fibroblasts into tumor-forming cells is a multi-stepped process and that at least one step, acquisition of anchorage independence, occurs as a mutagenic event. The data indicate that excision repair in these fibroblasts is essentially an error-free process and that the ability to excise potentially cytotoxic, mutagenic, or transforming lesions induced in DNA by UV radiation or by several classes of chemical carcinogens determines their ultimate biological consequences. Cells deficient in excision repair, such as those derived from cancer-prone xeroderma pigmentosum patients, are abnormally sensitive to the cytotoxic, mutagenic, and transforming effects of carcinogens. The data further suggest that there is a certain limited amount of time available between the initial exposure and the onset of the cellular events responsible for mutation induction, for cell transformation to anchorage independence, and for cell killing, and that the critical event for mutations and this partial transformation is DNA replication on a template containing unexcised lesions (photoproducts, adducts).

Acknowledgements—We wish to express our indebtedness to our colleagues, J. C. Ball, R. C. Corner, R. D. Curren, M. E. Dolan, J. Domoradzki, A. Grossmann, R. H. Heflich, J. N. Howell, B. Konze-Thomas, J. W. Levinson, E. M. Mahoney, L. M. Ouellette, J. D. Patton, A. E. Pegg, K. C. Silinskas, L. Simon, L. L. Yang, M. Watanabe and J. C. Wigle for their valuable contributions to the research summarized here. The excellent technical assistance of D. J. Dorney, R. M. Hazard, S. A. Kateley-Kohler, L. L. Lommel, A. L. Mendrala, L. D. Milam, M. M. Moon, L. A. Rowan and J. E. Tower is gratefully acknowledged. We thank Carol Howland for preparation of the manuscript. The labeled BPDE was provided by the Cancer Research Program of the National Cancer Institute and the labeled aromatic amines by the late Dr. John Scribner of the Pacific Northwest Research Foundation. The research summarized in this report was supported in part by Contract ES-78-4659 from the Department of Energy and by Grant CA21253, CA21289, and ES07076 from the Department of Health and Human Services, NIH. Additional financial assistance was provided by the Michigan Osteopathic College Foundation.

REFERENCES

ARLETT, C. F. and HARCOURT, S. A. (1983) Variation in response to mutagens amongst normal and repair-defective human cells. In: *Induced Mutagenesis Molecular Mechanisms and their Implications for Environmental Protection*, pp. 249–270. Lawrence, C. W. (ed.). Plenum Press, New York.
BELLETT, A. J. D. and YOUNGHUSBAND, H.B. (1979) Spontaneous, mutagen-induced and adeno-virus-induced anchorage independent tumorigenic variants of mouse cells. *J. Cell Physiol.* **101**: 33–48.
BERWALD, Y. and SACHS. (1965) *In vitro* transformation of normal cells to tumor cells of carcinogenic hydrocarbons. *J. natn. Cancer Inst.* **35**: 641–661.
BOREK, C. (1980) X-ray induced *in vitro* neoplastic transformation of human diploid cells. *Nature* **283**: 776–778.
BOREK, C. and SACHS, L. (1967) Cell susceptibility to transformation by x-irradiation and fixation of the transformed state. *Proc. natn. Acad. Sci. USA* **57**: 1522–1527.
CLEAVER, J.E. (1968) Defective repair replication of DNA in xeroderma pigmentosum. *Nature* **218**: 652–656.
DIPAOLO, J. A., NELSON, R. L. and DONOVAN, P. J. (1972) *In vitro* transformation of Syrian hamster embryo cells by diverse chemical carcinogens. *Nature* **235**: 270–280.
DOMORADZKI, J., PEGG, A. E., DOLAN, M. E., MAHER, V. M. and MCCORMICK, J. J. (1984) Correlation between transmethylase activity and resistance of familial polyposis coli, Gardner's syndrome, xeroderma pigmentosum and normal human cells to mutations by N-methyl-N'-nitro-N-nitrosoguanidine. *Carcinogenesis* **5**: 1641–1647.
GROSSMANN, A., MAHER, V. M. and MCCORMICK, J. J. (1985) The frequency of mutants in human fibroblasts UV-irradiated at various times during S-phase suggests that genes for thioguanine and diphtheria toxin resistance are replicated early. *Mutat. Res.* **152**: 67–76.
GROSOVSKY, A. J. and LITTLE, J. B. (1983) Mutagenesis and lethality following S phase irradiation of xeroderma pigmentosum and normal human diploid fibroblasts with ultraviolet light. *Carcinogenesis* **4**: 1389–1394.
HEFLICH, R. H., HAZARD, R. M., LOMMEL, L., SCRIBNER, J. D., MAHER, V. M. and MCCORMICK, J. J. (1980) A comparison of the DNA binding, cytotoxicity and repair synthesis induced in human fibroblasts by reactive derivatives of aromatic amide carcinogens. *Chem. Biol. Interactions* **29**: 43–56.
HEIDELBERGER, C. (1973) Chemical oncogenesis in culture. *Adv. Cancer Res.* **18**: 317–366.
HEIDELBERGER, C. and IYPE, P. T. (1967) Malignant transformation *in vitro* with carcinogenic hydrocarbons. *Science* **155**: 214–217.
HUMBERMAN, E. and SACHS, L. (1966) Cell susceptibility to transformation and cytotoxicity by the carcinogenic hydrocarbon benzo(a)pyrene. *Proc. natn. Acad. Sci. USA* **56**: 1123–1129.
KAKUNAGA, T. (1973) A quantitative system for assay of malignant transformation by chemical carcinogens using a clone derived from BALB/3T. *Int. J. Cancer* **12**: 463–473.
KAKUNAGA, T. (1974) Requirement for cell replication in the fixation and expression of the transformed state in mouse cells treated with 4-nitro-quinoline-1-oxide. *Int. J. Cancer* **14**: 736–742.
KAKUNAGA, T. (1975) The role of cell division in the malignant transformation of mouse cells treated with 3-methylcholanthrene. *Cancer Res.* **35**: 1637–1642.
KAKUNAGA, T. (1977) The transformation of human diploid cells by chemical carcinogens. In: *Origins of Human Cancer*, pp. 1537–1548. Hiatt, J. J., Watson, J. D., Winsten, J. A. (eds). Cold Spring Harbor Laboratory Press, Cold Spring Harbor.
KAKUNAGA, T. (1978) Neoplastic transformation of human diploid fibroblast cells by chemical carcinogens. *Proc. natn. Acad. Sci. USA* **75**: 1334–1338.
KONZE-THOMAS, B., HAZARD, R. M., MAHER, V. M. and MCCORMICK, J. J. (1982) Extent of excision repair before DNA synthesis determines the mutagenic but not the lethal effect of UV radiation. *Mutat. Res.* **94**: 421–434.
KONZE-THOMAS, B., LEVINSON, J. W., MAHER, V. M. and MCCORMICK, J. J. (1979) Correlation among the rates of dimer excision, DNA repair replication, and recovery of human cells from potentially lethal damage induced by ultraviolet radiation. *Biophys. J.* **28**: 315–326.
MAHER, V. M. and MCCORMICK, J. J. (1976) Effect of DNA repair on the cytotoxicity and mutagenicity of UV irradiation and of chemical carcinogens in normal and xeroderma pigmentosum cells. In: *Biology of Radiation Carcinogenesis*, pp. 129–145. Yuhas, J. M., Tennant, R. W. and Regan, J.D. (eds). Raven Press, New York.
MAHER, V. M., CURREN, R. D., OUELLETTE, L. M. and MCCORMICK, J. J. (1976a) Role of DNA repair in the

cytotoxic and mutagenic action of physical and chemical carcinogens. In: *In Vitro Metabolic Activation in Mutagenesis Testing*, pp. 313–336. de Serres, F. J., Fouts, J. R., Bend, J. R. and Philpot, R. M. (eds). Elsevier Scient. Publ. Co., Amsterdam.

MAHER, V. M., DORNEY, D. J., MENDRALA, A. L., KONZE-THOMAS, B. and MCCORMICK, J. J. (1979) DNA excision repair processes in human cells can eliminate the cytotoxic and mutagenic consequences of ultraviolet irradiation. *Mutat. Res.* **62**: 311–323.

MAHER, V. M., HAZARD, R. M., BELAND, F. J., CORNER, R., MENDRALA, A. L., LEVINSON, J. W., HEFLICH, R. H. and MCCORMICK, J. J. (1980) Excision of the deacetylated C-8-guanine DNA adduct by human fibroblasts correlates with decreased cytotoxicity and mutagenicity. *Proc. Am. Ass. Cancer Res.* **21**: 71.

MAHER, V. M., HEFLICH, R. H. and MCCORMICK, J. J. (1981). Repair of DNA damage induced in human fibroblasts by N-substituted aryl compounds. In: *Carcinogenic and Mutagenic N-Substituted Aryl Compounds*, pp. 217–222. Thorgeisson, S. S., Wiesberger, E. K., King, C. M. and Scribner, J. D. (eds). Monograph 58, National Cancer Institute.

MAHER, V. M., LESKO, JR, S. A., STRATT, P. A. and TS'O, P. O. P. (1971) Mutagenic action, loss of transforming activity, and inhibition of deoxyribonucleic acid template activity in vitro caused by chemical linkage of carcinogenic polycyclic hydrocarbons to deoxyribonucleic acid. *J. Bacteriol.* **108**: 201–212.

MAHER, V. M., MCCORMICK, J. J., GROVER, P. L. and SIMS, P. (1977) Effect of DNA repair on the cytotoxicity and mutagenicity of polycyclic hydrocarbon derivatives in normal and xeroderma pigmentosum human fibroblasts. *Mutat. Res.* **43**: 117–138.

MAHER, V. M., MILLER, E. C., MILLER, J. A. and SZYBALSKI, W. (1968) Mutations and decreases in density of transforming DNA produced by derivatives of the carcinogens 2-acetylaminofluorene and N-methyl-4-aminoazobenzene. *Mol. Pharmacol.* **4**: 411–426.

MAHER, V. M., MILLER, J. A., MILLER, E. C. and SUMMERS, W. C. (1970) Mutations and loss of transforming activity of *Bacillus subtilis* DNA after reaction with esters of carcinogenic N-hydroxy aromatic amides. *Cancer Res.* **30**: 1473–1480.

MAHER, V. M., OUELLETTE, L. M., CURREN, R. D. and MCCORMICK, J. J. (1976b) Frequency of ultraviolet light-induced mutations is higher in xeroderma pigmentosum variant cells than in normal human cells. *Nature* **261**: 593–595.

MAHER, V. M., ROWAN, L. A., SILINSKAS, K. C., KATELEY, S. A. and MCCORMICK, J. J. (1982) Frequency of UV-induced neoplastic transformation of diploid human fibroblasts is higher in xeroderma pigmentosum cells than in normal cells. *Proc. natn. Acad. Sci USA* **79**: 2613–2617.

MAHONEY, E. M., BALL, J. C., SWENSON, D. H., RICHMOND, D., MAHER, V. M. and MCCORMICK, J. J. (1984) Cytotoxicity and mutagenicity of aflatoxin dichloride in normal and repair-deficient diploid human fibroblasts. *Chem. Biol. Interactions* **50**: 59–76.

MCCORMICK, J. J. and MAHER, V. M. (1981) Measurement of colony-forming ability and mutagenesis in diploid human cells. In: *DNA Repair: A Laboratory Manual of Research Procedures*, Vol. 1, Part B, pp. 501–521. Friedberg, E. C. and Hanawalt, P. C. (eds). Marcel Dekker, Inc., New York.

MCCORMICK, J. J. and MAHER, V. M. (1983) Role of DNA lesions and DNA repair in mutagenesis and transformation of human cells. In: *Human Carcinogenesis*, pp. 401–420. Harris, C. E. and Autrup, H. N. (eds). Academic Press, New York.

MCCORMICK, J. J., KATELEY-KOHLER, S. and MAHER, V. M. (1985) Factors involved in quantitating induction of anchorage independence in diploid human fibroblasts by carcinogens. In: *Carcinogenesis—A Comprehensive Survey Volume 9. Mammalian Cell Transformation: Mechanisms of Carcinogenesis and Assays for Carcinogens*, pp. 233–247. Barrett, J. C. and Tennant, R. W. (eds). Raven Press, New York.

MCCORMICK, J. J., KATELEY-KOHLER, S., WATANABE, M. and MAHER, V. M. (1986) Fibroblasts from xeroderma pigmentosum variants are abnormally sensitive to UV-induced transformation to anchorage independence. *Cancer Res.* **46**: 489–492.

MCCORMICK, J. J., SILINSKAS, K. C. and MAHER, V. M. (1980) Transformation of diploid human fibroblasts by chemical carcinogens. In: *Carcinogenesis, Fundamental Mechanisms and Environmental Effects*, pp. 491–498. Pullman, B., Ts'o, P. O. P. and Gelboin, H. (eds). D. Reidel Publ. Co., Dordrecht.

MEHTA, J. R., LUDLUM, D. B., RENARD, A. and VERLY, W. G. (1981) Repair of O^6-ethylguanine in DNA by a chromatic fraction from rat liver: transfer of the ethyl group to an acceptor protein. *Proc. natn. Acad. Sci. USA* **78**: 6766–6770.

MILO, G. E., JR and DIPAOLO, J. A. (1978) Neoplastic transformation of human diploid cells in vitro after chemical carcinogen treatment. *Nature* **275**: 130–132.

MILO, G. E., OLDHAM, J. W., ZIMMERMAN, R., HATCH, G. G. and WEISBRODE, S. A. (1981) Characterization of human cells transformed by chemical and physical carcinogens in vitro. *In Vitro* **17**: 719–729.

PATTON, J. D., ROWAN, L. A., MENDRALA, A. L., HOWELL, J. N., MAHER, V. M. and MCCORMICK, J. J. (1984) Xeroderma pigmentosum (XP) fibroblasts including cells from XP variants are abnormally sensitive to the mutagenic and cytotoxic action of broad spectrum simulated sunlight. *Photochem. Photobiol.* **39**: 37–42.

PEREZ-RODRIGUES, R., CHAMBARD, J. C., VAN OBBERGHEN-SCHILLING, E., FRANCHI, A. and POUYSSEGUR, J. (1981) Emergence of hamster fibroblast tumors in nude mice-evidence for in vivo section leading to loss of growth factor requirement. *J. Cellul. Phys.* **190**: 387–396.

PEREZ-RODRIGUES, R., FRANCHI, A., DEYS, B. F. and POUYSSEGUR, J. (1982) Evidence that hamster fibroblast tumors emerge in selections leading to growth factor 'relaxation' and to immune resistance. *Int. J. Cancer* **29**: 309–314.

PETINGA, R. A., ANDREWS, A. D., TARONE, R. E. and ROBBINS, J. H. (1977) Typical xeroderma pigmentosum complementation group A fibroblasts have detectable ultraviolet light-induced unscheduled DNA synthesis. *Biochim. Biophys. Acta* **479**: 400–410.

PETO, R. (1977) Epidemiology, multistage models and short-term mutagenicity tests: In: *Origins of Human Cancer*, Book C, pp. 1403–1428. Hiatt, H. H., Watson, J. D. and Winsten, J. A. (eds). Cold Spring Harbor Laboratory, Cold Spring Harbor, New York.

REZNIKOFF, C. A., BRANKOW, D. W. and HEIDELBERGER, C. (1973) Establishment and characterization of a cloned line of C3H mouse embryo cells sensitive to post confluence inhibition of division. *Cancer Res.* **33**: 3231–3238.

ROBBINS, J. H., KRAEMER, K. H., LUTZNER, M. A., FESTOFF, B. W. and COON, H. G. (1974) Xeroderma pigmentosum an inherited disease with sun sensitivity, multiple cutaneous neoplasms, and abnormal DNA repair. *Ann. intn. Med.* **80**: 221–248.

ROPER, P. R. and DREWINKO, B. (1976) Comparison of *in vitro* methods to determine drug-induced cell lethality. *Cancer Res.* **36**: 2182–2188.

SILINSKAS, K. C., KATELEY, S. A., TOWER, J. E., MAHER, V. M. and MCCORMICK, J. J. (1981) Induction of anchorage independent growth in human fibroblasts by propane sultone. *Cancer Res.* **41**: 1620–1627.

SIMON, L., HAZARD, R. M., MAHER, V. M. and MCCORMICK, J. J. (1981) Enhanced cell killing and mutagenesis by ethylnitrosurea in xeroderma pigmentosum cells. *Carcinogenesis* **2**: 567–570.

SUTHERLAND, B. M., CEMINO, J. S., DELIHAS, N., SHIH, A. and OLIVER, R. (1980) Ultra-violet light-induced transformation of human cells to anchorage-independent growth. *Cancer Res.* **40**: 1934–1939.

SUTHERLAND, B. M., DELIHAS, N. C., OLIVER, R. O. and SUTHERLAND, J. C. (1981) Action spectra for ultra-violet light-induced transformation of human cells to anchorage-independent growth. *Cancer Res.* **41**: 2211–2214.

YANG, L. L., MAHER, V. M. and MCCORMICK, J. J. (1980) Error-free excision of the cytotoxic mutagenic N^2-deoxyguanosine DNA adduct formed in human fibroblasts by ($\pm$)-$7\beta,8a$-dihydrox-$9a$, $10a$-epoxy, 7,8,9,10-tetrahydrobenzo(a)pyrene. *Proc. natn. Acad. Sci. USA* **77**: 5933–5937.

YANG, L. L., MAHER, V. M. and MCCORMICK, J. J. (1982) Relationship between excision repair and the cytotoxic and mutagenic effect of the 'anti' 7,9-diol-9,10-epoxide of benzo(a)pyrene in human cells. *Mutat. Res.* **94**: 435–447.

ZIMMERMAN, R. J. and LITTLE, J. B. (1983) Characterization of a quantitative assay for the *in vitro* transformation of normal human diploid fibroblasts to anchorage independence by chemical carcinogens. *Cancer Res.* **43**: 2176–2182.

CHAPTER 5

THE INDUCTION AND REGULATION OF RADIOGENIC TRANSFORMATION *IN VITRO*: CELLULAR AND MOLECULAR MECHANISMS

CARMIA BOREK

Radiological Research Laboratory, Department of Radiology and Department of Pathology, Columbia University College of Physicians & Surgeons, New York, NY 10032, U.S.A.

Abstract—Rodent and *human* cells in *culture, transformed in vitro* by *ionizing radiation, ultraviolet light,* or *chemicals* into malignant cells afford us the opportunity to probe into early and late events in the *neoplastic process* at a cellular and molecular level. Transformation can be regarded as an abnormal expression of cellular genes. The initiating agents disrupt the integrity of the genetic apparatus altering *DNA* in ways that result in the activation of cellular transforming genes (*oncogenes*) during some stage of the neoplastic process. Events associated with *initiation* and *promotion* may overlap to some degree, but in order for them to occur, cellular permissive conditions must prevail. *Permissive factors* include thyroid and steroid *hormones*, specific states of *differentiation*, certain stages in the *cell cycle*, specific *genetic impairment*, and inadequate antioxidants. Genetically susceptible cells require *physiological states* conducive to transformation. These may differ with *age, tissue,* and *species* and in part may be responsible for the observed *lower sensitivity* of human cells to transformation.

FOREWORD

Radiation is a fact of life. It occurs in nature and pervades the environment. It can also be produced artificially and as such has been used for several decades. The scope of this article is limited to the *in vitro* oncogenic effects of radiation with emphasis on ionizing radiation. However, because some aspects of radiation induced transformation are similar to those induced by other carcinogens one must sometimes expand the horizon and generalize beyond the scope.

Although some basic aspects of radiobiology and radiation physics are described in detail elsewhere (Bacq and Alexander, 1955; Rossi and Kellerer, 1974; Hall, 1978), some facts should be mentioned here. Radiation is termed 'ionizing' when it possesses sufficient energy to remove electrons from their orbits in atoms constituting the irradiated material. This leads to breaking of chemical bonds and results in permanent changes.

In most cases, ionization occurs through electrically charged particles or nuclear components such as protons or alpha particles. These are directly ionizing radiations. They may originate from external or internal sources. They can also be generated inside the irradiated matter following exposure to indirectly ionizing radiation including electromagnetic quanta (or photons) such as X rays and γ rays, or electrically neutral particles such as neutrons.

Different types of ionizing radiations are characterized by their linear energy transfer (LET), i.e. the rate of energy deposition along their tracks. Low energy radiations such as X rays or γ rays have low LET radiations and are sparsely ionizing whereas α particles, neutrons and heavy ions release densely clustered ionizations along their tracks and are high LET radiations. These different patterns of ionizations account for many differences in the biological effects of these radiations. Comparison of various radiations in producing a particular effect is defined as the relative biological effectiveness (RBE). It is expressed as the ratio of the absorbed doses required to produce the same biological effect (e.g. cell killing or cell transformation). Ionization results in short-lived (10^{-10} sec) ion pairs which go on to produce free radicals of a somewhat longer life (10^{-5} sec).

The amount of ionizing radiation can be characterized in terms of exposure, as expressed by the unit roentgen (R), or in terms of absorbed dose expressed as rad or Gray (1 Gy = 100 rad). The exposure of water or soft tissue to 1 roentgen of X rays or γ rays results in an absorbed dose or approximately 1 rad or 0.01 Gy.

1. INTRODUCTION

The carcinogenic potential of X rays in humans was realized within the first decade after their discovery by Roentgen in 1895 (Brown, 1936). This was confirmed in later years through epidemiological data, the largest single source of information being from Hiroshima and Nagasaki (Rossi and Kellerer, 1974; BEIR, 1972; UNSCEAR, 1977). The data provided good evidence to suggest that various forms of cancer including leukemia represent the most significant late effect when human populations are exposed to substantial doses of radiation.

Work with animals established at the experimental level the ability of radiation to cause many types of malignancies and indicated the existence of dose-related effects (Fry and Ainsworth, 1977; Upton et al., 1970; Ullrich et al., 1977; Bond et al., 1960; Kaplan, 1967).

While epidemiology and animal studies have yielded much information on the subject, they have their limitations in studies concerned with the oncogenic effects of low doses of radiation and in studies addressing the possible underlying mechanism of radiation carcinogenesis.

2. CELL CULTURES

The development of cell culture systems has made it possible to study the cellular and molecular mechanisms involved in radiation transformation under defined conditions devoid of host-mediated homeostatic modulating factors. These *in vitro* systems allow qualitative and quantitative assessments of underlying mechanisms. They afford the opportunity to study dose-related and time-dependent interactions of radiation with single cells and to identify factors and conditions which may prevent or enhance cellular transformation by radiation.

3. CELL TRANSFORMATION *IN VITRO*

A most significant contribution, which served as a basis for the study of radiation oncogenesis *in vitro* was the development of the clonal assay by Puck and Marcus (1956) and their demonstration of a dose-related effect of radiation on the survival of single cells. These findings made it possible in later years to assess which surviving cells had been transformed into a neoplastic state following exposure to radiation, and to determine the incidence of transformation.

Thus in 1966 Borek and Sachs first reported the direct oncogenic effect of X rays by exposing diploid hamster embryo cells of 300 rad of X rays and transforming a fraction of them into cells which differed morphologically from untreated controls (Fig. 1). The transformed cells gave rise to tumors upon injection into hamsters, whereas untreated cells showed no spontaneous transformation (Borek and Sachs, 1966a, 1967). The work demonstrated that in order to fix radiation transformation as a hereditary property, cell divisions must take place soon after exposure, and that subsequent additional replications are essential for expression of the neoplastic state (Borek and Sachs, 1966a, 1967, 1968). The work also indicated that there exists among cells a differential physiological and genetic competence to be transformed and that surface-mediated cell recognition was modified in culture upon transformation (Borek and Sachs, 1966b).

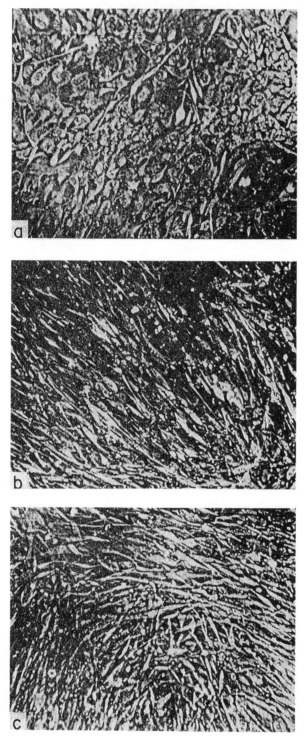

Fig. 1. A progression of neoplastic transformation *in vitro* in mass cultures of hamster embryo cells following exposure to 300 rad of X rays (Borek and Sachs, 1966a). (a) Appearance of single transformed fusiform cells 18 days after exposure on a background of flat untransformed cells. (b) The same culture 25 days after irradiation. (c) The same culture 58 days later. Note that the transformed cells dominate the culture. By this time the controls degenerated. Phase × 340.

Transformation of mammalian cells *in vitro* by radiation was later approached in mouse cell systems (Terzaghi and Little, 1976a; Little, 1979; Han and Elkind, 1979; Miller and Hall, 1978) and in human cells (Borek, 1980a), making it possible to evaluate the effects of radiation on cells across the lines of various species.

3.1. Culture Systems Currently used in Radiation Transformation Studies

There is a limited number of cell systems currently used in radiation transformation studies. These are composed of fibroblast-like cells, where morphological criteria serve well in quantitative assays of transformation. Because in the human the preponderance of carcinomas over sarcomas is unequivocal, there is a constant and urgent need to develop epithelial cultures to study transformation. A number of epithelial cell systems have been developed and used in studies on chemically induced transformation (reviewed in Borek, 1980a, 1982a, 1983; Upton, 1983), but so far not applied to studies in radiation carcinogenesis. There is a particular uniformity in fibroblast-like cells that does not exist in epithelial cells, whose susceptibility to radiogenic transformation may depend on the particular differentiated qualities and on the source and age of tissue from which they are derived. Human as well as animal data have indicated that in radiation carcinogenesis, latency, age and specific organ susceptibility determine the incidence of cancer. A further difficulty with epithelial cells arises from the fact that criteria for early stages in the neoplastic state of epithelial cells are expressed phenotypically in a less consistent manner than in fibroblasts (see Table 1).

Among the fibroblast lines there are two main cell systems used in radiogenic transformation studies: primary cultures and cell strains, and established cell lines.

3.1.1. Primary Cultures and Cell Strains

Primary cells are freshly derived from animal or human tissue. They descend from the cells *in situ* and consist of diploid cells. These cultures have a finite life span that differs from one cell type to another and is related to the longevity of the species from which they originate.

TABLE 1. *Characteristics of Fibroblasts and Epithelial Cells Malignantly Transformed* In Vitro *Distinguishing Them from Untransformed Parental Cells*

Property	Fibroblasts	Epithelial Cells
Morphology (light microscopy)	Pleomorphic, refractile crisscross orientation; irregular growth pattern	Often do not dramatically differ from normal, somewhat more pleomorphic in some cases (e.g. liver epidermal cells)
Topography (scanning electron microscope)	Surface features are enhanced	Inconsistent changes; sometimes an increase in microvilli
Cell density	Increased saturation density, loss of density-dependent inhibition of growth and multilayering	Inconsistent; depending on the cell line and the tissue of origin; in some cells lines piling up of cells, in others maintenance of monolayer growth pattern, some pleomorphism
Serum requirement for cellular growth	Lower in rodent cells; less pronounced feature in transformed human cells, because normal human cells can grow in lower serum levels	Low as in the normal; has yet to be studied in a variety of systems
Growth in medium with low calcium	Yes	Yes
Altered cell surface glycoproteins and glycolipids	Yes	Yes
Agglutinability by low concentrations of lectins	Yes	Yes
Increased protease production	Yes	Inconsistent
Changes in cytoskeleton	Pronounced	Inconsistent
Growth in agar	Yes	Yes
Tumorigenicity	Yes	Yes

IN VITRO CELL TRANSFORMATION

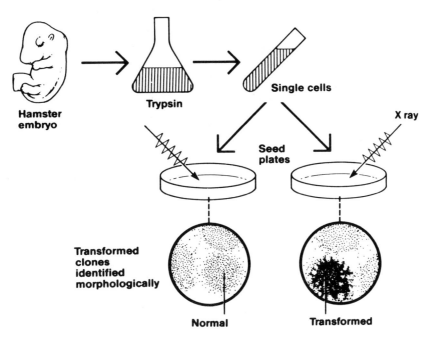

FIG. 2. Assay for hamster embryo cells transformed *in vitro* by radiation. Hamster embryos in midterm gestation are removed, minced and trypsinized progressively with 0.25% trypsin. After removal of the trypsin by centrifugation, cells are suspended in complete medium and seeded as single cells on feeder layers. They are exposed to radiation 24 hr later. After 8–10 days in incubation, cultures are fixed and stained. Transformed colonies are distinguished morphologically from controls.

The primary cultures used most commonly in radiation transformation studies are those of mixed cell populations derived from hamster embryo cells [Figs 1, 2, 3(a, b)] (Borek and Sachs, 1966a; Borek and Hall, 1973; DiPaolo *et al.*, 1976). Since these are diploid cells they have an advantage. They senesce upon continuous subculture, allowing 'immortal' transformants to emerge against a background of dying cells, thus confirming *in vitro* their distinctive transformed state. Cell survival and cell transformation can be scored simultaneously in the same dishes, and the rate of spontaneous transformation in these cells is less than 10^{-6}. Expression time for transformation is 8–10 days, a relatively short period, and the cells can be cryopreserved (Pienta *et al.*, 1977). Transformed colonies are identifiable by dense multilayered cells, random cellular arrangement, and haphazard cell-cell orientation accentuated at the colony edge [Figs 1, 3(a, b)]. Normal counterparts are usually flat, with an organized cell-cell orientation. Because of the mixed population of cells there exist untransformed cells which may possess a higher cell density than the usual flat colonies. These, however, do not exhibit the randomness at the colony edge just described.

Human primary cultures used in transformation studies are fibroblasts derived from adult human skin (Kakunaga, 1978; Borek, 1980a), human embryos (Sutherland *et al.*, 1981), or foreskin (Milo and DiPaolo, 1978; Silinskas *et al.*, 1981; Zimmerman and Little, 1983b). The assay is a focus assay in which the loss of cell density inhibition among the transformed fibroblasts renders them capable of proliferating over the untransformed sheets of cells, thus forming distinct recognizable foci (described later).

3.1.2. *Established Cell Lines*

Established cell lines have an unlimited life span. They represent cell populations which originated as primary cultures. Following a continuous and meticulously timed regime of subculturing a selected population emerges that had undergone a 'crisis', enabling the cells

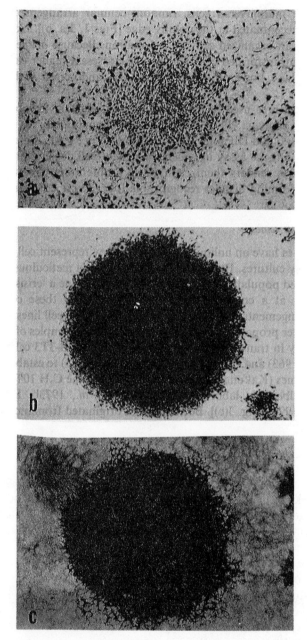

Fig. 3. (a) A normal colony of hamster embryo cells. (b) A colony of X-ray transformed hamster embryo cells. (c) A focus of C_3H 10T1/2 cells transformed by X ray growing over normal cells.

to grow indefinitely at a constant rate. The karyotype of these cells shows various chromosomal rearrangements and heteroploidy. Often these cell lines are cloned and the cloned cells are further propagated into large populations. Examples of cell lines that have been used extensively in transformation studies are the BALB-3T3 cell line developed by Todaro and Green (1963) and further cloned (Kakunaga, 1973) to establish susceptible and nonsusceptible cell lines (Kakunaga and Crow, 1980), and the C_3H 10T1/2 clone 8 cell line developed in Heidelberger's laboratory (Reznikoff et al., 1973b; Heidelberger, 1975; Heidelberger et al., 1978) [Fig. 3(c)]. Both cell lines originated from mouse embryos. They are transformable by a variety of oncogenic agents and used extensively in radiation transformation studies, in particular the 10T1/2 cells (Fig. 4). The advantage of these systems lies in the fact that they are 'immortal', so one can continuously utilize particular cell passages by maintaining 'banks' of frozen cells. The disadvantages are that the cells are not diploid, and if not treated meticulously as originally described they can give rise at high passage to spontaneous transformants. Transformation assay is a focus assay, thus survival must be scored in separate sets of dishes. In the C_3H 10T1/2 system three types of transformed foci are identifiable (Reznikoff et al., 1973a; Terzaghi and Little, 1976a): types I, II and III. Their morphology can be related to their oncogenic potential, type III being the most malignant [Fig. 3(c)].

3.1.3. In Utero–In Vitro Systems

Here exposure to radiation is carried out *in utero* and the tissues are cultured and assayed for transformation *in vitro* (Borek et al., 1977).

3.2. INITIATION AND PHENOTYPIC EXPRESSION OF TRANSFORMATION

One of the basic conundrums in cancer research evolves from our inability at the present time to distinguish unequivocally primary events associated with initiation of neoplastic transformation from those that function as secondary events. Although we aim to identify the process of initiation and consequently hope to modulate it, we are faced with the fact that at present we determine the occurrence of initiation by its phenotypic expression. Thus, although radiation carcinogenesis was recognized about 90 years ago, we are only now becoming aware of the mechanisms involved and still judge neoplastic transformation by a variety of phenomena associated with the neoplastic phenotype. These phenomena appear to be similar irrespective of the initiating oncogenic agent, whether it is a virus whose contribution is the introduction of new genetic material, a chemical carcinogen altering DNA, or radiation, whose initiating genetic damaging action on the cell is established and over within a fraction of a second.

We therefore strive to define various steps within the processes of transformation and try to associate cellular and molecular events with each step.

3.2.1. Sequence of Events in Transformation in vitro

(a) Initiation, i.e. exposure of cultured cells to the carcinogen and damage to DNA.
(b) Fixation of the transformed state requiring cell replication within hours after

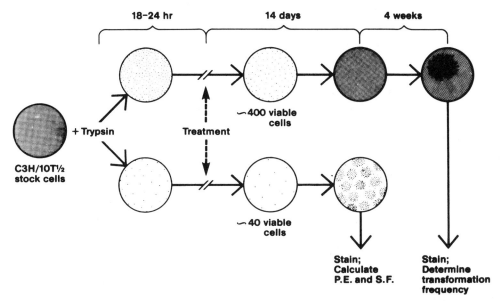

FIG. 4. Protocol for assay of cell transformation in C_3H 10T1/2 cells. Cells are allowed to attach for 24 hr before irradiation. For assessment of survival, cells are fixed and stained 2-weeks later. For transformation assays, cells are incubated for 6 weeks before staining and scoring for foci.

initiation (Borek and Sachs, 1966a, 1967, 1968; Reznikoff *et al.*, 1973a; Terzaghi and Little, 1976a; Kakunaga, 1974, 1975; Little, 1979).

(c) Expression of the transformed state of a single cell requiring several cell replications, depending on the cell type. The results are the growth of a focus or colony (depending on the assay), which in fibroblasts and some epithelial cells is morphologically distinct from control (Borek and Sachs, 1966b, 1967, 1968; Reznikoff *et al.*, 1973a; Kakunaga, 1974, 1975; Little, 1979) (Fig. 3). It should be added that there is little information on the neoplastic characteristic of exposed fibroblasts, which do not differ morphologically from the normal. So far, the assessment of the transformed state in fibroblasts has consistently adhered to the premise that the earliest observable phenotypic change in the process of transformation is morphological.

A variety of agents and factors can modify the events described above and enhance or inhibit the neoplastic process. These will be discussed in detail later. At this point it is important to state the following: at the time of initiation agents which have a damaging effect on DNA can act in concert with radiation and serve as cocarcinogens (Fig. 4). At stages of expression tumor promoters can act to enhance the frequency of neoplastic transformation by radiation (Borek, 1982a; Borek and Troll, 1983; Kennedy *et al.*, 1980b).

Agents or conditions which prevent initiation could serve as protective anticarcinogens while those which antagonize promotion could be defined as antipromoters.

3.2.2. *Methods Used for Transformation Studies*

(a) Exposure of mass cultures to radiation or other oncogenic agents and continuous subculturing for several weeks or months until transformed cells are selected out and form foci. Periodic clonings are made to assess the frequency of transformation (Borek and Sachs, 1966a, 1967).

(b) Treatment of mass cultures and cloning out at various periods of time after exposure. Transformation can then be scored in a colony or focus assay (Borek and Sachs, 1966a, 1967; Han and Elkind, 1979).

(c) Treatment of single cells seeded at a low-density clonal level sometimes on feeder cells (Puck and Marcus, 1956). Each cell is then allowed to proliferate into a distinct clone. Cultures are fixed and stained. Transformed clones are distinguished morphologically from normal by high cell density and random cell orientation. This method is used routinely in quantitative studies and has been applied in the hamster system (Borek and Hall, 1973, 1974; Borek *et al.*, 1978), where incubation time is 8–10 days (Fig. 3).

(d) Treatment of single cells seeded at low density. Cells are allowed to proliferate to high density. Transformed cells form foci clearly distinct against a background of flat density-inhibited cells. The characteristics of the foci are high cell density as well as randomness of cell organization, especially in the peripheral area of the focus invading the surrounding area. Incubation time for the appearance of these foci ranges from 4 to 8 weeks, depending on the cells employed (Fig. 2). The method is widely used, for the 3T3 and 10T1/2 transformation assays (4 and 6 weeks incubation time, respectively) (Reznikoff *et al.*, 1973b; Terzaghi and Little, 1976a; Miller and Hall, 1978; Borek *et al.*, 1979; Little, 1979) and can be applied in human transformation studies where incubation is approximately 8 weeks (Borek, 1980a; Kakunaga, 1978).

In all these transformation methods, especially in the treatment of single cells, an important prerequisite is knowledge of the cytotoxic effect of the oncogenic agent on the particular cells used. Thus a dose-response curve for cell survival is essential to evaluate the surviving fractions, following exposure to a specified radiation (Borek and Hall, 1973; Terzaghi and Little, 1976a; Miller *et al.*, 1979; Han and Elkind, 1979). Another factor of importance that can be applied from knowledge of the survival curve is the requirement for a particular number of viable cells. Transformation rate is decreased if the number of surviving cells exceeds a certain range, such as 200–400 viable cells in the 10T1/2 system (Terzaghi and Little, 1976a; Reznikoff *et al.*, 1973b) and approximately 10–30% viability in human cell transformation (Borek, 1980a). Though the mechanisms underlying this effect are not understood, an inhibitory action of normal cells or growth of transformed cells has been observed (Borek and Sachs, 1966a).

Quantitative analysis in the hamster cell system, where transformation rates are established from among the survivors in a single plate, can be dealt with statistically in several ways (Borek et al., 1978, 1983b). In the 10T1/2 cells, quantitation of transformation is different because scoring of survivors and transformants is carried out separately. These have been discussed in detail elsewhere (Reznikoff et al., 1973a, b; Terzaghi and Little, 1976a; Kennedy et al., 1980a; Fernandez et al., 1980; Mordan et al., 1983; Hall et al., 1982). For optimal quantitative analysis it is important to have no more than one transformed focus per plate because the lower adhesiveness of the transformed cells may lead to cell migration from a focus, forming secondary foci and introducing an error. Another factor that must be considered when comparing results from various laboratories is the method of irradiation. In some cases mass cultures are exposed to radiation and then trypsinized and cloned, thus introducing a variable of cellular plating efficiency *after* radiation. In other cases cells are cloned and then irradiated, thus plating efficiency is established *prior* to irradiation. Although this method is more cumbersome (requiring the irradiation of single cells contained in hundreds of plates), it eliminates the need for trypsinization after treatment, a procedure that introduces variables in the 10T1/2 system depending on the time of post irradiation at which cells are dissociated and cloned out.

3.3. Criteria for Transformation

The criteria for transformation are based on the characterization of transformed cells as compared to their untransformed parental cells. These are established using mass cultures of mixed cell populations as well as cultures grown from single isolated clones. Most of the characteristic alterations associations with the transformed state of fibroblasts hold true for epithelial cells, though in the latter with less consistency (Table 1).

Although we would like to evaluate early changes associated with transformation, we are limited by the low frequency of the event. Thus in order to be certain that we are indeed dealing with transformed cell populations, we must utilize cells that have expressed their transformed state by morphological alterations. The difficulties become evident. Critical evaluation of transformation is important at early stages following exposure and expression, yet for some assays, such as biochemical analysis, growth in agar or tumorigenicity, large cell populations are required. These depend on extensive propagation of the cells *in vitro* and the inevitable introduction of variables associated with continuous culture, such as karyotypic instability. Thus although the acquisition of phenotypic changes associated with neoplastic transformation has been considered a multistage process in rodent cells (Barrett and T'so, 1978a), the inability to assess some of the expressions until a late stage when further propagation is achieved complicates the issue.

3.3.1. *DNA and Chromosomes*

Because with ionizing radiation no specific DNA-repair enzymes have been identified in mammalian cells, exploration of the effects of radiation on transformation at the level of DNA damage and repair, and at a chromosomal level, is a study of associated phenomena. This is especially true because the frequency of transformation is low, so the relationship of chromosomal changes or DNA damage and repair, carried out on parallel cultures, can be inferred but not conclusively stated. This holds true for initiation by radiation, where DNA damage and repair (Painter and Cleaver, 1969; Painter and Young, 1972; Painter, 1978; Borek et al., 1984a, b, e) and sister chromatid exchanges (SCE) (Perry and Evans, 1975; Little et al., 1983; Miller et al., 1981; Geard et al., 1981; Borek et al., 1984e) have been investigated as well as studies on the modulation of transformation by a variety of agents (Nagasawa and Little, 1979; Kinsella and Radman, 1978; Miller et al., 1981; Geard et al., 1981; Borek et al., 1984a, b, e).

Observing a morphologically changed colony or a focus prepares the investigator to explore with more certainty the relationship between the karyotypic alteration and the phenotypic expression of the cells, though admittedly one can evaluate the neoplastic nature of these cells only later when progressive culture, and most probably karyotypic changes, have taken place. Another critical factor in studying chromosomal changes associated with transformation is the starting material. The use of the study of diploid cultures may differ from that of heteroploid cell lines where chromosomal imbalance has already taken place.

Subtle chromosomal alterations following exposure to oncogenic agents may denote genetic rearrangements and instability (Bloch-Schtacher and Sachs, 1976; Kinsella and Radman, 1978; Schimke et al., 1980; Klein, 1981), which may be associated with transformation. Specific genes may play a role in determining susceptibility to neoplastic transformation (Knudson, 1981) and point mutations may be associated with some stages in the neoplastic process. However, minimal changes in chromosome number (cells remaining near-diploid) and no changes in banding patterns were observed in diploid hamster or human cells transformed in vitro. These studies were carried out on cells within several passages after initiation and expression of transformation, i.e. at early stages after having expressed other criteria associated with their transformed state (Borek et al., 1977, 1978; DiPaolo et al., 1973; Borek, 1980b; Silinskas et al., 1981). Thus, chromosomal aberrations observed in neoplasms may be a secondary event and limited in kind and number (Mitelman, 1984).

3.3.2. *Loss of 'Contact Inhibition' and Changes in Cell Topography*

The most obvious phenotypic alterations seen in transformed fibroblasts are mediated via the cell surface. A loss of contact inhibition (Abercrombie, 1966) reduced density-dependent inhibition (Stoker and Rubin, 1967), irregular growth patterns, and ability to grow in multilayers (Berwald and Sachs, 1963; Borek and Sachs, 1966a, b; Reznikoff et al., 1973b; Terzaghi and Little, 1976a; Borek, 1980b; Kakunaga, 1978) are all features that characterize the transformed nature of fibroblasts derived from solid tissue and differentiate them from normal counterparts grown under the same conditions.

These morphological differences, so distinct in dense populations of cells, are not apparent at low density, when cells are not in contact with one another (Borek and Fenoglio, 1976). Changes at a single-cell level are seen when cell topography is evaluated using scanning electron microscopy (Borek and Fenoglio, 1976). At early stages following initiation, within 8 days after exposure to radiation and expression of transformation the relatively smooth and simple surface of normal hamster fibroblasts acquires a variety of excrescences and a marked cellular pleomorphism. These topographic changes, comprising ruffles and blebs, are present on the transformed cell surfaces throughout the cell cycle and remain an integral part of the transformed cell after many years of culture. The normal parental cells exhibited these complex features only during mitosis (Borek and Fenoglio, 1976), thus affirming other observations indicating that a variety of membrane associated properties characteristic of the neoplastic state are found in normal cells in mitosis (Borek, 1979a).

3.3.3. *Decreased Serum and Calcium Dependence*

Once rodent fibroblasts are initiated by radiation or other oncogenic agents and replicate to express their transformed state, their dependence on nutrients decreases and they can proliferate well in medium containing low serum concentrations, as low as 1% serum, whereas their untransformed counterparts remain essentially in a nonreplicating state (Borek et al., 1977; Borek, 1979a; Terzaghi and Little, 1976a). This feature, which serves well in selecting out low-frequency transformants from a background of normal cells, holds true for rodent cells but not necessarily for human fibroblasts, where some normal cells as well as the transformants grow at low serum concentrations (Borek, 1980a). Low calcium dependence, namely the ability to proliferate in medium containing less than 0.5% calcium, seems to be a feature common to transformed fibroblasts and epithelial cells of rodent origin (Swierenga *et al.*, 1978), as well as transformed human fibroblasts (Borek, 1980b). In all cases it has served as a selective feature for transformants, because normal cells die within days after exposure to maintenance medium containing a low calcium concentration, while the transformed cells thrive and can form distinct foci (Borek, 1980b).

3.3.4. *Membrane Structural Changes*

The cell membrane is a complex dynamic organelle which is thought to exert control over a variety of cellular patterns of behavior (Puck, 1977, 1979; Nielson and Puck, 1980). This control system may function by the coordination of interacting molecules of both surface receptors and submembranous fibrillar elements. This control appears to be transmembranous in nature, inextricably related to the structure of the cell membrane and to the molecular features of its surface receptors. These are glycoproteins, some of which traverse the matrix of the membrane as integral membrane proteins. Their hydrophilic portions project into the cytoplasm, where a number of interactions with cytoplasmic components such as cytoskeletal elements take place and can be affected by antimitotic drugs, hormones, and cyclic AMP (Puck, 1977).

Upon neoplastic transformation the cell membrane undergoes a variety of structural and functional changes. Neoantigens are observed (Embleton and Heidelberger, 1975) and glycoproteins decrease, disappear, or are no longer completely glycosylated (Gahnberg and Hakomori, 1973). Similarly, *sialoglycolipids* (*gangliosides*), a major group of membrane glycolipids, are incompletely glycosylated (Gahnberg and Hakomori, 1973), showing a reduction in higher gangliosides (Brady *et al.*, 1969; Borek *et al.*, 1977). The enzyme $Na^+/K^+/ATPase$, an Na-transport membrane-associated enzyme, is altered (Borek and Gurnsey, 1981) (Table 2). Intercellular communication is modified (Borek *et al.*, 1969) and cytoskeletal units are modified with the appearance of new elements (Leavitt and Kakunaga, 1980; Vandekerchkhove *et al.*, 1980; Hamada *et al.*, 1981). Some fibroblasts

TABLE 2. *Characteristics of Cloned Hamster Cells Transformed In Vitro by Neutrons*

Characteristic	Normal (Secondary Cultures)	Transformed (6th to 35th Passage)
Karyotype	Diploid	Near diploid
Growth in culture	Monolayers	Multilayers
Saturation density:		
No. of cells/cm², in 10% serum	8×10^4	2.2×10^5
No. of cells/cm², in 1% serum	Minimal growth	1.3×10^5
Doubling time (in 10% serum)	16 ± 2 hr	15 ± 2 hr
Morphology (scanning electron microscopy)	Flat and regular surface features except in mitosis	Pleomorphic with complex surface features throughout the cell cycle
Agglutinability by Con A (20 µg/ml)	—	+
Plating efficiency:		
In 10% serum	$3 \pm 1\%$	$65 \pm 5.5\%$
In 1% serum	$0.06 \pm 0.05\%$	$7.1 \pm 2.2\%$
Plasminogen activator	Low	High
Production of macrophage migration inhibitory-like factor	Absent	High
Growth in semisolid agar	—	+
Presence of C-type virus particles	—	+
Tumorigenicity	—	+

From Borek *et al.* (1978).

This table represents a study of transformed cells propagated from a transformed clone. Similar properties were found in all other clones tested, with the exception of one clone which differed by becoming malignant at a later stage.

and epithelial cells acquire an enhanced tendency to undergo agglutination following exposure to low levels of plant lectins (Borek et al., 1973), in contrast to normal counterparts, which exhibit this property in mitosis or following trypsinization.

This agglutinability could be inhibited by some protease inhibitors and enhanced by high levels of retinol (Borek, 1982b). The recovery of the membrane following trypsinization of the normal cells and their loss of agglutinability takes place within a period of approximately 6 hr, the same time scale required for restoration of ionic intercellular communication following trypsinization (Borek et al., 1969).

3.3.5. Proteolytic Enzymes Produced by Transformed Cells

The observation that a variety of surface properties characteristic of transformed cells can be mimicked in normal cells following trypsin treatment encouraged a major research effort to assess whether the process of neoplastic transformation is associated with an increased release of cellular proteases and whether protease inhibitors would modify transformation. Although information on increased protease activity (fibrinolysis) by neoplastic cells has been available since the early 1900s, when cells were grown on plasma clots, modern techniques with radioactive compounds have made it possible to analyze the processes involved (Wigler and Weinstein, 1976).

The amount of proteases produced by transformed fibroblasts depends on the origin of the cells studied and must be related to the production of these proteases by the untransformed counterparts. Such proteases include the plasminogen activator (Borek et al., 1977) and a serine protease (Borek, 1979a), an examination of 15 different acid hydrolases indicated that only acid phosphatases were elevated in the radiation-transformed cells (Borek et al., 1977).

3.3.6. Growth in Agar

The agar suspension assay serves as a selective assay for transformed cells. The underlying premise is the observation that normal cells derived from solid tissues cannot proliferate in suspension or in semisolid medium such as agar of methylcellulose. Thus the acquisition of the ability to grow in semisolid medium following exposure to oncogenic agents, including radiation (Borek and Sachs, 1966a; Borek and Hall, 1973, 1974; Lloyd et al., 1979), has been associated with the neoplastic state of the transformed cells for both rodent fibroblasts and epithelial cells (Fig. 4) (Borek, 1979a), as well as for human cells transformed in vitro (Borek, 1980a; Kakunaga, 1978; Sutherland et al., 1981; Silinskas et al., 1981; Milo and DiPaolo, 1978; Andrews and Borek, 1982; Borek and Andrews, 1983; Zimmerman and Little, 1983b).

Though growth in agar is an accepted criterion for transformation it is not proof of malignancy. There are instances where transformed cells do not grow in agar yet give rise to tumors (Borek and Sachs, 1966b), as well as other transformed cells of both rodent and human origin that grow in agar yet do not give rise to tumors (Borek, 1980b; Sutherland et al., 1981; Silinskas et al., 1981). Alternately, by manipulating growth conditions and fortifying the agar with high serum and hormonal factors, it is possible to observe colony formation by normal human fibroblasts, though these cells are unable to form tumors in the animal (Peehl and Stanbridge, 1981).

Thus growth in agar or on agar (Borek, 1980b; Zimmerman and Little, 1983a, b) can be used as a selective criterion for transformation when the employed conditions do not favor the growth of normal counterparts. To evaluate the neoplastic nature of the cells in an unequivocal way the inoculation of transformed cells into appropriate hosts should be carried out in addition to testing growth in agar.

It is of interest to note in this context that within the limited experience of human cell transformation in vitro, it seems that the ability of transformed human fibroblasts to grow in agar appears concomitantly with focus formation (Borek, 1980b; Andrews and Borek,

1982), and that often focus formation in culture can be circumvented (Silinskas *et al.*, 1981; Sutherland *et al.*, 1981; Milo and DiPaolo, 1978; Andrews and Borek, 1982; Zimmerman and Little, 1983a, b). By contrast, rodent diploid cells such as hamster cells, have been reported to grow in agar following transformation only at a later stage after morphological exhibition of transformation (Barrett and Ts'o, 1978a). This point has yet to be clarified because experiments carried out with the diploid rodent cells differ in protocol from those with the human cells. Furthermore, cell cycle time of the hamster cells is markedly shorter than that of the human cells, accounting in part for different patterns and temporal events associated with the progression of neoplastic transformation.

3.3.7. *Tumorigenicity*

The ultimate and unequivocal demonstration of malignancy is the induction of tumors in syngeneic inbred hosts or in immunosuppressed animals (Fig. 5). In the case of human cell transformation the animal of choice is the immunologically deficient nude mouse *nu/nu* (Kakunaga, 1978; Milo and DiPaolo, 1978; Borek, 1980b; Sutherland *et al.*, 1981; Silinskas *et al.*, 1981; Maher *et al.*, 1982). Because radiation transformation studies utilize fibroblasts, the injection of transformed cells gives rise to sarcomas with different degrees of differentiation. It is of interest to note that when hamsters were irradiated *in utero* and the embryos cultured *in vitro*, the transformed lines that arose induced both sarcomas and carcinomas (Borek *et al.*, 1977), indicating that though the neoplastic epithelial cells were not visible in culture, they developed in the host as epithelial tumors.

3.3.8. *DNA Transfection*

The ability of genomic high molecular DNA purified from *in vitro* transformed cells to transmit the transformed phenotype to normal cells by DNA mediated transfer (Wigler *et al.*, 1979; Perucho *et al.*, 1981) constitutes an important criterion for the neoplastic state of the cells exposed to the carcinogen (Shilo and Weinberg, 1981; Cooper, 1982; Borek *et al.*, 1984d; Borek *et al.*, 1987). It indicates that the transformed phenotype of the cells exposed

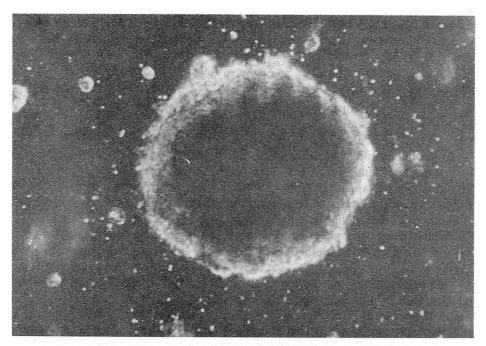

FIG. 5. Human cells from Bloom syndrome patient transformed by u.v. and growing in agar. (From Borek and Andrews, 1983.)

in vitro to the carcinogen is encoded in the DNA. This criterion more than all the others aids in the mechanistic studies of transformation by further analyzing the specific transforming genes which are activated as a result of exposure to the carcinogen and the elucidating nature of the genetic changes (Reddy *et al.*, 1982; Duesberg, 1983; Land *et al.*, 1983; Guerrero *et al.*, 1984; Borek, 1985b; Borek and Ong, 1984; Borek *et al.*, 1984d; Slamon *et al.*, 1984; Tabin *et al.*, 1982).

4. RADIOGENIC TRANSFORMATION *IN VITRO*

4.1. Ionizing Radiation

Though the criteria for transformation were described in the previous section the following will summarize these criteria as they have been applied by various investigators to radiogenic transformation studies.

(a) Formation of multilayered clones or foci under conditions where normal cells do not display this pattern (Borek and Sachs, 1966a, b, 1967, 1968; Borek and Hall, 1973, 1974; Borek *et al.*, 1977, 1978; Borek, 1980b; Terzaghi and Little, 1976a; Miller and Hall, 1978; Little, 1979; Han and Elkind, 1979; Lloyd *et al.*, 1979; Yang and Tobias, 1980).

(b) Alteration in cell topography as evaluated by scanning electron microscopy (Borek and Fenoglio, 1976).

(c) Chromosomal analysis (Borek and Hall, 1973; Borek *et al.*, 1977, 1978; Borek, 1980b; Miller *et al.*, 1981; Nagasawa and Little, 1979).

(d) Proliferation of transformed cells in low calcium medium (Borek, 1980b).

(e) Proliferation of transformed cells in medium containing low serum concentration (Borek and Hall, 1973; Borek *et al.*, 1977, 1978; Borek, 1980b; Terzaghi and Little, 1976a).

(f) Agglutination of transformants but not of normal cells by low concentrations of plant lectins (Borek and Hall, 1973; Borek *et al.*, 1978; Borek, 1980b).

(g) Alterations in ganglioside pattern (Brady *et al.*, 1969; Borek *et al.*, 1977).

(h) Increased proteolytic activity in transformed cells (Borek *et al.*, 1977; Borek, 1979a).

(i) Changes in intercellular communication (Borek *et al.*, 1969).

(j) Growth in soft agar under conditions where normal cells do not grow (Borek and Sachs, 1966a, b; Borek and Hall, 1973; Borek *et al.*, 1977, 1978; Borek, 1980b; Lloyd *et al.*, 1979).

(k) Tumor formation upon injection into appropriate hosts (Borek and Sachs, 1966a; Borek and Hall, 1973; Borek *et al.*, 1977, 1978; Borek, 1980b; Terzaghi and Little, 1976a; Han and Elkind, 1979).

(l) Transfection of DNA from radiation transformed cells into normal cells (Borek and Ong, 1984, Borek *et al.*, 1984d; Borek, 1985b; Borek *et al.*, 1986a).

4.1.1. *The Oncogenic Effects of Low-LET Radiation*

4.1.1.1. Single dose effects. (a) *Neoplastic transformation of diploid cell strains*

The successful induction of neoplastic cell transformation *in vitro* by X rays was first reported by Borek and Sachs (1966a). Short-term primary cultures derived from midterm golden hamster embryos were used as the source of normal cells. Mass cultures were irradiated with 300 rad and subcultured at low density 3 days later onto rat feeder layers. Further progressive subculturing without feeders of both irradiated cells and controls resulted in senescence in control cultures, and a gradual enhancement of mitotic rate in the X-ray-irradiated cells. Within 3 to 4 weeks after exposure, foci of fusiform cells began to pile up and overtake the culture. Quantitative evaluation of this transformation event was carried out by irradiating mass cultures and cloning them out immediately. A 0.7–0.8% transformation was observed in the treated cultures but none in the controls ($< 10^{-6}$). It is of interest to note that transformability of the embryonic cells by X rays declined with cell passage *in vitro*, and at passage three to four no transformation was observed following irradiation with 300 rad (Borek and Sachs, 1967). The ability of the transformed cells to grow in agar and to form tumors in 6-week-old hamsters was tested. Some of the cells did not grow in agar yet gave rise to tumors in the animals. Some of the tumors (fibrosarcomas) regressed (Borek and Sachs, 1966a); upon subsequent injection of cells, progressive tumor growth leading to the death of the animal was seen (Borek and Sachs, 1967).

Later studies to determine conditions and requirements for X-ray-induced (transformation showed that one to two cell replications were essential for the fixation of the transformed state (Borek and Sachs, 1967). Replication had to take place within 24 hr after exposure of log-phase cultures to X rays. Irradiation of mass cultures and trypsinization soon thereafter gave similar results. Inability to divide resulted in a loss of fixation of the transformed state (Borek and Sachs, 1967, 1968). This was indicated by maintaining cultures at plateau phase in liquid holding for 24 to 72 hr after irradiation. Transformation declined progressively and was inversely proportional to the duration of maintenance in plateau phase. Loss of fixation was inhibited by keeping the plateau-phase cells at 25°C for 24 to 72 hr. When cells were cloned after liquid holding at low temperature, transformation rates were fully restored (Borek and Sachs, 1967). This suggested that repair mechanisms were involved in the loss of fixation and that maintenance at low temperatures, where repair was delayed, prevented this loss. Further experiments indicated that expression of transformation could occur within 2 days after treatment. Irradiation of cloned cells at different stages of growth resulted in partial clonal transformation whose expression was related to the number of days in culture (Borek and Sachs, 1967). Once cells were transformed by X rays, they acquired surface properties that were specific for every transformed line individually derived. Thus cells transformed at different times by X rays were incapable of proliferating over other X-ray-transformed cell lines as well as on various chemically and virally transformed cell lines (Borek and Sachs, 1966a).

Surface-mediated changes following radiogenic transformation were also indicated by the loss of intercellular communication in some X-ray-transformed hamster embryo lines, as measured by ionic movement via permeable membrane junctions (Borek *et al.*, 1969) and in changes of cellular gangliosides (Brady *et al.*, 1969; Borek *et al.*, 1977).

The first dose-response relationship for radiation-induced *in vitro* malignant transformation was shown by Borek and Hall (1973) using the hamster clonal system. Single cells seeded onto syngeneic feeder cells were exposed 24 hr later to X ray doses ranging from 1 to 600 rad. The results shown in Fig. 6 indicated that transformation incidence per survivor was evident at doses as low as 1 rad and that the frequency rose with doses from 1 rad up to a plateau of 150–300 rad. At higher doses radiation toxicity appeared to be a competing factor in these asynchronous mixed cell populations and the transformation rate declined. Recent experiments using the same system indicate that cell transformation can be detected at doses as low as 0.3 rad of X rays (Borek and Hall, 1973) (Fig. 7). Other data indicate that at low doses the effectiveness of X rays in producing transformation is about twice as high as that induced by ^{60}Co γ rays (Borek *et al.*, 1983b) (Fig. 8). This could be predicted on microdosimetric grounds and is important in radiation protection.

(b) *Transformation of established cell lines.* Utilizing the C_3H 10T1/2 mouse embryo cell line (Reznikoff *et al.*, 1973b) and following the transformation assay described for chemically induced transformation (Reznikoff *et al.*, 1973b), Terzaghi and Little (1976b) carried out a study showing that these cells are transformable by X rays. A dose response for transformation indicated that transformation rate per viable cell rose exponentially from 100 rad up to 400 rad and remained constant at higher doses. Similar dose-response curves were reported by Han and Elkind (1979), Yang and Tobias (1980), Miller and Hall (1978) and Miller *et al.* (1979), though interlaboratory differences in transformation frequencies at the various doses may exist. It may be noted at this point that a variability in transformation in the 10T1/2 cells can vary with batches of serum.

Similar to the hamster embryo systems fixation of X-ray-induced transformation in the 10T1/2 cells required replication soon after exposure to radiation and further replications for expression of the transformed state (Terzaghi and Little, 1976a). As in the hamster

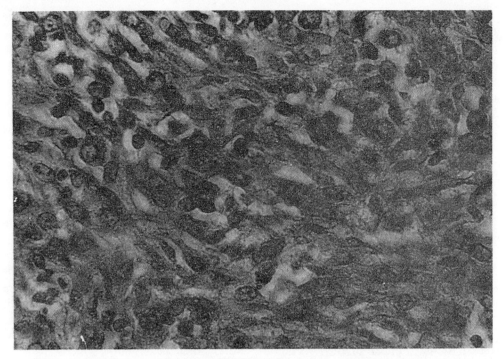

FIG. 6. Histological section of a sarcoma produced in a hamster following injection of X-ray transformed hamster embryo cells.

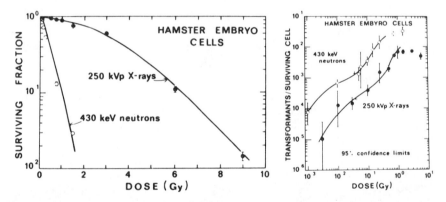

FIG. 7. (a) Pooled data for the survival of hamster embryo cells irradiated with 250 kVp X rays (open circles) or 430 keV monoenergetic neutrons (closed circles). Error bars show the estimated standard deviations. 1 Gy = 1 gray = 100 rad. (b) Pooled data for the hamster embryo cells of the number of transformants/surviving cell following irradiation with 250 kVp X rays (solid circles) or 430 keV monoenergetic neutrons (open circles) produced at the Radiological Research Accelerator Facility. (Error bars) 95% confidence intervals for the estimated value. The curves were obtained using B-spine fitting (in the least-squares sense) and with the additional constraint of nontoxicity upward; as such they shall be regarded only as a smooth representation of the shape of the data with a minimum of parametric-related bias. (Data from Borek et al., 1978; Borek and Hall, 1984.)

system (Borek and Sachs, 1967) liquid holding in plateau phase resulted here too in a loss of fixation (Terzaghi and Little, 1975). A reduced transformation rate was seen when cells were kept in a nondividing state for 24 hr or more after irradiation. Here, however, if cells were removed from liquid holding at 12 hr after exposure, transformation was enhanced.

The 10T1/2 cells transformed by X rays exhibited three different types of foci distinguished by their morphology, similar to those seen following treatment with chemical carcinogens (Reznikoff et al., 1973a), and these corresponded to different degrees of malignancy (Terzaghi and Little, 1976a). It was also observed that high cell density at treatment time resulted in a lower transformation frequency.

Another cloned cell line, the BALB/3T3 A31 (Kakunaga, 1973), has also been transformed by X rays (Little, 1979). Showing a dose-response curve similar in shape to that of the 10T1/2, this line requires a shorter time for expression of transformation. Here, too, the effect of cell density on transformation frequency was observed. Although the BALB/3T3 line could serve as an additional system to study radiation oncogenesis, several drawbacks were observed. Under certain conditions transformation varied from one experiment to another and a background of spontaneous transformation was observed. The complexity of this system may be in part explained by Kakunaga's finding that there exists in this cell line a range of cells whose competence to transformation by u.v. and perhaps by other agents ranges from highly susceptible cells to those that show resistance (Kakunaga and Crow, 1980).

4.1.1.2. *Split dose effects*

Transformation of cells directly *in vitro* by X rays has allowed detailed quantitative assessment of radiation effects over a wide range of doses. A logical extension of these studies has been to investigate the influence of the temporal distribution of radiation on transformation, in order to evaluate the oncogenic potential of radiation delivered as protracted doses or at low dose rates, the types of exposure that are of continuous concern to humans. Some unexpected information was obtained. Borek and Hall (1974) first reported with the hamster embryo cell system that fractionation of an X ray dose results in an elevated transformation incidence. In these experiments it was shown that two doses of 25 rad separated by 5 hr produced more transformation than a single dose of 50 rad, and that two doses of 37.5 rad were much more effective than a single exposure of 75 rad.

Splitting the dose resulted in a 70% enhancement in transformation rate as compared to the same dose delivered as a single exposure; but fractionation also produced a sparing effect on cell survival, indicating that cellular mechanisms involved in repair for survival differ from those associated with cell transformation.

Elkind and Han (1979) reported a similar observation, though they interpreted the observation differently. Experiments carried out by Terzaghi and Little (1976b), using the 10T1/2 cells and studying the effects of split doses at a higher dose of 150 to 800 rad, observed that splitting the dose resulted in a reduced transformation rate. Similar results were seen by Han and Elkind (1979). Miller and Hall (1978), utilizing the 10T1/2 cells and over a dose range from 25 to 800 rad, reported that although splitting the dose at the low-dose range resulted in enhanced transformation, a sparing effect on transformation was observed when doses above 150 rad were split into two equal fractions. Moreover, dividing the dose of 100 rad into two to four fractions over 5 hr resulted in a progressive

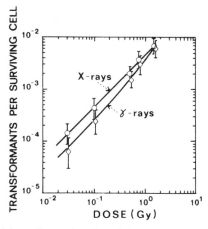

FIG. 8. Transformants/surviving cell as a function of dose for hamster embryo cells exposed to graded doses of 250 kVp X rays (open circles) or Cobalt-60 γ rays (open diamonds). Data pooled from several experiments. (Error bars) 95% confidence limits. (Data from Borek *et al.*, 1983b.)

enhancement of transformation. Borek, extending the split doses studies in the hamster system to a higher dose level (Borek, 1979b), reported that in this system too, at doses higher than 200 rad a lower transformation frequency was observed when the doses were split into two fractions.

Borek and Hall (1974), Miller and Hall (1978) and Borek (1979b) presented data in terms of the ratio of the transformation incidences produced by split to single doses. The agreement between the results obtained with the two different systems using different techniques is quite remarkable. These data, as well as similar observations in the 3T3 cells (Little, 1979), indicate that in three different cell systems the oncogenic effects of split doses of radiation differ with dose range. They clearly show that the use of linear extrapolation from high to low dose levels may lead to cancer risk estimates that are neither conservative nor prudent depending on the distribution of the dose in time.

4.1.2. *Hamster Cells Initiation* In Utero *and Assay* In Vitro

The available data on the subject of *in utero* carcinogenesis are limited. Experiments conducted by Borek *et al.* (1977) set out to compare transformation incidence in hamster embryo cells exposed *in utero* to 300 rad and then cultured *in vitro* immediately with cells of embryos cultured *in vitro* and later exposed to the same dose of radiation. The results indicated that transformation incidence induced *in utero* was ten-fold lower than that induced *in vitro*, thus closer to the frequency of oncogenesis *in vivo*. One cannot discount modifying effects of various cell populations and some differences in plating efficiency. However, these striking differences in incidence support some other factors that may be involved, such as host mediation, repair and loss of fixation at high density, inhibition of expression by cell-cell interaction, or other influences exerted by the tissue-specific organization present *in vivo*. When cells are transformed *in vitro* they are devoid of tissue-specific arrangements and are able to replicate under conditions where normally they may remain in a nonreplicating state. Thus the *in vitro* situation may yield an exaggerated rate of transformation because fixation of transformation can be carried out with ease in log-phase cultures.

A number of transformed cell lines were developed from these *in utero* experiments and studied for a variety of properties. The most striking finding was that injection of the mixed populations of transformed cells into hamsters yielded carcinomas as well as sarcomas. Although these epithelial transformed embryonic cells went undetected in cultures, because of their unaltered morphology they proliferated in the animals to form carcinomas.

4.1.3. *The Oncogenic Effects of High-LET Radiation*

As an oncogenic agent high-LET radiation is more effective than low-LET radiation. It is also more cytotoxic. The effectiveness of cell transformation is matched by the enhanced killing effect. There have been few reports on the oncogenic potential of high-LET radiation, and of these most of the available information is on the effectiveness of neutrons, a type of radiation that is not only used in therapy but is also a product of nuclear energy. Using the hamster embryo clonal system, Borek *et al.* (1978) reported the induction of cell transformation *in vitro* following exposure to 430 keV neutrons in the dose range of 0.1–150 rad. A 430-keV Van de Graaff accelerator was used, where protons accelerated onto a tritium source produced the spectrum of neutrons. The angle of the cells with respect to the source determined the neutron energy they received (Borek *et al.*, 1978). The results indicated that whereas neutrons were much more efficient than X rays in producing cell transformation, they were also more effective in cell killing (Fig. 7). Transformation was observed with neutron doses as low as 0.1 rad. It increased as a function of the dose to 150 rad, indicating a much higher frequency than that observed for X rays and resulting in an inability to evaluate RBE (ratio of doses which produce the same biological effect) at the dose level. However, when transformation frequency is expressed as the number of transformants per exposed cell, an evaluation more closely related to the *in vivo* situation, both neutrons and X-ray curves rise to the same peak value (Borek *et al.*, 1978). Thus whereas the RBE for survival varies and is inversely proportional to the square root of the neutron dose, RBE for transformation per exposed cell does not reflect this wide variation. Qualitatively, the hamster cells transformed by neutrons did not differ from those transformed by X rays (Borek *et al.*, 1978) (Table 2).

Transformation of the mouse 10T1/2 cells by single doses of 0.85 MeV fission neutrons as compared to X rays was reported by Han and Elkind (1979). The results were qualitatively similar to those reported by Borek *et al.* (1978) using the hamster system. Fractionation of the neutron dose spaced by an interval longer than 6 hr resulted in a reduced transformation rate (Han and Elkind, 1979).

The effect of low-energy α particles in transforming the 10T1/2 cells was evaluated by Lloyd *et al.* (1979). Transformation frequency per survivor increased as the cube of the dose, peaking at an α-particle fluence of 1.5×10^7 to 2.5×10^7 particles (205 342 rad). Maximum transformation frequency reached 4%. No parallel experiments were carried out with X rays; thus RBE for α radiation could not be determined. Recent experiments with α particles carried out by Robertson *et al.* (1983), in 3T3 cells, indicated an RBE of 2.0–2.3, when compared to X rays.

Data on the effects of other types of heavy ions on cell transformation are limited. Borek *et al.* (1978), using the hamster cell system, studied the transforming effects of argon ions, a type of radiation that has been considered for therapy. Cells were irradiated at the Bragg peak with high-energy argon ions (429 MeV/amu) at doses of 1 and 10 rad, resulting in transformation frequency of 0.26 and 0.70, respectively. Cell transformation by a 600 MeV amun ion beam has been investigated by Yang and Tobias (1980) in the C_3H 10T1/2 cells.

Studies by Little *et al.* (1983) on transformation induced by ^{125}I and [^3H]TdR indicated that ^{125}I was more efficient in producing transformation compared to ^3H, with both radioisotopes being more efficient transformants than X rays.

A summary of data on rodent cell transformation by various types of radiation is presented in Table 3.

4.1.4. Human Cell Transformation by X Rays

In 1980 Borek demonstrated the transformation of human skin fibroblasts by 400 rad X rays into cells which progressed *in vitro* to malignancy and were able to grow in agar and give rise to tumors when injected into nude mice (Borek, 1980b; Borek, 1981b). The cells used were a strain of diploid fibroblasts, the KD strain, previously used for studies on chemically induced transformation (Kakunaga, 1978).

Early passage cells were used and their diploid nature was ascertained by chromosome G-banding analysis. Their doubling time was 30–32 hr. Survival curves indicated that survival fraction following a dose of 400 rad was close to 12% of the total population. No shoulder was observed, indicating the different response of the human cells compared to that of the hamster (Borek and Hall, 1973).

The protocol used for transformation combined two methods employed for chemically induced human cell transformation, that of Kakunaga (1978), the focus assay; and that of Milo and DiPaolo (1978), where cells were synchronized prior to exposure to the oncogenic agent. Cells at concentrations of 2.6×10^5 were seeded into 75 mm² flasks and 2 days later medium was replaced by another medium containing 0.1% serum for a period of 24 hr, whereby cell proliferation was greatly reduced. Twenty-four hours later the medium was replaced by complete medium (10% serum), which contained either β-estradiol 1 μg/ml or the protease inhibitor antipain (6 μg/ml). Cells were then irradiated and subcultured 10 hr later, after which they were subcultured again upon reaching confluency.

The foregoing experimental protocol took advantage of the following: the cells were quiescent by serum deprivation, and thus entered a synchronous wave of DNA synthesis when released from quiescence by medium change. Treatment of the cells at this point

TABLE 3. Transformation of Rodent Cells In Vitro by Radiation

Radiation	Dose	Dose Rate	Lowest and Highest Rate of Transformation/Survivor	RBE	Cell System	Reference
X ray (250 or 50 kVp)	300 rad	60 or 280 rad/min	$7 \times 10^{-3} - 8 \times 10^{-3}$	1	Hamster Embryo	Borek and Sachs (1966a)
X ray (210 kVp)	1–600 rad	4.25 or 70.6 rad/min	$10^{-4} - 6 \times 10^{-3}$	1	Hamster Embryo	Borek and Hall (1973)
X ray (300 kVp)	0.3 rad	4.25 rad/min	10^{-5}	1	Hamster Embryo	Borek and Hall (1982)
X ray (100 kVp)	50–1200 rad	83.5 rad/min	$3 \times 10^{-5} - 3 \times 10^{-3}$	1	C3H 10T1/2	Terzaghi and Little (1976a)
X ray (50 kVp)	50–1200 rad	—	$8 \times 10^{-5} - 3 \times 10^{-3}$	1	C3H 10T1/2	Han and Elkind (1979)
X ray (300 kVp)	100–1000 rad	32 or 180 rad/min	$7 \times 10^{-5} - 2 \times 10^{-3}$	1	C3H 10T1/2	Miller and Hall (1978)
X ray (100 kVp)	10–400 rad	78 rad/min	$10^{-4} - 3 \times 10^{-3}$	1	3T3 A31	Little (1979)
Neutron (430 keV)	0.1–150 rad	10–80 rad/hr	$7.6 \times 10^{-3} - 3.2 \times 10^{-2}$	6–10	Hamster	Borek et al. (1978)
Particle (5.6 MeV)	205–342 rad	14 rad/sec	4×10^{-2}	—	C3H 10T1/2	Lloyd et al. (1979)
Particles	25–250 rad	24.2 rad/min	$2.3 \times 10^{-4} - 20 \times 10^{-4}$	2.2–3	3T3 A31	Robertson et al. (1983)
Fission neutron	25–60 rad	10.3–37.8 rad/min	$10^{-4} - 6 \times 10^{-3}$	2.6–10	C3H 10T1/2	Han and Elkind (1979)
Argon ion (429 MeV/amu)	1 and 10 rad	—	$2.7 \times 10^{-3} - 7 \times 10^{-3}$	—	Hamster Embryo	Borek et al. (1978)
Iron particles	10–90 rad	1 rad/min	$5 \times 10^{-4} - 6 \times 10^{-3}$	—	C3H 10T1/2	Yang and Tobias (1980)
u.v.	7.5–60 erg/mm²	0.76 erg/mm²/sec	$8 \times 10^{-3} - 7 \times 10^{-2}$	—	Fetal Hamster	DiPaolo and Donovan (1976)
u.v.	25–300 erg/mm²	4.5 erg/mm²/sec	$10^{-4} - 10^{-3}$	—	C3H 10T1/2	Chan and Little (1976)
u.v.	1.0–7.55/m³	0.38 J/m²/sec	$10^{-5} - 6 \times 10^{-4}$	—	3T3 A31	Little (1979)
u.v.	75 erg/min²	—	$1.6 \times 10^{-3} - 10^{-5}$	—	3T3 A31	Kakunaga and Crow (1980)

enabled the capturing of cells entering S phase. Within 60 to 80 days after treatment, foci appeared in treated cultures that were clearly distinguishable from the untreated controls (Fig. 9). They grew progressively when medium was changed to a low-calcium medium; the transformed morphology of the foci was enhanced, while the normal cells died within 24 hr. These clearly distinguishable foci were isolated and propagated *in vitro*. Chromosome G banding indicated a near diploid range of chromosomes (46–49), saturation density was two-fold compared to the normal, and the transformed but not the normal were agglutinable by 25 μg/ml of Con A. When seeded into 0.33% agar or grown on agar (0.5% concentration), the KD-transformed cells formed colonies. Cloning efficiency in agar was 2–3%. Neither the unirradiated nor the irradiated but untransformed cells formed colonies in this semisolid medium.

The ultimate proof of the neoplastic nature of the cells transformed *in vitro* is their ability to form tumors in an appropriate host. Swiss *nu/nu* mice were injected intradermally with 5×10^6 cells (suspended in 1 ml medium) into the subscapular region of animals X-ray irradiated 24 hr earlier with 450 rad, or unirradiated.

Whereas the five transformed lines tested formed colonies in agar, three gave rise within six weeks both in the irradiated and unirradiated mice to tumors that have been characterized as noninvasive fibrosarcomas. Cells cultured from the tumors possessed a human karyotype which will be described in detail elsewhere. None of the unirradiated or the irradiated but untransformed cells (seven samples of each) give rise to tumors in nude mice.

Cultures that were treated, i.e. irradiated but not allowed to replicate more than four or five times before reaching confluency, did not exhibit transformation. This indicated that, as in the rodent cells (Borek and Sachs, 1967; Terzaghi and Little, 1976a), replication is required following radiation for the fixation and expression of the transformed state.

At the present time quantitative assays for human cell transformation are not firmly

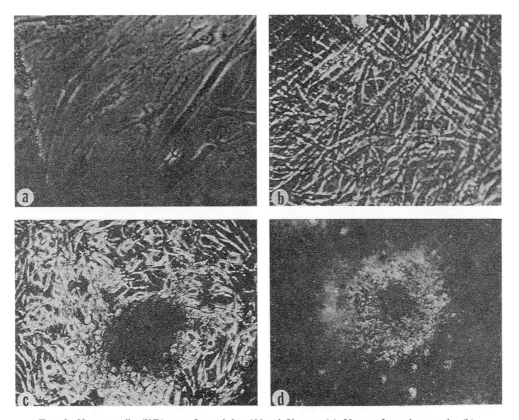

FIG. 9. Human cells (KD) transformed by 400 rad X-rays. (a) Untransformed controls. (b) Transformed cells. (c) Transformed cells growing in low Ca^{2+} medium. (d) Colonies of transformed KD cells growing in agar. (From Borek, 1980a).

established, though a number of laboratories have utilized growth in agar as an assay (Sunderland et al., 1981; Silinskas et al., 1981; Milo and DiPaolo, 1978; Andrews and Borek, 1982; Borek and Andrews, 1983; Zimmerman and Little, 1983a). The correlation between the malignant potential of the cells and their ability to proliferate as colonies in agar is clearly not established. On the one hand, as reported, normal human fibroblasts can give rise to colonies in semisolid medium (Peehl and Stanbridge, 1981), and on the other hand fibroblasts that showed anchorage independence could not form tumors in nude mice (Sutherland et al., 1981). Another problem with using agar as the sole endpoint for analysis of dose-response relationships in transformation studies is the inability to evaluate plating efficiency in agar. For example, do the number of clones growing in agar reflect the total number of cells that have undergone transformation or do they reflect only a fraction of transformants, those with a higher plating efficiency in this semisolid medium and an ability to proliferate under those conditions? Growth efficiency in agar can be enhanced by using a high-quality serum, and by washing the agar multiple times.

More recently human embryo cells have been transformed by X rays (Borek, 1985a). Their properties and transformability differ from those of the KD cells and are quite different from the hamster embryo cells (Table 4).

Though the quantitative aspects of radiogenic transformation of human cells are currently not precise, a number of observations using KD cells can be mentioned. The incidence of transformation in the human cells appears to be lower than that observed in the rodent cells, given the same dose of X rays. Whereas in the human cells the frequency per treated cells is closer to 10^{-6} at 400 rad, a frequency associated with mutational events, rodent cells show an incidence closer to 10^{-4} at that dose level. The number of doublings required for the expression of the transformed state of the human cells was approximately 10–13, similar to that observed in some rodent cells. It must be stressed that future experiments may indicate that the quantitative aspect of these endpoints may vary with the source of the human cells studied. Genetic and physiologic differences, as well as organ- and tissue-related variability in response may play important roles in the assay.

Other observations on the transformed human cells can be stressed. (a) Initial loss of contact inhibition is not as striking as that seen in rodent cells. (b) In contrast to rodent cells, the ability to proliferate in medium with low serum (%) is not confined to the transformed cells (Borek, 1980b); our normal KD cells as well as the transformed cells proliferated in medium containing low serum. (c) The transformed state is associated with membrane changes and, as in rodents cells, agglutinability by plant lectins can be used as a distinguishing probe. (d) Surface topography in the X-ray-transformed human cells is altered but not as dramatically as in the rodent cells. Microvilli found in abundance on rodent cells (Borek and Fenoglio, 1976) were not as abundant in transformed human cells (Borek, unpublished).

We cannot yet assess the roles played by growth inhibition and release from this arrest

TABLE 4. Radiogenic Transformation In Vitro of Diploid Cells

Parameters	Hamster Embryo	Human Embryo	Human Adult (KD)
Endpoint scored	Tr colonies*	Tr foci	Tr foci
Time of appearance (days)	10	28–38	30–60
Doubling time (no significant difference from normal)	16 ± 2	24 ± 2	32 ± 2
Growth in low Ca^{2+}	+ Tr	+ (Tr)	+ (Tr)
Growth in low serum	+ Tr	+ (Tr, N)†	+ (Tr, N)†
Growth in agar	+ Tr	+ (Tr)	+ (Tr)
Chromosomes (passage 5 after isolation)	Diploid or pseudo-diploid	Diploid or pseudo-diploid	Diploid or pseudo-diploid
Chromosomes (at time of tumor induction)	Heteroploid	Nt	Diploid or near diploid
Tumor	Sarcomas	Nt	Sarcomas
Frequency of transformation at 200 rad	5×10^{-3}	$\sim 3 \times 10^{-6}$	$\sim 8 \times 10^{-7}$

*Tr = Transformed.
†N = Normal.

and exposure in late G_1 in the process of X-ray-induced transformation. Recent experiments indicate that though these steps are not essential for observing transformation, without growth arrest transformation incidence is decreased and latency in appearance of transformed foci is lengthened (as long as 4 months). At present we are ignorant of the molecular mechanisms underlying the effects of growth inhibition and release from this arrest in the process of transformation. A tempting thought is provoked by reports of Woodcock and Cooper (1979) and Schimke et al. (1980). Their work indicates that following an inhibition of DNA, replication begins at the same origins (Woodcock and Cooper, 1979). Quiescence by serum deprivation and refeeding (Borek, 1980b) could lead to similar events. These may include disproportionate DNA replication and gene amplification (Schimke et al., 1980), which may play a role in potentiating the induction of radiogenic transformation and enhancing its frequency. Other studies (Namba et al., 1978) utilized high doses of ^{60}Co γ ray, delivered in multiple fractions to WI-38 fibroblasts. Exposure under these conditions resulted in marked and prominent chromosome aberrations in these cells. Morphologically transformed populations were observed 150 days after exposure, but no tumorigenicity in animals was observed.

5. NONIONIZING RADIATION

Of the nonionizing radiations, ultraviolet (u.v.) is most notorious for its oncogenic properties (Setlow, 1974; Setlow and Hart, 1974; Hart et al., 1977; Setlow, 1978).

5.1. U.V.-INDUCED ONCOGENIC TRANSFORMATION IN RODENT AND HUMAN CELLS

The effects of u.v. light are mediated via electron excitation and are exerted on cellular DNA (Setlow, 1966; Setlow et al., 1969; Wang, 1976; Cleaver, 1969; Cleaver and Bootsma, 1975). DNA absorbs light efficiently in the range of 240–300 nm. In so doing bases acquire excited energy states. The principal product of the photochemical reaction in DNA, at doses which are of biological significance, are cyclobutane pyrimidine dimers formed between adjacent pyrimidines (Setlow, 1966; Wang, 1976; Cleaver, 1978; Hanawalt et al., 1979; Haseltine, 1983).

Of all the lesions induced by u.v. in DNA, cyclobutane dimers are the most prevalent and have been implicated in lethality and mutagenesis (Maher et al., 1979) and in carcinogenesis (Hart et al., 1977). The precise role of other minor photoproducts has yet to be evaluated.

Studies on the repair of cyclobutane dimers demonstrated that they are removed from DNA in repair competent cells but not in repair deficient cells (Cleaver, 1969; Setlow et al., 1969).

The genetic loci controlling the removal of cyclobutane dimers have been identified and include loci that constitute the human u.v. sensitive cancer prone syndrome xeroderma pigmentosum (XP) (Cleaver, 1969, 1978; Setlow et al., 1969; Setlow, 1978). Mechanisms of u.v.-induced DNA damage and repair are quite different from those associated with ionizing radiation (Painter and Cleaver, 1969). Yet, similarly to ionizing radiation, the association between specific DNA damage and repair, and in vitro transformation is so far largely inferential. An association in vivo has been shown (Setlow and Hart, 1974). In rodent cells u.v.-induced transformation in vitro has been studied with the hamster cell system (DiPaolo and Donovan, 1976; Doniger et al., 1981), the C_3H 10T1/2 cell line (Chan and Little, 1976; Mondal et al., 1976), and the BALB/3T3 clone A31 cells (Little, 1979; Kakunaga, 1973).

Ultraviolet transformation has also been shown to be enhanced by X rays but not by chemicals (DiPaolo and Donovan, 1976), and by TPA (Mondal et al., 1976). The data from the various laboratories indicate that the frequency of u.v.-induced transformation is dose dependent. Cell susceptibility to u.v.-induced transformation may vary within the cell population (Kakunaga and Crow, 1980).

Action spectra for u.v.-induced transformation in rodent cells and its correlation with dimer production are reported using the hamster system and a morphological assay (Doniger et al., 1981). Using wavelengths between 240 and 313 nm, the most effective wavelengths in producing transformation were 265 and 270 nm. Relative sensitivities per quantum for transformation, pyrimidine dimer production and toxicity appeared the same at each of the wavelengths tested, thus implying that DNA may be the target for these processes (Doniger et al., 1981).

Data from various groups on the effectiveness of u.v.-irradiation in transforming rodent fibroblasts into morphologically identifiable colonies are presented in Table 3, where they are compared to data on transformation by ionizing radiation. Transformation in rodent cells has been largely based on morphological changes, growth in agar and tumorigenicity.

5.1.1. *Human Cells*

Transformation of human cells by u.v. into anchorage-independent cells which did not give rise to tumors has been induced in embryonic cells (Sutherland et al., 1981). Cell susceptibility to transformation by u.v. decreased with cell passage *in vitro* in the human embryo cells similar to that observed in experiments utilizing hamster embryo cells (Borek and Sachs, 1967). Transformation by u.v. radiation has also been reported in human foreskin cells, in both fibroblasts and epidermal cells (Milo et al., 1981) and in adult skin cells (Maher et al., 1982; Borek and Andrews, 1983). Growth in agar was reported and with the epidermal cells tumorigenicity was assessed in chick embryonic skin (Milo et al., 1981). The procedure for u.v. cell transformation, which generally followed that for chemical transformation (Milo and DiPaolo, 1978), was modified in the case of the epidermal cells, for it was found that the program of fixation and expression of transformation in the epidermal cells differed from that found in the fibroblasts (Milo et al., 1981).

More recently, adult skin fibroblasts from patients with xeroderma pigmentosum (Setlow et al., 1969; Cleaver, 1969) and Bloom syndrome cells (Chaganti et al., 1974) have been transformed *in vitro* by u.v. light into anchorage-independent cells (Borek and Andrews, 1983). The source of u.v. light was a UVB, a sunlamp type of irradiation relevant to human skin carcinogenesis. Single cells were exposed to wavelengths of 280 and 320 nm. Both morphological foci as well as ability to grow in agar were assessed. It appears, as mentioned earlier, that here, too, in human cells transformed by u.v., ability to grow in agar was exhibited concomitantly with the appearance of foci comprising piled-up, randomly oriented cells that were clearly distinguishable morphologically from the flat, orderly oriented control cells (Fig. 5). Transformation of XP cells by UVA has also been reported (Maher et al., 1982). A higher rate of transformation into anchorage-independent cells was observed in the repair-deficient XP cells as compared to the normal skin fibroblasts.

6. COCARCINOGENS AND MODULATORS OF RADIOGENIC TRANSFORMATION

In recent years it has become increasingly clear that environmental factors including diet play a crucial role in determining cancer incidence in humans (Higginson, 1979; Peto et al., 1981; Doll and Peto, 1981; Miller and Miller, 1979; Borek, 1982a, 1984a; Nagao et al., 1978; Cairns, 1981; Sugimura, 1982). Although radiation is a weak oncogenic agent compared to some chemicals, it is the most universal. Thus, in assessing the effects of radiation from the point of view of cancer risk to humans (BEIR, 1972; UNSCEAR, 1977), one cannot exclude the possibility that a multitude of genetic, physiological, and environmental factors influence cancer incidence initiated by radiation. The difficulty lies in identifying the carcinogens that act in an additive or synergistic manner. Once recognized, measures may be sought to alter exposure to these agents and to find ways to modify their effects and interactions.

Under defined conditions *in vitro* we can evaluate some of these interactions at a cellular level, although admittedly the absence of host-mediated effects or intact tissue organization give us a somewhat slanted view. Within the context of the accepted concept of cancer development (Farber, 1973; Berenblum, 1982; Pitot and Sirica, 1980), one thinks of neoplastic transformation as comprising the early phases of initiation and a later stage of expression by which we recognize initiation. Agents that modify expression can interact at early or at later stages, serving as promoters. Although initiation by radiation is irreversible, it can sometimes be prevented (Guernsey et al., 1980, 1981; Borek, 1981a; Borek et al., 1983a). Expression or promotion can be reversed (Borek et al., 1979; Miller et al., 1981; Kennedy and Little, 1978; Borek, 1981a, 1982b, 1984a, b; Borek and Troll, 1983; Borek et al 1986a).

6.1. Enhancement of transformation

The enhancement of viral transformation following a preexposure of cells to radiation was the first type of interaction observed, one example being the enhancement by radiation or SV40 transformation of 3T3 cells (Pollack and Todaro, 1968).

6.1.1. Chemicals

Preexposure of hamster embryo cell to X ray doses of 150 to 250 rad rendered them more responsive to transformation by benzo[a]pyrene as well as u.v. light (DiPaolo et al., 1971; DiPaolo and Donovan, 1976). Pretreatment with u.v. light did not have an enhancing effect. Maximum enhancement was found when radiation was administered 48 hr prior to the carcinogens, and it was directly related to the absorbed radiation. A synergistic interaction was reported between X rays and the food pyrolysate product 3-amino-1-methyl-5H-pyrido(4,3-b)indol (Trp-P-2) (Borek and Ong, 1981). The chemical, isolated in Japan from broiled meat and fish, foods widely consumed there, has been shown to be a mutagen and a carcinogen *in vivo* and *in vitro* (Nagao et al., 1978). Preexposure of hamster embryo cells to 50 or 150 rad of X rays following treatment by Trp-P-2 resulted in a dose dependent synergistic interaction between these two oncogenic and DNA damaging agents. The cocarcinogenesis between a dietary product highly consumed in Japan and radiation further compounds the interpretation of some of the data from Hiroshima and Nagasaki, the largest source of evidence on radiation carcinogenesis.

Radiogenic transformation can be markedly enhanced by ozone (O_3) a ubiquitous air pollutant and the key oxidant in photochemical smog. We find that the cocarcinogenic interaction of ozone and X rays are synergistic in nature and are associated with enhanced cellular levels of malonaldehyde, a product of lipid peroxidation and a known carcinogen (Borek et al., 1986b).

6.1.2. Cocarcinogenesis with Chemotherapeutic Agents

The long term survival of certain patients exposed to treatment with radiation and/or chemotherapeutic agents has brought about the realization that agents that effectively control cancer in human subjects also possess an oncogenic potential resulting in secondary malignancies (Arseneau et al., 1977).

Experiments using C_3H 10T1/2 cells have shown that a variety of chemotherapeutic agents are highly oncogenic (reviewed in Borek and Hall, 1982). Thus, the possible cocarcinogenic interaction between radiation and some chemotherapeutic agents should be taken into consideration in cancer treatment where combined modalities are used.

6.1.3. Promoters

Enhancing agents can also be part of the family of tumor promoters. The most widely studied promoter in conjunction with radiation transformation has been the phorbol ester derivative 12-*O*-tetradecanoylphorbol-13-acetate (TPA) (Hecker, 1971). Promoters in themselves are considered noncarcinogens and TPA has been considered a classic promoter. However, TPA does possess some initiating action (Kinsella and Radman, 1978; Emerit and Cerutti, 1981; Han and Elkind, 1982). The effects of TPA are pleotropic. In studies on radiation transformation, TPA shows an interaction with chromosomes as well as an effect on membrane-associated enzymes within the same cell system (Borek *et al.*, 1982; Borek, 1982a). The enhancement of X-ray-induced transformation by TPA has been studied in detail (Kennedy and Little, 1978; Borek, 1982a; Miller *et al.*, 1981; Borek, 1981b; Han and Elkind, 1982; Di-Paolo *et al.*, 1984). TPA exhibits a temporal enhancement of transformation consistent with the idea of promotion, namely that its effectiveness can be observed at various stages after initiation. While TPA is considered a true promoter, some of its actions have suggested that it may be a weak initiator (Han and Elkind, 1982). The interaction between X rays and TPA as cocarcinogens cannot be ruled out.

Another promoter, teleocidin-B (Fujiki *et al.*, 1979), has been shown to be one hundred times more potent than TPA in enhancing radiation transformation (Borek *et al.*, 1984a).

6.1.4. *Hormones*

6.1.4.1. *β-Estradiol*

The hormone β-estradiol potentiated X-ray-induced transformation in human cells (Borek, 1980a) but in itself did not transfer the human cells. In the C_3H 10T1/2 cells β-estradiol acted itself as a carcinogen as well as interacting with radiation to enhance transformation (Kennedy and Weichselbaum, 1981). These data alert one to the fact that in the process of evaluating the additive or synergistic interaction between agents to enhance carcinogenesis, we must be cognizant of the strain and species from which the cells are derived and to their response *in vivo*. For example, tumor incidence in the C_3H mouse strain is highly sensitive to the action of the hormone.

6.1.4.2. *Thyroid hormone*

Thyroid hormones play a critical role in cellular transformation by radiation (Guernsey *et al.*, 1980, 1981) as well as by chemicals (Borek *et al.*, 1983a) and viruses (Borek *et al.*, 1984c). Experiments conducted with hamster embryo cells and C_3H 10T1/2 cells indicated that the removal of thyroid hormones from the serum does not modify cell growth or cell survival but that these conditions inhibit transformation.

Under hypothyroid conditions no transformation was observed. When triiodothyronine was added to the medium at physiological levels over a range of 10^{-12} to 10^{10}, cell transformation was induced in a T_3 dose dependent manner (Guernsey *et al.*, 1981; Borek *et al.*, 1983a, 1984c) (Fig. 10). The induction of transformation took place within a confined window in time. Maximum transformation was observed when the hormone was added 12 hr prior to exposing the cells to the oncogenic agent and had no effect on transformation when added after exposure to radiation (Guernsey *et al.*, 1981), chemicals (Borek *et al.*, 1983a), or viruses (Borek *et al.*, 1984c). It is important to note that the action of T_3 was not mimicked by its inactive isomer reverse T_3. Cyclohexamide, an inhibitor of protein synthesis, inhibited T_3 action on transformation (Guernsey *et al.*, 1981).

While we do not know the mechanisms underlying thyroid hormone influence, two possibilities can be considered by our data. One suggests that T_3 may influence the induction of cellular transformation proteins associated with the initiation of the neoplastic process (Guernsey *et al.*, 1981; Borek *et al.*, 1983a). This possibility is underscored by the fact that the T_3 dose response relationship for radiation and chemical transformation is similar to the T_3 dose related induction of another cellular protein, that of Na^+/K^+ ATPase (Guernsey *et al.*, 1981) (Fig. 10). Another possibility, which is not mutally exclusive, resides in the fact that thyroid hormones affect the oxidative state of cells. As discussed later, the

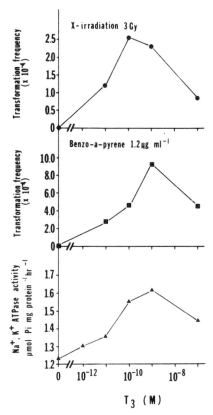

FIG. 10. The effect of varying concentrations of T3 on X-ray-induced (circle) and B[a]P-induced (square) transformation and on Na^+/K^+ ATPase activity (triangle) in stationary-phase C₃H 10T1/2 cells. For transformation experiments, cells were pretreated with various doses of T3 for 1 week prior to treatment with B[a]P (1.2 g/ml) or X rays (4 Gy) and maintained under the same conditions for the remainder of the experiment. (From Borek et al., 1983a.)

prooxidant state is of importance in transformation. We find that superoxide dismutase in part inhibits the effect of T_3 in potentiating radiogenic transformation and that it also counteracts the action of TPA and teleocidin which generate free radicals (Borek et al., 1984a).

The importance of thyroid hormone in radiogenic transformation is highlighted in the case of radiotherapy where a temporary state of hypothyroidism may be induced following treatment. This state may confer temporary protection by inhibiting the initiation of secondary malignancies by the radiation treatment. Recent evidence indicates that thyroid hormone modifies cellular genes. Under hypothyroid condition NIH-STS cells became refractory to transfection by gonomic DNA from X-ray-transformed cells (Borek, 1986).

6.1.5. The Protease Inhibitor Antipain, a Dual Action

Antipain (AP) (Umezawa et al., 1979) is a protease inhibitor that has been shown to have anticarcinogenic activity (Troll, 1976). Its action *in vitro* on influencing radiation-induced transformation has been dual (Borek et al., 1979; Geard et al., 1981). Antipain potentiated X-ray-induced transformation in human (Borek, 1980a), hamster, and 10T1/2 mouse cells, when added to the cells *prior* to radiation. It reduces X-ray-induced transformation when added *after* irradiation (Kennedy and Little, 1978; Borek et al., 1979; Geard et al., 1981; Han and Elkind, 1982). These dual actions are exerted without any effect on cell survival. The enhancing effects of AP are more pronounced if AP is removed after irradiation (Gerard et al., 1981), as compared to adding it before irradiation and keeping it on for the duration of the experiment, whereby its protective effect reduces the enhancing actions. The dual activity of AP was not reflected in chromosomal alterations as measured by sister chromatid analysis (Geard et al., 1981), or on DNA damage or replication (Borek and Cleaver, 1981). Another protease inhibitor, leupeptin, had the same effect on inhibiting transformation but not on DNA repair (Borek and Cleaver, 1981).

Clearly, the mechanisms of the protease inhibitor AP call for more inquiry. One can only speculate that the two diametrically opposed actions of AP on transformation are mediated via different mechanisms; some are associated with the direct cellular interactions with radiation at which time cascading events could occur, whereas in later events its inhibitory activity, acting in a temporal fashion (Borek et al., 1979), may be mediated via an effect on specific cellular proteases.

The effect of some proteases at a genetic level can be seen by their inhibitory action on poly(ADP)ribosylation (Cleaver et al., 1986).

6.1.6. Inhibition of Radiogenic Transformation

Although we try to identify agents that act as cocarcinogens or promoters, we aim at finding compounds or conditions that may inhibit the progression of transformation and, even better, prevent initiation.

6.1.6.1. Retinoids

Within the last decade analogs of vitamin A (retinoids) have been shown to modulate malignancy both in laboratory animals and in the clinic (Sporn et al., 1976; Lotan, 1980; Peto et al., 1981). The effectiveness in inhibiting X-ray-induced transformation in vitro was first shown using the 10T1/2 cells (Harisiadis et al., 1978), indicating that these compounds could act in vitro and that their action was not limited to an effect on neoplastic epithelial cells, as previously observed in vivo. The ability of retinoids to inhibit not only X-ray-induced transformation but also the enhancement of this transformation by TPA in both 10T1/2 and hamster embryo cells was reported more recently (Miller et al., 1981; Borek, 1981a). Retinoids were present in the medium at the time of irradiation and TPA was added after irradiation to some of the experimental plates. The retinoids were kept in contact with the cells for 4 days only and thereafter removed with an exchange of medium. TPA was maintained for 2 weeks (hamster) or 6 weeks (10T1/2) for the total length of the experiment. Thus inhibitory action of the retinoids had to be exercised within 4 days to override the enhancing effect of TPA. As seen in Table 5, the retinoids did indeed inhibit both X ray transformation and TPA enhancement of this transformation, thus indicating that their action on radiogenic transformation takes place within a short time and is irreversible. Studies to evaluate the mechanisms of action indicated that the inhibitory action of retinoids on transformation and on TPA action was not reflected in an inhibition of sister chromatid exchanges (SCE) (Miller et al., 1981; Borek, 1982a); TPA causes a slight enhancement of SCE, but so did the retinoids. Their action was reflected at the membrane level on the membrane-associated Na-transport enzyme Na^+/K^+ ATPase, but not on Mg^+ ATPase or on 5'-nucleotidase (Borek, 1982a; Borek and Guernsey, 1981). TPA enhanced the level of the enzyme; retinoids decreased it. When cells were exposed to both agents concomitantly the Na^+/K^+ ATPase returned to control level. Thus the retinoids appear to exercise their effect at the level of gene expression. Once cells are transformed and exhibit a neoplastic phenotype, their membrane Na^+/K^+ ATPase changes (Borek, 1982a). Its activity is no longer modulated by either retinoids or TPA. Although the mechanisms of action of the retinoids are not clear, the data lend support to the notion that these compounds are effective suppressors of carcinogen-induced neoplastic progression. They are agents that act on expression of oncogenesis. They increase cell adhesion to the plates and also alter the cellular morphology into fusiform enlongated cells. Although we considered their action on TPA as an effect on a promoter, we cannot rule out their action

TABLE 5. *Levels of Membrane Enzymes in Normal and Transformed Fibroblasts and the Effect of RA and TPA on the Enzyme Activities*

Treatment	Na^+/K^+ ATPase	Mg^+ ATPase	5'-Nucleotidase
Hamster embryo			
Control	1.21 ± 0.21	1.35 ± 0.17	1.73 ± 0.28
TPA (0.16 μM)	0.53 ± 0.43	1.44 ± 0.45	1.89 ± 0.31
Retinoid (7.1 μM)	0.78 ± 0.19	1.32 ± 0.36	1.75 ± 0.21
Retinoid, TPA	1.31 ± 0.27	1.19 ± 0.41	1.91 ± 0.32
C_3H 10T1/2			
Control	1.79 ± 0.32	1.26 ± 0.31	0.56 ± 0.08
TPA (0.16 μM)	2.18 ± 0.31	1.26 ± 0.26	0.60 ± 0.12
Retinoid (7.1 μM)	1.13 ± 0.20	1.32 ± 0.26	0.68 ± 0.05
Retinoid, TPA	1.73 ± 0.23	1.23 ± 0.28	0.64 ± 0.10
Transformed hamster embryo			
Control	2.46 ± 0.27	1.54 ± 0.31	5.75 ± 1.21
TPA (0.16 μM)	2.31 ± 0.21	1.38 ± 0.21	5.90 ± 1.38
Retinoid (7.1 μM)	2.50 ± 0.29	1.42 ± 0.21	5.83 ± 1.10
Retinoid, TPA	2.34 ± 0.31	1.50 ± 0.32	5.92 ± 1.23
Transformed C_3H 10T1/2			
Control	0.97 ± 0.09	0.87 ± 0.08	0.40 ± 0.07
TPA (0.16 μM)	1.21 ± 0.12	1.38 ± 0.22	0.46 ± 0.06
Retinoid (7.1 μM)	0.82 ± 0.21	8.99 ± 0.06	0.59 ± 0.02
Retinoid, TPA	1.04 ± 0.18	1.05 ± 0.10	0.46 ± 0.16

on X ray and TPA as an inhibition of cocarcinogenesis. Their inhibitory effect in cocarcinogenesis between radiation and pyrolysates of protein foods has been reported (Borek, 1982a). The effectiveness of retinoids in modifying gene expression has been shown in their action in inhibiting the expression of a cellular oncogene (Westin *et al.*, 1982). One must also remember that some forms of retinoids are radical scavengers (Leibovitz and Siegal, 1980) thus these dietary elements may have an effect in modifying free radicals produced by radiation and some chemicals (Borek *et al.*, 1986b).

6.1.7. *Free Radical Scavengers*

6.1.7.1. *SOD-catalase*

The generation of reactive oxygen species in living systems exposed to radiation has long been recognized, as well as the protective action by radical scavengers (Bacq and Alexander, 1955; Alexander and Lett, 1968). More recently, tumor promoters have been shown to produce free radicals. These include TPA (Goldstein *et al.*, 1981) and teleocidine (Fujiki *et al.*, 1979). Utilizing hamster embryo cells, the effects of superoxide dismutase (SOD) and catalase on X-ray-induced transformation (300 rad) and its enhancement by TPA were evaluated (Borek and Troll, 1983). SOD and catalase were added to the cells at seeding, or were combined during irradiation and removed immediately after exposure or at the end of the experiment. TPA was added after irradiation, and SOD and catalase were kept on (alone or in combination) for the full course of the experiment. The results indicate that SOD, and to a lesser extent catalase, inhibited both X-ray-induced transformation (inhibition of 50% by SOD) (Fig. 11) and the TPA-enhancing effect, while having a less marked effect in enhancing cell survival. Their effectiveness on lowering X-ray transformation was similar to that exhibited by *β*-all-*trans*-retinoic acid and antipain. A similar inhibition of transformation by bleomycin was observed (Borek and Troll, 1983).

The results suggest that bleomycin and X-ray-induced transformation as well as TPA action may be mediated in part via the action of free radicals. Superoxide dismutase (which was not detected in the serum used in these experiments) will convert O_2^- to H_2O_2, thus preventing it from forming toxic species such as HO_2^- and singlet oxygen (Fridovich, 1978). The prevailing H_2O_2, which is a substrate for many toxic peroxidases, is then converted by catalase. The lower effectiveness of catalase in inhibiting transformation may

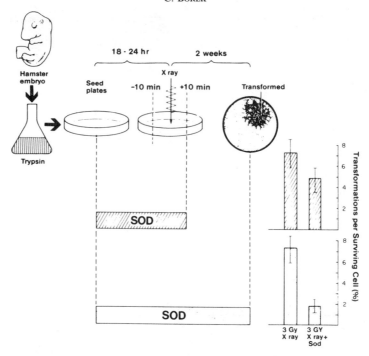

Fig. 11. The effect of SOD (10 units/ml) on the incidence of oncogenic transformation in hamster embryo cells produced by 300 rad X rays. hr = hours.

be due to the finding that the hamster embryo cells contain a high level of this enzyme, sufficient to remove the toxic H_2O_2.

Though we are relatively ignorant of the mechanisms by which free radicals may influence neoplastic transformation, one could speculate that their effect may involve membrane lipid peroxidation (Leibovitz and Siegal, 1980; Pryor, 1976), oxidation of proteins, and cross-linking of DNA, which may result in a cascade of events, with cell death and transformation being some of the consequences.

The efficiency of free radical scavengers in modifying transformation varies with species and probably tissue (Borek and Troll, 1983; Borek, 1984b). For example, hamster cells are rich in catalase and this enzyme had no effect on transformation by X rays or bleomycin in these cells (Borek and Troll, 1983). By contrast catalase did inhibit transformation in the C_3H 10T1/2 cells (Zimmerman and Cerutti, 1984).

6.1.7.2. Selenium and Vitamin E

The role of the prooxidant state of the cell in transformation is underscored by our findings that selenium inhibits radiogenic and chemically induced transformation (Borek, 1982a; Borek and Biaglow, 1984; Borek et al., 1986a).

Selenium (Se), a micronutrient in our diet, plays a critical role in protecting cellular systems from oxidative damage. Selenium is an essential component of the enzyme glutathione peroxidase which catalizes the breakdown of hydrogen peroxide and is necessary for the reduction of lipid peroxides to less reactive fatty acid alcohols (Tappel, 1973), thereby preventing the initiation and propagation of peroxidative chain reactions which are damaging to the cell (Pryor, 1976).

Our findings show that nontoxic levels of selenium significantly inhibit the oncogenic effects of X rays, benzo(a)pyrene, and tryptophane pyrolysates. We also find that pretreatment with selenium doubles the cellular capacity to destroy peroxides (Borek et al., 1986a). A marked enhancement in the levels of catalase and glutathione peroxidase was observed (Table 6). An increase in nonprotein thiols was also found, but no enhanced levels of glutathione-S-transferase were observed (Borek et al., 1986a).

TABLE 6. *Transformation, Glutathione Peroxidase (GSH). Catalase and Nonprotein Thiols (NPSH) in Selenium + Vitamin E Pretreated and Untreated C_3H 10T1/2 Cells*

	Untreated	Selenium Treated	Vitamin E Treated	Vitamin E + Selenium
Transformation by 400 rad X ray	9.8	1.8×10^{-4}	1.7×10^{-4}	0.9×10^{-4}
Transformed by B(a)P 1.2 µg/ml	11.6	1.6×10^{-4}	1.4×10^{-4}	0.8×10^{-4}
GSH px*	5.6	17.7	17.7	17.7
Catalase*	4.2	9.0	9.0	9.0
NPSH†				

*N moles H_2O_2 reduced/min/mg protein.
†N moles/mg protein.
Data from Borek et al. 1986b.

Results with Vitamin E (*a* tocophenol succinate), a known antioxidant, indicate that Vitamin E inhibits radiogenic and chemically induced transformation alone and in additive fashion with selenium. However, Vitamin E has no effect on altering the enzyme levels which are modified by selenium (Borek et al., 1986a).

These results further support the notion that radiation and chemically induced neoplastic transformation are mediated in part via the action of free radicals. The data indicate that selenium and Vitamin E act as radiopreventive and chemopreventive agents. Selenium confers protection by inducing or activating cellular enzymatic scavenging systems and enhancing peroxide destruction, Vitamin E acts by other mechanisms which remain to be determined in this system.

6.1.8. *Benzamides—Inhibitors of Poly(ADP-Ribose) Synthesis*

6.1.8.1. *Poly(ADP-ribose)*

Poly(ADP-ribose) is an important cellular regulatory molecule whose synthesis is stimulated by a variety of DNA damaging agents including radiation (Sugimura et al., 1980). The compounds 3-aminobenzamide (3-AB) and benzamide have been used as inhibitors of poly(ADP-ribose) in a variety of biological processes including DNA repair, gene expression, differentiation, mutagenesis and sister chromatid exchanges (Sugimura et al., 1980). We have investigated whether the benzamides modify neoplastic transformation induced by X rays or u.v. light. These DNA damaging agents are of particular interest since the benzamides, as inhibitors of poly(ADP-ribose), retard the ligation stage of DNA repair in cells damaged by alkylating chemicals and X rays but not in cells damaged in u.v. (Cleaver et al., 1983). This difference has made it possible to define a role for poly(ADP-ribosylation) in transformation that is distinct from its role in the ligation stage of DNA repair (Borek et al., 1984e, Borek and Cleaver, 1986).

We find that neoplastic transformation *in vitro* of hamster embryo cells and mouse C_3H 10T1/2 cells by X rays and u.v. light was suppressed by benzamide or 3-aminobenzamide, agents which inhibited poly(ADP-ribose) polymerization by about 75% and increased sister chromatid exchange frequencies, but had no influence on repair of X-ray and u.v. damage and had no detectable side effects on nucleotide precursor metabolism. Our findings indicate that the mechanisms regulating neoplastic transformation differ from those regulating mutagenesis in specific loci and sister chromatic exchanges and that they are mediated via alterations in poly(ADP-ribosylation), causing changes in gene control and expression (Fig. 12). Our recent findings showing that benzamides inhibit the enhancing action of tumor promoters indicate a role for poly(ADP)ribose in promotion as well as initiation (Borek and Cleaver, 1986).

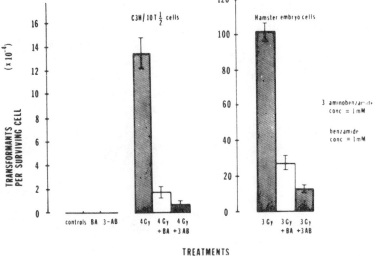

FIG. 12. (a) A comparison of the effects of X rays on transformation induction in the C_3H 10T1/2 cells and in hamster embryo cells in the absence or presence of benzamide (1 mM) or 3-AB (1 mM). (b) Inhibition of poly(ADP-ribose) synthesis (circle) and stimulation of sister chromatid exchanges (square) as a function of 3-AB concentration. Poly(ADP-ribose) polymerization was measured in C_3H 10T1/2 cells exposed to methyl methanesulfonate (5 mM) for 30 min immediately before permeabilization. SCEs were measured in Chinese hamster ovary cells grown for two cell cycles in 3-AB. (From Borek et al., 1984e.)

7. ONCOGENES IN RADIATION TRANSFORMATION

The role of DNA as a target in radiation transformation has been suggested by the requirement for DNA metabolism for fixation of the transformed state (Borek and Sachs, 1967). Yet, no direct proof was available.

As described earlier recent developments enabling DNA mediated gene transfer (transfection) have made it possible to address this question.

Hamster cells and C_3H 10T1/2 cells exposed to X rays and transformed *in vitro* yield populations of transformed cells from which high molecular DNA can be purified (Borek et al, 1987; Borek, 1985a, b; Borek et al., 1984d). The high molecular DNA is added to NIH/3T3 cells or the C_3H 10T1/2 cells using a modified calcium phosphate precipitation method (Shilo and Weinberg, 1981) in the presence or absence of a selective marker such as the Eco-gpt (Land et al., 1983). The ability of DNA from *in vitro* X-ray-transformed cells to confer

TABLE 7. *Transfection of DNA from Radiation* In Vitro *Transformed C_3H 10T1/2 and Hamster Embryo Cells onto NIH/3T3 and C_3H 10T1/2 Cells**

DNA Source	No. foci/μg DNA	
	On NIH/3T3	On C_3H 10T1/2
Normal hamster embryo	0	0
X-ray-transformed hamster embryo	0.27	0.20
Normal C_3H 10T1/2	0	0
X-ray-transformed C_3H 10T1/2	0.25	0.19
NIH 3T3	0	0
C_3H 10T1/2	0	0

*From Borek *et al.*, 1987.

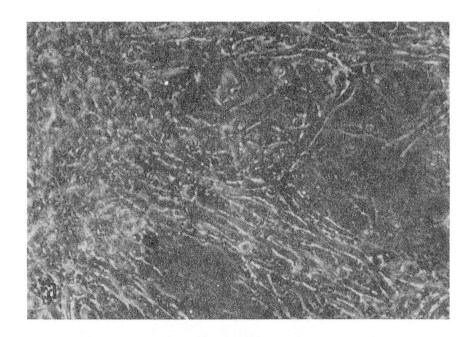

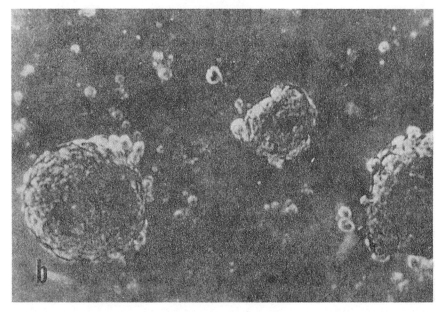

FIG. 13. (a) A focus of 3T3 cells transfected with X-ray-transformed hamster embryo cells. (b) Cells propagated from same population as in (a) growing in agar. (From Borek and Ong, 1984.)

the transformed phenotype on normal 3T3 or C_3H 10T1/2 cells by producing transformed foci (Table 7) which grow in agar, indicates that DNA codes for the radiation transformation phenotype following direct *in vitro* exposure to radiation. It also indicates that specific cellular oncogenes are activated. Restriction enzyme analysis indicated that specific segments of DNA are encoding the transformation phenotype and that these may differ in the *in vitro* transformed C_3H 10T1/2 and the hamster cells (Borek and Ong, in preparation). DNA isolated from the transfected foci can retransfect the 3T3 and 10T1/2 cells in a secondary and tertiary round of transfection and grow in agar (Fig. 13).

A Southern blot analysis carried out (Southern, 1975) on DNA from normal NIH/3T3 and from NIH/3T3 transformed by DNA from C_3H 10T1/2 or hamster cells indicated that no amplification or extra copies of any of the *ras* oncogenes were present in the transformants as compared to the control. The activated oncogenes in the radiation *in vitro* transformed cells are not of the *ras* family genes (Borek et al., 1984c; Borek et al., 1987; Borek, 1985a, b). This differs from observation *in vivo* by Guerrero et al. (1984), who observed an amplification of the Kirsten *ras* gene (Ki-*ras*) in thymic lymphomas induced by gamma radiation. While one must consider the different types of cells involved in these experiments, the results underscore the possibility that the activation of the *ras* oncogenes occurs late in the progression of the neoplastic process when cells have been established as progressive tumors.

The quick blot technique (Bresser et al., 1983a, b) was used to identify elevated transcripts of known oncogenes in the *in vitro* radiation transformed hamster and C_3H 10T1/2 cells. The C_3H 10T1/2 transformed cells showed elevated expression of three oncogenes, the c-*able, fms*, and Blym-1 as compared to the control (Borek et al., 1984e). None of the *ras* family gene transcripts appeared to be significantly elevated. Which cellular transforming genes are activated in radiogenic transformation *in vitro*, when, and how these events occur are some of the questions for the future.

8. DISCUSSION

We are constantly exposed to radiation. It occurs in nature and pervades the environment. Although a relatively weak carcinogen and mutagen compared to some chemicals, it is the most ubiquitous and measurable at low doses.

In recent years public interest and concern have focused on the potential biological hazards of low-dose radiation within the range of 0.1–1 rad. This is the dose level involved in public exposure from nuclear installations as well as from medical diagnostic X rays. We cannot discount the effect of low doses in the initiation of an event, which may later be amplified. Any carcinogens given at the right dose to a competent and specific target cell may serve as effective initiators and promoters. There is a need to develop suitable systems to assess directly the oncogenic potential of low-dose radiation. Epidemiologic studies do not lend themselves to evaluation of the carcinogenic effects of low doses of radiation. The contributions from epidemiology have been through extrapolation from incidents where a limited number of individuals received high doses of radiation delivered in most cases as single acute exposures. In practice, public exposure comprises multiple small doses. Animal systems, in which the induction of tumors by radiation serves to assess the oncogenic action of X rays, are limiting in studies in the low-dose range.

Cell cultures offer the best systems for evaluating the varied biological effects of radiation at a cellular level and investigating the mechanisms involved. Indeed, knowledge of basic aspects of cellular radiobiology was acquired within the last three decades. Clonal assays were developed, as well as survival curves, with the intriguing observation that some cells require 'company' provided by irradiated feeder cells (Puck and Marcus, 1956).

However, the use of cell culture to study radiation oncogenesis came later (Borek and Sachs, 1966a, 1967, 1968) and with it the opportunity to study cellular and molecular mechanisms associated with radiation oncogenesis as well as quantitative cancer risk estimates related to radiation quality, dose, dose rates, and cocarcinogenesis. Transformation *in vitro* was suggested as early as 1925 by Alexis Carrel, who stated that 'the best method of ascertaining the properties that characterize a malignant tissue would be to transform *in vitro* a strain of cells of a known type into cells capable of producing sarcomas or carcinomas and to study the changes undergone by the strain'.

The successful study of the direct oncogenic action of radiation on cells in culture was achieved in 1966 by Borek and Sachs (1966a), who showed that the exposure of mixed cultures of hamster embryo cells to 300 rad of X rays resulted in the neoplastic transformation of a defined low fraction of treated cells. At the same time, the unirradiated as well as the treated but not transformed cells, senesced and died. No spontaneous transformation was observed, demonstrating that the transforming effect resulted directly from the interaction of radiation with the exposed cells rather than from host effects present *in vivo*.

Although cell cultures offer defined systems that afford the opportunity to evaluate various aspects of transformation, they must be recognized as such; and their response to a variety of agents and conditions suggests rather than definitely establishes a condition comparable to that *in vivo*. 'Susceptibility' may vary among cell strains and lines, including cells from human origin where the genetic makeup of the donor cannot be excluded. We use *in vitro* cultures as simplified systems, yet these cells are derived of proliferating and nonproliferating tissues, 'forced' to grow freely *in vitro*. Though cell strains from freshly explanted cultures such as the hamster or human senesce *in vitro*, cell lines such as the 10T1/2 are populations of selected cells that are no longer subject to the control of 'time clocks' and finite life spans. Thus the radiation transformation process in the 10T1/2 established line (Fernandez *et al.*, 1980; Kennedy *et al.*, 1980a; Mordan *et al.*, 1983; Hall *et al.*, 1982) may differ from that observed in a cell strain like the hamster, consisting of normal diploid cells (Borek and Sachs, 1966a, 1967; Borek and Hall, 1973). Normal embryos explanted in culture give rise to cell populations in which some are competent to undergo induction by radiation (Borek and Sachs, 1967). The frequency of cells in this state of competence may vary not only with the cell population explanted and differ within a clonal population but also with the donor, because there exist differences in transformability among cells of different embryos (Borek, unpublished). The frequency of cells sensitive to initiation by X rays decreases with passage *in vitro* (Borek and Sachs, 1967) and this also holds true for cells from human embryos following u.v. treatment (Sutherland *et al.*, 1981). This is in contrast to the situation in the 10T1/2 and the 3T3 lines in which progressive culture *in vitro* may sometimes result in spontaneous transformation.

The clonal assay in the hamster cell system (Borek and Sachs, 1966b; Borek and Hall, 1973) does not require the lengthy cell-cell contact needed in the 10T1/2 cells suggested to be a major factor in determining the frequency of transformation in 10T1/2 (Kennedy *et al.*, 1980a). When hamster cells maintained under an identical environment are irradiated at various stages of clonal growth, only part of the clone may be transformed (Borek and Sachs, 1967), indicating a complex relationship related to genetic competence, sensitivity to transformation and physiological environment. Transformation by radiation can be compounded by viral transformation of these cells (Borek and Sachs, 1966a). This additional alteration modifies their surface properties and their interaction with parental X-ray-transformed cells as well as with normal cells which clearly inhibit the proliferation of the transformed (Borek and Sachs, 1966b). Change in cellular environment does not modify the inhibition of transformed cells by the normal ones, indicating that it is a contact phenomenon.

The success of a new field is measured by the unique and sometimes unexpected information contributed. We can therefore assess the success of cell cultures in studies related to radiation oncogenesis by the current acquisition of information unobtainable via other biological systems. Some of the following findings illustrate that, indeed, the *in vitro* systems have been useful and hold promise for the future.

(1) A number of assay systems based on rodent fibroblasts offer excellent tools for estimating quantitatively the incidence of radiation-induced oncogenic transformation under conditions where high and low toxicity prevail (Borek and Sachs, 1966a; Borek and Hall, 1973; Borek *et al.*, 1978; Terzaghi and Little, 1976a; Little, 1979; Han and Elkind, 1979; Lloyd *et al.*, 1979; Robertson *et al.*, 1983).

(2) Using these systems it is possible to obtain dose-response relationships over a wide range of doses and with a level of precision that cannot be rivaled by epidemiological studies of carcinogenesis in humans. (a) High-LET radiation is more oncogenic than X rays (Borek et al., 1978; Han and Elkind, 1979; Lloyd et al., 1979; Yang and Tobias, 1980; Robertson et al., 1983), but is also more toxic. (b) Transformation can be detected at doses as low as 0.3 rad of X rays. (c) The RBE for γ rays as compared to X rays at low-dose level is 0.5, which has important implications in medical radiation (Borek et al., 1983b). (d) The incidence of neutron-induced transformation peaks higher than X-ray transformation when assayed per cell survivor; but when evaluated on this basis of cells at risk, which is more relevant to the *in vivo* situation, X rays and neutrons rise to the same peak value (Borek et al., 1978; Han and Elkind, 1979). The incidence of transformation by a rays (Lloyd et al., 1979; Robertson et al., 1983) is higher than that induced by X rays.

(3) Assessment of transformation frequency can be made with *in vitro* assay systems at doses that are relevant to public health concern for exposure to medical radiation, with no extrapolation.

(4) Dose-response relationships established *in vitro* indicate that the data are poorly fitted by a simple linear relationship between dose and incidence. A linear extrapolation from data derived at high doses does not accurately predict transformation at low doses (Miller et al., 1979; Borek, 1979a).

(5) Data from several *in vitro* transformation systems indicate that fractionation of an X ray dose leads to an *enhanced* transformation for total doses less than about 150 rad. A linear extrapolation from single acute large doses therefore does not lead to cancer risk estimates that are either conservative or prudent for low doses as delivered as a series of fractions (Broek and Hall, 1974; Terzaghi and Little, 1976b; Miller and Hall, 1978; Miller et al., 1979; Borek, 1979b; Han and Elkind, 1979).

(6) The transforming effects of u.v. light are dose-dependent in rodent cells (DiPaolo and Donovan, 1976; Chan and Little, 1976; Little, 1979; Mondal et al., 1976) and in human cells (Sutherland et al., 1981; Borek and Andrews, 1983; Maher et al., 1982).

(7) Human fibroblasts can be transformed *in vitro* by X rays (Borek, 1980a) and u.v. (Sutherland et al., 1981; Milo et al., 1981; Borek and Andrews, 1983; Maher et al., 1982). The temporal process of transformation is different from that seen in rodent cells, as are also some aspects of the phenotypic expression of the cells. Potential to grow in agar appears concomitantly with morphological changes, and the frequency of human cell transformation by X rays is lower than that of rodent cells given the same dose of radiation (Borek, 1980a, 1981b).

(8) Agents that interact with radiation to enhance radiation transformation have been identified. These include TPA, chemicals, pyrolysate products, antipain added before radiation, and estradiol (Borek et al., 1979; Borek, 1980b; Kennedy and Little, 1978; Miller et al., 1981; Borek and Ong, 1981; Han and Elkind, 1982; DiPaolo et al., 1984; Borek et al., 1986a, Borek and Cleaver, 1986).

(9) We can successfully identify agents that suppress radiation-induced carcinogenesis and its cocarcinogenic interactions with other agents. These include the protease inhibitors antipain (added after irradiation), leupeptin (Kennedy and Little, 1978; Borek et al., 1979), retinoids (Miller et al., 1981; Borek et al., 1982; Borek, 1982a), inhibitors of poly(ADP-ribose) (Borek et al., 1984a, b, e).

(10) We can test conditions which modify cellular competence for transformation. Thus, thyroid hormones were found to be essential for the initiation of transformation (Guernsey et al., 1980, 1981; Borek et al., 1983a), while a hypothyroid state *in vitro* inhibited both X-ray- and chemically-induced transformation.

(11) Other agents that inhibit radiation-induced transformation and its enhancement by TPA and SOD and catalase, agents that scavenge free radicals, suggesting that X-ray-induced oncogenesis may be mediated in part via the effect of free radicals. The cellular content of catalase and SOD may determine to some extent cellular response to radiation (Borek and Troll, 1983; Borek, 1982b). Selenium, which induces catalase, nonprotein thiols and peroxidase is a powerful protective agent (Borek et al., 1986a; Borek and Biaglow,

1984) and enhanced peroxide breakdown. Vitamin E acts as an anticarcinogen by other mechanisms.

(12) It has been possible to evaluate underlying cellular and molecular mechanisms which regulate the effect of modulating compounds on radiation transformation. Thus the action of retinoids is not mediated by the type of damage inflicted on DNA, which can be monitored by SCE analysis (Borek, 1981a; Miller et al., 1981), but is mediated at the level of gene expression at the membrane level, expressed by altered adhesion and morphology and by changed levels of the Na-transport enzyme Na^+/K^+ ATPase (Borek, 1982a). The molecular effect of AP, which has diametrically opposed action on transformation depending on its temporal interaction with the cells being irradiated, does not alter DNA damage and replication (Borek and Cleaver, 1981). Antipain does not alter SCE in a direction parallel to its effect on transformation (Geard et al., 1981). Its effectiveness in the inhibition of carcinogen-induced chromosomal aberrations rather than SCE (Kinsella and Radman, 1978, 1980) suggests that perhaps aberrations are a more responsive assay to assess agents that modulate the action of carcinogens. The protective action of selenium is mediated via the induction of protective enzymes such as peroxidase and catalase (Borek and Biaglow, 1984).

(13) The surface expression of a variety of cellular features associated with the neoplastic state of X-ray-induced transformation can be identified within a week after exposure to radiation, thus allowing the identification of early transformants (Borek and Fenoglio, 1976). Such surface changes and other cytoskeletal modifications which may be associated could conceivably be intricately related to a cascade of events affecting the genetic apparatus (Puck, 1979). In chemically transformed cells a cytoskeletal element has been identified as a mutational product associated with transformation (Hamada et al., 1981).

(14) The role of chromosomal fine structure changes and instability associated with radiation-induced transformation can only be inferred, though it is compelling. Diploid strains of hamster and human origin remain near diploid or diploid even after transformation has taken place and other phenotypic changes have been expressed (Borek et al., 1977), but near diploid is perhaps sufficient for instability leading to malignancy.

(15) The suggestions of DNA damage repair and misrepair in the process of radiation transformation are again inferential. No DNA-repair enzyme has been identified in mammalian cells for ionizing radiation. Studies on u.v.-induced transformation in XP cells (Borek and Andrews, 1983; Maher et al., 1982) and on cells that are defective in DNA repair (Setlow et al., 1969; Cleaver, 1969) may bring us closer to the understanding of the exact role of DNA in transformation.

(16) Poly(ADP-ribose) appears to play a role in transformation though its action may differ in oncogenesis from that involving mutagenesis, sister chromatid exchanges and DNA repair (Borek et al., 1984a, b, e).

(17) Other experiments also suggest that mechanisms associated with repair or cell survival probably differ from those responsible for cell transformation. The increased survival associated with splitting the X ray doses is paralleled by higher transformation at low doses and lower transformation when the higher doses are split (Borek and Hall, 1974; Borek, 1979b; Miller et al., 1979). It has been suggested that subtransformation damage done by X rays is repaired more slowly than sublethal X ray damage (Elkind and Han, 1979).

(18) Although X rays are mutagenic (Cox and Masson, 1976) and some of the cell systems lend themselves to concurrent studies on both transformation and mutagenesis (Landolph and Heidelberger, 1979; Huberman et al., 1976; Barrett et al., 1978), transformation cannot be equated with mutagenesis (Barrett and Tso, 1978b; Borek et al., 1984e).

The underlying mechanisms associated within radiation transformation are still unclear. Normal cells contain various inherent growth factors which can serve as internal promoters and may vary from one cell type to another (Roberts et al., 1981). Chromosomal disturbances which may be amplified in diseased cells may be susceptible in the course of transformation. Events such as imbalance (Bloch-Schtacher and Sachs, 1976), disproportional DNA replication and gene amplification (Schimke et al., 1980), or involvement of specific genes (Knudson, 1981; Klein, 1981) following exposure to radiation cannot be excluded from playing a part in oncogenesis. Thus there is much to learn, in order to synthesize the information from various biochemical disciplines with our knowledge of radiation biology and to develop new techniques.

(19) The finding that radiation transformation *in vitro* of hamster embryo and C_3H 10T1/2 cells activates specific cellular transforming genes (oncogenes) (Table 7) (Borek and Ong, 1984; Borek *et al.*, 1984b; Borek *et al.*, 1987; Borek, 1985a, b) provides proof that DNA is the target in radiation malignant transformation. The fact that the activated oncogenes do not belong to the *ras* family oncogenes (Bishop, 1980; Guerrero *et al.*, 1984) suggests that a new non-*ras* transforming gene is associated with radiation transformation *in vitro* and that genes activated in cells transformed *in vitro* by radiation may differ from those in cells transformed by chemicals (Parada and Weinberg, 1983) and from those found in tumors induced by radiation (Guerrero *et al.*, 1984).

9. A FINAL WORD

One of the basic conundrums in carcinogenesis evolves from our inability unequivocally to distinguish primary events associated with initiation of malignant transformation from those which function as secondary events. Thus the role of oncogenes, mutations, gene rearrangements, amplification and other DNA alterations in transformation is yet unclear. The changes which take place give rise to abnormal expression of cellular genes.

We must always be cognizant of the fact that a variety of factors may modify the neoplastic process at its various stages of development. These constitute physiological permissive or protective factors (Borek, 1984a). When permissive factors prevail, such as genetic susceptibility, optimal stage in the cell cycle, optimal hormonal control or a particular stage in differentiation, initiation of transformation will take place. By contrast, if these permissive factors do not prevail, protective factors such as free radical scavengers will inhibit, to varying degrees, the onset and progression of the neoplastic process (Borek and Ong, in preparation). These may be inherent cellular factors (Borek, 1984a) or those added externally by dietary means acting as anticarcinogens (Ames, 1983; Borek *et al.*, 1986a) (Table 8). Thus, the interplay between inherent genetic and physiological factors

TABLE 8. *Permissive and Protective Factors in Transformation* In Vitro

	Parameter Observed
Permissive Factors:	
High genetic susceptibility	XP cells Bloom cells transform at higher rate than normal following u.v. exposure
Species' difference in susceptibility	Hamster > mouse > man
Thyroid hormones	Altered oxidative state and/or induction of 'cellular transformation associated proteins'
Stage in cell cycle	Some stages (e.g. G_1/S in human cells) are more permissive
Protective Factors:	
Dietary and cellular antioxidants	(Prevent the deleterious effects of reactive oxygen species)
1. Sellenium	Induces catalase and GSHpX, doubles cellular capacity to destroy peroxides
2. SOD	Scavenges superoxides formed by radiation and chemicals
NPSH (non-protein thiols)	Induced by Se scavenge reactive oxygen oxygen species
3. Retinoids	Various actions, some on cell membrane, e.g. induces Na^+/K^+ ATPase. Some derivatives scavenge free radicals. Affect oncogene expression (?)

and lifestyle influences (Sugimura et al., 1980; Ames, 1983; Arnott et al., 1982) which either enhance or inhibit the neoplastic process, are critical determinants in establishing the incidence of cancer.

Some important goals for the future include (1) the identification of transforming genes activated in cells transformed by radiation in vitro and their various forms of expression in early and late stages of transformation; (2) the study of age and tissue related sensitivity to radiation transformation by studying cellular and molecular events in various epithelial and fibroblast systems of several species including human; and (3) identify permissive and protective factors and their underlying mechanism of action.

It is of primary importance to improve culture conditions for human cells and thus encourage longer life span in vitro. Then, latency, expression and mechanisms of transformation to malignancy in human cells could be studied similar to rodent cells and one could even assess how these are related to the ones observed in rodent cells. Most of our data are derived from studies in rodent cells and we are still at a loss as to which of these cell systems matches most closely the human situation.

Acknowledgement—This article was supported by Grant No. CA-12536 by the National Cancer Institute awarded to the Radiological Research Laboratory at Columbia University and by a contract from the National Foundation for Cancer Research. I'd like to thank Ms. Michaela Delegianis, my secretary, who worked tirelessly and carefully to help to prepare this article.

REFERENCES

ABERCROMBIE, M. (1966) Contact inhibition: the phenomenon and its biological implications. *Natn. Cancer Inst. Monogr.* **26**: 249–277.
ALEXANDER, P. and LETT, J. (1968) *Comprehensive Biochemistry*, pp. 267–356, FLORKIN, M. and STOTS, E. (eds). Elsevier, Amsterdam.
AMES, B. N. (1983) Dietary carcinogens and anticarcinogens. Oxygen radicals and degenerative diseases. *Science* **211**: 1256–1264.
ANDREWS, A. D. and BOREK, C. (1982) Neoplastic transformation of human xeroderma pigmentosum and Bloom syndrome cells by UVB. *J. invest. Derm.* **78**: 355–356.
ARNOTT, M. S., VAN EYS, J. and WANG, Y. M. (1982) *Molecular Interactions of Nutrition and Cancer*, pp. 1–474, Raven Press, New York.
ARSENEAU, J., CANELLOS, G. P., JOHNSON, R. and DEVITA, V. T. (1977) Risk of new cancer in patients with Hodgkin's disease. *Cancer* **40**: 1912–1916.
BACQ, Z. M. and ALEXANDER, P. (1955) *Fundamentals of Radiobiology*, Academic Press, New York.
BARRETT, J.C. and Ts'o, P. O. P. (1978a) Evidence for the progressive nature of neoplastic transformation in vitro. *Proc. natn. Acad. Sci. U.S.A.* **71**: 3761–3765.
BARRETT, J. C. and Ts'o, P. O. P. (1978b) Relationship between somatic mutation and neoplastic transformation. *Proc. natn. Acad. Sci. U.S.A.* **75**: 3297–3301.
BARRETT, J. C., BIAS, N. E. and Ts'o, P. O. P. (1978) A mammalian cellular system for the concomitant study of neoplastic transformation and somatic mutation. *Mutat. Res.* **50**: 121–136.
BEIR (1972) Advisory Committee on the Biological Effects of Ionizing Radiations. The Effect on Populations of Exposure to Low Levels of Ionizing Radiations. National Research Council of the National Academy of Science.
BERENBLUM, I. (1982) Sequential aspects of chemical carcinogenesis: skin. In: *Cancer: A Comprehensive Treatise*, pp. 451–484, BECKER, F. F. (ed.) Plenum Press, New York.
BERWALD, Y. and SACHS, L. (1963) *In vitro* transformation with chemical carcinogens. *Nature* **200**: 1182–1184.
BISHOP, M. (1980) Retroviruses and cancer genes. *Adv. Cancer Res.* **37**: 1–29.
BLOCH-SCHTACHER, N. and SACHS, L. (1976) Chromosomal balance and the control of malignancy. *J. cell. Physiol.* **87**: 89–100.
BOND, V. P., CRONKITE, E. P., LIPPINCOTT, S. W. and SCHALLABARGER, C. F. (1960) Studies in radiation-induced mammary gland neoplasia in the rat. *Radiat. Res.* **12**: 276–285.
BOREK, C. (1979a) Malignant transformation in vitro: criteria, biological markers and application in environmental screening of carcinogens. *Radiat. Res.* **79**: 209–232.
BOREK, C. (1979b) Neoplastic transformation following split doses of X-rays. *Br. J. Radiol.* **52**: 845–848.
BOREK, C. (1980a) X-ray induced in vitro neoplastic transformation of human diploid cells. *Nature* **283**: 776–778.
BOREK, C. (1980b) Differentiation, metabolic activation and malignant transformation in cultured liver cells exposed to chemical carcinogens. In: *Advances in Modern Environmental Toxicology*, Vol. 1, pp. 297–318, MISHRA, N., DUNKEL, V. and MEHLMAN, M. A. (eds) Senate Press, Princeton, New Jersey.
BOREK, C. (1981a) Cellular transformation by radiation: induction, promotion and inhibition. *J. supramolec. Struct. cell. Biochem.* **16**: 311–336.
BOREK, C. (1981b) The induction, expression and modulation of radiation induced oncogenesis in vitro in diploid human and rodent cells. In: *Carcinogenesis: Fundamental Mechanisms and Environmental Effects*, pp. 509–516, PULLMAN, B., Ts'o, P. O. P. and BELBOIN, H. (eds). D. Reidel Publ. Co.

Borek, C. (1982a) Radiation oncogenesis in cell culture. *Adv. Cancer Res.* **37**: 159–232.
Borek, C. (1982b) Vitamins and micronutrients modify carcinogenesis and tumor promotion *in vitro*. In: *Molecular Interrelations of Nutrition and Cancer*, pp. 337–350, Arnott, M. S., van Eys, J. and Wang, Y. M. (eds). Raven Press, New York.
Borek, C. (1983) Epithelial *in vitro* cell systems in carcinogenesis studies. *Ann. N.Y. Acad. Sci.* **407**: 284–290.
Borek, C. (1984a) Permissive and protective factors in malignant transformation of cells in culture. In: *The Biochemical Basis of Chemical Carcinogenesis*, pp. 175–188. Greim, H., Juna, R., Kraemer, M., Marquardt, H. and Oesch, F. (eds). Raven Press, New York.
Borek, C. (1984b) *In vitro* cell cultures as tools in the study of free radicals and free radical modifiers in carcinogenesis. In: *Methods in Enzymology, Volume on Oxygen Radicals in Biological Systems*, pp. 465–479, Colowick, C. P., Kaplan, N. O. and Packer, L. (eds). Academic Press, New York.
Borek, C. (1985a) Cellular and molecular mechanisms in malignant transformation of diploid rodent and human cells by radiation. In: *Mammalian Cell Transformation. Mechanisms of Carcinogenesis and Assays for Carcinogens*, pp. 365–378. J. Carl Barrett and Raymond W. Tennant (ed). Raven Press, NY.
Borek, C. (1985b) Oncogenes and cellular controls in radiogenic transformation of rodent and human cells. In: *The Role of Chemicals and Radiation in the Etiology of Cancer*, pp. 303–316. Eliezer Huberman and Susan H. Barr (eds). Raven Press, NY.
Borek, C. (in press, 1986) Hormones and dietary factors controlling gene activation and expression in carcinogenesis. In: *Mechanisms of Antimutagenesis and Anticarcinogenesis*. D. Shankel *et al.* (eds). Plenum Press, NY.
Borek, C. and Andrews, A. (1983) Oncogenic transformation of normal, XP and Bloom syndrome cells by X-rays and ultraviolet irradiation. In: *Human Carcinogenesis*, pp. 519–541, Harris, C. C. and Antrup, H. (eds). Academic Press, New York.
Borek, C. and Biaglow, J. E. (1984) Factors controlling cellular peroxide breakdown: relevance to selenium protection against radiation and chemically induced carcinogenesis. *Proc. Am. Ass. Cancer Res.* **25**: 125 (abstract).
Borek, C. and Cleaver, J. E. (1981) Protease inhibitors neither damage DNA nor interfere with DNA in human cells. *Mutat. Res.* **82**: 373–380.
Borek, C. and Cleaver, J. E. (1986) Antagonistic action of a tumor promoter and a poly(adenosine diphosphoribose) synthesis inhibitor in radiation-induced transformation *in vitro*. *Biochemical and Physical Research Communication*, **134**: 1334–1341.
Borek, C. and Fenoglio, C. M. (1976) Scanning electron microscopy of surface features of hamster embryo cells transformed *in vitro* by x-irradiation. *Cancer Res.* **36**: 1325–1334.
Borek, C. and Guernsey, D. (1981) Membrane associated ion transport enzymes in normal and oncogenically transformed fibroblasts and epithelial cells. *Studia Biophysica* **84**(1): 53–54.
Borek, C. and Hall, E. J. (1973) Transformation of mammalian cells *in vitro* by low doses of X-rays. *Nature* **243**: 450–453.
Borek, C. and Hall, E. J. (1974) Effect of split doses of X-rays on neoplastic transformation of single cells. *Nature* **252**: 499–501.
Borek, C. and Hall, E. J. (1982) Oncogenic transformation produced by agents and modalities used in cancer therapy and its modulation. *Ann. N.Y. Acad. Sci.* **397**: 193–219.
Borek, C. and Hall, E. J. (1984) Induction and modulation of radiogenic transformation in mammalian cells. In: *Radiation Carcinogenesis: Epidemiology and Biological Significance*, pp. 291–302, Boice, J. D. and Fraumeni, J. F., Jr (eds). Raven Press, New York.
Borek, C. and Ong, A. (1981) The interaction of ionizing radiation and food pyrolysis products in producing oncogenic transformation *in vitro*. *Cancer Lett.* **12**: 61–66.
Borek, C. and Ong, A. (1984) Transfection of NIH/3T3 by DNA from x-ray transformed cells and assessment of expression of transforming genes in the donor cells, p. 124, *Proceedings of Radiation Research Society Meeting*, Orlando, Florida.
Borek, C. and Sachs, L. (1966a) *In vitro* cell transformation by X-irradiation. *Nature* **210**: 276–278.
Borek, C. and Sachs, L. (1966b) The difference in contact inhibition of cell replication between normal cells and cells transformed by different carcinogens. *Proc. natn. Acad. Sci. U.S.A.* **56**: 1705–1711.
Borek, C. and Sachs, L. (1967) Cell susceptibility to transformation by X-irradiation and fixation of the transformed state. *Proc. natn. Acad. Sci. U.S.A.* **57**: 1522–1527.
Borek, C. and Sachs, L. (1968) The number of cell generations required to fix the transformed state in X-ray induced transformation. *Proc. natn. Acad. Sci. U.S.A.* **59**: 83–85.
Borek, C. and Troll, W. (1983) Modifiers of free radicals inhibit *in vitro* the oncogenic actions of X-rays, bleomycin, and the tumor promoter 12-O-tetradecaoylphorbol 13-acetate. *Proc. natn. Acad. Sci. U.S.A.* **80**: 5749–5752.
Borek, C., Higashino, S. and Loewenstein, W. R. (1969) Intercellular communication and tissue growth —IV. Conductance of membrane junctions of normal and cancerous cells in culture. *J. Memb. Biol.* **1**: 274–293.
Borek, C., Grob, M. and Burger, M. M. (1973) Surface alterations in epithelial and fibroblastic cells in culture: a disturbance of membrane degradation versus biosynthesis. *Expl Cell Res.* **77**: 207–215.
Borek, C., Pain, C. and Mason, H. (1977) Neoplastic transformation of hamster embryo cells irradiated *in utero* and assayed *in vitro*. *Nature* **266**: 452–454.
Borek, C., Hall, E. J. and Rossi, H. H. (1978) Malignant transformation in cultured hamster embryo cells produced by X-rays. 430 keV monoenergetic neutrons, and heavy ions. *Cancer Res.* **38**: 2997–3005.
Borek, C., Miller, R., Pain, C. and Troll, W. (1979) Conditions for inhibiting and enhancing effects of the protease inhibitor antipain on X-ray-induced neoplastic transformation in hamster and mouse cells. *Proc. natn. Acad. Sci. U.S.A.* **76**: 1800–1803.
Borek, C., Miller, R. C., Geard, C. R., Guernsey, D. L. and Smith, J. E. (1982) *In vitro* modulation of oncogenesis and differentiation by retinoids and tumor promoters. In: *Carcinogenesis*. Vol. 7, pp. 277–284, Hecker, E. (ed.). Raven Press, New York.

BOREK, C., GUERNSEY, D. L., ONG, A. and EDELMAN, I. S. (1983a) Critical role played by thyroid hormone in induction of neoplastic transformation by chemical carcinogens in tissue culture. *Proc. natn. Acad. Sci. U.S.A.* **80**: 5749–5752.

BOREK, C., HALL, E. J. and ZAIDER, M. (1983b) X-rays must be twice as potent as gamma rays for malignant transformation at low doses. *Nature* **301** (5896): 156–158.

BOREK, C., CLEAVER, J. E. and FUJIKI, H. (1984a) Critical biochemical and regulatory events in malignant transformation and promotion *in vitro*. In: *Cellular Interaction by Environmental Tumor Promoters and Relevance to Human Cancer*, FUJIKI, H. *et al*. (eds). Japan Scientific Societies, Tokyo, in press.

BOREK, C., MORGAN, W. R., ONG, A. and CLEVER, J. E. (1984b) Malignant transformation *in vitro* is blocked by inhibitors of poly(ADP-ribose) synthesis. *Proc. natn. Acad. Sci. U.S.A.* **81**, 243–247.

BOREK, C., ONG, A. and RHIM, J. S. (1984c) Thyroid hormone modulates transformation induced by Kirsten Murine Sarcoma virus. *Cancer Res.* **45**: 1702–1706.

BOREK, C., ONG, A., BRESSER, J. and GILLESPIE, D. (1984d) Transforming activity of DNA of radiation transformed mouse cells and identification of activated oncogenes in the donor cells. *Proc. Am. Ass. Cancer Res.* **25**: 100.

BOREK, C., ONG, A., MORGAN, W. F. and CLEAVER, J. E. (1984e) Inhibition of X-ray and ultraviolet light-induced transformation *in vitro* by modifiers of poly(ADP-ribose) synthesis. *Radiat. Res.* **99**, 219–227.

BOREK, C., ONG, A., MASON, H., DONAHUE, L. and BIAGLOW, J. E. (1986a) Selenium and Vitamin E inhibit radiogenic and chemically induced transformation *in vitro* via different mechanisms of action. *Proc. Natn. Acad. Sci. (US)* **83**: 1490–1494.

BOREK, C., ZAIDER, M., ONG, A., MASON, H. and WITZ, G. (1986b) Ozone acts directly and synergistically with ionizing radiation to induce *in vitro* neoplastic transformation. Carcinogenesis 7 1611–1613

BOREK, C., ONG, A. and MASON, H. (1987) Distinctive transforming genes in x ray transformed mammalian cells. Proc. Nat. Acad. Sci (USA) **84** 794–798

BRADY, R. O., BOREK, C. and BRADLEY, R. M. (1969) Composition and synthesis of gangliosides in rat hepatocyte and hepatoma cell lines. *J. biol. Chem.* **244**: 6552–6554.

BRESSER, J., DOERING, J. and GILLESPIE, D. (1983a) Quickblot: Selective mRNA or DNA immobilization from whole cells. *DNA* **2**: 243–254.

BRESSER, J., HUBBEL, H. and GILLESPIE, D. (1983b) Biological activity of mRNA immobilized on nitrocellulose in NaI. *Proc. natn. Acad. Sci. U.S.A.* **80**: 6523–6527.

BROWN, P. (1936) *American Martyrs of Science Through Roentgen Ray*, Thomas Springfield, Illinois.

CAIRNS, J. (1981) The origin of human cancers. *Nature* **289**: 353–357.

CARREL, A. (1925) Essential characteristics of a malignant cell. *J. Am. Med. Ass.* **84**: 157–158.

CHAGANTI, R. S., SCHONBERG, S. and GERMAN, J. (1974) A manyfold increase in sister chromatid exchanges in Bloom's syndrome lymphocytes. *Proc. natn. Acad. Sci. U.S.A.* **71**: 4508–4512.

CHAN, C. and LITTLE, J. B. (1976) Induction of oncogenic transformation *in vitro* by ultraviolet light. *Nature* **264**: 442–444.

CLEAVER, J. E. (1969) Xeroderma pigmentosum: a human disease in which an initial stage of DNA repair is defective. *Proc. natn. Acad. Sci. U.S.A.* **63**: 428–435.

CLEAVER, J. E. (1978) DNA repair and its coupling to DNA replication in eukaryotic cells. *Biochim. biophys. Acta* **516**: 489–516.

CLEAVER, J. E. and BOOTSMA, D. (1975) Xeroderma pigmentosum: biochemical and genetic characteristics. *A. Rev. Genet.* **9**: 19–38.

CLEAVER, J. E., BANDA, M. J., TROLL, W. and BOREK, C. (1986) Short Communication: some protease inhibitors are also inhibitors of poly(ADP-ribose) polymerase. *Carcinogenesis* **7**: 323–325.

CLEAVER, J. E., BODELL, W. J., BOREK, C., MORGAN, W. F. and SCHWARTZ, J. L. (1983) Poly(ADP-ribose): spectator or participant in excision repair of DNA damage. In: *ADP-ribosylation, DNA Repair and Cancer*, pp. 139–205, HYASHI, O. *et al*. (eds). Japan Society Press, Tokyo.

COOPER, G. M. (1982) Cellular transforming genes. *Science* **218**: 801–806.

COX, R. and MASSON, W. K. (1976) X-ray-induced mutation to 6-thioguanine resistance in cultured human diploid fibroblasts. *Mutat. Res.* **37**: 125–136.

DIPAOLO, J. A. and DONOVAN, P. J. (1976) *In vitro* morphologic transformation of Syrian hamster cells by u.v.-irradiation is enhanced by X-irradiation and unaffected by chemical carcinogens. *Int. J. Radiat. Biol.* **30**: 41–53.

DIPAOLO, J. A., DONOVAN, P. J. and NELSON, R. L. (1971) X-irradiation enhancement of transformation by benzo(a)pyrene in hamster embryo cells. *Proc. natn. Acad. Sci. U.S.A.* **68**: 1734–1737.

DIPAOLO, J. A., DONOVAN, P. J. and POPESCU, N. C. (1976) Kinetic of Syrian hamster during X-irradiation enhancement of transformation *in vitro* by chemical carcinogens. *Radiat. Res.* **66**: 310–325.

DIPAOLO, J. A., EVANS, C. H., DEMARINIS, A. J. and DONIGER, J. (1984) Inhibition of radiation-induced anti-promoted transformation of hamster embryo cells by lymphotoxin. *Cancer Res.* **44**: 1465–1471.

DIPAOLO, J. A., POPESCU, N. C. and NELSON, R. L. (1973) Chromosomal banding patterns and *in vitro* transformation of Syrian hamster cells. *Cancer Res.* **33**: 3250–3258.

DOLL, R. and PETO, R. (1981) The causes of cancer: quantitative estimates of avoidable risks of cancer in the United States today. *J. natn. Cancer Inst.* **66**: 1191–1308.

DONIGER, J., JACOBSON, E. D., KRELL, K. and DIPAOLO, J. A. (1981) Ultraviolet light action spectra for neoplastic transformation and lethality of Syrian hamster embryo cells correlate with spectrum for pyrimidine dimer formation in cellular DNA. *Proc. natn. Acad. Sci. U.S.A.* **78**: 2378–2382.

DUESBERG, P. H. (1983) Retroviral transforming genes in normal cells? *Nature* **304**: 219–226.

ELKIND, M. M. and HAN, A. (1979) Neoplastic transformation and dose fractionation: does repair of damage play a role? *Radiat. Res.* **79**: 233–240.

EMBLETON, M. J. and HEIDELBERGER, C. (1975) Neoantigens on chemically transformed cloned C3H mouse embryo cells. *Cancer Res.* **35**: 2049–2055.

EMERIT, I. and CERUTTI, P. A. (1981) Tumor promoter phorbol-12-myrislate-13-acetate induces chromosomal damage via indirect action. *Nature* **293**: 144–146.

FARBER, E. (1973) Carcinogenesis—cellular evolution as a unifying thread: Presidential address. *Cancer Res.* **33**: 2537–2550.
FERNANDEZ, A., MONDAL, S. and HEIDELBERGER, C. (1980) Probabilistic view of the transformation of cultured C3H 10T1/2 mouse embryo fibroblasts by 3-methylcholanthrene. *Proc. natn. Acad. Sci. U.S.A.* **77**: 7272–7276.
FRIDOVICH, I. (1978) The biology of oxygen radicals. *Science* **201**: 875–880.
FRY, R. J. M. and AINSWORTH, E. J. (1977) Radiation injury: some aspects of the oncogenic effects. *Fedn Proc.* **36**: 1703–1707.
FUJIKI, H., MORI, M., NAKAYASU, M., TERADA, M. and SUGIMURA, T. (1979) A possible naturally occurring tumor promoter, teleocidin B from streptomyces. *Biochem. biophys. Res. Commun.* **90**: 976–983.
GAHNBERG, C. G. and HAKOMORI, S. I. (1973) Altered growth behavior of malignant cells associated with changes in externally labeled glycoprotein and glycolipid. *Proc. natn. Acad. Sci. U.S.A.* **70**: 3329–3333.
GEARD, C. R., FREEMAN, M. R., MILLER, R. C. and BOREK, C. (1981) Antipain and radiation effects on oncogenic transformation and sister chromatid exchanges in Syrian hamster embryo and mouse C3H 10T1/2 cells. *Carcinogenesis* **2**: 1229–1235.
GOLDSTEIN, B. D., WITZ, G., AMORUSO, M., STONE, D. S. and TROLL, W. (1981) Stimulation of human polymorphonuclear leukocyte superoxide anion radical production by tumor promoters. *Cancer Lett.* **11**: 257–262.
GUERNSEY, D. L., BOREK, C. and EDELMAN, I. S. (1981) Crucial role of thyroid hormone in X-ray induced transformation in cell culture. *Proc. natn. Acad. Sci. U.S.A.* **78**: 5708–5711.
GUERNSEY, D. L., ONG, A. and BOREK, C. (1980) Modulation of X-ray induced neoplastic transformation *in vitro* by thyroid hormone. *Nature* **288**: 591–592.
GUERRERO, I., CALZADA, P., MAYER, A. and PELLICER, A. (1984) A molecular approach to leukomogenesis. Mouse lymphoma contained an activated c-ras oncogene. *Proc. natn. Acad. Sci. U.S.A.* **81**: 202–205.
HALL, E. J. (1978) *Radiobiology for the Radiologist*, Harper and Row, New York.
HALL, E. J., ROSSI, H. H., ZAIDER, M., MILLER, R. C. and BOREK, C. (1982) The role of neutrons in cell transformation research. In: *Radiation Protection, Neutron Carcinogenesis*, pp. 381–406, BROERSE, J. J. and GERBER, G. B. (eds). Commission of European Communities, DCXII Biology, Radiation Protection and Medical Research and Radiobiology.
HAMADA, H., LEAVITT, J. and KAKUNAGA, T. (1981) Mutated beta-actin gene; coexpression with an unmutated allele in a chemically transformed human fibroblast cell line. *Proc. natn. Acad. Sci. U.S.A.* **78**: 3634–3638.
HAN, A. and ELKIND, M. M. (1979) Transformation of mouse C3H 10T1/2 cells by single and fractionated doses of X-rays and fission-spectrum neutrons. *Cancer Res.* **39**: 123–130.
HAN, A. and ELKIND, M. M. (1982) Enhanced transformation of mouse 10T1/2 cells by 12-O-tetradecanoyl-phorbol-13-acetate following exposure to X-rays or to fission spectrum neutrons. *Cancer Res.* **42**: 477–483.
HANAWALT, P. C., COOPER, P. K., GANESAN, A. K. and SMITH, C. A. (1979) DNA repair in bacteria and mammalian cells. *A. Rev. Biochem.* **48**: 783–836.
HARISIADIS, L., MILLER, R. C., HALL, E. J. and BOREK, C. (1978) A vitamin A analogue inhibits radiation-induced oncogenic transformation. *Nature* **274**: 486–487.
HART, R. W., SETLOW, R. B. and WOODHEAD, A. D. (1977) Evidence that pyrimidine dimers in DNA can give rise to tumors. *Proc. natn. Acad. Sci. U.S.A.* **74**: 5574–5578.
HASELTINE, W. A. (1983) Ultraviolet light repair and mutagenesis revisited. *Cell* **33**: 13–17.
HECKER, E. (1971) Isolation and characterization of the cocarcinogenic principles from croton oil. In: *Methods in Cancer Research*, Vol. 6, pp. 439–484, BUSCH, H. (ed.). Academic Press, New York.
HEIDELBERGER, C. (1975) Chemical carcinogenesis. *A. Rev. Biochem.* **44**: 79–121.
HEIDELBERGER, C., MONDAL, S. and PETERSON, A. R. (1978) Initiation and promotion in cell cultures. In: *Carcinogenesis, A Comprehensive Survey, Vol. 2—Mechanisms of Tumor Promotion and Co-Carcinogenesis*, pp. 197–220, SLAGA, T. J., SIVAK, A. and BOUTWELL, R. K. (eds). Raven Press, New York.
HIGGINSON, J. (1979) Perspectives and future developments in research on environmental carcinogenesis. In: *Identification and Mechanisms of Action*, 31st Annual Symposium on Fundamental Cancer Research, pp. 187–220, GRIFFIN, A. C. and SHAW, C. R. (eds). Raven Press, New York.
HUBERMAN, E., MAGER, R. and SACHS, L. (1976) Mutagenesis and transformation of normal cells by chemical carcinogens. *Nature* **264**: 360–361.
KAKUNAGA, T. (1973) A quantitative system for assay of malignant transformation by chemical carcinogens using a clone derived from BALB-3T3. *Int. J. Cancer* **12**: 463–473.
KARUNAGA, T. (1974) Requirement for cell replication in the fixation and expression of the transformed state in mouse cells treated with 4-nitroquinoline-1-oxide. *Int. J. Cancer* **14**: 736–742.
KAKUNAGA, T. (1975) The role of cell division in the malignant transformation of mouse cells treated with 3-methylcholanthrene. *Cancer Res.* **35**: 1637–1642.
KAKUNAGA, T. (1978) Neoplastic transformation of human diploid fibroblast cells by chemical carcinogens. *Proc. natn. Acad. Sci. U.S.A.* **75**: 1334–1338.
KAKUNAGA, T. and CROW, J. D. (1980) Cell variants showing differential susceptibility to ultraviolet light-induced transformation. *Science* **209**: 505–507.
KAPLAN, H. S. (1967) On the natural history of the murine leukemias: Presidential address. *Cancer Res.* **27**: 1325–1340.
KENNEDY, A. R. and LITTLE, J. B. (1978) Protease inhibitors suppress radiation-induced malignant transformation *in vitro*. *Nature* **276**: 825–826.
KENNEDY, A. R. and WEICHSELBAUM, R. R. (1981) Effects of 17 beta-estradiol on radiation transformation *in vitro*; inhibition of effects by protease inhibitors. *Carcinogenesis* **2**: 67–69.
KENNEDY, A. R., FOX, M., MURPHY, G. and LITTLE, J. B. (1980a) Relationship between X-ray exposure and malignant transformation in C3H 10T1/2 cells. *Proc. natn. Acad. Sci. U.S.A.* **77**: 7262–7266.
KENNEDY, A. R., MURPHY, G. and LITTLE, J. B. (1980b) Effect of time and duration of exposure to 12-O-tetradecanoylphorbol-13-acetate on X-ray transformation of C3H 10T1/2 cells. *Cancer Res.* **40**: 1915–1920.

Kinsella, A. R. and Radman, M. (1978) Tumor promoter induces sister chromatid exchanges: relevance to mechanisms of carcinogenesis. *Proc. natn. Acad. Sci. U.S.A.* **75**: 6149–6153.

Kinsella, A. R. and Radman, M. (1980) Inhibition of carcinogen-induced chromosomal aberrations by an anticarcinogenic protease inhibitor. *Proc. natn. Acad. Sci. U.S.A.* **77**: 3544–3547.

Klein, G. (1981) The role of gene dosage and genetic transpositions in carcinogenesis. *Nature* **294**: 313–318.

Knudson, A. G., Jr. (1981) Human cancer genes. In: *Symposium on Fundamental Cancer Research, Vol. 33—Genes, Chromosomes, and Neoplasia*, Houston, Texas, pp. 453–462, Arrighi, F. E., Rao, P. N. and Stubblefield, E. (eds). Raven Press, New York.

Land, H., Parada, L. F. and Weinberg, R. A. (1983) Tumorigenic conversion of primary embryo fibroblasts requires at least two cooperative oncogenes. *Nature* **304**: 596–602.

Landolph, J. R. and Heidelberger, C. (1979) Chemical carcinogens produce mutations to ouabain resistance in transformable C3H 10T1/2 Cl 8 mouse fibroblasts. *Proc. natn. Acad. Sci. U.S.A.* **76**: 930–934.

Leavitt, J. and Kakunaga, T. (1980) Expression of a variant form of actin and additional polypeptide changes following chemical-induced *in vitro* neoplastic transformation of human fibroblasts. *J. biol. Chem.* **255**: 1650–1661.

Leibovitz, B. E. and Siegal, B. V. (1980) Aspects of free radical reactions in biological systems: aging. *J. Geront.* **35**: 45–56.

Little, J. B. (1979) Quantitative studies of radiation transformation with the A31-11 mouse BALB/3T3 cell line. *Cancer Res.* **39**: 1474–1480.

Little, J. B., LeMotte, P. K. and Liber, H. L. (1983) Quantitative studies of cytotoxicity, mutagenesis and oncogenic transformation by radioisotopes incorporated into DNA. In: *Human Carcinogenesis*, pp. 545–559, Harris, C. C. and Antrup, H. N. (eds). Academic Press, New York.

Lloyd, E. L., Gemmell, M. A., Henning, C. B., Gemmell, D. S. and Zabransky, B. J. (1979) Transformation of mammalian cells by alpha particles. *Int. J. Radiat. Biol.* **36**: 467–478.

Lotan, R. (1980) Effects of vitamin A and its analogs (retinoids) on normal and neoplastic cells. *Biochim. biophys. Acta.* **605**: 33–91.

Maher, V. M., Dorney, E. J., Mendrola, A. L., Konze-Thomas, B. and McCormick, J. J. (1979) DNA excision repair processes in human cells can eliminate the cytotoxic and mutagenic consequences of ultraviolet irradiation. *Mutat. Res.* **62**: 311–323.

Maher, V. M., Rowan, L. A., Silinskas, K. C., Kateley, S. A. and McCormick, J. J. (1982) Frequency of u.v.-induced neoplastic transformation of diploid human fibroblasts is higher in xeroderma pigmentosum cells than in normal cells. *Proc. natn. Acad. Sci. U.S.A.* **79**: 2613–2617.

Miller, E. C. and Miller, J. A. (1979) Naturally occurring chemical carcinogens that may be present in foods. In: *International Review of Biochemistry, Vol. 27—Biochemistry of Nutrition*, pp. 123–166, Neuberger, A. and Jukes, T. H. (eds) University Park Press, Baltimore, Maryland.

Miller, R. C. and Hall, E. J. (1978) X-ray dose fractionation and oncogenic transformations in cultured mouse embryo cells. *Nature* **272**: 58–60.

Miller, R. C., Geard, C. R., Osmak, R. S., Rutledge-Freeman, M., Ong, A., Mason, H., Napholtz, A., Perez, N., Harisiadis, L. and Borek, C. (1981) Modified sister chromatid exchanges and radiation-induced transformation in rodent cells by the tumor promoter 12-O-tetradecanoyl-phorbol-13-acetate and two retinoids. *Cancer Res.* **41**: 655–659.

Miller, R. C., Hall, E. J. and Rossi, H. H. (1979) Oncogenic transformation of mammalian cells *in vitro* with split doses of x-rays. *Proc. natn. Acad. Sci. U.S.A.* **76**: 5755–5758.

Milo, G. E. and DiPaolo, J. A. (1978) Neoplastic transformation of human diploid cells *in vitro* after chemical carcinogen treatment. *Nature* **275**: 130–132.

Milo, G. E., Noyes, I., Donohue, J. and Weisbrode, S. (1981) Neoplastic transformation of human epithelial cells *in vitro* after exposure to chemical carcinogens. *Cancer Res.* **41**: 5096–5102.

Mitelman, F. (1984) Restricted number of chromosomal regions implicated in aetiology of human cancer and leukaemia. *Nature* **310**: 325–327.

Mondal, S., Brankow, D. W. and Heidelberger, C. (1976) Two-stage chemical oncogenesis in cultures of C3H 10T1/2 cells. *Cancer Res.* **36**: 2254–2260.

Mordan, L. J., Martner, J. E. and Betram, J. S. (1983) Quantitative neoplastic transformation of C3H 10T1/2 fibroblasts: dependence upon the size of the initiated cell colony at confluence. *Cancer Res.* **43**: 4062–4067.

Nagao, M., Sugimura, T. and Matsushima, T. (1978) Environmental mutagens and carcinogens. *A. Rev. Genet.* **12**: 117–159.

Nagasawa, H. and Little, J. B. (1979) Effect of tumor promoters, protease inhibitors, and repair processes on X-ray-induced sister chromatid exchanges in mouse cells. *Proc. natn. Acad. Sci. U.S.A.* **76**: 1943–1947.

Namba, M., Nishitani, K. and Kimoto, T. (1978) Carcinogenesis in tissue culture.—29. Neoplastic transformation of a normal human diploid cell strain, WI-38, with Co-60 gamma rays. *Jap. J. exp. Med.* **48**: 303–311.

Nielson, S. E. and Puck, T. T. (1980) Deposition of fibronectin in the course of reverse transformation of Chinese hamster ovary cells by cyclic AMP. *Proc. natn. Acad. Sci. U.S.A.* **77**: 985–989.

Painter, R. B. (1978) DNA synthesis inhibition in HeLa cells as a simple test for agents that damage human DNA. *J. Environ. Path. Toxicol.* **2**: 65–78.

Painter, R. B. and Cleaver, J. E. (1969) Repair replication, unscheduled DNA synthesis, and the repair of mammalian DNA. *Radiat. Res.* **37**: 451–466.

Painter, R. B. and Young, B. R. (1972) Repair replication in mammalian cells after X-irradiation. *Mutat. Res.* **14**: 225–235.

Parada, L. and Weinberg, R. A. (1983) Presence of a Kirsten Murine Sarcoma virus ras oncogene in cells transformed by 3-methylcholanthrene. *Molec. and Cell. Biol.* **3**: 2298–2301.

Peehl, D. M. and Stanbridge, E. J. (1981) Anchorage-independent growth of normal human fibroblast. *Proc. natn. Acad. Sci. U.S.A.* **78**: 3053–3057.

Perry, P. and Evans, H. J. (1975) Cytological detection of mutagen-carcinogen exposure by sister chromatid exchange. *Nature* **258**: 121–125.

PERUCHO, M., GOLDFARB, M., SHIMIZUK, M., FOGH, J. and WIGLER, M. (1981) Human tumor derived cell lines contain common and different transforming genes. *Cells* **27**: 467–476.

PETO, R., DOLL, R., BUCKLEY, J. D. and SPORN, M. B. (1981) Can dietary beta-carotene materially reduce human cancer rates? *Nature* **290**: 201–208.

PIENTA, R. J., POILEY, J. A. and LEBHERZ, III, W. B. (1977) Morphological transformation of early passage golden Syrian hamster embryo cells derived from cryopreserved primary cultures as a reliable *in vitro* bioassay for identifying diverse carcinogens. *Int. J. Cancer* **19**: 642–655.

PITOT, H. C. and SIRICA, A. E. (1980) The stages of initiation and promotion in hepatocarcinogenesis. *Biochim. biophys. Acta* **605**: 191–215.

POLLACK, E. J. and TODARO, G. J. (1968) Radiation enhancement of SV40 transformation in 3T3 and human cells. *Nature* **219**: 520–521.

PRYOR, W. A. (1976) The role of free radical reactions in biological systems. In: *Free Radicals in Biology*, Vol. 1, pp. 1–49, PRYOR, W. A. (ed.). Academic Press, New York.

PUCK, T. T. (1977) Cyclic AMP, the microtubule–microfilament system, and cancer. *Proc. natn. Acad. Sci. U.S.A.* **74**: 4491–4495.

PUCK, T. T. (1979) Studies on cell transformation. *Somatic Cell Genet.* **5**: 973–990.

PUCK, T. T. and MARCUS, P. I. (1956) Action of X-rays on mammalian cells. *J. exp. Med.* **103**: 653–666.

REDDY, E. P., REYNOLDS, R. K., SANTOS, E. and BARBACID, M. (1982) A point mutation is responsible for the acquisition of transforming properties by the T24 human bladder carcinoma oncogene. *Nature* **300**: 149–152.

REZNIKOFF, C. A., BERTRAM, J. S., BRANKOW, D. W. and HEIDELBERGER, C. (1973a) Quantitative and qualitative studies of chemical transformation of cloned mouse C3H 10T1/2 mouse embryo cells sensitive to post confluence inhibition of cell division. *Cancer Res.* **33**: 3239–3249.

REZNIKOFF, C. A., BRANKOW, D. W. and HEIDELBERGER, C. (1973b) Establishment and characterization of a cloned line of C3H mouse embryo cells sensitive to post confluence inhibition of division. *Cancer Res.* **33**: 3231–3238.

ROBERTS, A. B., ANZANO, M. A., LAMB, L. C., SMITH, J. M. and SPORN, M. B. (1981) New class of transforming growth factors potentiated by epidermal growth factor: isolation from non-neoplastic tissues. *Proc. natn. Acad. Sci. U.S.A.* **78**: 5339–5343.

ROBERTSON, J. B. KOEHLER, A., GEORGE J. and LITTLE, J. B. (1983) Oncogenic transformation of mouse BALB/3T3 cells by plutonium-238 alpha particles. *Radiat. Res.* **96**: 261–274.

ROSSI, H. H. and KELLERER, A. M. (1974) The validity of risk estimates of leukemia incidence based on Japanese data. *Radiat. Res.* **58**: 131–140.

SCHIMKE, R. T., BROWN, P. C., KAUFMAN, R. J., MCGROGAN, M. and SLATE, D. L. (1980) Chromosomal and extrachromosomal localization and amplified dihydrofolate reductase genes in cultured mammalian cells. *Symposium on Quantitative Biology*, Part 2, pp. 785–797, Cold Spring Harbor.

SETLOW, R. B. (1966) Cyclobutane-type pyrimidine dimers in polynucleotides. *Science* **153**: 379–386.

SETLOW, R. B. (1974) The wavelengths in sunlight effective in producing skin cancer: a theoretical analysis. *Proc. natn. Acad. Sci. U.S.A.* **71**, 3363–3366.

SETLOW, R. B. (1978) Repair deficient human disorders and cancer. *Nature* **271**: 713–717.

STELOW, R. B. and HART, R. W. (1974) Direct evidence that damaged DNA results in neoplastic transformation—a fish story. *Radiat. Res.* **59**: 73–77.

SETLOW, R. B., REGAN, T. D., GERMAN, J. and CARRIER, W. L. (1969) Evidence that Xeroderma pigmentosum cells do not perform the first step in the repair of ultraviolet damage to their DNA. *Proc. natn. Acad. Sci. U.S.A.* **64**: 1035–1040.

SHILO, B. Z. and WEINBERG, R. A. (1981) Unique transforming gene in carcinogen-transformed mouse cells. *Nature* **289**: 607–609.

SILINSKAS, K. C., KATELEY, S. A., TOWER, J. E., MAHER, V. M. and MCCORMICK, J. (1981) Induction of anchorage-independent growth in human fibroblasts by propane sultone. *Cancer Res.* **41**: 1620–1627.

SLAMON, D. J., DEKERNION, J. B., VERMA, I. M. and CLINE, M. J. (1984) Expression of cellular oncogenes in human malignancies. *Science* **224**: 257–262.

SOUTHERN, E. M. (1975) Detection of specific sequences among DNA fragments separated by gel electrophoresis. *J. molec. Biol.* **98**: 503–517.

SPORN, M. B., DUNLOP, N. M., NEWTON, D. L. and HENDERSON, W. R. (1976) Relationships between structure and activity of retinoids. *Nature* **263**: 110–113.

STOKER, M. G. and RUBIN, H. (1967) Density dependent inhibition of cell growth in culture. *Nature* **215**: 171–172.

SUGIMURA, T. (1982) Tumor initiators and promoters associated with ordinary foods. In: *Molecular Interrelations of Nutrition and Cancer*, pp. 3–24, ARNOTT, M. S., VAN EYS, J. and WANG, Y. M. (eds). Raven Press, New York.

SUGIMURA, T., MIWA, M., SAITO, H., KANAI, Y., IKEJIMA, M., TERADA, M., YAMADA, M. and UTAKOJI, T. (1980) Studies of nuclear ADP-ribosylation. *Adv. Enzyme Res.* **18**: 195–220.

SUTHERLAND, B. M., DELIHAS, N. C., OLIVER, R. P. and SUTHERLAND, J. C. (1981) Action spectra for ultraviolet light-induced transformation of human cells to anchorage-independent growth. *Cancer Res.* **41**: 2211–2214.

SWIERENGA, S. H., WHITFIELD, J. F. and KARASAKI, S. (1978) Loss of proliferative calcium dependence: simple *in vitro* indicator of tumorigenicity. *Proc. natn. Acad. Sci. U.S.A.* **75**: 6069–6072.

TABIN, C. J., BRADLEY, S. M., BARBMANN, C. I., WEINBERG, R. A., PAPAGEORGE, A. G., SCOLNICK, E. M., DHAR, R., LOWY, D. R. and CHANG, E. H. (1982) Mechanism of activation of a human oncogene. *Nature* **300**: 143–152.

TAPPEL, A. L. (1973) Lipid peroxidation damage to cell compounds. *Fedn Proc.* **32**: 1870–1874.

TERZAGHI, M. and LITTLE, J. B. (1975) Repair of potentially lethal radiation damage in mammalian cells is associated with enhancement of malignant transformation. *Nature* **253**: 548–549.

TERZAGHI, M. and LITTLE, J. B. (1976a) X-radiation-induced transformation in a C3H mouse embryo-derived cell line. *Cancer Res.* **36**: 1367–1374.

Terzaghi, M. and Little, J. B. (1976b) Letter: Oncogenic transformation *in vitro* after split-dose X-irradiation. *Int. J. Radiat. Biol.* **29**: 583–587.
Todaro, G. J. and Green, H. (1963) Quantitative studies of the growth of mouse embryo cells in culture and their development into established lines. *J. Cell Biol.* **17**: 299–313.
Troll, W. (1976) Blocking tumor promotion by protease inhibitors. In: *Fundamentals in Cancer Prevention*, pp. 41–50, Magee, P. N. *et al.* (eds). University Park Press, Baltimore, Maryland.
Ullrich, R. L., Jernigan, M. C. and Storer, J. C. (1977) Neutron carcinogenesis.—Dose and dose-rate effects in BALB/c mice. *Radiat. Res.* **72**: 487–498.
Umezawa, K., Sawamura, M., Matsuchima, T. and Sugimura, T. (1979) Inhibition of chemically induced sister chromatid exchanges by elastatinal. *Chem. Biol. Interact.* **24**: 107–110.
UNSCEAR (United Nations Scientific Committee on the Effects of Atomic Radiation) (1977) Report to the General Assembly on Sources and Effects of Ionizing Radiations. United Nations, New York.
Upton, A. C. (1983) Short term *in vitro* test for identifying carcinogens; transformation of mammalian cells in culture. In: *Short Term Tests for Environmentally Induced Chronic Health Effects*, pp. 38–68, Woodhead, A. (ed.). Washington, D.C.
Upton, A. C., Randolph, M. L. and Conklin, J. W. (1970) Late effects of fast neutrons and gamma rays in mice influenced by the dose rate of radiation: life shortening. *Radiat. Res.* **41**: 467–491.
Vandekerchkhove, J., Leavitt, J., Kakunaga, T. and Weber, K. (1980) Coexpression of a mutant beta-actin and the two normal beta- and gamma-cytoplasmic actins in a stably transformed human cell line. *Cell* **22**: 893–899.
Wang, S. Y. (1976) Pyrimidine bimolecular photoproducts. In: *Photochemistry and Photobiology of Nucleic Acids*, pp. 295–356, Wang, S. Y. (ed.). Academic Press, New York.
Westin, E. H., Gallo, R. C., Arya, S. K., Eva, A., Souza, L. M., Baluda, M. A., Aaronson, S. A. and Wong-Staal, F. (1982) Differential expression of the *amv* gene in human hematopoietic cells. *Proc. natn. Acad. Sci. U.S.A.* **79**: 2194–2198.
Wigler, M. and Weinstein, I. B. (1976) Tumor promoter induces plasminogen activator. *Nature* **259**: 232–233.
Wigler, M., Pellicer, A., Silverstein, S. and Axel, R. (1979) Biochemical transfer of single copy enkaryotic genes using total DNA as donor. *Cells* **14**: 725–731.
Woodcock, D. M. and Cooper, I. A. (1979) Aberrant double replication of segments of chromosomal DNA following DNA synthesis inhibition by cytosine arabinoside. *Exp. Cell Res.* **123**: 157–166.
Yang, T. C. and Tobias, C. A. (1980) Radiation and cell transformation *in vitro*. *Adv. Biol. Med. Phys.* **17**: 417–461.
Zimmerman, R. and Cerutti, P. (1984) Active oxygen as a promoter of transformation in mouse embryo C3H 10T1/2/C18 fibroblasts. *Proc. natn. Acad. Sci. U.S.A.* **81**: 2085–2087.
Zimmerman, R. and Little, J. B. (1983a) Characterization of a quantitative assay for the *in vitro* transformation of normal human diploid fibroblasts to anchorage independence by chemical carcinogens. *Cancer Res.* **43**: 2176–2182.
Zimmerman, R. and Little, J. B. (1983b) Characterization of human diploid fibroblasts transformed *in vitro* by chemical carcinogens. *Cancer Res.* **43**: 2183–2189.

CHAPTER 6

THE MECHANISM OF CELL TRANSFORMATION BY SV40 AND POLYOMA VIRUS

Claudio Basilico

Dept. of Pathology, New York University School of Medicine, New York, N.Y. 10016, U.S.A.

Abstract—The study of the small DNA tumor viruses Polyoma and SV40 has provided invaluable contributions to our knowledge of viral carcinogenesis as well as of the organization and expression of eukaryotic genes. This article reviews the molecular basis of the alterations of the cell's phenotype brought about by these viruses, their interaction with cells *in vivo* and *in vitro*, the association of the viral genes with the genome of the transformed cells, and discusses the mechanisms by which the expression of specific viral oncogenes may lead to the alterations of growth control typical of neoplastic cells.

1. INTRODUCTION: EARLY STUDIES

It is over twenty years since the pioneering studies of Eddy, Hilleman, Dulbecco, Stoker and several others focused their attention on the tumorigenic capacity of the small DNA viruses of the Papova group. These viruses, whose family name derives from papilloma, polyoma and vacuolating viruses, share many properties in addition to their oncogenic potential. They are icosahedrical in shape, contain double stranded closed circular DNA as their genetic material, and have a relatively limited amount of genetic information, since the size of their genome ranges from 5 to 10 Kbase-pairs. By far the best studied viruses of the group are Polyoma (a mouse virus) and Simina virus 40 (SV40, a monkey virus). They are quite similar in genomic organization, regulation of expression, etc., although they encode different proteins. It is undoubted that they must have derived from a common ancestor and analysis of their DNA sequence reveals regions of homology which were not evident in early studies of nucleic acid hybridization.

Much of the interest that these viruses have elicited results from the relative simplicity of their genome. Polyoma virus was isolated by Gross (1953) from cell-free extracts from AKR mice, and originally named parotid tumor agent. When it became clear that the virus was capable of inducing a variety of tumors, the name Polyoma was adopted. Early studies showed that the virus genetic material consisted of a DNA molecule of about 3×10^6 daltons. Since such a limited amount of genetic information was not likely to code for more than five to eight proteins, many investigators envisaged the possibility of easily identifying the viral functions responsible for neoplastic transformation. Similarly, in the case of SV40, which was originally discovered as a contaminant of polio vaccines (Sweet and Hilleman, 1960) and soon found to be tumorigenic in newborn hamsters (Eddy *et al.*, 1962; Girardi *et al.*, 1962), the ease of manipulation of the virus and the small size of the genome made researchers hopeful of quickly identifying the viral functions involved in tumorigenesis.

The use of cultured cells of rodents, and primate origin, and the demonstration that such cells could be converted *in vitro* to a new 'transformed' phenotype following infection with these viruses greatly facilitated their initial study. Polyoma and SV40 exhibit two types of interaction with cells *in vitro* and *in vivo*: a lytic interaction and a transforming one. In the lytic interaction, the virus is adsorbed and penetrates into the cell, where uncoating and expression of its early genes is followed by a high rate of viral DNA replication. Synthesis of late proteins and assembly of new virus particles in the nucleus then follows, with death and lysis of the infected cell. In the transforming interaction, while the early events are indistinguishable from those of the lytic interaction, there is no detectable viral DNA replication, no production of infectious virus and the cell is not killed. However, a small proportion of the infected cells acquires new properties which are characteristic of transformation *in vitro*.

The type of interaction that the virus enters with the host cell depends on the cell species: infection with polyoma virus of mouse cells results almost exclusively in viral multiplication, while infection of rat and hamster cells results in transformation. SV40 multiplies in African green monkey cells, and transforms hamster, mouse and rat cells. The molecular basis for permissiveness (the ability to support viral multiplication) and nonpermissiveness (here used in the sense of not supporting viral multiplication but being transformable) is unclear. It is known that nonpermissive cells lack one or more host factors which are necessary directly or indirectly for viral DNA replication. The main evidence supporting this conclusion is that hybrids or heterokaryons between permissive and nonpermissive cells generally support viral multiplication, provided they maintain a sufficient complement of permissive chromosomes (Watkins and Dulbecco, 1967; Koprowski et al., 1967; Basilico et al., 1970). The nature of these factor(s) is however still unknown.

Early studies focused on the characterization and quantitation of cell transformation *in vitro*, on the biological properties of the virus, and attempted to dissect the viral transforming functions by the use of genetics. They also discovered unexpected effects of viral infection on the cell metabolism, such as induction of host DNA synthesis, and attempted to understand whether transformation resulted from a hit and run mechanism, or required the continuous presence of all or part of the viral genome. These studies (reviewed in Tooze, 1981) revealed a number of basic aspects of transformation by these viruses.

(1) Transformed cells generally did not produce virus.

(2) Transformation followed single hit kinetics.

(3) Not all viral functions were necessary for transformation, but only those encoded in about 50% of the genome.

(4) Cells transformed by polyoma and SV40 contained new viral specific antigens, which were called T (for tumor) antigens.

(5) The genome of the viruses consisted of a double stranded closed circular DNA molecule of about 3×10^6 daltons, and the DNA molecule was infectious.

(6) Infection with polyoma and SV40 of both permissive and nonpermissive cells induced transiently host functions, including cellular DNA synthesis, enzymes, etc.

In addition, the early use of ts mutants (Fried, 1965) allowed the conclusion that some viral functions were necessary at least for the establishment of transformation. In 1966 Benjamin showed that Polyoma transformed cells contained virus-specific RNA, a result which fortified the belief held by many investigators in the field that transformation required the continuous expression of viral functions. These findings probably mark the end of the early period of Polyoma and SV40 research.

With the following revolution in molecular biology and genetics, of which these viruses were some of the most important protagonists, much more became known about their genetic organization, regulation of expression, cell interaction etc. Thus, although it would be intellectually of interest to follow the development of our knowledge in a chronological manner, I will instead discuss first the analysis of the viral genomes and that of their interaction with their host cells and eventually the properties and function of viral transforming proteins. While this should result in a more logical picture, it has to be pointed out that often this is not the sequence in which these findings have emerged.

2. GENOME ORGANIZATION AND EXPRESSION

Within the icosahedral capsid of Polyoma and SV40 is contained a single DNA molecule of about 5.3 Kb.p. The molecule is complexed with histones (of host cell origin) to form a nucleosome-DNA structure which has been called a minichromosome.

The DNA of Polyoma and SV40 has been totally sequenced (Fiers *et al.*, 1978; Soeda *et al.*, 1979, 1980) and extensively analyzed.

Polyoma. The functional and physical map of Polyoma DNA is shown in Fig. 1. The circular DNA molecule is divided in two approximately equal halves: an early region, encoding viral functions that in lytic infection are expressed prior to viral DNA synthesis, and a late region, which are separated by a noncoding regulatory region (see below). Through the use of restriction enzymes the genome has been divided in map units (m.u.) (0–100) starting at the unique Eco RI site. More recently the nomenclature using nucleotide numbers, starting at the putative origin of replication, has become widely used. A variety of studies which are beyond the scope of this review has revealed much about the organization and the expression of the Polyoma genome (see reviews by Fried and Griffin, 1977; Ito, 1980; Benjamin, 1982; De Pamphilis and Wassarman, 1982; Griffin and Dilworth, 1983). Transcription of the early and late region take place on opposite strands and in opposite polarity. The late region encodes the three viral structural proteins VP1, VP2 and VP3, whose three mRNAs are generated from a common precursor molecule. The generation of these mature transcripts is complex. All mRNAs have heterogenous 5′ ends mapping between nucleotide (n) 5170 and 5070, followed by three to four copies of a repeatedly spliced leader unit (n 5076–5020) and then by a further splice to the body segment of the specific mRNA (Flavell *et al.*, 1979, 1980; Legon *et al.*, 1979; Treisman, 1980). They are thought to be produced by the splicing together of long, head to tail multimeric transcripts of the entire polyoma late strand which are detected in the nucleus during the late phase of lytic infection (Acheson, 1978).

Several lines of evidence indicate that the polyoma late region is not involved in transformation: (1) late mRNAs are generally not expressed in transformed cells, which also do not produce virions; (2) late mutations do not affect the virus transforming ability; (3) the late region can be deleted from the viral DNA without affecting transformation. For

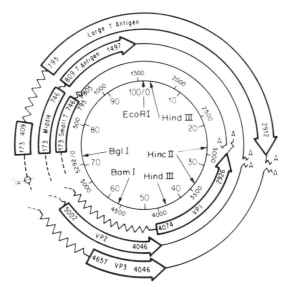

FIG. 1. Physical and functional map of Polyoma virus DNA (A2 strain). The physical map is divided into map units (starting clockwise from 0 at the EcoRI site) or nucleotide number (starting clockwise from 0 at the origin of replication). The mRNAs for the three T-antigens and the late proteins VP1, VP2, and VP3 are shown with their coding regions (□); 3′ noncoding regions (———); and intervening sequences (∿). The numbers within the coding regions represent initiation and termination codons and splice junctions. OR = Origin of viral DNA replication. Adapted from Tooze (1981).

this reason, that applies also to SV40, we will concentrate on the early and regulatory regions.

The early region is transcribed into a unique precursor RNA molecule which is then polyadenylated and spliced to generate three mature mRNA molecules, which have identical 5' and 3' ends but differ in the sequences removed by splicing (Treisman et al., 1981a). The reading frame of these mRNAs is initially identical (Fig. 1) but becomes different following splicing. Thus the shortest mature mRNA molecule, which is spliced from n 409 to n 795, has a long open reading frame which terminates near the poly A tail. Translation of this mRNA produces a protein at about 100 K daltons named large T-Antigen (Hunter et al., 1978). The two other mRNAs are spliced from a different donor site (at n 746) and thus contain sequences which are removed in the large T-mRNA. In one of these, splicing from this donor site to the same acceptor of the large T-mRNA (at n 795) leaves the translational reading frame unchanged. However, a termination codon at n 805 limits the coding capacity of this mRNA, which has the longest 3' untranslated region, to a protein of about 22 K daltons called small T-Antigen. The last mRNA molecule known is spliced from nucleotide 795 to 809 and encodes a protein of about 60 K daltons called middle T-Antigen. This method of RNA processing has the obvious result of greatly increasing the coding capacity of this relatively small DNA segment. It should also be mentioned that the early region of Polyoma DNA contains an alternative poly-A addition signal at 99 m.u. (Soeda et al., 1980). This site is normally not used, but becomes functional when the regular poly A site at 25 m.u. is deleted, a situation often found following integration (see later).

The regulatory region comprised between the late and the early contains the origin of viral DNA replication, characterized by palindromic structures observed also in SV40, from which replication starts bidirectionally to be concluded approximately at 180° from the start (reviewed by De Pamphilis and Wassarman, 1982). The early promoter region contains a TATA box and binding sites for large T-antigen. Between the origin and the late promoters are DNA sequences not yet fully characterized (they are better known in SV40) which constitute the so-called 'enhancer' sequences (Tyndall et al., 1981). These sequences, which have been found also in the LTR's of retroviruses, in immunoglobulin DNA etc. are essential for transcription in vivo, and their deletion renders the viral genome essentially transcriptionally inert (Khoury and Gruss, 1983). Enhancer sequences show some species specificity, although not absolute, function only in cis but in either polarity, and can be positioned also at some distance from the transcriptional unit they activate (Laimins et al., 1982; Khoury and Gruss, 1983). Their precise function is not known. They could represent a site for RNA pol II entry, or a 'hook' necessary for attachment to the nuclear matrix, a process which may be essential for gene expression.

SV40. As in the case of Polyoma, the circular DNA molecule is functionally divided in early and late regions, and noncoding regulatory sequences (Fig. 2). The strategy of SV40 gene expression is similar to that of Polyoma DNA. Late sequences are also not involved in transformation and code for the three viral structural proteins. The mRNAs for these proteins are generated also by differential splicing of a common precursor, but it is not known whether VP3 is translated from its own message or could be a processed form of VP2. A small protein (called agnoprotein) of about 8000 daltons is encoded in an open reading frame contained in the leader sequences of the late mRNA (Jay et al., 1981). This basic protein has been detected in cytoplasmic extracts from infected cells, and could play a role in packaging, or regulate late transcription as an attenuator (Hay et al., 1982; Nomura et al., 1983).

The early region of SV40 is transcribed from the opposite strand of the late and in counterclockwise polarity. Differential splicing of the common mRNA precursor yields two mRNA molecules (Aloni et al., 1977; Berk and Sharp, 1978). The shortest, with a splice of 346 nucleotides, encodes the SV40 large T protein of about 90 K daltons. The longest mRNA, with a large 3' untranslated region, codes for a protein of about 17 K daltons, named small T-Antigen. The N-terminal half of this protein is identical to the N-terminal of

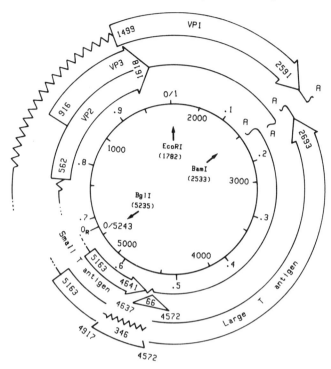

FIG. 2. Physical and functional map of SV40 DNA. The physical map is divided into map units (starting clockwise from 0 at the EcoRI site) or nucleotide numbers (1–5243, starting clockwise from 0 at the origin of replication). The mRNAs for the two T-Antigens and the late proteins VP1, VP2 and VP3 are shown with their coding regions (□), 3' noncoding regions (——) and intervening sequences (∿). The size (in nucleotides) and boundaries of the early intervening sequences are also indicated. OR = origin of viral DNA replication. Adapted from Tooze (1981).

large T-Antigen (Crawford et al., 1978). No third mRNA species (or protein) similar to Polyoma middle T-Antigen has been identified in SV40 infected or transformed cells. The origin-promoter regions also show similarities to that of Polyoma. Early and late promoters are positioned on both sides of the origin of viral DNA replication (Benoist and Chambon, 1981). On the late side of the origin lie the SV40 enhancer sequences, characterized by two 72 b.p. repeats (Gruss et al., 1981). The repetition is not essential for its enhancer function, but this arrangement, and the occurrence of a unique restriction enzyme site within the 72 b.p. repeat has made easier the precise mapping of the enhancer sequences of SV40 with respect to those of Polyoma. It is worth mentioning that 'enhancer' sequences from a murine retrovirus can replace those of SV40, yielding a functioning viral genome (Levinson et al., 1982). As for Polyoma, SV40 contains an additional polyadenylation signal in the middle of the early region, but its utilization has never been reported (Fitzgerald and Shenk, 1980).

While the complex regulation of transcription and RNA processing of SV40 and Polyoma is outside the scope of this review, some of the most important aspects of these phenomena are discussed below.

2.1. Early Phase of Infection—the Viral Early Proteins

Infection by Polyoma and SV40 of both permissive and nonpermissive cells is followed by a number of events which are similar in both cell species and thus will be treated together.

Following uncoating and penetration into the nucleus, the viral early region is transcribed as described above. Unspliced precursor RNAs can be detected in the nucleus, but the largest amount of stable viral transcripts is found capped, polyadenylated and processed in the cytoplasm. For reasons which are not clear, both in the case of Polyoma and SV40, splicing favors the large T-mRNA which is generally the most abundant species found. No evidence of stable polymeric early transcripts has been found, suggesting efficient polyadenylation or some form of transcription termination (reviewed in Tooze, 1981).

Early transcription is followed by synthesis of viral early proteins. In permissive cells, binding of large T-Antigen at the origin of DNA replication allows DNA replication to occur, followed by synthesis of late mRNA and late proteins (reviewed by De Pamphilis and Wasserman, 1982). It is not yet clear what mediates the early-late shift in Polyoma and SV40. Large T-Antigen binds to specific sites in the early promoter region and then represses early transcription (Reed et al., 1976). Concomitantly new minor early promoters (upstream from the normally used ones) become utilized (Ghosh and Lebowitz, 1981; Hansen et al., 1981; Fenton and Basilico, 1982a). It is not clear whether DNA replication itself is absolutely necessary for late gene expression, although recent results (Wasylyk et al., 1983; Keller and Alwine, 1984) seem to contradict this hypothesis.

In nonpermissive cells, viral DNA replication does not occur to an appreciable extent and the virus growth cycle is essentially interrupted here.

The early gene products of Polyoma and SV40 originally were called T-(tumour) antigens, since they were first recognized in virus-induced tumors or transformed cells, with the use of antisera raised in animals bearing SV40 or Polyoma induced tumors. It was eventually recognized (Hoggan et al., 1965) that their presence is not restricted to tumors, but to every situation in which the early region of the viral genome is expressed, i.e. they are the proteins encoded in the early region, but the name has stuck. While a detailed discussion of the known and possible biological activities of these proteins will be entertained in the following sections, it is important at this point to outline their most relevant physical and biochemical properties.

2.1.1. SV40 T-Antigens

The large T-Antigen of SV40 is a phosphorylated protein of an apparent molecular weight of 94,000 daltons as determined by gel electrophoresis. Genetic and biochemical analysis has identified this protein as the product of the SV40 gene A, and thus large T-Ag is sometimes referred to as the viral A gene product. In its native form, large T-Ag exists as a variety of multimeric forms (Greenspan and Carroll, 1981) which in infected or transformed cells are complexed with a cellular protein, named p53 (see later). Large T-Antigen is found in the nucleus of SV40 infected or transformed cells. However, a small fraction is found on the plasma membrane, where it probably represents the virus specific TSTA. SV40 large T-Ag also possesses an ATPase activity (Tjian and Robbins, 1979; reviewed by Rigby and Lane, 1983). In addition to its role in transformation, which will be discussed later, a number of other activities and properties of this protein are known (reviewed by Weil, 1978; Martin, 1981; Rigby and Lane, 1983).

(1) Large T-Ag is a DNA-binding protein (Carroll et al., 1974, Jessel et al., 1975), and it has been shown to bind specifically to three sites near or at the origin of viral DNA replication, and to protect them from nuclease digestion (Jessel et al., 1976; Reed et al., 1975; Tjian, 1981).

(2) Large T-Ag is the SV40 replication initiator, and the presence of a functional large T-protein is required for the initiation of each round of viral DNA replication (Tegtmeyer, 1972).

(3) SV40 large T-Ag regulates its own synthesis by repressing early transcription (Tegtmeyer et al., 1975; Reed et al., 1976). In addition it promotes also directly or indirectly efficient late transcription (Keller and Alwine, 1984).

(4) SV40 large T-Ag is involved in many cell responses to viral infection, particularly the induction of host DNA synthesis and DNA synthetic enzymes.

Much less is known about the small T-protein of about 17,000 daltons. Its role in viral multiplication is unclear and small-T defective mutants are viable (Sleigh et al., 1978). The localization of this protein is mostly cytoplasmic, where it appears to disrupt actin cables, thus having a profound effect on cytoskeletal organization (Pollack et al., 1975; Graessmann et al., 1980; Topp and Rifkin, 1980). Recent evidence (Ellmann et al., 1984) suggests that the protein is both nuclear and cytoplasmic, and binds *in vitro* to actin and tubulin.

2.1.2. Polyoma T-Antigens

As in SV40, Polyoma large T-Antigen is a phosphorylated protein of about 100 K which is encoded in most of the early region. Unlike SV40 T, however, it is not generally found associated with the p53 host protein, and its localization appears so far to be exclusively nuclear. Some of the functions of Polyoma large T-Ag are common to those of the analogous SV40 protein, with which it shows considerable sequence homology (for a detailed review, see Weil, 1978; Griffin and Dilworth, 1983).

(1) The Polyoma large T-Ag is also a DNA binding protein that will specifically protect viral DNA sequences near or at the origin of replication (Gaudray et al., 1981). Although the studies with Polyoma large T-Ag are less advanced than those with SV40, mainly because of greater difficulties in producing large amounts of pure protein, there is every reason to believe that the properties of Polyoma large T-Ag in this respect are very similar to those of SV40.

(2) Large T-Ag is the Polyoma replication protein, necessary for the initiation of each round of viral DNA synthesis (Francke and Eckhart, 1973).

(3) Polyoma large T-AG autoregulates its own synthesis by inhibiting early transcription (Cogen, 1978). It may also be involved in the early-late shift of transcription (Fenton and Basilico, 1982b).

(4) The role of Polyoma large T-Ag in induction of cellular DNA synthesis is less clear and will be discussed in the next section.

The middle T-protein has a molecular weight of $\sim 60{,}000$ daltons and has attracted considerable interest mainly because it is the virus' main transforming protein. It is found associated with the plasma membrane (Ito et al., 1977), a function for which the hydrophobic tail of the protein is essential (Carmichael et al., 1982). Studies with monoclonal antibodies also suggest a perinuclear distribution (Griffin and Dilworth, 1983). It has been reported to possess protein kinase activity specifically directed at the amino acid tyrosine (Smith et al., 1980), and at least two tyrosine residues in middle T are themselves phosphorylated (Eckhart et al., 1979; Schaffhausen and Benjamin, 1981; Segawa and Ito, 1982; Hunter et al., 1984; Courtneidge et al., 1984). Recent studies, however, support the hypothesis that the protein kinase activity is only associated with middle T, not being an integral part of the molecule. It has been noted (Courtneidge and Smith, 1983) that Polyoma middle T is generally found complexed with the pp60 c-Src protein (the cellular analog of the Rous Sarcoma virus transforming protein), itself a protein kinase (Collett and Erikson, 1978; Levinson et al., 1978). While the functional importance of this association is still unclear, it is very likely that middle T-Ag is phosphorylated by the host pp60 Src protein.

The specific role of the middle T-protein in viral replication is unknown. Middle T-mutants replicate, although not very efficiently. The protein could be involved in some activation of host cell functions, which may enhance viral replication but not be totally essential, especially in tissue culture.

Even less is known about the function of the small T polypeptide (MW 20,000) than for its SV40 counterpart. It is cytoplasmic (Griffin and Dilworth, 1983), probably involved in disruption of actin cables. For its role in replication, the same considerations made for middle T-Ag apply. It is worth noting that the entire sequence of Polyoma small T-Ag, with the exception of the last four amino acids, is contained within the N-terminal half of middle T-Ag (Smart and Ito, 1978).

3. EFFECTS OF VIRAL INFECTION ON THE HOST CELL METABOLISM

Following infection of susceptible cells with SV40 or Polyoma, a number of early effects on the host cell metabolism have been observed, which are sometimes grouped together under the name of abortive transformation, since they mimic temporarily the host of phenotypic changes displayed by virus-transformed cells. They are extensively described in Tooze (1981) and I will summarize them only briefly here.

3.1. Induction of Host DNA Synthesis, Enzymes

Infection of permissive and nonpermissive cells results in the induction of high levels of enzymes of DNA metabolism, such as DNA polymerase, thymidine kinase, DCMP deaminase, DNA ligase. This increase is followed by a wave of host DNA synthesis and, in the case of nonpermissive cells, by mitosis. It has to be mentioned that this phenomenon is observed only after infection of quiescent cells, i.e. cells which are arrested in G_1 (G_0) by density, serum starvation etc. It resembles the induction of G_1-S transition induced by serum and is accompanied by increased synthesis of mitochondrial DNA and histones. Exponentially growing cells are not induced into a higher rate of DNA synthesis by viral infection, in agreement with the conclusion that the normal temporal sequence of S phase events is not perturbed in this phenomenon. This event can therefore be viewed as a mitogenic stimulation of resting cells. It does not require viral DNA synthesis (occurs also in nonpermissive cells, in cells infected with mutants unable to replicate etc.).

3.2. Abortive Transformation Proper

Infection of nonpermissive cell which are unable to grow in agar suspension medium, results in the temporary acquisition by most cells of the ability to perform one or a few doublings in agar; a property which is then maintained in the small proportion of infected cells which became stably transformed.

3.3. Induction of Morphological Changes and Serum Independence

Most infected cells acquire very soon a new morphology, reminiscent of that of stably transformed cells, and are capable of growing for two or three cell doublings in medium with very low serum concentrations. Enhanced agglutinability by plant lectins (concanavalin A, wheat germ agglutinin) can also be detected in a large fraction of the cell population (Burger, 1969; Benjamin and Burger, 1970).

3.4. Induction of RNA Synthesis

There are several reports (Oda and Dulbecco, 1968; May et al., 1976; Salomon et al., 1977) on the induction of a higher rate of transcription by SV40 and Polyoma infection, but since the increases noted were difficult to define qualitatively, they did not receive a very wide attention. It appeared also possible that the increase in ribosomal RNA synthesis was not direct, but connected to the stimulation of cell cycle progression. This idea however is not in line with the discovery by Baserga's group (Soprano et al., 1979) of SV40 reactivation of ribosomal DNA synthesis in human-mouse hybrids. Somatic cell hybrids between mouse and human cells generally synthesize only mouse ribosomal RNA, while transcription of the human rDNA genes is suppressed. Infection with SV40 reactivates the expression of human ribosomal DNA and it can be shown that this effect is due to the viral large T-Antigen. These findings have taken new significance with the discovery that both SV40 and polyoma large T-Antigens can activate a number of promoters (including their own late promoter) in trans (Keller and Alwine, 1984; Brady and Khoury, 1985; Alwine, 1985; Kern *et al.*, 1986). Future work should be able to elucidate whether any of the effects these viruses have on the phenotype of infected cells may depend on the activation of expression of specific cellular genes.

What is the general significance of these phenomena which so closely resemble the acquisition of properties of stably transformed cells? The prevailing view is that they are the result of the expression of some viral functions by viral genomes which are only transiently associated with the infected cell (i.e. they are not integrated into the host DNA). In permissive cells, these events are followed by death of the infected cell. In nonpermissive cells the nonintegrated viral genome is diluted out by cell division or degraded, and the cells return to a normal behavior.

The second question concerns the specific viral gene product involved in these effects. For SV40, the viral large T-Antigen is clearly necessary for induction of host DNA synthesis and enzymes (reviewed by Martin, 1981), and an elegant demonstration of this conclusion is the fact that the microinjection of purified large T-polypeptide (or rather an adenovirus-SV40 hybrid protein which however has mostly SV40 sequences) into cultured cells is sufficient to induce host DNA synthesis (Tjian *et al.*, 1978). How does large T induce host DNA synthesis? One hypothesis is that it does so by recognizing cellular origins of replication and acting there much in the same way as in viral DNA replication. However, this mechanism does not explain the induction of enzymes and RNA synthesis. It is possible that large T-Ag acts by inducing the transcription of specific cellular genes, whose products then stimulate cell cycle progression.

In the case of Polyoma the situation is more complex. Polyoma tsA mutants (producing a thermolabile large T-Ag) induce cell DNA synthesis at the nonpermissive temperature (Fried, 1970), but this is also true (although the effect is less pronounced) of hrt (host-range/nontransforming) mutants, which do not produce functional middle and small T-Ags (Schlegel and Benjamin, 1978). Perhaps both the large and middle T-Ags of Polyoma are capable of eliciting some form of induction of DNA synthesis, the full expression being probably the result of the combined action of these two proteins.

Even less clear is the situation with respect to induction of classical abortive transformation (growth in agar), lectin agglutination and decreased serum requirements. Polyoma hrt mutants (lacking functional middle and small T-Ag) (see Benjamin, 1982) do not induce abortive transformation of lectin agglutinability (Fluck and Benjamin, 1979), suggesting that middle or small T-Antigen are involved. The same mutants however induce some DNA synthesis (Schlegel and Benjamin, 1978). In contrast mutants producing a nonfunctional large T-Ag are still capable of inducing temporary growth in agar medium (Stoker and Dulbecco, 1969). These data therefore indicate that middle T-Ag is the main viral product responsible for this phenomenon.

An attractive hypothesis could be that the polyoma viral middle T-Ag (or the SV40 large T-Ag) induce the production of transforming growth factors (TGF) (De Larco and Todaro, 1978), which are small polypeptides capable of promoting growth in agar and in low serum. These factors could create an autocrine system (temporary in abortive transformation, permanent in stably transformed cells). However, although it has been proved that Polyoma or SV40 transformed cells produce TGFs (Kaplan and Ozanne, 1982), the involvement of these factors in abortive transformation has not been verified. The important question of whether all these changes reflect one single basic phenomenon or can occur independently will be examined later.

4. CELL TRANSFORMATION BY POLYOMA AND SV40

4.1. Oncogenic Potential *In Vivo*

Although Polyoma virus was isolated from a mouse tumor, it is frequently present in mouse populations which bear no tumors. Thus its importance as an etiological agent of cancer in mice is likely to be low. Nevertheless, Polyoma virus readily produces a variety of tumors when injected into newborn mice, hamsters or rats. Adult animals are generally resistant and the mechanism is likely to be immunological. X-irradiated animals in fact develop tumors and adult nude mice are very susceptible to the virus, developing a variety of tumors, some of which (e.g. mammary carcinomas) are rarely seen in normal animals (McCormick, 1982). Polyoma DNA is also tumorigenic if injected into newborn hamsters or rats.

SV40 has an even more limited oncogenic potential than Polyoma. Not only has it never been associated with a tumor in monkeys (the natural host), but it is probably not tumorigenic even in mice. The only animal where SV40 readily causes tumors is the hamster, where the virus can cause a variety of neoplasia if the inoculum is high.

While these considerations may lessen the interest in these viruses as a natural cause of cancer, it has not decreased their usefulness as model systems.

4.2. Transformation *In Vitro*

A great deal of the impulse in this field of research came from the possibility of achieving in tissue culture a set of cellular changes analogous to neoplastic conversion *in vivo*. Thus the study of cell transformation has relied heavily on *in vitro* systems in which the cell, following virus infection, acquires new properties, which allow a distinction from its normal counterpart.

Early studies utilized mostly primary fibroblast cultures whose transformed derivatives were recognized because of their altered morphology, dense pattern of growth and indefinite growth in culture. Later the advent of 'normal' continuous cell lines allowed more precise quantitative studies since these populations were easily clonable and homogeneous. Such studies proved very useful even if they dealt with cells which had already acquired one aspect of transformation, i.e. immortalization.

The altered properties of Polyoma and SV40 transformed cells are listed in Table 1. Most are probably well known to the reader and will not be discussed in detail (for an extensive review, see Tooze, 1981). It is useful to remember, however, that several properties (e.g. high saturation density and lack of contact inhibition of growth) are clearly a consequence one of the other and do not represent independent parameters. Tumorigenicity *in vivo* in isologous animals is probably the only real proof of neoplastic conversion. However, this

TABLE 1. *Properties of Cells Transformed by Polyoma Virus or SV40*

Growth related
 Infinite or indefinite lifetime
 Reduced serum requirement
 Ability to grow in agar or Methocel suspension (anchorage independence)
 Lack of contact (density)-inhibition of movement, of DNA synthesis and of growth
 High saturation density
 Ability to grow on monolayers of normal cells
 Changes in the pattern of growth (less-oriented, piling up)
 Inability to reach a viable G_1 arrest
 Tumorigenicity in isologous animals

Surface Properties
 Increased agglutinability by plant lectins
 Changes in composition of glycolipids and glycoproteins
 Virus-specific transplantation antigen
 Increased rate of transport of nutrients
 Increased secretion of proteases or growth factors

Intracellular Properties
 Cytoskeletal disorganization
 Tendency towards aneuploidy

Virus-Diagnostic Properties
 Presence of viral DNA sequences
 Presence of viral mRNA
 Presence of virus-specific antigens
 Infectious virus can be rescued in some cases

statement should not be taken too religiously, because the identification of other altered properties, even when and probably because they are not sufficient to cause tumorigenicity, has helped tremendously in studies aimed at the dissection of the various facets of the transformed phenotype.

The properties of Polyoma or SV40 transformed cells which have been most useful for the isolation of transformed lines and the quantitation of the phenomenon are the ability to form dense foci over a monolayer of normal cells, and the ability to grow without anchorage when suspended in semisolid media (agar or methocel). These techniques allow the isolation of clonal populations of transformed cells very soon after infection (or transfection of viral DNA). Typically an agar colony or a focus can be recognized 7–10 days after plating and selectively picked a few days later.

4.3. Transformation of Permissive and Nonpermissive Cells

Most studies on SV40 or Polyoma transformation have been carried out with nonpermissive cells (typically rat or hamster for Polyoma, mouse or rat for SV40) for the following main reasons: (1) they are easier and more quantitative, since there is practically no cell killing and reinfection and (2) the number of viral functions one has to cope with is more limited. Between the fully permissive and nonpermissive cells lies the grey area of so-called semipermissive cells, a definition which unfortunately groups together cells with very different responses to viral infection, such as human cells for SV40, that are essentially permissive (burst size is high, but reinfection is probably low) and rat cells for Polyoma, in which virus production cannot be detected but the viral DNA probably performs a few rounds of replication. It is my personal bias that the totally nonpermissive cell probably does not exist, and the term nonpermissive is a useful definition for cells in which viral DNA synthesis and production of virus particles do not take place to any appreciable extent. In studies of transformation this distinction is probably not so important as permissive cells are also susceptible to transformation. It is in fact clear that permissive cells are not resistant to transformation, but simply cannot be easily transformed because they are killed as a result of viral replication and maturation. Permissive Polyoma and SV40 transformed cells have existed for a while (Polyoma causes mouse tumors after all) but with the advent of recombinant DNA technology it has become increasingly easier to produce them. By and large three mechanisms of transformation of permissive cells can be envisaged:

(1) Transformation of a rare nonpermissive variant by wild type virus. Surprisingly, the existence of this class of transformants has never been conclusively demonstrated (Wilson et al., 1976).

(2) Transformation by defective viruses which are impaired in the ability to replicate but not in the ability to transform. A very useful class of this type of transformants are monkey cells transformed by origin-defective SV40 molecule (COS cells) (Gluzman et al., 1980; Gluzman, 1981) or mouse cells similarly transformed by origin-defective Polyoma genomes (Tyndall et al., 1981). These molecules cannot undergo replication because of the deletion of large T-Ag binding sites at the origin, but can express early functions, integrate and transform. Permissive cells harboring such molecules provide a very useful system of complementation of T-Antigen defective molecules and can drive the replication of practically any DNA molecule containing the origin of replication (Gluzman, 1981; Conrad et al., 1982).

(3) In rare cases, fully permissive cells transformed by wt virus can be isolated. These cells generally shed large amounts of virus and cannot be maintained easily in culture unless reinfection is blocked, e.g. by the use of antiviral serum in the medium. These transformants probably originate by the integration of viral DNA molecules before the virus infectious cycle has been completed. Integrated molecules can excise, replicate and kill the cells. However, if the rate of excision is not too high, the cells can be maintained as a line, since they multiply at a faster rate than they are killed. A way of controlling this phenomenon consists of infecting mouse cells with the tsA mutants of Polyoma virus, and

shifting the cells to high temperature 2–3 days after infection (Bourgaux et al., 1978; Bourgaux, 1981; Liboi and Basilico, 1984). Since the viral large T-Ag is necessary for excision, this procedure facilitates the isolation of transformed cells which do not shed virus as long as they are maintained at the nonpermissive temperature.

4.4. Transformation of Primary Cultures and Continuous Cell Lines

As previously mentioned, a large number of studies on Papovavirus transformation have been conducted using continuous cell lines, which although more or less 'normal' from the point of view of behavior in culture, tumorigenicity in animals etc., are obviously abnormal in having acquired already one property of transformed cells, i.e. an indefinite capacity to proliferate (immortalization). Although this caveat was always kept in mind, until recently most studies seemed to fortify the assumption that this reservation was not extremely important, as most oncogenic viruses (including Papovaviruses, several retro-viruses, etc.) seemed equally capable of transforming cells of primary cultures as of continuous cell lines. The dissection of the viral transforming genes has however provided evidence that immortalization is an important aspect of cell transformation, and that specific viral genes seem to affect this property without conferring on the cell the rest of the phenotypic changes characteristic of transformation. This will be discussed in more detail when dealing with the role of each viral gene product in transformation. This finding however does not invalidate the studies conducted with continuous cell lines, and particularly the studies on the molecular events of this process, which are likely to be indistinguishable in all cell types.

5. THE STATE OF THE VIRAL GENOME IN POLYOMA OR SV40 TRANSFORMED CELLS

Following infection by Polyoma and SV40 or nonpermissive cells (this has been the system most studied, but there is no reason to think that the main molecular steps would be different in permissive cells), the viral genome is transcribed and expresses its early proteins. This causes the phenomena of abortive transformation described above. Eventually a small percentage of the cells stably acquires transformed growth properties. These cells generally do not produce infectious virus or virions, but express virus-specific antigens.

Several lines of evidence suggested that at least part of the viral genome persisted in transformed cells, but the direct demonstration of this hypothesis had to wait for the modern techniques of nucleic acid hybridization. While early studies (Westphal and Dulbecco, 1968; Sambrook et al., 1968; Gelb et al., 1971) established that viral DNA sequences were present in all transformed cell lines, that they were associated and linked to chromosomal DNA, and that their amounts varied from less than one to several genome equivalents/cell, it was the advent of restriction endonuclease analysis, Southern (1975) blotting and molecular cloning that produced more detailed information about the state of the viral DNA sequences in transformed cells, their stability and expression.

The viral DNA molecules are colinearly inserted into the chromosomal DNA without any precise rule as to the viral or cellular site of integration. Most often the left and right joint of the viral insertion differ, i.e. the insertion does not correspond to an integral number of viral DNA molecules (1, 2, 3 etc.) (Kettner and Kelly, 1976; Botchan et al., 1976; Steinberg et al., 1978; Birg et al., 1979; Basilico et al., 1979; Gattoni et al., 1980; Lania et al., 1980; Sambrook et al., 1980). There is no evidence of special structures (e.g. inverted repeats) near the viral-host DNA junctions. The number of independent insertions into the host DNA varies from one to several (even 10–20).

There is only one rule, i.e. the integrity of all or most of the early region of the viral DNA is preserved. This however does not reflect the specificity of an integration process that can only take place in the viral late region, but simply the selection of transformation, which depends on the expression of the viral early functions. This is clearly demonstrated by the fact that the most common type of integration is that of a tandem, head or tail repeat of up to 3–4 viral DNA molecules (Steinberg et al., 1978; Birg et al., 1979; Campo et al., 1979; Gattoni et al., 1980; Basilico et al., 1980; Lania et al., 1980; Della Valle et al., 1981; Clayton and Rigby, 1981; Mendelsohn et al., 1982). While the mechanism leading to this type of integration will be discussed later, tandem insertion has the obvious result of preserving an intact early region even if integration has occurred in the same sequences.

With respect to the host site of insertion, in spite of some reports suggesting chromosomal specificity (Croce and Koprowski, 1975), the overwhelming evidence favors random or semirandom integration. Restriction enzyme analysis shows that in *independent* transformed lines, the viral-host DNA junctions are generally different, and sequence analysis of several cloned viral insertions also showed no similarities (Botchan et al., 1980; Clayton and Rigby, 1981; Stringer, 1981; Hayday et al., 1982; Neer et al., 1983). Total randomness or totally illegitimate recombination cannot be demonstrated. Indeed, the analysis of some cloned host-viral DNA junctions suggest that some degree of partial homology may have played a role in the recombination of the host DNA with the viral sequences (Bourgaux et al., 1982; Stringer, 1982a; Chowdhury et al., 1984), but this is not a generalized finding (Savageau et al., 1983). There is instead good agreement on the fact that viral insertions are generally haploid, i.e. they do not tend to become homozygous in the transformed cell. In addition, in all cases examined, the cloning of viral insertions and use of the flanking host DNA sequences as probes has allowed the conclusion that integration does not happen by a precise insertion between host sequences which then become separated by the viral DNA, but is accompanied by a deletion of the host DNA at the site of insertion (Stringer, 1982a; Hayday et al., 1982; Neer et al., 1983; Basilico et al., 1983). Only in one case, to my knowledge, have the boundaries of these deletions been mapped, and they flank a deleted DNA fragment of about 3 Kb (Neer et al., 1983).

While integration obviously ensures the hereditary transmission of the viral DNA sequences, it could represent a mutagenic event. Some evidence of this effect has been presented (reviewed by Martin, 1981), but SV40 or Polyoma are not efficient mutagens, presumably also because viral insertions will only inactivate one allele of any autosomal gene. In addition to the integrated viral DNA molecules, polyoma or SV40 transformed cells often contain a small number of free viral DNA molecules. The origin of these will be discussed in the following sections.

5.1. Mechanism of Integration

The mechanism of integration of foreign genes into animal cell DNA has attracted considerable interest also because of its implications for studies of genetic engineering. For this and other reasons investigators have tried to elucidate the mechanism by which Papovaviruses integrate their DNA and the viral and host functions involved. Studies have centered on the existence of viral functions involved in integration and on the role of specific integration precursors.

We have discussed above the evidence which supports the idea that integration generally occurs by illegitimate recombination. It is also known that practically any DNA sequence can be integrated, with an efficiency which may depend on the cells used, the method of introduction of the DNA, etc. The question one can then ask is whether Polyoma and SV40 possess specific mechanisms which promote their integration, and whether the type of integration generally observed is related to specific events. Early studies (Fried, 1965; Di Mayorca et al., 1969; Eckhart, 1969; Fluck and Benjamin, 1979) suggested that the Polyoma A gene-product (large T-Ag) was involved in the establishment, but not in the maintenance, of transformation. These studies showed that tsA Polyoma mutants could transform cells at 33°C. At 39°C their efficiency of transformation was severely impaired but not totally abolished. However, cells transformed at 33°C maintained the transformed phenotype when shifted to 39°C, suggesting that the A function was necessary only transiently and thus could have well consisted of promoting integration. This prediction was borne out by the experiments of Della Valle et al. (1981) who, using tsA viral DNA or restriction fragments capable of encoding only truncated large T-Ag, showed that the

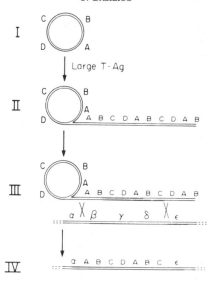

Fig. 3. A model of the mechanism by which Polyoma DNA integrates in a tandem arrangement. The circular viral DNA molecule (I) initiates replication (requiring large T-Antigen) through a rolling circle mechanism. This produces tandem repeats of viral DNA sequences in a head-to-tail arrangement (II). A double-crossing over event (III) between these oligomers and cellular DNA leads to tandem integration with concomitant loss of the host DNA sequences comprised between the cross-over sites (IV).

frequency of transformation was about 30-fold higher in the presence of large T-Ag function than in its absence. Furthermore, while cells transformed in the presence of a functional large T-protein almost invariably displayed integration of tandemly arrayed molecules, cells transformed in the absence of this viral function generally had integrations of less than single-copy molecules.

This suggested a model in which large T-Ag promoted integration by allowing the formation of oligomeric DNA forms which then integrate by a double crossing over, causing a deletion of host DNA (Fig. 3). A likely way to create these oligomers would be a rolling circle mechanism of replication (Bjursell, 1978; Robberson et al., 1976), and this hypothesis is in line with the finding of Chia and Rigby (1981) of linear oligomeric SV40 forms shortly after infection of mouse cells. This model does not contradict the finding that a variety of nonreplicating molecules can also integrate in tandem, and thus tandem insertion can also be produced by recombination (either before integration, or by integration within a previously inserted molecule) (Fluck et al., 1983), since in rare cases Polyoma molecules incapable of producing large T-Ag can also be found in tandem arrangement. However the frequency of transformation obtained with these molecules is always much lower than with T-Ag competent molecules (Della Valle et al., 1981; Dailey et al., 1984). In addition, recent experiments (Dailey et al., 1984) show that origin-defective Polyoma molecules transform with a low efficiency which is not altered by the presence or absence of a functional large T-Ag, and that they generally integrate in single copy. These experiments strongly suggest that the role of large T-Ag in promoting integration is mainly a replicative one, rather than recombination-promoting. Thus the model depicted in Fig. 3 is in substantial agreement with all the available data, even if it does not have to be operative in *all* transforming events.

5.2. Stability of Integrated Genomes

It was originally thought that Polyoma and SV-40 integrations were very stable, since in general transformed cells did not produce virus or contain detectable amounts of nonintegrated viral DNA. In addition, several studies directed at isolating revertants of virus-transformed cells had generally failed to obtain cells which owed their loss of the transformed phenotype to loss of integrated viral genomes (see Section 9). Only one situation in which the stability of integration could be disrupted was well known. In 1967, Watkins and Dulbecco, and at the same time Koprowski et al. (1967), found that fusion of nonpermissive SV40 transformed mouse cells with permissive monkey cells resulted in virus production in the heterokaryons. It was however unclear whether this process involved physical excision of the viral DNA (prior to replication promoted by permissive cell factors) or if the first free DNA template could have been produced by other mechanisms.

In the case of Polyoma virus, the demonstration that fusion with permissive cells results in virus rescue took much longer (Prasad et al., 1976), and originally it was reported that most Polyoma transformed cells did not produce virus following fusion (Watkins and Dulbecco, 1967). The reason for this observation was probably the use of cells which contained defective molecules.

It was later noted that Polyoma transformed rat cells contained, in addition to integrated viral DNA sequences, a small number (10–50/cell, on average) of free viral DNA molecules (Prasad et al., 1976). The presence of these molecules was restricted to a small proportion (0.1–1%) of the cell population at any given time and was under the control of the viral large T-Ag. Rat fibroblasts transformed by a tsA polyoma mutant, which codes for a temperature sensitive large T-antigen, produce free viral DNA at the permissive temperature whereas they do not express it at the nonpermissive temperature. However, even after months of propagation at a high temperature, the shifting of tsA transformants to 33°C results in the rapid reappearance of free viral DNA molecules (Zouzias et al., 1977). This showed that free viral DNA production was under the control of the A gene function and that the free molecules derived from the integrated ones, in agreement with the finding that the free viral DNA produced in a number of transformed lines faithfully reflected the main species of integrated viral DNA, which were generally present in a head-to-tail tandem arrangement (Gattoni et al., 1980).

It was suggested (Botchan et al., 1979) that the same mechanism that promotes 'excision' following fusion with permissive cells could have been responsible for the spontaneous production of free viral DNA in rat cells transformed by Polyoma virus. The requirement for a functional large T-antigen could have been explained merely by the fact that such protein is necessary for replication of the excised viral DNA; however, it appears that the A-gene function is also necessary for excision itself. Induction of free viral DNA synthesis in transformed rat cells is accompanied by loss of integrated viral DNA molecules as occurs during 'curing' of lysogenic bacteria (Basilico et al., 1979). My laboratory studied whether the Polyoma virus A-gene product was necessary for the excision event by comparing the extent of curing taking place in Polyoma virus tsA-transformed cells that had been propagated at the permissive or the nonpermissive temperature. The strategy used to isolate putative cured cell lines was based on the selection of cells that had regained a normal phenotype, and then to investigate whether such cells had lost integrated viral genomes. It was found that the rate of reversion was quite high and dependent on the presence of a functional large T-Ag. Thus tsA transformed cells propagated at high temperature gave rise to almost no revertants, while at low temperature reversion occurred at high rate (Basilico et al., 1979, 1980). Analysis of integrated viral DNA sequences revealed that: (1) all the original transformed lines had tandem insertions of viral DNA molecules, (2) revertants had lost an integral number of viral DNA molecules from the original tandem insertion, while the host–viral joints appeared unchanged, (3) the degree of reversion was correlated with the viral DNA sequences retained, i.e. whether they could still code for transforming proteins or not.

At about the same time other studies revealed that the instability of Polyoma integration could not only lead to excision, but also to amplification of integrated viral DNA sequences (Colantuoni et al., 1980). Although several instances of amplification have been found, by far the most common leads to the generation, from insertions of 1–2 tandem copies of viral DNA, to longer tandem repeats of 3–5 genome equivalent. Like in excision, the host–viral DNA joints are generally unchanged, and thus the amplification is internal. The modality of amplification, which generally does not involve random fragments of the viral genome, but an integral number of viral genome equivalents, also suggests the involvement of a step of homologous recombination. Amplification is very rapid (with the appropriate cell line it can be shown to have involved 20–30% of the cell population after ~ 10 division cycles) and also requires the presence of a functional large T-Ag (Colantuoni et al., 1980, 1982).

The similar requirements of excision and amplification suggest a common mechanism, which involves large T-Ag and homologous recombination (Fig. 4). The latter requirement is further shown by the finding that insertions of less than one copy of viral DNA, thus lacking regions of homology, do not excise or amplify to any significant extent, even in the presence of a functional large T-Antigen. Occasional events of amplification observed in these cells seemed to involve flanking host DNA sequences, perhaps because of partial homology with the viral sequences (Colantuoni et al., 1982). Basically three possible mechanisms have been considered, accounting for the known requirements of these phenomena.

(1) A replication–recombination model. In this mechanism, Py large T-Antigen binds sporadically to the integrated viral origin of replication, inducing replication *in situ*. When duplicated regions of Py genomes are present, homologous sequences can recombine, producing complete viral molecules that can be removed and circularized as free DNA. If the parental strands are involved in this recombination event, this would result in physical excision of integrated viral DNA and generate 'cured' cells. Finally the newly replicated molecules could recombine with homologous sequences of the parental strands, leading to an increase in the copy number of integrated viral genomes (Botchan et al., 1979; Colantuoni et al., 1982). The process of replication itself may allow the 'exposure' to recombination of existing homologous sequences by introducing transient nicks and gaps in these regions.

(2) It is also possible that recombination occurs without replication, implying that Py large T-Antigen can have a recombination-promoting activity, similar, for example, to that of the Rec A protein, independent of its role in viral DNA replication. In this case, the regions of homology would determine the size of the DNA segment that will become the 'unit' involved in excision or, after replication, in amplification.

(3) Another possible mechanism for the generation of excision and amplification (but not free viral DNA production) in polyoma-transformed rat cells with tandem viral integrations could be that of unequal sister chromatid exchange, resulting from the out-of-register pairing of the repeated regions of the viral DNA insertions. This mechanism predicts that for any excision event taking place, a parallel amplification event should occur.

Strong support for the first model comes from recent experiments involving the use of origin-defective tsA Polyoma mutants (Pellegrini et al., 1984). A partial tandem of these molecules was created *in vitro*, and following integration and transformation, the ability of these origin-defective tandem insertions to undergo excision or amplification was studied.

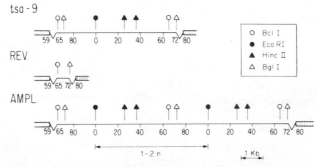

FIG. 4. An example of amplification and excision of integrated Polyoma viral DNA sequences in a line of transformed rat cells (tsA-9). The figure shows the restriction maps of the integrated viral DNA of tsA-9 and derivatives of tsA-9 after propagation at 33C. REV, a tsA-9 revertant; AMPL, a derivative of tsA-9 in which the integrated Py DNA has undergone amplification. The cross-hatched areas represent flanking host DNA and the thin lines integrated viral DNA sequences. Numbers below these lines indicate Polyoma map units (see Fig. 1). The cleavage sites of some restriction enzymes used to construct this map are also shown. From Colantuoni et al. (1982).

It was found that the origin defective molecules did not undergo any detectable rearrangement even in the presence of intramolecular homology and a functional large T-Ag. This did not result from an inability of the cells to support amplification, since when a plasmid containing the Polyoma origin of replication and the bacterial sequences coding for G418 resistance (Colbere-Garapin et al., 1981) were introduced into the same cells, these latter sequences readily amplified under the control of the large T-molecule produced by the origin defective genomes. It appears therefore that the origin of replication is a strong requirement for excision and amplification, a finding which strongly supports the replication–recombination model.

Using an inducible Polyoma-transformed line of rat cells, Manor's group has provided direct evidence of *in situ* replication (Baran et al., 1983). In the LPT line treatment with mitomycin C results in a high rate of production of free viral DNA and infectious virus. Concomitantly, the integrated viral DNA sequences and flanking cellular DNA increase in their relative concentration about ten-fold. Unless this result is an artefact of the usage of mitomycin C, it provides clear evidence that excision and free viral DNA production are initiated by multiple rounds of replication of the integrated viral DNA.

Less is known about SV40 excision and amplification than for Polyoma, probably because in polyoma transformed cells the role of large T-antigen in promoting gene rearrangements can be more easily separated from its role in transformation. Most experiments on SV40 excision have been carried out using fusion of SV40 transformed cells with permissive monkey cells and analysis of rescued free viral DNA. The results obtained are generally in line with those of Polyoma, even if in this system it has never been possible to demonstrate physical excision. Production of free DNA in fusion experiments is not so dependent on tandem insertions as in 'spontaneous' excision; however the amounts of free viral DNA are much higher following fusion of lines with tandem insertions (Botchan et al., 1979, 1980). Probably here it is a matter of detecting even lower probability excision, since the fusion amplifies the number of molecules produced. It is important to note here that free SV40 molecules derived from tandem insertions are generally homogeneous, while those derived from single copy insertions are heterogeneous, and contain flanking host DNA which often shows partial homology to the vital sequences (Bullock and Botchan, 1982), in line with the hypothesis (see above) that even in these cases excision involves homologous recombination.

When human cells transformed by a plasmid containing the SV40 origin of replication linked to the chicken TK gene, were fused to monkey cells transformed by origin-defective SV40 molecules (COS cells), extrachromosomal replication of the SV40 origin-TK plasmid was observed (Conrad et al., 1982). Thus also in this system excision can result from the interaction of large T-Ag with a functional origin of replication, and such interaction can take place in trans.

In addition to these phenomena, less understood rearrangements occur also relatively frequently in Polyoma and SV40 transformed cells, and they will be discussed in Section 7.

6. EXPRESSION OF INTEGRATED GENOMES

Once the viral genome of Polyoma or SV40 is integrated into the host DNA, it behaves in many ways as any other cellular gene, in that it is replicated synchronously with the host DNA and its expression depends on host enzymes. However, the fine regulation of transcription seems to be still mainly dependent on the viral regulatory elements.

It has been generally believed that transcription of integrated SV40 or Polyoma genomes is limited to the early region, or more precisely, that stable polyadenylated cytoplasmic RNAs represent only early region transcripts (see Tooze, 1981). The reason why Polyoma or SV40 transformed cells do not express stable late transcripts, even when it can be shown that the corresponding DNA sequences are integrated intact, are largely unknown. There could be a block to transcription or to RNA processing. Sensitive methods (Lange et al., 1981; Lange and May, 1983) revealed low, but measurable levels of cytoplasmic late transcripts in SV40 transformed cells in absence of free viral DNA. The block to late expression is therefore probably not complete, but still quite severe. Estimates of the ratios of late to early mRNAs vary from 3/100 to 1/10 (Lange and May, 1983). Excision and production of free DNA lead to a rapid increase in the amount of late mRNAs (Manor and Kamen, 1978), showing that this finding is not dependent on the cell type studied, but rather on the physical state of viral DNA. We will return to this point later.

6.1. Early Transcripts

Viral early transcripts are present in all cells transformed by Polyoma and SV40. By a variety of techniques these transcripts have been compared to those detected in lytic infection (that were characterized earlier) to establish whether integration resulted in a different control of transcription.

Both in Polyoma and SV40 transformed cells it appears that, when a complete early region is present, its transcripts are identical to those found in early lytic infection (Bacheler, 1977; May et al., 1980; Kamen et al., 1980; Fenton and Basilico, 1981). This is particularly true of polyadenylated cytoplasmic RNAs. Both in Polyoma and SV40 transformed cells 5' ends, splice junctions and mRNA 3' ends are indistinguishable from those found early in lytic infection, and therefore Polyoma transformed cells contain three main species and SV40 two main species of viral mRNAs.

A number of unusual mRNAs have however been described. These fall into two categories: (a) larger than normal transcripts, with normal 5' and 3' ends. These generally result from reiteration of integrated viral DNA sequences. A case in point is the mRNA coding for 'super' T-Ag, which has been observed in some SV40 transformed cells (May et al., 1981; Chen et al., 1981; Clayton et al., 1982a; Lovett et al., 1982), (b) mRNAs, whose 5' ends and splice junctions are normal but with unusual 3' ends. These transcripts are quite common in Polyoma transformed cells (Kamen et al., 1980; Fenton and Basilico, 1981), where deletions of the distal portion of the early region (coding for the C-terminal portion of large T-Ag) are quite frequent. In fact most if not all of these unusual 3' transcripts derive from the interruption of the early region (in general by host DNA sequences, but in some cases by the deletions of the viral DNA) upstream from the usual polyadenylation signal at 25.8 m.u. (Fenton and Basilico, 1981). This generates fused viral–host transcripts which are polyadenylated at host poly A signals, or, in the case of viral deletions, long polymeric viral transcripts that include DNA homologous to anti-late sequences (Fenton and Basilico, 1981). In all of these cases the poly A signal at 99 m.u., which is normally utilized with low efficiency, is used much more frequently, generating RNAs of approxi-

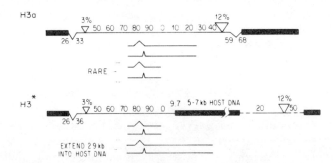

FIG. 5. Schematic representation of the transcription pattern of integrated Polyma DNA in two related transformed cell lines (H3A and H3*). H3* retains the left end of the H3A proviral insertion, but a 5 to 7 kb translocation of host DNA (not drawn to scale) has occurred, displacing the right end of the viral DNA insertion. ▽ indicates a deletion. The position of the 3' end of the H3* transcripts extending into host DNA has not been precisely mapped. Heavy lines represent host sequences flanking proviral DNA. Numbers above the proviral insertion are viral DNA map units; the approximate positions of the host–virus DNA junctions are also shown. Spliced sequences present in the small and middle T-Ag mRNAs are not distinguished in the diagram, but are both represented by the short early splice near 86 m.u. From Fenton and Basilico (1981).

mately 1100 and 1450 nucleotides (Kamen et al., 1980; Fenton and Basilico, 1981). The reasons behind the preferential utilization of the 25.8 polyadenylation signal in Polyoma are not known. It is possible that (i) viral DNA near 25.8 may encode signals for transcription termination of (ii) sequences from this region, when present in primary transcripts, may signal very efficient cleavage and polyadenylation of viral RNA. When these recognition signals are deleted, transcription either continues downstream, or cleavage near the alternative poly A addition site at 99 m.u. occurs (Fig. 5).

In any case, it is interesting to note that the poly A signal at 99 m.u. directs the addition of poly A residues at two positions 10 and 30 bases downstream, which bracket the termination codon for middle T-Ag (Fenton and Basilico, unpublished). Thus RNAs polyadenylated in this area can encode a complete small and middle T-Ags, although only a truncated large T-protein.

With respect to the regulation of viral transcription, available data support the idea that it depends mainly on viral controlling elements. Thus the 5′ ends of viral mRNA are generally identical to those found in lytic infection and no evidence of transcription initiating from the host DNA has been detected, in line with the finding that viral molecules which are interrupted just upstream of the early coding regions or are missing the viral enhancer sequences are not transforming. This does not of course prove that in rare cases integration next to host promoter or enhancer may not be essential to viral expression, but clearly shows that this is not the case in the great majority of transformed cells. In this respect it is worth mentioning that host sequences can replace viral regulatory sequences needed for transcription. Fried et al. (1983) transfected rat cells with a Polyoma DNA fragment lacking sequences from the BamHI site (58 m.u.) to the HaeII site (at 72 m.u.) and thus missing promoter, origin and enhancer sequences, after random ligation to mouse DNA. They could obtain rare transformants from which they were able to clone the Polyoma insertion and flanking host DNA. The host DNA 5′ to the Polyoma sequences is necessary for the biological activity of the new molecules, which although replication defective, can now transform cells quite efficiently.

In spite of these curiosities it is still clear that the main regulation of Polyoma and SV40 transcription is viral. Large T-Ag, known to regulate early transcription in the lytic cycle (Reed et al., 1976; Khoury and May, 1977; Rio et al., 1980; Cogen, 1978), can do so also in virus transformed cells. While the extent of transcription inhibition mediated by large T-Ag in SV40 transformed cells may be somewhat less than what is observed in the lytic cycle (Alwine et al., 1977), in the case of Polyoma transformants the increase in transcriptional activity obtained when large T-Ag is nonfunctional appears to be of the same order of magnitude (5–10 fold) as in the lytic cycle (Fenton and Basilico, 1982b).

A higher order of host control at the level of chromatin structure, integration site, etc. cannot however be excluded and is in fact suggested by several experiments. In 1974, Swetly and Watanabe showed that in a line of SV40 transformed rat cells viral transcription took place preferentially during the S-phase. Taking advantage of host mutations, which somehow interfere with the expression of the transformed phenotype, it has been possible to show that in a few SV40 transformed mouse cell lines, transcription of integrated SV40 molecules is strongly cell cycle regulated, such that G_1-arrested cells do not transcribe integrated SV40 DNA and do not express viral antigens (Basilico and Zouzias, 1976). Transcription of other cellular genes proceeds unimpaired (La Bella et al., 1983). Whether this cell cycle control results from the availability of host factors necessary for SV40 transcription, or from the site of viral DNA integration is not known. In the latter case, it would have to be postulated that SV40 preferentially integrates in regions of the host DNA that are not accessible to transcription in G_1 because of chromatin conformation or other undefined factors. Clearly further work is needed to resolve this issue and this work may be useful also for understanding the factors regulating the expression of animal cell genes. This work could also help answering the old question of whether 'cryptic' transformants really exist, in other words whether an appreciable number of cells may carry integrated, nondefective viral genomes which are not expressed.

6.2. Late Transcripts

As mentioned above, Polyoma and SV40 late transcripts are not present or present in very low amounts in transformed cells. This could be due to the fact that viral late promoters are not active in an integrated state, perhaps because they need independent viral DNA replication to become available (Lange and May, 1983), or to a defective processing of the late transcripts.

Recent results however strongly suggest that the latter mechanism is the most likely. In recombinant plasmids, a DNA fragment corresponding to the origin-promoter region of SV40 can activate transcription of the bacterial guanine phosphoribosyl transferase (gpt) gene in transfected cells following integration. This is true but only when the gpt gene is inserted downstream from the early, but also from the late promoter (Bourachot et al., 1982; Cereghini et al., 1983). The late promoter seems to be somewhat less active (~ 3 fold) than the early and this effect is more pronounced (~ 10 fold) when the Polyoma late promoter strength is compared to the early promoter (Bourachot et al., 1982).

The relative weakness of the late promoter sequences in these experiments could be due however to the construction of the plasmids. We have introduced (Kern and Basilico, 1985) two foreign genes into the Polyoma late region at the BclI site (just upstream of the initiation codon of VP2): a promoter-less Herpes Thymidine Kinase (TK) gene, and the coding sequences conferring resistance to the antibiotic G418. In both cases, the AUG of the foreign gene was within 5–10 nucleotide of the Polyoma BclI site. Rat cells were transfected with these hybrid molecules, selected for morphological transformation, and then tested for the acquisition of the TK^+ or G418 resistant phenotype. Surprisingly, in both cases more than 90% of the transformants had acquired the TK^+ or G418 resistance phenotype and expressed mRNAs homologous to these sequences in amounts similar to those of the viral early transcripts. The 5' ends of these mRNAs map within the Polyoma late sequences. This finding suggests that Polyoma late promoters are as active as the early ones in transformed cells. Thus the explanation why late transcripts are generally absent in transformed cells could be one of the following: (a) since the foreign genes tested were all unspliced, some splicing defect may be responsible for the inefficient processing of viral late-RNAs; (b) expression of Polyoma late mRNA could be selected against, since it is lethal to the cells; (c) The stability of Polyoma late m-RNAs may require the presence of specific 5' structures which cannot be generated in transformed cells, such as reiterated leader sequences. Novel structures occurring in our chimeric m-RNAs may obviate the need for such 5' structures.

In any case, these experiments show that Polyoma late promoters can be active in transformed cells and that probably transcription itself is not blocked.

6.3. Expression of Viral Proteins

The modalities of transcription of the viral DNA are the main determinants of the viral proteins found in transformed cells. There is no evidence of translational control, i.e. by and large all major viral proteins found can be accounted for by their mRNA structure. Late viral proteins are generally not detected except in the presence of detectable amounts of free viral DNA (Prasad et al., 1976).

The great majority of cells transformed by SV40 express both the small and large viral T-Ags. Nonpermissive transformed cells do not synthesize truncated or defective large T-molecules, as frequently as is the case in Polyoma transformants. On the other hand, *permissive* (human, monkey) transformed cells often produce defective T-Antigens which because of mutations, small deletions, or host substitutions are incapable of supporting viral DNA replication, but have transforming ability (reviewed by Rigby and Lane, 1983). This finding can probably be ascribed to a selection against viral replication occurring in permissive cells.

The occurrence of a larger form of T-Ag (called super T) has been reported in a few transformed cell lines (May et al., 1981; Chen et al., 1981; Clayton et al., 1982a; Lovett et *al.*, 1982). As discussed above, it appears that super T-Ag is encoded in integrated SV40 DNA molecules with tandem duplications of parts of the early region. In addition to the nuclear form of large T-Ag, SV40 transformed cells express a surface form of this protein, which probably represents the TSTA of these cells (reviewed by Mora, 1982).

In the case of Polyoma transformants, the only two proteins which are found invariably expressed in transformed cells are the small and middle T-Antigen (Griffin and Dilworth, 1983). Full-size large T-Ag molecules are often expressed, but a large number of transformants (especially if carried in culture for long periods of time) produce truncated forms of this protein, whose structure is reflected in the mRNAs encoding it. Thus the shorter forms are encoded in large T-mRNA polyadenylated at 99 m.u., as described above, and the larger forms generally consist of fused viral–host proteins, which are terminated at host sequences near the viral–host DNA junctions of the Polyoma DNA insertion (Lania *et al.*, 1981; Dailey *et al.*, 1982). All of these large T-Ag forms are generally replication-incompetent, and accordingly large T-Ag molecules which are defective in replication because of point mutations or in frame deletions have also been identified (Hayday *et al.*, 1983). The frequent occurrence of these defective large T-Ag forms can be explained probably in two ways. Large T-Ag, or at least its C-terminus, is not necessary for the expression of the transformed phenotype, and in addition, a selection against the expression of this protein exists in transformed cells, as will be discussed in the next section.

7. EVOLUTION OF TRANSFORMED CELL LINES

The pattern of integration of Polyoma and SV40 DNA in transformed cells is not totally stable. In addition to the phenomena of amplification and excision previously mentioned, other less understood rearrangements of integrated viral DNA often take place. These rearrangements are sometimes difficult to characterize and their mechanism is not so clear, even if in some cases they probably also involve recombination and may be promoted by large T-Ag (Bender and Brockman, 1981). In any case, they can lead to reversion of the transformed phenotype (Steinberg *et al.*, 1978; Bender *et al.*, 1983), increase in tumorigenicity and changes in expression of viral proteins (Hiscott *et al.*, 1980; Sager *et al.*, 1981; Chen *et al.*, 1984).

In the case of Polyoma tranformants, a clear evolution leading to lack of expression of a full-size large T-Ag molecule can often be demonstrated. Israel *et al.* (1980) found that tumors caused in hamsters by the injection of complete viral DNA molecules generally did not produce a functional large T antigen. This and the frequent finding of Polyoma-transformed lines which produced defective large T antigen molecules, or did not contain any recognizable form of this protein (Hutchinson *et al.*, 1978; Lania *et al.*, 1980), suggested the possibility that an actual selection against the expression of this viral protein took place during the propagation (*in vitro* or *in vivo*) of Polyoma-transformed lines. Accordingly, injection into rats of T-Ag positive transformed lines resulted in the emergence of tumorigenic cells which through deletions, rearrangement and mutations had lost the ability to produce a functional (at least for replication) large T-molecule, while maintaining production of small and middle T-Ag (Lania *et al.*, 1981). Long term propagation in tissue culture resulted in similar changes (Dailey *et al.*, 1982).

These data lead to the conclusion that large T-Ag production is not only unnecessary for the maintenance of the Polyoma-transformed phenotype, but is also somewhat detrimental to its full expression. One of the hypotheses first proposed to explain these observations was that in animals the synthesis of this protein would lead to immunological rejection of the tumor cells (Israel *et al.*, 1980). This explanation is made unlikely by the experiments, which show that selection for large-T negative cells occurs also in tissue culture and only in conditions in which T antigen is present as a functional protein (Dailey *et al.*, 1982). The most likely explanation is in my view that large T-Ag negative cells may be selected merely because their integration pattern is stable, and these cells, while still transformed, do not undergo processes of excision, amplification and production of free DNA. The selective disadvantage conferred by these phenomena to a transformed cell population may consist not only in the fact that they may cause reversion (e.g. by excision), but also in that they may lead to high frequency of rearrangements, mutations, etc., which may be lethal for the cells or suppress their transformed phenotype.

Similar phenomena cannot be so easily observed in SV40 transformed populations, probably because in this case the transforming protein is large T-Ag itself, whose elimination would lead to loss of the transformed phenotype. However, the production of large T-Ag molecules which are transformation competent, but replication defective, is not uncommon (see Rigby and Lane, 1983) and will be discussed in the next section. This phenomenon could be considered a case analogous to the evolution of Polyoma trans formed cells, although in this case it is not known whether it results from cell transformation by mutated SV40 molecules or evolved during propagation *in vitro*.

8. ROLE OF THE VIRAL PROTEINS IN TRANSFORMATION

A considerable body of evidence supports the idea that SV40 and Polyoma cell transformation is caused by the expression of the viral early proteins. Some of this evidence was reviewed before, and additional data will be presented in this section. The sustained expression of certain viral gene-products in transformed cells suggest that the same products are responsible for the transformed phenotype, but does not answer the question of which ones are the most important or by which mechanism they act. In addition, some viral proteins could be found in transformed cells by coincidence, such as being contained in the same transcriptional unit of the transforming protein, but without playing any crucial role in the process of transformation.

The genetic dissection of transformation has progressed further than the biochemical studies. Thus, while the transforming genes of Polyoma and SV40 have been identified, the precise mechanism by which their gene-products so drastically subvert the host cell behavior is largely unknown. Although the mechanism of transformation by Polyoma and SV40 may be ultimately similar, there is enough difference between their main transforming proteins to justify treating them separately.

8.1. SV40 Proteins and Transformation

SV40 large T-Ag was identified quite early as the essential transforming protein of this virus (reviewed by Martin, 1981). The main evidence for this conclusion is the following: temperature sensitive mutants of the A complementation group (shown to produce a thermolabile large T-Ag) are inhibited in transforming ability at high temperature, i.e. cells infected with tsA mutants at 39–40°C are not transformed, while the same experiment yields a normal frequency of transformants at 32–33°C (Osborn and Weber, 1975; Brugge and Butel, 1975; Tegtmeyer, 1975; Martin and Chou, 1975). This type of result, however, does not distinguish between a viral function necessary for establishment or initiation of transformation (a term which could include many early events of which the only one clearly identified is integration) and transformation maintenance, i.e. the capacity to confer to the cells an altered phenotype. It was therefore very important that several authors (Brugge and Butel, 1975; Martin and Chou, 1975; Osborn and Weber, 1975; Tegtmeyer, 1975; Anderson and Martin, 1976; Noonan et al., 1976; Brockman, 1978; Seif and Cuzin, 1977; Seif and Martin, 1979a; O'Neill et al., 1980) could also show that the phenotype of cells transformed by SV40 tsA mutants at 33°C reverted to a more or less normal one when the cells were incubated at the nonpermissive temperature for large T-Ag function, since this provided direct evidence that the functional state of this protein was continually required for transformation.

While this and similar experiments prove quite conclusively that the large T-protein is necessary for the expression of the SV40 transformed phenotype, a number of observations in this field still need to be explained. The most important is probably the frequent occurrence of SV40 tsA transformed lines which are totally temperature-independent for the expression of transformation (reviewed by Martin, 1981). The simplest explanation, i.e. that this behavior is due to back-mutants, which now produce a wt large T-Ag, is contradicted by the relatively high frequency of temperature independent transformants, and in some cases by direct experimentation showing that these cells still harbored tsA genomes (Brockman, 1978; Tenen et al., 1977). Leakiness of the mutants used probably plays a role in the phenomenon, particularly since the transforming ability may be more resistant to temperature inactivation than the ability to support viral DNA replication (Chou and Martin, 1975), and this is more easily measured. Gene-dosage effects due to rearrangements and amplification of integrated viral DNA sequences have also been implicated (Hiscott et al., 1980). Synthesis of very large amounts of a somewhat defective protein could in some cases be responsible for its retained ability to modify the cell's behavior (Brockman, 1978; Tenen et al., 1977). All of these explanations probably play a role; it is also possible that in some cases the transformed phenotype has become independent of the presence of this viral protein, either because of the constitutive expression of some cellular oncogenes, or of some unspecified mechanism of transformation, that, once primed by large T-Ag, does not require it any longer.

The involvement of small T-Antigen in transformation remains more elusive and its role does not seem essential. Viral DNA molecules encoding almost exclusively small T-Ag (i.e. carrying large deletions of the early region starting immediately downstream from the small T-terminator codon) do not transform cells in culture (Rubin et al., 1982; Chang et al., 1984). Several studies have been carried out on deletion mutants, which remove sequences of the viral DNA unique to the C-terminal portion of small T-Ag, and do not affect large T-Ag since they are part of large T-Ag intervening sequences (see Fig. 2). These mutants can transform cells in culture, but with a somewhat reduced efficiency with respect to the wild type (Bouck et al., 1978; Sleigh et al., 1978; Rubin et al., 1982). Most important, the phenotype of these transformants is 'weak' so that if transformation is scored by the ability to form foci, the mutants are only slightly defective, but their ability to promote growth in agar suspension is more impaired, and is complemented by small T-Ag (Bouck et al., 1978; Rubin et al., 1982; Chang et al., 1984). These data would assign to small T-Ag, a secondary, but essential role in transformation, but there is some disagreement as to the relative 'weight' of small T contribution to the transformed phenotype. The efficiency of transformation of small T deletion mutants may depend on the state of the cells at the time of infection and the type of cells infected (Seif and Martin, 1979b). In addition, their tumorigenicity in hamsters is not greatly reduced (Lewis and Martin, 1979; Topp et al., 1981). Since SV40 small T-Ag is involved in the disorganization of the cell cytoskeleton found in most transformed cells, it is likely that this facet of transformation is more essential in certain cell types than in others, and thus the contribution of this protein to malignancy varies with the experimental system used. Alternatively, small T-Ag may contribute functions which can also be provided by serum, a hypothesis suggested by the observation that the ability of cells transformed by small T deletion mutants to grow in agar is greatly influenced by the type and amounts of serum present in the medium (Martin et al., 1979).

The genetic identification of SV40 large T-Ag as the essential transforming protein of this virus does not unfortunately provide specific clues as to its mechanism, except for suggesting that one or more of its better known functions may be directly related to its ability to transform. The first of such functions is its ability to promote viral DNA replication, mediated by its specific binding to the origin region of SV40 DNA. In addition, its ATPase activity may be strictly involved in replication, since mutations which inactivate this enzymatic activity also affect the replicative ability of large T-Ag (Manos and Gluzman, 1984). A strong case has been made for the hypothesis that SV40 T-Ag could maintain transformation by inducing cellular DNA synthesis at host origins (Martin, 1981), but a number of experimental data suggest that the ability of SV40 large T-Ag to drive viral DNA replication and by inference DNA replication in general is not directly related to its transforming potential.

As mentioned before, the finding of replication-defective large T-Ag molecules, although not as frequent as in Polyoma transformed cells, is fairly common in SV40 transformed cells, and was only more difficult to recognize because these replication defective forms of SV40 large T-Ag are generally not grossly reduced in size. The first type of such T-Antigen molecules have generally a MW in excess of 100,000 and have been called super T-Antigens (see Section 6). It is not clear whether these super T-Ags are in some way enhanced in their ability to induce transformation. It is however clear that they are defective in replication, since transfection of the molecules encoding them into cells yields transformants, but the molecules do not replicate (Chen et al., 1981; Clayton et al., 1982a; Lovett et al., 1982).

Other examples of transformation competent, replication defective T-Antigen molecules, have been discovered. They include mutations and deletions. As already mentioned, they have been found not only in permissive, but also in nonpermissive cells (Chaudry et al., 1982; Gluzman and Ahrens, 1982; Stringer, 1982b; Prives et al., 1983; Manos and Gluzman, 1984). Gluzman's group has studied several of these molecules and found that they generally fall into two classes: (1) mutant proteins which are defective in binding to the SV40 origin of DNA replication (Prives et al., 1983), (2) T-Antigens which carry mutations, which somehow decrease or suppress ATPase activity (Manos and Gluzman, 1984). These mutations generally fall into different regions of the protein, and provide a plausible basis for the replication defect of these molecules. They also show, as already discussed, that large T-Ag is a multifunctional protein, and some domains are not necessary for transformation.

The data discussed above do not support the hypothesis that SV40 large T-Ag transform cells by driving uncontrolled, cellular DNA replication by interacting with cellular origins or with integrated viral origins, but do not rule it out altogether. The capacity to bind calf thymus DNA is preserved in these mutant proteins (Prives et al., 1982, 1983) and indeed the sequences contributing to origin-binding map between 0.5 and 0.54 m.u., whereas 'nonspecific' cell DNA binding seems to require sequences between 0.39 and 0.44 m.u. (Prives et al., 1982; Morrison et al., 1983). It could thus be argued that these replication-deficient, transformation competent large T-proteins can still induce host DNA synthesis (Fig. 6). It is my personal bias that the replication function of large T is not intrinsic to transformation, even if it may play a role in tumor evolution, etc., but clearly further experiments are necessary to settle this point.

Another property of large T-Ag may be very important in transformation. It has been shown that a small fraction of large T-molecules are always found associated with the plasma membrane on the cell surface. The evidence for this conclusion is of two types: (1) immunological (Chang et al., 1979; Flyer and Tevethia, 1982; Tevethia et al., 1980; Reddy et al., 1982), showing basically that injection of purified large T-Ag is able to protect mice

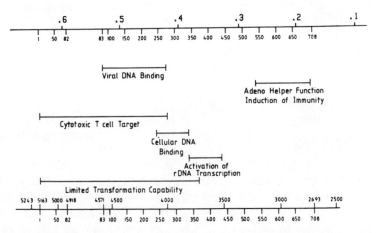

FIG. 6. Functional map of SV40 large T-Antigen. The upper line maps the amino acid number (1–708) of large T-Ag against viral map units, the lower against nucleotide numbers. The location of the functional domains should be considered tentative. Relevant references are in the text.
Adapted from Rigby and Lane (1983).

from SV40 tumors, and the demonstration that cytotoxic T cells, primed by large T-Ag, can selectively attack SV40 tumor cells, (2) the combined use of monoclonal and monospecific antibodies and biochemical procedures (Deppert et al., 1980; Gooding et al., 1984; Henning et al., 1981; Santos and Butel, 1982; Walter et al., 1980), has shown that SV40 transformed cells express specific parts of large T-Ag on their surface, generally not including the central area of the protein. This evidence is still somewhat controversial, mainly because of the lack of any detectable biochemical difference (so far) between surface and nuclear T-Ag. Yet there is enough evidence to conclude that a portion of T-Ag is expressed on the cell surface, and so ask what its role may be. The answer to this question is difficult: while it is clear that it must represent the SV40 TSTA (Mora, 1982), its physiological role in viral growth or transformation is unclear. However the fact that many transforming proteins are membrane associated suggests it may be important. In this respect it is worth mentioning that two SV40 large T-Ag mutants which affect transport into the nucleus have been recently described. One mutant carries an amino acid substitution of lysine 128 (Lanford and Butel, 1984), the other a deletion in frame removing amino acids 110 to 152 (Fischer-Fantuzzi and Vesco, 1985). Both mutants can transform cells from continuous lines, such as NIH3T3, but are somewhat defective in transformation of primary embryo cells. The study of these mutant T-Ags should provide important information as to the relative roles of nuclear and surface T in transformation.

Another aspect of T-Ag which has received wide attention is its interaction with p53 (Lane and Crawford, 1979), a host protein which the virus induces and with which T-Ag complexes in a rather stable manner (reviewed by Crawford, 1983). The p53 is probably a growth related cellular protein (Crawford, 1983), but its function is still unknown. The p53-TAg complex stabilizes p53, and this could provide a mechanism by which both the inducer (T-Ag) and the growth promoting protein (p53) are activated (Greenspan and Carroll, 1981; Linzer and Levine, 1979; Oren et al., 1981; Lane et al., 1982). However this mechanism is not suggested by any firm data, and will require a much better knowledge of p53 function to be examined in depth.

Although this function of SV40 large T-Ag has been studied less than others, there is mounting evidence that this protein can activate the transcription of cellular genes (Schutzbank et al., 1982; Scott et al., 1983; Brickell et al., 1983). It is therefore tempting to speculate that this could be its most important activity vis à vis transformation, since clearly the activation of growth related genes could be a way of inducing transformed growth behavior. There is however no evidence of substance yet linking this function to cell transformation.

Since SV40 large T-Ag is a multifunctional protein, each function could reside in distinct domains (Fig. 6). A considerable amount of data exists on the ability of viral molecules deleted in various regions to induce cell transformation, cell DNA synthesis or to activate RNA transcription. All of these studies suffer from certain limitations, mainly the fact that deletion of a specific protein sequence may affect its conformation, stability etc. and thus its function, even when the sequence deleted is not directly involved in the function tested. With all these limitations, the available data suggest that the C-terminal portion of SV40 large T-Ag (from nucleotide 3500 approximately) is not necessary for induction of cellular DNA and rRNA synthesis (Galanti et al., 1981; Mueller et al., 1978; Soprano et al., 1981), while it is required for viral DNA replication. In addition, Sompayrac and Danna (1983a, b) and Clayton et al. (1982b) have shown that DNA molecules truncated even further upstream (nucleotides 4000 and 3700, respectively) can still transform continuous rat cell lines, although with very low efficiency. Colby and Shenk (1982) in addition have shown that similar molecules, although not fully transforming, can immortalize primary rat embryo cultures, giving them at least an aspect of the transformed phenotype. A complex picture of SV40 large T emerges from these studies: that of a multifunctional protein (Fig. 6) which can drive viral DNA replication, a function for which its C-terminus is clearly an essential domain, can in a still unknown way be partitioned between the nucleus and the plasma membrane, and is uncommonly 'plastic' in maintaining residual biological activities even when deleted of a large part of its sequences.

8.2. Polyoma Proteins and Transformation

Differently from SV40, a number of experiments support the idea that Polyoma large T-Ag does not play a direct role in transformation at least in certain cell types. Polyoma tsA mutants are generally not blocked in the maintenance of transformation, but only in some 'establishment' functions, that is cells transformed by tsA mutants at 33°C remain transformed by 39°C (Fried, 1965; Di Mayorca et al., 1969; Eckhart, 1969), and as already discussed, at least part of this 'establishment' function consists of promoting integration (Della Valle et al., 1981; Dailey et al., 1984). In addition, a very common finding in Polyoma transformed lines is that of truncated large T-Ags, in some cases resulting from deletion of practically all unique large T-DNA sequences, downstream from the EcoRI site (Fig. 1). Accordingly, Polyoma DNA fragments extending from the regulatory region to the EcoRI site (a few base-pairs downstream from the middle T-Ag termination codon) can transform cells with a somewhat reduced efficiency (Hassell et al., 1980; Novak et al., 1980; Chowdhury et al., 1980; Della Valle et al., 1981), but resulting in a phenotype indistinguishable from that of wild type transformed cells.

While the majority of studies carried out using rat and hamster cell lines strongly suggest that Polyoma large T-Ag is not involved in the maintenance of transformation (see Griffin and Dilworth, 1983), it has been reported (Seif and Cuzin, 1977; Rassoulzadegan et al., 1978) that Polyoma tsA transformants of the FR3T3 line fall into two classes: A (isolated as agar colonies) and N (isolated as dense foci). A transformants are temperature independent, while N transformants are ts for the expression of the transforming phenotype. The exact difference between A and N transformants is difficult to assess, and it has to be mentioned that host cell transformation mutants exist, although their frequency is probably too low to explain the N transformants. The result may depend on the cell types used. It is possible that transformation of FR 3T3 cells requires viral functions that other cell lines express constitutively (see later).

Irrespective of whether Polyoma large T-Ag contributes to the transformed phenotype, it is not sufficient for transformation. hrt Polyoma mutants, carrying a deletion within the large T-intron, and producing no functional small and middle T-Ag (Schaffhausen et al., 1978; Silver et al., 1978, and reviewed by Benjamin, 1982), can integrate, and express large T-Ag but the cell is not transformed (Lania et al., 1979). Similarly, one of the very useful constructions of Treisman et al. (1981b), a recombinant plasmid encoding only large T-Ag since the entire intervening sequence is removed, can be integrated and expressed but the cell remains phenotypically 'normal' (Rassoulzadegan et al., 1982). In conclusion, most of the data described show that a replication competent large T-molecule is not required for transformation, but they do not rule out that its N-terminal half (up to the EcoRI site) may not play a cooperative role.

On the other hand, there is no doubt that the essential transforming protein of Polyoma virus is middle T-Ag. The evidence for this conclusion has been mentioned before and can be summarized as follows:

(1) Polyoma virus transformed cells invariably produce full size middle T-Ag (Ito et al., 1980; Smith and Ely, 1983).

(2) HR-T mutations, which affect middle T production abolish transforming ability and tumorigenicity (Benjamin, 1982).

(3) Two mutations, mapping in the DNA sequences common to middle and large T-Ag, and that introduce a termination colon within the middle T-reading frame, but only an amino acid substitution into the large T-reading frame, are viable but cannot transform (Carmichael et al., 1982; Templeton and Eckhart, 1982).

(4) A Polyoma mutant which is cold-sensitive for the maintenance of transformation contains a single amino acid change in the middle T-Ag two residues upstream from the C-terminal hydrophobic region, and no changes in large T-Ag (Templeton and Eckhart, 1984).

(5) An intron-less recombinant plasmid encoding exclusively middle T-Ag can transform

continuous rat cell lines and the transformants are tumorigenic in animals (Treisman et al., 1981b).

All of these observations suggest that the middle T protein is necessary and sufficient to maintain cell transformation, and that the role of the other two early proteins in transformation may only be that of favoring integration, or contributing to the transformed phenotype in a manner not essential to tumorigenicity.

In line with this conclusion are observations which fail to attribute to Polyoma small T-Ag an essential role in transformation: mutations affecting both middle and large T-Ag may or may not transform, yet they all produce full size small T-Ag (Griffin and Dilworth, 1983), and infection with the small T-only recombinant plasmid does not induce a transformed phenotype (Rassoulzadegan et al., 1982). The Polyoma small T-Ag has striking homology to SV40's and thus also probably contributes to disruption of cytoskeletal organization. In most cases however this does not seem to be an essential requisite for expression of the transformation.

This relatively simple picture has been recently made much more complex (and interesting) by the finding that if transformation is assayed on primary embryo fibroblasts (generally rat), rather than on continuous lines, the role of small and large T-Ag appears much more important than previously thought.

Transfection into primary rat embryo fibroblasts of the Py MT only plasmid does not result in any detectable transformation (Rassoulzadegan et al., 1982), and the same molecule is not tumorigenic if injected into young rats (Asselin et al., 1983). Thus in this system middle T alone is not sufficient to induce transformation. Since one obvious difference between primary and continuous cell lines is that the latter are immortal, Cuzin's group investigated the ability of molecules encoding large T-Ag to elicit the formation of immortal cell lines and indeed both the large T-only recombinant plasmid and hrt mutants can induce the formation of continuous cell lines from primary cultures. These lines are phenotypically normal, and can now be transformed if superinfected with the middle T-only plasmid (Rassoulzadegan et al., 1983). Most important, this effect can be produced by plasmids encoding only the N-terminal half of large T-Ag (up to the EcoRI site) (Rassoulzadegan et al., 1982, 1983), again separating this 'establishment' function from the ability of this molecule to drive viral DNA replication.

These results strongly suggest that the full expression of transformation requires at least two Polyoma functions, one of which is dispensable if the cell is already 'immortalized', but still leave a number of questions to be answered.

(1) The inability of Polyoma middle T to transform primaries could still be due to problems of expression of this plasmid in primary cells, a problem which is relieved in an unknown way by large T-Ag. This possibility is unlikely, but it has to be kept in mind that middle T alone can induce tumors in hamsters, albeit with a long latency (Asselin et al., 1984).

(2) Co-transfection of the middle T and large T plasmids into primary cells does not seem to lead to transformation, while if the large T, middle T and small T plasmids are mixed transformation is obtained in one step (Rassoulzadegan et al., 1983).

(3) In some cases complementation of middle T-Ag for transformation and tumorigenicity seems to require small T-Ag, not large T-Ag (Asselin et al., 1983).

(4) In many cases the synthesis of the relevant proteins has not been demonstrated in the cell lines examined.

Another point that has to be considered is that immortalization of primary cells by large T-Ag encoding molecules has a low efficiency, less than expected from the frequency of transformation using wt genomes. Some of this is probably due to lethal effects of replication-competent large T-Ag encoding molecules (Land et al., 1983), as origin-defective LT plasmids function better in this assay (Pellegrini and Basilico, unpublished), but the possibility remains that large T can immortalize only cells that have already spontaneously undergone some change. Finally, it is not yet entirely clear whether large T promoted immortalization requires the continuous expression of this viral protein, or whether cells may become independent of it (Rassoulzadegan et al., 1983). In this latter case, the function of large T could well be that of activating, in a probabilistic manner, some cellular gene and could explain the low frequency of this phenomenon. We should remember that rodent primary cultures typically can be passaged about six to seven times before growth stops.

When transformation is assayed in primary cultures the role of small T-Ag appears to be more important. As mentioned before, only the transfection of plasmids encoding all three early proteins causes typical transformation. Transfection of small and middle T-Ag plasmids results in our hands (Liboi and Basilico, in preparation) consistently in the formation of agar colonies, which grow large enough to be isolated. Upon isolation however, most of these cells reveal themselves incapable to sustained growth. While this finding should probably be interpreted as showing that these cells lack the 'immortalized' phenotype, it also shows that small T-Ag contributes something to transformation, as cells transfected with the middle T-plasmid only never gave us any agar colony. Furthermore, the lack of tumorigenicity of middle T-plasmids in rats can be complemented by small T-Ag (Asselin et al., 1983, 1984). Perhaps there the number of targets is so high that some cells having undergone spontaneous establishment changes can be reached by the viral molecules.

The role of the Polyoma early proteins in transformation can therefore provisionally be summarized as follows: middle T-Ag is necessary for cell transformation and tumorigenicity, and also sufficient in continuous cell lines. Middle T-Ag is associated with the plasma membrane, is phosphorylated at tyrosine residues and is generally found associated with the product of the C-Sarc gene (Courtneidge et al., 1984). As in the case of the SV40 T-Ag p53 complex it is not known whether this association has functional implications, but the association with the product of a cellular oncogene is unlikely to have no functional meaning.

The C-terminal half of large T-Ag as well as its replication promoting ability are involved in establishment functions, the most important of which is probably integration. A full size large T-Ag is not necessary for the expression of transformation, but the information encoded in its N-terminal half can apparently favor the establishment of primary cells into lines, and thus complement middle T for transformation of these cells. In addition it may play a role in decreasing the serum requirements of transformed cells (Rassoulzadegan et al., 1982). It is interesting to note that another nuclear protein, the product of the v-myc retroviral gene, has been implicated in establishment functions (Land et al., 1983).

Small T-AG seems clearly not necessary for the transformation of continuous lines, but contributes to the transformed phenotype of primary cells and tumors, perhaps by causing changes in cytoskeletal organization which are required for anchorage independent growth. Thus Polyoma virus appears to possess three transforming genes with specialized functions, all of which may, depending on the target cell, play a role in transformation.

9. HOST CONTROL OF VIRAL TRANSFORMATION

While it is clear that in Polyoma or SV40 transformed cells, the transformed phenotype is maintained by the presence of the viral early proteins, it is legitimate to ask whether it can also be controlled by host functions. It seems somewhat obvious that some form of cellular control of the expression of viral transformation must exist. For example, if transformation ultimately depended on the autocrine production and utilization of TGF (Sporn and Todaro, 1980), a cell lacking receptors for such TGF's could not behave as transformed. More generally, the absence or defectiveness of a specific cellular protein with which viral proteins have to interact, would also inhibit the expression of transformation.

A number of presumed host cell mutants which somehow seem to fall into these categories do indeed exist. Revertants have been isolated from pure populations of SV40 or Polyoma transformants following treatments (Vogel and Pollack, 1974) which would selectively kill transformed cells and thus enrich the populations in cells with normal growth behavior. Under conditions in which normal cells should not grow but remain viable, such as in agar suspension, serum starvation or confluency, the addition to the culture medium of FudR, BudR and light, etc., results in the preferential killing of transformed cells, which continue to proliferate, and the surviving population is enriched in revertants, which can then be clonally isolated on the basis of their normal morphology. Most of this work was carried out some years ago (reviewed in Tooze, 1981), and in many cases the aim was that of isolating revertants which had lost the viral genome. Surprisingly however, the great majority of revertants isolated by these methods had not lost viral specific antigens, and still contained the viral genome, and thus were called 'phenotypic'. Their degree of reversion varied from a reversion of all the phenotypic traits of transformed cells, including tumorigenicity, to partial or limited reversion, sometimes depending on the selection system used (Pollack et al., 1968; Culp and Black, 1972; Kelly and Sambrook, 1975; Pollack and Vogel, 1973; Rabinowitz and Sachs, 1968; Wyke, 1971).

The fact that reversion resulted from host cell rather than viral mutations was suggested by two observations. In most cases tested, the virus rescued from these revertants by fusion with permissive cells was wild type (Pollack et al., 1968; Culp and Black, 1972; Ozanne et al., 1973; Toniolo and Basilico, 1975), and the revertants could not be transformed by superinfection with the endogenous virus (Wyke, 1971; Ozanne, 1973), a result consistent with the hypothesis that their behavior results from some cellular changes which somehow interfere with the expression of transformation. Still it has to be remembered that most of these results were obtained before the modern techniques of analysis of integrated genomes were available. Thus it is possible that many of the 'phenotypic' revertants could have contained many viral insertions, and the virus rescued was not the one controlling the phenotype. Indeed a more recent selection using an SV40 cell line with a single, well characterized viral insertion resulted almost exclusively in the isolation of revertants which owed their phenotypic changes to alteration or deletion of the viral genome (Steinberg et al., 1978). In Polyoma transformed cells with tandem insertions, FUDR selection results almost exclusively in the isolation of cells which have excised integrated viral genomes (Basilico et al., 1979), thus losing DNA sequences encoding the transforming proteins.

An interesting class of ts transformants are the ts-SV40 transformed 3T3 cells isolated and studied in our laboratory (Renger and Basilico, 1972, 1973). These cells were isolated by a selection procedure aimed at cells expressing the transformed phenotype at 33°C but not at 39°C, and they behave accordingly. They are considered host mutants because they express SV40 T-Ags and early mRNAs, the virus rescued from three independent lines is wild-type (Renger and Basilico, 1972, 1973), and we have determined that one of these lines carries a single insertion of viral DNA (La Bella et al., 1983). They cannot be retransformed by SV40 at 39°C, but are susceptible to transformation by another unrelated virus, Murine Sarcoma virus (Renger, 1972). Thus the nature of the ts mutation is quite specific, since it does not affect transformation by a different transforming gene. As mentioned before, these cells also have the unusual property of extinguishing viral DNA transcription and T-Antigen expression if arrested in G_1 at 39°C. However this property is unlikely to be directly related to the ts SV3T3 mutation, since *growing* cells express SV40 gene products at 39°C as well as at 33°C (Basilico and Zouzias, 1976).

In conclusion, there is good evidence that host cell functions can control the expression of the transformed phenotype, but unfortunately, the nature of the mutations which alter the cell's response to the viral gene products has remained so far elusive. It is hoped that when our knowledge of the chain of events involved in malignant transformation will become more extensive, these cell mutants will prove advantageous at identifying cellular molecules which are the targets of transforming proteins, or interact in any way with these proteins to activate them, stabilize them, etc.

10. MODELS FOR TRANSFORMATION BY POLYOMA AND SV40

The identification of specific gene or gene products as responsible for the changes in cellular phenotype that we call neoplastic transformation, does not allow unfortunately any precise conclusion as to their mode of action. The molecular events which are responsible for cell transformation are still largely elusive, and one is forced to build over analogies, and sometimes weak indications offered by a biochemical property of a given protein or another. In addition, since neoplastic transformation only defines an end-point, it is possible that several diverse mechanisms can lead to this ill defined terminal phenotype.

Probably the most exciting recent development in this field is the number of data lending credit to the theory that carcinogenesis may be a multistep process, and the tentative identification of at least two classes of transforming genes responsible for distinct events in transformation.

Although this hypothesis was first put forward by Vogt and Dulbecco (1963) about 20 years ago, it represents to an extent a change of mind in the field of viral carcinogenesis, where the prevailing idea was probably until recently that one single transforming protein could have been sufficient to achieve one-step transformation. While this idea may still be correct in some cases, the evidence that many proteins encode more than one function renders it in a sense passé. On the contrary, it is the study of very specialized transforming proteins, or subsets thereof, which will eventually answer the fundamental questions of whether transformation results from a single primary change, leading sequentially to a cascade of alterations, or influencing simultaneously many independent metabolic pathways, or whether in fact it is caused by the independent activation of several pathways, each contributing one aspect of the transformed phenotype. The evidence supporting one mechanism or another has been so far mostly circumstantial and indirect. The dissection of the properties of the Polyoma transforming genes, and those of the Myc, and Ras genes and Adenovirus EIA however has recently provided strong evidence in favor of at least two classes of transforming genes, each devoted to quite distinct tasks.

The first class includes the Polyoma middle T gene, the Ras gene(s), and the EIB gene(s) of Adenovirus (Rassoulzadegan et al., 1982, 1983; Land et al., 1983; Ruley, 1983). While not much is known about the Adeno EIB protein, both the Ras and middle T-Ag are membrane proteins, one (ras) possessing a protein kinase activity, the other being itself phosphorylated at tyrosine residues (probably by the associate c-src protein). Both proteins are essential for transformation, their presence being associated with growth in agar, tumorigenicity, and possibly the production of TGF's (Kaplan and Ozanne, 1982). Their main mechanism of action could be at the level of the plasma membrane, altering their lipids fluidity, permeability, or possibly the presence of activity of receptors for a variety of growth factors. This coupled with the frequent production of TGF by the same cells could be responsible for their altered growth control, and 'unresponsiveness' to stimuli which normally limit cell growth. The well-documented involvement of protein kinases in receptors–hormone interaction (Cohen, 1982) could be related to the action of protein kinases which are found to be intrinsic activities or associated with transforming proteins.

An alternative possibility, which applies however only to Polyoma middle T-Ag, is that its association with the c-src protein is essential for transformation, since it could lead to activation or modification of a protein involved in the control of cell proliferation (Courtneidge et al., 1984). While the hypothesis that the highly conserved c-src protein is involved in some aspects of cell proliferation is certainly tenable, this model does not answer the final question of what determines transformation and I consider it unlikely that the only function of middle T-Ag is to activate the c-src protein.

The changes brought about by membrane-associated proteins however are apparently not sufficient to originate a totally transformed phenotype in fully normal cells, but only in cell types which have already performed some step towards transformation, one of them that of having acquired an indefinite lifetime.

The second set of transforming genes would be responsible for these other changes, and include Polyoma large T-Ag (or its N-terminal half), the myc protein, the EIA protein(s) of

Adenovirus (Rassoulzadegan et al., 1982, 1983; Land et al., 1983; Ruley, 1983). The first two are nuclear, DNA binding proteins, and the EIA Adenovirus region as well as Polyoma large T-Ag can activate transcription of specific genes in trans (Green et al., 1983; Kern et al., 1986. They do not cause fibroblast transformation per se, but are capable at least in some cases of converting primary cells to an indefinite lifetime. They could act by activating cellular genes, resulting in the constitutive expression of a gene-product which is normally switched off during differentiation or senescence, and this could be achieved directly or by promoting gene rearrangements. Large T-Ag may also be involved in determining the decreased serum requirements of transformed cells. If this is the case they could have a regulatory effect on the distribution of receptors on the cell surface, or mimic some growth hormones which act on DNA. Alternatively these gene products could induce DNA synthesis directly.

The emerging picture, that of a transformed cell acquiring the ability to grow indefinitely under the influence of one transforming protein, and lacking the response to stimuli which would keep growth in check under another is attractive, since clearly tumor cells possess both these properties. However it is probably simplistic and almost certainly incomplete.

This model can be accommodated also to SV40. According to this view, nuclear SV40 large T-Ag would perform the same function as Polyoma large T-Ag, mainly immortalization and some evidence suggests that, like in Polyoma, this function is encoded in its N-terminal portion. In addition, SV40 large T-Antigen clearly is capable of activating transcription in trans (Keller and Alwine, 1984; Brady and Khoury 1985; Alwine, 1985). The function of middle T would be carried out by the membrane associated T-Ag, either in a manner similar to middle T or in a different fashion, and would be encoded yet in a different protein domain. This lack of specialization could account for SV40 being a less competent tumor virus than Polyoma, possibly more reliant on spontaneous cellular alterations, and yet paradoxically requiring fewer genes to perform its full action.

This oversimplified picture does not provide an essential role for small T-Ag, that both in Polyoma and SV40 clearly contributes to the transformed phenotype, the most likely way being that of altering cytoskeletal organization, thus decreasing cell adhesion, cell to cell contact etc., processes which may be very critical in certain cell types and not others. Other cellular changes may be required. Immortalization is certainly a simplistic term, and may not be all that is needed to make a cell susceptible to a single oncogene, as apparently some continuous cell lines cannot be transformed in one step by the transforming genes that encode membrane proteins. Clearly the challenge for the future is the elucidation of the precise mechanism by which transforming proteins act, and the seemingly simple system of Papovaviruses may in this context reserve more surprises and provide invaluable knowledge to the students of the basic mechanisms of carcinogenesis.

Acknowledgements—I wish to thank all of my colleagues who made available to me their manuscripts before publication and contributed much useful material. I am indebted to Drs C. Ceccarini, R. Rappuoli, G. Ratti and M. Rossini for helpful criticism of this manuscript. Finally I am particularly grateful to Ms Susan Weemys and Ms Vanna Pieri of the Sclavo Research Center, Siena, Italy for much help in editing and typing this manuscript.

REFERENCES

ACHESON, N. H. (1978) Polyoma giant RNAs contain tandem repeats of the nucleotide sequence of the entire viral genome. *Proc. natn. Acad. Sci. U.S.A.* **75**: 4754–4758.

ALONI, Y., DHAR, R., LAUB, O., HOROWITZ, M. and KHOURY, G. (1977) Novel mechanism for RNA maturation: the leader sequences of Simian virus 40 mRNA are not transcribed adjacent to the coding sequences. *Proc. natn. Acad. Sci. U.S.A.* **74**: 3686–3690.

ALWINE, J. C. (1985) Transient gene expression control: effects of transfected DNA stability and transactivation by viral early Proteins. *Mol. Cell. Biol.* **5**: 1034–1042.

ALWINE, J. C., REED, S. I. and STARK, G. R. (1977) Characterization of the autoregulation of Simian virus 40 gene A. *J. Virol.* **24**: 22–27.

ANDERSON, J. L. and MARTIN, R. G. (1976) SV40 transformation of mouse brain cells: critical role of gene A in maintenance of the transformed phenotype. *J. cell. Physiol.* **88**: 65–76.

ASSELIN, C., GELINAS, C. and BASTIN, M. (1983) Role of the three polyoma virus early proteins in tumorigenesis. *Molec. cell. Biol.* **3**: 1451–1459.

ASSELIN, C., GELINAS, C., BRANTON, P. E. and BASTIN, M. (1984) Polyoma middle T antigen requires collaboration from another gene to express the malignant phenotype *in vitro*. *Molec Cell. Biol.* **4**: 755–760.

BACHELER, L. T. (1977) Virus-specific transcription in 3T3 cells transformed by the tsA mutant of polyoma virus. *J. Virol.* **22**: 54–64.

BARAN, N., NEER, A. and MANOR, H. (1983) 'Onion skin' replication of integrated polyoma virus DNA and flanking sequences in polyoma-transformed rat cells: termination within a specific cellular DNA segment. *Proc. natn. Acad. Sci. U.S.A.* **80**: 105–109.

BASILICO, C., MATSUYA, Y. and GREEN, H. (1970) The interaction of polyoma virus with mouse–hamster somatic hybrid cells. *Virology* **41**: 295–305.

BASILICO, C. and ZOUZIAS, D. (1976) Regulation of viral transcription and tumor antigen expression in cells transformed by Simian virus 40. *Proc. natn. Acad. Sci. U.S.A.* **73**: 1931–1935.

BASILICO, C., GATTONI, S., ZOUZIAS, D. and DELLA VALLE, G. (1979) Loss of integrated viral DNA sequences in Polyoma transformed cells is associated with an active viral A function. *Cell* 645–659.

BASILICO, C., ZOUZIAS, D., DELLA VALLE, G., GATTONI, S., COLANTUONI, V., FENTON, R. and DAILEY, L. (1980) Integration and excision of polyoma virus genomes. *Cold Spr. Harb. Symp. quant. Biol.* **44**: 611–620.

BASILICO, C., DAILEY, L., PELLEGRINI, S., FENTON, R. G. and LA BELLA, F. (1983) Integration and expression of the polyoma virus oncogenes in transformed cells. In: *Application of Biological Markers to Carcinogen Testing*, pp. 441–452, Plenum Publishing Corporation, New York.

BENDER, M. A. and BROCKMAN, W. W. (1981) Rearrangement of integrated viral DNA sequences in mouse cells transformed by Simian virus 40'. *J. Virol.* **38**: 872–879.

BENDER, M. A., CHRISTENSEN, L. and BROCKMAN, W. W. (1983) Characterization of a T-antigen-negative revertant isolated from a mouse cell line which undergoes rearrangement of integrated Simian virus 40 DNA. *J. Virol.* **47**: 115–124.

BENJAMIN, T. L. (1966) Virus-specific RNA in cells productively infected or transformed by polyoma virus. *J. molec. Biol.* **16**: 359–373.

BENJAMIN, T. L. (1982) The hr-t gene of polyoma virus. *Biochim. biophys. Acta* **695**: 69–95.

BENJAMIN, T. L. and BURGER, A. M. (1970) Absence of a cell membrane alteration function in non-transforming mutants of polyoma virus. *Proc. natn. Acad. Sci. U.S.A.* **67**: 929–934.

BENOIST, C. and CHAMBON, P. (1981) *In vivo* sequence requirements for the SV40 early promotor region. *Nature* **290**: 304–309.

BERK, A. J. and SHARP, P. A. (1978) Spliced early mRNAs of Simian virus 40. *Proc. natn. Acad. Sci. U.S.A.* **75**: 1274–1278.

BIRG, F., DULBECCO, R., FRIED, M. and KAMEN, R. (1979) State and organization of polyoma virus DNA sequences in transformed rat cell lines. *J. Virol.* **29**: 633–648.

BJURSELL, G. (1978) Effects of 2′-deoxy-2′-azidocytidine on polyoma virus DNA replication: evidence for rolling circle-type mechanism. *J. Virol.* **26**: 136–142.

BOTCHAN, M., STRINGER, J., MITCHISON, R. and SAMBROOK, J. (1980) Integration and excision of SV40 DNA from the chromosome of a transformed cell. *Cell* 143–152.

BOTCHAN, M., TOPP, W. and SAMBROOK, J. (1976) The arrangement of Simian virus 40 sequences in the DNA of transformed cells. *Cell* **9**: 269–287.

BOTCHAN, M., TOPP, W. and SAMBROOK, J. (1979) Studies on SV40 excision from cellular chromosomes. *Cold Spr. Harb. Symp. quant. Biol.* **43**, 709–719.

BOUCK, N., BEALES, B., SHENK, T., BERG, P. and DI MAYORCA, G. (1978) New region of Simian virus 40 genome required for efficient viral transformation. *Proc. natn. Acad. Sci. U.S.A.* **75**: 2473–2477.

BOURACHOT, B., JOUANNEAU, J., GIRI, I., KATINKA, M., CEREGHINI, S. and YANIV, M. (1982) Both early and late control sequences of SV40 and polyoma promote transcription of *Escherichia coli* gpt gene in transfected cells. *EMBO J.* **1**: 895–900.

BOURGAUX, P. (1981) Murine polyoma virus transformation: integration and excision of the viral genome. *Can. J. Microbiol.* **27**: 559–562.

BOURGAUX, P., DELBECCHI, L., YU, K., HERRING, E. and BOURGAUX-RAMOISY, D. (1978) A mouse embryo cell line carrying an inducible, temperature-sensitive, polyoma virus genome. *Virology* **88**: 348–360.

BOURGAUX, P., SYLLA, B. and CHARTRAND, P. (1982) Excision of polyoma virus DNA from that of a transformed mouse cell: identification of a hybrid molecule with direct and inverted repeat sequences at the viral–cellular joints. *Virology* **122**: 84–97.

BRADY, J. and KHOURY, G. (1985) Transactivation of the SV40 late transcription unit by T-Antigen. *Mol. Cell. Biol.* **5**: 1391–1399.

BRICKELL, P., LATCHMAN, D., MURPHY, D., WILLISON, K. and RIGBY, P. (1983) Activation of a Qa/Tla class I major histocompatibility antigen gene is a general feature of oncogenesis in the mouse. *Nature* **306**: 756–760.

BROCKMAN, W. W. (1978) Transformation of Balb/c-3T3 cells by tsA mutants of Simian virus 40-temperature sensitivity of transformed phenotype and retransformation of wild-type virus. *J. Virol.* **25**: 860–870.

BRUGGE, J. S. and BUTEL, J. S. (1975) Role of Simian virus 40 gene A function in maintenance of transformation. *J. Virol.* **15**: 619–635.

BULLOCK, P. and BOTCHAN, M. (1982). Molecular events in the excision of SV40 DNA from the chromosomes of cultured mammalian cells. In: *Gene Amplification*, pp. 215–224, SCHIMKE, R. T. (ed.) Cold Spring Harbor Lab., New York.

BURGER, M. M. (1969) A difference in the architecture of the surface membrane of normal and virally transformed cells. *Proc. natn. Acad. Sci.* **62**: 994–1001.

CAMPO, M. S., CAMERON, I. R. and ROGERS, M. E. (1979) Tandem integration of complete and defective SV40 genomes in mouse–human somatic-cell hybrids. *Cell* **15**: 1411–1426.

CARMICHAEL, G., SCHAFFHAUSEN, B. S., DORSKY, D. I., OLIVER, D. B. and BENJAMIN, T. (1982) Carboxy terminus of polyoma middle-sized tumor antigen is required for attachment to membranes, associated protein kinase activities and cell transformation. *Proc. natn. Acad. Sci. U.S.A.* **79**: 3579–3583.

CARROLL, R. B., HAGER, L. and DULBECCO, R. (1974) Simian virus 40 T-antigen binds to DNA. *Proc. natn. Acad. Sci. U.S.A.* **71**: 3754–3757.

CEREGHINI, S., HERBOMEL, P., JOUANNEAU, J., SARAGOSTI, S., KATINKA, M., BOURACHOT, B., DE CROMBRUGGHE, B. and YANIV, M. (1983) Structure and function of the promoter-enhancer region of polyoma and SV40. *Cold Spr. Harb. Symp. quant. Biol.* **47**: 935–944.

CHANG, C., MARTIN, R. G., LIVINGSTON, D. M., LUBORSKY, S. W., HU, C-P. and MORA, P. T. (1979) Relationship between T-antigen and tumor-specific transplantation antigen in Simian virus 40-transformed cells. *J. Virol.* **29**: 69–75.

CHANG, L. S., PATER, M., HUTCHINSON and DI MAYORCA, G. (1984) Transformation by purified early genes of SV40. *Virology* **133**: 341–353.

CHAUDRY, F., HARVEY, R. and SMITH, A. E. (1982) Structure and biochemical functions of four Simian virus 40 truncated large T-antigens. *J. Virol.* **44**: 54–66.

CHEN, S., BLANCK, G. and POLLACK, R. (1984) Reacquisition of a functional early region by a mouse transformant containing only defective Simian virus 40 DNA. *Molec. cell. Biol.* **4**: 666–670.

CHEN, S., VERDERAME, M., LO, A. and POLLACK, R. (1981) Nonlytic Simian virus 40-specific 100K phosphoprotein is associated with anchorage-independent growth in Simian virus 40-transformed and revertant mouse cell lines. *Molec. cell. Biol.* **1**: 994–1004.

CHIA, W. and RIGBY, P. W. (1981) Fate of viral DNA in nonpermissive cells infected with Simian virus 40. *Proc. natn. Acad. Sci. U.S.A.* **78**: 6638–6642.

CHOU, J. Y. and MARTIN, R. G. (1975) DNA infectivity and the induction of host DNA synthesis with temperature-sensitive mutants of Simian virus 40. *J. Virol.* **15**: 145–150.

CHOWDHURY, K., GARON, C. F. and ISRAEL, M. A. (1984) Structural analysis of integrated polyoma viral DNA in a polyoma-induced hamster tumor cell line. *J. Virol.*, in press.

CHOWDHURY, K., LIGHT, S. E., GARON, C. F., ITO, Y. and ISRAEL, M. A. (1980) A cloned polyoma DNA fragment representing the 5' half of the early gene region is oncogenic. *J. Virol.* **36**: 566–574.

CLAYTON, C. E. and RIGBY, P. W. J. (1981) Cloning and characterization of the integrated viral DNA from three lines of SV40-transformed mouse cells. *Cell* **25**: 547–559.

CLAYTON, C. E., LOVETT, M. and RIGBY, P. W. J. (1982a) Functional analysis of a Simian virus 40 super T-antigen. *J. Virol.* **44**: 974–982.

CLAYTON, C. E., MURPHY, D., LOVETT, M. and RIGBY, P. W. J. (1982b) A fragment of the SV40 large T-antigen gene transforms. *Nature* **299**: 59–61.

COGEN, B. (1978) Virus-specific early RNA in 3T6 cells infected by a ts mutant of polyoma virus. *Virology* **85**: 222–230.

COHEN, P. (1982) The role of protein phosphorylation in neural and hormonal control of cellular activity. *Nature* **296**: 613–620.

COLANTUONI, V., DAILEY, L. and BASILICO, C. (1980) Amplification of integrated viral DNA sequences in polyoma virus transformed cells. *Proc. natn. Acad. Sci. U.S.A.* **77**: 3850–3854.

COLANTUONI, V., DAILEY, L., DELLA VALLE, G. and BASILICO, C. (1982) Requirements for excision and amplification of integrated viral DNA molecules in polyoma virus-transformed cells. *J. Virol.* **43**: 617–628.

COLBERE-GARAPIN, F., HORODNICEANU, F., KOURILSKY, P. and GARAPIN, A. C. (1981) A new dominant hybrid selective marker for higher eukaryotic cells. *J. molec. Biol.* **150**: 1–14.

COLBY, W. W. and SHENK, T. (1928) Fragments of the Simian virus 40 transforming gene facilitate transformation of rat embryo cells. *Proc. natn. Acad. Sci. U.S.A.* **79**: 5189–5193.

COLLETT, M. S. and ERIKSON, R. L. (1978) Protein kinase activity associated with the avian sarcoma virus src gene product. *Prod. natn. Acad. Sci. U.S.A.* **75**: 2021–2024.

CONRAD, S. E., LIU, C. and BOTCHAN, M. R. (1982) Fragment spanning the SV40 replication origin is the only DNA sequence required in cis for viral excision. *Science* **218**: 1223–1225.

COURTNEIDGE, S. A. and SMITH, A. (1983) Polyoma virus transforming protein associates with the product of the c-src cellular gene. *Nature* **303**: 435–439.

COURTNEIDGE, S. A., OOSTRA, B. and SMITH, A. E. (1984) Tyrosine phosphorylation and polyoma virus middle T. *Cold Spr. Harb. Conf. Cell Proliferation and Cancer* **11**: in press.

CRAWFORD, L. (1983) The 53,000 dalton cellular protein and its role in transformation. *Int. Rev. exp. Path.* **25**: 1–50.

CRAWFORD, L. V., COLE, C. N., SMITH, A. E., PAUCHA, E., TEGTMEYER, P., RUNDELL, K. and BERG, P. (1978) Organization and expression of early genes of Simian virus 40. *Proc. natn. Acad. U.S.A.* **75**: 117–121.

CROCE, C. M. and KOPROWSKI, H. (1975) Assignment of gene(s) for cell transformation to human chromosome 7 carrying the Simian virus 40 genome. *Proc. natn. Acad. Sci. U.S.A.* **72**: 1658–1660.

CULP, L. A. and BLACK, P. H. (1972) Contact-inhibited revertant cell lines isolated from Simian virus 40 transformed cells. III. Concanavalin-A selected revertant cells. *J. Virol.* **9**: 611–620.

DAILEY, L., COLANTUONI, V., FENTON, R. G., LA BELLA, F., ZOUZIAS, D., GATTONI, S. and BASILICO, C. (1982) Evolution of polyoma transformed rat cell lines during propagation *in vitro*. *Virology* **116**: 207–220.

DAILEY, L., PELLEGRINI, S. and BASILICO, C. (1984) Deletion of the origin of replication impairs the ability of polyoma virus DNA to transform cells and form tandem insertions. *J. Virol.* **49**: 984–987.

DELLA VALLE, G., FENTON, R. G. and BASILICO, C. (1981) Polyoma large T antigen regulates the integration of viral DNA sequences into the genome of transformed cells. *Cell* **23**: 347–355.

DELARCO, J. E. and TODARO, G. J. (1978) Growth factors from murine sarcoma virus-transformed cells. *Proc. natn. Acad. Sci. U.S.A.* **75**: 4001–4005.

DE PAMPHILIS, M. and WASSARMAN, P. R. (1982) Organization and replication of papovavirus DNA. In: *Organization and Replication of Viral DNA*, pp. 37–114, KAPLAN, A. S. (ed.) CRC Press.

DEPPERT, W., HANKE, K. and HENNING, R. (1980) Simian virus 40 T-antigen-related cell surface antigen: serological demonstration on Simian virus 40-transformed monolayer cells *in situ*. *J. Virol.* **35**: 505–518.

DI MAYORCA, G., CALLENDER, J., MARIN, G. and GIORDANO, R. (1969) Temperature-sensitive mutants of polyoma virus. *Virology* **38**: 126–133.

ECKHART, W. (1969) Complementation and transformation by temperature-sensitive mutants of polyoma virus. *Virology* **38**: 120–125.
ECKHART, W., HUTCHINSON, A. and HUNTER, T. (1979) An activity phosphorylating tyrosine in polyoma T antigen immunoprecipitates. *Cell* **18**: 925–933.
EDDY, B. E., BORMAN, G. S., GRUBBS, G. E. and YOUNG, R. D. (1962) Identification of the oncogenic substance in rhesus monkey kidney cell cultures as Simian virus 40. *Virology* **17**: 65–75.
ELLMAN, M., BIKEL, J., FIGGE, J., ROBERTS, T., SCHLOSSMAN, R. and LIVINGSTON, D. M. (1984) Localization of the SV40 small T-Antigen in the nucleus and cytoplasm of monkey and mouse cells. *J. Virol.* **50**: 623–628.
FENTON, R. G. and BASILICO, C. (1981) Viral gene expression in polyoma transformed rat cells and their cured revertants. *J. Virol.* **40**: 150–163.
FENTON, R. G. and BASILICO, C. (1982a) Changes in the topography of early region transcription during polyoma virus lytic infection. *Proc. natn. Acad. Sci. U.S.A.* **79**: 7142–7146.
FENTON, R. G. and BASILICO, C. (1982b) Regulation of Polyoma virus early transcription in transformed cells by large T-Antigen. *Virology* **121**: 384–392.
FIERS, W., CONTRERAS, R., HAEGEMAN, G., ROGIERS, R., VAN DE VOORDE, A., VAN HEUVERSWYN, H., VAN HERREWEGHE, J., VOLCKAERT, G. and YSEBAERT, M. (1978) Complete nucleotide sequence of SV40 DNA. *Nature* **273**: 113–120.
FISCHER-FANTUZZI, L. and VESCO, C. (1985) Deletion of 43 amino acids in the N-terminal half of the large T-Ag of SV40 results in a non-caryophylic protein capable of transforming established cells. *Proc. natn. Acad. Sci. U.S.A.* **82**: 1891–1895.
FITZGERALD, M. and SHENK, T. (1980) The site at which late mRNAs are polyadenylated is altered in SV40 mutant d1882. *Ann. N.Y. Acad. Sci.* **354**: 53–59.
FLAVELL, A. J., COWIE, A., ARRAND, J. R. and KAMEN, R. (1980) Localization of three major capped 5'-ends of polyoma virus late mRNAs within a single tetranucleotide sequence in the viral genome. *J. Virol.* **33**: 902–908.
FLAVELL, A. J., COWIE, A., LEGON, S. and KAMEN, R. (1979) Multiple 5'-terminal cap structures in late polyoma virus RNA. *Cell* **16**: 357–371.
FLUCK, M. M. and BENJAMIN, T. L. (1979) Comparisons of two early gene functions essential for transformation in polyoma virus and SV40. *Virology* **96**: 205–228.
FLUCK, M. M., SHAIKH, R. and BENJAMIN, T. L. (1983) An analysis of transformed clones obtained by co-infections with Hr-t and Ts-a mutants of polyoma virus. *Virology* **130**: 29–43.
FLYER, D. C. and TEVETHIA, S. S. (1982) Biology of Simian virus 40 (SV40) transplantation antigen (TrAg). VIII. Retention of SV40 TrAg sites on purified SV40 large T antigen following denaturation with sodium dodecyl sulfate. *Virology* **117**: 267–270.
FRANCKE, B. and ECKHART, W. (1973) Polyoma gene function required for viral DNA synthesis. *Virology* **55**: 127–135.
FRIED, M. (1965) Cell transforming ability of temperature-sensitive mutants of polyoma virus. *Proc. natn. Acad. Sci. U.S.A.* **53**: 486–491.
FRIED, M. A. (1970) Characterization of a temperature-sensitive mutant of polyoma virus. *Virology* **40**: 605–617.
FRIED, M. and GRIFFIN, B. E. (1977) Organization of the genomes of polyoma virus and SV40. *Adv. Cancer Res.* **24**: 67–113.
FRIED, M., GRIFFITHS, M., DAVIES, B., BJURSELL, G., LA MANTIA, G. and LANIA, L. (1983) Isolation of cellular DNA sequences that allow expression of adjacent genes. *Proc. natn. Acad. Sci. U.S.A.* **80**: 2117–2121.
GALANTI, N., JONAK, G. J., SOPRANO, K. J., FLOROS, J., KACZMAREK, L., WEISSMAN, S., REDDY, V. B., TILGHMAN, S. M. and BASERGA, R. (1981) Characterization and biological activity of cloned Simian virus 40 DNA fragments. *J. biol. Chem.* **256**: 6469–6474.
GATTONI, S., COLANTUONI, V. and BASILICO, C. (1980) Relationship between integrated and nonintegrated viral DNA in rat cells transformed by polyoma virus. *J. Virol.* **34**: 615–626.
GAUDRAY, P., TYNDALL, C., KAMEN, R. and CUZIN, F. (1981) The high affinity binding site on polyoma virus DNA for the viral large-T protein. *Nucleic Acids Res.* **9**: 5697–5710.
GELB, L. D., KOHNE, D. E. and MARTIN, M. A. (1971) Quantitation of Simian virus 40 sequences in African green monkey, mouse and virus-transformed cell genomes. *J. molec. Biol.* **57**: 129–145.
GHOSH, P. K. and LEBOWITZ, P. (1981) SV-40 early mRNAs contain multiple 5'-termini upstream and downstream from a Hogness–Goldberg sequence. A shift in 5'-termini during the lytic cycle is mediated by large T-antigen. *J. Virol.* **40**: 224–240.
GIRARDI, A. J., SWEET, B. H., SLOTNICK, V. B. and HILLEMAN, M. R. (1962) Development of tumors in hamsters inoculated in the neonatal period with vacuolating virus, SV40. *Proc. Soc. exp. Biol. Med.* **109**: 649–660.
GLUZMAN, Y. (1981) SV40-transformed Simian cell support the replication of early SV40 mutants. *Cell* **23**: 175–182.
GLUZMAN, Y. and AHRENS, B. (1982) SV40 early mutants that are defective for viral DNA synthesis but competent for transformation of cultured rat and Simian cells. *Virology* **123**: 78–92.
GLUZMAN, Y., FRISQUE, R. J. and SAMBROOK, J. (1980) Origin defective mutants of SV40. *Cold Spr. Harb. Symp. quant. Biol.* **44**, 293–300.
GOODING, L. R., GEIB, R. W. O., CONNELL, K. A. and HARLOW, E. (1984) Antibody and cellular detection of SV40 T-antigenic determinants on the surfaces of transformed cells. *Cancer Cells* **1**: 263–270.
GRAESSMANN, M., GRAESSMANN, A. and MUELLER, C. (1980) Monkey cells transformed by SV40 DNA fragments: flat revertants synthesize large and small T antigens. *Cold Spr. Harb. Symp. quant. Biol.* **44**: 605–610.
GREEN, M. R., TREISMAN, R. and MANIATIS, T. (1983) Transcriptional activation of cloned human β-globin genes by viral immediate-early gene products. *Cell* **35**: 137–148.
GREENSPAN, D. S. and CARROLL, R. B. (1981) Complex of Simian virus 40 large tumor antigen and 48,000-dalton host tumor antigen. *Proc. natn. Acad. Sci. U.S.A.* **78**: 105–109.
GRIFFIN, B. E. and DILWORTH, S. M. (1983) Polyoma virus: an overview of its unique properties. *Adv. Cancer Res.* **39**: 183–268.
GROSS, L. (1953) A filterable agent, recovered from Ak leukemic extracts, causing salivary gland carcinomas in C3H mice. *Proc. Soc. exp. Biol. Med.* **83**: 414–421.

Gruss, P., Dhar, R. and Khoury, G. (1981) Simian virus 40 tandem repeated sequences as an element of the early promoter. *Proc. natn. Acad. Sci. U.S.A.* **78**: 943–947.

Hansen, U., Tenen, D. G., Livingston, D. M. and Sharp, P. A. (1981) T antigen repression of SV40 early transcription from two promoters. *Cell* 603–612.

Hassell, J. A., Topp, W. C., Rifkin, D. B. and Moreau, P. (1980) Transformation of rat embryo fibroblasts by cloned polyoma virus DNA fragments containing only part of the early region. *Proc. natn. Acad. Sci. U.S.A.* **77**: 3978–3982.

Hay, N., Skolnik-David, H. and Aloni, Y. (1982) Attenuation in the control of SV40 gene expression. *Cell* **29**: 183–193.

Hayday, A., Chaudry, F. and Fried, M. (1983) Loss of polyoma virus infectivity as a result of a single amino acid change in a region of polyoma virus large T antigen when it has an extensive amino acid homology and Simian virus 40 large T antigen. *J. Virol.* **45**: 693–699.

Hayday, A., Ruley, H. E. and Fried, M. (1982) Structural and biological analysis of integrated polyoma virus DNA and its adjacent host sequences cloned from transformed rat cells. *J. Virol.* **44**, 67–77.

Henning, R., Lange-Mutschler, J. and Deppert, W. (1981) SV40-transformed cells express SV40 T antigen-related antigens on the cell surface. *Virology* **108**: 325–337.

Hiscott, J., Murphy, D. and Defendi, V. (1980) Amplification and rearrangement of integrated SV40 DNA sequences accompany the selection of anchorage-independent transformed mouse cells. *Cell* **22**: 535–543.

Hoggan, M. D., Rowe, W. P., Black, P. H. and Huebner, R. J. (1965) Production of 'tumor-specific' antigens by oncogenic viruses during acute cytolytic infections. *Proc. natn. Acad. Sci. U.S.A.* **53**: 12–19.

Hunter, T., Hutchinson, M. A. and Eckhart, W. (1978) Translation of polyoma virus T antigens *in vitro*. *Proc. natn. Acad. Sci. U.S.A.* **75**: 5917–5921.

Hunter, T., Hutchinson, M. A. and Eckhart, W. (1984) Polyoma middle-sized T antigen can be phosphorylated on tyrosine at multiple sites *in vitro*. *EMBO J.* **3**: 73–79.

Hutchinson, M. A., Hunter, T. and Eckhardt, W. (1978) Characterization of T antigens in polyoma-infected and transformed cells. *Cell* **15**: 65–77.

Israel, M. A., Vanderryn, D. H., Meltzer, M. L. and Martin, M. A. (1980) Characterization of polyoma viral DNA sequences in polyoma-induced hamster tumor cell lines. *J. biol. Chem.* **255**: 3798–3805.

Ito, Y. (1980) Organization and expression of the genome of polyoma virus. In: *Viral Oncology*, pp. 447–480, Klein G. (ed.) Raven Press, New York.

Ito, Y., Brocklehurst, J. R. and Dulbecco, R. (1977) Virus-specific proteins in the plasma membrane of cells lytically infected or transformed by polyoma virus. *Proc. natn. Acad. Sci. U.S.A.* **74**: 4666–4670.

Ito, Y., Spurr, N. and Griffin, B. (1980) Middle T antigen as primary inducer of full expression of the phenotype of transformation by polyoma virus. *J. Virol.* **35**: 219–232.

Jay, G., Nomura, S., Anderson, C. W. and Khoury, G. (1981) Identification of the SV40 agnogene product: a DNA binding protein. *Nature* **291**: 346–349.

Jessel, D., Hudson, T., Landau, T., Tenen, D. and Livingston, D. M. (1975) Interaction of partially purified SV40 T antigen with circular viral DNA molecules. *Proc. natn. Acad. Sci. U.S.A.* **72**: 1960–1964.

Jessel, D., Laundau, T., Hudson, J., Lalor, T., Tenen, D. and Livingston, M. (1976) Identification of regions of the SV40 genome which contain preferred SV40 T antigen-binding sites. *Cell* **8**: 535–545.

Kamen, R., Favaloro, J., Parker, J., Treisman, R., Lania, L., Fried, M. and Mellor, A. (1980) A comparison of polyoma virus transcription in productively infected cells and transformed rodent cell lines. *Cold Spr. Harb. Symp. quant. Biol.* **44**: 63–75.

Kaplan, P. L. and Ozanne, B. (1982) Polyoma virus-transformed cells produce transforming growth factor(s) and grow in serum-free medium. *Virology* **123**: 372–380.

Keller, J. and Alwine, J. (1984) Activation of the Simian virus 40 late promoter: direct effects of T antigen in the absence of viral DNA replication. *Cell* **36**: 381–389.

Kelly, F. and Sambrook, J. (1975) Variants of SV40-transformed mouse cells resistant to cytochalasin B. *Cold Spr. Harb. Symp. quant. Biol.* **39**: 345–353.

Kern, F. G. and Basilico C. (1985) Transcription from the polyoma late promoter in cells stably transformed by chimeric plasmids. *Mol. Cell. Biol.* **5**: 797–807.

Kern, F. G., Pellegrini, S., Cowie, A. and Basilco, C. (1986). Regulation of polymerims late promoter activity by viral early proteins *J. Virol.* **60**: 275–285.

Kettner, G. and Kelly, T. J., Jr. (1976) Integration of SV40 sequences in transformed cell DNA: analysis using restriction endonucleases. *Proc. natn. Acad. Sci. U.S.A.* **73**: 1102–1107.

Khoury, G. and Gruss, P. (1983) Enhancer elements. *Cell* **33**: 313–314.

Khoury, G. and May, E. (1977) Regulation of early and late Simian virus 40 transcription: overproduction of early viral RNA in the absence of a functional T-antigen. *J. Virol.* **23**: 167–176.

Koprowski, H., Jensen, F. C. and Steplewski, Z. S. (1967) Activation of production of infectious tumor virus SV40 in Heterokaryon cultures. *Proc. natn. Acad. Sci. U.S.A.* **58**: 127–133.

La Bella, F., Brown, E. H. and Basilico, C. (1983) Changes in the levels of viral cellular gene-transcripts in the cell cycle of SV40-transformed mouse cells. *J. cell. Physiol.* **117**: 62–68.

Laimins, L. A., Khoury, G., Gorman, C., Howard, B. and Gruss, P. (1982) Host-specific activation of transcription by tandem repeats from Simian virus 40 and Moloney murine sarcoma virus. *Proc. natn. Acad. Sci. U.S.A.* **79**: 6453–6457.

Land, H., Parada, L. and Weinberg, R. (1983) Tumorigenic conversion of primary embryo fibroblasts requires at least two cooperating oncogenes. *Nature* **304**: 596–601.

Lane, D. P. and Crawford, L. V. (1979) T antigen is bound to a host protein in SV40-transformed cells. *Nature* **278**: 261–263.

Lane, D. P., Gannon, J. and Winchester, G. (1982) The complex between p53 and SV40 T antigen. *Adv. Viral Oncol.* **2**: 23–39.

Lanford, R. E. and Butel, J. S. (1984) Construction and characterization of an SV40 mutant defective in nuclear transport of T-antigen. *Cell* **37**: 801–813.

Lange, M., May, E. and May, P. (1981) Ability of non-permissive mouse cell to express a Simian virus 40 late function(s). *J. Virol.* **38**: 940–951.

LANGE, M. and MAY, E. (1983) Evidence of transcription from the late region of the integrated Simian virus 40 genome in transformed cells: location of the 5' ends of late transcripts in cells abortively infected and in cells transformed by Simian virus 40. *J. Virol.* **46**: 756–767.

LANIA, L., GRIFFITHS, M., COOKE, B., ITO, Y. and FRIED, M. (1979) Untransformed rat cells containing free and integrated DNA of polyoma non-transforming mutant. *Cell* **18**: 793–802.

LANIA, L., GRIFFITHS, M., COOKE, B., ITO, Y. and FRIED, M. (1980) The polyoma virus 100K large T antigen is not required for the maintenance of transformation. *Virology* **101**: 217–232.

LANIA, L., HAYDAY, A. and FRIED, M. (1981) The loss of functional large T-antigen and free viral genomes from cells transformed *in vitro* by polyoma virus after passage *in vivo* as tumor cells. *J. Virol.* **39**: 422–431.

LEGON, S., FLAVELL, A. J., COWIE, A. and KAMEN, R. (1979) Amplification in the leader sequence of late polyoma virus mRNAs. *Cell* **16**: 373–388.

LEVINSON, A. D., OPPERMANN, H., LEVINTOW, L., VARMUS, H. and BISHOP, J. M. (1978) Evidence that the transforming gene of avian sarcoma virus encodes a protein kinase associated with a phosphoprotein. *Cell* **15**: 561–572.

LEVINSON, B., KHOURY, G., VANDE WOUDE, G. and GRUSS, P. (1982) Activation of SV40 genome by 72-base pair tandem repeats of Moloney sarcoma virus. *Nature* **295**: 568–572.

LEWIS, A. M., JR. and MARTIN, R. G. (1979) The oncogenicity of Simian virus 40 deletion mutants that induce altered 17K t-proteins. *Proc. natn. Acad. Sci. U.S.A.* **76**: 4299–4302.

LIBOI, E. and BASILICO, C. (1984) Inhibition of polyoma gene expression in transformed mouse cells by hypermethylation. *Virology* **135**: 440–451.

LINZER, D. I. H. and LEVINE, A. J. (1979) Characterization of a 54K dalton cellular SV40 tumor antigen present in SV40-transformed cells and uninfected embryonal carcinoma cells. *Cell* **17**: 43–52.

LOVETT, M., CLAYTON, C. E., MURPHY, D., RIGBY, P. W. J., SMITH, A. E. and CHAUDRY, F. (1982) Structure and synthesis of a Simian virus 40 super T-antigen. *J. Virol.* **44**: 963–973.

MANOR, H. and KAMEN, R. (1978) Polyoma virus-specific RNA synthesis in an inducible line of polyoma virus-transformed rat cells. *J. Virol.* **25**: 719–729.

MANOS, M. M. and GLUZMAN, Y. (1984) Simian virus 40 large T-antigen point mutants that are defective in viral DNA replication but competent in oncogenic transformation. *Molec. cell. Biol.* in press.

MARTIN, R. G. (1981) The transformation of cell growth and transmogrification of DNA synthesis by Simian virus 40. *Adv. Cancer Res.* **34**: 1–68.

MARTIN, R. G. and CHOU, J. Y. (1975) Simian virus 40 functions required for the establishment and maintenance of malignant transformation. *J. Virol.* **15**: 599–612.

MARTIN, R. G., STELOW, V. P., EDWARDS, C. A. F. and VEMBU, D. (1979) The roles of the Simian virus 40 tumor antigens in transformation of Chinese hamster lung cells. *Cell* **17**: 635–643.

MAY, P., KRESS, M., LANGE, M. and MAY, E. (1980) New genetic information expressed in SV40-transformed cells: characterization of the 55K proteins and evidence for unusual SV40 mRNAs. *Cold Spr. Harb. Symp. quant. Biol.* **44**: 189–200.

MAY, E., KRESS, L., DAYA-GROSJEAN, L., MONIER, R. and MAY, P. (1981) Mapping of the viral mRNA encoding a super-T-antigen of 115,000 daltons expressed in Simian virus 40-transformed rat cell lines. *J. Virol.* **37**: 24–35.

MAY, P., MAY, E. and BORDÉ, J. (1976) Stimulation of cellular RNA synthesis in mouse-kidney cell cultures infested with SV40 virus. *Exp. Cell Res.* **100**: 433–436.

MCCORMICK, K. (1982) Oncogenic viruses in the nude mouse. In: *The Nude Mouse in Experimental and Clinical Research*, Vol. 2, pp. 39–66, FOGN J. and GIOVANNELLA, B. C. (eds).

MENDELSOHN, E., BARAN, E., NEER, A. and MANOR, H. (1982) Integration site of polyoma virus DNA in the inducible LPT line of polyoma-transformed rat cells. *J. Virol.* **41**: 192–209.

MORA, P. T. (1982) The immunopathology of SV40-induced transformation. *Springer Semin. Immunopath.* **5**: 7–32.

MORRISON, B., KRESS, M., KHOURY, G. and JAY, G. (1983) Simian virus 40 tumor antigen: isolation of the origin-specific DNA-binding domain. *J. Virol.* **47**: 106–114.

MUELLER, C., GRAESSMANN, A. and GRAESSMANN, M. (1978) Mapping of early SV40-specific functions by microinjection of different early viral DNA fragments. *Cell* **15**: 579–585.

NEER, A., BARAN, B. and MANOR, H. (1983) Integration of polyoma virus DNA into chromosomal DNA in transformed rat cells causes deletion of flanking cell sequences. *J. gen. Virol.* **64**: 69–82.

NOMURA, S., KHOURY, G. and JAY, G. (1983) Subcellular localization of the Simian virus 40 agnoprotein. *J. Virol.* **45**: 428–433.

NOONAN, C. A., BRUGGE, J. S. and BUTEL, J. S. (1976) Characterization of Simian cells transformed by temperature-sensitive mutants of Simian virus 40. *J. Virol.* **18**: 1106–1119.

NOVAK, Y., DILWORTH, S. M. and GRIFFIN, B. E. (1980) Coding capacity of a 35% fragment of the polyoma virus genome is sufficient to initiate and maintain cellular transformation. *Proc. natn. Acad. Sci. U.S.A.* **77**: 3278–3282.

ODA, K. and DULBECCO, R. (1968) Induction of cellular mRNA synthesis in BSC-1 cells infected by Simian virus 40. *Virology* **35**: 439–444.

O'NEILL, F. J., COHEN, S. and RENZETTI, L. (1980) Temperature dependency for maintenance of transformation in mouse cells transformed by Simian virus 40 tsA mutants. *J. Virol.* **35**: 233–245.

OREN, M., MALTZMAN, W. and LEVINE, A. J. (1981) Post-translational regulation of the 54K cellular tumor antigen in normal and transformed cells. *Molec. cell. Biol.* **1**: 101–110.

OSBORN, M. and WEBER, K. (1975) Simian virus 40 gene A function and maintenance of transformation. *J. Virol.* **15**: 636–644.

OZANNE, B. (1973) Variants of Simian virus 40-transformed 3T3 cells that are resistant to concanavalin A. *J. Virol.* **12**, 79–89.

OZANNE, B., SHARP, P. A. and SAMBROOK, J. (1973) Transcription of Simian virus 40. II. Hybridization of RNA extracted from different lines of transformed cells to the separated strands of Simian virus 40 DNA. *J. Virol.* **12**: 90–98.

Pellegrini, S., Dailey, L. and Basilico, C. (1984) Amplification and excision of integrated polyoma DNA sequences require a functional origin of replication. *Cell* **36**: 943–949.

Pollack, R. and Vogel, A. (1973) Isolation and characterization of revertant cell lines. II Growth control of a polyploid revertant line derived from SV40-transformed 3T3 mouse cells. *J. cell. Physiol.* **82**: 93–100.

Pollack, R. E., Green, H. and Todaro, G. J. (1968) Growth control in cultured cells. Selection of sublines with increased sensitivity to contact inhibition and decreased tumor-producing ability. *Proc. natn. Acad. Sci. U.S.A.* **60**: 126–133.

Pollack, R. E., Osborn, M. and Weber, K. (1975) Patterns of organization of actin and myosin in normal and transformed cultured cells. *Proc. natn. Acad. Sci. U.S.A.* **72**: 994–998.

Prasad, I., Zouzias, D. and Basilico, C. (1976) State of the viral DNA in rat cells transformed by polyoma virus. I. Virus rescue and the presence of nonintegrated viral DNA molecules. *J. Virol.* **18**: 436–444.

Prives, C. L., Barnet, B., Scheller, A., Khoury, G. and Jay, G. (1982) Discrete regions of Simian virus 40 large T-antigen are required for nonspecific and viral origin-specific DNA binding. *J. Virol.* **43**: 73–82.

Prives, C., Covey, L., Scheller, A. and Gluzman, Y. (1983) DNA-binding properties of Simian virus 40 T-antigen mutants defective in viral DNA replication. *Molec. cell. Biol.* **3**: 1958–1966.

Rabinowitz, Z. and Sachs, L. (1968) Reversion of properties in cells transformed by polyoma virus. *Nature* **220**: 1203–1206.

Rassoulzadegan, M., Seif, R. and Cuzin, F. (1978) Conditions leading to establishment of N (A gene dependent) and A (A gene independent) transformed states after polyoma virus infection of rat fibroblasts. *J. Virol.* **28**: 421–426.

Rassoulzadegan, M., Cowie, A., Carr, A., Glaichenhaus, N., Kamen, R. and Cuzin, F. (1982) The roles of individual polyoma virus early proteins in oncogenic transformation. *Nature* **300**: 713–718.

Rassoulzadegan, M., Naghashfar, Z., Cowie, A., Carr, A., Grisoni, M., Kamen R. and Cuzin, F. (1983) Expression of the large T protein of polyoma virus promotes the establishment in culture of 'normal' rodent fibroblast cell lines. *Proc. natn. Acad. Sci. U.S.A.* **80**: 4354–4358.

Reddy, V. B., Tevethia, M. J., Tevethia, S. S. and Weissman, S. M. (1982) Molecular dissection of MHC complex and of SV40 induced surface antigen. *Ann. N.Y. Acad. Sci.* **397**: 229–237.

Reed, S. I., Ferguson, J., Davis, R. W. and Stark, G. R. (1975) T-Antigen binds to Simian virus to 40 DNA at the origin of replication. *Proc. natn. Acad. Sci. U.S.A.* **72**: 1605–1609.

Reed, S. I., Stark, G. R. and Alwine, J. C. (1976) Autoregulation of Simian virus 40 gene A by T antigen. *Proc. natn. Acad. Sci. U.S.A.* **73**: 3083–3087.

Renger, H. (1972) Retransformation of temperature-sensitive SV40 transformed line at nonpermissive temperature. *Nat. New Biol.* **240**: 19–21.

Renger, H. C. and Basilico, C. (1972) Mutation causing temperature-sensitive expression of cell transformation by a tumor virus. *Proc. natn. Acad. Sci. U.S.A.* **69**: 109–114.

Renger, H. C. and Basilico, C. (1973) Temperature-sensitive Simian virus 40-transformed cells. Phenomena accompanying transition from transformed to normal state. *J. Virol.* **11**: 702–708.

Rigby, P. W. J. and Lane, D. P. (1983) The structure and function of the Simian virus 40 large T-antigen. *Adv. Viral Oncology* **3**: 31–57.

Rio, D., Robbins, A., Myers, R. and Tjian, R. (1980) Regulation of Simian virus 40. Early transcription *in vitro* by a purified tumor antigen. *Proc. natn. Acad. Sci. U.S.A.* **77**: 5705–5710.

Robberson, D. L., Crawford, L. V., Syrett, C. and James, A. W. (1976) Unidirectional replication of a minority of polyoma virus and SV40 DNAs. *J. gen. Virol.* **26**: 56–69.

Rubin, H., Figge, J., Bladon, M. T., Chen, L. B., Ellman, M., Bikel, I., Farrell, M. and Livingston, D. M. (1982) Role of small T-antigen in the acute transforming activity of SV40. *Cell* **30**: 469–480.

Ruley, H. E. (1983) Adenovirus early region 1A enables viral and cellular transforming genes to transform primary cells in culture. *Nature* **304**: 602–606.

Sager, R., Anisowicz, A. and Howell, N. (1981) Genomic rearrangements in a mouse cell line containing integrated SV40 DNA. *Cell* **23**: 41–50.

Salomon, C., Türler, H. and Weil, R. (1977) Polyoma-induced stimulation of cellular RNA synthesis is paralleled by changed expression of the viral genome. *Nucleic Acids Res.* **4**: 1483–1503.

Sambrook, J., Westphal, H., Srinivasan, P. R. and Dulbecco, R. (1968) The integrated state of viral DNA in SV40-transformed cells. *Proc. natn. Acad. Sci. U.S.A.* **60**: 1288–1295.

Sambrook, J., Greene, R., Stringer, J., Mitchison, T., Hu, S.-L. and Botchan, M. (1980) Analysis of the sites of integration of viral DNA sequences in rat cells transformed by adenovirus 2 or SV40. *Cold Spr. Harb. Symp. quant. Biol.* **44**: 569–584.

Santos, M. and Butel, J. S. (1982) Association of SV40 large tumor antigen and cellular proteins on the surface of SV40 transformed mouse cells. *Virology* **120**: 1–17.

Savageau, M. A., Metter, R. and Brockman, W. W. (1983) Statistical significance of partial base-pairing potential: implications for recombination of SV40 DNA in eukaryotic cells. *Nucleic Acids Res.* **11**: 6559–6570.

Schaffhausen, B. S. and Benjamin, T. L. (1981) Comparison of phosphorylation of two polyoma virus middle T antigens *in vivo* and *in vitro*. *J. Virol.* **40**: 184–196.

Schaffhausen, B. S., Silver, J. E. and Benjamin, T. L. (1978) T-antigen(s) in cells productively infected by wild-type polyoma virus and mutant NG-18. *Proc. natn. Acad. Sci. U.S.A.* **75**: 79–83.

Schlegel, R. and Benjamin, T. L. (1978) Cellular alterations dependent upon the polyoma virus Hr-T function: separation of mitogenic from transforming capacities. *Cell* **14**: 587–599.

Schutzbank, T., Robinson, R., Oren, M. and Levine, A. J. (1982) SV40 large tumor antigen can regulate some cellular transcripts in a positive fashion. *Cell* **30**: 481–490.

Scott, M., Westphal, K. and Rigby, P. W. (1983) Activation of mouse genes in transformed cells. *Cell* **34**: 557–567.

Segawa, K. and Ito, Y. (1982) Differential subcellular localization of *in vivo*-phosphorylated and nonphosphorylated middle-sized tumor antigen of polyoma virus and its relationship to middle-sized tumor antigen phosphorylating activity *in vitro*. *Proc. natn. Acad. Sci. U.S.A.* **79**: 6812–6816.

Sief, R. and Cuzin, F. (1977) Temperature-sensitive growth regulation in one type of transformed rat cells induced by TsA mutant of polyoma virus. *J. Virol.* **24**: 721–728.

Seif, R. and Martin, R. G. (1979a) Growth state of the cell early after infection with SV40 determines whether the maintenance of transformation will be A-gene dependent or independent. *J. Virol.* **31**: 350–359.

Seif, R. and Martin, R. G. (1979b) Simian virus 40 small T antigen is not required for the maintenance of transformation but may act as a promoter (cocarcinogen) during establishment of transformation in resting rat cells. *J. Virol.* **32**: 979–988.

Silver, J., Schaffhausen, B. and Benjamin, T. (1978) Tumor antigens induced by nontransforming mutants of polyoma virus. *Cell* **15**: 485–496.

Sleigh, M., Topp, W. C., Hanich, R. and Sambrook, J. (1978) Mutants of SV40 with an altered small T protein are reduced in their ability to transform cells. *Cell* **14**: 79–88.

Smart, J. E. and Ito, Y. (1978) Three species of polyoma virus tumor antigens share common peptides probably near the amino termini of the proteins. *Cell* **15**: 1427–1437.

Smith, A. E. and Ely, B. K. (1983) The biochemical basis of transformation by polyoma virus. *Adv. Viral Oncology* **3**: 3–30.

Smith, A. E., Fried, M., Ho, Y., Spurr, N. and Smith, R. (1980) Is polyoma virus middle T antigen a protein kinase? *Cold Spr. Harb. Symp. quant. Biol.* **44**: 141–147.

Soeda, E., Arrand, J. R., Smolar, N. and Griffin, B. E. (1979) Polyoma virus DNA. I. Sequence from the early region that contains the origin of replication and codes for small, middle and (part of) large T-antigens. *Cell* **17**: 357–370.

Soeda, E., Arrand, J. R., Smolar, N. and Griffin, B. E. (1980) Coding potential and regulatory signals of the polyoma virus genome. *Nature* **283**: 445–453.

Sompayrac, L. and Danna, K. J. (1983a) Simian virus 40 deletion mutants that transform with reduced efficiency. *Molec. cell. Biol.* **3**: 484–489.

Sompayrac, L. and Danna, K. J. (1983b) Simian virus 40 sequences between 0.168 and 0.424 map units are not required for abortive transformation. *J. Virol.* **46**: 475–480.

Soprano, K. J., Dev, V. G., Croce, C. M. and Baserga, R. (1979) Reactivation of silent rRNA genes by Simian virus 40 in human-mouse hybrid cells. *Proc. natn. Acad. Sci. U.S.A.* **76**: 3885–3889.

Soprano, K. J., Jonak, G. J., Galanti, N., Floros, J. and Baserga, R. (1981) Identification of an SV40 DNA sequence related to the reactivation of silent rRNA genes in human mouse hybrid cells. *Virology* **109**: 127–136.

Southern, E. (1975) Detection of specific sequences among DNA fragments separated by gel electrophoresis. *J. Molec. Biol.* **98**: 503–515.

Sporn, M. B. and Todaro, G. J. (1980) Autocrine secretion and malignant transformation. *New Engl. J. Med.* **303**: 878–880.

Steinberg, B., Pollack, R., Topp, W. and Botchan, M. (1978) Isolation and characterization of T-antigen-negative revertants from a line of transformed rat cells containing one copy of the SV40 genome. *Cell* **13**: 19–32.

Stoker, M. and Dulbecco, R. (1969) Abortive transformation by the TsA mutant of polyoma virus. *Nature* **223**: 397–398.

Stringer, J. R. (1981) Integrated Simian virus 40 DNA: nucleotide sequences at cell-virus recombinant junctions. *J. Virol.* **38**: 671–679.

Stringer, J. R. (1982a) DNA sequence homology and chromosomal deletion at a site of SV40 DNA integration. *Nature* **296**: 363–366.

Stringer, J. R. (1982b) Mutant of Simian virus 40 large T-antigen that is defective for viral DNA synthesis, but competent for transformation of cultured rat cells. *J. Virol.* **42**: 854–864.

Sweet, B. H. and Hilleman, M. R. (1960) The vacuolating virus, SV40. *Proc. Soc. exp. Biol. Med.* **105**: 420–427.

Swetly, P. and Watanabe, Y. (1974) Cell cycle dependent transcription of SV40 DNA in SV40-transformed cells. *Biochemistry* **13**: 4122–4126.

Tegtmeyer, P. (1972) Simian virus 40 DNA synthesis: the viral replicon. *J. Virol.* **10**: 591–598.

Tegtmeyer, P. (1975) Function of Simian virus 40 gene A in transforming infection. *J. Virol.* **15**: 613–618.

Tegtmeyer, P., Schwartz, M., Collins, J. K. and Rundell, K. (1975) Regulation of tumor antigen synthesis by SV40 gene A. *J. Virol.* **16**: 168–178.

Templeton, D. and Eckhart, W. (1982) Mutation causing premature termination of the polyoma virus medium T antigen blocks cell transformation. *J. Virol.* **41**: 1014–1024.

Templeton, D. and Eckhart, W. (1984) Characterization of viable mutants of polyoma virus cold-sensitive for the maintenance of cell transformation. *J. Virol.* **49**: 799–805.

Tenen, D., Martin, R., Anderson, J. and Livingston, D. (1977) Biological and biochemical studies of cells transformed by Simian virus 40 temperature-sensitive gene A mutants and A mutant revertants. *J. Virol.* **22**: 210–218.

Tevethia, M. J., Flyer, D. C. and Tjian, R. (1980) Biology of Simian virus 40 (SV40) transplantation antigen (TrAg). VI. Mechanism of induction of SV40 transplantation immunity in mice by purified SV40 T antigen (D2 protein). *Virology* **107**: 13–23.

Tjian, R. (1981) T antigen binding and control of SV40 gene expression. *Cell* **26**: 1–2.

Tjian, R., Fey, G. and Graessmann, A. (1978) Biological activity of purified Simian virus 40 T antigen proteins. *Proc. natn. Acad. Sci. U.S.A.* **75**: 1279–1283.

Tjian, R. and Robbins, A. (1979) Enzymatic activities associated with a purified Simian virus 40 T antigen-related protein. *Proc. natn. Acad. Sci. U.S.A.* **76**: 610–614.

Toniolo, T. and Basilico, C. (1975) SV40-transformed cells with temperature-dependent serum requirements. *Cell* **4**: 255–262.

Tooze, J. (1981) *DNA Tumor Viruses* (2nd edn. revised) Cold Spring Harbor Lab., New York.

Topp, W. C. and Rifkin, D. B. (1980) The small-T protein of SV40 is required for loss of actin cable networks and plasminogen activator synthesis in transformed rat cells. *Virology* **106**: 282–291.

TOPP, W. C., RIFKIN, D. B. and SLEIGH, M. J. (1981) SV40 mutants with an altered small-T protein are tumorigenic in newborn hamsters. *Virology* **111**: 341–350.
TREISMAN, R. (1980) Characterization of the polyoma late mRNA leader sequences by molecular cloning and DNA sequence analysis. *Nucleic Acids Res.* **8**: 4867–4888.
TREISMAN, R., COWIE, A., FAVALORO, J., JAT, P. and KAMEN, R. (1981a). The structures of the spliced mRNAs encoding polyoma virus early region proteins. *J. molec. Appl. Genet.* **1**: 83–92.
TREISMAN, R., NOVAK, U., FAVALORO, J. and KAMEN, R. (1981b) Transformation of rat cells by an altered polyoma virus genome expressing only the middle T protein. *Nature* **292**: 595–600.
TYNDALL, C., LA MANTIA, G., THACKER, C. M., FAVALORO, J. and KAMEN, R. (1981) A region of the polyoma virus genome between the replication origin and late protein coding sequences is required in cis for both early gene expression and viral DNA replication. *Nucleic Acids Res.* **9**: 6231–6250.
VOGEL, A. and POLLACK, R. (1974) Methods of obtaining revertants of transformed cells. *Methods Cell Biol.* **8**: 75–92.
VOGT, M. and DULBECCO, R. (1963) Steps in the neoplastic transformation of hamster embryo cells by polyoma virus. *Proc. natn. Acad. Sci. U.S.A.* **49**: 171–179.
WALTER, G., SCHEIDTMANN, K.-H., CARBONE, A., LAUDANO, A. P. and DOOLITTLE, R. F. (1980) Antibodies specific for the carboxy- and amino-terminal regions of Simian virus 40 large tumor antigen. *Proc. natn. Acad. Sci. U.S.A.* **77**: 5197–5200.
WASYLYK, B., WASYLYK, C., MATTHES, H., WINTZERITH, M. and CHAMBON, P. (1983) Transcription from the SV40 early-early and late-early overlapping promoters in the absence of DNA replication. *EMBO J.* **2**: 1605–1611.
WATKINS, J. F. and DULBECCO, R. (1967) Production of SV40 virus in heterokaryons of transformed and susceptible cells. *Proc. natn. Acad. Sci. U.S.A.* **58**: 1396–1403.
WEIL, R. (1978) Viral 'tumor antigens'. A novel type of mammalian regulator protein. *Biochim. biophys. Acta* **516**: 301–388.
WESTPHAL, H. and DULBECCO, R. (1968) Viral DNA in polyoma and SV40-transformed cell lines. *Proc. natn. Acad. Sci. U.S.A.* **59**: 1158–1165.
WILSON, J. H., DEPAMPHILIS, M. and BERG, P. (1976) Simian virus 40 permissive cell interactions—selection and characterization of spontaneously arising monkey cells that are resistant to Simian virus 40 infection. *J. Virol.* **20**: 391–399.
WYKE, J. (1971) Phenotypic variation and its control in polyoma-transformed BHK21 cells. *Exp. Cell Res.* **66**: 209–233.
ZOUZIAS, D., PRASAD, I. and BASILICO, C. (1977) State of the viral DNA in rat cells transformed by polyoma virus. II. Identification of the cell containing non-integrated viral DNA and the effect of viral mutations. *J. Virol.* **24**: 142–150.

CHAPTER 7

CELLULAR TRANSFORMATION BY ADENOVIRUSES

S. J. FLINT

Department of Molecular Biology, Princeton University, Princeton, New Jersey 08544, U.S.A.

Abstract—The examination of integrated viral DNA sequences present in adenovirus transformed rodent cells and genetic studies have established that products of the viral, early E1A and E1B transcription units are necessary and sufficient for adenovirus transformation of rodent cells in culture. The viral gene products encoded by these complex transcription units are described. The roles of the individual E1A and E1B gene products in induction and maintenance of transformation and in determining tumorigenicity of transformed cells and of the virus are discussed, as is our present knowledge of the molecular and biochemical properties of these adenoviral transforming proteins.

1. INTRODUCTION

Some ten years after the discovery of the adenovirus group, Trentin and colleagues announced that human adenovirus type 12 (Ad12) induced tumors with high efficiency when inoculated into newborn rodents (Trentin et al., 1962). This observation was quickly reproduced and extended to additional human serotypes, as well as adenoviruses isolated from other animals, such as dogs and monkeys (Huebner et al., 1962, 1963; Yabe et al., 1962; Girardi et al., 1964; Rabson et al., 1964; Darbyshire, 1966; Hull et al., 1965; Sarma et al., 1965, 1967). Moreover, all human adenovirus serotypes tested were shown to transform non- or semi-permissive rodent cells in culture (Pope and Rowe, 1964; Freeman et al., 1967; McAllister et al., 1969a, b), regardless of whether their oncogenic potential was manifested in animals (Huebner et al., 1965). Such properties focused attention on a previously obscure group of viruses: attempts to understand both the mechanisms whereby adenoviruses transform cells in culture and the fascinating differences in oncogenicity in animals displayed by the human serotypes, which can be classified into highly-, weakly- or non-oncogenic subgroups (Huebner et al., 1965), have comprised an important thrust in adenovirus research for the last 20 years.

The study of adenovirus transformation has progressed from investigations of virus-cell interactions and the properties of transformed cells to the identification of the adenovirus gene products necessary and sufficient to induce and maintain the transformed state and, most recently, to detailed genetic and biochemical analyses of the products of such transforming genes. A great deal has been learnt, particularly of the contributions made by different adenovirus gene products to the transformation process. Nevertheless, we have not yet gained a great deal of insight into the molecular mechanisms whereby adenovirus gene products transform mammalian cells.

The pioneering work on adenovirus transformation has been reviewed in detail by several authors (see, for example, Tooze, 1973, 1980; Casto, 1973; Flint, 1982) and will be mentioned only briefly in this article, which will concentrate on the identification, functions and properties of the adenovirus gene products that participate in transformation.

2. IDENTIFICATION OF ADENOVIRUS TRANSFORMING GENES

Transformation by human adenoviruses is a very inefficient process, even when a particular serotype infects cells that are completely non-permissive for its replication, for example, adenovirus type 12 (Ad12) in hamster cells (Doerfler, 1968, 1969; Zur Hansen and Sokol, 1969; Doerfler and Lundholm, 1970). It has been estimated that one focus-forming (*i.e.* transformation) unit corresponds to anywhere from 10^4 to 10^7 plaque-forming units of virus in the various adenovirus-rodent cell systems that have been studied (Pope and Rowe, 1964; Freeman *et al.*, 1967; McAllister *et al.*, 1969a, b; Castro, 1973; Gallimore, 1974). Consequently, direct study of the induction of transformation is impossible: only a few cells in an infected culture will give rise to descendents that display the characteristic properties of transformed cells and such potential transformants cannot be identified until these properties become manifest, that is, until the transformation process is complete. Foci of adenovirus transformed cells do, however, display typical transformed phenotypes, including oncogenicity in rodents (see Castro, 1973; Tooze, 1973) and can be readily established as immortal cell lines. The first attempts to identify adenoviral gene products that contribute to transformation therefore relied on such cell lines.

Fortunately, as it turned out, detailed attention was first paid to lines of rat or hamster embryonic cells transformed by the subgroup C (non-oncogenic) serotypes Ad2 or Ad5, respectively. The ability of total DNA purified from independent lines of transformed cells to accelerate the rate of renaturation of denatured, restriction endonuclease fragments of labeled Ad2 or Ad5 DNA (Gelb *et al.*, 1971) was employed, both to map the adenoviral DNA sequences integrated into the cellular genome and to estimate their concentration (Gallimore *et al.*, 1974; Sambrook *et al.*, 1975). None of the twelve or so transformed cell lines examined retained an intact copy of the adenoviral genome, a finding in complete agreement with previous failures to rescue infectious virus from transformed cell lines (Landau *et al.*, 1966; Larsson *et al.*, 1966). The most striking result to emerge from this analysis was, however, the discovery of only one set of adenoviral DNA sequences common to all lines examined, those homologous to the left-hand 14% of the viral genome (Gallimore *et al.*, 1974; Sambrook *et al.*, 1975). Some lines contained only these sequences, whereas others retained sequences homologous to many other regions of the adenoviral genome. Nevertheless, the invariable presence of sequences comprising the left-hand 14% of the adenoviral genome, which includes the early regions E1A and E1B (see next section) permitted the conclusion that the products of these genes were sufficient to maintain the complete transformation phenotype exhibited by the cell lines in question (Williams, 1973; Gallimore, 1974). This conclusion has been amply confirmed by the results of subsequent, similar investigations of other subgroup C adenovirus-transformed cell lines (Flint *et al.*, 1976b; Johansson *et al.*, 1977, 1978; Frolova *et al.*, 1978; Frolova and Zalmanzon, 1978; Visser *et al.*, 1979; Dorsch-Häsler *et al.*, 1980).

Had attempts to identify adenovirus transforming genes by virtue of their ubiquitous integration into the genome of transformed rodent cells first been made with cells transformed by subgroup A (oncogenic) serotypes, such as Ad12, no such clear answer would have emerged: the great majority of Ad12 transformed or tumor cell lines whose integrated viral DNA sequences have been examined, by the methods used in the experiments of Sambrook and colleagues or by the blotting technique of Southern (1975), contain integrated, colinear copies of the entire Ad12 genome (Fanning and Doerfler, 1976; Sutter *et al.*, 1978; Mak *et al.*, 1979; Doerfler *et al.*, 1979; Ibelgaufts *et al.*, 1980; Dewing *et al.*, 1981; Eick and Doerfler, 1982; Starzinski-Powitz *et al.*, 1982). An alternative approach to identification of transforming genes, the introduction of viral DNA or restriction endonuclease fragments into appropriate cells by calcium phosphate-co-precipitation (Graham and van der Eb, 1973) has therefore proved equally valuable. Transfection of individual, purified restriction endonuclease fragments of subgroup C adenovirus DNA into normal rodent cells in culture confirmed that the viral functions both necessary and sufficient, under these conditions, to induce and maintain transformation are encoded within the extreme left-hand segment of the viral genome (Graham *et al.*, 1975a; van der Eb *et al.*, 1979; Rowe *et al.*, 1984). Similar results were obtained when fragments generated from Ad12 DNA or the genome of Ad7, a member of the weakly-oncogenic subgroup B, were tested in this way (Shiroki *et al.*, 1977; Seikikawa *et al.*, 1978; Sawada *et al.*, 1979; van der Eb *et al.*, 1979; Dijkema *et al.*, 1979), observations of little surprise in light of the similar genetic organization exhibited by members of the three subgroups (see, for example, Ortin *et al.*, 1976; Smiley and Mak, 1978; Section 3).

The majority of adenovirus mutants whose lesions lie within the transforming region identified by these methods, that is, early regions E1A and E1B, show defects in transformation activity when tested in appropriate cell types, as will be discussed in detail in Section 4. Interestingly, however, genetic analysis of adenovirus transformation has identified additional viral genes whose products influence transformation by the virus. Temperature-sensitive mutants of adenovirus type 5, such as H5ts125 (Ensinger and Ginsberg, 1972), bearing lesions in early region E2A, which encodes a 72kd single-stranded DNA-binding protein (DBP) (Levine *et al.*, 1975; Lewis *et al.*, 1976; Grodzicker *et al.*, 1977) essential for viral DNA replication (Wilkie *et al.*, 1973; Levine *et al.*, 1975), transform secondary rat embryo cells three- to eight-fold more efficiently at a temperature non-permissive for replication than at a permissive temperature (Ginsberg *et al.*, 1975; Meyer and Ginsberg, 1977). The transformation frequency at permissive temperatures is the same as the temperature-independent frequency exhibited by wild-type Ad5. Thus, the H5ts125 mutation induces an increase in transformation efficiency (Ginsberg *et al.*, 1975; Meyer and Ginsberg, 1977). The genomes of cell lines established after H5ts125 transformation at a non-permissive temperature contain integrated, colinear copies of the entire Ad5 genome, whereas cells transformed at a permissive temperature resemble their counterparts transformed by wild-type Ad5 and usually retain only sequences homologous to the left-hand end of Ad5 DNA (Meyer and Ginsberg, 1977; Dorsch-Häsler *et al.*, 1980). Rat cells are quite permissive for subgroup C adenovirus replication (Gallimore, 1974, for example), but a completely non-permissive system is established when rat cells infected by H5ts125 are maintained at a non-permissive temperature. In this sense a non-permissive H5ts125-infection is analogous to an Ad12 infection of hamster cells, a situation that might account for the integration of complete copies of the adenoviral genome in both instances.

The non-permissiveness of an H5ts125-infection at a high temperature cannot, however, account for the increased transformation frequency, compared to a permissive H5ts125 or a wild-type Ad5 infection of the same cells: other adenovirus mutants in which replication of the virus is blocked and which complement H5ts125 fail to exhibit increased transformation frequencies under non-permissive conditions. The group N temperature-sensitive mutants of Ad5 such as H5ts36 or H5ts149, fall into this category (Wilkie *et al.*, 1973; Ginsberg *et al.*, 1975; Williams *et al.*, 1975, 1979). These mutations lie in the gene encoding the viral DNA polymerase (Stillman *et al.*, 1982; Friefeld *et al.*, 1983b). Thus, viral DNA replication can never begin in semi-permissive cells infected by group N mutants at a non-permissive temperature. By contrast, the 72kd DBP mutated in H5ts125 is essential to elongation (van der Vliet *et al.*, 1977; Friefeld *et al.*, 1983a, van Bergen and van der Vliet, 1983), although it may play some role in early replication steps (van der Vliet and Sussenbach, 1975). It is, therefore, possible that viral DNA molecules present in H5ts125-infected cells at a non-permissive temperature adopted a configuration that is more favorable to the recombination event(s) that must mediate integration of viral DNA sequences, for example, structures containing single-stranded ends as a result of initiation of viral DNA replication (see Tooze, 1980). Alternatively, the increased transformation frequency exhibited by H5st125 might reflect the effects of this mutation upon viral early gene expression: the steady-state levels of all viral early mRNA sequences are increased some five- to ten-fold in H5ts125-infected human cells at a non-permissive temperature (Carter and Blanton, 1978a, b; Babich and Nevins, 1981). Assuming that viral gene expression in rodent cells is also elevated as a consequence of the H5st125 mutation, then enhanced expression of the transforming E1A and E1B regions could readily account for the more efficient transformation by H5ts125 at high temperature. Regardless of the mechanism whereby the mutated E2A 72kd DBP actually influences the frequency of transformation, it is clear that this viral product is not a necessary participant in the transformation process.

Genetic studies have, however, identified a viral gene product other than those encoded by early regions E1A or E1B that does appear to be essential to transformation when the adenoviral genome is introduced into rodent cells by virus infection rather than by DNA transfection. The Ad5 group N mutants mentioned previously have been reported to be completely deficient in transformation of secondary rat embryo fibroblasts at a non-permissive temperature (H5st149) or to display greatly reduced transformation frequencies (H5ts36, H5ts69) (Williams *et al.*, 1975; 1979). In experiments similar in use both of embryonic rat cells and a focus-formation assay for transformation, Ginsberg *et al.* (1975) failed to observe temperature-sensitive transformation by H5ts149. The reasons for these differences have never been explained, but temperature-sensitive transformation by Ad5 group N mutants has been observed by Williams and colleagues in many independent experiments. When rat embryo cells infected by any of the three group N Ad5 mutants were held at 32.5°C for two days before shift to a non-permissive temperature, close to normal transformation frequencies were observed (Williams *et al.*, 1975, 1979). Thus, the product altered by the group N mutants must be required to mediate an early step in the transformation process, during what is generally termed initiation of transformation.

The group N mutations lie within the segment 18.5 to 22.0 map units in the Ad5 genome (Williams *et al.*, 1979; Galos *et al.*, 1979), a region that has been shown to be expressed as rare, viral early mRNA species (E2B) (Galos *et al.*, 1979; Stillman *et al.*, 1981), one of which encodes the adenoviral DNA polymerase (Stillman *et al.*, 1982; Friefeld *et al.*, 1983b). The transformation properties of the Ad5 group N mutants might appear difficult to reconcile with the findings, discussed previously in this section, that viral DNA fragments that comprise as little as the region left-hand 12 to 14% of the genome are both necessary and sufficient for complete transformation. The resolution to this apparent paradox must surely lie in the forms in which viral DNA has been introduced into rodent cells, as a natural viral genome packaged in its core proteins on the one hand and naked viral DNA coprecipitated with calcium phosphate and irrelevant carrier DNA on the other: it is well established that transfected DNA is efficiently ligated to form complex molecular structures (Pellicer *et al.*, 1978; Perucho *et al.*, 1980; Scangos *et al.*, 1980) and recent work suggests that such ligation occurs by unusual mechanisms (Kopchick and Stacey, 1984). The failure of group N mutants to transform appears, therefore, to suggest that such unnatural forms are more suitable substrates for an early transformation step, presumably integration, than the double-stranded, viral DNA molecules present when replication is absolutely blocked, as it is under non-permissive conditions of group N infection (Wilkie *et al.*, 1973; Ginsberg *et al.*, 1975). The implication of this hypothesis that in rodent cells infected by the virus, replicating viral DNA molecules serve as substrates for integration is certainly consistent both with the role of the N-gene products in initiation of transformation and the partially single-stranded character of typical adenoviral replicating intermediates (see Tooze, 1980), but has not been tested directly.

In this context, it is of interest to consider more carefully the efficiencies of transformation by virus and purified viral DNA fragments. A typical transformation frequency for wild-type Ad2 or Ad5 virus is 5 foci per 10^6 cells, following an infection at 10 pfu/cell. It has been reported that close to one molecule of viral DNA can be detected in the nucleus of infected cells for each plaque-forming unit of virus to which the cells were exposed (Flint *et al.*, 1976a; Thomas and Mathews, 1980). Thus, in the best possible case, it can be calculated that 5 foci of transformed cells are produced for each 10^7 molecules of viral DNA that enter the cells. The particle: pfu ratio of subgroup C adenovirus preparations ranges from 20 to 100 to 1, so that in the worst possible case, in which each virus particle defective in a plaque assay nevertheless contained a complete copy of transforming sequences, 10^9 viral DNA molecules would result in 5 transformed foci. Low as such frequencies are, those following introduction of purified viral DNA fragments are even worse. About the best transformation efficiencies reported following DNA transfection are of the order of 12 foci per microgram equivalent of viral DNA, a value that translates to 5 foci per 10^{10} DNA molecules. Thus, transformation by purified adenoviral DNA molecules is from twenty-five to perhaps as much as two thousand-fold less efficient than that induced following virus infection of similar (rodent embryo) cells. Such a difference is certainly consistent with the notion that alternative molecular mechanisms mediate early steps in transformation, presumably integration of genes whose products induce and maintain the transformed phenotype, when rodent cells are presented with adenoviruses of deproteinized viral DNA.

In summary, then, it seems clear that the adenovirus DNA polymerase is required during initiation of transformation, shortly after infection of rodent cells by virus. Equally clearly, this viral gene product cannot participate in maintenance of the transformed phenotype: more often than not the E2B DNA sequences are not retained in fully-transformed cells. Moreover, cells transformed by Ad5 group N mutants at a permissive temperature maintain a fully-transformed phenotype when shifted to a non-permissive temperature (Williams et al., 1975, 1979). Thus, only the gene products encoded by the extreme left-hand segment of the adenoviral genome must be needed to maintain the transformed phenotype, a conclusion that has focused a great deal of attention upon these regions, E1A and E1B, and their products.

3. THE ADENOVIRUS E1A AND E1B REGIONS AND THEIR PRODUCTS

Early studies of Ad2 and Ad5 gene expression established that the transforming region identified by the methods discussed in Section 2 contained two transcriptional units encoded in the r-strand, termed E1A and E1B (Sharp et al., 1975; Flint et al., 1975; Pettersson et al., 1976; Wilson et al., 1979a) and that both were expressed in all lines of transformed cells examined (Flint et al., 1975, 1976b). Each unit, however, encodes at least two protein products specified by differentially spliced mRNA species (Berk and Sharp, 1978; Chow et al., 1979; Kitchingman and Westphal, 1980). The nucleotide sequences of the E1A and E1B genes of Ad2 (Gingeras et al., 1982), Ad5 (van der Eb et al., 1979; Maat et al., 1980; van Ormondt et al., 1980a, b, 1983; Bos et al., 1981), Ad7 (Dijkema et al., 1980) and Ad12 (Sugisaki et al., 1980; Bos et al., 1981) have been determined, as have those of E1A and E1B mRNA species of Ad2 and Ad12 cloned as cDNA copies (Perricaudet et al., 1979, 1980a, b; Aleström et al., 1980; Virtanen et al., 1982; Virtanen and Pettersson, 1983; 1985).

Figure 1 illustrates the detailed picture of the E1A region that has emerged from such intense scrutiny, in this case that of Ad2. The transcription of E1A DNA initiates at nucleotide 498 (Baker and Ziff, 1979, 1981) (where the first nucleotide in the r-strand is designated 1) and the site of polyadenylation lies at nucleotide 1608, preceded by a translational termination codon at nucleotide 1540. The three mRNA species processed

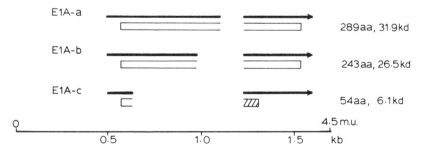

FIG. 1. Adenovirus region E1A products. A. The mRNA and protein encoded by Ad2 E1A are shown as horizontal arrows, drawn in the direction of transcription in which gaps represent sequences removed during splicing. Coding sequences are shown as boxes below the mRNA species. The different reading frames employed in translation are indicated by open and cross-hatched areas within the boxes. The mRNA species are designated according to the nomenclature of Chow et al. (1979) and to the right of each is listed the predicted length and molecular weight of its protein product. The sources of these data are given in the text.

from E1A transcripts, first distinguished by nuclease S1 mapping (Berk and Sharp, 1978) and examination of DNA:RNA hybrids by electron microscopy (Chow et al., 1979; Kitchingman and Westphal, 1980), share sequences from a 3′ splice site at nucleotide 1226 to the poly(A)-addition site (Perricaudet et al., 1979), but are made by splicing from three different 5′ splice sites, at nucleotides 1111, 973 and 636 to create species E1A-a, E1A-b and E1A-c, respectively (in the nomenclature of Chow et al., 1979). As illustrated in Fig. 1, all three E1A mRNA species must share 5′ coding sequences from the initiation codon at nucleotide 558 to the 5′ splice site of E1A-c mRNA. Species E1A-a and E1A-b also share sequences from this site, nucleotide 636, to the E1A-b 5′ splice site at nucleotide 973, whereas coding sequences from this point to nucleotide 1111 are unique to species E1A-a. Splicing introduces a two nucleotide shift in reading frame in the 3′ compared to the 5′ coding segments, in both the E1A-a and E1A-b mRNA species. Thus, their protein products differ only in the number of amino-acids expressed from the internal portion of the region, 46 amino acids uniquely encoded in E1A-a mRNA (see Fig. 1).

The amino acid sequences of proteins encoded by E1A-a and E1A-b mRNA have been predicted from the complete genomic and mRNA sequence (Perricaudet et al., 1979; Gingeras et al., 1982). The polypeptides would be expected to exhibit molecular weights of 32,000 daltons (289 amino acids) and 26,000 daltons (243 amino acids), respectively. The E1A-encoded proteins made in Ad2, or Ad5, infected or transformed cells have been examined in numerous laboratories by such methods as immunoprecipitation of labeled extracts with sera from tumor-bearing animals or from rabbits innoculated with synthetic peptides predicted by the DNA sequence and in vitro translation of hybridization-selected E1A mRNA (Lewis et al., 1976; Gilead et al., 1976; Harter and Lewis, 1978; Levinson and Levine, 1977a, b; Johansson et al., 1978; Halbert et al., 1979; Lewis et al., 1979; Green et al., 1979; Lassam et al., 1979; Schrier et al., 1979; Brackmann et al., 1980; Ross et al., 1980; Esche et al., 1980; Lupker et al., 1981; Lewis and Mathews, 1981; Smart et al., 1981; Esche, 1982; Halbert and Raskas, 1982; Rowe et al., 1983b; Yee et al., 1983). The values reported for the apparent molecular weights of the E1A proteins have varied somewhat, but all workers agree in estimates that are substantially larger than those predicted, 38–53kd and 34–47kd, respectively. Moreover, four to six E1A polypeptides have been routinely observed in such experiments. In vitro translation of size fractionated E1A mRNA species has assigned polypeptides of some 53 and 44kd apparent molecular weight to E1A-a mRNA and those estimated to be 47 and 35kd to species E1A-b (Halbert et al., 1979; Esche et al., 1980; Smart et al., 1981), in agreement with results of peptide mapping that place the four E1A polypeptides commonly observed into two highly-related pairs (Harter and Lewis, 1978; Lewis et al., 1979; Green et al., 1979; Halbert and Raskas, 1982). Most recently, the E1A proteins (immunoprecipitated from infected cell extracts with monoclonal antibodies) have been examined in two-dimensional gels, an approach that has separated more than 60 E1A polypeptides (Harlow et al., 1985). At least forty of these were identified as products of the two larger E1A mRNA species (Harlow et al., 1985). The far larger number of species detected in this way was attributed by Harlow et al. (1985) to several factors. These include the nature of the separation system employed, the longer labelling period, four hours compared to two in most earlier experiments and the use of monoclonal antibodies, which yield cleaner immunoprecipitates and thus permit longer exposures and the detection of minor species. The origins of such a variety of forms of the E1A polypeptides are not known, but pulse-chase experiments indicated that they arise because of post-translational modifications (cited in Harlow et al., 1985). The E1A polypeptides are phosphorylated (Lassam et al., 1979; Gaynor et al., 1982; Yee et al., 1983; Spindler et al., 1984; Lucher et al., 1984), but it seems likely that additional modifications account for the variety of E1A polypeptides observed. The aberrant mobility displayed by E1A proteins during migration in conventional SDS-acrylamide gels is generally ascribed to their high proline content.

The E1A-a and E1A-b mRNA species or the proteins they encode have been detected in all lines of Ad2 or Ad5 transformed cells examined (Flint et al., 1976b; Graham et al., 1977; Johansson et al., 1978; Levinson and Levine, 1977a, b; Sambrook et al., 1979; Schrier et al.,

1979; Flint and Beltz, 1979; Lassam *et al.*, 1979; Wilson *et al.*, 1979b; Schrier *et al.*, 1979; Flint and Beltz, 1979; Lassam *et al.*, 1979; Wilson *et al.*, 1979b; Lewis and Mathews, 1981; Esche, 1982; Jochemsen *et al.*, 1981; Green *et al.*, 1983a, b; Rowe *et al.*, 1984), suggesting that these viral gene products are important in transformation. It is not known whether transformed cells also routinely express the E1A-c mRNA and its product. This mRNA species, preferentially expressed during the late phase of productive infection (Spector *et al.*, 1978; Chow *et al.*, 1979), was initially reported to encode a protein of apparent molecular weight 28kd (Halbert *et al.*, 1979; Esche *et al.*, 1980). Such an E1A-encoded protein could not be detected in several lines of Ad2 transformed cells (Lewis and Mathews, 1981; Esche, 1982), although an E1A-c mRNA species was observed in one line of Ad2 transformed rat embryo fibroblasts (Sambrook *et al.*, 1979). The significance of failures to find an E1A 28kd protein is difficult to assess for the sequence of E1A-c mRNA indicates that it can encode only a 6.1kd polypeptide (Virtanen and Pettersson, 1983): it is not clear at present whether the 28kd protein is in some way related to the predicted 6.1kd E1A-c mRNA product or whether additional, small E1A mRNA species exist.

The different oncogenic properties displayed by human adenoviruses have stimulated comparisons of the organization and products of the transforming genes of different serotypes. The E1A region of Ad12 (sub-group A) does display some differences when compared to Ad2 E1A, but these are minor. Thus, Ad12-infected or transformed cell mRNA preparations contain four species complementary to E1A rather than the two most abundant Ad2 or Ad5 E1A-a and E1A-b mRNAs discussed in previous paragraphs (Sawada and Fujinaga, 1980; Segawa *et al.*, 1980; Saito *et al.*, 1983). The two major Ad12 mRNAs are indeed analogous to these two subgroup C mRNA species (Perricaudet *et al.*, 1980b; Sawada and Fujinaga, 1980; Saito *et al.*, 1983), whereas the two more minor species are variants that carry longer 5' untranslated regions. It should be noted that an alternative, upstream E1A initiation site is also utilized, albeit rarely, in subgroup C-infected cells (Chow *et al.*, 1979; Osborne *et al.*, 1982; Osborne and Berk, 1983). Transcription from such upstream sites does not alter the coding capacity of the E1A mRNA species for the first open reading frame lies downstream from the second, more frequently used transcription initiation site (see Fig. 1 for Ad2 E1A). Thus, it is not surprising that the polypeptides encoded by the Ad12 E1A region are analogous to those listed previously for subgroup C adenoviruses (van der Eb *et al.*, 1979; Ribeiro and Vasconcelos-Costa, 1981; Jochemsen *et al.*, 1980, 1982; Segawa *et al.*, 1980; Saito *et al.*, 1983).

Comparison of the Ad5 E1A sequence to those of Ad12 (van Ormondt *et al.*, 1980a) and the subgroup B serotype Ad7 (Dijkema *et al.*, 1980) reveals an overall homology of 31–35% among all three strains, or 50–55% homology between any pair. Even greater homology is observed when the predicted amino-acid sequences are compared (van Ormondt *et al.*, 1980a; Perricaudet *et al.*, 1980b): blocks of remarkably conserved sequences are interspersed with segments that are not conserved.

Figure 2A depicts the early and transformed cell E1B mRNA species specified by Ad2 (Berk and Sharp, 1978; Spector *et al.*, 1978; Chow *et al.*, 1979; Kitchingman and Westphal, 1980; Bos *et al.*, 1981; Perricaudet *et al.*, 1980a). Transcription of the Ad2 E1B region is initiated at nucleotide 1699 or 1701 (Baker and Ziff, 1979, 1981). The two major mRNA species have in common sequences from this point to the E1B-b mRNA 5' splice site at nucleotide 2249, which is joined to a 3' splice site at nucleotide 3589. The larger E1B-a mRNA includes sequences from nucleotide 2249 to its 5' splice site at nucleotide 3504 which is ligated to the same 3' splice site. Thus, these two E1B mRNA species, like those complementary to the E1A gene, have both 5' and 3' terminal segments in common. By contrast to E1A, however, the polypeptides encoded by these two mRNA species have no common sequences.

An open reading frame lies near the N-terminal end of the Ad2 E1B region, nucleotides 1711–2236, of sufficient size to encode a polypeptide of 21kd. A protein exhibiting an apparent molecular weight of 15–19kd (Lewis *et al.*, 1976; Levinson and Levine, 1977a, b; Harter and Lewis, 1978; van der Eb *et al.*, 1979; Brackmann *et al.*, 1980; Green *et al.*, 1982) has, on the basis of its N-terminal amino acid sequence (Anderson and Lewis, 1980), indeed

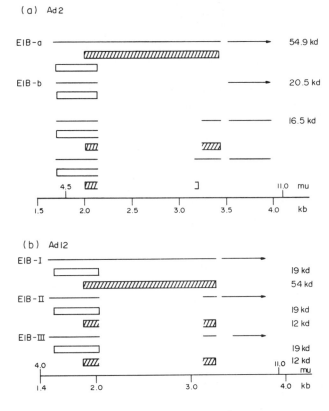

FIG. 2. Adenovirus region E1B products. The E1B mRNA and protein species encoded by Ad2 (part A) or Ad12 (part B) are illustrated as described in the legend to Fig. 1. The nomenclature of the Ad12 mRNA species is that of Virtanen *et al.* (1982). The sources of these data are given in the text.

assigned to this coding region. It is the only product specified by E1B-b mRNA, but can also be translated *in vitro* from the E1B-a mRNA species (Halbert *et al.*, 1979; Esche *et al.*, 1980; Bos *et al.*, 1981). The termination codon of the 19kd E1B polypeptide at nucleotide 2236 precedes the E1B-b mRNA 5' splice site at 2249. Thus, none of the C-terminal portion of the smaller E1B mRNA species beyond the 3' splice site is expressed.

A second major E1B-encoded polypeptide, apparent molecular weight 53 to 65kd (Lewis *et al.*, 1976; Levinson and Levine, 1977a, b; Harter and Lewis, 1978; van der Eb *et al.*, 1979; Brackmann *et al.*, 1980) must be expressed from a different reading frame or from a coding region that does not overlap with that specifying the 19kd protein: the two proteins share no methionine-containing tryptic peptides (Bos *et al.*, 1981; Green *et al.*, 1982) and antibodies directed against synthetic peptides predicted from the sequence of one of the E1B proteins fail to recognize the other (Green *et al.*, 1983a; Yee *et al.*, 1983; Lucher *et al.*, 1984). A long open reading frame, predicted coding capacity 55kd, spans nucleotides 2016 to 3501 (Fig. 2). Although the N-terminal segment of this overlaps the 19kd protein coding sequence in the region nucleotides 2016–2236, the larger, more downstream reading frame is shifted by two nucleotides relative to the first (Bos *et al.*, 1981). As shown in Fig. 2, the second reading frame can be expressed only in E1B-a mRNA, which can indeed be translated into the 55kd polypeptide *in vitro* (Halbert *et al.*, 1979; Esche *et al.*, 1980; Bos *et al.*, 1981). It must therefore be concluded that an internal AUG codon of the E1B-a mRNA, at nucleotide 2016, can be efficiently recognized, an inference confirmed by the ability of antibodies raised against the N-terminal sixteen amino acids predicted from the sequence beginning at nucleotide 2016 to immunoprecipitate the Ad2 55kd polypeptide (Lucher *et al.*, 1984).

A third E1B protein, apparent molecular weight 20kd, has been identified in Ad2-

infected and transformed cells. Unlike the 15–19kd polypeptide, the Ad2 20kd protein shares methionine-containing peptides with the 55kd protein (Green et al., 1982; Matsuo et al., 1982) and can be immunoprecipitated by 55kd-specific antibodies (Lucher et al., 1984). Moreover, the use of antibodies against synthetic peptides corresponding to the predicted N- or C-terminal sequences of the Ad2 E1B 55kd protein has established that the 20kd protein is also produced by initiation of translation at nucleotide 2016 (Lucher et al., 1984). What may be the same protein has been observed when E1B-selected mRNA from infected cells harvested during the late phase was translated in vitro (Esche et al., 1980). The subgroup C mRNA species encoding the 20kd polypeptide, as well as a second new E1B mRNA species, were identified only recently (Virtanen and Pettersson, 1985). These two, less abundant Ad2 E1B mRNA species each contain not only the intron present in E1B-a (22S) mRNA, but also an additional intron, which removes sequences of the 55kd protein. The mRNA species designated 14S by Virtanen and Pettersson (Fig. 2) encodes a protein of predicted molecular weight (16.5kd) which shares N- and C-terminal sequences with the E1B 55kd protein and must correspond to the 20kd protein described previously. The 14.5S mRNA could encode a 9.2k polypeptide, with a unique C-terminus (Fig. 2A).

All Ad2 or Ad5-transformed cells containing a complete copy of the E1B region (see Section 4.1) that have been examined synthesize the E1B proteins described in previous paragraphs (see, for example van der Eb et al., 1979; Schrier et al., 1979; Ross et al., 1980; Lewis and Mathews, 1981; Jochemsen et al., 1981; Esche, 1982; Green et al., 1982, 1983a; Rowe et al., 1984). Although rodent cells transformed by Ad12 have been reported to synthesize analogous E1B proteins, exhibiting apparent molecular weights of 50–60kd, and 17–19kd (for example, van der Eb et al., 1979; Shiroki et al., 1979; Mak et al., 1979; Jochemsen et al., 1982; Saito et al., 1983; Fujinaga et al., 1979) the organization of the Ad12 E1B gene and its expression are not identical to those displayed by Ad2 or Ad5. Cloning and sequencing of c-DNA copies of Ad12 E1B mRNA (Virtanen et al., 1982) has identified three abundant E1B mRNA species. The structures of these has been described precisely following comparison of these cDNA sequences to that of the Ad12 E1B DNA sequence (Kimura et al., 1981; Bos et al., 1981). These are summarized in Fig. 2B, for comparison with the Ad2 E1B mRNA species depicted in Fig. 2A. The highly oncogenic Ad12 encodes two small doubly-spliced mRNA species, each of which can in principle express the unique E1B 19kd protein and a 54kd protein related-protein of predicted molecular weight 12kd. Small E1B proteins exhibiting apparent molecular weights of 14 and 14.5kd have been observed in Ad12 transformed cell lines (Shiroki et al., 1981; Segawa et al., 1980; Saito et al., 1983). The two small Ad12 E1B mRNA species possess identical coding capacities (Fig. 2B), so it is not clear why two differently spliced forms should be made. Virtanen et al. (1982) have suggested that the different 3'-terminal segments carried by the two mRNA species might influence their transport or stability or the frequency of initiation codon usage and thus the nature of the protein made from them, but these ideas remain to be tested experimentally.

All three Ad12 E1B mRNA species contain, and could in principle express, the open reading frame spanning nucleotide 1541 to nucleotide 2030, coding capacity 19kd, exactly analogous to the N-terminal reading frame of the Ad2 E1B gene (compare Fig. 2A and 2B). Similarly, a second open reading frame of some 54kd capacity, nucleotide 1846 to nucelotide 3292, could also be expressed from the largest Ad12 E1B mRNA.

It is also unclear whether the differences between the third Ad12 and Ad2 E1B gene products, 12kd and 20kd, respectively, have any significance in relation to the transforming and oncogenic properties of these viruses. It has been reported that the two smaller Ad12 E1B mRNA species are not, in fact, expressed in certain lines of Ad12-transformed cells that contain an integrated copy of the complete E1B gene and apparently display a fully-transformed phenotype (Saito et al., 1983). This observation implies that the Ad12 12kd protein is either completely irrelevant to the transformation process or important by virtue of its absence. The former interpretation appears more reasonable at present, but the latter is difficult to exclude in the absence of detailed information about the biological characteristics of Ad12-transformed cell lines that do produce the smaller E1B mRNA species and their unique product.

4. THE ROLES OF E1A AND E1B GENE PRODUCTS IN TRANSFORMATION

The multiplicity of products encoded by the adenoviral E1A and E1B transcriptional units immediately raises the question of what roles the individual proteins play in the initiation or maintenance of transformation, indeed whether all products of these regions are important in transformation. The answers to this question provide an essential basis for attempts to elucidate the molecular mechanisms of adenovirus transformation. Consequently, they have been sought assiduously, by both introduction into rodent cells of adenoviral DNA segments that ressect the region containing the E1A and E1B regions and genetic analysis of the transforming functions of the E1A and E1B gene products.

As we have seen, transfection does not necessarily reveal the normal process by which adenoviruses transform cells: the results discussed in Section 2 argue strongly that early steps ('initiation') occur by different molecular mechanisms when viral DNA sequences are introduced by transfection or as part of the genomes of infectious particles. Such potential differences must therefore be considered when comparing results obtained by DNA tranfection to those from genetic analysis using infectious virus. This point might seem obvious, but has not always been considered fully by workers in this area. Interpretation of the very large body of information now available is also hampered by other experimental variables, notably the use of different types of cell and of different transformation assays. Although all experiments to be considered in this section have analyzed transformation of rodent cells, the cells have in some cases been truly normal and in others, partially transformed, primary or secondary cultures and established lines of immortal cells, respectively. Some functions that may be essential to transformation of normal cells might appear to be dispensable for transformation of immortal cells. The most common criterion of transformation employed has been the formation of foci of morphologically transformed cells. However, the precise conditions used have not been identical in all experiments, and in some important experiments only focus-formation in soft-agar, generally regarded as a more stringent assay, has been measured.

4.1. Functions Ascribed to E1A Gene Products

In some of the earliest attempts to separate the E1A and E1B transformation functions, purified restriction endonuclease fragments, spanning increasing lengths of the E1A + E1B region, were introduced into baby rat kidney cells by transfection and transforming activity assayed as the appearance of foci of morphologically transformed cells. Purified, or cloned, restriction endonuclease fragments that comprise only the E1A region, for example, *Hpa*1 fragment E, 0 to 4.36 mu, of Ad2 or Ad5 DNA, are sufficient to induce partial transformation, regardless of the serotype of origin of the DNA (van der Eb *et al.*, 1977, 1979; Seikikawa *et al.*, 1978; Yano *et al.*, 1977; Dijkema *et al.*, 1979; Houweling *et al.*, 1980; Shiroki *et al.*, 1979, 1980; van den Elsen *et al.*, 1982). Such E1A-transformed cells display a recognizably altered morphology and, unlike their normal parents, are able to grow continuously in culture. On the other hand, it has often proved difficult to establish clonal lines from such E1A-transformed cell foci (see, for example, van der Eb *et al.*, 1977). More strikingly, E1A transformed cells display only a limited set of transformation phenotypes: they grow more slowly and to lower saturation densities than do fully transformed cells, and retain a more fibroblastic-like morphology than typical adenovirus transformants, which are generally described as epitheloid (van der Eb *et al.*, 1979; Houweling *et al.*, 1980; for example). These properties indicate that, upon transfection, E1A gene products are sufficient to induce and maintain the immortal state and some changes in growth and morphology. Additional viral products, must, however, be present to permit acquisition of the fully transformed phenotype. The ability of purified or cloned adenoviral DNA fragments that contain complete copies of *both* the E1A and the E1B gene to induce complete transformation (Graham *et al.*, 1975a; van der Eb *et al.*, 1977, 1979; Yano *et al.*, 1977; Shiroki *et al.*, 1977; van den Elsen *et al.*, 1982) fully supports this conclusion.

Mutants of Ad5 from which most of either the E1A or the E1B region has been deleted fail to transform primary rat embryo cells (Jones and Shenk, 1979a; Shenk et al., 1979). However, H5dl313, which carries an E1A gene but no functional E1B sequences (Jones and Shenk, 1979a) does induce transformed foci in an established line of rats cells, 3Y1 (Shiroki et al., 1981). Cell lines established from such H5dl313 transformed foci, unlike those established after transformation by wild-type virus, cannot form colonies in soft-agar, that is, they have not acquired the phenotype of anchorage-independent growth (Shiroki et al., 1981). They are, therefore, partially transformed. Nevertheless, the E1A region of H5dl313 was sufficient to induce the formation of recognizably transformed 3Y1 cell foci, a property arguing strongly that an E1A gene product(s) maintains certain aspects of morphological transformation.

This conclusion receives compelling support from the behavior of certain mutants that are temperature dependent for transformation. The Ad5 mutant H5hr1, whose lesion lies within the E1A region (Frost and Williams, 1978), originally isolated as a host-range mutant no longer capable of growth on HeLa cells (Harrison et al., 1977) and defective in transformation of rat embryo or brain cells at 37°C (Graham et al., 1978), has subsequently been shown to be cold sensitive for transformation of these cells (Ho et al., 1982). Several similar mutants belonging to the same complementation group, for example, Hrhr^{cs}11 and H5hr^{cs}12, have been isolated (Ho et al., 1982). These mutants transform primary rat cells as well as, or better than, wild-type Ad5 at 38.5°C, but at 32.5°C either are completely defective (H5hr1 and H5hr^{cs}12) or display severely impaired transformation activity (H5hr^{cs}11) (Ho et al., 1982). H5hr1, as well as additional Ad5 mutants engineered to alter the expression of products of the largest E1A mRNA, such as H5dl1101 (Babiss et al., 1984a), have also been reported to be cold-sensitive for transformation of a clonal line of established (i.e. immortal), rat fibroblasts, CREF cells (Babiss et al., 1983). These properties are quite in line with those described previously, emphasizing the role of an E1A product(s) in initiation of transformation. Equally interesting, however, are the results of temperature-shirt experiments: when shifted from a permissive to a non-permissive temperature, both CREF and primary rat embryo cells transformed by H5hr1 or H5dl1101, unlike wild-type transformants, lost the ability to grow efficiently in soft agar (Babiss et al., 1983; 1984b cited in Ho et al., 1982). Some lines also displayed temperature-dependent growth to high saturation densities (Babiss et al., 1983), whereas others formed colonies at much reduced efficiency when plated upon mouse 3T3 cells at 32.5°C (cited in Ho et al., 1982).

Because an E1A product potentiates expression of other viral early genes (Jones and Shenk, 1979b; Berk et al., 1979) it could be argued that the apparent requirement for an E1A function in maintenance of these phenotypes is merely a reflection of temperature-dependent E1A-expression and, therefore, temperature-dependent expression of an E1B gene product that performs the maintenance functions. The E1A protein that potentiates expression of viral early genes is that encoded by the E1A-a mRNA: (Berk et al., 1979; Ricciardi et al., 1981; Montell et al., 1982). The H5hr1 mutant carries a frameshift mutation, the result of a single base-pair deletion at nucleotide 1055, which is followed by a termination codon, 11 codons beyond the frame shift (Ricciardi et al., 1981). The H5hr1 mutation lies in the unique portion of the E1A-a mRNA (see Fig. 3) and thus alters only the protein(s) specified by this species. Production of viral early mRNA species is indeed severely depressed in H5hr1-infected cells (Berk et al., 1979), but alternative mutations that have less radical effects upon E1B expression have been introduced into E1A. Carlock and Jones (1981), for example, inserted an octanucleotide EcoR1 linker into the Sma1 site at nucleotide 1010 in Ad5 E1A DNA, a site that is expressed solely in E1A-a mRNA (Fig. 1 and 3). The resulting mutant, H5in500, cannot express normal E1A-a mRNA products because the frame into which translation is shifted by the insertion terminates at nucleotide 1086. In this respect, H5in500 resembles H5hr1 (Fig. 3). HeLa cells infected by H5in500 produce close to wild-type levels of E1B mRNA species, although expression of the E2A and E4 genes is substantially reduced (Carlock and Jones, 1981). Nevertheless, H5in500 displayed no transforming activity whatsoever when tested on rat embryo cells and

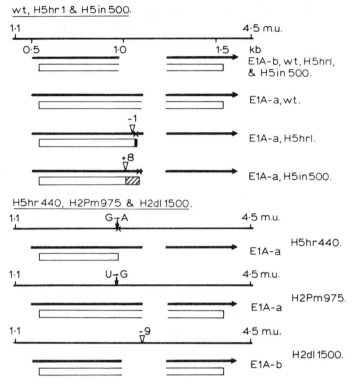

Fig. 3. Sites and consequences of subgroup C adenovirus E1A mutations. In both parts of the figure, the E1A mRNA and protein species are depicted as described in the legend to Fig. 1. The species encoded by wild-type virus are redrawn immediately below the horizontal line representing the viral genome in the top part of the figure. In both sections, the sites of nucleotide additions or deletions are shown by ▽, the introduction of new termination codons by X and the introduction of single base changes by vertical arrows. Changes in reading frame, compared to wild-type, are shown by the solid or cross-hatched boxes.

substantially depressed frequencies of transformation of BRK cells (Carlock and Jones, 1981). An independent mutant with similar properties was isolated by Solnick (1981) following nitrous acid mutagenesis of E1A DNA. This isolate, H5hr440 contains point mutations that lead to synthesis of a truncated E1A-a mRNA product (Fig. 3), and inhibit production of E1A-b mRNA. H5hr440 is also defective for transformation of BRK cells, despite the synthesis of normal levels of E1B mRNA species and polypeptides in H5hr440-infected HeLa cells (Solnick and Anderson, 1982). Similar results have been obtained with Ad12 mutant plasmids in which expression of the largest E1A mRNA species was specifically impaired (Bos et al., 1983). Thus, the function of product(s) specified by E1A-a mRNA in transformation cannot be merely to induce expression of the E1B gene.

Direct evidence that E1A gene products influence the phenotype of adenovirus transformed cells has also come from experiments in which pairs of cloned, Ad5 or Ad12 E1A and E1B regions have been introduced into rodent cells by transfection. Baby rat kidney cells transformed by Ad5 (subgroup C) can be distinguished morphologically from Ad12 (subgroup A) transformants: the former appear as well-defined foci in which the cells are very densely packed, whereas Ad12 transformed cells form less closely packed foci that have no sharp boundaries. The morphology of BRK cells transformed by the combinations Ad5E1A + Ad12E1B and Ad2E1A + Ad5E1B correlated with the transfected E1A gene (van den Elsen et al., 1983a), and this same correlation was obtained even when E1B expression was rendered independent of E1A gene products in plasmids in which the Ad5 or Ad12 E1B region was placed under the control of SV40 early promoter and enhancer elements (van den Elsen et al., 1983a).

The properties of transformation by hybrid plasmids also indicate that E1A gene products influence the efficiency of transformation. Cloned Ad5 DNA fragments comprising the region 0–15.5 mu transform BRK cells at a frequency of the order of 12 foci per μg equivalent DNA. The corresponding, cloned Ad12 DNA fragments display approximately a 20-fold lower transformation efficiency (Bernards *et al.*, 1982). Furthermore, Ad5 transformed foci become visible in 10–12 days, whereas those induced by Ad12 do not become visible for 3–4 weeks (Bernards *et al.*, 1982). The efficiencies of transformation by Ad5/Ad12 E1 hybrid plasmids were found to be determined by the source of the E1A gene: the Ad12E1A + Ad5E1B and Ad5E1A + Ad12E1B hybrid plasmids transformed BRK cells at low and high frequency, respectively (Bernards *et al.*, 1982).

The body of evidence reviewed in previous paragraphs provides compelling evidence that adenovirus E1A gene products participate in both early events in the induction of transformation and maintenance of certain aspects of the transformed phenotype. Moreover, the failure of H5in500 and H5hr440, to transform rat cells suggests that the 289 amino acid protein encoded by E1A-a mRNA (see Fig. 3) plays an essential role in transformation (Carlock and Jones, 1981; Solnick and Anderson, 1982). The cold-sensitive phenotype of H5hr1 and H5dl1101 transformation and the resulting transformed cells suggests that the 289aa protein is important not only in initiation, but also in maintenance of certain aspects of transformation (Ho *et al.*, 1982; Babiss *et al.*, 1983, 1984b). Temperature-dependent phenotypes displayed by H5hr1 transformed primary or established rat cells include anchorage independent growth (Babiss *et al.*, 1983, 1984a; cited in Ho *et al.*, 1982). The morphology of H5hr1 transformed cells has also been observed to be less epitheloid than is typical of adenovirus transformants, regardless of whether the cells transformed were the established CREF line (Babiss *et al.*, 1983), BRK cells (Ruben *et al.*, 1982) or primary rat embryo cells (J. Williams, personal communication). However, Babiss *et al.* (1983) reported an additional flattening of the cells when H5hr1 transformed CREF cells were shifted from a permissive to a non-permissive temperature, whereas Ho *et al.* (1982) discerned no temperature-dependent alterations in morphology. This difference is unlikely to represent variable expression of E1B functions, for H5hr1 is not temperature-dependent for growth on HeLa cells (Ho *et al.*, 1982), that is, it fails to express normal levels of early mRNA species at any temperature. And, as we have seen E1A gene products can influence morphology in a manner that is independent of any effects on E1B gene expression (van den Elsen *et al.*, 1983a). These observations are therefore difficult to reconcile, but might result from subtle differences in the permissive and non-permissive conditions employed, or perhaps the use of established cell lines compared to primary cells.

Be that as it may, the most obvious interpretation of the temperature-dependent phenotypes reproducibly displayed by H5hr1 transformed rat cells, such as anchorage-independent growth, is that the truncated version of the 289aa E1A protein encoded by this mutant can provide the necessary function at a high but not a low temperature: the H5hr1 mutation does not impair expression, or alter the protein product of E1A-b mRNA (see Fig. 3). It is, of course, formally possible that this latter protein mediates these E1A transformation functions and is inherently cold-sensitive. Unlikely as this interpretation might seem, it has recently received support from the properties displayed by several Ad2 E1A mutants. Two were constructed initially to prevent expression of either the E1A-a (H2dl1500) or the E1A-b (H2Pm975) mRNA species by the introduction of point mutations at the appropriate 5' splice sites (see Fig. 3, Montell *et al.*, 1982, 1984). In the construction of Pm975, advantage was taken of the degeneracy of the triplet code to introduce a T→G transversion at the E1A-b mRNA 5' splice site without alteration of the E1A-a mRNA coding sequence (Montell *et al.*, 1982). The mutant which expresses only the 243aa protein, H2dl1500, like H5hr1 (Graham *et al.*, 1978; Ho *et al.*, 1982; Babiss *et al.*, 1983), displayed temperature-dependent transformation: it transformed CREF cells at 37°C but displayed no transformation activity at 32°C (Montell *et al.*, 1984). These authors therefore concluded that the cold-sensitivity of H5hr1 for initiation of transformation can be ascribed to inherent cold-sensitivity of the 243aa protein.

At this point, it is worth recalling that there is no direct assay for initiation: because the transformants must be recognized by some criterion that distinguishes them from their untransformed parents all assays of transformation require some 'maintenance' functions. If cells transformed at a permissive temperature by a temperature-dependent mutant remain recognizably transformed when transferred to a non-permissive, temperature, but the virus shows no or reduced transformation under non-permissive conditions, then conclusions about function of the mutated product in initiation are justified. Just such findings have been reported for H5hr1 (Ho et al., 1982; Babiss et al., 1983, 1984a). Temperature-shift experiments with H2dl1500 transformed cells were not reported by Montell et al. (1984). However, rat cells (both CREF and BRK) transformed by a second virus constructed to express only the 12S E1A-b mRNA, H5dl1520, have been reported to be cold-sensitive, growing very poorly at 32°C (Haley et al., 1984).

Several additional complications arise when considering the roles of the 243 and 289aa proteins in transformation: all portions of the former are included within the latter, so that the two proteins might be at least partially interchangeable. The efficient transformation induced by H5dl1500 and H5dl1520 (Montell et al., 1984; Haley et al., 1984) might imply that the 289aa protein is not essential for initiation and the maintenance functions assayed by focus formation. Nevertheless, H2Pm975, which can express only this protein, is not defective for transformation either. Rather it induces typical transformed foci in CREF cells at about 20% the frequency of wild-type Ad2 and the absolute frequency is even higher, close to 60% of the wild-type level, when the microfoci that this mutant induces are included in the count (Montell et al., 1984). Thus, the E1A-a and E1A-b proteins routinely observed in transformed cells appear to be functionally equivalent, to some degree, at least when it comes to transformation of immortalized cells. There is, on the other hand, some evidence to suggest that the 289 and 243aa E1A proteins are not perfect functional homologues: although H2dl1500, which cannot express the 289aa protein, transforms CREF cells with a greater efficiency than wild-type virus, H5hr1, which encodes a truncated N-terminal version of this protein as well as the 243aa protein (see Fig. 3), is even more efficient (Montell et al., 1984).

The 'supertransformation' induced by these mutants whose expression of the E1A-a mRNA or its normal protein products is impaired emphasizes that transformation by infectious virus may be the outcome of several 'competing' virus–host cell interactions, particularly in the case of subgroup C adenoviruses, for which all rodent cells are semipermissive (see Section 2). It seems that the ability of H5hr1 or H2dl1500 to transform at higher efficiency than wt Ad2 or Pm975 can be attributed to their decreased cytopathogenicity to rat cells (Montell et al., 1984), presumably the result of inefficient expression of other, potentially toxic, viral early gene products: a greater fraction of the infected cells survive the virus infection to manifest a transformed phenotype.

Whether the two E1A proteins are also partially interchangeable for transformation of normal cells is not clear. H2Pm975, which cannot express the 243aa protein, appears to be completely defective for transformation of primary rat embryo cells (Montell et al., 1984). Although the activity of H2dl1500 in primary rat cells was not reported, the analogous mutant of Ad5, H5dl1520, transformed baby rat kidney cells more efficiently than did wild-type Ad5, the same behavior as shown when the mutant was tested for transformation of established CREF cells (Haley et al., 1984). An additional mutant of Ad2, H2dl1231 that transforms rat embryo fibroblasts at 20–50% the efficiency of wild-type Ad2 can express normal quantities of E1A-a mRNA, but reduced quantities of a defective E1A-b mRNA (Hurwitz and Chinnadurai, 1985). Ad5 cDNA viruses whose genomes encode only E1A-a (dl348) or E1A-b (dl347) mRNA both display reduced abilities to transform rat embryo fibroblasts (Winberg and Shenk, 1984). The two mutants H5hr1 and H5in500 that can express a normal 243aa protein, but truncated, N-terminal portions of the 289aa protein were reported to be severely impaired for transformation of primary rat embryo cells at 37°C (Graham et al., 1978; Carlock and Jones, 1981; Solnick and Anderson, 1982). These observations suggest that both E1A proteins are necessary to the transformation of cells that have not previously been immortalized. Transformation was assayed by focus-formation in all experiments cited in this paragraph, a procedure which, as discussed previously, requires that phenotypes such as altered morphology and growth properties are manifest. It will, therefore, be of considerable interest to compare the properties of the numerous E1A mutant virus in more subtle assays.

Certain phenotypes displayed by cells transformed by the E1A mutants discussed in previous paragraphs have also been examined. CREF cells transformed by H2Pm975 or H2d11500 were examined by Montell et al. (1984). Although the latter formed somewhat smaller colonies in soft agar than wild-type transformed cells, a far more dramatic reduction in colony size was shown by H2Pm975 transformants (Montell et al., 1984). The 243aa protein, not expressed in H2Pm975-infected cells, therefore appears to be the more important determinant of anchorage-independent growth. A similar conclusion was reached by Hurwitz and Chinnadurai (1985), who observed that rat embryo fibroblasts transformed by H2d1231, which can express only small quantities of aberrant E1A-b mRNA, were defective in the ability to form colonies in soft agar. In addition, either CREF or normal baby rat kidney cells transformed by H5d11520, which cannot express E1A-a mRNA and its products, displayed a more fibroblastic morphology than normal adenovirus transformants (Haley et al., 1984), suggesting that 289R E1A proteins are important to induction of this phenotype.

The results of genetic analysis of the roles of E1A proteins in adenovirus transformation leave several issues to be resolved. Nevertheless, certain conclusions can be justified by the available data. It seems clear, for example, that the two E1A proteins are partial functional homologues, at least during transformation of established cell lines: as discussed, neither H2d11500 nor H2Pm975 is absolutely defective for transformation of these cells and both alter the ability, when compared to wild-type Ad2, of transformed CREF cells to grow in an anchorage-independent fashion (Montell et al., 1984). This conclusion cannot be too surprising in view of the close relationship between the two proteins (Figure 1). On the other hand, it can be argued that the 289aa protein plays a more important role in transformation of primary rodent cells. Quite a large number of mutant viruses expressing only the E1A-a or the E1A-b mRNA normally have now been constructed and further study of their properties, for example, using assays designed to examine immortalization of primary cells in the absence of other phenotypic changes, will undoubtedly improve our understanding of the roles played by these two proteins during transformation by adenoviruses.

4.2. FUNCTION ASCRIBED TO E1B GENE PRODUCTS

The early DNA transfection experiments discussed in Section 4.1 indicated that E1A DNA was sufficient to induce partial transformation, but that E1B sequences were required if the fully transformed phenotype was to be displayed. The failure of a plasmid containing only the Ad5 E1B gene to transform BRK cells (van den Elsen et al., 1982) is consistent with the conclusion that E1A gene products are both necessary and sufficient to 'initiate' transformation. It is, however, well established that 289aa product of E1A-a mRNA permits efficient expression of all other viral early genes (Berk et al., 1979; Jones and Shenk, 1979b; Nevins, 1981; Ricciardi et al., 1981; Montell et al., 1982). The failure of the cloned E1B gene to manifest transforming activity might, therefore, simply reflect very inefficient expression of the E1B gene in the absence of an E1A gene. To test this possibility, van den Elsen et al. (1983b) placed the Ad5 E1B region under the control of the SV40 early transcriptional control region and enhancer elements. Rodent cells into which such a hybrid E1B region was introduced by co-transfection with a dominant, selectable marker synthesized E1B polypeptides at levels comparable to those made in cells transformed by a cloned Ad5 DNA fragment containing both the E1A and E1B genes. Nevertheless, no foci of morphologically transformed cells were observed unless the BRK cells also received a plasmid containing the E1A region (Bernards et al., 1983c), indicating that elevation of expression of E1B in the absence of E1A gene products was not sufficient to induce morphological transformation of normal BRK cells.

The negative results obtained in these experiments suggest the E1B region might provide functions acting 'later' in transformation than those that mediate immortalization, which, as we have seen can be achieved by the transfected E1A gene alone. One might, therefore, expect E1B DNA to be sufficient to transform immortal, but otherwise normal rodent cells.

Such a separation of transformation functions would be in line with results obtained with other oncogenes: both certain viral genes, for example, cloned cDNA expressing only the polyoma middle T-antigen (Rassoulzadegan et al., 1982), and cellular genes, such as activated members of the ras gene family (see for example, Der et al., 1982; Parada et al., 1982),can morphologically transform established cell lines, such as NIH/3T3, but fail to transform cultures of primary or secondary cells (Rassoulzadegan et al., 1982, 1983; Land et al., 1983a, b; Newbold and Overall, 1983). In either case, co-transfection of the adenovirus E1A region permits transformation of such normal cells (Ruley, 1983, Land et al., 1983a, b). By contrast, the Ad5 E1B gene, cannot morphologically transform rat cells of the immortal 3Y1 line, even when expressed from SV40 transcriptional control elements, (van den Elsen et al., 1983b). This failure distinguishes the Ad5 E1B gene from members of the ras class of oncogenes (see, for example, Land et al., 1983b). It also illustrates that a simple two-step model, in which the E1A region would supply early immortalization functions whereas the E1B region would mediate complete morphological transformation and the acquisition of other phenotypes is too simple to explain adenovirus transformation. Rather, the results discussed in Section 4.1 establish that E1A gene products must do more than immortalize normal cells. And genetic analysis of the roles of the E1B gene products in transformation has demonstrated that at least one provides not only maintenance but also initiation functions.

Several mutants of Ad5 or Ad12 in which expression of the E1B region is compromised by insertion or deletion of sequences do not transform primary rat embryo fibroblasts (Jones and Shenk, 1979a; Bos et al., 1983; Fukui et al., 1984). Similar properties are displayed by host-range or host-range, cold sensitive E1B mutants (group II mutants) of Ad5 obtained following nitrous acid mutagenesis, which are defective in transformation of rat embryo fibroblasts, rat brain and BRK cells (Graham et al., 1978; Ho et al., 1982). The inability of such E1B mutants to transform appears to be at odds with the partial transformation induced by DNA of the E1A region (see Section 4.1). As the same cell types and similar transformation assays (focus formation in liquid media) have been used when viral genes were introduced by transfection or following infection by mutant viruses, the resolution to this apparent paradox would appear to lie in the way in which E1A DNA was presented to the cell, as deproteinized DNA on the one hand or as part of a viral genome on the other. The properties of E1B mutants suggest that E1A DNA carried as part of a viral genome that cannot express functional E1B product(s) is not sufficient to initiate transformation of cells that are not immortal, whereas transfected E1A DNA is sufficient. Bizarre as this conclusion might appear, no other obvious explanation can account for the discrepant observations made with transfected DNA and mutant viruses. Moreover, Rowe and Graham (1983) have reported that the DNA of two host-range, transformation-defective E1B mutants, H5hr6 and H5hr60, when transfected into baby rat kidney cells transforms them as efficiently as wild-type DNA, just as the explanation given above would predict. Thus, we are forced to the inevitable conclusion that any picture of adenovirus transformation based solely upon the results of DNA transfection experiments will be incomplete: this method of introduction of viral genes permits several steps essential in transformation by the virus to be bypassed. Moreover, it is clear that under normal circumstance, that is, following adeno*virus* infection of semi- or non-permissive cells, E1B functions contribute to initiation of transformation of primary cells.

Considerable progress has been made recently toward the elucidation of the roles of the E1B 55kd and 19kd proteins in transformation. These might be expected to perform unrelated functions for they share no coding sequences (Section 3). The lesions carried by two of the Ad5 group II mutants mentioned previously, H5hr6 and H5hr^{cs}13, have been mapped to the region 6.1 to 8.0 mu by marker rescue (Williams et al., 1979) and recently located to sites expressed only in the largest E1B protein (Williams et al, 1986). These Ad5 mutants transform rat embryo cells not at all, or very poorly, at all temperatures tested in the range 32.5°C to 38.5°C (Graham et al., 1978; Ho et al., 1982), demonstrating the importance of the largest E1B protein in transformation of primary cells: as discussed previously at least some E1B mutants have been shown to induce a partially-transformed phenotype in established rat cell lines. Although the Ad5 group II mutants transform very poorly, H5hr^{cs}13 did give rise to a few foci of transformed rat embryo cells at 38.5°C, from which cell lines were established (Ho et al., 1982). Such cell lines displayed a cold-sensitive phenotype, similar to that discussed in Section 4.1 for H5hr1 cold-sensitive transformants, (cited in Ho et al., 1982). Thus, the E1B 55kd protein of Ad5 has been implicated in maintenance of certain transformation phenotypes, a well as initiation of transformation by infectious virus.

Several lines of evidence have been cited in support of the view that the E1B 19kd protein is also essential to the induction of complete transformation, a view initially based on results of DNA transfection experiments. Fragments of subgroup A or subgroup C adenovirus DNA comprising the left-hand 7.2 or 7.9%, respectively, were found to induce complete transformation of primary rat cells with an efficiency comparable to that obtained with DNA fragments containing an intact E1B (Graham et al., 1975a; Shiroki et al., 1977; van der Eb et al., 1979; Dijkema et al., 1979). It can be seen in Fig. 2 that these DNA fragments contain all coding sequences of the E1B 19kd protein, but cannot express the normal E1B 55kd protein. It was therefore concluded that the 19kd protein was essential to transformation. Additional evidence cited in support of this conclusion has come from the properties of cytopathic mutants of Ad2. Two such large-plaque (lp) mutants obtained after nitrous acid mutagenesis of Ad2, or DNA comprising the left-hand 15% of the viral genome, have been examined in detail (Chinnadurai, 1983). One isolate, H21p5, carries two G→T transversions, at nucleotides 1954 and 2237. The former changes a tyr to an asp codon near the N-terminal end of the 19kd protein, whereas the latter affects all known E1B proteins: it alters an ile to a met codon near the N-terminus of the 55kd and 20kd proteins but also substitutes a leu for the normal 19kd stop codon. This mutant should therefore encode a larger 19kd-related protein and such a polypeptide, of 21kd, was indeed observed in H21p5-infected cells (Chinnadurai, 1983). The second mutant, H21p3, carries a single C→T transition at nucleotide 1718, changing a val to an ala codon in the 19kd protein. H21p3 transformed 3Y1 cells with only 10% the efficiency of wild-type Ad2, whereas H21p5 was completely defective (Chinnadurai, 1983). At first sight, these observations indicate that the E1B 19kd protein, the only product mutated in H21p3, is necessary to achieve efficient transformation, even of established cells. However, Chinnadurai (1983) assayed transformation as the appearance of foci when infected cells were maintained in semisolid medium, a more stringent assay than typically used. Thus, it was not clear from these experiments whether the 19kd protein was actually required early in transformation or induced the phenotype assayed by Chinnadurai. In this context, it should be recalled that Shiroki et al. (1981) have shown that an Ad5 mutant lacking all E1B sequences will transform 3Y1 cells by the criterion of focus formation in liquid medium. More recently, a large number of mutants of Ad2 or Ad5 bearing lesions restricted to E1B sequences encoding the 19kd protein have been constructed and assayed for transforming activity (Subramanian et al., 1984; Babiss et al., 1984b; Pilder et al., 1984; Takemori et al., 1984). All such mutant viruses show severe transformation defects when assayed by the ability to induce focus formation upon infection of either established cells of the CREF or 3Y1 lines (Subramanian et al., 1984; Babiss et al., 1984b; Pilder et al., 1984; Takemori et al., 1984) or rat embryo fibroblasts (Subramanian et al., 1984; Pilder et al., 1984). Similar mutants of Ad12, constructed by insertion or deletion of sequences at a restriction endonuclease site near the N-terminal end of the 19kd coding sequences to induce shifts in reading frame and premature termination of translation, show a decrease in efficiency of transformation of 3Y1 cells, but are not absolutely defective (Fukui et al., 1984). It can, therefore, be concluded that the E1B 19kd protein is essential to focus formation, regardless of the state of the infected rodent cells. Moreover, all the cell lines established from foci of cells transformed by such mutants that have been examined show little ability to grow in soft agar, that is, are deficient for anchorage-independent growth (Fukui et al., 1984; Subramanian et al., 1984; Pilder et al., 1984). These observations, and the initial results of Chinnadurai (1983) emphasise the importance of the E1B 19kd protein to induction of this phenotype. In addition, foci of cells transformed by E1B 19kd mutants occasionally exhibit atypical morphology (Fukui *et al.*, 1984; Subramanian *et al.*, 1984).

Results obtained with mutants carrying lesions in the 19kd E1B protein should be considered in the context of the increased cytopathogenicity of such viruses. The E19 19kd mutants of Chinnadurai (1983), for example, were isolated by virtue of their ability to induce huge plaques in permissive cells and most of the E1B 19kd mutants discussed in the previous paragraph display this phenotype and/or induce degradation of viral and cellular DNA (Subramanian *et al.*, 1984; Pilder *et al.*, 1984; Takemori *et al.*, 1984; White *et al.*, 1984). A similar class of Ad12 mutants, the *cyt* mutants first described by Takemori *et al.* (1968), destroy infected cells much more efficiently than wild-type Ad12. The majority of those tested transform primary rat kidney cells at reduced efficiencies compared to wild-type Ad12 (Mak and Mak, 1983), and the mutants have long been known to be less oncogenic in animals than wild-type Ad12 (Takemori *et al.*, 1968; Yamamoto *et al.*, 1972). The *cyt* mutations have been located to the E1B region and probably affect the 19kd protein (Lai Fatt and Mak, 1982), although they have not yet been precisely placed by DNA sequence analysis. It could, therefore, be argued that mutations within the E1B 19kd coding sequences reduce transformation efficiency by virtue of their increased ability to kill infected cells.

Indeed, several E1B 19kd mutants examined by White and colleagues (1984) induced such severe cell killing of primary rate cells that transformation could not be examined. Moreover, the one *cyt* mutation, present in a virus defective by virtue of a second mutation, that did not induce such a dramatic cytopathic effect did not alter transformation efficiency (White *et al.*, 1984). On the other hand, Chinnadurai (1983) reported that wild-type and large-plaque mutant Ad5-infected RY1 cells displayed similar cloning efficiencies and the dependence of transformation frequency upon multiplicity of infection was found to be very similar for wild-type Ad12 and several *cyt* mutants (Mak and Mak, 1983). Similarly, Takemori *et al.* (1984) did not observe excessive killing of rat 3Y1 cells when testing the transforming activity of a large number of *cyt* mutants of Ad2 or Ad5. It therefore seems clear that, while the cytocidal and transforming activities of mutants carrying lesions in the E1B 19kd protein-coding sequences vary somewhat with the nature of the mutation, the impaired ability to transform displayed by these mutant viruses is not simply a trivial result of their cytotoxicity.

The results discussed in this section raise an obvious paradox: if the largest E1B protein were important both early in the transformation process and in the maintenance of certain phenotypes, then viral DNA fragments that include only E1A-proximal part of the E1B region would be expected to be defective in at least some aspect(s) of transformation. Yet, such fragments have been routinely reported to transform efficiently and fully (Graham *et al.*, 1975b, 1977; Shiroki *et al.*, 1977; Dijkema *et al.*, 1979; Jochemsen *et al.*, 1982; Rowe *et al.*, 1984). A reasonable explanation of this apparent contradiction is not too difficult to find, that it is the N-terminal portion of the larger E1B protein that carries transformation functions. Thus, the lesions carried by group II host-range or host-range, cold-sensitive Ad5 E1B mutants lie with this region (Ho *et al.*, 1982; B. Karger and J. Williams, personal communication). Moreover, the efficiency of transformation is progressively impaired as deletions within the E1B 55kd coding region approach the N-terminus of the protein (Babiss *et al.*, 1984b). It can therefore be argued that the smallest DNA fragments capable of inducing complete transformation, 0–7.2 mu or 0–7.9 mu of Ad12 of Ad5 DNA, respectively, encode a sufficient portion of this important N-terminal region of the larger E1B protein. Indeed, such fragments include some 25kd worth of coding capacity starting from the 55kd internal initiation codon (see Fig. 2). It is therefore quite possible that cells transformed by such DNA fragments synthesize a truncated protein that is related to a 55kd or 20kd E1B proteins (see Section 3). Low molecular weight, E1B-encoded proteins in addition to the 19kd protein have been observed in cells transformed both by Ad5 and Ad12 DNA fragments (van der Eb *et al.*, 1979; Jochemsen *et al.*, 1982) but whether any are related to the 55kd E1B protein has not been established. Clearly further work on the nature of the E1B products synthesized in cells transformed by DNA fragments that contain N-terminal portions of the E1B region, in conjunction with detailed comparisons of the biological properties exhibited by such cells to those shown by cells transformed by an intact E1B region, will be required to settle the question of whether the truncated E1B 55kd or 20kd proteins can indeed substitute for their larger relatives.

The most striking feature of our present picture of the roles of the E1A and E1B gene products in adenovirus transformation is its complexity: each of the four proteins encoded by these regions and routinely expressed in transformed cells (as well as the DNA polymerase) appears to be necessary if normal cells are to be stably converted to a fully-transformed state following virus infection. In this sense, adenoviruses stand in marked contrast to other transforming viruses, whose transforming functions reside in one or two proteins, for example, acute retroviruses and polyomavirus, respectively.

While the work described in this section represents substantial progress in our understanding of the roles played by the adenovirus E1A and E1B gene products in transformation, the definitions of these roles has not yet progressed beyond terms that are purely descriptive, conveying little information about the underlying molecular changes. Our ignorance in this area reflects, at least in part, our incomplete picture of the molecular and biochemical properties of the adenovirus transforming proteins themselves (see Section 6). Moreover, the very low efficiency with which non- or semi-permissive cells become transformed following adenovirus infection raises an almost impenetrable barrier to the direct elucidation of the early virus-cell interactions that will lead to transformation. On the other hand, it should eventually be possible to obtain a molecular description of the phenotypes of adenovirus transformed cells and of the functions of the individual E1A and E1B proteins: the methods applied to the separation of adenoviral transforming functions discussed in this section in conjunction with molecular analysis of transformed phenotypes and the properties of the transforming proteins should prove valuable in this regard.

5. TUMORIGENICITY OF ADENOVIRUSES AND ADENOVIRUS TRANSFORMED CELLS

The past year or two has brought much new information concerning the determinants of the ability of adenovirus transformed cells to induce tumors and some insights into the molecular mechanisms that might mediate the differences in oncogenicity displayed by human adenoviruses. Once again, both genetic analyses and experiments with interserotypic hybrid plasmids have made important contributions.

Baby rat kidney cells transformed by hybrid plasmid carrying an Ad12 E1B region exhibited the higher tumorigenicity (very close to 100%) characteristic of Ad12 transformed cells, whereas transformants containing an Ad5 E1B region induced tumors with a frequency 0–20%, even lower than that displayed by Ad5 transformants, 50% (Bernards et al., 1982). Similarly, rodent cells transformed by recombinant viruses that express the Ad12 E1A region but not the Ad12 E1B region were unable to induce tumors in newborn, syngenic rats, in this respect resembling wild-type Ad5 transformants (Shiroki et al., 1982). These results therefore suggest that adenovirus E1B gene products determine the tumorigenic potential of adenovirus transformed cells. Interestingly, however, the Ad12 E1B region alone does not appear to be sufficient to establish the oncogenic potential of the virus upon direct inoculation into animals: Ad5 recombinant viruses in which the E1B region has been replaced by that of Ad12, while faithfully expressing the Ad12 E1B region, were found to be incapable of tumor induction (0 in 90 days) when inoculated into newborn hamsters, a result typical of wild-type Ad5 (Bernards et al., 1983c). Adenovirus type 12, by contrast, induced tumors in 90% of the animals in 20–50 days. It would therefore seem that viral products other than those encoded by the E1B region influence the oncogenicity of the virus in a serotype-specific fashion.

The question of whether one, other or both the E1B proteins are important determinants of transformed cell tumorigenicity has been explored using additional hybrid plasmids. Two were constructed by Bernards et al. (1983a) such that each could express the E1A region of both Ad5 and Ad12, but psT12 expressed an Ad12 19kd E1B protein and an Ad5

55kd E1B protein, whereas pLT12 encoded an Ad5 19kd E1B protein and an Ad12 54kd protein. A third construct that contained the Ad12 E1A region but only the 54kd coding sequences of E1B induced transformed BRK cells that were completely non-tumorigenic in nude mice (Bernards et al., 1983b). This result demonstrates the necessity of the E1B 19kd protein to the ability of transformed cells to induce tumors in this system, but appears in conflict with the greatly reduced frequency of transformation displayed by virus carrying mutations in the E1B 19kd coding sequences (see Section 4.2). In this case, the discrepancy does not seem to lie in the method of introduction of viral DNA sequences into recipient cells, transfection of a mutant plasmid or infection by mutant viruses: one set of E1B 19kd mutants were also tested by transfection of deproteinized viral DNA into CREF cells and found to be as defective for transformation as when cells were infected with mutant viruses (Babiss et al., 1984b). By contrast, the plasmid containing the Ad12 E1A region but only E1B 54kd coding sequences examined by Bernards et al. (1983a) transformed BRK cells by the criterion of focus formation in liquid medium nearly as efficiently as wild-type Ad12 E1A + E1B DNA (Bernards et al., 1983a). Whether the observed differences are the result of transfection of plasmid DNA (Bernards et al., 1983a) as opposed to complete viral DNA molecules (Babiss et al., 1984b) is not known.

Although synthesis of an E1B 19kd protein is necessary if adenovirus transformed cells are to be tumorigenic in nude mice, the different oncogenic potentials displayed by Ad5 compared to Ad12 transformed cells appear to be determined by the larger E1B protein. Thus, pST12 (Ad5 E1B 55kd protein) transformed cells induced tumors in 8–25% of innoculated nude mice with a latent period of 50–100 days, whereas those transformed by pLTAd12 (Ad12 54kd E1B protein) were tumorigenic in all the animals in 20–40 days and thus display the high oncogenicity typical of cells transformed by a plasmid containing only Ad12 E1B coding sequences (Bernards et al., 1983a).

The E1B 19kd protein appears to be a critical determinant of the tumorigenicity of adenoviruses as well as adenovirus transformed cells in nude mice: the Ad12 E1B 19kd insertion and deletion mutants of Fukui et al. (1984) induced no tumors whatsoever when mutants virus was inoculated into baby hamsters. Wild-type Ad12 at similar doses caused tumors to appear in at least 80% of the infected animals. Whether the largest E1B protein also influences virus oncogenicity has not been established. While these two E1B gene products appear to be necessary and sufficient to confer the subgroup-specific, tumorigenic properties of adenovirus transformed cells in nude mice they cannot, as discussed previously, be the sole determinants of their oncogenicity in newborn hamsters (Bernards et al., 1983a).

Some recent comparisons of the tumorigenicity of BRK cells transformed by hybrid plasmids in immunocompetent, syngeneic rats or nude mice may shed some light on this discrepancy. In nude mice, the transformed cell lines tested exhibited the high or low tumorigenicity characteristic of the virus from which the E1B region was derived (Bernards et al., 1983b), as in the experiments mentioned in previous paragraphs. However, only transformed cell lines that contained an Ad12 E1A region induced tumors in immunocompetent rats: those containing an Ad5 E1A and an Ad12 E1B region were not tumorigenic at all under the conditions employed (Bernards et al., 1983b). These results imply that cells expressing an Ad5, but not an Ad12, E1A gene can be eliminated by the animals' thymus-dependent immune response, which is of course functional in syngeneic rats but not nude mice. Indeed, transformed cells that were tumorigenic in immunocompetent rats were found to be less susceptible to in vitro killing by allogenic, cytotoxic T lymphocytes (CTLs) (Bernards et al., 1983b).

The molecular basis of the ability of these transformed cells expressing an Ad12 E1A region to escape immune surveillance and induce tumors in immunocompetent animals appears to lie in suppression of expression of class I major histocomptability antigens: a 45kd rat class I MHC RT1. A heavy chain, as well as a protein of 32kd, was found to be absent from cells transformed by Ad12 or hybrid plasmids that contained an Ad12 E1A region, but was normally produced in untransformed BRK cells or those transformed by Ad5 (Schrier et al., 1983). Inhibition of expression of these proteins in transformed cells was independent of the identity of the E1B region (Schrier et al., 1983). Only plasmids that could express the Ad12 protein encoded by E1A-a mRNA were able to inhibit expression of the 45kd protein in transformed cells and such cells were tumorigenic in rats (Bernards et al., 1983b). Thus, in addition to its ability to induce expression of other viral early genes (Bos and ten Wolde-Kraamwinkel 1983), the largest Ad12 E1A protein seems to possess the unique capacity (compared to its subgroup C counterparts) to suppress expression of certain cellular genes. The molecular basis of this phenomenon is not yet known. However, the Ad5 E1A-a mRNA product can block such inhibition of production of the cellular 45kd antigen, when both Ad5 and Ad12 E1A regions are expressed in transformed cells (Bernards et al., 1983b).

These workers have therefore concluded that the differences in oncogenicity displayed by subgroup A and subgroup C human adenoviruses are mediated by the unique ability of the largest Ad12 (subgroup A) E1A protein to block expression of class I MHC genes (Bernards et al., 1983b). It has recently been shown that expression of at least the N-terminal portion of the E1B gene (i.e. cells transformed by a minimum of Ad12 HindIII G, 0–7.2 mu) was necessary if Ad12 transformed cells were either to be killed efficiently by CTLs raised against syngeneic transformed cells or to elicit such CTLs when innoculated into animals (Föhring et al., 1983), suggesting that the E1B 19kd protein, a cell-surface protein at least in the case of Ad12 (see Section 6), is the viral antigen recognized by cytotoxic T lymphocytes. However, CTLs recognized viral antigens only in the context of 'self' class I MHC antigens (Zinkernagel and Doherty, 1979), a fact that appears to explain the dependence of in vitro killing on both the E1B gene product (Föhring et al., 1983) and on the nature of the E1A region expressed (Bernards et al., 1983b). Moreover, data presented by Bernards et al. (1983b) suggest that the identity of the E1B protein expressed is not important to CTL killing in vitro or to tumorigenicity: cells transformed by Ad12 E1A and Ad5 E1B regions were, for example, found to be as susceptible as those transformed by Ad12 E1A and E1B regions.

Satisfying as this picture might be, it cannot account for the results of several other studies of host defence systems evaded by oncogenic adenoviruses or cells transformed by them. Moreover, the relevance of reduced expression of MHC class I antigens to tumorigenicity of transformed cells has been questioned. Thus, for example, Mellow et al. (1984) examined several lines of syngeneic Ad12 transformed rat cells that displayed varying degrees of tumorigenicity in newborn rats and syngeneic, non-oncogeneic Ad2 transformed lines for the presence of MHC class I antigens. While a somewhat lower concentration of these antigens were indeed observed on Ad12 transformed cells, confirming the results of Schrier et al. (1983), no correlation was observed between this parameter and sensitivity of the transformed cells to killing with allogenic CTL's induced in vivo and in vitro, nor between sensitivity to killing in vitro and tumorigenicity in vivo (Mellow et al., 1984). Mellow and colleagues (1984) suggested that the most likely explanations for these differences in cell killing were the less pronounced differences in concentration of MHC class I antigens among the different cell lines examined than in those observed by Bernards et al. (1983b) and Schrier et al. (1983), or the different nature of the control experiments performed. Be that as it may, these more recent results suggest that inhibition of expression of MHC class I proteins does not provide a universal, or complete, explanation of the tumorigenicity of adenovirus transformed cells. Indeed, while such observations as the tumorigenicity displayed by certain lines of Ad2 transformed cells in athymic animals, but not in immuno-competent, syngeneic animals (Harwood and Gallimore, 1975; Gallimore et al., 1977) implicate T-cells in the host's defense against adenovirus transformed cells, other experiments point to an equally, if not more important role for natural killer (NK) cells. A direct, and absolute correlation of the tumorigenity of adenovirus transformed cells and their susceptibility to in vitro lysis by non-T, NK cells (Cook et al., 1982; Raskă and Gallimore, 1982; Sheil et al., 1984; Cook et al., 1982) or macrophages (Cook et al., 1982) from normal animals has been established. This cytotoxic activity has been shown to be independent of MHC class I antigens (Raskă and Gallimore, 1982). Similar differences in sensitivity to NK lysis are also exhibited by hamster cells infected by Ad12 (more resistant) and Ad2 (more sensitive) (Cook and Lewis, 1984). In the case of Ad2-infected cells, susceptibility to NK lysis increased throughout the period examined, in parallel with the production of viral early proteins (Cook and Lewis, 1984), leading to the suggestion that some difference in expression of viral early genes leads to an increased ability of Ad12-infected cells to survive the NK defence system (Cook and Lewis, 1984). The relevant molecular mechanism is not known and it has also to be established that mechanisms that can contribute to an explanation of how transformed *cells* evade immune surveillance in an animal also apply to cells infected *in vivo* following inoculation of *virus*.

6. PROPERTIES OF E1A AND E1B POLYPEPTIDES

Despite the progress towards identification of at least some of the roles played by the polypeptides encoded by adenovirus E1A and E1B transcriptional units in transformation, very little is known of the molecular mechanisms whereby these products alter normal cellular phenotypes, or indeed of their molecular properties. None of the proteins participating in adenovirus transformation is made in more than small quantities in infected or transformed cells and attempts to obtain them from other sources have begun only recently (see, for example, Ferguson *et al.*, 1984; Ko and Harter, 1984). The locations of the various E1A and E1B-encoded proteins within infected or transformed cells have been studied in some detail, with antibodies of increasing specificity and we possess, in certain cases, information after their functions in productive infections. As the human adenovirus E1A and E1B regions did not evolve to permit transformation of rodent cells, a property that has no relevance to the natural life history of the virus, it seems reasonable to consider their transforming activities as unusual manifestations of their functions in a productive infection.

6.1. The E1A Polypeptides

The E1A polypeptides, labelled in a one hour period, were found to be distributed equally between the nucleus and cytoplasm of infected cells by biochemical fractionation followed by immuno-precipitation with sera from animals bearing adenovirus-induced tumors (Rowe *et al.*, 1983a). None of the tumor sera employed in these experiments distinguish the two major forms of the E1A proteins, nor do those raised against an E1A-related protein synthesized in *E. coli* (Spindler *et al.*, 1984). However, Feldman and Nevins (1983) synthesized a 13 amino acid peptide predicted from the sequence of the unique portion of Ad2 E1A-a mRNA (Fig. 1). Rabbit antibodies raised against this peptide located the 289 amino acid polypeptide to both the nucleus and the cytoplasm, the protein being slightly more concentrated in the nucleus. The nuclear form could not be released from material that sedimented to the bottom of 10–30% glycerol gradients in low salt buffer and was also retained in the fraction surviving DNAase I digestion and extraction with 2M NaCl (Feldman and Nevins, 1983). These authors therefore concluded that the 289 amino acid E1A protein is associated with the nuclear matrix. By contrast, Spindler *et al.* (1984) reported that at least 90% of nuclear E1A protein could be solubilized in moderate concentrations of ammonium sulphate. The precision location of E1A proteins within the nucleus is, therefore, controversial. Cytoplasmic E1A polypeptides were found in both soluble and cytoskeletal fractions (Rowe *et al.*, 1983a).

Both E1A proteins are phosphorylated (Lassam *et al.*, 1979; Gaynor *et al.*, 1982; Yee *et al.*, 1983; Spindler *et al.*, 1984) and, as discussed in Section 3, may also be modified in other ways. The functional significance of phosphorylation is not presently understood.

The E1A 289 amino acid protein enhances transcription of other viral genes in productively infected cells (Berk *et al.*, 1979; Jones and Shenk, 1979b; Ricciardi *et al.*, 1981; Nevins, 1981). The precise molecular mechanism of action of this protein is not understood (Nevins, 1981; Gaynor and Berk, 1983; see Flint, 1984 for discussion). The ability to induce transcription might seem to be an ideal property for a protein involved in induction and maintenance of transformation. Indeed, this same E1A protein has been shown to induce expression of cellular genes, including that encoding a human 72kd heat-shock protein, (Nevins, 1982; Kao and Nevins, 1983), a human β-globin gene (Green et al., 1983b), a rat preproinsulin gene (Gaynor et al., 1984) and human histone and tubulin genes (Stein and Ziff, 1984). Thus, it is easy to imagine that continued expression of the E1A 289 amino acid protein in a transformed cell could permanently activate expression of inappropriate cellular genes whose products regulate the growth rate, metabolism or other phenotypes of the cell. Attractive as such a model might be, it has no direct experimental support at present. Furthermore, the experiments discussed in Section 4.2 seem to indicate that the 243 amino acid E1A protein, which cannot induce transcriptional activation of viral genes efficiently (Montell et al., 1982; Haley et al., 1984; Winberg and Shenk, 1984; Leff et al., 1984), is more important in transformation than the 289 amino acid protein. Clearly, further work will be required to establish whether the well-known ability of the 289 amino acid protein to activate transcription contributes to the molecular foundation of adenovirus transformation.

Recently, both the 289R and 243R E1A proteins have been shown to repress transcription of sequences under the control of papovaviral enhancer elements (Borelli et al., 1984; Velich and Ziff, 1985). The molecular basis of such repression is not understood, although it does appear to involve the enhancer directly (Borelli et al., 1984) and is mediated by E1A proteins in trans (Borelli et al., 1984; Velich and Ziff, 1985). The ability of E1A proteins to repress enhancer-mediated transcription is of obvious interest to the roles of these proteins in transformation (see Velich and Ziff, 1985, for discussion).

Interestingly, the 243 amino acid polypeptide, although not essential for adenovirus replication in growing cells, is required for efficient growth of the virus in strictly-arrested permissive cells (Spindler et al., 1985). It has been known for some time that adenovirus infection of growth arrested human or rodent cells can induce cellular DNA synthesis and abnormal cell-cycle progression (see Flint, 1984, for a review). The viral function responsible has been mapped to the E1A gene and shown to be independent of the transcriptional activation of other viral early genes by the 289 amino acid E1A protein (Braithewaite et al., 1983; Bellett et al., 1985).

The Ad2 pm975 mutant, which cannot produce the 243R E1A protein, could not alter control of cell cycle progression of rat cells to the same extent as wild-type virus, whereas H5hr1 and H5in500, producing the 243R protein and truncated 289R polypeptides (see Figure 3) were completely defective for this activity (Bellett et al., 1985). Thus, the 289R protein is essential for adenovirus-induced alterations in cell-cycle progression but its activity is augmented by the 243R protein, which alone is inactive. These observations are particularly interesting in light of the conclusions that both E1A proteins play important roles during transformation (section 4.1). The mitogenic activity of E1A proteins may be important, because the abnormal cell cycles induced by the E1A region are characterized by the presence of chromosomal aberrations (see Flint, 1984). These, and the induction of cellular DNA synthesis, could be well involved in integration of viral early genes, an essential prerequisite to the development of stable transformants.

6.2. The E1B 55kd Protein

The Ad2 or Ad5 E1B 55kd proteins are those predominantly recognized by antibodies of sera from tumor-bearing animals (see, for example, Levinson and Levine, 1977a, b; van der Eb et al., 1979; Ross et al., 1980). Immunofluorescence studies with such sera initially located this antigen predominantly to the nucleus of adenovirus transformed cells, where it appears as characteristic flecks and granules, a result confirmed with antibodies raised against synthetic, 55kd-specific peptides (Yee et al., 1983).

This protein, like the E1A polypeptides, is phosphorylated (Levinson and Levine, 1977a, b; van der Eb et al., 1979) and has also been reported to the phosphorylated *in vitro* by a protein kinase activity present in immunoprecipitates of infected-cell proteins obtained with sera from animals bearing tumors induced by an Ad5 transformed hamster cell line expressing only the E1A and E1B regions (Branton et al., 1981). This kinase phosphorylates serine and threonine residues, but can be distinguished from kinases present in uninfected cell extracts, by, for example, its mode of phosphorylation of histone H3 (Branton et al., 1981). Whether the kinase is an intrinsic activity the E1B 55kd protein or merely associated with it has not, however, been established.

A potentially interesting property of the large E1B protein is its association in transformed, but not productively-infected, cells with a cellular protein exhibiting an apparent molecular weight of 53kd: monoclonal antibodies directed against either the Ad5 55kd E1B protein or the cellular 53kd protein immunoprecipitate both proteins (Sarnow et al., 1982). The 53kd cellular protein recovered in such complexes is that found in association with the viral large T-antigen in SV40-transformed cells (Sarnow et al., 1982).

The E1B 55kd and cellular p53 proteins have recently been shown to exhibit the same intracellular distribution in transformed cells, most of each protein being found in an unusual filamentous organelle located close to the nucleus (Zantema et al., 1985). The interaction of products of transforming genes of different viruses with this cellular protein suggests that its association with viral proteins is likely to be of importance to the molecular mechanisms whereby transformed phenotypes are induced or maintained. Particularly exciting in this context are recent results implicating the cellular p53 protein in growth regulation in normal cells. When growth-arrested NIH/3T3 cells were stimulated to enter the cell cycle by addition of serum, the levels of the p53 protein and its mRNA were observed to increase prior to cellular DNA synthesis (Reich and Levine, 1984). Moreover, microinjection of species-specific antibodies against the p53 protein into arrested cells inhibited serum-stimulated DNA synthesis in a species-specific manner, but had no effect on the ability of cells to traverse the G_1 phase (Mercer et al., 1982, 1984). Most excitingly, the mouse p53 protein expressed at high levels from viral transcriptional control regions will cooperate with the c-ras oncogene to mediate transformation of primary rodent cells (Eliyahau et al., 1984; Parada et al., 1984) or immortalize primary rat chondrocytes (Jenkins et al., 1984).

Clearly, further studies of the functional consequences of the interaction of the p53 protein with viral transforming proteins such as the E1B 55kd polypeptide are likely to be rewarding.

The 55kd-related 20kd protein has been reported to be predominantly cytoplasmic in location in Ad2-transformed cells (Green et al., 1982) and a major virus-specific product in the lines examined (Green et al., 1982; Matsuo et al., 1982), but little else is known about it.

6.3. The E1B 19kd Protein

The E1B 19kd protein is associated with the plasma membrane of infected and transformed cells. Persson et al. (1982), for example, purified a virus-specific protein exhibiting an apparent molecular weight of 15kd from the membrane fraction of Ad2-infected HeLa cells. Monospecific antibodies raised against the purified protein immunoprecipitated the smaller product of in vitro translation of E1B-selected mRNA and the identity of the membrane protein to the E1B 19kd protein was confirmed by tryptic fingerprinting (Persson et al., 1982). Membrane-associated E1B 19kd protein was also found in all lines of adenovirus transformed cells examined (Persson et al., 1982). This viral protein behaves as if it were firmly associated with the membrane, being released only by detergents (Persson et al., 1982). However, its immunoprecipitation by antibodies raised against extreme N-terminal (and C-terminal) peptides (Green et al., 1983a) indicates that, unlike many membrane proteins, the E19 19kd protein is not subjected to proteolytic processing in vivo, nor was it upon membrane insertion in vitro (Persson et al., 1982).

Interestingly, two forms of membrane associated E1B protein were found when membrane-fractions of Ad5-infected cells were analyzed with relatively mono-specific tumor sera: these polypeptides, seen after a one hour pulse label, exhibited apparent molecular weights of 18.5 and 19kd and appeared identical by partial proteolytic mapping (Rowe et al., 1983a). Upon longer labeling, the two forms apparently changed to species of 17 and 17.5kd (Rowe et al., 1983a). The E1B 19kd protein is not labeled by [^{3}H]mannose of [^{3}H]glucosamine, nor does it bind to lentil lecithin (Persson et al., 1982), suggesting that it does not carry carbohydrate chains. The structural relationships among the various E1B 19kd-related species reported by Rowe et al. (1983a) is therefore not understood at present, nor is their functional significance. Several E1B encoded, low molecular proteins have also been described in Ad12-transformed cells (Jochemsen et al., 1980; Esche and Siegmann, 1982; Föhring et al., 1983, for example), but their relationships to either the 19kd or the 54kd proteins have not been established.

Although it seems to be established that the E1B 19kd protein is a plasma membrane protein, investigations of whether it is exposed on the outer surfaces of infected or transformed cells have yielded equivocal results. When tumor sera that had a strong avidity for the 19kd, but not the 55kd, Ad5 E1B protein were employed to examine whole infected cells, only a small fraction, less than 1%, displayed strong fluorescence over their entire surfaces (Rowe et al., 1983a). These authors also cited failures to detect the 19kd protein on the surface of fixed, but impermeable cells by immunofluorescence or cell-surface iodination. Similar studies of adenovirus transformed cells were not, unfortunately, reported. The surfaces of Ad12-transformed cells that express at least the N-terminal portion of the E1B region have, by contrast, been reported to be stained by antibodies raised against transformed cells expressing the E1B region (Föhring et al., 1983). Furthermore, an E1B encoded protein, apparent molecular weight 18kd and unrelated to the E1B 54kd protein, has been localized to the surface of Ad12 transformed cells (Grand and Gallimore, cited in Föhring et al., 1983). More detailed studies of cell lines transformed by members of the subgroup A and C adenoviruses will be required to establish whether cell-surface expression of the E1B 19kd polypeptide does indeed display subgroup specificity. This question is of considerable interest in light of the potential role(s) of this protein in cytotoxic T cell targeting and tumorigenicity, discussed in Section 5.

The properties of the adenovirus E1A and E1B proteins, therefore, provide interesting clues about the molecular mechanisms that might underlie their roles in transformation. Particularly intriguing are the potential functions of the E1A proteins in induction and/or repression of cellular gene expression and of entry of quiescent cells into the cell cycle and the interaction of the E1B 55kd protein with the cellular p53 protein which we now know can display transforming activity under certain circumstances. Obviously, much further work will be needed before the molecular basis of the contributions of these proteins to adenovirus transformation is established. Nevertheless, the rapid pace at which genetic and biochemical analyses of these proteins have progressed in recent years presents an encouraging prospect.

Acknowledgements—I thank colleagues who have communicated results prior to publication and Noël Mann for her patient work in preparing this manuscript.

REFERENCES

ALESTRÖM, P., AKUSJÄRVI, G., PERRICAUDET, M., MATHEWS, M. B., KLESSIG, D. and PETTERSSON, U. (1980) The gene for polypeptide IX of adenovirus type 2 and its unspliced messenger RNA. *Cell* **19**: 671–681.

ANDERSON, C. W. and LEWIS, J. B. (1980) Amino-terminal sequence of Ad2 proteins: hexon, fiber, component IX and early protein 1B-15K. *Virology* **104**: 27–41.

BABICH, A. and NEVINS, J. R. (1981) The stability of early adenovirus RNA is controlled by the viral 72kd DNA binding protein. *Cell* **26**: 371–379.

BABISS, L. E., GINSBERG, H. S. and FISHER, P. B. (1983) Cold-sensitive expression of transformation by a host-range mutant of type 5 adenovirus. *Proc. natn. Acad. Sci. U.S.A.* **80**: 1352–1356.

BABISS, L. E., FISHER, P. B. and GINSBERG, H. S. (1984a) Deletion and insertion mutations of early region 1A of type 5 adenovirus that produce cold-sensitive or defective phenotypes for transformation. *J. Virol.* **49**, 731–740.

BABIS, L. E., FISHER, P. B. and GINSBERG, H. S. (1984b) Effect on transformation of mutations in the early region 1B encoded 21- and 55-kilodalton proteins of adenovirus 5. *J. Virol.* **52**, 389–395.

BAKER, C. C. and ZIFF, E. B. (1979) Biogenesis, structure and sites encoding the 5′ termini of adenovirus 2 mRNAs. *Cold Spring Harb. Symp. quant. Biol.* **44**: 415–428.

BAKER, C. C. and ZIFF, E. B. (1981) Promoters and heterogeneous 5′ termini of the messenger RNAs of adenovirus 2. *J. molec. Biol.* **148**: 189–222.

BELLETT, A. J. D., LI, P., DAVID, E. T., MAKEY, E. J., BRAITHEWAITE, A. W. and CUTT, J. R. (1985) Control functions of adenovirus transforming region E1A gene products in rat and human cells. *Mol. Cell. Biol.* **5**, 1933–1939.

BERK, A. J., LEE, F., HARRISON, T., WILLIAMS, J. F. and SHARP, P. A. (1979) Pre-early adenovirus 5 gene product regulate synthesis of viral early messenger RNAs. *Cell* **17**: 935–944.

BERK, A. J. and SHARP, P. A. (1978) Structure of the adenovirus 2 early mRNAs. *Cell* **14**: 695–711.

BERNARDS, R., HOUWELING, A., SCHRIER, P. J., BOS, J. L. and VAN DER EB, A. J. (1982) Characterization of cells transformed by Ad5/Ad12 hybrid early region 1 plasmids. *Virology* **120**: 422–432.

BERNARDS, R., SCHRIER, P. I., BOS, L. and VAN DER EB, A. J. (1983a) Roles of adenovirus types 5 and 12 early region 1B tumor antigens in oncogenic transformation. *Virology* **127**: 45–53.

BERNARDS, R., SCHRIER, P. I., HOUWELING, A., BOS, J. L., VAN DER EB, A. J., ZIJLSTRA, M. and MELIEF, C. J. M. (1983b) Tumorigenicity of cells transformed by adenovirus type 12 by evasion of T-cell immunity. *Nature* **305**: 767–776.

BERNARDS, R., VAESSEN, M. J., VAN DER EB, A. J. and SUSSENBACH, J. C. (1983c) Construction and characterization of an adenovirus type 5 adenovirus type 12 recombinant virus. *Virology* **131**: 30–38.

BORELLI, E., HEN, R. and CHAMBON, R. (1984) The adenovirus 2-E1a products repress stimulation of transcription by enhancers. *Nature* **312**, 608–612.

BOS, J. L. and TEN WOLDE-KRAAMWINKEL, H. C. (1983) The E1B promoter of Ad12 in mouse L tk-cells is activated by adenovirus region E1A. *EMBO J.* **2**: 73–76.

BOS, J. L., JOCHEMSEN, A. G., BERNARDS, R., SCHRIER, P. I., VAN ORMONDT, H. and VAN DER EB, A. J. (1983) Deletion mutants of region E1A of Ad12 plasmids: effect on oncogenic transformation. *Virology* **129**: 393–400.

BOS, J. L., POLDER, L. J., BERNARDS, R., SCHRIER, P. I., VAN DER ELSEN, P. J., VAN DER EB, A. J. and VAN ORMONDT, H. (1981) The 2.2kb E1B mRNA of human Ad12 and Ad5 codes for two tumor antigens starting at different AUG triplets. *Cell* **27**: 121–131.

BRACKMANN, K. H., GREEN, M., WOLD, W. S. M., CARTAS, M., MATSUO, T. and HASTIMOTO, S. (1980) Identification and peptide mapping of human adenovirus 2-induced early polypeptides isolated by two-dimensional gel electrophoresis and immunoprecipitation. *J. biol. Chem.* **255**: 6772–6779.

BRAITHEWAITE, A. W., CHEETHAM, B. F., LI, P., PARISH, C. R., WALDRON-STEVENS, L. K. and BELLETT, A. J. D. (1983) Adenovirus-induced alterations of the cell growth cycle: a requirement for expression of E1A but not E1B. *J. Virol.* **45**, 192–199.

BRANTON, P. E., LASSAM, N. J., DOWNEY, J. F., YEE, S-P., GRAHAM, F. L., MAK, S. and BAYLEY, S. T. (1981) Protein kinase activity immunoprecipitated from adenovirus-infected cells by sera from tumor-bearing animals. *J. Virol.* **37**: 601–608.

CARLOCK, L. R. and JONES, N. C. (1981) Transformation defective mutant of adenovirus type 5 containing a single altered E1A mRNA species. *J. Virol.* **40**: 657–664.

CARTER, T. H. and BLANTON, R. A. (1978a) Possible role of the 72,000 dalton DNA binding protein in regulation of adenoviral type 5 early gene expression. *J. Virol.* **25**: 664–674.

CARTER, T. H. and BLANTON, R. A. (1978b) Autoregulation of adenovirus type 5 early gene expression II. Effect of temperature-sensitive early mutations on virus RNA accumulation. *J. Virol.* **28**: 450–456.

CASTRO, B. C. (1973) Biologic parameters of adenovirus transformation. *Prog. exp. Tumor Res.* **18**: 166–198.

CHINNADURAI, G. (1983) Adenovirus 2 lp$^+$ locus codes for a 19kd tumor antigen that plays an essential role in cell transformation. *Cell* **33**: 759–766.

CHOW, L.T., BROKER, T. R. and LEWIS, J. B. (1979) Complex splicing patterns of RNA from the early regions of adenovirus 2. *J. molec. Biol.* **134**: 265–303.

COOK, J. L. and LEWIS, A. M. (1984). Differential NK and macrophage killing of hamster cells infected with nononcogenic or oncogenic adenovirus. *Science* **224**: 612–615.

COOK, S. L., HIBBS, J. B. and LEWIS, A. M. (1982) DNA virus transformed hamster cell-host effector cell interactions: level of resistance to cytolysis is correlated with tumorigenicity. *Int. J. Cancer* **30**: 795–803.

DARBYSHIRE, J. H. (1966) Oncogenicity of bovine adenovirus type 3 in hamsters. *Nature* **211**: 102–104.

DER, C. J., KRONTIRIS, T. G. and COPPER, G. M. (1982) Transforming genes of human bladder and lung carcinoma cell lines are homologous to the *ras* genes of Harvey and Kirsten sarcoma viruses. *Proc. natn. Acad. Sci. U.S.A.* **79**: 3637–3640.

DEWING, R., WINTERHOFF, U., TAMANOI, F., STABEL, S. and DOERFLER, W. (1981) Site of linkage between adenovirus type 12 and cell DNA in hamster tumor line CLAC 3. *Nature* **293**: 81–84.

DIJKEMA, R., DEKKER, B. M. M. and VAN ORMONDT, H. (1980) The nucleotide sequence of the transforming BglII-H fragment of adenovirus 7 DNA. *Gene* **9**: 141–156.

DUKEMA, R., DEKKER, B. M. M., VAN DER FELTZ, J. M. and VAN DER EB, A. J. (1979) Transformation of primary rat cells by DNA fragments of weakly oncogenic adenoviruses. *J. Virol.* **32**: 943–950.

DOERFLER, W. (1968) The fate of the DNA of adenovirus 12 in baby hamster kidney cells. *Proc. natn. Acad. Sci. U.S.A.* **60**: 636–643.

DOERFLER, W. (1969) Non-productive infection of baby hamster kidney cells with adenovirus type 12, *Virology* **38**: 587–606.

DOERFLER, W. and LUNDHOLM, U. (1970) Absence of the replication of the DNA of adenovirus type 12 in BHK 21 cells. *Virology* **40**: 754–756.

DOERFLER, W., STABEL, S., IBELGAUFTS, H., SUTTER, D., NEUMANN, R., GRONEBERG, J., SCHEIDTMANN, K. H., DEURING, R. and WINTERHOFF, U. (1979) Selectivity in integration sites of adenoviral DNA. *Cold Spring Harb. Symp. Quant. Biol.* **44**: 551–564.

DORSCH-HASLER, K., FISHER, P. B., WEINSTEIN, I. B. and GINSBERG, H. S. (1980) Patterns of viral DNA integration of cells transformed by wild-type or DNA-binding mutants of adenovirus type 5 and effects of chemical carcinogens on integration. *J. Virol.* **34**: 305–314.

EICK, D. and DOERFLER, W. (1982) Integrated adenovirus type 12 DNA in the transformation hamster cell line T637: sequence arrangements at the termini of viral DNA and mode of amplification. *J. Virol.* **42**: 317–321.

ELIYAHAU, D., RAZ, A., GRUSS, P., GIVOL, D. and OREN, M. (1984) Participation of the p53 cellular tumor antigen in transformation of normal embyotic cells. *Nature* **312**, 646–649.

ENSINGER, M. and GINSBERG, H. S. (1972) Selection and preliminary characterization of temperature-sensitive mutants of type 5 adenovirus. *J. Virol.* **10**: 323–339.

ESCHE, H. (1982) Viral gene products in adenovirus 2 transformed hamster cells. *J. Virol.* **41**: 1076–1082.

ESCHE, H. and SIEGMANN, B. (1982) Expression of early viral gene products in adenovirus type 12-infected and transformed cells. *J. gen. Virol.* **60**: 99–113.

ESCHE, H., MATHEWS, M. B. and LEWIS, J. B. (1980) Proteins and messenger RNAs of the transforming region of wild-type and mutant adenovirus. *J. molec. Biol.* **142**: 399–417.

FANNING, E. and DOERFLER, W. (1976) Intracellular forms of adenovirus DNA V. Viral DNA sequences in hamster cells abortively infected and transformed with human adenovirus type 12. *J. Virol.* **20**: 373–383.

FLEDMAN, L. T. and NEVINS, J. R. (1983) Localization of the adenovirus E1A-a protein, a positive-acting transcriptional factor, in infected cells. *Molec. Cell. Biol.* **3**: 829–838.

FERGUSON, B., JONES, N., RICHTER, J. and ROSENBERG, M. (1984). Adenovirus E1a gene product expressed in *E. coli* is functional. *Science* **224**, 1342–1346.

FLINT, S. J. (1982) Organization and expression of viral genes in adenovirus transformed cells. *Int. Rev. Cytol.* **76**: 47–66.

FLINT, S. J. (1984) Adenovirus Cytopathology. In *Comprehensive Virology*, FRAENKEL-CONRAT, H. and WAGNER, R. R. (eds) Plenum Press, NY., in press.

FLINT, S. J. and BELTZ, G. A. (1979) Expression of transforming viral genes in semipermissive cells transformed by SV40 or adenovirus type 2 or type 5. *Cold Spring Harb. Symp. quant. Biol.* **44**: 89–102.

FLINT, S. J., BERGET, S. M. and SHARP, P. A. (1976a) Characterization of the single-stranded viral DNA sequences present during replication of adenovirus types 2 and 5. *Cell* **9**: 559–571.

FLINT, S. J., GALLIMORE, P. H. and SHARP, P. A. (1975) Comparison of viral RNA sequences in adenovirus 2 transformed and lytically infected cells. *J. molec. Biol.* **96**: 47–68.

FLINT, S. J., SAMBROOK, J., WILLIAMS, J. F. and SHARP, P. A. (1976b) Viral nucleic acid sequences in transformed cells. IV. A study of the sequences of adenovirus DNA and RNA in four lines of adenovirus 5 transformed rodent cells using specific fragments of the viral genome. *Virology* **72**: 456–470.

FÖHRING, B., GALLIMORE, P. H., MELLOW, G. H. and RAŠKA, K. (1983) Adenovirus type 12 specific cell surface antigen in transformed cells is a product of the E1B early region. *Virology* **131**: 463–472.

FREEMAN, A. E., BLACK, P. H., VANDERPOOL, J. H., HENRY, P. H., AUSTIN, J. B. and HUEBNER, R. J. (1967) Transformation of primary rat embryo cells by adenovirus type 2. *Proc. natn. Acad. Sci. U.S.A.* **58**: 1205–1212.

FRIEFELD, B. R., KREVOLIN, M. D. and HOROWITZ, M. S. (1983a) Effects of the adenovirus H5ts125 and H5ts107 DNA binding proteins on DNA replication *in vitro*. *Virology* **124**: 380–389.

FRIEFELD, B. R., LICHY, J. H., HURWITZ, J. and HORWITZ, M. S. (1983b) Evidence for an altered adenovirus DNA polymerase in cells infected with the mutant H5ts149. *Proc. natn. Acad. Sci. U.S.A.* **80**: 1589–1593.

FROLOVA, E. I. and ZALMANZON, E. S. (1978) Transcription of viral sequences in cells transformed by adenovirus type 5. *Virology* **89**: 347–359.

FROLOVA, E. I., ZALMANZON, E. S., LUKANIDIN, E. M. and GEORGIEV, G. P. (1978) Studies of the transcription of viral genome in adenovirus 5 transformed cells. *Nucleic Acids Res.* **5**: 1–11.

FROST, E. and WILLIAMS, J. F. (1978) Mapping temperature-sensitive and host-range mutants of adenovirus type 5 by marker rescue. *Virology* **91**: 39–50.

FUJINAGA, K., SAWADA, Y., UEMIZU, Y., YAMASHITA, T., SHIMOJO, H., SHIROKI, K., SUGISAKI, H., SUGIMOTO, K. and TAKANAMI, M. (1979) Nucleotide sequences, integration and transcription of the adenovirus 12 transforming genes. *Cold Spring Harb. Symp. quant. Biol.* **44**: 519–532.

FUKUI, Y., SAITO, I., SHIROKI, K. and SHIMOJO, H. (1984) Isolation of transformation-defective, replication non-defective mutants of adenovirus 12. *J. Virol.* **49**: 154–161.

GALLIMORE, P. H. (1974) Interactions of adenovirus type 2 with rat embryo cells: permissiveness, transformation and *in vitro* characterization of adenovirus 2 transformed rat embryo cells. *J. gen. Virol.* **25**: 263–272.

GALLIMORE, P. H., MCDOUGALL, J. and CHEN, L. B. (1977). *In vitro* traits of adenovirus transformed cell lines and their relevance to tumorigenicity in nude mice. *Cell* **10**, 669–678.

GALLIMORE, P. H., SHARP, P. A. and SAMBROOK, J. (1974) Viral DNA in transformed cells. II. A study of the sequences of Ad2 DNA in nine lines of transformed rat cells using specific fragments of the viral genome. *J. molec. Biol.* **89**: 49–72.

GALOS, R. B., WILLIAMS, J. F., BINGER, M-H. and FLINT, S. J. (1979) Location of additional early gene sequences in the adenoviral chromosome. *Cell* **17**: 945–956.

GAYNOR, R. B. and BERK, A. J. (1983) *Cis*-acting induction of adenovirus transcription. *Cell* **33**: 683–693.

GAYNOR, R. B., HILLMAN, D. and BERK, A. J. (1984) Adenovirus early region 1A protein activates transcription of a non viral gene introduced into mammalian cells by infection or transfection. *Proc. natn. Acad. Sci. U.S.A.* **81**: 1193–1197.

GAYNOR, R. B., TSUKUMOTO, A., MONTELL, C. and BERK, A. J. (1982) Enhanced expression of adenovirus transforming proteins. *J. Virol.* **44**: 276–285.

GELB, L. D., KOHNE, D. E. and MARTIN, M. A. (1971) Quantitation of SV40 sequences in African green monkey, mouse and virus transformed cells. *J. molec. Biol.* **57**: 129–145.

GILEAD, Z., JENG, I-H., WOLD, W. J. M., SUGAWARA, K., RHO, H. M., HARTER, M. L. and GREEN, M. (1976) Immunological identification of two adenovirus 2 induced early proteins possibly involved in cell transformation. *Nature* **264**: 263–265.

GINGERAS, T. R., SCIAKY, D., GELINAS, R. E., BING-DONG, J., YEN, C. E., KELLY, M. E., BULLOCK, P. A., PARSON,

B. L., O'NEILL, K. E. and ROBERTS, R. J. (1982) Sequences from the adenovirus 2 genome. *J. biol. Chem.* **257**: 13475–13491.

GINSBERG, H. S., ENSINGER, M. J., KAUFMAN, R. S., MAYER, A. J. and LUNDHOLM, O. (1975) Cell transformation: a study of regulation with types 5 and 12 adenovirus temperature-sensitive mutants. *Cold Spring Harb. Symp. quant. Biol.* **39**: 419–426.

GIRARDI, A. S., HILLEMAN, M. R. and ZWICKEY, R. E. (1964) Tests in hamsters for the oncogenicity of ordinary virus including adenovirus type 7. *Proc. Soc. exp. Biol. Med.* **115**: 1141–1150.

GRAHAM, F. L. and VAN DER EB, A. J. (1973) Transformation of rat cells by DNA of human adenovirus 5. *Virology* **54**: 536–539.

GRAHAM, F. L., ABRAHAMS, P. J., MULDER, C., HEIJNEKER, H. L., WARNAAR, S. O., DE FRIES, F. A. J., FIERS, W. and VAN DER EB, A. J. (1975a) Studies on *in vitro* transformation by DNA and DNA fragments of human adenovirus type 5. *J. gen. Virol.* **36**: 59–72.

GRAHAM, F. L., ABRAHAMS, P. J., MULDER, C., HEIJNEKER, H. L., WARNAAR, S. O., DE FRIES, F. A. J., FIERS, W. and VAN DER EB, A. J. (1975b) Studies on *in vitro* transformation by DNA and DNA fragments of human adenoviruses and SV40. *Cold Spring Harb. Symp. quant. Biol.* **39**: 637–650.

GRAHAM, F. G., HARRISON, T. J. and WILLIAMS, J. F. (1978) Defective transforming capacity of adenovirus type 5 host-range mutants. *Virology* **86**: 10–21.

GRAHAM, F. L., SMILEY, J., RUSSELL, W. C. and NAIRN, R. (1977) Characteristics of a human cell line transformed by DNA from human adenovirus. *J. gen. Virol.* **36**: 59–72.

GREEN, M., BRACKMANN, K. H., CARTAS, M. A. and MATSUO, T. (1982) Identification and purification of a protein encoded by the human adenovirus type 2 transforming region. *J. Virol.* **42**: 30–41.

GREEN, M., BRACKMANN, K. H., LUCHER, L. A., SYMINGTON, J. S. and KRAMER, T. A. (1983a) Human adenovirus 2 E1B-19K and E1B-53K tumor antigens: antibodies targeted to the NH_2 and COOH termini. *J. Virol.* **48**: 604–615.

GREEN, M., WOLD, W. S. M., BRACKMANN, K. H. and CARTAS, M. A. (1979) Identification of families of overlapping polypeptides encoded by early (transforming) gene region 1 of human adenovirus type 2. *Virology* **97**: 275–286.

GREEN, M. R., TREISMAN, R. and MANIATIS, T. (1983b) Transcriptional activation of cloned human β-globin genes by viral immediate-early gene products. *Cell* **35**: 137–148.

GRODZICKER, T., ANDERSON, C. W., SAMBROOK, J. and MATHEWS, M. B. (1977) The physical location of structural genes in adenovirus DNA. *Virology* **80**: 111–126.

HALBERT, D. N. and RASKAS, H. J. (1982) Tryptic and chymotryptic methionine peptide analysis of the *in vitro* translation products specified by the transforming region of adenovirus type 2. *Virology* **116**: 406–418.

HALBERT, D. N., SPECTOR, D. J. and RASKAS, H. J. (1979) *In vitro* translation products specified by the transforming region of adenovirus type 2. *J. Virol.* **31**: 621–629.

HALEY, K. P., OVERHAUSER, J., BABISS, L. E., GINSBERG, H. S. and JONES, N. C. (1984) Transformation properties of type 5 adenovirus mutants that differentially express the E1A gene products. *Proc. Nat. Acad. Sci. U.S.A.* **81**, 5734–5738.

HARLOW, E., FRANZA, B. R. and SCHLEY, C. (1985). Monoclonal antibodies specific for adenovirus early region 1A proteins: extensive heterogeneity in early region 1A products. *J. Virol.* **55**, 533–546.

HARRISON, T. J., GRAHAM F. and WILLIAMS, J. F. (1977) Host range mutants of adenovirus type 5 defective for growth on HeLa cells. *Virology* **77**: 319–329.

HARTER, M. L. and LEWIS, J. B. (1978) Adenovirus type 2 early proteins synthesized *in vitro* and *in vivo*: identification in infected cells of the 38,000 to 50,000 molecular weight protein encoded by the left end of the adenovirus type 2 genome. *J. Virol.* **26**: 736–749.

HARWOOD., L. M. and GALLIMORE, P. H. (1975) A study of the oncogenicity of adenovirus type 2-transformed rat embryo cells. *Int. J. Cancer* **16**, 498–508.

HO, Y-S., GALOS, R. and WILLIAMS, J. F. (1982) Isolation of type 5 adenovirus mutants with a cold-sensitive phenotype: genetic evidence of an adenovirus transformation maintenance function. *Virology* **112**: 109–124.

HOUWELING, A., VAN DER ELSEN, P. J. and VAN DER EB, A. J. (1980) Partial transformation of primary rat cells by the left most 4.5% fragment of adenovirus 5 DNA. *Virology* **105**: 537–550.

HUEBNER, R. J., CASEY, M. J., CHANOCK, R. N. and SCHELL, K. (1965) Tumors induced in hamsters by a strain of adenovirus type 3: sharing of tumor antigens and 'neoantigens' with those produced by adenovirus type 1 tumors. *Proc. natn. Acad. Sci. U.S.A.* **54**: 381–388.

HUEBNER, R. J., ROWE, W. P. and LANE, W. T. (1962) Oncogenic effects in hamsters of human adenovirus types 12 and 18. *Proc. natn. Acad. Sci. U.S.A.* **48**: 2051–2058.

HUEBNER, R. J., ROWE, W. P., TURNER, H. C. and LANE, W. T. (1963) Specific adenovirus complement fixing antigens in virus-free hamster and rat tumors. *Proc. natn. Acad. Sci. U.S.A.* **50**: 379–389.

HULL, R. N., JOHNSON, I. S., CULBERTSON, C. Q., REIMER, C. B. and WRIGHT, H. F. (1965) Oncogenicity of the simian adenoviruses. *Science* **150**: 1044–1046.

HURWITZ, D. R. and CHINNADURAI, G. (1985). Evidence that a second tumor antigen encoded by adenovirus early region E1A is required for efficient cell transformation. *Proc. Nat. Acad. Sci. U.S.A.* **82**, 163–167.

IBELGAUFTS, H., DOERFLER, W., SCHEIDTMANN and WECHSLER, W. (1980) Adenovirus type 12 induced rat tumor cells of neuroepithelial origin: persistence and expression of the viral genome. *J. Virol.* **33**: 423–437.

JENKINS, J. R., RUDGE, K. and CURRIE, G. A. (1984) Cellular immortalization by a cDNA clone encoding the transformation-associated phosphoprotein p53. *Nature* **312**, 651–654.

JOCHEMSEN, H., DANIELS, G. S. G., HERTOGHS, J. J. L., SCHRIER, P. I., VAN DER ELSEN, P. J. and VAN DER EB, A. J. (1982) Identification of adenovirus 12 gene products involved in transformation and oncogenesis. *Virology* **122**: 15–28.

JOCHEMSEN, H., DANIELS, G. S., LUPKER, J. H. and VAN DER EB, A. J. (1980) Identification and mapping of the early gene products of adenovirus type 12. *Virology* **105**: 551–563.

JOCHEMSEN, H., HERTOGHS, J. J. L., LUPKER, J. H., DAVIS, A. and VAN DER EB, A. J. (1981) *In vitro* synthesis of adenovirus type 5 T antigens II. Translation of virus-specific RNA from cells transformed by fragments of adenovirus type 5 DNA. *J. Virol.* **37**: 530–534.

JOHANSSON, K., PETTERSSON, U., LEWIS, A. M., PERSSON, H., TIBBETTS, C. and PHILIPSON, L. (1978) Viral DNA sequences and genes products in hamster cells transformed by adenovirus type 2. *J. Virol.* **27**: 628–639.

JOHANSSON, K., PETTERSSON, U., PHILIPSON, L. and TIBBETTS, C. (1977) Reassociation of complementary strand-specific adenovirus type 2 DNA with sequences of transformed cells. *J. Virol.* **33**: 29–35.

JONES, N. and SHENK, T. (1979a) Isolation of adenovirus type 5 host-range deletion mutants defective for transformation of rat embryo cells. *Cell* **17**: 683–689.

JONES, N. and SHENK, T. (1979b) An adenovirus type 5 early gene function regulates expression of other early genes. *Proc. natn. Acad. Sci. U.S.A.* **76**: 3665–3669.

KAO, H-T. and NEVINS, J. R. (1983) Transcriptional activation and subsequent control of the human heat shock gene during adenovirus infection. *Molec. Cell. Biol.* **3**: 2058–2065.

KARGER, B. et al. (in press) *Cancer Cells* **4**: Cold Spring Harbour, N.Y.

KIMURA, T., SAWADA, Y., SHINAGAWA, M., SHIMIZU, Y., SHIROKI, K., SHIMOJO, H., SUGISAKI, H., TAKANAMI, M., UEMIZU, Y. and FUJINAGA, K. (1981) Nucleotide sequence of the transforming early region E1b of adenovirus type 12 DNA: structure of gene organization and comparison with those of adenovirus type 5. *Nucleic Acids Res.* **9**: 6571–6589.

KITCHINGMAN, G. R. and WESTPHAL, H. (1980) The structure of adenovirus 2 early nuclear and cytoplasmic RNAs. *J. molec. Biol.* **137**: 23–48.

KO, J-L. and HARTER, M. L. (1984) Plasmid-directed synthesis of genuine adenovirus 2 early region 1A and 1B proteins. *Mol. Cell. Biol.* **4**, 1427–1439.

KOPCHICK, J. J. and STACEY, D. W. (1984) Differences in intracellular DNA ligation after microinjection and transfection. *Molec. cell. Biol.* **4**: 240–246.

LAI FATT, R. B. and MAK, S. (1982) Mapping of an adenovirus function involved in the inhibition of DNA degradation. *J. Virol.* **42**: 969–977.

LAND, H., PARADA, L. F. and WEINBERG, R. A. (1983a) Tumorigenic conversion of primary embryo fibroblasts requires at least two cooperating oncogenes. *Nature* **304**: 596–602.

LAND, H., PARADA, L. F. and WEINBERG, R. A. (1983b) Cellular oncogenes and multistep carcinogenesis. *Science* **222**: 771–778.

LANDAU, B. J., LARSSON, V. M., DEVERS, G. A. and HILLEMAN, M. R. (1966) Studies on induction of virus from adenovirus and SV40 tumors I. Chemical and physical agents. *Proc. Soc. exp. Biol. Med.* **122**: 1176–1181.

LARSSON, V. M., GOSNELL, M. A. and HILLEMAN, M. R. (1966) Studies on induction of virus from adenovirus and SV40 tumors II. 'Helper' virus. *Proc. Soc. exp. Biol. Med.* **122**: 1182–1191.

LASSAM, N. J., BAYLEY, S. T. and GRAHAM, F. L. (1979) Tumor antigens of adenovirus 5 in transformed cells and in cells infected with transformation-defective host-range mutants. *Cell* **18**: 781–791.

LEFF, T., ELKAIM, R., GODIN, C. R., JALINOT, R., SASSONE-CORSI, P., PERRICAUDET, M., KÉDINGER, C. and CHAMBON, P. (1984) Individual products of the adenovirus 12S and 13S E1A mRNAs stimulate viral E11a and E111 expression at the transcriptional level. *Proc. Nat. Acad. Sci. U.S.A.* **81**, 4381–4385.

LEVINE, A. J., VAN DER VLIET, P. C., ROSENWIRTH, B., RABEK, J., FRENKEL, G. and ENSINGER, M. (1975) Adenovirus infected cell-specific DNA binding proteins. *Cold Spring Harb. Symp. quant. Biol.* **39**: 559–566.

LEVINSON, A. and LEVINE, A. J. (1977a) The isolation and identification of the adenovirus group C tumor antigens. *Virology* **76**: 1–11.

LEVINSON, A. and LEVINE, A. J. (1977b) The group C adenovirus tumor antigens: identification in infected and transformed cells and a peptide map analysis. *Cell* **11**: 871–879.

LEWIS, J. B. and MATHEWS, M. B. (1981) Viral messenger RNAs in six lines of adenovirus transformed cells. *Virology* **115**: 345–360.

LEWIS, J. B., ATKINS, J. F., BAUM, P. R., SOLEM, R., GESTELAND, R. and ANDERSON, C. W. (1976) Location and identification of the genes for adenovirus type 2 early polypeptides. *Cell* **7**: 121–151.

LEWIS, J. B., ESCHE, H., SMART, J. E., STILLMAN, B. W., HARTER, M. L. and MATHEWS, M. B. (1979) Organization and expression of the left third of the genome of adenovirus. *Cold Spring Harb. Symp. quant. Biol.* **44**: 493–508.

LUCHER, L. A., BRACKMANN, K. H., SYMINGTON, J. S. and GREEN, M. (1984) Antibody directed to a synthetic peptide encoding the NH_2-terminal 16 amino acids of the adenovirus type 2 E1B-53k tumor antigen recognizes the E1B-20k tumor antigen. *Virology* **132**: 217–221.

LUPKER, J.H., DAVIS, A., JOCHEMSEN, H. and VAN DER EB, A. J. (1981) In vitro synthesis of adenovirus type 5 T antigens. I. Translation of early region 1-specific RNA from lytically infected cells. *J. Virol.* **37**: 524–529.

MAAT, J., VAN BEVEREN, C. P. and VAN ORMONDT, H. (1980) The nucleotide sequence of adenovirus type 5 early region E1: the region between map positions 8.0 (*Hind* III site) and 11.8 (*Sma* 1 site). *Gene* **10**: 27–38.

MAK, S. and MAK, I. (1983) Transformation of rat cells by *cyt* mutants of adenovirus type 12 and mutants of adenovirus type 5. *J. Virol.* **45**: 1107–1117.

MAK, S., MAK, I., SMILEY, J. R. and GRAHAM, F. L. (1979) Tumorigenicity and viral gene expression in rat cells transformed by Ad12 virions or by the *Eco*R1 C fragment of Ad12 DNA. *Virology* **98**: 456–460.

MATSUO, T., WOLD, W. S. M., HASHIMOTO, S., RANKIN, A., SYMINGTON, J. and GREEN, M. (1982) Polypeptides encoded by transforming region E1B of human adenovirus 2: immunoprecipitation from transformed and infected cells and cell-free translation of E1B-specific mRNA. *Virology* **118**: 456–465.

MCALLISTER, R. M., NICOLSON, M. O., LEWIS, A. M., MACPHERSON, I. and HUEBNER, R. J. (1969a) Transformation of rat embryo cells by adenovirus type 1. *J. gen. Virol.* **4**: 29–36.

MCALLISTER, R. M., NICOLSON, N. O., REED, G., KERN, J., GILDEN, R. V. and HUEBNER, R. J. (1969b) Transformation of rodent cells by adenovirus 19 and other group D adenoviruses. *J. natn. Cancer Soc.* **43**: 917–923.

MELLOW, G. H., FÖHRING, B., DOUGHERTY, J., GALLIMORE, P. H. and RASKA, K. (1984) Tumorigeneity of adenovirus-transformed rat cells and expression of class I major histocompatibility antigens. *Virology* **134**, 460–465.

MERCER, W. E., NELSON, D., DELEO, A. B., OLD, L. J. and BASERGA, R. (1982) Microinjection of monoclonal antibody to protein p53 inhibits serum-induced DNA synthesis in 3T3 cells. *Proc. natn. Acad. Sci. U.S.A.* **79**: 6309–6312.

MERCER, W. E., AUIGNOLO, C. and BASERGA, R. (1984) Role of the p53 protein in cell proliferation as studied by microinjection of monoclonal antibodies. *Molec. cell. Biol.* **4**: 276–281.

MEYER, A. J. and GINSBERG, H. S. (1977) Persistence of type 5 adenovirus DNA in cells transformed by a temperature-sensitive mutant, H5ts125. *Proc. natn. Acad. Sci. U.S.A.* **74**: 785–788.

MONTELL, C., COURTOIS, G., ENG, C. and BERK, A. (1984) Complete transformation by adenovirus 2 requires both E1A proteins. *Cell* **36**: 951–961.

MONTELL, C., FISHER, E. F., CARUTHERS, M. H. and BERK, A. J. (1982) Resolving the functions of overlapping viral genes by site-specific mutagenesis of a mRNA splice site. *Nature* **295**: 380–384.

NEVINS, J. R. (1981) Mechanisms of activation of early viral transcription by the adenovirus E1A gene product. *Cell* **26**: 213–220.

NEVINS, J. R. (1982) Induction of the synthesis of a 70,000 Dalton mammalian heat shock protein by the adenovirus E1A gene product. *Cell* **29**: 913–919.

NEWBOLD, R. F. and OVERALL, R. W. (1983) Fibroblast immortality is a prerequisite for transformation by EJ-c-Ha-ras oncogene. *Nature* **304**: 648–651.

ORTIN, J., SCHEIDTMANN, K. H., GREENBERG, R., WESTPHAL, M. and DOERFLER, W. (1976) Transcription of the genome of adenovirus type 12-III. Transcription maps in productively infected human cells and in abortively infected and transformed hamster cells. *J. Virol.* **201**: 355–372.

OSBORNE, T. F. and BERK, A. J. (1983) Far upstream initiation sites for adenovirus early region 1A transcription are utilized after the onset of viral DNA replication. *J. Virol.* **45**: 594–599.

OSBORNE, T. F., GAYNOR, R. B. and BERK, A. J. (1982) The TATA homology and the mRNA 5' untranslated sequence are not required for expression of essential adenovirus E1A functions. *Cell* **29**: 139–148.

PARADA, L. F., TABIN, C. J., SHIH, C. and WEINBERG, R. A. (1982) Human EJ bladder carcinoma oncogene is homologous to the Harvey sarcoma virus gene. *Nature* **297**: 474–479.

PARADA, L. F., LAND, H., WEINBERG, R. A., WOLF, D. and ROTTER, V. (1984) Cooperation between gene encoding p53 tumor antigen and ras in cellular transformation. *Nature* **312**: 649–651.

PELLICER, A., WIGLER, M., AXEL, R. and SILVERSTEIN, S. (1978) The transfer and stable integration of the HSV thymidine gene into mouse cells. *Cell* **14**: 133–141.

PERRICAUDET, M., AKUSJÄRVI, G., VIRTANEN, A. and PETTERSSON, U. (1979) Structure of the two spliced mRNAs from the transforming region of human subgroup C adenoviruses. *Nature* **281**: 694–696.

PERRICAUDET, M., LEMOULLEC, J. M. and PETTERSSON, U. (1980a) Predicted structure of two adenovirus tumor antigens. *Proc. natn. Acad. Sci. U.S.A.* **77**: 3778–3782.

PERRICAUDET, M., LEMOULLEC, J. M., TIOLLAIS, P. and PETTERSSON, U. (1980b) Structure of two adenovirus type 12 transforming polypeptides and their evolutionary implications. *Nature* **288**: 174–176.

PERSSON, H., KATZE, M. G. and PHILIPSON, L. (1982) Purification of a native membrane-associated adenovirus tumor antigen. *J. Virol.* **42**: 905–917.

PERUCHO, M., HANAHAN, D. and WIGLER, M. (1980) Genetic and physical linkage of exogenous sequences in transformed cells. *Cell* **22**: 309–317.

PETTERSSON, U., TIBBETTS, C. and PHILIPSON, L. (1976) Hybridization maps of early and late mRNA sequences on the adenovirus type 2 genome. *J. molec. Biol.* **101**: 479–502.

PILDER, S., LOGAN, J. and SHENK, T. (1984). Deletion of the gene encoding the adenovirus 5 early region 1B 21,000-molecular weight polypeptide leads to degradation of viral and host cell DNA. *J. Virol.* **52**, 664–671.

POPE, J. H. and ROWE, W. P. (1964) Immunofluorescent studies of Ad12 tumors and of cells transformed or infected by adenovirus. *J. exp. Med.* **120**, 577–587.

RABSON, A. S., KIRSCHSTEIN, R. L. and PAUL, F. J. (1964) Tumors produced by adenovirus 13 in mastomys and mice. *U.S. Natn. Cancer Inst. J.* **32**: 77–87.

RAŠKA, K. and GALLIMORE, P. H. (1982) An inverse relationship of the oncogenic potential of adenovirus-transformed cells and their sensitivity to killing by syngeneic natural killer cells. *Virology* **123**: 8–18.

RASSOULZADEGAN, M., COWIE, A., CARR, A., GLAICHENHAUS, N., KAMEN, R. and CUZIN, F. (1982) The roles of individual polyoma virus early proteins in oncogenic transformation. *Nature* **300**: 713–718.

RASSOULZADEGAN, M., NAGHASHFAR, Z., COWIE, A., CARR, A., GRISONI, M., KAMEN, R. and CUZIN, F. (1983) Expression of the large T protein of polyma virus promotes the establishment in culture of 'normal' rodent fibroblast cell lines. *Proc. natn. Acad. Sci. U.S.A.* **80**: 4354–4358.

REICH, N. C. and LEVINE, A. J. (1984) Growth regulation of a cellular tumor antigen, p53, in non-transformed cells. *Nature* **308**: 199–201.

RIBEIRO, G. and VANCONCELOS-COSTA, J. (1981) Identification of polypeptide components of adenovirus type 12 tumor antigen from productivity and abortively infected cells and from tumor cells. *Virology* **112**: 775–779.

RICCIARDI, R. P., JONES, R. L., CEPKO, C. T., SHARP, P. A. and ROBERTS, B. E. (1981) Expression of early adenovirus genes requires a viral encoded acidic polypeptide. *Proc. natn. Acad. Sci. U.S.A.* **78**: 6121–6125.

ROSS, S., FLINT, S. J. and LEVINE, A. J. (1980) Identification of the adenovirus early proteins and their genomic map positions. *Virology* **100**: 419–432.

ROWE, D. T. and GRAHAM, F. L. (1983) Transformation of rodent cells by DNA extracted from transformation-defective adenovirus mutants. *J. Virol.* **46**: 1039–1044.

ROWE, D. T., BRANTON, P. E., YEE, S-P., BACCHETTI, S. and GRAHAM, F. L. (1984) Establishment and characterization of hamster cell lines transformed by restriction endonuclease fragments of adenovirus 5. *J. Virol.* **49**: 162–170.

ROWE, D. T., GRAHAM, F. L. and BRANTON, L. E. (1983a) Intracellular localization of adenovirus type 5 tumor antigens in productively infected cells. *Virology* **129**: 456–468.

ROWE, D. T., YEE, S-P., OTIS, J., GRAHAM, F. L. and BRANTON, P. E. (1983b) Characterization of human Ad5 early region 1A polypeptides using antitumor sera and an antiserum specific for the carboxy terminus. *Virology* **127**: 253–271.

RUBEN, M., BACCHETTI, S. and GRAHAM, F. L. (1982) Integration and expression of viral DNA in cells transformed by host-range mutants of adenovirus type 5. *J. Virol.* **41**: 674–685.

RULEY, H. E. (1983) Adenovirus early region 1A enables viral and cellular transforming genes to transform primary cells in culture. *Nature* **304**: 602–606.

SAITO, I., SHIROKI, K. and SHIMOJO, H. (1983) mRNA and proteins of adenovirus 12 transforming regions: identification of proteins translated from multiple coding stretches in 2.2 kb region 1B mRNA *in vitro* and *in vivo*. *Virology* **127**: 272–289.
SAMBROOK, J., BOTCHAN, M., GALLIMORE, P., OZANNE, B., PETTERSON, U., WILLIAMS, J. and SHARP, P. A. (1975) Viral DNA sequences in cells transformed by SV40, adenovirus type 2 and adenovirus type 5. *Cold Spring Harb. Symp. quant. Biol.* **39**: 615–630.
SAMBROOK, J., GREENE, R., STRINGER, J., MITCHISON, T., HU, S-L. and BOTCHAN, M. (1979) Analysis of the sites of integration of viral DNA sequences in rat cells transformed by adenovirus 2 or SV40. *Cold Spring Harb. Symp. quant. Biol.* **44**: 569–584.
SARMA, P. S., HUEBNER, R. J. and LANE, W. T. (1965) Induction of tumors in hamsters with an avian adenovirus (CELO). *Science* **149**: 1108.
SARMA, P. S., VASS, W., HUEBNER, R. J., IGEL, H., LANE, W. T. and TURNER, H. C. (1967) Induction of tumors in hamsters with infectious canine hepatitis virus. *Nature* **215**: 293–294.
SARNOW, P., HO, Y-S., WILLIAMS, J. and LEVINE, A. J. (1982) Adenovirus E1B 58kd tumor antigen and SV40 large T-antigen are physically associated with the same 54kd cellular protein in transformed cells. *Cell* **28**: 387–394.
SAWADA, Y. and FUJINAGA, K. (1980) Mapping of adenovirus 12 mRNAs transcribed from the transforming region. *J. Virol.* **36**: 639–651.
SAWADA, Y., OJIMA, S., SHIMOJO, H., SHIROKI, K. and FUJINAGA, K. (1979) Transforming DNA sequences in rat cells transformed by DNA fragments of highly oncogenic human adenovirus type 12. *J. Virol.* **32**: 379–385.
SCANGOS, G. A., HUTTNER, K. M., JURICEK, D. K. and RUDDLE, F. H. (1980) DNA-mediated gene transfer in mammalian cells: molecular analysis of unstable transformants and their progression to stability. *Molec. cell. Biol.* **1**: 111–120.
SCHRIER, P. I., BERNARDS, R., VAESSEN, M. J., HOUWELING, A. and VAN DER EB, A. J. (1983) Expression of class I major histocompatibility antigens switched off by highly oncogenic adenovirus 12 in transformed rat cells. *Nature* **305**: 771–775.
SCHRIER, P. I., VAN DER ELSEN, P. J., HERTOGHS, J. L. and VAN DER EB, A. J. (1979) Characterization of tumor antigens in cells transformed by fragments of adenovirus type 5 DNA. *Virology* **99**: 372–385.
SEGAWA, K., SAITO, I., SHIROKI, K. and SHIMOJO, H. (1980) *In vitro* translation of adenovirus type 12 specific mRNA complementary to the transforming gene. *Virology* **107**: 61–70.
SEIKIKAWA, K., SHIROKI, K., SHIMOJO, H., OJIMA, S. and FUJINAGA, K. (1978) Transformation of a rat cell line by an adenovirus 7 DNA fragment. *Virology* **88**: 1–7.
SHARP, P. A., GALLIMORE, P. H. and FLINT, S. J. (1975) Mapping of adenovirus 2 RNA sequences in lytically infected cells and transformed cell lines. *Cold Spring Harbor Symp. quant. Biol.* **39**: 457–474.
SHEIL, J. M., GALLIMORE, P. H., ZIMMER, S. G. and SOPORI, M. L. (1984). Susceptibility of adenovirus 2-transformed rat cell lines to natural killer (NK) cells: a direct correlation between NK resistance and *in vivo* tumorigenesis. *J. Immunol.* **132**, 1578–1582.
SHENK, T., JONES, N., COLBY, W. and FOWLKES, D. (1979) Functional analysis of adenovirus 5 host-range deletion mutants defective for transformation of rat embryo cells. *Cold Spring Harb. Symp. quant. Biol.* **44**: 367–376.
SHIROKI, K., HANDA, H., SHIMOJO, H., YANO, S., OJIMA, S. and FUJINAGA, K. (1977) Establishment and characterization of rat cell lines transformed by restriction endonuclease fragments of adenovirus 12 DNA. *Virology* **82**: 462–471.
SHIROKI, K., MARUYAMA, K., SAITO, I., FUKUI, Y. and SHIMOJO, H. (1981) Incomplete transformation of rat cells by a deletion mutant of adenovirus type 5. *J. Virol.* **38**: 1048–1054.
SHIROKI, K., MARUYAMA, K., SAITO, I., FUKUI, Y., YAZAKI, K. and SHIMOJO, H. (1982) Dependence of tumor forming capacities of cells transformed by recombinants between adenovirus types 5 and 12 on expression of early region 1. *J. Virol.* **42**: 708–718.
SHIROKI, K., SEGAWA, K. and SHIMOJO, H. (1980) Two tumor antigens and their polypeptides in adenovirus type 12 infected and transformed cells. *Proc. natn. Acad. Sci. U.S.A.* **77**: 2274–2278.
SHIROKI, K., SHIMOJO, H., SAWADA, Y., UEMIZU, Y. and FUJINAGA, K. (1979) Incomplete transformation of rat cells by a small fragment of adenovirus 12 DNA. *Virology* **95**: 127–136.
SMART, J. E., LEWIS, J. B., MATHEWS, M. B., HARTER, M. L. and ANDERSON, C. W. (1981) Adenovirus type 2 early proteins assignment of the early region 1A protein synthesized *in vivo* and *in vitro* to specific mRNAs. *Virology* **112**: 703–713.
SMILEY, J. R. and MAK, S. (1978) Transcription maps for adenovirus type 12 DNA. *J. Virol.* **28**: 227–239.
SPINDLER, K. R., ENG, C. Y. and BERK, A. J. (1985) An adenovirus early region 1A protein is required for maximal viral DNA replication in growth-arrested human cells. *J. Virol.* **53**, 742–750.
SOLNICK, D. (1981) An adenovirus mutant defective in splicing RNA from early region E1A. *Nature* **291**: 508–516.
SOLNICK, D. and ANDERSON, M. (1982) Transformation-deficient adenovirus mutant defective in expression of region 1A but not region 1B. *J. Virol.* **42**: 106–113.
SOUTHERN, E. (1975) Detection of specific sequences among DNA fragments separated by gel electrophoresis. *J. molec. Biol.* **98**: 504–517.
SPECTOR, D. J., MCGROGAN, M. and RASKAS, H. J. (1978) Regulation of appearance of cytoplasmic RNAs from region 1 of the adenoviral 2 genome. *J. molec. Biol.* **126**: 395–414.
SPINDLER, K. R., ROSSER, D. S. E. and BERK, A. J. (1984) Analysis of adenovirus transforming proteins from early regions 1A and 1B with sera to inducible fusion antigens produced in *Escherichia coli*. *J. Virol.* **49**: 132–141.
STARZINSKI-POWITZ, A., SCHULZ, M., ESCHE, H., MUKAI, N. and DOERFLER, W. (1982) The adenovirus type 12-mouse cell system: permissivity and analysis of integration patterns of viral DNA in tumor cells. *EMBO J.* **1**: 493–497.
STEIN, R. and ZIFF, E. (1984) HeLa cell β-tubulin gene transcription is stimulated by adenovirus 5 in parallel with viral early genes by an E1a dependent mechanism. *Mol. Cell. Biol.* **4**: 2792–2801.
STILLMAN, B. W., LEWIS, J. B., CHOW, L. T., MATHEWS, M. B. and SMART, J. E. (1981) Identification of the gene and mRNA for adenovirus terminal protein precursor. *Cell* **23**: 497–508.
STILLMAN, B. W., TAMANOI, F. and MATHEWS, M. B. (1982) Purification of an adenovirus-coded DNA polymerase that is required for initiation of DNA replication. *Cell* **31**: 613–632.

SUBRAMANIAN, T., KUPPUSWAMY, M., MAK, S. and CHINNADURAI, G. (1984) Adenovirus cyt⁺ lucis, which controls cell transformation is an allele of 1P+ locus, which codes for a 19-kilodalton tumor antigen. *J. Virol.* **52**: 336–343.

SUGISAKI, H., SUGIMOTO, K., TAKANAMI, J., SHIROKI, K., SAITO, I., SHIMOJO, H., SAWADA, Y., UEMIZU, Y., UESUGI, S. and FUJINAGA, K. (1980) Structure and gene organization in the transforming *Hind*III G fragment of Ad12. *Cell* **30**: 777–786.

SUTTER, D., WESTPHAL, M. and DOERFLER, W. (1978) Patterns of integration of viral DNA sequences in the genomes of adenovirus type 12 transformed hamster cells. *Cell* **14**: 569–585.

TAKEMORI, N., CLADARAS, C., BHAT, B., CONLEY, A. J. and WOLD, W. S. M. (1984) Cyt gene of adenovirus 2 and 5 is an oncogene for transforming function in early region 1B and encodes the E1B 19,000-molecular weight polypeptide. *J. Virol.* **52**, 793–805.

TAKEMORI, N., RIGGS, J. L. and ALDRICH, C. (1968) Genetic studies with tumorigenic adenoviruses 1. Isolation of cytocidal (*cyt*) mutants of adenovirus type 12. *Virology* **36**: 575–586.

THOMAS, G. P. and MATHEWS, M. B. (1980) DNA replication and the early to late transition in adenovirus infection. *Cell* **22**: 523–534.

TOOZE, J. Ed. (1973) *The Molecular Biology of Viruses*, First Edition, Cold Spring Harbor Laboratory, N.Y.

TOOZE, J. Ed. (1980) *The Molecular Biology of Tumor Viruses—2 DNA Tumor Viruses*, Second Edition, Cold Spring Harbor Laboratory, N.Y.

TRENTIN, J. J., YABE, Y. and TAYLOR, G. (1962) The quest for human cancer viruses. *Science* **137**: 835–841.

VAN BERGEN, B. G. M. and VAN DER VLIET, P. C. (1983) Temperature-sensitive initiation and elongation of adenovirus DNA replication with nuclear extracts from H5ts36-, H5ts149- and H5ts125-infected HeLa cells. *J. Virol.* **46**: 642–648.

VAN DEN ELSEN, P. J., DE PATER, S., HOUWELING, A., VAN DER VREER, J. and VAN DER EB, A. J. (1982) The relationship between region E1A and E1B of human adenoviruses in cell transformation. *Gene* **18**: 175–185.

VAN DEN ELSEN, P. J., HOUWELING, A. and VAN DER EB, A. J. (1983a) Morphological transformation of human adenoviruses is determined to a large extent by gene products of region 1A. *Virology* **131**: 242–246.

VAN DEN ELSEN, P. J., HOUWELING, A. and VAN DER EB, A. J. (1983b) Expression of region 1B of human adenoviruses in the absence of region 1A is not sufficient for complete transformation. *Virology* **128**: 377–390.

VAN DER EB, A. J., MULDER, C., GRAHAM, F. L. and HOUWELING, A. (1977) Transformation with specific fragments of adenovirus DNAs 1. Isolation of specific fragments of adenovirus DNAs 1. Isolation of specific fragments with transforming activity of adenovirus type 5. *Gene* **2**: 115–132.

VAN DER EB, A. J., VAN ORMONDT, H., SCHRIER, P. I., LUPKER, J. H., JOCHEMSEN, H., VAN DEN ELSEN, P. J., DELAYS, R. J., MAAT, J., VAN BERESEN, C. P., DIJKEMA, R. and DE WAARD, A. (1979) Structure and function of the transforming genes of human adenoviruses and SV40. *Cold Spring Harb. Symp. quant. Biol.* **44**: 383–399.

VAN DER VLIET, P. C. and SUSSENBACH, J. J. (1975) An adenovirus 5 gene function required for initiation of viral DNA replication. *Virology* **67**: 415–426.

VAN DER VLIET, P. C., ZANDBERG, J. and JANSZ, H. S. (1977) Evidence for a function of the adenovirus DNA binding protein in initiation of DNA synthesis as well as in elongation on nascent DNA chains. *Virology* **80**: 98–110.

VAN ORMONDT, H., MAAT, J. and DIJKEMA, R. (1980a) Comparison of the nucleotide sequences of the early region E1A regions for subgroups A, B and C of human adenoviruses. *Gene* **12**: 63–76.

VAN ORMONDT, H., MAAT, J. and VAN BEVEREN, P. (1980b) The nucleotide sequences of the transforming early region E1 of adenovirus type 5 DNA. *Gene* **11**: 299–309.

VELICH, A. and ZIFF, E. (1985) Adenovirus E1A proteins repress transcription from the SV40 early promoter. *Cell* **40**, 705–716.

VIRTANEN, A. and PETTSERSON, U. (1983) Molecular structure of the 9S mRNA from early region 1A of adenovirus sero type 2. *J. molec. Biol.* **165**: 496–499.

VIRTANEN, A. and PETTERSSON, U. (1985). Organization of early region 1B of human adenovirus type 2: identification of four differentially spliced mRNAs. *J. Virol.* **54**: 383–391.

VIRTANEN, A., PETTERSSON U., LEMOULLEC, J. M., TIOLLAIS, P. and PERRICAUDET, M. (1982) Different mRNAs from the transforming region of highly oncogenic and non-oncogenic human adenoviruses. *Nature* **295**: 705–707.

VISSER, L., VAN MAARSCHALKERWEERD, M. W., ROZIJN, T. H., WASSENAAR, A. D. C., REEMST, A. M. C. B. and SUSSENBACH, J. S. (1979) Viral DNA sequences in adenovirus-transformed cells. *Cold Spring Harb. Symp. quant. Biol.* **44**: 541–550.

WHITE, E., GRODZICKER, T. and STILLMAN, B. W. (1984). Mutations in the gene encoding the adenovirus early region 1B 19,000-molecular-weight tumor antigen cause the degradation of chromosomal DNA. *J. Virol.* **52**, 410–419.

WILKIE, N. M., USTACELABI, S. and WILLIAMS, J. F. (1973) Characterization of temperature-sensitive mutants of adenovirus type 5: nucleic acid synthesis. *Virology* **51**: 499–503.

WILLIAMS, J. F. (1973) Oncogenic transformation of hamster embryo cells *in vitro* by adenovirus type 5. *Nature* **243**: 162–163.

WILLIAMS, J. F., GALOS, R. S., BINGER, M-H. and FLINT, S. J. (1979) Location of additional early regions within the left quarter of the adenoviral genome. *Cold Spring Harb. Symp. quant. Biol.* **44**: 353–366.

WILLIAMS, J. F., YOUNG, H. and AUSTIN, P. (1975) Genetic analysis of human adenovirus type 5 in permissive and non-permissive cells. *Cold Spring Harb. Symp. quant. Biol.* **39**: 427–432.

WILLIAMS,, J. F., KARGER, B. D., HO, Y. S., CASTIGLIA, C. L., MANN, T. and FLINT, S. J. (1986). The adenovirus E1B 495R protein plays a role in regulating the transport and stability of viral late messages. *Cancer Cells* **4**, 275–284.

WILSON, M. C., FRASER, N. W. and DARNELL, J. E. (1979a) Mapping of DNA initiation sites by high doses of uv-irradiation: evidence for three independent promoters within the left-hand 11% of the Ad2 genome. *Girology* **94**: 175–184.

WILSON, M. C., NEVINS, J. R., BLANCHARD, J-M., GINSBERG, H. S. and DARNELL, J. E. (1979b) Metabolism of mRNA from the transforming region of adenovirus 2. *Cold Spring Harb. Symp. quant. Biol.* **44**: 447–455.

WINBERG, G. and SHENK, T. (1984) Dissection of overlapping functions within the adenovirus type 5 E1A gene. *The EMBO J.* **3**: 1907–1912.

YABE, Y. TRENTIN, J. J. and TAYLOR, G. (1962) Cancer induction in hamsters type 12 adenovirus: effects of age and of virus dose. *Proc. Soc. exp. Biol. Med.* **111**: 343–344.

YAMAMOTO, H., SHIMOJO, H. and HAMADA, C. (1972) Les tumorigenic (*cyt*) mutants of adenovirus 12 defective in induction of cell surface change. *Virology* **50**: 743–752.

YANO, S., OJIMA, S., FUJINAGA, K., SHIROKI, K. and SHIMOJO, H. (1977) Transformation of a rat cell line by an adenovirus type 12 DNA fragment. *Virology* **82**: 214–220.

YEE, S. P., ROWE, D. T., TREMBLAY, M. L., McDERMOTT, M. and BRANTON, P. E. (1983) Identification of human adenovirus early region 1 products by using antisera against synthetic peptides corresponding to the predicted carboxy termini. *J. Virol.* **46**: 1003–1013.

ZANTEMA, A., FRANSEN, J. A. M., DAVIS-OLIVER, A., RAMAEKERS, F. C. S., VOOIJS, G. P., DELEYS, B. and VAN DER EB, A. J. (1985) Localization of the E1B proteins in adenovirus 5 transformed cells as revealed by interaction with monoclonal antibodies. *Virology* **142**: 44–58.

ZINKERNAGEL, R. M. and DOHERTY, P. C. (1979) MHC—restricted cytotoxic T cells: studies on the biological role of polymorphic major transplantation antigens determining T-cell restriction—specificity, function, and responsiveness. *Adv. Immun.* **27**: 51–177.

ZUR HAUSEN, H. and SOKOL, F. (1969) Fate of adenovirus type 12 genome in non-permissive cells. *J. Virol.* **4**: 255–263.

CHAPTER 8

CELLULAR TRANSFORMATION BY THE HERPESVIRUSES AND ANTIVIRAL DRUGS

JUNG-CHUNG LIN* and JOSEPH S. PAGANO†

Departments of Biochemistry and Nutrition, and Medicine, Microbiology and Immunology,†
Lineberger Cancer Research Center, University of North Carolina at Chapel Hill, Chapel Hill,
North Carolina 27514, U.S.A.

Abstract—This article is intended (i) to discuss broadly the etiological role played by herpesviruses in the development of human cancers, and (ii) to provide some detailed information on the mode of action of antiviral agents. The association between virus and malignancy rests on two general grounds: *in vivo* evidence coming from epidemiologic, serologic and molecular association of virus and malignancy; and *in vitro* cellular transforming effects of the virus. Other cofactors such as genetic disposition, environmental or nutritional carcinogens, chromosomal translocations with oncogene activation and suppression of cell-mediated immune functions are also discussed. The selectivity of most antiherpes drugs depends essentially on two virus-encoded enzymes, dThd kinase and DNA polymerase. From a consideration of the mechanisms of action of both the drugs and the viruses in the cells that they infect, we can attempt to predict whether such drugs would have any impact in the transforming process.

1. INTRODUCTION

The case for virus causation of cancer has strengthened in recent years, not through direct proof which has remained elusive, but through increasing associations on many levels ranging from the molecular to the epidemiologic. Among the most durable in their association with human malignancy have been several of the human herpes-group viruses. Other prime candidates include hepatitis B virus, suspected as a causal agent of hepatocellular carcinoma, and human T-cell leukemia virus, suspected as the cause of a form of leukemia and lymphoma found in southern Japan, the Caribbean basin and central Africa. The association between virus and malignancy rests on two general grounds: in vivo evidence coming from epidemiologic, serologic and molecular association of virus and malignancy; and in vitro cellular transforming effects of the virus.

There have been many, probably too many, reviews written about viruses and cellular transformation. Our aim here is to take a different point of view of this topic guided by a perspective provided by a consideration of antiviral drugs. The dawn of antiviral drugs for use in human beings has arrived. If viruses do cause cancer would these drugs be effective, not only against viruses as inciting causes, but also against the cancer itself and under what circumstances? We will focus on the Epstein–Barr virus which provides an abundance of *in vivo* and *in vitro* data on transformation. From a consideration of the mechanisms of action of both the drugs and the viruses in the cells that they infect, we will attempt to predict whether such drugs would have any impact in the transforming process.

1.1. VIRUSES AND CANCER

The number of human viruses now associated with malignant disease has increased. Human T-cell leukemia virus type 1 (HTLV-1) is a retrovirus isolated from patients with T-cell lymphomas characterized by an aggressive course and infiltration of the skin (Poiesz *et al.*, 1980). Among the DNA viruses many types of papillomaviruses, the causative agents of warts, have now been distinguished by molecular cloning of their genomes. Some warts undergo malignant degeneration, in particular, the multiple flat warts known as epidermodysplasia verruciformis. In addition there is now clear evidence that selected types of papillomavirus genomes are present in some vulvar cancers and cervical carcinomas (Green et al., 1982; zur Hausen, 1982).

Papoviruses such as Simian Virus 40 (SV40) in monkeys and BK virus in human beings cause latent infections particularly in the brain and in the kidney. Although papovaviruses can transform human cells *in vitro* into malignant-appearing cells, the only disease with which the virus has been associated is the rare progressive multifocal leukoencephalopathy (PML) and not with cancer. PML has been described in patients with Acquired Immunodeficiency Syndrome (AIDS) (Miller et al., 1982). The human papovavirus, BK, is reactivated and excreted with high frequency in immunosuppressed persons and in pregnant women without apparent injury to the host (Coleman et al., 1980).

Human adenoviruses like SV40 can readily transform human cells *in vitro*, and several adenovirus types also produce tumors when inoculated into newborn hamsters. However, there is no evidence linking adenoviruses with human malignancy.

Of the human herpesviruses, herpes simplex type 2 has been associated for many years with carcinoma of the cervix chiefly by suggestive but far from conclusive epidemiologic evidence. The virus may prove to be an initiating factor in the disease, but it does not appear to be the sole cause of this malignancy (Galloway and McDougall, 1983). The role of human cytomegalovirus has been scrutinized in a number of cancers including cervical carcinoma and prostatic cancer, but the strongest evidence for an oncogenic relation comes from studies of CMV in the endemic form of Kaposi's sarcoma in Africa and in the United States (Giraldo et al., 1980; Boldogh et al., 1981, 1983).

Finally, the association of hepatitis B virus with hepatocellular carcinoma is now well established, and it is likely that the virus has a key role as an initiator in the pathogenesis of this malignancy (Blumberg and London, 1981).

TABLE 1. *Herpesviruses Associated with Cancer in Human Beings*

Virus	Strong association	Weak association
Epstein–Barr virus	African Burkitt's lymphoma Nasopharyngeal carcinoma B-lymphocytic immunoblastic sarcomas	Supraglottic carcinoma of larynx Palatine tonsilar carcinoma Parotid tumors
Cytomegalovirus	Kaposi's sarcoma	Cervical carcinoma Prostatic cancer
Herpes simplex type 2	None	Cervical cancer
Herpes simplex type 1	None	Oral cancer
Varicella-zoster	None	None

1.1.1. Clues to Viral Etiology

Certain features of malignancies suggest causation or initiation by virus infection. Most of the cancers suspected of a viral etiology are not common malignancies in Western nations. Diseases such as Burkitt's lymphoma and unusual forms of leukemias, Kaposi's sarcoma and hepatocellular carcinoma are prominently linked to virus infection, whereas gastrointestinal and lung tumors are not. Except for carcinoma of the posterior nasopharynx and cervical carcinoma, most of the virus-associated cancers are not carcinomas. However, from a world-wide perspective, virus-associated cancers account for a large percentage of malignant disease.

Viruses linked to cancers generally produce, like most viruses, silent primary infection. The viruses are common infectious agents with the oncogenic outcome a rare event. The cancer is a late manifestation appearing after a period of latency following early infection, ranging from an average of seven years in the case of Burkitt's lymphoma, to 40–60 years for nasopharyngeal carcinoma and hepatomas in some endemic regions (de-Thé, 1982). The appearance of the malignancy may be tied to reactivation of virus infection (for example, Burkitt's lymphoma, nasopharyngeal carcinoma) or to persistent active infection (hepatoma).

The hallmark of a possible viral etiology is an endemic pattern of incidence of the cancer, usually in only certain parts of the world. However, virus-associated malignances exhibit both sporadic and endemic patterns of incidence depending upon where they occur.

In the endemic regions, cofactors are suspected of playing a role to explain the high incidence. Postulated cofactors cover a broad range of agents and conditions including exposure to virus infection early in life, genetic disposition, environmental or nutritional carcinogens, chromosomal translocations with oncogene activation and suppression of cell-mediated immune functions.

1.1.2. *Properties of Oncogenic Viruses*

Both RNA- and DNA-containing viruses have now been associated with a variety of malignancies. All of the viruses identified so far have the capability of forming circular DNA genomes or proviral genomes. This physical property may promote or facilitate integration of viral genomes into chromosomal DNA. The viral genomes are retained, usually *in toto*, either integrated into cellular DNA by covalent bonding or as plasmid or episomal forms. In most cases only a very limited amount of the viral genetic information is expressed in the tumor tissue. However, since whole genomes are retained, it might be possible to recover complete viruses, at least from explanted tumor tissues. Exceptions to retention of complete genomes in transformed tissue seem to be provided by herpes simplex type 2 virus and human cytomegalovirus; in cells transformed *in vitro* by these viruses, viral genetic material seems to be progressively lost and can finally no longer be detected by hybridization techniques in cells transformed *in vitro* (Rapp and Westmoreland, 1976; Huang *et al.*, 1983).

However, the *in vitro* findings do not necessarily correlate with results of analyses of human tissues such as cervical carcinoma and Kaposi's sarcoma. In the case of cervical carcinoma, there is some evidence of continued presence of viral genes coming from detection of HSV RNA in carcinoma *in situ* (Galloway and McDougall, 1983). The significance of these observations is taken up later in this article. In Africa Kaposi's sarcoma viral DNA has been detected directly in tumor tissue in some specimens but not others. Finally, in adenovirus experimental systems, only a small fraction of the genome may be retained and expressed in the *in vitro* transformed cells (Sharp *et al.*, 1974).

As might be expected, the dominant effects of infection on the host cell are nonlytic and stimulatory of cellular functions. These effects culminate in tissue proliferation. However, cultivation of tumor cells *in vitro* may permit derepression of viral genes with fuller expression and replication of virus.

1.2. ANTIVIRAL DRUGS

The first drugs proposed for use against herpesvirus infections were only marginally effective. They did not discriminate between cellular and viral macromolecular synthesis nor preferentially inhibit virus-controlled activities in the cell. A halogenated pyrimidine such as iododeoxyuridine when used for the treatment of herpes simplex infection of the eye exhibited a slight selectivity based on the differential kinetics of synthesis of viral vs cellular DNA. In the infected eye viral DNA synthesis is maximal whereas cellular DNA synthesis is at a low level. Furthermore, the accessibility of the site made it possible to achieve local concentrations of the drug that would be toxic if given systemically. These advantages were sufficient to permit limited efficacy of the drug for herpes simplex keratitis. The first drug used successfully to treat herpesvirus infection systemically was adenine arabinoside, which

TABLE 2. *Effects of Antiviral Drugs on the Herpesviruses*

Drug	Virus	Infection state		
		Active replication	Latent infection	Transformed cells
PFA*	HSV-1, -2, CMV, EBV	Strongly inhibitory	No effect	Nonspecific effect in EBV-transformed lymphocytes
ACV†	HSV-1, -2, EBV, VZV	Strongly inhibitory	No effect	No effect
DHPG‡	HSV-1, -2, CMV, EBV, VZV, some ACV-resistant HSV	Strongly inhibitory	No effect	Not tested
BVDU	HSV-1, EBV (other?)	Strongly inhibitory	No effect	Not tested
FMAU	HSV-1, -2, CMV, EBV, VZV	Strongly inhibitory	No effect	Not tested
FIAC	HSV-1, -2, CMV, EBV, VZV	Strongly inhibitory	No effect	Not tested

*Phosphonoformic acid.
†Proven clinical utility against HSV-1, -2, VZV; preliminary evidence of utility against EBV.
‡Preliminary evidence of clinical utility against CMV.

reduced mortality and morbidity of herpesvirus encephalitis (Whitley *et al.*, 1977). The milestone advance in the development of antiviral therapy came with the introduction of Acyclovir, subsequently also used to treat herpes encephalitis (Sköldenberg *et al.*, 1984). The difference in the strategy of this drug compared with IUDR is fundamental in that the action of Acyclovir depends upon virus-specific enzymatic processes, and thus the action of the drug is highly discriminatory affecting viral synthetic processes and sparing cellular DNA synthesis. Acyclovir is a particularly interesting drug because it is, in the case of herpes simplex infection, both activated through phosphorylation by a virus-specific enzyme and has as its target of action another virus-specific enzyme.

Acyclovir [9-(2-hydroxyethoxymethyl)guanine], the first potentially, clinically useful drug effective against replication of EBV, is without effect against latent or persistent EBV infection (Colby *et al.*, 1982). Acyclovir is relatively ineffective against human cytomegalovirus (CMV). Three halogenated nucleoside analogs, E-5-(2-bromovinyl)-2'-deoxyuridine (BVDU), 1-(2-deoxy-2-fluoro-7-D-arabinofuranosyl)-5-iodocytosine (FIAC) and 1-(2-deoxy-2-fluoro-7-D-arabinofuranosyl)-5-methyluracil (FMAU), are potent inhibitors of EBV replication *in vitro* (Lin *et al.*, 1983). Moreover in contrast to the reversibility of viral inhibition by Acyclovir, these three drugs have prolonged effects in suppressing viral replication even after the drugs are removed from persistently infected cell cultures. These drugs, which have been used experimentally *in vitro*, are also active against CMV and HSV. Finally an Acyclovir-like guanosine analog, 9-(1,3-dihydroxy-2-propoxymethyl)guanine (BW 759U, DHPG) inhibits herpes simplex virus type 1 (HSV-1), HSV-2, CMV and EBV replication at concentrations that do not inhibit cell growth in culture. The potency of the drug against all of these viruses is greater than that of ACV (Cheng *et al.*, 1983b; Lin *et al.*, 1984). Unfortunately, the clinical utility of DHPG will be limited because of toxicity.

2. THE HERPESVIRUSES

2.1. Virology*

HSV, VZV, CMV and EBV particles appear to be identical by electron mircoscopy. They are large viruses with a diameter of 150–200 nm. An inner core consisting of DNA interwined with protein is surrounded by a protein capsid of symmetrical structure

*Adapted from Pagano and Lemon, 1984.

composed of 162 capsomeres. This nucleocapsid is covered by an amorphous envelope that is derived from the nuclear and plasma membrane of the host cell. The nucleocapsid is assembled in the cell nucleus and is enveloped as it exits from the nucleus. Loss of the envelope largely abolishes virus infectivity. There are at least 33 polypeptides involved in the structure of the HSV virion including some that are included in the envelope. Other polypetides are in the core of the virion intimately associated with the viral DNA and may affect regulation of viral gene expression. Except possibly for a protein kinase, these viruses do not appear to contain polypetides with enzymatic activities as part of their structure. Their genomes do code for several virus-specific enzymes, including DNA polymerases, thymidine kinases and ribonucleotide reductases that are distinct from those found in mammalian cells.

Herpesvirus genomes consist of linear double-stranded DNA ranging from approximately 90×10^6 daltons (140 kilobases) for certain VZV strains to 150×10^6 daltons (235 kilobases) for CMV. Herpesvirus genomes possess terminal and internal stretches of repeated sequences in addition to the unique sequences. In some herpesvirus genomes, nucleotides sequences found at the ends of the genome are also situated internally in the genome. A variety of such arrangements, ranging from simple to complex permuted structures, exist in the herpes group of viruses. Such sequence arrangements, the significance of which is unknown, appear to exist only in herpesvirus genomes. These novel arrangements may relate to the generation of defective genomes.

Defective or incomplete virus particles are common among the herpesviruses. These virions possess less than the full complement of viral DNA. For example, in the case of CMV, many more virions contain only 100×10^6 daltons of DNA than the full genome complement of 150×10^6 daltons. Herpes simplex virions may possess a genome that is of normal unit length but is still defective in that unique sequences are missing and are replaced by reiterations of a single segment of the genome. These defective forms may be important in some of the biologic effects of the herpesviruses such as persistent infection.

In addition to the linear form, the EBV genome can assume an entirely novel closed circular form within the cell. This supercoiled genome, called the EBV episome or plasmid, is not encapsulated by virion structural protein. It is found in the nuclei of latently infected cells within a nucleosomal structure similar to that of cellular chromatin. During latency the EBV episome is replicated as if it were a cellular constituent by host-cell DNA polymerase. This is the only episomal or plasmid DNA form known to exist in eucaryotic cells save for certain yeasts. Other herpesvirus genomes may be able to assume a similar episomal form, but thus far such closed circular genomes have been detected only in cells infected with an oncogenic herpesvirus of monkeys, *Herpesvirus saimiri*. In the laboratory, however, linear HSV genomes can be circularized by exonuclease digestion of the ends of the DNA strands. This digestion exposes homologous nucleotide sequences at either end of the genome, thereby permitting circularization. Some EBV DNA may also be directly integrated by covalent bonding into cellular DNA in transformed lymphoblasts.

With the exception of HSV type 1 and HSV type 2, the genomes of the human herpesviruses are completely or almost completely different as shown by cross-hybridization experiments. HSV type 1 and HSV type 2 share approximately 50% of their genome content. A lesser degree of homology between, for example, the Epstein–Barr virus and cytomegalovirus—in the range of 5% or less—has not yet been excluded. With the exception of certain simian EBV-like agents, the many herpes-group viruses of other animal species bear little if any genetic homologous relation to human herpes-group viruses. A few of the animal herpesviruses are capable of infecting man—herpesvirus simiae and perhaps simian CMV—but do so rarely.

Since only the EBV genome has been entirely sequenced, sequence homology of DNA virus genomes is analyzed by digestion with restriction endonucleases. These nucleases digest only at certain sites on different genomes, and thus generate specific DNA fragments which when examined by electrophoresis show characteristic patterns of sizes. Such patterns are a direct and precise reflection of virus strain differences that may be difficult to detect otherwise. HSV and CMV and to a lesser extent, EBV and VZV genomes isolated from various sources exhibit great diversity, especially in some regions of the genomes. Such analyses have demonstrated epidemiologic usefulness in tracing transmission of specific strains of virus. Homology of genomes can be proved further by cross-hybridization of fragments of the genomes.

2.1.1. *Antigenic Composition*

The large herpesvirus genome contains sufficient genetic information to encode more than 100 different proteins, placing the herpesviruses among the most complex viruses. Some of these proteins, many of which are not present in the intact virion, are efficient antigens. However, despite the identical morphology of the herpesviruses, there is little antigenic relatedness among most members of this group.

Of the human herpesviruses only HSV type 1 and HSV type 2 have a significant degree of antigenic relatedness. Although these two viruses are distinguishable by serologic techniques, antigenically related polypeptides common to both HSV type 1 and HSV type 2 have been identified by immunoprecipitation. This is not surprising in view of the high degree of nucleic acid homology (approximately 50%) between these viruses. Heterologous neutralizing antisera are cross-reactive, which indicates that related antigenic sites exist on the surface of the viral envelope. Cross-reacting internal and nonstructural antigens have been identified as well. The antigenic diversity is not restricted to differences between type 1 and type 2 viruses; there is considerable variation among strains of the same type. Such variations may explain so-called 'intermediate' types between type 1 and 2.

VZV isolates are relatively homogeneous in terms of their antigenic composition. There is no cross-reactivity between VZV and HSV antigens despite the heterologous antibody rises that are occasionally observed following infection with one of the viruses.

Human cytomegalovirus strains, while antigenically unrelated to other human herpesviruses, are related to some degree to CMV-like agents isolated from nonhuman primates. As with HSV, strains of CMV demonstrate considerable antigenic diversity, but they have not been grouped by these means. Recent studies have identified 'early' nonstructural CMV antigens that are distinct from 'late' CMV antigens including virion components. Early and late refer to approximate time of appearance in the viral replicative cycle, with 'early' antigens being synthesized before, and 'late' antigens after, viral DNA synthesis begins.

EBV has several distinctive antigens, none of which cross-reacts with the other human herpesviruses, reflecting the general lack of nucleic acid homology among these agents. By fluorescent antibody technique viral capsid antigen (VCA) activity has been recognized as part of the virion structure. However, there are additional antigenic activities that are nonstructural. EBV 'early' antigen (EA) is found in cells that are abortively infected with virus and unable to produce mature infectious virions as well as in those producing virus. 'Membrane' antigen (MA) appears on the surface of some EBV-producing cells grown *in vitro* from neoplasms associated with EBV. Such cells as well as latently infected cells also contain EBV nuclear antigen (EBNA), the function of which is still unknown, although it may be involved in maintenance of the EBV episome in latently infected cells. Recently a second form of EBNA (EBNA-2) has been described that may be involved in the transforming function of EBV. The appearance of antibodies to each of these antigens follows a characteristic temporal sequence after primary infection with the virus, and they are therefore of diagnostic significance.

2.2. Epstein–Barr virus associated diseases

EBV has a well established association with two malignancies, carcinoma of the posterior nasopharynx (NPC), which is common and distributed world-wide, and Burkitt's lymphoma (BL), which is rare and largely confined to a midcontinental belt of Africa (Pagano and Henry, 1983; Epstein and Achong, 1979). These diseases, one an epithelial cell malignancy and the other a B-cell lymphoma, are so different as to seem to defy assignment of an etiologic role for EBV to both. Moreover, the virus is the principal cause of a third disease, infectious mononucleosis (IM), in its acute, less defined chronic, and rare lethal forms. A strong association with yet another group of diseases, usually classified as immunoblastic sarcomas, has emerged recently (for review, see Sixbey and Pagano, 1984).

2.2.1. Burkitt's Lymphoma

The idea that Burkitt's lymphoma might be caused by a virus came primarily from the epidemiology of the disease. The epidemiologic features include age-related incidence, geographic distribution and case-clustering, which resemble the classical pattern of a childhood viral illness with largely subclinical infection, acquired immunity and possible vector transmission. The virus was discovered (Epstein et al., 1964) in an explanted specimen of Burkitt's lymphoma.

Patients with Burkitt's lymphoma have EBV antibodies including a response to early antigen (EA). However, the virus itself is not found in tumor tissue, nor even always in explanted tissue. Molecular hybridization analyses show that EBV DNA is present in almost all specimens of African Burkitt's lymphoma; over 98% contain hybridizable EBV DNA. In the United States only about one-seventh of BL contain hybridizable EBV DNA.

2.2.2. Nasopharyngeal Carcinoma

Carcinoma of the posterior nasopharynx (NPC) is also relatively common in Africa. Unlike BL, NPC is an epithelial cell malignancy although it has been classed as a lymphoepithelioma because of infiltration of lymphocytes in the malignant tissue. The anaplastic form of NPC was associated with EBV by the high antibody titers to VCA and EA in some patients in a pattern reminiscent of BL. As with BL, NPC tissue does not appear to contain virus; however, NPC tissue can not be cultivated *in vitro* and so the consistent association between virus and disease required the demonstration that NPC's throughout the world contain hybridizable EBV DNA. EBV genomes in NPC are in the transformed epithelial elements of the tumor as shown by *in situ* cytohydridization studies and also by cultivating the tumors in nude mice.

NPC also has a relatively high incidence in North Africa. However, the most striking incidence is in the Chinese, especially in southeast China, as well as in persons of Chinese extraction worldwide. In south China, the disease is endemic in middle-aged persons, especially males. There is also a high incidence of NPC in American Eskimos. In other areas the disease is sporadic and much lower in frequency. Of the three histologic types of NPC, the anaplastic of type III form has the strongest association with EBV, but EBV DNA and RNA have also been detected in type I and type II disease (Raab-Traub et al., 1983). Type I disease is a differentiated, keratinized, squamous cell type without infiltrating lymphocytes.

The endemicity of BL and of NPC in certain areas, coupled with the worldwide incidence of EBV infection in all peoples, point to the existence of local cofactors which may have an etiologic role, but none has been conclusively identified. Genetic predisposition has been implicated but not proven in NPC as suggested by the prevalence of certain HLA types in the patients. Immunosuppression and chronic antigenic stimulation produced by malaria might predispose to Burkitt's lymphoma. Recently HTLV-3 antibodies have been detected in as many as 70% of persons tested in equatorial Africa. Translocations from chromosome 8 to chromosomes 14, 22 or 2 are invariably demonstrable in BL. The cellular oncogene, *c-myc*, is transposed and activated by these translocations (Dalla-Favera et al., 1982).

2.2.3. Acute EBV Lymphoproliferative Syndromes in Allograft Recipients

Lymphoproliferative disorders have been increasingly recognized in kidney transplant recipients (Hanto et al., 1981). Patients can be divided into two groups, the first with a diffuse lymphoproliferative disorder developing soon after transplantation (mean age, 11 months). All of the patients have EBV antibodies. The second group are older patients (mean age, 53 years). The onset of the disease after transplantation was a mean of 42 months. The lymphoproliferative disorder was more localized. There was a 70% mortality within nine months of diagnosis. The lymphoid process was classified as polymorphic diffuse B-cell lymphoma with invasive proliferation involving follicular center cells, lymphocytes with plasmacytic differentiation and large immunoblasts with atypical nuclei. Proliferating lymphocytes with monoclonal Ig surface markers may emerge in these conditions.

All these diseases may originate similarly but manifest themselves variably: diffuse or localized, invasive, or merely reactive and hyperplastic lymphoproliferative. The basic pathologic cell type is the B-lymphocyte and there may be either polyclonal or monoclonal cell proliferation. The diffuse form of the disease tends to be acute in onset and produced by primary or reactivated infection with EBV. The localized form may be more insidious and accompanied by reactivated EBV infection. The pathologic cell, whether polyclonal or monoclonal, contains the EBV nuclear antigen and EBV genomes. The common denominator appears to be impairment of cell-mediated immune mechanisms, centering on T-lymphocytes, produced either iatrogenically or naturally occurring. Such immune dysfunction is sometimes mirrored in the distinctive EBV antibody patterns characterizing these syndromes.

2.2.4. Kaposi's Sarcoma

If a virus plays a role in Kaposi's sarcoma as suspected, the virus is probably a common one since Kaposi's sarcoma occurs in such diverse subjects as male homosexuals and allograft recipients. Kaposi's sarcoma in allograft recipients antedated the appearance of AIDS. By analogy with EBV, infection with this agent probably occurs earlier in life, with the infection then becoming latent. Virus infection, perhaps when reactivated by immunosuppression, may trigger proliferation of the target cell of Kaposi's sarcoma, vascular endothelial stem cells. For infection itself, the primary target cell type may be respiratory or oropharyngeal mucosal with the endothelial cells as secondary targets reached by virus in blood or blood cells. Initial infection sites in the oropharynx or respiratory tract seem more likely than a genital site if the infection is a common one. Occurrence of Kaposi's sarcomas in allograft recipients before sexual maturity might indicate a nonsexual mode of transmission.

Cytomegalovirus remains a prime suspect in the pathogenesis of Kaposi's sarcoma. CMV infects early in life, may infect *in utero* or perinatally, causes silent infections, enters a latent phase and is readily reactivated by immunosuppression. There is intense CMV replicative activity in essentially all allograft recipients and most patients with AIDS around the time that Kaposi's sarcoma appears. Kaposi's sarcoma is one of the earliest malignancies to appear in allograft recipients, and it is the most frequent malignancy in AIDS—again, perhaps an indicator of close temporal association with CMV replication as well as a short latent period. CMV has striking tissue pleiotropism as well as tissue-proliferative stimulatory capabilities *in vitro*. Finally, CMV is capable of infecting blood cells, so that the virus could gain access to vascular endothelium or stem cells. When released from immunologic control mechanisms, latent CMV infection might trigger proliferation of vascular endothelial cells and the appearance of Kaposi's sarcoma. The multifocal character of Kaposi's lesions suggests a blood-borne dissemination of the virus.

2.2.5. Epstein–Barr Virus Infection in AIDS

Many patients with AIDS have high antibody titers to EBV antigens, produced probably through reactivation of EBV but also through EBV-induced lymphoproliferation. The nonendemic EBV lymphotropic diseases all seem to arise in a setting of deficiency of cell-mediated immune mechanisms, either primary or acquired. X-linked lymphoproliferative syndrome is a sex-linked dominant genetic disease which leads to fatal infectious mononucleosis upon primary infection in males within the kindred or to death later from B-cell lymphomas. More recently the Wiscott–Aldrich syndrome, ataxia telangiectasia and combined immunodeficiency disease, has been associated with a high incidence of B-cell lymphomas, which contain Epstein–Barr virus DNA (Sixbey and Pagano, 1984). American Burkitt's lymphomas have been described in AIDS, and Burkitt's lymphoma may arise through immunosuppression produced by malaria or possibly by subclincal HTLV-3 infection.

In the pathogenesis of EBV-associated lymphoproliferative diseases (reviewed in Sixbey and Pagano, 1984), there is evidence that the Epstein–Barr virus primarily infects epithelial cells in the oropharynx with virus replicating in these cells and excreted in the saliva (de-Thé, 1982; Lemon et al., 1977). The virus, once replicated, causes secondary infection of B-lymphocytes via specific EBV receptors. In vivo infection of B-lymphocytes is essentially latent and virus is not produced in these cells. The cells are, however, stimulated to proliferate and polyclonal B-cell proliferation results. The latently infected B-lymphocytes carry EBV genomes and EBNA. They also probably have virus-specified alterations of the plasma membrane which promote the recognition of these cells by reactive cytotoxic lymphoctyes. Finally, after the initial wave of EBV-induced B-cell proliferation and destruction, a steady state is reached and a few latently infected B-cells persist for life.

In the absence of a normal cellular immune response, which seems to be needed to maintain latent EBV infection, it is possible that the viral genomes in B-lymphocytes may be expressed and become new centers of virus production in vivo. Consequently, virus infection, primary or reactivated, may stimulate rises in EBV antibody titers. The polyclonal EBV lymphoproliferation, which is ordinarily a benign outcome of EBV infection, becomes progressive and invasive with behavior like malignant disease. In some cases, an actual monoclonal malignant cell line may emerge. The biology of polyclonal and monoclonal forms of these conditions probably differs and has implications for therapy.

3. THE ANTIVIRAL DRUGS

The goal of antiviral chemotherapy is to develop an agent that selectively inhibits the replication of virus in infected cells without affecting the metabolic processes of host cells. Many compounds have been synthesized and reported to have selective inhibitory effects on the replication of herpesviruses. In this section we focus on in vitro effects of some of the promising anti-herpesvirus agents that have been developed.

5-Iodo-2'-deoxyuridine (IdUrd) was first synthesized as an anticancer drug (Prusoff, 1959). Soon after its synthesis, Herrmann (1961) reported that the compound inhibited the replication in cultured cells of several DNA-containing animal viruses including herpes simplex virus. Since then several studies have indicated that IdUrd is a potent inhibitor of the replication of almost all DNA viruses (Prusoff and Goz, 1975). Kaufman and his coworkers (1960) observed that IdUrd was effective in the treatment of human herpetic keratitis, and the compound was approved by the U.S. Food and Drug Administration as the first antiviral drug for clinical use. It is interesting that IdUrd is an inducer rather than an inhibitor of replication of the Epstein–Barr virus (Glaser et al., 1973; Lin and Pagano, 1980a,b).

1-β-D-Arabinofuranosylcytosine (ara-C) and 9-β-D-arabinofuranosyladenine (ara-A) have shown antiviral activity in human beings. Ara-C was reported to be a potent antiviral agent against various herpetic infections (Prusoff and Goz, 1973) but is not used clinically because of its marked toxicity for host cells. However, it has proven effective in the treatment of certain leukemias. Ara-A, which is relatively nontoxic, has been shown to inhibit herpes simplex virus DNA replication in vitro (Müller et al., 1977) and to cure cases of herpes encephalitis in clinical trials. In vitro ara-A also inhibits productive EBV infection in P3HR-1 cells (Coker-Vann and Dolin, 1977) and viral DNA synthesis in superinfected Raji cells (Benz et al., 1978).

Ara-A was approved in 1978 for treatment of herpes encephalitis in man. However, ara-

A is very extensively deaminated, and only a small fraction is converted intracellularly to the active compound, ara-ATP. The deamination of ara-A and ara-C limits the availability of these aranucleosides for conversion to their respective nucleotides. However, the antiherpes activity of ara-A and ara-C can be potentiated by combination with their corresponding inhibitors of deaminase such as erythro-9-(2-hydroxy-3-nonyl)adenine (EHNA) and tetrahydrouridine (THU). The antiviral properties of ara-A and ara-C and other aranucleosides have been reviewed recently (North and Cohen, 1979, 1982).

The two pyrophosphate analogs, phosphonoformic acid (Foscarnet, PFA) and phosphonoacetic acid (PAA), are potent antiviral agents *in vitro*. PAA reversibly inhibits the multiplication of a number of herpesviruses. These drugs are specific inhibitors of the herpesvirus-induced DNA polymerases *in vivo* and *in vitro* (Shipkowitz *et al.*, 1973; Bolden *et al.*, 1975; Datta and Hood, 1981). PAA (50 µg/ml) reversibly inhibits cytomegalovirus DNA replication in tissue culture (Huang, 1975). The productive synthesis of EBV DNA is inhibited by PAA (100 µg/ml) in producer cells (B95-8, P3HR-1), but the resident EBV DNA in Raji was not sensitive to PAA (Summers and Klein, 1976; Yajima *et al.*, 1976). PAA (100 µg/ml) also inhibited EBV DNA synthesis in superinfected Raji cells (Yajima *et al.*, 1976).

FIG. 1. Structures of the pyrophosphate analogs, PFA and PAA.

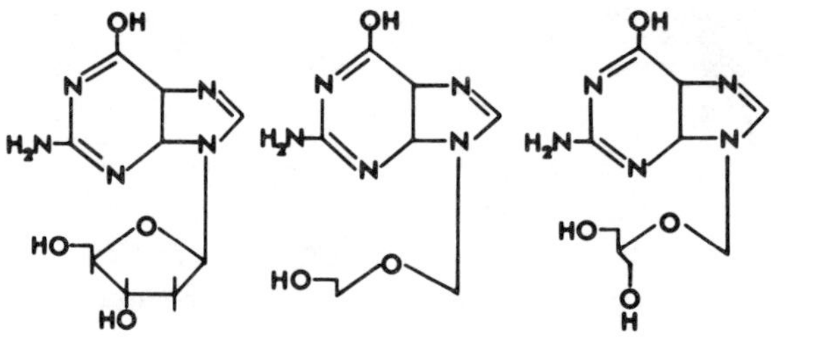

FIG. 2. Structures of the analogs of 2'-deoxyguanosine.

Recently several new nucleoside analogs which are highly selective, potent inhibitors of replication of herpesviruses have been developed. Among these are 9-(2-hydroxyethoxymethyl)guanine (Acyclovir or ACV), 1-(2-fluoro-2-deoxy-β-D-arabinofuranosyl)-5-iodocytosine (FIAC), E-5-(2-bromovinyl)-2'-deoxyuridine (BVDU) and 9-(1,3-dihydroxy-2-propoxymethyl)guanine (BW759U, DHPG).

In cell culture, BVDU inhibits the replication of HSV-1 (DeClercq *et al.*, 1979), and varicella-zoster virus (VZV) (for a review see DeClercq, 1982). BVDU is virtually inactive against dTK$^-$ (deoxythymidine kinase-deficient) mutants of HSV-1 or against VZV (DeClercq *et al.*, 1980; DeClercq, 1982). This finding indicates that the drug must be phosphorylated by the virus-induced thymidine kinase to exert its antiviral effects. The activity of BVDU against human cytomegalovirus (CMV) and EBV has not been fully assessed. However, we have recently shown that BVDU is a potent and selective inhibitor of replication of EBV *in vitro*. At 20 µM BVDU effectively inhibits replication of EBV DNA both in a virus-producing cell line (P3HR-1) and in superinfected Raji cells (Lin *et al.*, 1983). The inhibitory effect of BVDU (20 µM) on EBV genome replication appears to be irreversible for 10 days after removal of the drug from the cultured medium (Lin *et al.*, 1983). The 50% inhibitory dose for viral replication (ED_{50}) is 0.06 µM and for cell growth (ID_{50}) in P3HR-1 cells it is 390 µM (Lin *et al.*, 1985). The therapeutic index for BVDU (ID_{50}/ED_{50}) thus calculated is 6500.

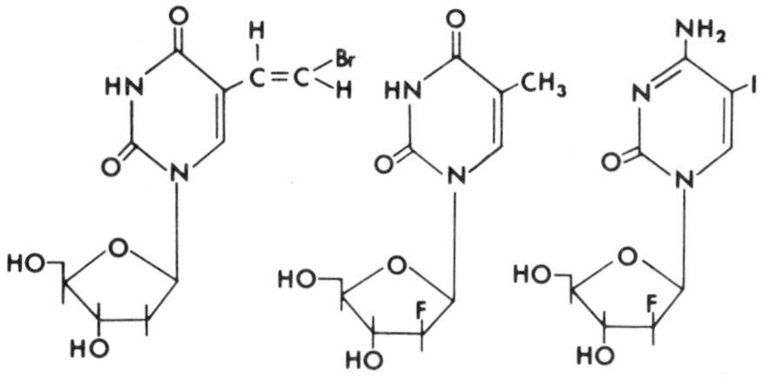

FIG. 3. Structures of the halogenated nucleoside analogs.

Recently, Watanabe and coworkers (1979) have synthesized a new series of nucleoside analogs. One of these compounds, FIAC, was found to have potent anti-HSV-1, HSV-2 and VZV activity with minimal cytotoxicity for uninfected cells (Lopez et al., 1980). Furthermore, FIAC has potent anti-CMV (Lopez et al., 1979; Mar et al., 1983) and EBV (Lin et al., 1983) activity. The EBV DNA replication in P3HR-1 cells remained suppressed for more than one month after removal of the drug (Lin et al., 1983; Lin and Pagano, unpublished data). The ED_{50} of FIAC for EBV is 0.005 μM and the ID_{50} for P3HR-1 cells is 5 μM (Lin et al., 1985), giving a calculated in vitro therapeutic index of 1,000.

ACV is one of the most potent antiviral agents against HSV replication (Elion et al., 1977; Schaeffer et al., 1978). In contrast, it has limited antihuman CMV activity (Crumpacker et al., 1979; Mar et al., 1982; Plotkin et al., 1982). ACV inhibits EBV DNA replication in a virus-producing cell line (P3HR-1) but it is essentially without effect on the nonproducer Raji cells (Colby et al., 1980; Lin et al., 1983). Replication of latent EBV DNA in Burkitt somatic hybrid cells (D98/HR-1) induced by IdUrd is inhibited by ACV (Lin and Pagano, 1980a). Furthermore, both the spontaneously induced and TPA-induced EBV DNA synthesis are inhibited by ACV (Lin et al., 1981).

Although ACV is a potent antiviral agent, potential disadvantages of this drug are its reversibility of inhibition of EBV replication (Colby et al., 1980; Lin et al., 1983, 1984) and rapid development in vitro of drug-resistant HSV mutants either at the TK (Field et al., 1981; Coen and Schaffer, 1980; Crumpacker et al., 1980) or DNA polymerase loci (Furman et al., 1981; Coen et al., 1982), usually the former.

ACV is widely used clinically for various herpetic infections by intravenous (Corey et al., 1983; Meyers et al., 1982), oral (Bryson et al., 1983) or topical (Corey et al., 1982a,b) administration. In oral administration less than 20% of the dose is absorbed in human beings (de Miranda and Blum, 1983). A new form of Acyclovir, 6-deoxyacyclovir, has recently been developed (Krenitsky et al., 1984). This prodrug is 18 times more water soluble than ACV and is readily oxidized to ACV by xanthine oxidase (Krenitsky et al., 1984). The rapid absorption and metabolic conversion of 6-deoxyacyclovir to Acyclovir makes this prodrug clinically useful for oral administration.

DHPG, a congener of ACV, was at first reported to have anti-HSV-1 and HSV-2 activity but only marginal activity against EBV and none against CMV (Smith et al., 1982). However, we have demonstrated that DHPG is active not only against replication of HSV-1 and HSV-2 but also against CMV and EBV (Cheng et al., 1983b). In addition, some but not all variants of HSV with altered virally encoded thymidine kinase and DNA polymerase that are resistant to ACV were still as sensitive to DHPG as the parental virus (Cheng et al., 1983b).

The antiviral activity of DHPG against human CMV was further documented by more recent studies (Mar et al., 1983; Tocci et al., 1984).

In a comparative study we found (Lin et al., 1984) that both ACV and DHPG inhibited EBV DNA replication in P3HR-1 cells and in superinfected Raji cells, but neither inhibited replication of the plasmid form of the EBV genome in latently infected Raji cells. The kinetics of inhibition and reversibility of EBV replication indicated that the inhibitory effect of ACV was readily reversed within 11 days after removal of the drug, in contrast to

6-deoxyacyclovir acyclovir

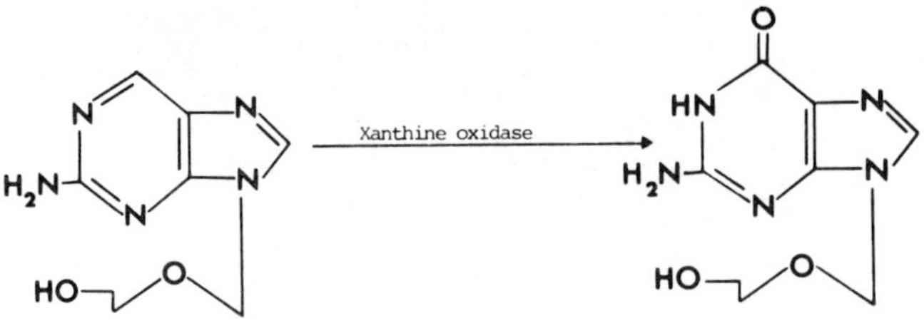

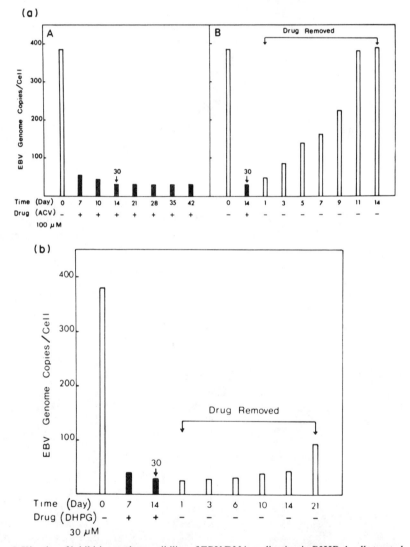

FIG. 4. Acyclovir and its prodrug (reproduced from Krenitsky *et al.*, 1984).

FIG. 5. Kinetics of inhibition and reversibility of EBV DNA replication in P3HR-1 cells treated with ACV (a) and DHPG (b) (reproduced from Lin *et al.*, 1984).

the more prolonged effect exerted by DHPG, which persisted for more than 21 days. The viral 50% effective dose (ED_{50}) of inhibition of DHPG was 0.05 μM. The 50% inhibitory dose of DHPG for cell growth (ID_{50}) was 200 μM. Thus, the *in vitro* therapeutic index (ID_{50}/ED_{50}) was 4,000. Synthesis of EBV-associated polypeptides was also affected (Lin *et al.*, 1984) by both drugs, presumably secondary to exhaustion of template DNA.

TABLE 3. *Relative Efficacy of Antiviral Drugs* in vitro

Drug*	Antiviral effect ED_{50} (μM)			Anticellular effect ID_{50} (μM)			Therapeutic index ID_{50}/ED_{50}		
	HSV	CMV	EBV	HSV	CMV	EBV	HSV	CMV	EBV
ACV	0.07[a]	86–100[b]	0.3[c]	15[a]	250[b]	250[c]	214[a]	<3[b]	833[c]
DHPG	0.05[a]	1–4.8[b]	0.05[c]	150[a]	300[b]	200[c]	3000[a]	<300[b]	4000[c]
BVDU	0.01–0.1[d,f]	—	0.06[c]	150–140[d,f]	—	390[c]	>4400[d,f]	—	6500[c]
FMAU	—	0.1–0.2[c]	0.0065[c]	—	6[e]	1	—	<60[e]	154[c]
FIAC	0.02[d]	0.3–0.7[e]	0.005[c]	64[d]	16[e]	5	3200[d]	<50[e]	1000[c]
FIAU	—	0.4–0.6[e]	0.005[c]	—	10[e]	1	—	<25[e]	200[c]

*ACV, 9-[2-(hydroxymethoxy)methyl]guanine
 DHPG, 9-(1,3-dihydroxy-2-propoxymethyl)guanine
 BVDU, E-5-(2-bromovinyl)-2'-deoxyuridine
 FMAU, 1-(2-deoxy-2-fluoro-β-D-arabinofuranosyl)-5-methyl uridine
 FIAC, 1-(2-deoxy-2-fluro-β-D-arabinofuranosyl)-5-iodocytosine
 FIAU, 1-(2-deoxy-2-fluro-β-D-arabinofuranosyl)-5-iodouridine
[a]Smee *et al.*, 1983.
[b]Mar *et al.*, 1983.
[c]Lin *et al.*, 1985.
[d]Larsson and Öberg, 1981.
[e]Mar *et al.*, 1984.
[f]DeClercq *et al.*, 1979.

4. MECHANISM OF ACTION OF DRUGS

4.1. Toxicity from Point of View of Mechanism of Action

The cytotoxic action of most nucleoside analogs centers on the replicative viral DNA polymerases which are encoded by all herpesviruses, human and nonhuman. The triphosphate derivatives of the drugs inhibit competitively the utilization of the corresponding deoxyribonucleoside triphosphate in the DNA polymerase reaction. Alternatively, the analog is incorporated into DNA and this lesion impedes synthesis by providing a poor primer for further chain elongation. Some drugs make use of both mechanisms. The phosphono compounds, which are not base analogs, have a different mechanism in that they interact directly with viral DNA polymerase at the pyrophosphate-binding site and do not require phosphorylation (Leinbach *et al.*, 1976).

4.2. General Target of Action of Antiviral Drugs

The selection of a suitable antiviral drug for treating virus diseases has been hampered by the fact that most antiviral compounds are capable of interfering with the molecular processes of host cell functions. Thus, the ideal antiviral drug would interfere only with a virally encoded process and not with processes in uninfected cells. On the basis of mode of action antiviral drugs can be divided into five categories: (a) antiviral drugs which interfere with cellular processes required by the virus for its replication; (b) antiviral drugs which selectively bind to, or interfere with, virus-coded enzymes and thus inhibit their function; (c) antiviral drugs that bind to the virus nucleic acid and therefore inhibit its expression and function; (d) antiviral drugs which prevent the processing of the viral precursor polypeptides; (e) antiviral drugs which intervere with the virus assembly and inhibit the formation of the virus progency; (f) antiviral drugs which modify the viral proteins on the surface of the viral envelope and thus prevent the virus from infecting new cells.

4.3. Virus-Coded Enzymes as Targets for Antiviral Drugs

Among the virus-coded enzymes, viral DNA polymerase has received a great deal of attention for the development of synthetic and natural antiviral compounds. Overby and coworkers (1974) reported that the inhibition of HSV-1 replication by PAA appeared to be a result of an inhibition of the viral DNA syntheses. The molecular mechanism by which PAA exerts its antiherpetic effect was later demonstrated by Mao and coworkers (1975). PAA was observed to inhibit herpesvirus DNA polymerase activity in a noncompetitive manner with respect to the dNTPs and in an uncompetitive manner with respect to the varied amounts of activated DNA at a saturating level. Later Leinbach and coworkers (1976) proposed that PAA, by interacting at the pyrophosphate binding site, blocks the formation of the 3'-5'-phosphodiester bond between the primer and substrate and thus prevents further DNA chain elongation. The mechanism of inhibition by foscarnet (PFA) is similar to that of PAA.

The lethal activity of IdUrd results from first, the initial phosphorylation of the nucleoside and second, the incorporation of the nucleoside triphosphate into viral DNA. Both of these critical events depend upon the activity of herpes viral enzymes, dThd kinase and the viral DNA polymerase. Lin and Riggs (1976) showed that IdUrd substitution in DNA increases the binding of regulatory proteins such as the *lac* repressor to the *lac* operator and nonhistone chromosomal proteins to the DNA.

The primary target of action of ara-A and ara-C is presumably DNA replication. Both ara-C and ara-A are competitive inhibitors of herpes-induced DNA polymerase (Furth and Cohen, 1967) and this activity requires formation of the 5'-triphosphate. The triphosphate derivative has been reported to be incorporated into DNA (Rashbaum and Cozzarelli, 1976).

Inhibitors which bind to, and inhibit, the viral DNA-synthesizing enzymes would be highly effective as antiviral agents, provided they are selective and inhibit only virus-coded enzymes. However, when inhibitors of the viral DNA polymerase were studied, it became evident that drug-resistant virus mutants are easily selected in the presence of the inhibitor. In this connection, it may be necessary to consider the use of two antiviral drugs in treating virus infections, so as to overcome mutation or selection of resistant virus strains.

4.4. Phosphorylation, Incorporation and Chain Termination

The selectivity of most antiherpes agents depends essentially on two virus-induced enzymes, dThd kinase and DNA polymerase. For instance, Acyclovir is selectively phosphorylated by an HSV-coded thymidine kinase (Elion *et al.*, 1977), and the resultant monophosphate, acyclo-GMP, is phosphorylated to acyclo-GDP by host cell GMP kinase (Miller and Miller, 1980). The acyclo-GDP is further converted by cellular kinases to the active form, acyclo-GTP (Miller and Miller, 1982), which is a potent inhibitor of the DNA polymerases of herpesviruses. The resulting drug triphosphate in turn serves as a substrate for viral DNA polymerase and is incorporated into the viral DNA primer-template, causing chain termination and inactivation of the polymerase (Elion *et al.*, 1977). Uninfected cells do not phosphorylate ACV to any significant extent and they have DNA polymerases which are much less sensitive to ACV triphosphate than are the herpesvirus DNA polymerases.

The preferential phosphorylation of BVDU (Cheng *et al.*, 1981) or FIAC (Lopez *et al.*, 1984) by the viral dThd kinase was also demonstrated in herpesvirus-infected cells. Thus, BVDU or FIAC is readily converted to its 5'-monophosphate in HSV-1 infected cells but poorly in uninfected cells. The 5'-triphosphate of BVDU or FIAC inhibits the HSV-1 DNA

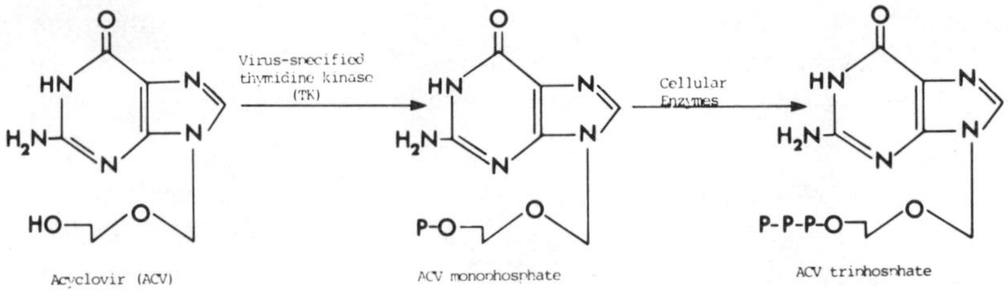

Fig. 6. Enzymic conversion of Acyclovir to its phosphorylated forms (reproduced from Elion, 1983).

polymerase to a considerably greater extent than the cellular DNA polymerases α, β and γ (Allaudeen et al., 1981). Phosphorylation of ACV in EBV-infected cells is 100-fold less efficient than that of its congener DHPG (Lin et al., 1985, 1986). The triphosphate of DHPG persists in cells at a significant level for 5 days after drug removal which may account for its prolonged inhibitory effect observed in vitro (Lin et al., 1984, 1986).

Although CMV and EBV do not seem to code for their own dThd kinase (Estes and Huang, 1977; Datta et al., 1980), nucleoside analogs that are inhibitors of HSV replication are also potent inhibitors against EBV and CMV (Lin et al., 1983, 1984; Cheng et al., 1983b; Mar et al., 1983). Since all herpesvirus-induced DNA polymerases share common features and may have similar behaviors toward these analog triphosphates, it is conceivable that the amount of analog triphosphate required to inhibit EBV or CMV DNA replication is less than that required to inhibit DNA synthesis in uninfected cells.

4.5. Reversibility

We recently reported that BVDU, FIAC, FMAU (Lin et al., 1983) and DHPG (Lin et al., 1984) have more prolonged inhibitory effects on EBV replication than ACV. Furthermore, their inhibition of EBV DNA replication appears to be irreversible whereas the effect of ACV is readily reversible which could be due to the action of ACV as chain terminator of EBV DNA replication, which could be easily corrected by endonucleases. ACV is incorported only at the terminus. Endonuclease, which can remove the terminal fragments of DNA containing ACV, could reverse the action of ACV; thus, the action of ACV was reversible once the compound was removed from the culture medium, whereas BVDU, FIAC, FMAU and DHPG (Cheng et al., 1983a) may incorporate into the internucleotide region of DNA. Such mass incorporation of the drug would be difficult to excise and the action would therefore be quite irreversible.

In vitro analysis of interaction of EBV-associated DNA polymerase with ACV-TP indicates that the triphosphorylated drug behaves as a classical competitive inhibitor of viral DNA polymerase with respect to dGTP (Datta et al., 1980). Furthermore, the same authors demonstrated that binding of EBV DNA polymerase to DNA chains terminated with ACV monophosphate residues, is reversible in terms of inhibition of viral DNA synthesis. Similarly, a reversible inhibition of viral DNA synthesis by the ACV-MP-bound primer-template was reported by Derse and coworkers (1981) in purified HSV-1 DNA polymerase. Recently, by a more careful kinetic analysis, Furman and coworkers (1984) reported that the inhibition of HSV-1 DNA polymerase activity by ACV was caused by the irreversible binding of ACV-MP-terminated primer-template to polymerase. Thus, ACV-TP is called a suicide inactivator of HSV DNA polymerase.

5. A PERSPECTIVE ON TREATMENT OF EBV INFECTION STATES

Treatment of herpesvirus infections is complicated by the different forms and states of infection that these viruses may bring about. EBV exemplifies the problem in that it can cause at least five types of infection: (1) acute (both primary and reactivated), (2) persistent active, (3) latent, (4) oncogenic and (5) abortive. Most of these virologic states have their counterparts in infections produced by one or more of the other herpesviruses. EBV mimicks these infection states in cell culture, which makes it possible to study the effects of antiviral drugs in vitro and to predict effects in vivo.

5.1. Epstein–Barr Virus Cell Systems

EBV exists in HR1 B-lymphoblastoid cells in a chronic virus-producing state equivalent to persistent infection. Only a minority of cells produce virus at any given time. If such a cell culture is treated with inducing agents such as the phorbol ester, TPA, or the base analog, Budr, then the nonvirus-producing cells in the culture, which contain latent viral genomes, are induced to produce virus, mimicking virus reactivation. Acute infection is mimicked *in vitro* by infecting Raji cells with EBV harvested from P3HR-1 cells; exogenous virus is used for infection and an infectious cycle ensues. However, Raji cells already have endogenous viral genomes which in some way contribute to the outcome. Raji cells provide a cellular model of latent EBV infection; the molecular basis for the latent state of the infection seems to be the plasmid or episomal form of the EBV genome. Cell models also exist for the oncogenic relation, captured in malignant cells explanted from Burkitt's lymphomas. Such cell lines, which are monoclonal, may contain covalently integrated EBV DNA sequences. However, these cells also contain latent viral genomes in the form of EBV plasmids. A B-lymphocytic line exists that contains only integrated viral sequences without episomal forms (Namalwa). Finally, there is a fifth virologic state which is suspected, both in nature and in *in vitro* models, namely, abortive infection. *In vitro* this state is produced by exposing Raji cells to TPA which causes activation of EBV gene expression with production of early antigen, but without replication of EBV genomes. *In vivo*, there is evidence of a corresponding virologic state in nasopharyngeal carcinoma in which activated EBV gene expression appears to begin but apparently stops short of virus replication in the tumor tissue.

5.2. Summary of Mechanism of Acyclovir

In HSV infection, Acyclovir is monophosphorylated by virus-encoded thymidine kinase and then di- and triphosphorylated by cellular enzymes. The triphosphorylated drug which is the active form specifically interacts with HSV DNA polymerase; ACV-triphosphate has at least 100-fold greater affinity for the viral than the normal a cellular polymerase. In EBV-infected cells, monophosphorylation of ACV is apparently accomplished by cellular kinases rather than by EBV-encoded TK since such an enzyme has not been identified in EBV-infected cells. Moreover, phosphorylation of ACV in EBV-infected cells is very inefficient (Pagano and Datta, 1982; Colby *et al.*, 1980). In general, therefore, the specificity of action of ACV and hence its relative nontoxicity depends on selective phosphorylation of the drug, preferential affinity of the drug-triphosphate for the viral polymerase, or both. The relative contributions of these steps in the action of the drug differ with the different herpesviruses (Pagano and Datta, 1982). The nontoxicity of Acyclovir is assisted by the localization of the active form of the drug in infected tissue inasmuch as ACV-triphosphate cannot permeate normal cell membranes. Phosphorylated drug is in highest concentration in cells where phosphorylation is carried out, namely, infected cells, which have the viral polymerase with its affinity for the triphosphate. The nature of the interaction with herpesvirus polymerases is as a competitive inhibitor of dGTP. There is evidence for both reversible and nonreversible aspects of this interaction. In HSV infection, ACV triphosphate is incorporated into the DNA; the incorporation causes immediate chain termination and binding of the viral polymerase. Incorporation into EBV DNA probably also occurs. Even though small amounts of the drug are phosphorylated in normal tissue, the amount depending on cell type and cellular metabolism, the likelihood of incorporation into viral DNA is much greater than it is into cellular DNA because of the preference of drug for the viral polymerase and the greater concentrations of triphosphorylated drug in virus-infected tissue in HSV-infected cells.

5.3. Effects on Protein Synthesis

DHPG is more efficiently phosphorylated in EBV-infected cells than is ACV, which may account for its superior inhibitory effect (Lin *et al.*, 1985; 1986). The affinity of EBV DNA polymerase for DHPG-triphosphate is under study. ACV and DHPG also have secondary effects on EBV polypeptide synthesis in infected cells. HSV polypeptides have been classified by Roiziman, as γ_1, which are independent of, but amplified by, viral DNA synthesis, and γ_2, which are stringently dependent on viral DNA synthesis. Late EBV polypeptides that appear not to be synthesized and others which are synthesized in reduced amounts in the presence of ACV have been identified; DHPG has similar but greater effects than ACV (Lin et al., 1984). However, the majority of EBV polypeptides are synthesized normally which suggests that viral polypeptides arising in EBV-cell states that do not involve viral DNA replication will be unaffected by these drugs.

5.4. Reversal of Drug Action

Because all herpetic infection states involve persistent replication or persistence of viral genomes, what happens after removal of drug from infected cells is relevant to treatment outcome. When ACV is applied to an EBV-producing cell line, free viral DNA replication is abolished and virus production ceases. However, a small persistent fraction of viral DNA remains in the treated cell cultures. When the drug is removed, viral DNA replication and virus production resume rapidly and reach the pretreatment levels. Virus replication is suppressed as long as the drug is present. This reversibility of inhibition suggests either that the drug triphosphate is reversibly bound to viral DNA polymerase or that new polymerase molecules are generated and become functional upon removal of the drug. There is some evidence to suggest that viral DNA polymerase, which is probably an early polypeptide, is synthesized even in the presence of the drug but remains inactive. The possibility that the interaction between ACV triphosphate and EBV DNA polymerase is reversible is suggested also by studies of the kinetics of the interaction of the triphosphate with EBV DNA polymerase *in vitro* (Datta et al., 1980). On the other hand, in the case of HSV DNA replication, ACV triphosphate interacts irreversibly with viral DNA polymerase, forming a complex with DNA into which the drug incorporates, terminating chain elongation (Furman et al., 1984). New studies are under way to ascertain whether the interaction of the triphosphate with EBV DNA polymerase is also irreversible.

Different inhibitors behave differently with respect to the duration of the drug effect after removal. DHPG is more strongly inhibitory than ACV and FMAU even more so. Moreover, after removal of DHPG from the virus-producing cell line, the kinetics of recovery of virus production are slower, taking about 21 days for a restoration to 25% of predrug exposure levels (Lin et al., 1984). In the case of FMAU, the persistent drug effect after removal is remarkable—more than 58 days (Lin et al., 1983). These different kinetics of recovery of virus replication point to different modes of action of the three drugs. The differences in persistent effects are more likely to be related to incorporation of drug into DNA than to differences in drug metabolism.

5.5. Latent Infection

None of the antiviral drugs has any effect on replication of the episomal form of the EBV genome in Raji cells, regardless of the differences in potency and mode of action of the various drugs. The lack of effect is independent of drug phosphorylation; neither Acyclovir, which is poorly phosphorylated in lymphoblastoid cells, nor DHPG, which is better phosphorylated, has any effect on the latent infection. Furthermore, phosphonoformate does not require phosphorylation and is without effect on EBV plasmid replication. The residual EBV genomes that persist in the presence of high inhibitory concentrations of the drugs are present in the form of EBV episomes or plasmids. Thus, two forms of the EBV genome are replicated in P3HR-1 cells: linear genomes that become encapsidated and circular genomes that remain intracellular. The circular episomal genomes, which are present in a nucleosomal arrangement in the host-cell DNA, are evidently replicated by host DNA polymerase rather than by the viral polymerase, as inferred from the effects of the inhibitors. Since Acyclovir spares host DNA polymerase activity while inhibiting polymerization by the viral enzyme, it is likely that FMAU has a similar dichotomy of effect on the two classes of polymerase. It appears that only cytotoxic drugs are liable to have any effect on the maintenance of EBV episomal forms, but at the cost of general, nonselective effects on cellular replication and viability.

5.6. Transformation

Transformation or immortalization of lymphocytes by EBV proceeds readily in the presence of inhibitory levels of Acyclovir ($>ED_{90}$) (Sixbey and Pagano, 1985). Therefore, transformation is not mediated by the viral DNA polymerase producing a round of viral DNA replication. However, EBV genomes in the form of episomes do become established in the immortalized cells. Relatively few EBV episomes (2–10/cell) become fixed in the cells with or without the drug and the immortalized cells do not make the virus. Already transformed cell lines including Burkitt's lymphoma cells are also unaffected by these antiviral drugs *in vitro*. The only effects of transformed cells are dose-dependent cytoxic effects that are probably nonselective.

5.7. EBV Pathogenesis and Predictions about Therapy

One view of the pathogenesis of EBV infection is that EBV enters the body through oropharyngeal contact and primarily infects epithelial cells in the oropharynx (Sixbey *et al.*, 1983; Sixbey and Pagano, 1984). These cells located in the oro- or naso-pharynx support the replication of the virus which is excreted in the saliva; the virus also replicates in epithelial cells in the parotid gland or its duct. The secondary cell target is the B-lymphocyte; these cells, which bear EBV receptors, are infected early in the course of infection. However, the B-lymphocytes seem not to be sites of active replication of the virus in normal hosts; rather a small percentage of circulating B-lymphocytes harbor latent EBV genomes and display expression only of the EBV antigens not tied to virus replication. Although these cells do not support virus replication, they are induced to proliferate; the resulting lymphoproliferation is polyclonal and limited by the host immune responses. These events occur in acute infectious mononucleosis in silent EBV infection.

We might, therefore, predict that the drugs discussed would interfere with virus replication in epithelial cells and shedding in the oropharynx, with suppression of replication as long as the drug is administered. Since infection of B-lymphocytes occurs early, the drug would be without effect on the already latently infected B-lymphocytes. Preliminary results of a trial of Acyclovir in patients with acute infectious mononucleosis have confirmed these predictions (Pagano *et al.*, 1983). Administration of Acyclovir transiently suppressed virus excretion but did not abolish it, nor was there any effect on the ability to establish EBV-infected B-lymphocyte lines from the peripheral blood of patients being given Acyclovir. However, suppression of virus production might reduce the number of B-lymphocytes that become newly infected with EBV during infection, perhaps with favorable effects. Moreover, a respite in virus replication might tip the balance in favor of the host immune mechanisms and aid in recovery even though virus replication and excretion is eventually resumed. Shedding of virus in the oropharynx continues asymptomatically for years in untreated patients. Some manifestations of acute EBV infection in mononucleosis such as the Guillan–Barré syndrome, hepatitis and suppression of hemopoiesis may be manifestations of secondary immune responses to latently infected EBV-containing lymphocytes. These immunologically based phenomena should be indifferent to the effect of the antiviral drugs except through interruption of continuing amplification of the population of EBV-infected lymphocytes. Burkitt's lymphoma seems to be a consequence of one or more additional steps in this pathogenetic scheme with chromosomal translocations producing critical activation of a cellular oncogene (*c-myc*) which leads to a monoclonal B-lymphocytic malignancy. None of the cytogenetic, molecular or cellular changes is likely to be susceptible to the action of Acyclovir or the other presently available drugs.

In nasopharyngeal carcinoma tissue, EBV episomes are in the epithelial elements of the neoplasm. Presumably, the transformed epithelial cells arise from rare epithelial cells infected many years earlier but not lyzed by the virus. NPC tissues do not seem to contain antigens associated with viral replication. An activated transcriptional state described in some NPC tissues (Raab-Traub *et al.*, 1983) is believed to stop short of virus replication, and therefore it is unlikely that Acyclovir would have any effect on this process. Nor would the drug be expected to have an effect on the transformed epithelial cells bearing EBV episomes. One conceivable point of action might be synthesis of EBV polypeptides equivalent to the γ_2 class of HSV polypeptides. If these late polypeptides are among those that begin to be synthesized during the activated transcriptional state found in some NPCs, their synthesis might be limited in so far as it depends upon amplification of viral DNA templates. However, it is not yet known whether any late polypeptides are in fact produced in NPC, and if they are, what their pathologic consequences might be.

In the case of NPC, there is some evidence to suggest that reactivation of EBV precedes and may trigger the onset of the disease. This evidence rests primarily on IgA responses to EA and VCA that precede (but do not necessarily lead to) the appearance of a significant percentage of NPCs (de-Thé, 1982). Whether these antibody responses signify full-fledged virus replication, abortive replication or merely an activated transcriptional state and where these antigenic stimuli arise at the cellular level are unknown. However, if reactivation of viral replication is a prelude to NPC, then antiviral drugs given prophylactically to high-risk patients in NPC endemic areas might interfere with appearance of the malignancy. Drugs that can be given orally are nontoxic and a relatively long duration of action would be needed for this purpose.

In invasive B-lymphocytic proliferation that occurs in immunocompromised hosts, there are anecdotal reports of an apparent efficacious effect of Acyclovir on polyclonal B-cell proliferation (Hanto et al., 1982). However, in two patients with a congenital immune defect Acyclovir appeared to have no effect (Sullivan et al., 1984). We would not expect that either polyclonal or monoclonal B-lymphocytic proliferation based on already infected cells would be affected by the antiviral drugs. However, if lymphoproliferation in such patients is based on continuing infection of additional B-lymphocytes, then antiviral drugs might indirectly curb polyclonal lymphoproliferation. The intensity and nature of the immune defect would also determine outcome, with lymphoproliferation in milder acquired immune defects being more susceptible to treatment.

In conclusion, we now have at least one drug available for testing in acute herpetic infection states, either primary or reactivated, that has therapeutic effects in human beings. Other drugs still confined to the laboratory may point the way to therapeutic agents useful for treatment of persistent infection states. No drugs that have been identified until now specifically inhibit latent EBV infection or growth transformation of EBV-infected lymphocytes. The same lack of effect on analogous infection states with the other herpesviruses is predictable. The treatment of these crucial infection states is for now beyond our reach and will require deeper understanding of the mechanism of establishment and maintenance of the latent herpesvirus genomes and transformed states before further progress. Nevertheless, the prospects for treatment of EBV, as well as other herpesvirus infections, have brightened in the interval since the effect of Acyclovir on EBV replication was first described in 1980 (Colby et al., 1980), following the description of the antiherpetic effects of ACV (Elion et al., 1977; Schaeffer et al., 1978).

6. GENERAL DISCUSSION AND SUMMARY: HERPESVIRUSES AND HUMAN NEOPLASIA

Do, then, virus-associated human cancers cause cancer and do they retain viral genomic sequences? In experimental systems—virus-transformed cell lines and virus-induced tumors in laboratory animals—involving both RNA- and DNA-containing viruses, viral genomes or portions of them are retained in the transformed tissue, usually in an integrated state. Furthermore in actual human tumors there are persistent viral genomes, usually without virus production, in the tumor tissue. There may, however, be other human tumors that are not thought to contain viral genomes because a suspect virus has not been identified, so that it is not possible to probe for viral sequences.

Another realm of possibility is that viruses may initiate changes in cells that lead to their transformation, but the viral genome is not retained. Experimentalists would prefer it to be otherwise, because to trace and dissect the track of a virus that has come and gone is a baffling enterprise. Perhaps cancers that do *not* exhibit the hallmarks of virus-associated malignancies are those whose existence has been triggered by virus-cell interactions remote in time and later coupled with events in the cell that promote its eventual transformation.

An intermediate situation is suggested by cellular oncogenes which may be activated by a variety of factors including probably some of the herpesviruses.

Often most suggestive of a viral etiology have been the epidemiologic features, especially endemicity. Of the cancers suspected of having a viral etiology, only carcinoma of the cervix does not exhibit such a pattern. On the other hand had we looked from the epidemiologic perspective at human T-cell leukemia virus infection only in the United States, HTLV-1 might not have been suspected of causing subacute T-cell leukemia. The endemic pattern of incidence of the disease in southern Japan was crucial in fueling suspicions. Is there also a class of malignancies, carcinoma of the cervix included, that may at least be initiated by virus infection, but the linkage to malignancy is too remote and indirect to generate florid endemicity? The next generation of research focusing on viruses and cancer may deal with such subtle virus-cell interactions. The growing evidence of a link between human papillomavirus infection and cervical carcinoma may conform to conventional patterns of sexually transmitted disease, or it may come to suggest an actual causative role for these viruses.

Recent workers have essayed a fresh look at the question of whether herpesvirus genes persist in cells transformed by herpes simplex virus and concluded that specific viral genes are not retained either in cells transformed *in vitro* or in human neoplastic tissue. Nor are any specific RNAs or viral gene products consistently expressed in transformed or tumor tissue. In coming to their hypothesis, the investigators take into account that HSV may require the expression of several genes for transformation and that the viral genes themselves may still be difficult to detect in transformed tissue even with the refined probes

TABLE 4. *Evidence Linking Herpesviruses with Cellular Transformation*

	Condition	In vitro	In vivo	Epidemiologic association
EBV	Burkitt's lymphoma	Transforms human lymphocytes	EBV DNA, RNA, EBNA in tumor cells	Strong
	Nasopharyngeal carcinoma	Infects human epithelial cells*	EBV DNA, RNA, EBNA in tumor cells	Strong
	Immunoblastic sarcoma	Transforms human B-lymphocytes	EBV DNA, RNA, EBNA in cells	Strong
	Supraglottic laryngeal carcinoma	Infects human epithelial cells*	EBV DNA, EBNA in tumor cells	Weak
CMV	Kaposi's sarcoma	Transforms human fibroblasts†	CMV DNA, RNA, EBV in some tumors	Strong
	Cervical carcinoma	Transforms human fibroblasts*	CMV DNA, virus in some tumors normal tissues	Weak
	Prostatic cancer	Transforms human fibroblasts	CMV DNA, virus in some tumors, normal tissues	Weak
HSV 2	Cervical carcinoma	Transforms rodent fibroblasts‡	HSV RNA in some carcinomas *in situ*	Weak
HSV 1	Oral carcinoma	Transforms rodent fibroblasts‡	None	Weak

*Transformation of human epithelial cells not demonstrated.
†Transformation of human endothelial cells not demonstrated.
‡Transformation of human cells not demonstrated.

now becoming available—and knowing that the interplay of expression of such genes may be difficult to capture technically and recreate temporally. These interpretations are complicated by the fact that we still do not really know the HSV gene products necessary to the transformed state. Nevertheless, taking everything into consideration, these investigators come down on the side of the oft-cited but little-loved hit-and-run theory of viral oncogenesis. They open Pandora's box, knowing full well that what they will find in it is negative evidence: 'the failure to detect a consistent set of viral sequences or a specific protein in transformed cells or human tumors' (Galloway and McDougall, 1983).

However there are flaws in the arguments. First the experimental: no one has yet succeeded in transforming human cells with HSV. At the same time, the true biologic meaning of the ability to transform rodent cells, especially hamster cells, *in vitro* has yet to be understood. In short we do not know whether sequences identified as transforming by transfection of rodent cells are in fact the same sequences needed to transform human cells *in vitro*, let alone the sequences active in actual human neoplasia.

Second, there is the problem of latent infection. HSV can exist in cervical tissue in a latent infection state without ever causing cancer. However, latent viral genomes, unlike viral genes in transformed tissue, may be present in only a few cells in a given tissue rather than in every cell. Hence it might be difficult to detect latent viral genomes in the cervix even with present-day technology, and furthermore the results might be inconsistent because of the very biologic nature of the latent virus relation. Further, such latent genomes may, quite conceivably, be expressed, with viral RNA and gene products characteristic of latency found in the infected cells. Neither RNA nor specific proteins peculiar to latency have yet been identified—probably because of the paucity of valid *in vitro* cellular models for latent HSV infection. Nor in our preoccupation with the oncogenic relation have we deliberately looked for the RNA and glycoproteins expressed in latently infected tissue. We should now be in a position to begin to deduce latent viral functions.

From this perspective there is no compelling evidence establishing HSV-2 as a causative agent in carcinoma of the cervix. The sero-epidemiologic evidence implicating HSV-2 in carcinoma of the cervix has never seemed as cogent as it is for the other virus-associated malignancies such as Burkitt's lymphoma, nasopharyngeal carcinoma, B-cell lymphomas in allograft recipients and in patients with AIDS, hepatocellular carcinoma, human T-cell leukemia and perhaps Kaposi's sarcoma. This is not to say that HSV-2 has nothing to do with carcinoma of the cervix, but offers the possibility that HSV-2 may be merely one link in a chain of causality—the question being whether this link is indispensible. CMV may be as good a candidate as HSV with respect to cervical carcinoma, but the needed sero-epidemiologic studies have not been done. CMV methodology, both serology and virus detection, is more difficult than for HSV. Had the same studies been done for CMV in relation to cervical carcinoma that were done years ago with HSV, we might now be chasing different viral genes and their products. The consequences of this line of reasoning is that failure to find specific HSV genes in cervical carcinoma may not favor a hit-and-run hypothesis, but be telling us that HSV is not an essential participant in the pathogenesis of cervical carcinoma.

6.1. CYTOMEGALOVIRUS

Typically CMV infects us harmlessly, persists for the rest of our lives quietly in a perfectly adapted latent infection and reactivates silently. Of the five human herpesviruses CMV is the only one that can transform human fibroblasts *in vitro* into cells with unlimited growth potential. This is a property shared by few viruses, among them Simian Virus 40 and human adenovirus. Rapp and his colleagues (Albrecht and Rapp, 1973) first showed the ability of CMV to transform human cells *in vitro*, and recently Huang and colleagues (1983) have succeeded in transforming normal diploid human fibroblasts into fast-growing abnormal-appearing cells that produce large tumors when inoculated into nude mice. The transformation was accomplished by transfection of the cells with CMV DNA which led to focus formation, but not to the establishment of cell lines, which seemed to depend upon the addition of a tumor-promoting substance, the phorbol ester, TPA, to the *in vitro* system. The transformed cells contained CMV DNA, antigens and polypeptides. Gradually with passage of time the cells lost detectable CMV DNA, but continued to express some CMV polypeptides, a result which suggests that a functional residuum of the CMV genome persisted in the cells. The design of this experiment leads to the notion that transformation of cells may be initiated by CMV and then facilitated by a tumor promoter in a two-step concept of carcinogenesis.

The relevance of those experimental findings to human experience is open to question. The best-characterized human fibroblast-transforming viruses, SV40 and adenovirus, have never been associated with any human malignancy despite surveillance of persons of all ages exposed to these viruses either by natural or artificial means. Nevertheless the experimental findings are likely to be significant inasmuch as they disclose the basic biologic ability of CMV at the cellular level to alter human cells in the direction of malignant behavior. In an intact organism other factors certainly need to operate to facilitate the full expression of such a cellular alteration by the formation of a tumor. There is much to support the contemporary view that cancer is a multifactorial process and it is reasonable to hold for the moment that CMV under the right circumstances is capable of initiating such a process.

The *in vivo* investigations linking CMV to Kaposi's sarcoma do not prove causation, but they provide crucial associative data that may lay the basis for proof of an etiologic relation later. At this point, however, it is not possible to say whether the CMV DNA and transcripts found in some KS tissue indicate secondary infection or primary oncogenic relation. It is not even clear whether intact virus is in the tumor tissue, or only portions of the CMV genome are present and expressed. Finally it is puzzling why a virus that is as pleiotrophic as CMV, affecting many tissues throughout the body, should have its strongest association until now only with as rare and unusual a cancer as Kaposi's sarcoma.

6.2. Epstein–Barr Virus

Is Epstein–Barr virus oncogenic? The involvement of EBV in lymphomas that arise in allograft recipients and now in AIDS may be the most persuasive evidence yet as to the oncogenicity of EBV. Human lymphocytes growth-transformed *in vitro* by EBV into polyclonal lines correspond closely to the transformation of B-lymphocytes now recognized in immunosuppressed patients. These cells, also polyclonally transformed, in and of themselves are capable of invasive behavior and may cause lethal disease, in every sense true malignancy even though polyclonal. EBV alone seems to be entirely capable of mediating this crucial change in cell behavior. Sometimes a transition from polyclonal to monoclonal pathologic cell type also transpires with perhaps a corresponding biologic change, namely, increased resistance to therapy. The only other factor needed for this oncogenic process, apparently triggered by EBV, is whether or not the immunosuppression hits or misses. Immunosuppression, broad and intense as in AIDS, permits release of the invasive phenotypes of EBV-infected polyclonal B-cells because AIDS destroys immunosurveillance for EBV-transformed cells. In an intermediate position are allograft recipients who may or may not develop EBV-induced lymphomas, presumably depending upon whether or not the immunosuppression happens to hit immune mechanisms crucial in the EBV process. At the other extreme is the precise targeting of a defect in cell-mediated immunity that affects only EBV-related mechanisms, as in Duncan's syndrome. These patients do not have other immune defects initially. From all this it seems that EBV-associated lymphomas in the immunosuppressed display perhaps the least multifactorial, most direct process for any of the oncogenic herpesviruses. The strength, consistency and variety of evidence linking EBV to B-cell lymphomas may revive the moribund Koch–Henle postulates and give them a new lease on life!

However as with HSV before there is a central problem in dealing with EBV. In all of the cell systems and in the pathologic tissues the entire EBV genome is invariably present. It is not possible, therefore, to discern whether latency or the oncogenic relation is being addressed because the EBV-transformed cell lines are both transformed and latently infected. It then becomes crucial to try once again to establish whether or not there are integrated EBV genes as well as the EBV plasmids. There has recently been headway on this critical question. The distinction between a latent infectious state and an oncogenic relation may come to rest on the fact of, and perhaps sites of, integrated EBV genes. The role of the EBV plasmid in the oncogenic process may be relegated to a role as a vehicle for the transposition, if that occurs, of viral genes into an integrated state. An *in vitro* model for the generation of plasmid forms of the EBV genome and testing whether insertion of integrated EBV sequences follows in some sequential fashion would be invaluable.

The question of integrated viral genes raises the question then, what of oncogenes? It is fashionable to hold that activation of cellular oncogenes may be an important and perhaps even necessary step in viral oncogenesis for a number of viruses. In the case of Burkitt's lymphoma the 8; 14 translocations described for so many years, but not understood, seem now to make some sense because of the linkage of these translocations to *c-myc*. However in the immunoblastic sarcomas of the B-cell type found in allograft recipients no such chromosomal translocations are described. Either *c-myc* is not the necessary final common pathway for lymphoma induction by EBV, or another oncogene is involved.

Finally, what of NPC—that remarkable malignancy arising near the primary site of EBV replication in the body? The linkage of NPC to EBV reactivation especially in south China is compelling. de-Thé and Chinese colleagues have estimated that up to 2% of persons 40 years of age or older in south China will have an IgA response to EBV antigens annually. Approximately 1–2% of the persons who have this distinctive antibody response will manifest NPC within twelve to eighteen months. Since the period of high risk in south China is between the ages of 40 and 60 years, and the annual risk of NPC is 0.1% or more, then the cumulative risk in this area is in the neighbourhood of 1–2% over the 20 year period. These observations are nothing short of phenomenal.

But what is EBV doing in epithelial cells since EBV is supposed to be a lymphotrophic virus? Recent work indicates that EBV can infect epithelial cells not only *in vivo* but also *in vitro* (Pagano and Lemon, 1984). Some years ago Glaser working with de-Thé (Glaser *et al.*, 1976) was able to infect NPC cells *in vitro* which indicates that NPC's do bear receptors for EBV. NPC could represent a clonal expansion of a subpopulation of epithelial cells bearing EBV receptors. In any case the new *in vitro* studies seem to show quite clearly that EBV can infect epithelial cells *in vitro*, albeit only a small percentage. The question of whether EBV can transform epithelial cells *in vitro* is unanswered.

Returning to the problem of latency vs oncogenesis, and since recalling that the whole genome is present in all EBV diseases, it seems reasonable to consider EBV expression in the different EBV diseases, asking whether there are favored expressions that are disease-related. Examination of transcription in NPC tissues should prove to be a promising line of study. Not only is such work relevant to a dissection of the oncogenic relation of EBV to NPC, but a similar line of pursuit should be capable of delineating transcription peculiar to latency. The main goal would be to distinguish between the latent and the oncogenic relations by a precise mapping of RNA transcripts instead of by determining viral gene content. The identification of integrated viral genes should contribute to the distinction. For these studies we need veritable latent infection systems whose cells are not also transformed and transformation systems uncontaminated by latent infection.

Thus, the breadth and depth of information linking viruses to cancer is increasing dramatically at a time when at last there are effective antiviral drugs available for study and use. At present for the most part there is little to indicate that such drugs will be effective in virus-transformed cells, but timely intervention may have an effect.

Acknowledgements—We thank Etsuyo I. Choi for the artwork and Laura T. Ginsberg for typing the manuscript.

REFERENCES

ALBRECHT, T. and RAPP, F. (1973) Malignant transformation of hamster embryo fibroblasts following exposure to ultraviolet-irradiated human cytomegalovirus. *Virology* **55**: 53–61.

ALLAUDEEN, H. S., KOZARICH, J. W., BERTINO, J. R. and DECLERCQ, E. (1981) On the mechanism of selective inhibition of herpesvirus replication by (E)-5-(2-bromovinyl)-2′-deoxyuridine. *Proc. natn. Acad. Sci. U.S.A.* **78**: 2698–2702.

BENZ, W. C., SIEGEL, P. J. and BAER, J. (1978) Effects of adenine arabinoside on lymphocytes infected with Epstein–Barr virus. *J. Virol.* **27**: 475–482.

BLUMBERG, B. S. and LONDON, W. T. (1981) Hepatitis B virus and prevention of primary hepatocellular carcinoma. *New Engl. J. Med.* **304**: 782–784.

BOLDEN, A., AUCKER, J. and WEISSBACH, A. (1975) Synthesis of herpes simplex virus, vaccinia virus and adenovirus DNA in isolated HeLa cell nuclei. I. Effect of virus-specific antisera and phosphonoacetic acid. *J. Virol.* **16**: 1584–1592.

BOLDOGH, I., BETH, E., HUANG, E. S., KYALWAZI, S. K. and GIRALDO, G. (1981) Kaposi's sarcoma: IV. Detection of CMV DNA, CMV RNA and CMNA in tumor biopsies. *Int. J. Cancer* **28**: 469–474.

BOLDOGH, I., BASKAR, J. F., MAR, E. C. and HUANG, E. S. (1983) Human cytomegalovirus and herpes simplex types 2 virus in normal and adenocarcinomatous prostate glands. *J. natn. Cancer Inst.* **70**: 819–826.

BRYSON, Y. J., DILLON, M., LOVETT, M., ACUNA, G., TAYLOR, S., CHERRY, J. D., JOHNSON, B. L., WIESMEIER, E., GROWDON, W., CREAGH-KIRK, T. and KEENEY, R. (1983) Treatment of first episodes of genital herpes simplex virus infection with oral acyclovir. A randomized double-blind controlled trial in normal subjects. *New Engl. J. Med.* **308**: 916–921.

CHENG, Y.-C., DUTSCHMAN, G. E., DECLERCQ, E., JONES, A. S., ROHIM, S. G., VERHELST, G. and WALKER, R. T. (1981) Differential affinities of 5-(2-halogenovinyl)-2′-deoxyuridines for deoxythymidine kinases of various origins. *Molec. Pharmac.* **20**: 230–233.

CHENG, Y.-C., GRILL, S. P., DUTSCHMAN, G. E., NAKAYAMA, K. and BASTOW, K. F. (1983a) Metabolism of 9-(1,3-dihydroxy-2-propoxymethyl)guanine, a new anti-herpes virus compound, in herpes simplex virus-infected cells. *J. biol. Chem.* **258**: 12460–12464.

CHENG, Y.-C., HUANG, E.-S., LIN, J.-C., MAR, E. C., PAGANO, J. S., DUTSCHMAN, G. E. and GRILL, S. P. (1983b) Unique spectrum of activity of 9-[(1,3-dihydroxy-2-propoxy)methyl]-guanine against herpesvirus *in vitro* and its mode of action against herpes simplex virus type. 1. *Proc. natn. Acad. Sci. U.S.A.* **80**: 2767–2770.

COEN, D. M. and SCHAFFER, P. A. (1980) Two distinct loci confer resistance to acycloguanosine in herpes simplex virus type 1. *Proc. natn. Acad. Sci. U.S.A.* **77**: 2265–2269.

COEN, D. M., FURMAN, P. A., GELEP, P. T. and SCHAFFER, P. A. (1982) Mutation in herpes simplex virus DNA polymerase gene can confer resistance to 9-β-D-arabinofuranosyladenine. *J. Virol.* **41**: 909–918.

COKER-VANN, J. and DOLIN, R. (1977) Effect of adenine arabinoside on Epstein–Barr virus *in vitro*. *J. infect. Dis.* **135**: 447–453.

COLBY, B. M., SHAW, J. E., ELION, G. B. and PAGANO, J. S. (1980) Effect of acyclivor [9-(2-hydroxyethoxymethyl)-guanine] on Epstein–Barr virus DNA replication. *J. Virol.* **34**: 560–568.

COLBY, B. M., SHAW, J. E., DATTA, A. K. and PAGANO, J. S. (1982) Replication of Epstein–Barr virus DNA in lymphoblastoid cells treated for extended periods with Acyclovir. *Am. J. Med.* **73**(1A): 77–81.

COLEMAN, D. V., WOLFENDALE, M. R., DANIEL, R. A., DHANJAL, N. K., GARDNER, S. D., GIBSON, P. E. and FIELD, A. M. (1980) A prospective study of human polyomavirus infection in pregnancy. *J. infect. Dis.* **142**(1): 1–8.

COREY, L., BENEDETTI, J. K., CRICHLOW, C. W., REMINGTON, M. R. WINTER, C. A., FAHNLANDER, A. L., SMITH, K., SALTER, D. L., KEENEY, R. E., DAVIS, L. G., HINTZ, M. A., CONNOR, J. D. and HOMES, K. K. (1982a) Double-blind controlled trial of topical acyclovir in genital herpes simplex virus infections. *Am. J. Med.* **73**(1A): 326–334.

COREY, L., NAHMIAS, A. J., GUINAN, M. E., BENEDETTI, J. K., CRITCHLOW, C. W. and HOLMES, K. K. (1982b) A trial of topical acyclovir in genital herpes simplex virus infections. *New Engl. J. Med.* **306**: 1313–1319.

COREY, L., BENEDETTI, J. K., FAHNLANDER, A., HINTZ, M. A., FIFE, K. H., WINTER, C. A., CONNOR, J. D. and HOLMES, K. K. (1983) Intravenous acyclovir for the treatment of primary genital herpes. *Ann. intern. Med.* **98**: 914–921.

CRUMPACKER, C. S., SCHNIPPER, L. E., ZAIA, J. A. and LEVIN, M. J. (1979) Growth inhibition by acycloguanosine of herpesvirus isolated from human infections. *Antimicrob. Ag. Chemother.* **15**: 642–645.

CRUMPACKER, C. S., CHARTRAND, P., SUBAK-SHARPE, J. H. and WILKIE, N. M. (1980) Resistance of herpes simplex virus to acycloguanosine-genetic and physical analysis. *Virology* **105**: 171–184.

DALLA-FAVERA, R., BREGNI, M., ERIKSON, J., PATTERSON, D., GALLO, R. C. and CROCE, C. M. (1982) Human c-myc onc gene is located on the region of Chromosome 8 that is translocated in Burkitt lymphoma cells. *Proc. natn. Acad. Sci. U.S.A.* **79**: 7824–7827.

DATTA, A. K. and HOOD, R. E. (1981) Mechanism of inhibition of Epstein–Barr virus replication by phosphonoformic acid. *Virology* **114**: 52–59.

DATTA, A. K., COLBY, B. M., SHAW, J. E. and PAGANO, J. S. (1980) Acyclovir inhibition of Epstein–Barr virus replication. *Proc. natn. Acad. Sci. U.S.A.* **77**: 5163–5166.

DECLERCQ, E. (1982) Specific targets for antiviral drugs. *Biochem. J.* **205**: 1–13.

DECLERCQ, E., DESCAMPS, J., DESOMER, P., BARR, P. J., JONES, A. S. and WALKER, R. T. (1979) (E)-5-(2-bromovinyl)-2′-deoxyuridine: a potent and selective anti-herpes agent. *Proc. natn. Acad. Sci. U.S.A.* **76**: 2947–2951.

DECLERCQ, E., DESCAMPS, J., VERHELST, G., WALKER, R. J., JONES, A. S., TORRENCE, P. F. and SHUGAR, D. (1980) Comparative efficacy of different antiherpes drugs against different strains of herpes simplex virus. *J. infect. Dis.* **141**: 563–574.

DEMIRANDA, P. and BLUM, M. R. (1983) Pharmacokinetics of acyclovir after intravenous and oral administration. *J. Antimicrob. Chemother.* **12**(Suppl.B): 29–37.

DERSE, D., CHENG, Y. C., FURMAN, P. A., ST. CLAIR, M. H. and ELION, G. B. (1981) Inhibition of purified human and herpes simplex virus-induced DNA polymerase by 9-(2-hydroxyethoxymethyl)guanine triphosphate. *J. biol. Chem.* **256**: 11447–11451.

DE-THÉ, G. (1982) Epidemiology of Epstein–Barr virus and associated diseases in man. In: *The Herpesviruses*, pp. 25–103, ROIZMAN, B. (ed.) Plenum Press.

ELION, G. B. (1983) The biochemistry and mechanism of action of acyclovir. *J. Antimicrob. Chemother.* **12**(Suppl. B): 9–17.
ELION, G. B., FURMAN, P. A., JAMES, A. F., DEMIRANDA. P., BEAUCHAMP, L. and SCHAEFFER, H. J. (1977) Selectivity of action of an antiherpetic agent, 9-(2-hydroxyethoxymethyl)guanine. *Proc. natn. Acad. Sci. U.S.A.* **74**: 5716–5720.
EPSTEIN, M. A. and ACHONG, B. G. (1979) In: *The Epstein–Barr Virus*, EPSTEIN, M. A. and ACHONG, B. G. (eds) Springer-Verlag, New York.
EPSTEIN, M. A., ACHONG, B. G. and BARR, Y. M. (1964) Virus particles in cultured lymphoblasts from Burkitt's lymphoma. *Lancet* **i**: 702–703.
ESTES, J. and HUANG, E.-S. (1977) Stimulation of cellular thymidine kinase by human cytomegalovirus. *J. Virol.* **24**: 13–21.
FIELD, H. J., MCMILLAN, A. and DARBY, G. (1981) The sensitivity of acyclovir-resistant mutants of herpes simplex virus to other antiviral drugs. *J. infect. Dis.* **143**: 281–285.
FURMAN, P. A., COEN, D. M., ST. CLAIR, M. H. and SCHAEFFER, P. A. (1981) Acyclovir-resistant mutants of herpes simplex virus type 1 express altered DNA polymerase or reduced acyclovir phosphorylating activities. *J. Virol.* **40**: 936–938.
FURMAN, P. A., ST. CLAIR, M. H. and SPECTOR, T. (1984) Acyclovir triphosphate is a suicide inactivator of the herpes simplex virus DNA polymerase. *J. biol. Chem.* **259**: 9575–9579.
FURTH, J. J. and COHEN, S. S. (1967) Inhibition of mammalian DNA polymerase by the 5'-triphosphate of 9-β-D-arabinofuranosyl adenine. *Cancer Res.* **27**: 1528–1533.
GALLOWAY, D. A. and MCDOUGALL, J. K. (1983) The oncogenic potential of herpes simplex viruses: evidence for a 'hit-and-run' mechanism. *Nature* **302**: 21–24.
GIRALDO, G., BETH, E. and HUANG, E. S. (1980) Kaposi's sarcoma and its relationship to cytomegalovirus (CMV). III. CMV DNA and CMV early antigens in Kaposi's sarcoma. *Int. J. Cancer* **26**: 23–29.
GLASER, R., NONOYAMA, M., DECKER, B. and RAPP, F. (1973) Synthesis of Epstein–Barr virus antigens and DNA in activated Burkitt somatic cell hybrids. *Virology* **55**: 62–69.
GLASER, R., DE-THÉ, G., LENOIR, G. and HO, J. H. C. (1976) Superinfection of epithelial nasopharyngeal carcinoma cells with Epstein–Barr virus. *Proc. natn. Acad. Sci. U.S.A.* **73**: 960–963.
GREEN, M., BRACKMANN, K. H., SANDERS, P. R., LOWENSTEIN, P. M., FREEL, J. H., EISINGER, M. and SWITLYK, S. A. (1982) Isolation of a human papillomavirus from a patient with epidemodysplasia verruciformis: presence of related viral DNA genomes in human urogenital tumors. *Proc. natn. Acad. Sci. U.S.A.* **79**: 4437–4441.
HANTO, D. W., FRIZZERA, G., PURTILO, D. T., SAKAMOTO, K., SULLIVAN, J. L., SAEMUNDSEN, A. K., KLEIN, G., SIMMONS, R. L. and NAJARIANA, J. S. (1981) Clinical spectrum of lymphoproliferative disorders in renal transplant recipients and evidence for the role of Epstein–Barr virus. *Cancer Res.* **41**: 4253–4261.
HANTO, D. W., FRIZZERA, G., GAJL-PECZALSKA, K. J., SAKAMOTO, K., PURTILO, D. T., BALFOUR, H. H., SIMMONS, R. L. and NAJARIAN, J. S. (1982) Epstein–Barr virus-induced B-cell lymphoma after renal transplantation: Acyclovir therapy and transition from polyclonal to monoclonal B-cell proliferation. *New Engl. J. Med.* **306**: 913–918.
HERRMANN E. C., JR. (1961) Plaque inhibition test for detection of specific inhibitors of DNA containing viruses. *Proc. Soc. exp. Biol. Med.* **107**: 142–145.
HUANG, E. S. (1975) Human cytomegalovirus. IV. Specific inhibition of virus-induced DNA polymerase activity and viral DNA replication of phosphonoacetic acid. *J. Virol.* **16**: 1560–1565.
HAUNG, E. S., BOLDOGH, I. and MAR, E. C. (1983) Cytomegalovirus: evidence associated with human cancer. In: *Viruses Associated with Human Cancer*, pp. 161–194, PHILLIPS, L. (ed.) Marcel-Dekker, New York.
KAUFMAN, H. E., MAROLA, E. and DOHLMAN, C. (1960) Use of 5-iodo-2'-deoxyuridine (IDU) in treatment of herpes simplex keratitis. *Archs Ophthal.* **68**: 235–239.
KRENITSKY, T. A., HALL, W. W., DEMIRANDA, P., BEAUCHAMP, L. M., SCHAEFFER, H. J. and WHITEMAN, P. D. (1984) 6-Deoxyacyclovir: a xanthine oxidase-activated prodrug of acyclovir. *Proc. natn. Acad. Sci. U.S.A.* **81**: 3209–3213.
LARSSON, A. and ÖBERG, B. (1981) Selective inhibition of herpesvirus deoxyribonucleic acid synthesis by acycloguanosine, 2'-fluoro-5-iodo-aracytosine, and (E)-5-(2-bromovinyl)-2'-deoxyuridine. *Antimicrob. Ag. Chemother.* **19**: 927–929.
LEINBACH, S. S., RENO, R. M., LEE, L. F., ISBELL, A. F. and BOLZI, J. A. (1976) Mechanism of phosphonoacetate inhibition of herpesvirus-induced DNA polymerase. *Biochemistry* **15**: 426–430.
LEMON, S. M., HUTT, L. M., SHAW, J. E., LI, J. L. and PAGANO, J. S. (1977) Replication of EBV in epithelial cells during infectious mononucleosis. *Nature* **268**: 268–270.
LIN, J.-C. and PAGANO, J. S. (1980a) Synthesis of chromosomal proteins and Epstein–Barr virus DNA in activated Burkitt somatic cell hybrids. *Virology* **106**: 50–58.
LIN, J.-C. and PAGANO, J. S. (1980b) Effect of 5-iodo-2'-deoxyuridine on physical properties and nonhistone chromosomal proteins of chromatin from Burkitt somatic cell hybrids. *Archs Biochem. Biophys.* **200**: 567–574.
LIN, J.-C., SMITH, M. C. and PAGANO, J. S. (1981) Induction of replication of Epstein–Barr virus DNA by 12-O-tetradecanoyl-phorbol-13-acetate. II. Inhibition by retinoic acid and 9-(2-hydroxyethoxymethyl)guanine. *Virology* **111**: 294–298.
LIN, J.-C., SMITH, M. C., CHENG, Y. C. and PAGANO, J. S. (1983) Epstein–Barr virus: inhibition of replication by three new drugs. *Science* **221**: 578–579.
LIN, J.-C., SMITH, M. C. and PAGANO, J. S. (1984) Prolonged inhibitory effect of 9-(1,3-dihydroxy-2-propoxymethyl)guanine against replication of Epstein–Barr virus. *J. Virol.* **50**: 50–55.
LIN, J.-C., SMITH, M. C. and PAGANO, J. S. (1985) Comparative efficacy and selectivity of some nucleoside analogs against Epstein–Barr virus. *Antimicrob. Ag. Chemother.* **27**: 971–973.
LIN, J.-C., NELSON, D. J., LAMBE, C. U., CHOI, E. I. and PAGANO, J. S. (1985) Effects of nucleoside analogs in inhibition of Epstein–Barr virus. In: *Proceedings of the International Symposium on Pharmacological and Clinical Approaches to Herpes Viruses and Virus Chemotherapy, Oiso, Japan, 1984.* Exerpta Medica International Congress Series ICS 667. Elsevier Science Publishers, Amsterdam: pp 225–227.

LIN, J.-C., NELSON, D. J., LAMBE, C. U., and CHOI, E. I. (1986) Metabolic activation of 9([2-hydroxy-1-(hydroxymethyl)ethoxy]methyl) guanine in human lymphoblastoid all lines infected with Epstein-barr virus. *J. Virol* **60**: 569–573.

LIN, S. Y. and RIGGS, A. D. (1976) The binding of lac repressor and the catabolic gene activator protein to halogen-substituted analogues of poly d(A-T). *Biochim. biophys. Acta* **432**: 185–191.

LOPEZ, C., LIVELLI, T., WATANABE, K. A., REICHMAN, U. and FOX, J. J. (1979) 2′-Fluoro-5-iodo-aracytosine: a potent anti-herpesvirus nucleoside with animal toxicity to normal cells. *Proc. Am. Ass. Cancer Res.* **20**: 183.

LOPEZ, C., WATANABE, K. A. and FOX, J. J. (1980) 2′-Fluoro-5-iodo-aracytosine, a potent and selective anti-herpesvirus agent. *Antimicrob. Ag. Chemother.* **17**: 803–806.

LOPEZ, C., CHOU, T.-C., WATANABE, K. A. and FOX, J. J. (1984) 2′-Fluoro-5-iodo-1-β-D-arabinofuranosylcytosine [FIAC]: synthesis and mode of anti-herpesvirus activity. In: *Antiviral Drugs and Interferon: The Molecular Basis of Their Activity*, pp. 105–115, BECKER, Y. (ed.) Martinus Nijhoff, Boston.

MAO, J. C.-H., ROBISHAW, E. E. and OVERBY, L. R. (1975) Inhibition of DNA polymerase from herpes simplex virus-infected WI-38 cells by phosphonacetic acid. *J. Virol.* **15**: 1281–1283.

MAR, E. C., PATEL, P. C. and HUANG, E. S. (1982) Effect of 9-(2-hydroxyethoxymethyl)guanine on viral-specific polypeptide synthesis in human cytomegalovirus-infected cells. *Am. J. Med.* **3**: 82–85.

MAR, E. C., CHENG, Y. C. and HUANG, E. S. (1983) Effect of 9-(1,3-dihydroxy-2-propoxymethyl)guanine on human cytomegalovirus replication *in vitro*. *Antimicrob. Ag. Chemother.* **24**: 518–521.

MAR, E. C., PATEL, P. C., CHENG, Y. C., FOX, J. J., WATANABE, K. A. and HUANG, E. S. (1984) Effects of certain nucleoside analogues on human cytomegalovirus replication *in vitro*. *J. gen. Virol.* **64**: 47–53.

MEYERS, J. D., MITCHELL, C. D., LIETMAN, P. S., LEVIN, M. J., BALFOUR, H. H., WADE, H. C., SARAL, R., DURACK, D. T. and SEGRETI, A. C. (1982) Multicenter collaborative trial of intravenous acyclovir for treatment of mucocutaneous herpes simplex virus infection in the immunocompromised host. *Am. J. Med.* **73**(1A): 229–235.

MILLER, J. R., BARRETT, R. E., BRITTON, C. B., TAPPER, M. L., BAHR, G. S., BRUNO, P. J., MARQUARDT, M. D., HAYS, A. P., MCMURTRY, J. G., WEISSMAN, J. B. and BRUNO, M. S. (1982) Progressive multifocal leukoencephalopathy in a male homosexual with T-cell immune deficiency. *New Engl. J. Med.* **307**: 1436–1438.

MILLER, W. H. and MILLER, R. L. (1980) Phosphorylation of acyclovir (acycloguanosine) monophosphate by GMP kinase. *J. biol. Chem.* **255**: 7204–7207.

MILLER, W. H. and MILLER, R. L. (1982) Phosphorylation of acyclovir diphosphate by cellular enzymes. *Biochem. Pharmac.* **31**: 3879–3884.

MÜLLER, W. E. G., ZAHN, R. K., BEYER, R. and FALKE, D. (1977) 9-β-D-Arabinofuranosyladenine as a tool to study herpes simplex virus DNA replication *in vtiro*. *Virology* **76**: 787–796.

NORTH, T. W. and COHEN, S. S. (1979) Aranucleosides and aranucleotides in viral chemotherapy. *Pharmac. Ther.* **4**: 81–108.

NORTH, T. W. and COHEN, S. S. (1982) In: *The International Encylopedia of Pharmacology and Therapeutics*, SHUGAR, D. (ed.) Pergamon Press, Oxford.

OVERBY, L. K., ROBISHAW, E. E., SCHLEICHER, J. B., RUETER, A., SHIPKOWITZ, N. L. and MAO, J. C.-H. (1974) Inhibition of herpes simplex virus replication by phosphonoacetic acid. *Antimicrob. Ag. Chemother.* **6**: 360–365.

PAGANO, J. S. and DATTA, A. K. (1982) Perspectives of the interaction of acyclovir with Epstein–Barr virus and other herpesviruses. *Am. J. Med.* **73A**: 18–26.

PAGANO, J. S. and HENRY, B. E. (1983) The Epstein–Barr virus: biochemistry and relation to human malignancy. In: *Viruses Associated with Human Cancer*, pp. 125–160, PHILLIPS, L. (ed.) Marcel Dekker, New York.

PAGANO, J. S. and LEMON, S. M. (1984) The Herpesviruses. In: *Medical Microbiology and Infectious Diseases* (2nd Edn), pp. 541–549, BRAUDE, A. (ed.) W. B. Saunders, Philadelphia.

PAGANO, J. S., SIXBEY, J. W. and LIN, J.-C. (1983) Acyclovir and Epstein–Barr virus infection. *J. Antimicrob. Chemother.* **12**(Suppl. B): 113–121.

PLOTKIN, S. A., STAN, S. E. and BRYAN, C. K. (1982) *In vitro* and *in vivo* responses of cytomegalovirus to acyclovir. *Am. J. Med.* **73**: 257–261.

POIESZ, B. F., RUSCETTI, F. W., GAZDAR, A. F., BUNN, P. A., MINNA, J. D. and GALLO, R. C. (1980) Detection and isolation of type C retrovirus particles from fresh and cultured lymphocytes of a patient with cutaneous T-cell lymphoma. *Proc. natn. Acad. Sci. U.S.A.* **77**: 7415–7419.

PRUSOFF, W. H. (1959) Synthesis and biological activities of iododeoxyuridine, an analog of thymidine. *Biochim. biophys. Acta* **32**: 295–296.

PRUSOFF, W. H. and GOZ, B. (1973) Chemotherapy—molecular aspects. In: *The Herpes Viruses*, pp. 641–663, KAPLAN, A. S. (ed.) Academic Press, New York.

PRUSOFF, W. H. and GOZ, B. (1975) Halogenated pyrimidine deoxyribonucleosides. In: *Antineoplastic and Immunosuppression Agents*, Vol. 2 pp. 272–347, SARTORELLI, A. C. and JOHNS, D. G. (eds) Springer, Berlin.

RAAB-TRAUB, N., HOOD, R., YANG, C. S., HENRY, B. and PAGANO, J. S. (1983) Epstein–Barr virus transcription in nasopharyngeal carcinoma. *J. Virol.* **48**: 580–590.

RAPP, F. and WESTMORELAND, D. (1976) Cell transformation by DNA-containing viruses. *Biochim. biophys. Acta.* **458**: 167–211.

RASHBAUM, S. and COZZARELLI. (1976) Mechanism of DNA synthesis inhibition by arabinosylcytosine and arabinosyladenine. *Nature* **264**: 679–680.

SCHAEFFER, H. J., BEAUCHAMP, L. M., DEMIRANDA, P., ELION, G. B., BAUER, D. J. and COLLINS, P. (1978) 9-(2-hydroxyethoxymethyl)guanine activity against viruses of the herpes group. *Nature* **272**: 583–585.

SHARP, P. A., PETTERSSON, U. and SAMBROOK, J. (1974) Virus DNA in transformed cells. I. A study of the sequences of adenovirus 2 DNA in a line of transformed rat cells using specific fragments of the viral genome. *J. molec. Biol.* **86**: 709–726.

SHIPKOWITZ, N. L., BOWER, R. R., APELL, R. N., NORDEEN, C. W. and OVERBY, L. R. (1973) Suppression of herpes simplex virus infection by phosphonoacetic acid. *Appl. Microbiol.* **26**: 264–267.

SIXBEY, J. W. and PAGANO, J. S. (1984) New perspectives on the Epstein–Barr virus in the pathogenesis of lymphoproliferative disorders. In: *Current Clinical Topics in Infectious Diseases*, pp. 146–176, REMINGTON, J. and SCHWARZ, M. (eds) McGraw-Hill, New York.

SIXBEY, J. W. and PAGANO, J. S. (1985) Epstein–Barr virus transformation of human B lymphocytes despite inhibition of viral polymerase. *J. Virol.* **53**: 299–301.

SIXBEY, J. W., VESTERINEN, E. H., NEDRUD, J. G., RAAB-TRAUB, N., WALTON, L. A. and PAGANO, J. S. (1983) Replication of Epstein–Barr virus in human epithelial cells infected *in vitro*. *Nature* **306**: 480–483.

SKÖLDENBERG, B., FORSGREEN, M., ALESTIG, K., BERGSTROM, T., BURMAN, L., DAHLQUIST, E., FORKMAN, A., FRYDEN, A., LOVGREN, K., NORLIN, K., NORRBY, R., STENKVIST, E. O., STIERNSTEDT, G., UHNOO, I. and VAHL, K. (1984) Acyclovir versus vidarabine in herpes simplex encephalitis. *Lancet* **i**: 707–711.

SMEE, D. F., MARTIN, J. C., VERHEYDEN, J. P. H. and MATTHEWS, T. R. (1983) Anti-herpesvirus activity of the cyclic nucleoside 9-(1,3,-dihydroxy-2-propoxymethyl)guanine. *Antimicrob. Agents Chemother.* **23**: 676–682.

SMITH, K. O., GALLOWAY, K. S., KENNELL, W. L., OGILVIE, K. K. and RADATUS, B. K. (1982) A new nucleoside analog, 9-[2-hydroxy-1-(hydroxymethyl)ethoxymethyl]guanine, highly active *in vitro* against herpes simplex virus types 1 and 2. *Antimicrob. Ag. Chemother.* **22**: 55–61.

SULLIVAN, J. L., MEDVECZKY, P., FORMAN, S. J., BAKER, S. M., MONROE, J. E. and MULDER, C. (1984) Epstein–Barr virus induced lymphoproliferation: implications for antiviral chemotherapy. *New Engl. J. Med.* **311**: 1163–1167.

SUMMERS, W. C. and KLEIN, G. (1976) Inhibition of Epstein–Barr virus DNA synthesis and late gene expression by phosphonoacetic acid. *J. Virol.* **18**: 151–155.

TOCCI, M. J., LIVELLI, T. J., PERRY, H. C., CRUMPACKER, C. S. and FIELD, A. K. (1984) Effects of the nucleoside analog 2'-nor-2'-deoxyguanosine on human cytomegalovirus replication. *Antimicrob. Ag. Chemother.* **25**: 247–252.

WATANABE, K. A., REICHMAN, A., HIROTA, K., LOPEZ, C. and FOX, J.J. (1979) Nucleosides. 110. Synthesis and antiherpes virus activity of some 2'-fluoro-2'-deoxyarabinofuranosylpyrimidine nucleosides. *J. med. Chem.* **22**: 21–24.

WHITLEY, R. J., SOONG, S. J. and DOLIN, R. (1977) Adenine arabinoside therapy of biopsy-proved herpes simplex encephalitis. *New Engl. J. Med.* **297**: 289–294.

YAJIMA, Y., TANAKA, A. and NONOYAMA, M. (1976) Inhibition of productive replication of Epstein–Barr virus DNA by phosphonoacetic acid. *Virology* **71**: 352–354.

ZUR HAUSEN, H. (1982) Human genital cancer: synergism between the two virus infections or synergism between a virus infection and initiating events? *Lancet* **ii**: 1370–1372.

CHAPTER 9

THE ROLE OF HEPADNAVIRUSES IN HEPATOCELLULAR CARCINOMA

WILLIAM S. ROBINSON, ROGER H. MILLER and PATRICIA L. MARION
Stanford University School of Medicine, Stanford, California 94301, USA

1. INTRODUCTION

Members of several virus families have the capacity to transform animal cells in culture and to induce tumors *in vivo* (Weiss *et al.*, 1982; Tooze, 1982). Transformation of cells in culture by several such viruses has been studied intensively leading to considerable understanding of molecular mechanisms of viral transformation. Although some of the tumor viruses studied most intensively appear to induce tumors only under experimental conditions, natural infection with other viruses are associated with naturally occurring tumors in certain species including man. The strongest evidence for a causal role for virus in tumor formation is in subprimate species, although certain herpes viruses (reviewed in Aurelian, 1983, and in Pagano and Henry, 1983), papova viruses (reviewed in Howley, 1983) and retroviruses (Poiesz *et al.*, 1980, 1981; Miyoshi *et al.*, 1981; Yoshida *et al.*, 1982) have been implicated in human neoplasia. Among the strongest human tumor virus candidates is hepatitis B virus (HBV). Although the acute viral hepatitis associated with primary HBV infection was recognized for many years before HBV was physically identified and characterized (called serum hepatitis because it was recognized to be commonly transmitted by virus-containing human serum), the association between HBV infection and hepatocellular carcinoma (HCC) has been recognized much more recently (Szmuness, 1978). The development of serologic tests for HBV led to the recognition that many HBV infections become persistent with continued virus replication in the liver and the presence of viral forms in the blood for many years. Persistent HBV infections were found to be unexpectedly common with prevalences exceeding 10% to 20% in some populations in certain geographic areas of the world and it has been estimated that there are more than 170 million such infections worldwide (Szmuness, 1978). Such persistent infections are sometimes associated with chronic hepatitis and cirrhosis. It is in this setting that the association of HBV infection and HCC was first noted (Sherlock *et al.*, 1970) and in which a great majority of hepatocellular carcinomas (HCC) appear to occur in man (Szmuness, 1978; Beasley *et al.*, 1978). HBV may be the single most common cause of chronic liver disease and HCC in the world, and HCC is one of the most common cancers in the world.

Although, when first characterized HBV was considered to be a unique virus, very similar viruses have been more recently discovered in three different animal species. Woodchuck hepatitis virus (WHV), the first of these to be discovered, was looked for and found in a colony of captive woodchucks (*Marmota monax*) in the eastern United States because the animals were observed commonly to develop HCC (Summers *et al.*, 1978b). Ground squirrel hepatitis virus (GSHV) was next found (Marion *et al.*, 1980a) in California in Beechey ground squirrels (*Spermophilus beecheyi*) which are closely related to woodchucks. Duck hepatitis B virus (GHBV) was first found in Pekin ducks with HCC in China (J. Summers *et al.*, unpublished with permission of author), and more recently in Pekins in the United States (Mason *et al.*, 1980) and Japan (Omata *et al.*, 1983). The family name hepadnaviridae has been suggested for viruses of this kind (Robinson, 1980; Robinson *et al.* 1982; Gust *et al.*, in press). Among the features which define this new virus family are unique virion ultrastructure; characteristic polypeptide and antigenic composition; and common genome size, structure and mechanism of replication. Although these are DNA viruses, they appear to utilize a reverse transcriptase step in replication of their genome. Incompletely replicated (partially single stranded) circular DNA molecules and occasionally RNA-DNA hybrid molecules appear to be packaged in virions along with DNA polymerase activity which synthesizes viral DNA minus strands on an RNA template and DNA plus strands of the partially single stranded circular DNA molecules to make them completely double stranded molecules of approximately 3200 bp. The reverse transcriptase step is reminiscent of retrovirus replication and recent studies showing hepadnavirus genome nucleotide sequence homology with retroviruses (Toh *et al.*, 1983; Miller and Robinson, 1986), suggests that these two virus families are related. This relatedness is of particular interest since infections with members of both virus families are associated with neoplasis. Interesting biological features of hepadnaviruses include a striking tropism for hepatocytes, the common occurrence of persistent infections with complete and incomplete viral forms in high concentrations in the blood and lower concentrations in other body fluids continuously for years. Naturally occurring infections with hepadnaviruses are associated with acute and chronic hepatitis, sometimes immune complex mediated disease, and hepatocellular carcinoma (HCC).

Here we will review recent knowledge of the virion structure, the molecular structure and mechanism of replication of the genomes of these viruses, their relation to retroviruses, evidence for the association of hepadnavirus infection and HCC, the physical state of the virus in tumor cells, and consider whether hepadnaviruses could function by any of the mechanisms of tumor induction known for other viruses. It is of interest to define similarities and differences between hepadnaviruses and retroviruses in order to understand their evolutionary relationship and to determine whether they share a common oncogenic mechanism since infection with members of both virus families is associated with neoplastic disease.

2. VIRUS MORPHOLOGY AND ANTIGENIC STRUCTURE

The virions of HBV are spherical particles approximately 42–47 nm in diameter with an electron dense spherical inner core with diameter of approximately 22–25 nm and an outer shell or envelope approximately 7 nm in thickness (Dane *et al.*, 1970; Summers *et al.*, 1978b; Marion *et al.*, 1980a; Mason *et al.*, 1980). The lipid-containing envelope bears the viral surface antigen to which virus neutralizing antibody is directed. The inner core particles which can be released from virions by detergent treatment bear the viral core antigen (Almeida *et al.*, 1971) and contain the viral DNA (Robinson *et al.*, 1974; Summers *et al.*, 1978b; Marion *et al.*, 1980a; Mason *et al.*, 1980), a DNA polymerase activity (Kaplan *et al.*, 1973; Robinson and Greenman, 1974; Summers *et al.*, 1978b; Marion *et al.*, 1980a; Mason *et al.*, 1980) and a protein kinase activity (Albin and Robinson, 1980; Feitelson *et al.*, 1982a) which phosphorylates the viral genome specified major polypeptide of the core. The virion ultrastructures of WHV (Summers *et al.*, 1978b), GSHV (Marion *et al.*, 1980a), and DHBV (Mason *et al.*, 1980) are similar but not identical to that of HBV. The virion morphologies of the three mammalian viruses are more similar to each other than to the duck virus, the core of which is covered with spike-like projections (Mason *et al.*, 1980) not apparent on the cores of the mammalian viruses.

In addition to virions, there are more numerous particulate forms which bear viral surface antigen in serum of infected animals. In the case of the three mammalian viruses, these are small (16–25 nm diameter) spherical particles and filamentous forms 22 nm in width and varying in length (Dane *et al.*, 1970; Summers *et al.*, 1978b; Marion *et al.*, 1980a; Mason *et al.*, 1980). The filamentous forms in ground squirrel serum are particularly long and numerous (Marion et al., 1980a). In the well-studied HBV, these particles have been shown to be composed of lipid, protein and carbohydrate, and lack virion core components. Thus they are considered to be incomplete viral forms. Only spherical forms and no filamentous forms have been described for the DHBV (Mason et al., 1980). The spherical particles associated with DHBV are larger and more pleomorphic than those of the mammalian viruses.

No antigenic variation of the surface antigens as known for HBsAg (Courouce-Pauty and Soulier, 1974; Courouce et al., 1976) has been described for the other hepadnaviruses although this question has not yet been investigated in detail.

The third HBV antigen, the hepatitis Be antigen (HBeAg), was first detected as a soluble antigen in serum of HBV infected patients (Magnius and Espmark, 1972). The soluble HBeAg is antigenically and physically distinct from HBsAg particles described above. HBeAg also appears to be present in a cryptic form in the virion core and can be detected only after disruption of core particles as for example by detergent treatment (Takahashi et al., 1979). The major polypeptide of virion cores appears to manifest HBeAg specificity when isolated from cores (Takahashi et al., 1979). HBeAg activity has not been reported in association with other hepadnaviruses but has not been extensively investigated.

Several studies have been done to determine the relatedness of the virion antigens of the different hepadnaviruses but it is difficult to compare results because different methods have been used in different laboratories to quantitate antigenic relatedness. Antibody to HBsAg (anti-HBs) appears to bind to both WHsAg and GSHsAg with reported cross reactivity of 1% with WHsAg (Werner et al., 1979) and 48% with GSHsAg (Gerlich et al., 1980) using different assay methods. These two studies showed less binding of HBsAg by anti-GSHs and anti-WHs respectively than by anti-HBs, using the same methods. Tryptic peptide mapping suggests that a common amino acid sequence is present in polypeptides of HBsAg of all antigenic subtypes, GSHsAg and WHsAg (Feitelson et al., 1983) suggesting this could represent the group specific *a* determinant of HBsAg. Comparison of GSHsAg and WHsAg have shown them to be highly related, but not identical (Feitelson et al., 1983).

A high degree of cross reactivity between WHcAg and HBcAg using counter electrophoresis (12.5–25%) has been reported (Werner et al., 1979). It was suggested that since the cross reactivity between human and animal virus core antigens was apparently much greater than between the surface antigens of these species (1%), the hepadnavirus core antigens might serve as 'group specific' antigens. In support of this, DNA sequence data (Galibert et al., 1982) indicates that there is a greater amino acid sequence homology between virion core polypeptides (73%) than surface antigen polypeptides (62%) of HBV and WHV.

3. VIRION POLYPEPTIDES

Multiple polypeptides have been found in purified surface antigen particles of HBV, WHV and GSHV. A major pair of polypeptides with apparent molecular weights of approximately 24,000 ($P-24^S$) and 27,000 ($GP-27^S$) daltons and minor components with apparent sizes of 31,000, 33,000, 39,000 and 42,000 daltons (designated $P-31^S$, $GP-33^S$, $P-39^S$ and $GP-42^S$, respectively) have been repeatedly observed in HBsAg preparations (reviewed in Tiollais et al., 1985). Most of these react with antibody raised against intact HBsAg particles and against the major polypeptides indicating that they share at least some amino acid sequences. This conclusion has been confirmed by tryptic peptide mapping of the different polypeptides (Feitelson et al., 1983). $P-24^S$ and $P-27^S$ share amino acid sequences at their carboxy and amino termini (Peterson et al., 1977, 1978) but only P-27 is glycosylated, suggesting that the two have identical primary sequences coded by the S-region of the S open reading frame (see *Viral Genome*, below) and differ in electrophoretic mobility only because of the carbohydrate in P-27. $P-31^S$ and $GP-33^S$ appear to represent the nonglycosylated and glycosylated polypeptides coded by the $preS_2$ plus S sequence of the S open reading frame, and $P-39^S$ and $GP-42^S$ the nonglycosylated and glycosylated polypeptides coded by the $preS_1$, $preS_2$ plus S sequences of the S open reading frame (reviewed by Tiollais et al., 1985). Sodium dodecylsulfate polyacrylamide gel electrophoresis (SDS-PAGE) has shown that the two major polypeptides of WHsAg and GSHsAg each have a lower electrophoretic mobility than the respective polypeptides of HBsAg, and both share a similar tryptic peptide map homology with the HBsAg polypeptides (Feitelson et al., 1981). The apparent size difference in the polypeptides of HBsAg and WHsAg by SDS-PAGE is not substantiated by the theoretical molecular weights calculated from the nucleotide sequence of HBV and WHV DNA (Galibert et al., 1982), suggesting that post-translational processing, or some other mechanism, may be responsible for the difference in electrophoretic mobility. Evidence for a close relatedness of HBsAg, WHsAg and GSHsAg comes not only from their serologic cross reactivity but also from tryptic peptide mapping of their polypeptides. More than 50% of the tryptic peptides of the major nonglycosylated polypeptides of WHsAg and GSHsAg are identical (Feitelson et al., 1981). 25% of the tryptic peptides of the major nonglycosylated polypeptides of HBsAg are subtypes adw, adr and ayw; GSHsAg and WHsAg are identical indicating that a significant region of this polypeptide has been conserved during the evolution of these viruses. A 66% homology has been found in the nucleotide sequences of the genes for the major surface antigen polypeptides of HBV and WHV (Galibert et al., 1982). Analysis of purified DHBsAg has revealed that it is composed of single predominant polypeptide of 17,500 daltons instead of the major pair of polypeptides seen with the mammalian surface antigens and it lacks the abundant minor polypeptides of the latter (Marion et al., 1983a). This observation is supported by the published nucleotide sequence of DHBV DNA in which an open reading frame of 501 nucleotides in the region of the genome corresponding to the location of the surface antigen gene of the mammalian viruses (Mandart et al., 1984), would specify a polypeptide of 18,204 daltons.

Virion cores of HBV, WHV and GSHV all contain a single major polypeptide of approximately 22,000 daltons (P-22^C) (although the polypeptides of the three viruses are not identical in electrophoretic mobility) and several minor polypeptides with larger apparent sizes (Feitelson et al., 1982b). Purified cores from both DHBV virions and infected liver contain a 37,000 dalton polypeptide (Newbald et al., 1984), close to the 34,986 dalton molecule predicted by the nucleic acid sequence (Mandart et al., 1984). This greater size of the duck virus core polypeptide apparently results from fusion of the open reading frame of the duck genome which corresponds to the core polypeptide gene of the mammalian viruses with a second smaller open reading frame corresponding to the X-gene of the mammalian viruses as described below under *Physical and Genetic Structure of the Viral Genome*. The antigenic relatedness of the polypeptides of GSHV and DHBV cores has not been studied. Approximately 56% of the tryptic peptides of the major core polypeptides of HBV and GSHV are identical as might be expected by serologic relatedness of HBcAg and GSHcAg (Feitelson et al., 1982b). The approximate 73% amino acid sequence homology calculated from the nucleotide sequences between the major core polypeptides of HBV and WHV (Galibert et al., 1982), further confirms the close relatedness of these two viruses.

4. PHYSICAL AND GENETIC STRUCTURE OF THE VIRAL GENOME

Hepadnaviruses have among the smallest genomes of all known viruses and the form of the genome in virions from the blood of infected individuals is a small circular DNA molecule (Robinson et al., 1974) that is partially single stranded (Summers et al., 1975; Landers et al., 1977; Hruska et al., 1977) (shown schematically in Fig. 1). The DNAs of the three mammalian viruses contain single stranded regions which vary in length from approximately 15–60% of the circle length in different molecules (Summers et al., 1978a; Marion et al., 1980a; Summers et al., 1975; Landers et al., 1977; Hruska et al. 1977). The single stranded region of DHBV is apparently much smaller and many full length molecules are packaged in virions (Mason et al., 1980). Thus these DNAs consist of a long strand (this

HBV DNA (HBsAg, adw$_2$)

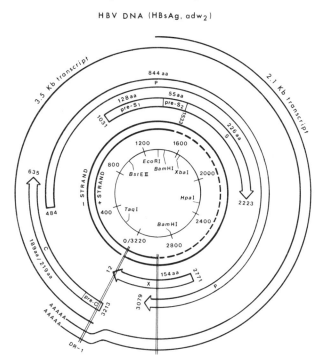

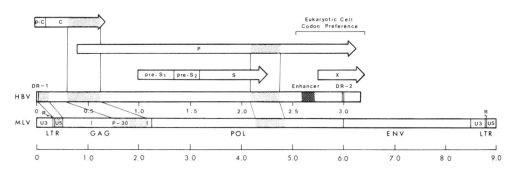

Fig. 1. Physical and genetic map of HBV DNA and murine leukemia virus (MLV). In the circular map of HBV DNA the broken line in the short (+) DNA strand represents the region within which the 3' end of the + strand may occur in different molecules, and the corresponding region of the long − strand is that which may be single stranded in different molecules. The restriction sites and locations of the nick in the − strand, the 5' end of the + strand, and the location of the single-stranded region are reported by Siddiqui et al. (1979), and the open reading frames (large arrows) and 11 bp direct repeat sequences (DR-1 and DR-2) are found in the nucleotide sequence published by Valenzuella et al. (1980). The RNA transcripts are as mapped by Moroy et al. (1985) and Enders etc al. (1985). In the HBV genome represented as a linear sequence, the transcriptional enhancer location is that reported by Shaul et al. (1985) and Tognoni et al. (1985). The MLV proviral DNA structure is as reviewed by Coffin (1982). The regions of shared nucleotide sequence homology (stipled regions) are those described by Miller and Robinson (1986) and Toh (1983).

is the minus DNA strand since its sequence is complementary to that of viral messenger RNA) of constant length (between 3000 to 3300 bases in different viruses) in all molecules and a short (or plus) strand which varies in length between 1700 and 2800 bases in different molecules. A DNA polymerase activity in the virion (Summers et al., 1978a; Marion et al., 1980a; Mason et al., 1980; Kaplan et al., 1973; Robinson and Greenman, 1974) repairs the single stranded region in the viral DNA to make fully double-stranded molecules. DNA synthesis is initiated for this reaction at the 3' end of the short strand, which occurs at different sites within a specific region (50%) of the DNA in different molecules. DNA synthesis is terminated when the uniquely located 5' end of the short strand is reached. The virus strand is not a closed circle but a nick exists at a unique site approximately 225 bp from the 5' end of the plus strand in mammalian viruses (Sattler and Robinson, 1979,

Siddiqui et al., 1979; Seeger et al., 1986), and 69 bp in DHBV DNA (Molnar-Kimber et al., 1984). An approximately nine nucleotide terminal repetition (r) has been shown in the minus DNA strand of DHBV (Lien et al., 1986) and GSHBV (Seeger et al., 1986) which may be important in circularizing the DNA and in template switching during synthesis of the plus DNA strand (see *Genome Replication* below). The circular DNA of the mammalian viruses can be converted to a linear form with single-stranded cohesive ends by selectively denaturing the 225 bp region between the 5' ends of the short and long strands by heating under appropriate conditions (Sattler and Robinson, 1979). The resulting linear form can be recircularized by reassociation of the complementary single-stranded ends.

The 5' ends of both the long and short strands of HBV DNA appear to be blocked in a manner which prevents phosphorylation with polynucleotide kinase (Gerlich and Robinson, 1980). A polypeptide appears to be covalently attached to the 5' end of the long strand of HBV (Gerlich and Robinson, 1980), DHBV (Molnar-Kimber et al., 1984) and GSHV (Ganem et al., 1982) DNAs isolated from virions and this undoubtedly prevents phosphorylation of this strand. Evidence with DHBV (Molnar-Kimber et al., 1983) and GSHV (Ganem et al., 1982) suggests that this protein functions as a primer for synthesis of the viral minus DNA strand. Recently a nineteen nucleotide capped oligoribonucleotide covalently attached to the 5' end of the plus DNA strand of DHBV (Lien et al., 1986) and GSHV (Seeger et al., 1986) which appears to function as a primer for synthesis of this DNA strand has been described. This capped oligoribonucleotide undoubtedly accounts for the inability to phosphorylate the 5' end of the plus DNA strand. The unusual virion DNA structure of hepadnaviruses results from packaging into virions which are released from cells incompletely replicated viral DNA molecules consisting of complete minus strands (the first strand to be synthesized) and variably completed plus strands, and the primer for each strand has remained covalently attached to the 5' end (see below under *Virus Replication*).

The complete nucleotide sequences of the cloned DNAs of nine HBV isolates (Fujiyama et al., 1983; Ono et al., 1983; Kobayashi and Koike, 1984; Gan et al., 1984; Valenzuela et al., 1980; Galibert et al., 1979; Bichko et al., 1985; Pasek et al., 1979), two DHBV isolates (Mandart and Galibert, 1984; Sprengel et al., 1985), and one each of WHV (Galibert et al., 1982) and GSHV (Seeger et al., 1984) have been reported. The genomes of the three mammalian viruses have four long open reading frames in the complete or long (minus) DNA strand and these have similar locations in each virus with respect to the cohesive ends of the DNAs (Fig. 1). The genes for the two major virion polypeptides have been identified with certainty. The C gene specifies the major viral core or nucleocapsid polypeptide ($P-22^c$) and the polypeptide with HBeAg specificity (reviewed in Tiollais et al., 1985), and this open reading frame sometimes includes a short preC (precore) sequence. The S gene including $preS_1$, $preS_2$, and S regions specifies the viral surface antigen reactive polypeptides ($P-24^S$, $GP-27^S$, $P-31^S$, $GP-33^S$, $P-39^S$ and $GP-42^S$) in the virion (Dane particle) envelope and in incomplete viral forms (surface antigen particles) found in serum and liver of infected individuals. These represent the glycosylated and non-glycosylated forms of three polypeptides coded respectively by S alone, by $preS_2$ and S, and by $preS_1$, $preS_2$ and S regions of the S open reading frame (reviewed in Tiollais et al., 1985). The P gene is thought to specify the virion associated DNA polymerase (or reverse transcriptase). The small X-gene appears to specify a polypeptide of unknown function detected in liver of some infected patients (Moriarty et al., 1985; Kay et al., 1985). The C, P, $preS_2$ and S genes are approximately the same size and occupy the same relative genomic positions in all mammalian hepadnaviruses. The $preS_1$ varies somewhat in size in different viruses suggesting that the polypeptide sequence specified by this DNA sequence is less functionally critical than that specified by the $preS_2$ and S regions. The size of the X-gene varies by up to 50% among nine different HBV isolates and the carboxyl terminal 90% of the X open reading frame has been deleted in DHBV. Several other functionally important elements have been identified in hepadnavirus genomes. These include eleven bp direct repeat sequences (5'TTCACCTCTGC3') designated DR1 and DR2 which are approximately 225 bp apart in the mammalian viruses and appear to play a critical role in viral DNA replication (see below under *Virus Replication*). The 5' end of the minus DNA strand occurs within DR1 and the 5' end of the plus DNA strand at the 3' boundary of DR2 (Seeger *et al.*, 1986; Molnar-Kimber *et al.*, 1984). Transcriptional control elements that have been identified include two promoter sequences. One in the preS$_1$ region (Cattaneo *et al.*, 1983) appears to direct synthesis of a 2.1 kb RNA transcript that probably functions as messenger for preS$_2$ and S gene specified polypeptides, and another upstream from the start of the C-gene (Moroy *et al.*, 1985; Enders *et al.*, 1985) appears to direct synthesis of a greater than genome length (3.5 kb) RNA transcript, which may function as messenger for synthesis of C and P gene specified polypeptides, and appears to serve as a template for synthesis of the virus minus DNA strand by reverse transcription (see below under *Virus Replication*). These are the major viral transcripts which appear to be unspliced, and to have different 5' ends but colinear, polyadenylated 3' ends. A transcriptional enhancer element has been localized to a region approximately 450 bp upstream from the C-gene promoter (Shaul *et al.*, 1985; Tognoni *et al.*, 1985) and occurs either immediately upstream from the start of the X-gene in viruses with a short (typical) X-gene or within the 5'end of that open reading frame in viruses with a long X-gene (e.g. in HBV subtype adr). Glucocorticoids increase levels of HBsAg expression in HBV infected patients *in vivo* and in HCC cell lines expressing HBsAg in culture. Recently dexamethasone has been shown to stimulate expression of chloramphenicol amino transferase (CAT) driven by the promoter when a region of HBV DNA was inserted into plasmid pA$_{10}$CAT (Tur-Kaspa *et al.*, 1986). The glucocorticoid responsive region of HBV DNA was localized to the S region of the S open reading frame, and a fifteen nucleotide sequence (5' NCAANNTGTYCT 3') similar to other known glucocorticoid responsive DNA elements was identified in that region (approximately 2097–2112 on the map in Fig. 1). This glucocorticoid responsive enhancer-like sequence is distinct from the enhancer sequence described above. A polyadenylation signal that appears to be used by both transcripts described above lies within the beginning of the C-gene and approximately 20 bp upstream from the 3' end of the minus DNA strand (Cattaneo *et al.*, 1983, 1984). Finally, a highly conserved 60–70 bp sequence with a high degree of homology with the U-5 sequence of the retrovirus LTR (see U-5-homologous sequence in *Genome Homology*, below) is present just downstream from DR-1 and within the pre-C region of the genome (Miller and Robinson, 1986).

Fig. 1 also shows the linear arrangement of the four genes of a typical murine leukemia virus for comparison with the hepadnavirus genome. The gag, pol and env gene products are functionally analogous to the C, P and S genes respectively of hepadnaviruses. The terminal regions of retrovirus RNA are duplicated at the ends of the proviral DNA synthesized in infected cells by reverse transcription of the virion RNA. This long terminal repeat or LTR sequence contains a short sequence, U-5, derived from the 5'D end of genomic RNA and a U3 region derived from the 3'D end of the genomic RNA. U3 appears to contain transcriptional and other viral control sequences. Several retroviruses including HTLV I, II and III and bovine leukemia virus contain another open reading frame located between the env gene and the 3'D LTR in HTLV I and II and separated in two parts in HTLV III, and these have been called the long open reading frame (LOR) or pX genes which appear to specify a protein involved in transactivation of transcription directed by the LTR (reviewed in Wong-Staal and Gallo, 1985). Some other retroviruses contain viral oncogenes that are often inserted into the env gene (Coffin, 1982).

5. GENOME HOMOLOGY OF HEPADNAVIRUSES AND RETROVIRUSES

Comparison of the nucleotide sequences of eleven cloned mammalian hepadnavirus isolates [(nine HBV isolates of subtypes adw (Ono *et al.*, 1983; Valenzuela *et al.*, 1980), adr (Fujiyama *et al.*, 1983; Ono *et al.*, 1983; Kobayashi and Koike, 1984; Gan *et al.*, 1984), ayw (Galibert *et al.*, 1979; Bichko *et al.*, 1985), and adyw (Pasek *et al.*, 1979), and single isolates of GSHV (Seeger *et al.*, 1984) and WHV (Galibert *et al.*, 1982)] which have been completely sequenced with sequences of retrovirus genomes has revealed regions of homology between specific regions of the genomes of viruses in the two families (Miller and Robinson, 1986). The most highly conserved sequence in hepadnaviruses was found to be a 111 nucleotide sequence extending from the carboxyl terminal end of the X-gene and the 3' end of DR-1 into the C-gene (shown in Fig. 1). The sequence of the first fifty nucleotides of this 111 nucleotide sequence was found to be 99% conserved, and the entire sequence more than 95% conserved among the eleven sequenced mammalian hepadnaviruses. The greatest homology between hepadnavirus and retrovirus genomes was found between the first sixty-six nucleotides of this highly conserved hepadnavirus sequence and almost the entire U-5 sequence of certain mammalian retroviruses. The short inverted repeat sequence at the 3' end of U-5 which is essential for integration of proviral DNA of retroviruses into cellular DNA (Panganiban and Temin, 1983) was the only region of U-5 without a high degree of homology in the hepadnavirus U-5-homologous sequence.

The retroviruses with the highest degree of UV5 homology with mammalian hepadnaviruses are the C-type murine leukemia/sarcomaviruses and the endogenous retrovirus-like elements found in human and simian chromosomal DNA (O'Connell et al., 1984; Bonner et al., 1982a; Repaske et al., 1983, 1985; Bonner et al., 1982b; O'Connell and Cohen, 1984; Paulson et al., 1985). Although the U-5 sequences of retroviruses represent the most conserved region of the retroviral LTR, the function of this sequence is not well understood and whether the U-5-homologous sequence in hepadnaviruses has a specific function is not known. The fact that this sequence is highly conserved in hepadnaviruses (and in retroviruses) suggests it has some essential function. Its location near DR1 which is the site where hepadnavirus minus DNA strand synthesis begins suggests that the U-5-homologous sequence in the intracellular RNA template could play a role such as binding the protein primer or reverse transcriptase for initiation of viral minus DNA strand synthesis. Such possibilities have yet to be investigated. Hepadnaviruses do not contain a genomic region which is physically and functionally identical to the retrovirus LTR. The LTR is a complex terminally repeated sequence in the retroviral pregenomic DNA derived from sequences at both ends of genomic RNA, and containing transcriptional control elements and controls for orderly integration of the pregenomic DNA into chromosomal DNA (Varmus, 1982). Hepadnaviruses do not undergo such orderly and regular integration (hepadnavirus integration occurs but appears to be an infrequent event, is associated with genomic rearrangements, and is more similar to integration by viruses such as SV-40 than to retrovirus integration) and thus appears to have lost critical functional elements for integration. The hepadnavirus pregenomic RNA template possesses a terminal repetition with sequences corresponding to certain control elements of genomic DNA (e.g. the polyA addition signal, DR-1 and the U-5-homologous sequence) but as RNA, these sequences do not function like LTRs in retroviral pregenomic DNA.

The C-gene is the hepadnavirus gene whose sequence is most highly conserved among the eleven mammalian hepadnavirus sequences studied. Comparison of amino acid sequences predicted from the C-gene nucleotide sequence with predicted amino acid sequences of retroviral genes revealed 41% homology between a ninety-eight amino acid sequence at carboxyl terminal end of the hepadnavirus C-gene product and the carboxyl terminal ninety-eight amino acids of the P-30 gag protein of the C-type murine leukemia/sarcoma viruses, a similar homology with endogenous retroviral elements of primates, and slightly lower homology with gag P-30 of other mammalian retroviruses. The P-30 sequence is highly conserved in mammalian retroviruses (Hunter et al., 1978; Tamura and Takano, 1982; O'Rear and Temin, 1982; Oroszlan et al., 1981a,b; Cohen et al., 1981) and mutations in this region of the gag gene are lethal (Goff, 1984) indicating the essential nature of this sequence for retrovirus replication. The C-terminus of gag P-30 and the hepadnavirus C-gene product contain highly basic amino acids suggesting that these polypeptides may bind nucleic acids as might be expected for nucleocapsid proteins.

Comparison of the predicted amino acid sequences of the putative polymerase (P) gene of mammalian hepadnaviruses and the pol gene of mammalian retroviruses has revealed a 40% homology in a ninety-four amino acid sequence in the mid portion of these genes

(Miller and Robinson, 1986; Toh et al., 1983), a region that is highly conserved among retroviruses (Hunter et al., 1978; Tamura and Takano, 1982; O'Rear and Temin, 1982; Oroszlan et al., 1981a,b; Cohen et al., 1981). Similar homology has been found in the pol gene of cauliflower mosaic virus (Toh et al., 1983), copia-like element 17.6 of *Drosophila* (Varmus, 1985), and mobile Ty elements of *Sacharomyces* (Varmus, 1983). Hepadnaviruses also have other similarities with cauliflower mosaic virus which is a DNA (8 kb) plant virus utilizing a reverse transcriptase step in its replication (reviewed in Varmus, 1983). Hepadnavirus genomes do not have homology with retroviral pol gene region (3' end) which specifies the endonuclease function required for efficient integration of the retroviral DNA provirus (reviewed in Varmus, 1985). The absence of such a function in hepadnaviruses could account for the failure of these viruses to integrate in the orderly and efficient manner demonstrated by retroviruses.

The presence of significant homology in genes with comparable functions (e.g. nucleocapsid and polymerase) and control regions (e.g. U-5) suggest that hepadnaviruses and retroviruses (and cauliflower mosaic virus) are related and probably evolved from a common ancestor. No such homology was found between hepadnaviruses and other virus families. The preservation of sequences in a major portion of the P-30 region of the gag gene and a middle region of the pol gene of retroviruses with functionally and structurally analogous genes of hepadnaviruses (and the pol gene of cauliflower mosaic virus) suggest that the conserved sequences specify functionally important peptide regions within the protein gene products that have little tolerance for change. Similarly the highly conserved U-5 sequence in retroviruses and the U-5-homologous sequence in hepadnaviruses must be an essential sequence with an important function for viruses in these two families.

6. CODON PREFERENCES IN HEPADNAVIRUS GENES

There is degeneracy of the genetic code in that any of two to four different bases may occur in the third position for codons specifying specific amino acids. Genes of different classes of organisms have been found preferentially to contain different bases in the third position of codons (Table 1). The base C occurs more often than T, and G more often than A in this position for genes of eukaryotic cells, and the reverse is true of genes of eukaryotic viruses (Wain-Hobson et al., 1981). Analysis of third base preferences for codons of mammalian hepadnavirus genes reveals that the C, P and S genes contain codon frequencies typically found in eukaryotic viral genes and the X-gene codon usage is like that of eukaryotic cell genes (Miller and Robinson, 1986) (Table 1). Similar analysis of retroviral genes has shown that the gag, pol and env genes exhibit codon usage patterns typical of eukaryotic viral genes. Retroviral oncogenes as well as the pX genes of certain retroviruses described above, on the other hand, exhibit codon preferences of eukaryotic cell genes. It is intriguing that the X-gene of hepadnaviruses, and oncogenes and pX genes of retroviruses are located in the same genomic position relative to the other viral genes and all exhibit codon preferences like those of cellular genes. There is good evidence that retroviral oncogenes have been recently derived from cellular genes (Coffin, 1982) and the

TABLE 1 *Codon 3rd Base Preference*

	euk cell[1] genes	euk virus[1] genes	phage and prok: & cell[1] genes	hepadnavirus genes[2]			
				C	P	S	X
$\dfrac{A+U}{G+C}$	$\dfrac{1}{50.7}$	$\dfrac{9.6}{1}$	$\dfrac{1}{1.2}$	$\dfrac{17}{1}$	$\dfrac{2.5}{1}$	$\dfrac{1.7}{1}$	$\dfrac{1}{2.5}$
Number of sequences analyzed	16	10	10	12	12	12	12

1. Wain-Hobson et al., Cell **13**:355 (1981).
2. Miller and Robinson, *PNAS* (1986).

data described here suggest this may also be true for the pX genes of retroviruses and the X-gene of hepadnaviruses. In the case of hepadnaviruses, the DNA region exhibiting cell-like codon preferences always includes the transcriptional enhancer element suggesting that like the enhancer of BK virus (Rosenthal et al., 1983), the hepadnavirus enhancer may be of cellular origin. The function of the X-gene of hepadnaviruses is not known but it seems clearly not to be an oncogene like those of retroviruses which appear to lead to rapid transformation of cells in which they are expressed. Whether the X-gene has a function analogous to that of the pX gene of retroviruses is an intriguing possibility but has not yet been investigated.

7. THE MECHANISM OF HEPADNAVIRUS REPLICATION

The earliest studies of HBV in infected hepatocytes revealed that HBcAg could be detected only in nuclei of hepatocytes by immunofluorescent staining, and HBsAg in cytoplasm and on cell surfaces (Barker et al., 1973; Gudat et al., 1975; Ray et al., 1976). Consistent with this, electron microscopy demonstrated particles with the appearance of virion cores exclusively in hepatocyte nuclei (Almeida et al., 1970; Huang, 1971; Camamia et al., 1972). Particles resembling HBsAg forms have not been readily detected in cells and the morphogenesis of these particles, including complete virions, is unclear. During persistent infection, a variable number of cells contain detectable viral antigens by immunofluorescent staining (from less than 1% to virtually all hepatocytes in different patients) (Barker et al., 1973; Gudat et al., 1975; Ray et al., 1976). Interestingly, the pattern of viral antigen expression appears to be different in different cells of the same chronically infected liver. Commonly, most positive cells stain only for HBsAg, fewer have only detectable HBcAg, and even fewer cells contain both HBsAg and HBcAg. In the liver of some chronic carriers, HBsAg is the only detectable viral antigen. In all chronic carriers producing relatively high concentrations of viral DNA and DNA polymerase containing virions, significant numbers of HBcAg-positive cells can be found. The different patterns of viral antigen synthesis in individual cells of the same chronically infected liver indicate that individual viral genes are expressed differently in different cells. There is much less evidence concerning the cellular location of viral antigens for the hepadnaviruses of subprimate species but a recent study (Ponzetto et al., 1984) has demonstrated that both the core and surface antigens of WHV are detected by immunofluorescence in the cytoplasm of hepatocytes in liver biopsies of WHV infected woodchucks. Thus the viral core antigens of HBV and WHV appear to have different cellular sites of accumulation when detected by immunofluorescence.

More recently forms of viral RNA and DNA and synthesis of viral DNA strands have been studied and results indicate that hepadnavirus DNAs replicate in a unique and interesting way. These studies have been done in liver tissue of infected animals because no infected cell culture systems have been available. Viral DNA is present in infected liver cells in several distinct forms. The predominant form is 3200 bp closed circular (form 1) DNA which is found exclusively in a free form in the cell nucleus (Mason et al., 1982; Miller and Robinson, 1984a). Several viral DNA forms are present in liver cell cytoplasm and these appear to be contained in particles with DNA polymerase activity (Miller and Robinson, 1984a; Summers and Mason, 1982; Miller et al., 1984b). In DHBV infected duck liver these particles have been shown to have properties of viral cores (Summers and Mason, 1982). These particles contain 3200 bp relaxed circular (form II) and linear (form III) viral DNA, 3200 nucleotide single (predominantly minus) viral DNA strands, and viral DNA-RNA hybrid molecules (Miller and Robinson, 1984b; Summers and Mason, 1982; Miller et al., 1984b). The hybrid molecules appear to contain the longer than genome length plus strand RNA polyadenylated transcript with 200–300 terminal repeat (R) (Miller and Robinson, 1984b; Summers and Mason, 1982; Miller et al., 1984b). The endogenous DNA polymerase activity in the particles catalyzes the incorporation of nucleoside triphosphates into viral DNA minus strands of the DNA-RNA hybrid molecules in the presence or absence of actinomycin D which inhibits DNA dependent DNA synthesis. This suggests that the minus DNA strand is synthesized on an RNA template by a reverse transcriptase mechanism analogous to that known for retroviruses. In the DHBV (Summers and Mason, 1982) the RNA strand of the DNA-RNA hybrid molecules appeared to be degraded as the DNA strand length increased reminiscent of the RNase H activity of retroviruses. Nucleoside triphosphates are also incorporated into the plus strand of the relaxed circular DNA molecules in a DNA-dependent DNA synthetic reaction in the liver particles of HBV and GSHV as well as DHBV. Direct evident that the newly synthesized viral DNA minus strand of the DNA-RNA hybrid molecules in particles from GSHV infected ground squirrel liver is a precursor of the relaxed circular DNA molecules in these particles was obtained in pulse-chase experiments (Miller et al., 1984b). After a 5-min incubation of liver particles in a DNA polymerase reaction with $[a-^{32}P]$-dNTP, approximately equal amounts of ^{32}P were detected in minus DNA strands of DNA-RNA hybrid molecules and in relaxed circular DNA molecules. When the reaction was continued for 30 min in the presence of unlabeled dNTP, ^{32}P-DNA disappeared from the DNA-RNA hybrid and appeared in the relaxed circular DNA forms suggesting that minus DNA strands in the DNA-RNA molecules become components of relaxed circular DNA molecules during the DNA polymerase reaction taking place in the liver particles. Freezing of the liver or storage of particles before the experiment led to marked reduction in the polymerase activity catalyzing the incorporation of dNTP into DNA of the DNA-RNA hybrid but not that involved in incorporation into pure DNA suggesting the two activities have different stabilities.

Much greater detail of the replication mechanism of these genomes has been deduced from experiments identifying the primers for synthesis of both DNA strands and from nucleotide sequence analysis precisely identifying the ends of the two DNA strands. Evidence in the DHBV (Molnar-Kimber et al., 1983) and GSHV (Ganem et al., 1982) systems suggests that a protein serves as a primer for minus DNA strand synthesis and growing minus DNA strands of the DNA-RNA hybrid (and minus DNA strands in mature virions) are found to be covalently attached to the protein primer. The 5' end of the minus DNA strand of DHBV (Lien et al., 1986; Molnar-Kimber et al., 1984) and GSHV (Seeger et al., 1986) has been shown to correspond to the origin of reverse transcription and occurs within the DR1 sequence. The 3' ends of the minus DNA strands of DHBV (Lien et al., 1986) and GSHV (Seeger et al., 1986) have been localized to sites approximately five nucleotides beyond the DR-1 sequence indicating an approximately nine nucleotide terminal redundancy (r) in this strand. Recent studies have demonstrated the 5' ends of the plus DNA strands of DHBV (Lien et al., 1986; Molnar-Kimber et al., 1984) and GSHV (Seeger et al., 1986) to be the last nucleotide (3' end) of DR-2 or the following nucleotide. A nineteen or twenty base oligoribonucleotide with a sequence corresponding to that of the 5' end of the long RNA transcript serving as a template for minus DNA strand synthesis was found to be covalently attached to the 5' end of the plus DNA strand (Seeger et al., 1986; Lien et al., 1986). This oligoribonucleotide contains the DR-sequence and is thought to serve as the primer for plus DNA strand synthesis and it remains covalently attached to that strand in virion DNA.

These findings reported in most detail for DHBV (Lien et al., 1986; Molnar-Kimber et al., 1983, 1984; Mason et al., 1982; Summers and Mason, 1982), and more recently for GSHV (Seeger et al., 1986) and substantiated in some respects for HBV infected human liver (Miller and Robinson, 1984b; Miller et al., 1984a, 1984b) suggest the model shown in Fig. 2 for replication of hepadnaviruses which is based on that proposed by Summers and Mason (Summers and Mason, 1982) and Lien et al. (Lien et al., 1986) for DHBV replication and by Seeger et al. (Seeger et al., 1986) for GSHV. Following virus entry into liver cells, the infecting viral genome is converted to 3200 bp closed circular viral DNA which is found in abundance in the cell nucleus. This conversion requires removal of the oligoribonucleotide from the 5' end of the plus strand, and the protein and one nine nucleotide terminal redundancy (r) from the minus strand of virion DNA, and ligation of the ends of both strands to form closed circular (cc) molecules. This ccDNA found in the cell nucleus probably functions as a template for viral messenger RNA synthesis and synthesis of the longer than genome length transcript with the DR-2 sequence near the polyadenylated 3' end, and the DR-1 sequence and U-5-homologous sequence at both ends within the 200–300 nucleotide terminal redundancy. This transcript then serves as a template for minus strand DNA synthesis. The long RNA transcript, newly synthesized viral DNA polymerase (reverse transcriptase), and the protein primer for minus DNA strand synthesis are assembled with the major structural polypeptide of the viral core into core particles or viral nucleocapsids. Viral minus-strand DNA synthesis by reverse transcriptase within the core particles in the cell cytoplasm is initiated within the DR-1 sequence (probably that near the 3' end of the RNA template) utilizing the protein primer. If the DR-1 sequence at the same (e.g., 3') end of the RNA template is always the site for initiation of DNA minus strand synthesis, the protein primer and/or reverse transcriptase must recognize not only the DR-1 or some neighboring sequence such as the U-5-homologous sequence, but must also distinguish the two ends of the RNA template perhaps by recognizing a sequence unique to one end of the RNA (i.e. a sequence not within R). The RNA template appears to be degraded by RNase H-like activity as DNA synthesis proceeds. The primer for synthesis of the viral DNA plus strand synthesis is a nineteen or twenty oligoribonucleotide with a sequence corresponding to that of a 5' terminal fragment of the RNA template containing the DR-1 sequence. DNA plus strand synthesis is initiated at the last nucleotide of DR-2 near the 5' end of the minus DNA strand template. This suggests that the oligoribonucleotide primer is generated from the 5' end of the RNA template by ribonuclease action, that RNA fragment is then dissociated from the completed 3' end of the newly synthesized minus DNA strand and translocated by an unknown mechanism to the 5' end of the minus DNA strand where it base pairs with the DR-2 sequence in the minus strand for initiation of DNA plus strand synthesis at the boundary of the DR-2 sequence. When the 3' end of the elongating DNA plus strand reaches the 5' end of the DNA minus strand template, it must switch templates to the 3' end of the DNA minus strand forming a circular molecule. Template switching and circle formation could be facilitated by local denaturation of the newly formed double stranded terminus at the 5' end of the minus strand. The 3' end of the new plus strand dissociated from the 5' end of the minus strand would contain a sequence complimentary to the short terminal redundancy (r) of the minus strand. This complimentary sequence in the plus strand could then base pair with r at the 3' end of the minus DNA strand resulting in circularization of the DNA and positioning the 3' end of the minus strand for use as template for continued elongation of the DNA plus strand. It has been pointed out that the high AT content of the r sequence might facilitate local denaturation of a double stranded terminal r sequence (Seeger et al., 1986).

Core particles are assembled into complete virions with HBsAg and cell membrane lipid containing envelopes. In the case of HBV, virus formation and release from the cell can apparently take place at almost any step after intracellular assembly of the core particle, since virions (Dane particle) can be found in the blood which contain DNA-RNA hybrid molecules as well as virions containing partly single stranded circular DNA molecules, and endogenous DNA polymerase activity in the virions catalyzes the incorporation of nucleotides into minus DNA strands of the former and plus DNA strands of the latter (Miller et al., 1984b).

Results of restriction endonuclease digestion of DNA from HBV infected liver and Southern blot analysis suggest that viral DNA sequences are also integrated in cellular DNA in at least some HBV infected livers (and in hepatocellular carcinomas (HCC) to be discussed in *State of Hepadna Viruses in HCC*, below). Evidence for viral integration in some studies of chronically HBV infected human liver (Brechot et al., 1981a; Kam et al., 1982) was the finding of one or more DNA fragments containing viral DNA sequences that are larger than unit length viral DNA (3200 bp) after but not before digestion of cell DNA with a restriction enzyme (e.g. HindIII) for which no recognition sites exist in the viral DNA. The specific high-molecular weight-HindIII DNA fragments containing viral sequences have been found to be different in liver of different chronically infected patients.

The ability to detect such DNA fragments by Southern blot analysis has been interpreted to mean that viral DNA is integrated in the same site in many different cells of the liver of each chronically infected patient, but the site is different in different patients. Direct evidence such as cloning and sequence analysis of cellular DNA fragments containing HBV DNA sequences to prove integration is not yet available. Other viruses such as retroviruses which readily integrate in cellular DNA appear to do so at many and possibly random sites in the cellular DNA and specific integration sites are not detected in infected tissue DNA by the experimental strategy just described for HBV unless the cells are of clonal origin (e.g. as are cells in most viral-induced tumors) (reviewed in Varmus and Swanstrom, 1982).

Other studies (Koshy et al., 1981; Miller et al., 1985) have obtained evidence for random integration by detecting subgenomic sized-DNA fragments with HBV DNA sequences after digestion of infected liver DNA with a restriction enzyme which cleaves HBV DNA at more than one site, and no fragments of any specific size after HindIII digestion can be detected by Southern blot analysis. Southern blot analysis of livers of many infected humans, woodchucks, ground squirrels, and ducks has failed to demonstrate high molecular weight viral DNA forms suggesting integration. This would suggest that hepadna viral integration probably does not occur in all infected cells and under some conditions of infection integration may be present in few, if any, cells. High molecular weight forms of viral DNA have been found in productively infected woodchuck (Ogston et al., 1982) and ground squirrel (Marion et al., 1982) liver when the much more sensitive method of DNA cloning in the lambdoid vector Charon 30 was used. The high molecular weight viral DNA forms that have been isolated in this way and analyzed to date appear to be oligomeric or longer than unit length viral DNA with deletions and rearrangements. In the woodchuck system, at least, these oligomeric forms have been found integrated into host DNA sequences. The finding of these forms in infected ground squirrel and woodchuck liver raises a question whether they also exist in HBV infected human liver and whether the high molecular weight DNA bands containing viral sequences detected in Southern blots of HBV infected liver represent such oligomeric forms alone rather than integrated viral DNA. Proof of the nature of the high molecular weight fragments in infected human liver will require a detailed analysis that is most feasible only after the fragments are cloned. Similar non-integrated oligomeric forms of viral DNA have recently been found in lymphocytes from the blood of chronic HBV carrier humans (86) and chimpanzees, and WHV infected woodchucks (Korba et al., 1986).

Integrated HBV DNA is usually present in much smaller amounts (e.g. < 1 copy per cell) than the free viral DNA forms (e.g. > 500 copies per cell) in liver in which HBV is replicating (Kam et al., 1982; Miller et al., 1985). Although the role of integrated viral DNA in virus replication has not been established, HBsAg has been shown to be expressed in cells in which the only detectable viral DNA is apparently integrated in cellular DNA (Kam et al., 1982; Miller et al., 1985; Marion et al., 1980b). Expression of no other viral gene in an integrated form has been observed.

8. RELATIONSHIP OF HEPADNAVIRUSES AND RETROVIRUSES

Common features such as the number and arrangement of genes, the nucleotide sequence homology in several essential genome regions, the unusual reverse transcription mechanism for viral genome replication, and the use of an oligoribonucleotide fragment of the RNA template as primer for second DNA strand synthesis indicate that the hepadnavirus and retrovirus families are phylogenetically related. The murine type C retroviruses and endogenous retroviral sequences found in human and other primate genomes share the greatest homology with mammalian hepadnaviruses (Miller and Robinson, 1986) suggesting that they have diverged most recently from a common ancestor. The high degree of homology of hepadnaviruses with human endogenous retroviral elements is in contrast to the much lower homology with the known exogenous human T-lymphotropic retroviruses (HTLV-I, HTLV-II, and HTLV-III/LAV) indicating a much more distant relationship.

The significant homology between regions of the HBV genome and endogenous retroviral elements in the human genome raises the question of whether these sequences can be detected by hybridization of human DNA with an HBV DNA probe under some conditions and thus account for some of the sequences detected with HBV DNA probes by Southern blot hybridization in DNA from liver and other tissues of patients with no serologic evidence of current or past HBV infection (Brechot et al., 1981a, 1982, 1984; Laure et al., 1985).

Although homology in certain genes and control (e.g. U-5) sequences, and features of replication sited above have been retained by members of the two virus families (and by cauliflower mosaic virus, Pfeiffer and Holu, 1983) during their evolution (and these features are not found in other virus families) viruses of the two families have diverged in certain important ways. Notable among the differences are that: (1) Hepadnaviruses package viral DNA in virions and retroviruses package RNA. (2) Hepadnaviruses use a protein and retroviruses t-RNA as primer for minus (first) strand DNA synthesis. (3) During replication retroviruses form in their pregenomic DNA a terminally repeated control sequence (LTR) derived from both ends of genomic RNA and hepadnaviruses do not. The RNA transcript that serves as template for hepadnavirus DNA synthesis is longer than the viral DNA genome; and retroviruses have the opposite length relationship, that is the intracellular DNA pregenome exceeds the length of virion RNA which serves as template for viral DNA synthesis. In both cases the longer genome form owes its greater length to terminally repeated sequences. The long terminal repeat (LTR) of the retroviral DNA pregenome is derived from specific sequences at both the 3' (U-3) and 5' (U-5) ends of the viral RNA template (reviewed in Seeger et al., 1986). The order of sequences in the LTR is generated because synthesis of the first DNA strand is initiated near the 5' end of the RNA template and after copying the U-5 sequence the template is switched to the 3' end of the RNA and DNA synthesis continues by copying the U-3 sequence of the RNA. The resulting sequences at the 5' end of the first DNA strand (U-5, R and U-3) are then duplicated at the 3' end of the same strand by displacement synthesis during completion of that strand. Such an LTR is not formed during synthesis of hepadnavirus DNA because synthesis of the first DNA strand is initiated at the 3' end of the RNA template (Fig. 2) so that sequences from both ends of the RNA template are not copied during formation of the 5' end of that DNA strand. In addition, there is no evidence of displacement synthesis in hepadnavirus replication which is the retroviral mechanism for duplicating the 5' sequence at the 3' end of the first DNA strand. (4) Retroviruses have the capacity for precise and orderly integration into cell DNA that occurs in all infected cells and is an integral part of virus replication, and hepadnaviruses do not (possibly related to absence of the pol gene sequences specifying the retroviral endonuclease function and/or absence of the inverted repeat at the 3' end of retroviral U-5 sequence, both of which are essential for retrovirus integration). Hepadnavirus integrations appear to be sporadic and may increase with longer duration of infection, and possibly with liver injury and regeneration, and rearrangements and deletions (at least in HCC) appear to be the rule in the integrated viral genomes. Thus hepadnavirus integrations are more similar to those of DNA viruses such as SV40 than to retroviruses. Although integration is a regular, ordered event in retrovirus infection and the integrated proviral DNA is the template for most viral transcription, integration is not essential for retroviral gene expression and virus replication (Panganiban and Temin, 1983) as it is not for hepadnaviruses. A common behavior of viruses of both groups is expression of just the envelope gene *in vivo* by cells containing only integrated viral DNA (Kam et al., 1982; Miller et al., 1985; Robinson et al., 1981). Hepadnavirus genomes are approximately one-third the length of typical retroviral genomes. However, unusual retrovirus recombinants (O'Rear and Temin, 1982) and endogenous retrovirus-like elements in some human cells (Repaske et al., 1983) have been described with sizes and gene arrangements that are quite similar to those of hepadnaviruses. These may have arisen through evolutionary mechanisms including selective deletions that have been utilized in the evolution of hepadnaviruses.

The genes specifying the major virion envelope proteins of hepadnaviruses (S gene) and retroviruses (env gene) do not share sequence homology and have different relative

HEPADNAVIRUS GENOME REPLICATION

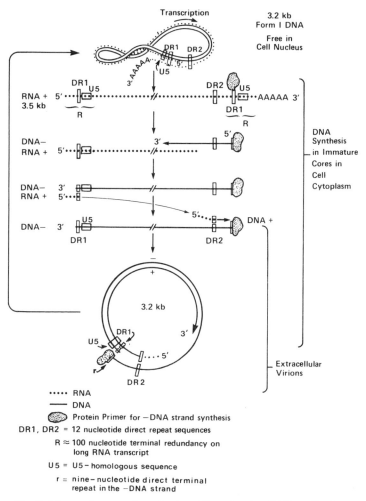

Fig. 2. Scheme of proposed mechanism of hepadnavirus DNA replication.

locations in the respective genomes indicating a significant evolutionary divergence of this viral function. The env gene of typical large retroviruses is a non-overlapping open reading frame located between the pol gene and the 3' LTR and it appears to be frequently inactivated by deletion (Coffin, 1982; Paulson et al., 1985) or insertion of new (e.g. oncogene) DNA within the env gene (Coffin, 1982). The resulting defective retroviruses often survive by phenotypic mixing with coinfecting 'helper viruses' which provide an env protein. In contrast, the S gene of hepadnaviruses consists of an open reading frame entirely within the large open reading frame thought to specify the viral polymerase protein (and translated in a different reading frame).

A mechanism by which new overlapping genes can arise at any time as alternate open reading frames in DNA with oligomeric repeats without termination codons has been described and analysis of the DNA encoding the HBV P and S genes suggests that these overlapping genes have arisen by that mechanism (Ohno, 1984). This is a mechanism by which a new env gene could arise within another open reading frame of a retrovirus whose non-overlapping env gene had been deleted or interrupted, for example by insertion of an oncogene sequence. In fact, overlapping open reading frames have been found in small or truncated endogenous retrovirus elements in human cells which are transcribed as episomal and integrated forms (Paulson et al., 1985). Thus there are evolutionary mechanisms by which the differences in the envelope genes of hepadnaviruses and retroviruses could have arisen as the viruses evolved from a common ancestor.

Two interesting but unanswered questions about these viruses are whether the hepadnavirus X-gene may have a function similar to any of the retroviral genes of probable cellular origin (e.g. pX-gene of HTLV) that occur in a similar genomic location in some retroviruses, and whether members of the two virus families may share common oncogenic mechanisms.

9. THE ASSOCIATION OF HEPADNAVIRUS INFECTION AND HEPATOCELLULAR CARCINOMA (HCC)

HCC in man has a worldwide distribution and numerically is one of the major cancers in the world today. Although HCC is rare in many parts of the world, it occurs commonly in sub-Saharan Africa, eastern Asia, Japan, Oceania, Greece and Italy. In certain areas of Asia and Africa, it is probably the most common cancer. Geographic areas with the highest incidence of HCC are also areas where persistent HBV infections occur at the highest known frequencies. Within the limits of the data available, there appears to be a good correlation between geographic distribution of HCC and active HBV infection with the highest frequency of both being sub-Saharan Africa and eastern Asia (Szmuness, 1978). In addition, in many retrospective studies hepatitis B surface antigen (HBsAg), a marker of active HBV infection, has been found four to six times more frequently in serum of patients with HCC than in tumor negative controls in both high-HCC-incidence and low-HCC-incidence geographic areas (Szmuness, 1978). An ongoing prospective study of 22,707 male government workers in Taiwan, 15% of which were HBsAg positive, revealed the incidence of HCC to be more than 300 times higher in HBsAg positive than HBsAg negative individuals and provided evidence that HBV infection precedes the development of HCC (Beasley et al., 1978). 3% of HBsAg patients 50 years or older developed HCC per year, and 43% of all deaths in HBsAg carriers 40 years or older were due to HCC.

The high incidence of persistent HBV infection in mothers of HCC patients, in contrast to that in fathers (Larouze et al., 1976), suggests that transmission from mothers to newborn or infant children may be a frequent mode and the time of HBV infection in HCC patients. The finding of low HBsAg titers, together with the rare occurrence of HB core antigen (HBcAg) and HB e antigen (HBeAg) in most patients (Szmuness, 1978; Nishioka et al., 1973), also suggests that the persistent infections in HCC patients are of long duration. If HBV infection does occur frequently at very early ages in HCC patients, the age distribution of patients when the tumors were clinically recognized in high-incidence areas (Steiner, 1960) would suggest that tumors appear after a mean duration of approximately 35–40 years of HBV infection. Very few cases of HCC occur in children (Steiner, 1960). Between 60% and 90% of HCC patients have coexisting cirrhosis (Beasley et al., 1978; Steiner, 1960; Trichopoulos et al., 1975; Peters, 1976), and in the prospective study by Beasley et al. (1978), more than 12% per year of HBsAg carriers with cirrhosis developed HCC, an incidence more than ten times that in carriers without evidence of cirrhosis. These findings suggest that cirrhosis in association with persistent HBV infection may predispose to HCC, although clearly the presence of cirrhosis is not an absolute requirement. Epidemiological data refining the association of HCC and persistent HBV infection have been reviewed in detail (Szmuness, 1978) and these data represent strong evidence for an important role of HBV in HCC formation in man.

Persistent WHV infection in wild-caught woodchucks is common in regions of the Eastern United States (Tyler et al., 1981) and HCC appears to occur more frequently in such infected captive animals than in HBV infected humans. In two colonies of woodchucks in this country approximately one-third of animals persistently infected with WHV have been observed to develop HCC per year (Summers et al., 1978b; Tyler et al., 1981). All such animals have histological findings of moderately severe active hepatitis and (unlike man) never cirrhosis, and they have moderately high levels of DNA containing virions and WHsAg in serum. No hepatomas have been reported in uninfected animals. A histological study by Snyder of the livers of 139 wild-caught woodchucks that died in captivity showed that hepatitis was present in 75% of the animals, although only 35% had markers of active WHV infection (Snyder and Summers, 1980), suggesting that factors other than WHV infection may be responsible for at least some mild hepatitis in these animals. Severe hepatitis was observed in 28% of the animals. Although the hepatitis in woodchucks may be severe and progressive, it does not appear to lead to cirrhosis.

GSHV infected ground squirrels have a different disease response. While the prevalence of persistent infection is very high in some endemic areas (more than 50%) and the titer of virus (measured by Virion DNA polymerase assay) is unusually high (Marion et al., 1980a), little or no hepatitis occurs in the infected animals (Marion et al., 1983b). None of 25 infected captive animals followed for more than four years in this lab have developed significant hepatitis or cirrhosis. Only a mild form of hepatitis has been detected in some infected as well as uninfected animals (Marion et al., 1983b). Some animals with high titers of virus in their sera show no histological abnormalities in their livers. In two colonies of captive ground squirrels used for GSHV research, no HCC were seen in chronically infected animals for several years of prospective study. In one colony, HCC have recently begun to appear exclusively in animals that have been continuously infected or have had evidence of past infection for 4.4 years or longer. In this colony, HCC has developed in four of eleven such GSHsAg positive animals; two of three GSHsAg negative, anti-GSHc positive animals, and none of eight animals with no markers of GSHV infection (Marion and Robinson, unpublished results). GSHV DNA has been found in some of those tumors. Thus HCC appears to be associated with GSHV infection of ground squirrels but there is clearly a much longer incubation period than in WHV infected woodchucks and whether the lifetime incidence will turn out to be the same or different remains to be determined. The incidence would appear to be equal to or higher, however, than the incidence of HCC in chronic HBV carrier humans.

DHBV has been found in domestic ducks in parts of China and in up to 10% of Pekin ducks in many commercial flocks in this country. Histologic studies of livers of ducks from Chi-tung county in China (Marion et al., 1984; Omata et al., 1983) revealed some degree of hepatitis in most and no correlation between severity of hepatitis and presence of virus in serum or viral DNA in liver suggesting that nonviral factors are involved in at least some hepatitis in this duck population. Inoculation of DHBV into chorioallantoic veins of embryonated duck eggs led to moderate to severe hepatitis in 6-week-old hatchlings (Marion et al., 1984) indicating the DHBV infection can cause hepatitis in young ducks. Clearly HCC occurs in domestic ducks in China but the limited data now available do not suggest that active DHBV infection is as closely associated with HCC in ducks as is hepadnavirus infection in humans, woodchucks or ground squirrels with HCC. Of two HCC examined in one study (Marion et al., 1984), only one contained viral DNA at the lower limit of detection. No HCC have been observed in DHBV infected or uninfected Pekin ducks from California followed as long as 2 years in this lab, and none has been reported in the commercial flocks in which DHBV has been studied in this country. Further study will be needed better to define the association of DHBV infection and HCC in ducks in China and to rule out a role for aflatoxins and other etiologic factors in the formation of these tumors.

In summary, persistent infection is common for all known members of the hepadnavirus family, although the range of virus titers in serum varies among the four virus-host systems. The severity of liver injury (hepatitis) during acute and chronic infections vary from minimal disease in ground squirrels to occasional severe disease in woodchucks and man. Cirrhosis, a significant sequelum of chronic HBV infection in man, has been observed in the ducks (Omata et al., 1983), but it has not yet been correlated with virus infection. No cirrhosis has been found in ground squirrels or woodchucks. The differences in hepatitis and HCC formation associated with the different hepadnavirus infections of their natural hosts are striking. Whether these differences are due to differences in the pathogenicity of these closely related viruses, genetic differences in the hosts, or to environmental factors remains to be determined. The hepatitis associated with hepadnavirus infections in the three subprimate species occurs on a background of similar disease apparently caused by other factors, suggesting that a role for non-viral factors in association with the virus could be important in the pathogenesis of hepadnavirus associated liver disease observed in these animals.

10. THE STATE OF HEPADNAVIRUSES IN HCC

Several studies have evaluated the state of virus in HCC tissue from humans, woodchucks, ground squirrels and ducks, and in tissue culture lines from human HCC. Immunofluorescent and immunoperoxidase staining of tumor tissue have demonstrated that in patients with HBsAg in the blood and in whom nontumorous liver cells are positive for HBsAg and/or HBcAg, tumor cells appear most often to be negative, although some studies have reported small numbers of HBsAg-positive cells in tumors (reviewed in Kew, 1978). HBcAg has been detected even more rarely. Thus few tumor cells appear to express either viral gene product in amounts that can be detected by immunofluorescent staining. Among cell lines isolated from human HCC, some clearly express HBsAg in cell culture and others do not (Chang et al., 1985).

The earliest studies of viral DNA in tumors (Lutwick and Robinson, 1977; Summers et al., 1978a) utilized virion DNA radiolabeled by the endogenous virion DNA polymerase reaction as a probe for hybridization in solution with DNA extracted from HCC tissue. Viral DNA sequences were found in some but not all tumors with a lower limit of detection of approximately one viral genome copy per cell. Alkaline sucrose gradient centrifugation of the tumor DNA suggested that some viral DNA sequences were integrated in high molecular weight cell DNA (Lutwick and Robinson, 1977).

Southern blot analysis has revealed the presence of replicating viral DNA forms in some HCC and evidence for integrated viral DNA in some but not all human HCC (Brechot et al., 1980; Koshy et al., 1981; Shafritz et al., 1981; Shafritz and Kew, 1981; Brechot et al., 1981b, 1982; Shafritz, 1982; Chen et al., 1982; Hino et al., 1984; Miller et al., 1985), in woodchuck HCC (Summers et al., 1980; Ogston et al., 1981; Mitamura et al., 1982), ground squirrel HCC (Marion and Robinson, unpublished results), and in human cell lines isolated from HCC (Marion et al., 1980b; Chakraborty et al., 1980; Brechot et al., 1980; Edman et al., 1980; Twist et al., 1981; Miller and Robinson, 1983). Restriction endonuclease HindIII (an enzyme which cleaves cellular DNA at specific sites but not most HBV DNAs so far examined) digestion of tumor DNA and Southern blot analysis has revealed DNA fragments (usually one to four) apparently containing viral DNA sequences that are larger than unit length (3200 bp) viral DNA similar to those described above for nontumorous infected liver. This suggests that viral DNA may be integrated at a few specific cellular DNA sites in such HCC tissue but always at different sites in different tumors. The presence of unique integration sites and the rough correspondence of the viral genome copy number per cell and the number of integration sites in tumors without replicating viral DNA (Miller et al., 1985) suggests that the cells of these tumors are of clonal origin and when multiple sites of integration exist, they probably arise through DNA rearrangements of the original integration.

HBV DNA sequences appear to be integrated in at least eight specific cellular DNA sites in HCC cell line PLC/PRF/5 in culture and these have been shown to be extensively methylated, unlike HBV DNA forms in virions and non-tumorous liver in which there was no detectable methylation (Miller and Robinson, 1983). The more extensive methylation of the coding sequences for the core polypeptide than sequences coding for the HBsAg polypeptide correlated with the expression of HBsAg but not HBcAg or other viral gene products by this cell line. This finding raises the possibility that methylation of viral DNA may be involved in regulation of viral gene expression in HCC.

Cloning and sequence analysis of integrated viral DNA with flanking cellular DNA sequences has proven that viral sequences are integrated in host DNA of HCC. In almost all cases of woodchuck (Ogston et al., 1982) and human tumors (Fun et al., 1984, 1985; Matsubara et al., 1985) and tumor cell lines (Standring et al., 1983; Dejean et al., 1983) studied to date, the viral DNA contains extensive deletions and rearrangements which are different for each integrated viral sequence. The site in viral DNA that joins cellular DNA among thirty or more cloned HBV integrations that have been analyzed is near the end of the X-gene in approximately 50% and at other (possibly random) viral genome sites in the others (Matsubara et al., 1985). The apparently more frequent integration near the end of the X-gene and between DR-1 and DR-2 (i.e. between the 5' ends of the two virion DNA strands) suggests the cohesive ends of virion DNA or a nucleotide sequence in that region plays a role in many hepadnavirus integrations. There is no evidence to date that the clone flanking cellular sequences correspond to known proto-oncogene sequences (Fung et al., 1984, 1985) or that transcription of known proto-oncogene sequences is altered in HCC cells. At this time there is no demonstrated difference between the state of integrated viral DNA in HCC and in infected nontumorous liver.

A finding to be emphasized is that some tumors in woodchucks (Marion et al., 1982) and humans (Barker et al., 1973; Gudat et al., 1975; Kam et al., 1982; Varmus and Swanstrom, 1982; Popper et al., 1981; Miller et al., 1985; Hino et al., 1984) contain no detectable viral DNA with sensitivity down to 0.01 genome copy per haploid cell genome equivalent in one study (Miller et al., 1985). A significant number of patients without detectable HBV DNA in their tumor have serum HBsAg indicating their liver is infected with HBV. The absence of detectable viral DNA in some HCC suggests either that virus was not involved in development of those tumors or that the continuing presence of viral DNA sequences are not necessary for growth of those tumors, although the presence of a very small viral DNA sequence (i.e., <100 bp per cell) has not been excluded. Thus although there is a strong association between long standing hepadnavirus infection and HCC formation in humans and woodchucks, and viral DNA can be shown to be integrated in the cellular DNA of many tumors, exactly how these viruses are involved in HCC formation is not clear.

Viral integration is a critical event in neoplastic transformation by some viruses. Retroviral LTR sequences when integrated in appropriate cellular DNA sites appear to influence directly expression of cellular oncogenes resulting in cell transformation (called the 'promoter-insertion' mechanism) (Hayward et al., 1981). The findings to date that HBV and WHV DNA are integrated at different cellular DNA sites in different tumors, that cellular DNA flanking viral inserts have not been found to contain cellular oncogene sequences, that the sites in viral DNA which join cellular DNA are frequently different in different integrations and that a significant number of tumors contain no detectable HBV DNA, appear to be inconsistent with such a promoter-insertion mechanism. Integration at cellular DNA sites adjacent to new and as yet undescribed cellular oncogenes would not be detected by hybridization with known oncogene probes. Failure to detect viral sequences by the hybridization employed does not exclude the presence of a small viral sequence (i.e. <100 bp). It is not known whether all HCC arise by the same mechanism and it cannot be excluded that those without detectable viral DNA (or even those containing viral DNA) arose through a mechanism unrelated to a direct transforming action by HBV.

Infection of cells with retroviruses (Varmus and Swanstrom, 1982) and other viruses (Galloway and McDougall, 1983) can result in point mutations, translocations, deletions or other rearrangements of cellular DNA. It has been postulated (Varmus and Swanstrom, 1982) that such virus effects on cellular DNA could alter the structure or expression of cellular oncogenes resulting in cell transformation without an apparent lasting viral DNA insert (a 'hit-and-run' mechanism) or with residual viral sequences. Rearrangements of cellular DNA in conjunction with HBV integrations have been described. Inverted duplication of HBV and cellular sequences (Mizusawa et al., 1985), and amplification and transposition of integrated viral plus flanking cellular DNA (Koch et al., 1984; Zeimer et al., 1985) have been reported in HCC cell lines. Whether these cellular DNA rearrangements were present in the tumors from which the cell lines were derived or occurred during passage of cells in culture is unclear. More recently Rogle et al. have described a large deletion of cellular DNA at chromosome position $11p^{13}$ in association with an HBV integration (Rogler et al., 1985), and a cellular DNA translocation between chromosome 17 and 18 accompanying a deletion of 1.3 Kb of chromosome 18 DNA at the translocation site

(Hino et al., 1987) in different HCC containing single HBV integrations. Thus HBV integrations can lead to significant rearrangements in cellular DNA and such rearrangements could play a role in HCC formation even when a viral insert is not retained. Such cellular DNA changes resulting from hepadnavirus integrations would appear to be a promising lead for further investigation of the role of HBV in HCC formation.

Certain retroviruses such as Rous sarcoma viruses (RSV) and murine sarcoma viruses (MSV) (Bishop and Varmus, 1982), and other viruses such as SV40 (Topp et al., 1980) and adenoviruses (Flint, 1980) contain genes whose expression leads to rapid transformation of cells in culture and to tumor induction in vivo. The continued presence and expression of these genes is essential to maintain the neoplastic phenotype of cells transformed by this mechanism. Cells in culture can usually be transformed by transfection with the DNAs of such viruses.

Recent experiments in this laboratory (Klote and Robinson, unpublished results) have shown that transfection of N1H3T3 cells with an HBV DNA head-to-tail dimer within plasmid vector pBR322 and with a BamHI fragment of HBV DNA containing DR-1, DR-2, the distal two-thirds of the X-gene, the preC and C gene, the proximal third of the P gene and the $preS_1$ region of the S gene results in approximately a tenfold higher rate of cell transformation (i.e. transformants able to grow as colonies in soft agar and form tumors in nude mice) than results from transfection with pBR322 or salmon sperm DNA alone, although a 100-fold lower rate than with SV40 DNA. Analysis of DNA extracted from clones resulting from HBV dimer DNA transfection revealed HBV DNA sequences. The mechanism and the significance of this apparent transformation by DNAs containing HBV sequences is not clear. The low rate of transformation and the failure to find HBV DNA in the transformants suggests that this phenomenon is not a response to a viral oncogene but more likely analogous to the transformation related to specific sequences in herpes simplex and cytomegaloviruses (Galloway et al., 1984). At this time there is no evidence that expression of any HBV or WHV gene is directly involved in initiating or maintaining the transformed state as is the case with viruses such as RSV.

Further investigation is needed to test more completely whether HBV and WHX may induce HCC by any of the mechanisms known for other viruses, or whether they may have some other direct transforming effect. The evidence to date does not exclude the possibility that HBV and WHV do not have a direct transforming effect on cells but trigger HCC formation in a more indirect way such as through chronic liver injury and regeneration. Such a possibility is suggested by the evidence that liver injury leading to cirrhosis over many years greatly increases the risk of HCC in patients with chronic HBV infection, compared with carriers having little or no liver damage (Szmuness, 1978; Beasley et al., 1978), and although WHV and GSHV are very similar viruses and their natural hosts are related rodents, chronic WHV infection of woodchucks is associated with moderately severe hepatitis with evidence of liver regeneration and a high incidence of HCC (Summers et al., 1978b; Popper et al., 1981) and, in contrast, chronic GSHV infection of Beechey ground squirrels is associated with no apparent liver injury and HCC formation is apparently delayed in comparison with the occurrence of HCC in WHV infected woodchucks (Marion et al., 1980a, 1983b). Comparison of these animal models should provide an opportunity to analyze the viral and/or host factors that are important in HCC formation. It has been speculated (Popper, 1978) that some other environmental factor such as a chemical carcinogen, in addition to virus, may be necessary for HCC formation, but there is no direct or experimental evidence for such a mechanism at this time. However, recent epidemiologic evidence shows that HCC appears to occur at a much higher frequency in HBsAg carriers in Chi Dong Island near Shanghai than in Shanghai, China, although the prevalence of HBsAg carriers is approximately the same in the two geographic areas.

Acknowledgement—Part of the work described here was supported by NIH grant AI 13526.

REFERENCES

ALBIN, C. and ROBINSON, W. S. (1980) Protein kinase activity in hepatitis B virus. *J. Virol.* **34**: 297–302.
ALMEIDA, J. D., RUBENSTEIN, D. and STOTT, E. J. (1971) New antigen antibody system in Australia antigen positive hepatitis. *Lancet* **II**: 1225–1227
ALMEIDA, J. D., WATTERSON, A. P., TROWEL, J. M. and NEALE, G. (1970) The finding of virus-like particles in two Australia-antigen-positive human livers. *Microbios.* **2**: 145–153.
AURELIAN, L. (1983) Herpes viruses and cervical cancer. In: *Viruses Associated with Human Cancer*, pp. 79–123, Phillips, L. A. (ed.), Marcel Dekker, Inc., New York.
BARKER, L. F., CHISARI, F., MCGRATH, P. P., DALGARD, D. W., KIRSCHSTEIN, R. L., ALMEIDA, J. D., EDGINGTON, T. S., SHARP, D. C. and PETERSON, W. R. (1973) Transmission of type B viral hepatitis to chimpanzees. *J. Inf. Dis.* **127**: 648–662.
BEASLEY, R. P., LIN, C. C., HWANG, L. Y., and CHIEN, C. S. (1978) Hepatocellular carcinoma and hepatitis B virus: a prospective study of 22,707 men in Taiwan. *Lancet* **2**: 1129–1133.
BICHKO, V., PUSHKO, P., DREILINA, D., PUMPEN, P. and GREN, E. (1985) Subtype ayw variant of hepatitis B virus. DNA primary structure analysis. *FEBS Letters* **185**: 208–212.
BISHOP, J. M. and VARMUS, H. (1982) Functions and origins of retroviral transforming genes. In: *RNA Tumor Viruses* pp. 999–1108, Weiss, R., N. Teich, H. Marmus and J. Coffin (eds), Cold Spring Harbor Press.
BONNER, T .I., BIRKENMEIER, E. H., GONDA, M .A., MARK, G. E., SEARFOSS, G. H., and TODARO, G. J. (1982a) Molecular cloning of a family of retroviral sequences found in chimpanzee but not human DNA. *J. Virol.* **43**: 914–924.
BONNER, T. I., O'CONNELL, C. and COHEN, M. (1982b) Cloned endogenous retroviral sequences from human DNA. *Proc. Natl Acad. Sci. USA* **79**: 4709–4713.
BRECHOT, C., HADCHOUEL, M., SCOTTO, J., FONCK, M., POTET, F., VYAS, G N. and TIOLLAIS, P. (1981a) State of hepatitis B virus DNA in hepatocytes of patients with hepatitis B surface antigen-positive and -negative liver disease. *Proc. Natl Acad. Sci.* **78**: 3906–3910.
BRECHOT, C., LUGASSY, A. D., DEJEAN, A., PONTISSO, P., THIERS, V., BERTHELOT, P. and TIOLLAIS, P. (1984) Hepatitis B virus in infected human tissues. In: *Viral Hepatitis and Liver Disease*, pp. 395–409, Vyas, G. N., Dienstag, J. L. and Hoofnagle, J. H. (eds), Grune and Stratton, New York.
BRECHOT, C. POURCEL, C., LOUISE, A., RAIN, B. and TIOLLAIS, P. (1980) Presence of integrated hepatitis B virus DNA sequences in cellular DNA of human hepatocellular carcinoma. *Nature (London)* **286**: 533–535.
BRECHOT, C., POURCEL, C., LOUISE, A., RAIN, B. and TIOLLAIS, P. (1981b) Detection of hepatitis B virus DNA sequences in human hepatocelluar carcinoma in an integrated form. *Prog. Med. Virol.* **27**: 99–102.
BRECHOT, C., POURCEL, C., HADCHOUEL, M., DEJEAN, A., LOUISE, A., SCOTTO, J. and TIOLLAIS, P. (1982) State of hepatitis B virus DNA in liver diseases. *Hepatology* **2**: 27S–34S.
CAMAMIA, F., DEBAC, C. and RICCI, G. (1972) Virus-like particles within hepatocytes of Australia antigen carriers. *Am. J. Dis. Child.* **123**: 309.
CATTANEO, R., WILL, H. and SCHALLER, H. (1984) Hepatitis B virus transcription in the infected liver. *EMBO J.* **3**: 2192–2196.
CATTANEO, R., WILL, H., HERNANDEZ, N. and SCHALLER, H. (1983) Signals regulating hepatitis B surface antigen transcription. *Nature* **305**: 336–338.
CHAKRABORTY, P., RUIZ-OPAZO, N., SHOUVAL, D. and SHAFRITZ, D. (1980) Identification of integrated hepatitis B virus DNA and expression of viral DNA in an HBsAg producing human hepatocellular carcinoma cell line. *Nature* **286**: 531–533.
CHANG, C., CHOU, C. K., TING, L. P., SU, C. S., HAN, S. H. and HU, C. P. (1985) The cellular and antigenic properties of human hepatoma cell lines. In: *Molecular Biology of Hepatitis B Viruses*, 84, Cold Spring Harbor Press.
CHEN, D. S., HOYER, B. H., NELSON, J., PURCELL, R. H. and GERIN, J. L. (1982) Detection and properties of hepatitis B virus DNA in liver tissue from patients with hepatocellular carcinoma. *Hepatology* **2**: 425–465.
COFFIN, J. C. In: Structure of the retroviral genome. (1982) *RNA Tumor Viruses* pp. 261–368. Weiss, R., Teich, N., Varmus, H. and Coffin, J. (eds), Cold Spring Harbor Press.
COHEN, M., REIN, A., STEPHENS, R. M., O'CONNELL, C. O., GILDEN, R. V., SHURE, M., NICOLSON, M. O., MCALLISTER, R. M. and DAVIDSON, N. (1981) Baboon endogenous virus genome: molecular cloning and structural characterization of nondefective viral genomes from DNA of a baboon cell strain. *Pro. Natl Acad. Sci. USA* **78**: 5207–5211.
COUROUCE, A. M., HOLLAND, P. V., MULLER, J. Y. and SOULIER, J. P. (1976) HBs antigen subtypes. Proceedings of the International Workshop on HBs antigen subtypes. *Bibliotheca Haematol.* **42**: 1–158.
COUROUCE-PAUTY, A. M. and SOULIER, J. P. (1974) Further data on HBs antigen subtypes-geographical distribution. *Vox Sang* **27**: 533–549.
DANE, D. S., CAMERON, C. H. and BRIGGS, M. (1970) Virus-like particles in serum of patients with Australia antigen associated hepatitis. *Lancet* **II**, 695–698.
DEJEAN, A., BRECHOT, C., TIOLLAIS, P. and WAIN-HOBSON, S. (1983) Characterization of integrated hepatitis B viral DNA cloned from a human hepatoma and the hepatoma-derived cell line PLC/PRF/5. *Proc. Natl Acad. Sci. USA* **80**: 2505–2509.
EDMAN, J., GRAY, P., VALENZUELA, P., RALL, L. B. and RUTTER, W. J. (1980) Integration pattern of hepatitis B virus DNA sequences in human hepatoma cell lines. *Nature (London)* **286**: 535–538.
ENDERS, G. H., GANEM, D. and VARMUS, H. (1985) Mapping the major transcripts of ground squirrel hepatitis virus: the presumptive template for reverse transcriptase is terminally redundant. *Cell* **42**: 297–308.
FEITELSON, M. A., MARION, P. L. and ROBINSON, W. S. (1981) Antigenic and structural relatonships of the surface antigens of hepatitis B virus, ground squirrel hepatitis virus and woodchuck hepatitis virus. *J. Virol.* **39**: 447–454.
FEITELSON, M. A., MARION, P. L. and ROBINSON, W. S. (1982a) The core particles of HBV and GSHV. II.

Characterization of the protein kinase reaction associated with ground squirrel hepatitis virus and hepatitis B virus. *J. Virol.* **43**: 741–748.

FEITELSON, M. A., MARION, P. L. and ROBINSON, W. S. (1982b) The core particles of HBV and GSHV. I. Relationship between HBcAg and GSHcAg associated polypeptides by SDS-PAGE and tryptic peptide mapping. *J. Virol.* **43**: 687–696.

FEITELSON, M. A., MARION, P. L. and ROBINSON, W. S. (1983) The nature of polypeptides larger in size than the major surface antigen components of hepatitis B and like viruses in ground squirrels, woodchucks and ducks. *Virology* **130**: 76–90.

FLINT, S. J. (1980) Transformation by adenoviruses. In: *DNA Tumor Viruses*, pp. 547–576, Tooze, J. (ed.), New York, Cold Spring Harbor Press.

FUJIYAMA, A., MIYANOHARA, A., NOZAKI, C., YONEYAMA, T., OHTOMO, N. and MATSUBARA, K. (1983) Cloning and structural analyses of hepatitis B virus DNAs, subtype adr. *Nuclei Acids Res.* **11**: 4601–4610.

FUNG, G.-K. T., LAI, C. L., TODD, D., GANEM, D. and VARMUS, H. E. (1984) An amplified domain of cellular DNA containing a subgenomic insert of hepatitis B virus DNA in a human hepatoma. In: *The 1984 International Symposium on Viral Hepatitis*, Vyas, Alter and Hoofnagle (eds), (in press).

FUNG, Y. K., LAI, C. L., LOK, A., TODD, D. and VARMUS, H. E. (1985) Analysis of HBV-associated human hepatocellular carcinoma for oncogene expression and structure rearrangement. In: *Molecular Biology of Hepatitis B Viruses*, 80, Cold Spring Harbor Press.

GALIBERT, F., CHEN, T. N. AND MANDART, E. (1982) Nucleotide sequence of a cloned woodchuck hepatitis virus genome: comparison with the hepatitis B virus sequence. *J. Virol.* **41**: 51–65.

GALIBERT, F., MANDART, E., FITOUSSI, F., TIOLLAIS, P. and CHARNAY, P. (1979) Nucleotide sequence of hepatitis B virus genome (subtype ayw) cloned in *E. coli*. *Nature (London)* **281**: 646–650

GALLOWAY, D. A. and MCDOUGALL, J. K. (1983) The oncogenic potential of herpes simplex viruses: evidence for a 'hit and run' mechanism. *Nature* **302**: 21–24.

GAN, R-B., CHU, M-J., SHEN, L-P. and LI, Z-P. (1984) *Acta Biochimica et Biophysica Sinica* **16**: 316–319.

GANEM, D., GREENBAUM, L. and VARMUS, H.E. (1982) Virion DNA of ground squirrel hepatitis virus: structural analysis and molecular cloning. *J. Virol.* **44**: 374–383.

GERLICH, W.H. and ROBINSON, W. S. (1980) Hepatitis B virus contains protein attached to the 5' terminus of its complete DNA strand. *Cell* **21**: 801.

GERLICH, W. H., FEITELSON, M. A., MARION, P. L., ROBINSON, W. S. (1980) Structural relationships between the surface antigens of ground squirrel hepatitis virus and human hepatitis B virus. *J. Virol.* **336**: 787–795.

GOFF, S. P. (1984) The genetics of murine leukemia viruses. *Current Topics in Microbiology and Immunology* **112**: 45–71.

GUDAT, F., BIANCHI, O. and SONNABEND, W. (1975) Pattern of core and surface expression in liver tissue reflects state of specific immune response in hepatitis B. *Lab. Invest.* **32**: 1.

GUST, I. D., BURRELL, C. J., COULEPIS, A. G., ROBINSON, W. S. and ZUCKERMAN, A. J. (1986) Taxonomic classification of hepatitis B virus. *Intervirol.* in press.

HAYWARD, W., NEEL, B. G. and ASTRIN, S. M. (1981) Activation of a cellular *onc* gene by promoter insertion in ALV-induced lymphoid leukosis. *Nature* **290**: 475–480.

HINO, O., KITAGAWA, T., KOIKE, K., KOBAYASHI, M., HARA, M., MORI, W., NAKASHIMA, T., HATTORI, N. and SUGANO, H. (1984) Detection of hepatitis B virus DNA in hepatocellular carcinoma in Japan. *Hepatology* **4**: 90–95.

HINO, O. SHOWS, T. B. and ROGLER, C. E. (1986) Hepatitis B virus integration site in hepatocellular carcinoma at chromosome 17:18 translocation. *Proc. Natl Acad. Sci. USA* in press.

HOWLEY, P. M. (1983) Papovaviruses: search for evidence of possible association with human cancer. In: *Viruses Associated with Human Cancer*, pp. 253–306, Phillips, L. A. (ed.), Marcel Dekker, Inc., New York.

HRUSKA, J. F., CLAYTON, D. A., RUBENSTEIN, J. L. R. and ROBINSON, W. S. (1977) Structure of hepatitis B Dane particle DNA before and after the Dane particle DNA polymerase reaction. *J. Virol.* **21**: 666–672.

HUANG, S. A. (1971) Hepatitis associated antigen hepatitis: an electronmicroscopic study of virus-like particles in liver cells. *Am. J. Pathol.* **64**: 783.

HUNTER, E., BROWN, A. S. and BENNETT, J. C. (1978) Amino terminal amino acid sequence of the major structural polypeptides of avian retroviruses: sequence homology between reticulo and othaliosis virus p30 and p30s of mammalian retroviruses. *Proc. Natl Acad. Sci. USA* **75**: 2708–2712.

KAM, W., RALL, L., SMUCKLER, E., SCHMID, R. and RUTTER, W. (1982) Hepatitis B viral DNA in liver and serum of asymptomatic carriers. *Proc. Natl Acad. Sci. USA* **79**: 7522–7526.

KAPLAN, P. M., GREENMAN, R. L., GERIN, J. L., PURCELL, R. H. and ROBINSON, W. S. (1973) DNA polymerase associated with human hepatitis B antigen. *J. Virol.* **12** : 995–1005.

KAY, A., MANDART, E., TREPO, C. and GALIBERT, G. (1985) The HBV HBX gene expressed in *E. coli* is recognised by sera from hepatitis patients. *EMBO J.* **4**: 1287–1292.

KEW, D. M. (1978) In: *Viral Hepatitis: A Contemporary Assessment of Etiology, Epidemiology, Pathogenesis and Prevention*, p. 439, Vyas, G. N., et al. (eds), Philadelphia, Franklin Institute Press.

KOBAYSHI, M. and KOIKE, K. (1984) Complete nucleotide sequence of hepatitis B virus DNA of subtype adr and its conserved gene organization. *Gene* **30**: 227–232.

KOCH, S, VONLOUGHOVEN, A. F., HOFSCHNEIDER, P. F. and KOSHY, R. (1984) Amplification and rearrangement in hepatoma cell DNA associated with integrated hepatitis B virus DNA. *EMBO J.* **3**: 2185–2189.

KORBA, B. D., WELLS, F., TENNANT, B. C., YOAKUM, G. H., PURCELL, R. H. and GERIN, J. L. (1986) Hepadnavirus infection of peripheral blood lymphocytes *in vivo*: woodchuck and chimpanzee models of viral hepatitis. *Journal of Virology* **58**: 1–8.

KOSHY, R., MAUPAS, P., MULLER, R. and HOFSCHNEIDER, P. H. (1981) Detection of hepatitis B virus-specific DNA in the genomes of human hepatocellular carcinoma and liver cirrhosis tissues. *J. Gen. Virol.* **57**: 95–102.

LANDERS, T. A., GREENBERG, H. B. and ROBINSON, W. S. (1977) Structure of hepatitis B Dane particle DNA and nature of the endogenous DNA polymerase reaction. *J. Virol.* **23**: 368–376.

LAROUZE, B., LONDON, W. T., SAIMOT, B. G., et al., (1976) *Lancet* **2**: 534–538.

LAURE, F., ZAGURY, D., SAIMOT, A. G., GALLO, R. C., HAN, H. B. and BRECHOT, C. (1985) Hepatitis B virus DNA sequence in lymphoid cells from patients with AIDS and AIDS-related complex. *Science* **229**: 561–563.
LIEN, J-M., ALDRICH, C. E. and MASON, W. S. (1986) Evidence that a capped oligoribonucleotide is the primer for duck hepatitis B virus plus-strand DNA synthesis. *J. Virol.* **57**: 229–236.
LUTWICK, L. I. and ROBINSON, W. S. (1977) DNA synthesis in the hepatitis B Dane particle DNA polymerase reaction. *J. Virol.* **21**: 96.
MAGNIUS, L. O. and ESPMARK, J. A. (1972) New specificities in Australia antigen-positive sera distinct from Le Bouvier determinants. *J. Immunol.* **109**: 1017–1021.
MANDART, E., KAY, A. and GALIBERT, F. (1984) Nucleotide sequence of a cloned duck hepatitis B virus genome: comparison with woodchuck and human hepatitis B virus sequences. *J. Virol.* **49**: 782–792.
MARION, P. L., KNIGHT, S. S., FEITELSON, M. A., OSHIRO, L. S. and ROBINSON, W. S. (1983a) Major polypeptide of duck hepatitis B surface antigen particles. *J. Virol.* **48**: 534–541.
MARION, P. L., KNIGHT, S. S., HO, B-K., GUO, Y. G. and ROBINSON, W. S. (1984) Liver disease associated with duck hepatitis B virus infection of domestic ducks. *Proc. Natl Acad. Sci. USA* **81**: 898–902.
MARION, P. L., KNIGHT, S. S., SALAZAR, F. H., POPPER, H. and ROBINSON, W. S. (1983b) Ground squirrel hepatitis virus infection. *Hepatology* **3**: 519–527.
MARION, P. L., OSHIRO, L., REGNERY, D. C., SCULLARD, G. H. and ROBINSON, W. S. (1980a) A virus in Beechey ground squirrels that is related to hepatitis B virus of man. *Proc. Natl Acad. Sci. USA* **77**: 2941–2945.
MARION, P. L., ROBINSON, W. S., ROGLER, C. E. and SUMMERS, J. (1982) High molecular weight GSHV-specific DNA in chronically-infected ground squirrel liver. *J. Cellular Biochem. Suppl.* 6, p. 203.
MARION, P. L., SALAZAR, F. H., ALEXANDER, J. J. and ROBINSON, W. S. (1980b) The state of hepatitis B viral DNA in a human hepatoma cell line. *J. Virol.* **33**: 795–806.
MASON, W. S., ALDRICH, C., SUMMERS, J. and TAYLOR, J. M. (1982) A symmetric replication of duck hepatitis B virus DNA in liver cells (free minus-strand DNA). *Proc. Natl Acad. Sci. USA* **79**: 3997–4001.
MASON, W. S., SEAL, G. and SUMMERS, J. (1980) Virus of Pekin ducks with structural and biological relatedness to human hepatitis B virus. *J. Virol.* **36**: 829–836.
MATSUBARA, K., NAGAYA, A., FUKUSHIGE, FUJIYAMA, A., TSURIMOTO, T., CHISAKA, O., SHINYA, T. and NAKAMURA, T. (1985) Studies with junctions between integrated HBV-DNA and host chromosomal DNA. In: *Molecular Biology of Hepatitis B Viruses.* Cold Spring Harbor Press 82.
MILLER, R H. and ROBINSON, W. S. (1983) Integrated hepatitis B virus DNA sequences specifying the major viral core polypeptide are methylated in PLC/PRF/5 cells. *Proc. Natl Acad. Sci. USA* **80**: 2534–2538.
MILLER, R. H. and ROBINSON, W. S. (1984a) Hepatitis B viral DNA forms in nuclear and cytoplasmic fractions of infected human liver. *Virology* **137**: 390–399.
MILLER, R. H. and ROBINSON W. S. (1984b) Hepatitis B virus particles of plasma and liver contain viral DNA-RNA hybrid molecules. *Virology* **139**: 53–63.
MILLER, R. H. and ROBINSON, W. S. (1986) Common evolutionary origin of hepatitis B virus and retroviruses. *Proc. Natl Acad. Sci.* **83**: 2531–2535.
MILLER, R. H., LEE, S. C., LIAW, Y. F. and ROBINSON, W. S. (1985) Hepatitis B viral DNA in infected human liver and in hepatocellular carcinoma. *J. Infect. Dis.* **151**: 1081–1092.
MILLER, R. H., MARION, P. L. and ROBINSON, W. S. (1984a) Hepatitis viral DNA-RNA hybrid molecules in particles from infected liver are converted to viral DNA molecules during an endogenous DNA polymerase reaction. *Virology* **139**: 64–72.
MITAMURA, K., HOYER, B., PONZETTO, A., NELSON, J., PURCELL, R. H. and GERIN, J. L. (1982) Woodchuck hepatitis virus DNA in woodchuck liver tissue. *Hepatology* **2**: 47S–50S.
MIYOSHI, I., KUBONISHI, I., YOSHIMOTO, S., *et al.*, (1981) Type C virus particles in a cord T-cell line derived by cocultivating normal human cord leukocytes and human leukemic T cells. *Nature* **294**: 770–771.
MIZUSAWA, H., MASANORI, T., YAGINUMA, K., KOBYASHI, M., YOSHIDA, E. and KOIKE, K. (1985) Inversely repeating integrated hepatitis B virus DNA and cellular flanking sequences in the human hepatoma-derived cell line huSP. *Proc. Natl Acad. Sci. USA* **82**: 208–212.
MOLNAR-KIMBER, K. L., SUMMERS, J. and MASON, W. S. (1984) Mapping of the cohesive overlap of duck hepatitis B virus DNA and of the site of initiation of reverse transcription. *J. Virol.* **51**: 181–191.
MOLNAR-KIMBER, K., SUMMERS, J., TAYLOR, J. and MASON, W. (1983) Protein covalently bound to minus-strand DNA intermediates of duck hepatitis B virus. *J. Virol.* **45**: 165–172.
MORIARTY, A. M., ALEXANDER, H. and LERNER, R. A. (1985) Antibodies to peptides detect new hepatitis B antigen: serological correlation with hepatocellular carcinoma. *Science* **227**: 429–433.
MOROY, T., ETIEMBLE, J., TREPO, C., TIOLLAIS, P. and BUENDIA, M. A. (1985) Transcription of woodchuck hepatitis virus in the chronically infected liver. *EMBO J.* **4**: 1507–1515.
NEURATH, A. R., KENT, S. B. and STRICK, N. (1984) Location and chemical synthesis of a pre-S gene coded immunodominant epitope of hepatitis B virus. *Science* **224**: 392–395.
NEWBALD, J. E., MASON, W. S. and SUMMERS, J. (1984) Purification and characterization of DHBV liver cores. Abstracts: *The International Symposium on Viral Hepatitis,* p. 86. San Francisco.
NISHIOKA, K., HIRAYAMA, T., SEKINE, T., *et al.* (1973) *Gann. Monogr. Can. Res.* **14**: 167.
O'CONNELL, C. D. and COHEN, M. (1984) The long terminal repeat sequences of a novel human endogenous retrovirus. *Science* **226**: 1204–1206.
O'CONNELL, C., O'BRIEN, S., NASH, W. G. and COHEN, M. (1984) ERV3, a full-length human endogenous provirus: chromosomal localization and evolutionary relationships. *Virology* **138**: 225–235.
OGSTON, C. W., JONAK, G. J., ROGLER, C. E., ASTRIN, S. M. and SUMMERS, J. (1982) Cloning and structural analysis of integrated woodchuck hepatitis virus sequences from hepatocellular carcinomas of woodchucks. *Cell* **29**: 385–394.
OGSTON, C. W. JONAK, G. J., TYLER, G. V., SNYDER, R. L., ASTRIN, S. M. and SUMMERS, J. W. (1981) Integrated woodchuck hepatitis virus DNA in hepatocellular carcinomas from woodchucks. In: *Proceedings of the 1981 Symposium on Viral Hepatitis,* pp. 809–810, Alter, H., Maynard, J. and Szmuness, W. (eds.).

Ohno, S. (1984) Sequential homology and internal repetitiousness identified in putative nucleic acid polymerase and human hepatitis B surface antigen of human hepatitis B. *Proc. Natl Acad. Sci. USA* **81**: 3781–3785.

Omata, M., Uchiumi, K., Ito, Y., Yokosuka, O., Mori, J., Terao, K., Wei-Fa, Y., O'Connell, A. P., London, W. T. and Okuda, K. (1983) Duck hepatitis B virus and liver diseases. *Gastroenterology* **85**: 260–267.

Ono, Y., Onda, H., Sasada, R., Igarahi, K., Sugino, Y. and Nishioka, K. (1983) The complete nucleotide sequences of the cloned hepatitis B virus DNA; subtype adr and adw. *Nucleic Acids Res.* **11**: 1747–1757.

O'Rear, J. J. and Temin, H. M. (1982) Spontaneous changes in nucleotide sequence in proviruses of spleen necrosis virus, an avian retrovirus. *Proc. Natl Acad. Sci. USA* **79**: 1230–1234.

Oroszlan, S., Barbacid, M., Copeland, T. D., Aaronson, S. A. and Gilden, R. V. J. (1981a) Chemical and immunological characterization of the major structural protein (p20) of MMC 1, a rhesus monkey endogenous type C virus: homology with the major structural protein of avian reticuloendotheliosis virus. *Virol.* **39**: 845–854.

Oroszlan, S., Copeland, T. D., Gilden, R. V. and Todaro, G. J. (1981b) Structural homology of the major internal proteins of endogenous type C viruses of two distantly related species of Old World monkeys: *Macaca arctoides* and *Colobus polykomos*. *Virology* **115**: 262–271.

Pagano, J. S. and Henry, B. E. (1983) Epstein–Barr Virus: biochemistry and Relation to Human Malignancy. In: *Viruses Associated with Human Cancer*, pp. 125–160, Phillips, L. A. (ed.), Marcel Dekker, Inc., New York.

Panganiban, A. T. and Temin, H. M. (1983) The terminal nucleotides of retrovirus DNA are required for integration but not for virus production. *Nature (London)* **306**: 155–160.

Pasek, M., Goto, T., Gilbert, W., Zink, B., Schaller, H., MacKay, P., Leadbetter, G. and Murray, K. (1979) Hepatitis B virus genes and their expression in *E. coli*. *Nature (London)* **285**: 575–579.

Paulson, K. E., Deka, N., Schmid, C. W., Misra, R., Schindler, C. W., Rush, M. G., Kadyk, L. and Leinwand, L. (1985) A transposon-like element in human DNA. *Nature (London)* **316**: 359–361.

Peters, R. L. (1976) Pathology of hepatocellular carcinoma. In: *Hepatocellular Carcinoma*, p. 107. Okuda, K. and Peters, R. L. (eds), Wiley, New York.

Peterson, D. L. Chien, D. Y., Vyas, G. N., *et al.* (1978) Characterization of polypeptides of HBsAg for the proposed 'UC Vaccine' for hepatitis B. In: *Viral Hepatitis: a Contemporary Assessment of Etiology, Epidemiology, Pathogenesis and Prevention*, p. 569, Vyas, G. N., Cohen, S. N. and Schmid, R. (eds), Franklin Institute, Philadelphia.

Peterson, D. L., Roberts, I. M. and Vyas, G. N. (1977) Partial amino acid sequence of two major component polypeptides of hepatitis B surface antigen. *Proc. Natl Acad. Sci. USA* **74**: 1530.

Pfeiffer, P. and Holu, T. (1983) *Cell* **33**: 781; Thomas, C. M., Hull, R., Brynt, J. A. and Maule, A. J. (1985) Isolation of a fraction from cauliflower mosaic virus-infected protoplasts which is active in the synthesis of (+) and (−) strand viral DNA and reverse transcription of primed RNA templates. *Nucl. Acids Res.* **13**: 4557–4576.

Poiesz, B. J., Ruscetti, F. W., Gazdur, A. F., *et al.* (1980) Detection and isolation of type C retrovirus particles from fresh and cultured lymphocytes of a patient with cutaneous T-cell lymphoma. *Proc. Natl Acad. Sci. USA* **77**: 7415–7419.

Poiesz, B. J., Ruscetti, F. W., Reitz, M. S., *et al.* (1981) Isolation of a new type C retrovirus (HTLV) in primary uncultured cells of a patient with Sezary T-cell leukemia. *Nature* **294**: 268–272.

Popper, H. (1978) Considerations on an association between hepatocellular carcinoma and hepatitis B. In: *Viral Hepatitis*, pp. 451–454, Vyas, G., Cohen, S. and Schmid, R. (eds), Franklin Institute Press, Philadelphia.

Popper, H., Shih, J. W. K., Gerin, J. L., Wong, D. C., Hoyer, B. H., London, W. T., Sly, D. L. and Purcell, R. H. (1981) Woodchuck hepatitis and hepatocellular carcinoma: correlation of histologic with virologic observations. *Hepatology* **1**: 91.

Ponzetto, A., Cote, P. J., Ford, E. C., Purcell, R. H. and Gerin, J. L. (1984) Woodchuck hepatitis virus core antigen and antibody in woodchucks following infection with the woodchuck hepatitis virus. *J. Virol.* (in press).

Ray, M. B., Desmet, V. I. and Bradburne, A. F. (1976) Distribution patterns of hepatitis B surface antigen (HBsAg in liver of hepatitis patients). *Gastroenterol.* **71**: 462.

Repaske, R., O'Neill, R. R., Steele, P. E. and Marin, M. A. (1983) Characterization and partial nucleotide sequence of endogenous type C retrovirus segments in human chromosomal DNA. *Proc. Natl Acad. Sci. USA* **80**: 678–682.

Repaske, R., Steele, P. E., O'Neill, R. R., Rabson, A. B. and Martin, M. A. (1985) Nucleotide sequence of a full-length human endogenous retroviral segment. *J. Virol.* **54**: 764–772.

Robinson, H. L., Astrin, S. M., Senior, A. M. and Salazar, F. H. (1981) Host susceptibility to endogenous viruses: defective, glycoprotein-expressing proviruses interfere with infections. *J. Virol.* **40**: 745–751.

Robinson, W. S. (1980) Genetic variation among hepatitis B and related viruses. *Ann. New York. Acad. Sci.* **354**: 371–378.

Robinson, W. S. and Greenman, R. L. (1972) DNA polymerase in the core of the human hepatitis B virus candidate. *J. Virol.* **13**: 1231–1236.

Robinson, W. S., Clayton, D. A. and Greenman, R. L. (1974) DNA of a human hepatitis B virus candidate. *J. Virol.* **14**: 384–391.

Robinson, W. S., Marion, P. L., Feitelson, M. and Siddiqui, A. (1982) The hepadna virus group: hepatitis B and related viruses. In: *Viral Hepatitis*, pp. 57–68, Szmuness, W., Alter, H. J. and Maynard, J. E. (eds), Philadelphia: Franklin Institute Press.

Rogler, C. E., Sherman, M., Su, C. Y., Shafritz, D. A., Summers, J., Shows, T. B., Henderson, A. and Kew, M. (1985) Deletion in chromosome 11p associated with a hepatitis B integration site in hepatocellular carcinoma. *Science* **230**: 319–322.

Rosenthal, N., Kress, M., Gruss, P. and Khoury, G. (1983) BK viral enhancer element and a human cellular homolog. *Science* **222**: 749–755.

Sattler, F. and Robinson, W. S. (1979) Hepatitis B viral DNA molecules have cohesive ends. *J. Virol.* **32**: 226–233.

SEEGER, C., GANEM, D. and VARMUS, H. (1984) The nucleotide sequence of an infectious molecularly cloned genome of the ground squirrel hepatitis B virus. *J. Virol.* **51**: 367–375.

SEEGER, C., GANEM, D. and VARMUS, H. E. (1986) Biochemical and genetic evidence for the hepatitis B virus replication strategy. *Science* **232**: 477.

SHAFRITZ, D. A. (1982) Hepatitis B virus DNA molecules in the liver of HBsAg carriers: mechanistic considerations in the pathogenesis of hepatocellular carcinoma. *Hepatology* **2**: 35S–41S.

SHAFRITZ, D. and KEW, M. (1981) Identification of integrated hepatitis B virus DNA sequences in human hepatocellular carcinoma. *Hepatology* **1**: 1–8.

SHAFRITZ, D. A., SHOUVAL, D., SHERMAN, H., HADZIYANNIS, S. and KEW, M. (1981) Integration of hepatitis B virus DNA into the genome of liver cells in chronic liver disease and hepatocellular carcinoma. *New Engl. J. Med.* **305**: 1067–1073.

SHAUL, Y., RUTTER, W. J. and LAUB, O. (1985) A human hepatitis B viral enhancer element. *EMBO J.* **4**: 427–430.

SHERLOCK, S., FOX, R. A., NIAZI, S. P. *et al.* (1970) Chronic liver disease and primary liver-cell cancer with hepatitis-associated (Australia) antigen in serum. *Lancet* **1**: 1243–1247.

SIDDIQUI, A., SATTLER, F. R. and ROBINSON, W. S. (1979) Restriction endonuclease cleavage map and location of unique features of the DNA of hepatitis B virus, subtype adw_2. *Proc. Natl Acad. Sci. USA* **76**: 4664–4668.

SNYDER, R. L. and SUMMERS, J. (1980) Woodchuck hepatitis virus and hepatocellular carcinoma. In: *Viruses in Naturally Occurring Cancers*, Cold Spring Harbor Conferences on Cell Proliferation, vol. 7, pp. 447–457. Cold Spring Harbor, New York.

SPRENGEL, R., KUHN, C., WILL, H. and SCHALLER, H. (1985) Cloned duck hepatitis B virus DNA is infectious in Pekin ducks. *J. Med. Virol.* **15**: 323–333.

STANDRING, D., RALL, L., LAUB, O. and RUTTER, W. (1983) Hepatitis B virus encodes a DNA polymerase III transcript. *Mol. and Cell Bio.* **3**: 1774–1782.

STEINER, P. E. (1960) Cancer of the liver and cirrhosis in Trans-Saharan Africa and the United States of America. *Cancer* **13**: 1085–1166.

SUMMERS, J. and MASON, W. S. (1982) Replication of the genome of a hepatitis B-like virus by reverse transcription of an RNA intermediate. *Cell* **29**: 403–415.

SUMMERS, J. A., O'CONNELL, A. and MILLMAN, I. (1975) Genome of hepatitis B virus: restriction enzyme cleavage and structure of DNA extracted from Dane particles. *Proc. Natl Acad. Sci. USA* **72**: 4597–4601.

SUMMERS, J., O'CONNELL, A., MAUPAS, P., GOUDEAU, A., COURSAGET, P. and DRUCKER, J. (1978a) Hepatitis B virus in primary hepatocellular carcinoma tissue. *J. Med. Virol.* **2**: 207.

SUMMERS, J., SMOLEC, J. M. and SNYDER, R. (1978b) A virus similar to human hepatitis B virus associated with hepatitis and hepatoma in woodchucks. *Proc. Natl Acad. Sci. USA* **75**: 4533–4537.

SUMMERS, J., SMOLEC, J. M., WERNER, B. G., *et al.* (1980) Hepatitis B virus and woodchuck hepatitis virus are members of a novel class of DNA viruses. In: *Viruses in Naturally Occurring Tumors*, pp. 459–470. Cold Spring Harbor Conference on Cell Proliferation VII. New York: Cold Spring Harbor Press.

SZMUNESS, W. (1978) Hepatocellular carcinoma and the hepatitis B virus: evidence for a causal association. *Prog. Med. Virol.* **24**: 40–69.

TAKAHASHI, K., AKAHANE, Y., GOTANDA, T., MISHIRO, T., IMAI, M., MYAKAWA, Y. and MAYUMI, M. (1979) Demonstration of hepatitis B e antigen in the core of Dane particles. *J. Immunol.* **122**: 275–279.

TAMURA, R. and TAKANO, T. (1982) Long terminal repeat (LTR)-derived recombination of retroviral DNA: sequence analyses of an aberrant clone of baboon endogenous virus DNA which carries an inversion from the LTR to the *gag* region. *Nucleic Acids Res.* **10**: 5333–5343.

TIOLLAIS, P., POURCEL, C. and DEJEAN, A. (1985) The hepatitis B virus. *Nature* **317**: 489–95.

TOGNONI, A., CATTANEO, R., SERTLING, E. and SCHAFFNER, W. (1985) A novel expression selection approach allows precise mapping of the hepatitis B virus enhancer. *Nucl. Acid Res.* **13**: 7457–7472.

TOH, H., HAYASHIDA, H. and MIYATA, T. (1983) Sequence homology between retroviral reverse transcriptase and putative polymerases of hepatitis B virus and cauliflower mosaic virus. *Nature* **305**: 827–829.

TOOZE, J. (ed.) (1982) *DNA Tumor Viruses*, Cold Spring Harbor.

TOPP, W. C., LANE, D. and POLLACK, R. (1980) Transformation by SV40 and polyoma virus. In: *DNA Tumor Viruses*, pp. 205–296. Touze, T. (ed.), Cold Spring Harbor Press.

TRICHOPOULOS, D., VIOLAKI, M., SPARROS, L., *et al.*, (1975) *Lancet* **II**: 1038.

TUR-KASPA, R., BURK, R. D., SHAUL, Y. and SHAFRITZ, D. A. (1986) Hepatitis B virus DNA contains a glucorticoid responsive element. *Proc. Natl Acad. Sci. USA* **83**: 1627–1631.

TWIST, E. M., CLARK, H. F., ADEN, A. P., KNOWLES, B. B. and PLOTKIN, S. A. (1981) Integration pattern of hepatitis B virus DNA sequences in human hepatoma cell lines. *J. Virol.* **37**: 239–243.

TYLER, G. V., SUMMERS, J. W. and SNYDER, R. L. (1981) Woodchuck hepatitis virus in natural woodchuck populations. *J. Wildlife Dis.* **17**: 297–301.

VALENZUELA, P., QUIROGA, M., ZALDIVAR, J., GRAY, P. and RUTTER, W. J. (1980) The nucleotide sequence of the hepatitis B genome and the identification of the major viral genes. In: *Animal Virus Genetics*, pp. 57–70. Fields, B., Jaenisch, R. and Fox, C. F. (eds), New York, Academic Press.

VARMUS, H. E. (1982) Form and function of retroviral proviruses. *Science* **216**: 812–820.

VARMUS, H. E. (1983) Reverse transcriptase in plants? *Science* **304**: 116–117.

VARMUS, H. E. (1985) Reverse transcriptase rides again [news]. *Nature* (London) **314**: 583–584.

VARMUS, H. E. and SWANSTROM, R. (1982) Replication of retroviruses. In: *RNA Tumor Viruses*, pp. 369–512. Weiss, R., Teich, N., Varmus, H. and Coffin, J. (eds). New York, Cold Spring Harbor Press.

WAIN-HOBSON, S., NUSSINOV, R., BROWN, R. J. and SUSSMAN, J. L. (1981) Preferential codon usage in genes. *Gene* **13**: 355–364.

WEISS, R., TEICH, N., VARMUS, H. and COFFIN, J. (eds). *DNA Tumor Viruses*. Cold Spring Harbor.

WERNER, B. G., SMOLEC, J. M., SNYDER, R. and SUMMERS, J. (1979) Serologic relationship of woodchuck hepatitis virus and human hepatitis B virus. *J. Virol.* **32**: 314–322.

WONG-STAAL, F. and GALLO, R. C. (1985) Human T-lymphotropic retroviruses. *Science* **317**: 395.

Yoshida, M., Miyoshi, I. and Hinuma, Y. (1982) Isolation and characterization of retrovirus from cell lines of human adult T-cell leukemia and its implication in the disease. *Proc. Natl Acad. Sci. USA* **79**: 2031–2035.

Zeimer, M., Garcia, P., Shaul, Y. and Rutter, W. J. (1985) Sequence of hepatitis B virus DNA incorporated into the genome of a human hepatoma cell line. *J. Virol.* **53**: 885–892.

CHAPTER 10

TRANSFORMATION BY FELINE RETROVIRUSES

CHARLES J. SHERR

Department of Tumor Cell Biology, St. Jude Children's Research Hospital, 332 North Lauderdale, Memphis, TN 38101, U.S.A.

Abstract—The genome organization, transmission, epidemiology, and spectrum of diseases induced by feline leukemia (FeLV) and sarcoma (FeSV) viruses are discussed. FeLV can act as an insertional mutagen, and in certain lymphoid tumors, integrates into the vicinity of the c-*myc* proto-oncogene. Recombination of FeLV with other proto-oncogenes (including c-*fes*, c-*fms*, c-*abl*, c-*sis* and c-*fgr*) has led to the formation of acutely transforming FeSVs containing transduced oncogene sequences. With the exception of v-*sis* (which is related to the gene encoding the B chain of the platelet-derived growth factor), the other viral oncogenes specify tyrosine kinases. Of these, the v-*fms* gene encodes a glycoprotein related to the receptor for the macrophage growth factor (M-CSF or CSF-1).

1. INTRODUCTION

Since the discovery of the first RNA tumor virus by Peyton Rous in 1911 (Rous, 1911), retroviruses capable of inducing tumors have been isolated from a wide variety of avian and mammalian species (reviewed in Teich, 1982). These viruses are distinguished by a unique replication cycle, involving the synthesis of a DNA proviral intermediate that integrates into a host cell chromosome and then encodes new viral RNA (Temin and Baltimore, 1972; Varmus, 1983). In general, oncogenic retroviruses can be classified into one of two categories: (1) the majority of isolates which produce tumors (generally leukemia) only after a long latency period and (2) a relatively few isolates which rapidly induce neoplasias, most commonly sarcomas or acute hematologic malignancies. The elucidation of the detailed genomic structure of both classes of retroviruses has shown that the acutely transforming viruses are genetic recombinants that have acquired DNA sequences from the genomes of their hosts (Stehelin *et al.*, 1976a,b; Scolnick *et al.*, 1973; Frankel *et al.*, 1976, 1979; Roussel *et al.*, 1979). The transduced 'viral oncogenes' confer the properties of morphological transformation *in vitro* and tumor formation *in vivo* by encoding products that regulate the control of cell proliferation. Their discovery has, in turn, provided a means of identifying homologous 'proto-oncogenes' in the DNA of normal cells and studying their role in controlling normal cell growth and differentiation (reviewed in Bishop, 1983). Originally, it was difficult to understand how retroviruses which lack viral oncogenes could induce leukemias. However, recent results in several different systems now suggest that this class of retroviruses can integrate into cellular proto-oncogene sequences and thereby function as insertional mutagens (Hayward *et al.*, 1981; Payne *et al.*, 1982; Nusse and Varmus, 1982; Fung *et al.*, 1983). It seems likely, then, that each class of tumor-inducing retrovirus, whether chronic or acute, exerts its effect through the expression of oncogenes.

The feline retroviruses represent one family of mammalian viral agents responsible for a variety of malignancies in domestic cats (Hardy, 1980a,b). Like other retroviral families, they can be divided into the feline leukemia viruses (FeLV) which produce chronic anemias, leukemias, and lymphosarcomas, and the feline sarcoma viruses (FeSV) which induce acute fibrosarcomas. FeLV is a naturally transmitted leukemogenic agent in domestic cats and many outbred cat populations have been exposed to FeLV even though only a minority of infected animals actually develop viremia or frank disease (Essex, 1975). The relatively high incidence of FeLV exposure increases the probability that the virus can integrate into proto-oncogene loci and subsequently recombine with host cellular sequences to form an acutely transforming FeSV containing a viral oncogene. As predicted, novel viral strains have been identified which have acquired six of the approximately twenty known viral oncogenes. Studies of the products encoded by these different genes have helped to provide insights into the role of 'transforming proteins' and the manner by which they and their normal cellular homologues regulate growth.

2. GENETIC STRUCTURE AND REPLICATION OF FeLV

Like other retroviruses, FeLV contains a diploid RNA genome which is copied by viral RNA-dependent DNA polymerase ('reverse transcriptase') to form double-stranded proviral DNA. Each genomic RNA strand is presumed to be identical in sequence and has the same polarity as viral messenger RNA (Billeter et al., 1974; Beemon et al., 1974; Quade et al., 1974; Beemon et al., 1976). The RNA genome contains three genes which are required for viral replication: the *gag* gene which codes for structural proteins of the virion core, the *pol* gene which codes for viral polymerase and endonuclease, and the *env* gene which specifies the glycoproteins of the virion envelope. The organization of these genes in viral RNA is 5'-R-U5-*gag-pol-env*-U3-R-3', where U5 and U3 represent untranslated 5' and 3' sequences, respectively, and R is a short region of direct terminal redundancy (reviewed in Coffin, 1982).

Soon after infection, genomic RNA is copied into a DNA provirus which integrates at one of many sites (and possibly at random) in host cellular DNA. The various steps in proviral DNA synthesis are mediated entirely by proteins encoded by the virus, since disrupted virions can be used to synthesize linear DNA intermediates *in vitro* (Gilboa et al., 1979). The linear DNA differs from viral RNA at both termini where it contains long terminal repeats (LTRs), each composed of sequences from the ends of viral RNA in the order U3-R-U5 (Hsu et al., 1978; Shank et al., 1978). Circularization of the linear DNA intermediate and integration of the circular form generates a nonpermuted, integrated provirus which is colinear with viral RNA and has the gene order U3-R-U5-*gag-pol-env*-U3-R-U5. The U3 region of the LTR contains regulatory sequences required for RNA transcription, including enhancer and promotor signals; a polyadenylation sequence is found within the R region (Temin, 1981). The viral promotor in U3 positions the cap site for RNA transcripts at the beginning of the R region in the 5' LTR, and full length transcripts are polyadenylated at the end of the R region in the 3' LTR. Hence, the formation of unspliced, polyadenylated RNA transcripts from proviral DNA regenerates genomic RNA molecules.

Both spliced and full length RNA transcripts are used as mRNAs for the synthesis of viral structural proteins (Weiss et al., 1977; Hayward, 1977; Mellon and Duesberg, 1977; Rothenberg et al., 1978). Viral transcription is under the control of cellular polymerase II, and the processing, transport and translation of viral mRNAs occurs in a manner similar to that of other cellular genes (Temin, 1981; Varmus, 1983). Assembly of virions containing viral genomic RNA and core proteins occurs near the plasma membrane at sites where envelope gene products accumulate. Newly assembled virus particles bud from the cell surface taking portions of the membrane as their envelope. FeLV is an enveloped 'C type' retrovirus, based on the morphology of virus buds and mature extracellular virus particles (Teich, 1982).

3. EXPOSURE TO FELINE LEUKEMIA VIRUS

FeLV is a contagious virus (Jarrett et al., 1964a,b; Hardy et al., 1969; Rickard et al., 1969), and its spread can be prevented by eliminating contacts between infected and uninfected animals (Essex, 1975; Hardy, 1980a,b). Because the most common vehicle of infection is the saliva (Gardner *et al.*, 1971; Francis *et al.*, 1977), FeLV initially infects lymphoid organs of the head and neck (Rojko *et al.*, 1979). Neutralizing antibodies produced at early stages after infection are effective in eliminating the virus, and animals thereby develop immunity to viruses of the same serotype (Sarma *et al.*, 1974; Hardy *et al.*, 1976; Russell and Jarrett, 1978b). The failure to restrict primary infection can result in the systemic spread of the virus, primarily to rapidly dividing cells of the bone marrow and lymphoid organs where it can establish a nidus of chronic infection. FeLV then enters the blood stream either in peripheral leukocytes or as free viral particles and can then spread to other organs (Hardy, 1980b).

Many cats are exposed to FeLV, as evidenced by the presence of antibodies to the virus (Essex, 1975). Most do not become persistently infected and, in general, less than 1% of healthy stray animals harbor detectable virus (Hardy, 1980a). However, continued exposure to FeLV-infected animals in multiple cat households can increase the frequency to as high as 30%. Animals persistently infected with FeLV are very susceptible to viral induced lymphosarcoma although some animals remain resistant. In viremic cats, resistance is associated with humoral antibodies to the feline oncornaviral cell membrane antigen (FOCMA) (Essex *et al.*, 1971; 1975), an antigen of ill-defined specificity which is detected on the plasma membranes of FeLV-induced tumor cells. Whether the FOCMA antigen is defined by an unusual FeLV serotype or recombinant, or is a host cellular antigen expressed at high levels in tumor cells remains unclear (Teich *et al.*, 1982)

It seems likely that the spread of FeLV among cats preceded domestication of the animals by cat fanciers. At least three other species of *Felis*, including the European wild cat (*F. sylvestris*), the jungle cat (*F. chaus*), and the sand cat (*F. margarita*) contain gene sequences related to FeLV in their cellular DNA whereas other species of Felidae lack FeLV-related genes (Benveniste *et al.*, 1975). These 'endogenous' retroviral sequences are genetically transmitted and possibly represent the residue of FeLV proviral integrations occurring since the divergence of these four *Felis* species from a common ancestor. When cellular DNA from other mammals was examined for FeLV-related genes, sequences distantly related to FeLV were detected only in rodents, suggesting that FeLV may have itself evolved from ancestral retroviruses of rodent origin. The presence of endogenous FeLV-related sequences raises the possibility that recombination can occur between contagiously transmitted (exogenous) viruses and viral sequences preexistent in cat cellular DNA. Indeed, animals experimentally inoculated with viruses that efficiently replicate only in cat cells (FeLV subgroup A) frequently produce viruses with greatly expanded host ranges (FeLV subgroups B and C) able to replicate in cells of several other species (Jarrett *et al.*, 1973; Sarma and Log, 1973; Russell and Jarrett, 1978a). Since viral host range is determined by the product of the envelope gene, it seems likely that endogenous *env* sequences contribute to the formation of FeLV-B and FeLV-C genomes, possibly conferring an increased oncogenic potential (Hardy, 1984).

4. FeLV-INDUCED DISEASE

FeLV induces both neoplastic and nonneoplastic diseases (Hardy, 1980a,b; Teich *et al.*, 1982). Because of its predilection for lymphoid and hematopoietic cells, it can produce thymic atrophy and myelodegenerative diseases of both the red and white cell lineages and, in fact, the incidence of these syndromes exceeds the frequency of FeLV-induced neoplasia. Of the FeLV-induced neoplasms, lymphosarcoma is most common, accounting for about 30% of all cat tumors (Dorn *et al.*, 1968). Although there is a clear epidemiologic association between FeLV exposure and lymphosarcoma development, paradoxically, about one-third of feline lymphosarcomas lack evidence of FeLV antigens (Hardy *et al.*, 1980). This suggests that disease induction may not depend on the continued replication of the virus in tumor cells, and is consistent with the idea that the mechanism of FeLV-induced transformation is indirect. Several forms of feline lymphosarcoma can be distinguished and most commonly include multicentric presentations with tumor at several sites as well as localized thymic and alimentary forms (Crighton, 1969). Rare varieties, including cutaneous lymphoma or disease of the central nervous system, involve presentations in nonlymphoid organs. Most lymphosarcomas are T cell tumors except for the alimentary form which can involve B lymphoid cells. FeLV also induces myeloproliferative disorders affecting different hematopoietic cell types: these include erythro- and granulocytic leukemias, myelofibrosis and osteosclerosis. FeLV-induced anemias can either be associated with myeloproliferative diseases of the bone marrow or can occur independently (Hardy, 1980a).

The fact that FeLV itself lacks viral oncogene sequences suggests that it might act as an insertional mutagen. Like other retroviruses, the structure of the DNA provirus is similar to that of a transposon, and proviral insertion can occur at many sites in cellular DNA (reviewed in Varmus, 1983). Integration events occurring in susceptible target cells of viremic animals could eventually lead to insertions into specific cellular proto-oncogene sequences and subsequent tumor development. Indeed, the isolation of several different strains of feline sarcoma viruses (see below) indirectly argues that FeLV insertions in, or adjacent to, proto-oncogene sequences can occur. However, due to the presence of endogenous FeLV-related sequences in the genomes of domestic cats (Quintrell et al., 1974; Benveniste et al., 1975), it was not previously possible to demonstrate that proviral insertions occurred at specific sites in FeLV-induced tumor cells (Koshy et al., 1980; Casey et al., 1981).

The avian leukosis viruses can induce B-cell leukemia in birds by insertional mutagenesis of c-*myc* proto-oncogene sequences (Hayward et al., 1981; Payne et al., 1982). Recently, several groups of investigators searched for c-*myc* rearrangements in FeLV-induced lymphosarcoma cells (Neil et al., 1984; Levy et al., 1984; Mullins et al., 1984). In thymic lymphosarcomas, a small percentage of cases were found in which the feline c-*myc* gene had undergone rearrangement in a manner suggesting that FeLV proviral integration had occurred in the vicinity of the gene (Neil et al., 1984). Unexpectedly, a higher percentage of thymic tumors contained novel proviruses that included 'processed' *myc* sequences lacking the introns characteristic of the feline c-*myc* proto-oncogene (Neil et al., 1984; Levy et al., 1984; Mullins et al., 1984). The *myc*-containing viruses were produced by several tumor cell lines, and could be experimentally transmitted to other cells, including those of nonfeline origin. Thus, transduction of c-*myc* sequences may be a relatively frequent event in the pathogenesis of FeLV-induced thymic neoplasms, or alternatively, *myc*-containing viruses might themselves be horizontally transmitted among cats. The relatively high frequency of such viruses in T cell tumors coupled with the failure to demonstrate them in lymphosarcomas of the alimentary type (primarily B cell neoplasms) indicates that these mechanisms for FeLV-induced tumorigenesis are different. To date, no active oncogene capable of transforming fibroblasts has been recovered by DNA-mediated gene transfer from FeLV-induced tumors, nor do the *myc*-containing viruses themselves exhibit fibroblast transforming activity. Hence, these viruses differ from the previously described feline sarcoma viruses that are able to transform morphologically cultured fibroblasts and rapidly induce neoplasias in animals.

5. ISOLATION OF FELINE SARCOMA VIRUSES

The first feline sarcoma virus was isolated by Snyder and Theilen (1969) from a two year old cat with multiple subcutaneous fibrosarcomas. Since then, several other FeSV strains have been isolated, each from multicentric tumors in young animals (Gardner et al., 1970; McDonough et al., 1971; Irgens et al., 1973; Hardy et al., 1982). By contrast, solitary fibrosarcomas, which occur in older animals and represent the most frequent variety of feline fibrosarcoma, are FeSV-negative. FeSVs isolated from tumor cell homogenates were shown rapidly to induce fibrosarcomas after inoculation into newborn kittens or puppies (reviewed in Hardy, 1981). In addition, the sarcoma viruses, unlike FeLV, can morphologically transform fibroblasts and epithelial cells in culture, generating foci of transformed

TABLE 1. *Oncogenes of Feline Sarcoma Viruses*

Viral Oncogene	FeSV Strain	Viral Poloyprotein	Transforming Function	Similar Isolates
v-*fes*	Snyder-Theilen (ST) Gardner-Arnstein (GA) Hardy-Zuckerman 2 (HZ-1)	P85*gag-fes* P110*gag-fes* P95*gag-fes*	Tyrosine-specific protein kinase	Fujinami and PRCII strains of avian sarcoma virus (v-*fps*)
v-*abl*	Hardy-Zuckerman (HZ-2)	P95*gag-abl*	Tyrosine-specific protein kinase	Abelson strain of murine leukemia virus
v-*fgr*	Gardner-Rasheed (GR)	P70*gag-fgr*	Tyrosine-specific protein kinase; contains portion of actin gene	None known
v-*fms*	Susan McDonough (SM) Hardy–Zuckerman (HZ-5)	gP180*gag-fms*	Glycoprotein with *in vitro* tyrosine-specific protein kinase activity	None known
v-*sis*	Parodi-Irgens (PI)	P75*gag-sis*	Homologous to A chain of platelet-derived growth factor (PDGF)	Simian sarcoma virus

cells which are tumorigenic in animals of the same species. The incidence of multicentric fibrosarcomas in pet cats is very low, consistent with the conclusion that FeSVs are rare genetic recombinants that have transduced oncogene sequences (see below). Table 1 lists the best characterized FeSV strains.

Although FeLV is spread contagiously, there is no evidence for horizontal transmission of FeSV. Hence, the different FeSV strains appear to have arisen *de novo* in FeLV infected animals. Experimental inoculation of the different FeSV strains listed in Table 1 generally gives rise to progressive fibrosarcomas which yield transforming virus (Snyder and Theilen 1969; Gardner *et al.*, 1970; Theilen *et al.*, 1970; McDonough *et al.*, 1971). The latent period for tumor induction is as short as two to three weeks in animals inoculated at a young age. Multicentric fibrosarcomas induced by FeSV are rapidly growing invasive tumors which contain less collagen than solitary fibrosarcomas, and have a high mitotic index. The tumors are pleomorphic, highly vascular, and have central necrotic areas characteristic of cells which outgrow their blood supply (Hardy, 1981). Malignant melanomas of the skin have been obtained in kittens inoculated subcutaneously with the Gardner–Arnstein strain of FeSV (McCullough *et al.*, 1972). Recently, ocular melanomas were also induced after inoculation of the same strain into the eye (Niederkorn *et al.*, 1981). Hence, at least some strains of FeSV can transform cells of both mesodermal and ectodermal origin.

6. STRUCTURE AND REPLICATION OF FELINE SARCOMA VIRUSES

All FeSV strains are genetic recombinants formed between FeLV and proto-oncogene sequences derived from cat cellular DNA. In each case, the viral oncogene (v-*onc*) sequences have recombined into the viral genome at the expense of *gag, pol*, and *env* genes necessary for viral replication. Because FeSV genomes lack sequences required to complete their replication cycle, they can only be propagated in the presence of a helper virus which provides the complimenting functions. In the case of natural FeSV isolates, the sarcoma virus genome is replicated in a complex with replication-competent FeLV, and the defective genome is packaged into virions encoded by the coinfecting helper virus. Cells infected with FeSV alone acquire the FeSV provirus and express viral RNA, but cannot produce infectious viral particles (Sarma and Log, 1971; Henderson *et al.*, 1974). Because such cells will nevertheless undergo morphological transformation as a result of expression of v-*onc* sequences, they are referred to as 'transformed nonproducers'. Reinfection of nonproducer cells with a second helper virus again provides the missing replicative functions and gives rise to extracellular FeSV genomes packaged in envelopes of the helper virus. If a mammalian retrovirus other than FeLV is used to 'rescue' FeSV genomes from transformed nonproducer cells—for example, a murine leukemia virus—the transforming virus produced by the cells is referred to as a pseudotype. The host range of sarcoma viruses is dependent on envelope gene functions provided by the helper virus, whereas their ability to transform cells is determined by v-*onc* sequences in the sarcoma virus genome. Since the various FeLV subgroups differ in their viral host range (Sarma and Log 1973; Jarrett *et al.*, 1973), some stocks of FeSV (FeLV) will only replicate and transform feline cells, whereas others can also be propagated in cells of heterologous species, including carnivores and primates.

All FeSV genomes characterized to date have the gene order 5'-*gag-onc-pol/env*-3', where the acquired v-*onc* sequences have been inserted into the viral *gag* gene (Sherr *et al.*, 1980b; Fedele *et al.*, 1981; Donner *et al.*, 1982; Besmer *et al.*, 1983a,b; Naharro *et al.*, 1984). In almost all cases, virtually all of the viral *pol* gene and part of the *env* gene have been deleted. Figure 1 illustrates the schematic structure of the FeLV and FeSV proviruses. Because the *gag* and v-*onc* sequences are fused in frame with one another, the full length RNA transcript defines a single open reading frame beginning at an initiation codon in 5' *gag* sequences and ending at a termination codon within the v-*onc* gene. This sequence encodes a fusion polyprotein with aminoterminal amino acid residues defined by *gag* sequences and carboxylterminal residues specified by the v-*onc* gene. The polyproteins are produced at relatively high levels in FeSV-infected cells and are responsible for transformation (Stephenson *et al.*, 1977; Sherr *et al.*, 1978; Barbacid *et al.*, 1980; Ruscetti *et al.*, 1980).

Table 1 lists the genetically characterized FeSV strains, defines the oncogenes transduced in each FeSV strain, and indicates the apparent molecular weight of each of the FeSV-coded polyproteins. Three independent isolates contain the oncogene v-*fes*: the Snyder-Theilen (ST), Gardner-Arnstein (GA), and Hardy-Zuckerman 1 (HZ-1) strains. This oncogene is a cognate of another, originally designated v-*fps*, found in several strains of avian sarcoma viruses. Recently characterized FeSV strains have been shown to carry three other oncogenes: v-*sis*, v-*abl*, and v-*fgr* (Besmer *et al.*, 1983a,b; Naharro *et al.*, 1984). The v-*sis* and v-*abl* genes were also acquired by acutely transforming retroviruses of monkeys and mice, respectively, whereas v-*fgr* is a partial homologue of the avian oncogene, v-*yes*. The Susan McDonough strain of FeSV contains the oncogene, v-*fms*, which has not been transduced by other acutely transforming retroviruses of mammals or birds (Donner *et al.*, 1982).

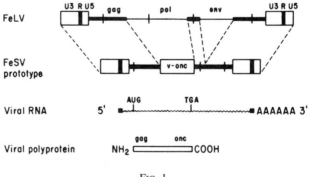

Fig. 1.

7. REGULATION OF FeSV TRANSCRIPTION

Although transformation by FeSV is dependent on the product of the viral oncogene, regulation of FeSV transcription determines the levels of transforming polyprotein synthesized in infected cells. The regulation of RNA transcription is mediated primarily through the promoter and enhancer signals in the U3 region of the 5' LTR (see, for example, Tsichlis and Coffin, 1980). Enhancer signals are known differentially to affect promoter 'strength' in different cell types or in cells of different species (Khoury and Gruss, 1983). Therefore, it seemed reasonable to assume that FeSV LTRs had enhancer elements which were most effective in the context of the cat genome but could be attenuated in cells of other species. Indeed, as a general rule, transcription of the FeSV provirus proceeds more efficiently in feline as compared to rodent cells. By manipulation of cloned DNA molecules in which murine retroviral LTRs were substituted for those of GA-FeSV, chimeric viruses were obtained with LTRs either of rodent or feline origin but with identical transforming genes. The different stocks of sarcoma virus, rescued with the same helper leukemia virus, had different transforming efficiencies in rodent and feline cells. The efficiency of transformation was reciprocally optimized when the LTRs were of a viral genetic background similar to that of the host cell (Even et al., 1983). Hence, FeSV-induced transformation is influenced by the action of host cell molecules which recognize viral LTR elements in a species-specific sense.

These observations imply that some proviral integration events will not lead to levels of transcription sufficient to transform different cell types. In fact, when rat cells were infected with wild type ST-FeSV, phenotypically normal cell clones were identified which contained silently integrated proviruses that did not express detectable FeSV transcripts (L. P. Turek and C. J. Sherr, unpublished). On prolonged passage in culture, these cells 'spontaneously' gave rise to foci of transformed cells at a frequency several orders of magnitude higher than the spontaneous transformation rate. The frequency of transformation could be further increased by treatment of the cells with 5-azacytidine (L. P. Turek, personal communication). The proviruses in transformed subclones were not detectably rearranged or amplified, but were found to be transcribing high levels of viral RNA. Presumably, factors from the host cell can differentially regulate the transcription of a provirus inserted at a single site in cellular DNA.

8. ONCOGENES ENCODING TYROSINE-SPECIFIC PROTEIN KINASES (v-*fes*, v-*abl*, v-*fgr*)

Of the three known FeSV strains containing the v-*fes* oncogene, two [ST- and GA-FeSV] have been molecularly cloned (Sherr et al., 1980b; Fedele et al., 1981) and the regions coding for the transforming polyprotein have been subjected to nucleotide sequence analysis (Hampe et al., 1982). The v-*fes* gene was found to be a member of a family of viral oncogenes which encode tyrosine-specific protein kinases. The latter include v-*src*, v-*abl*, v-*yes*, v-*ros*, v-*fps*, v-*fgr* and v-*ros*, all of which specify proteins which induce phosphorylation of heterologous proteins in tyrosine and are themselves phosphorylated, generally at preferred sites (reviewed in Bishop, 1983). As in other systems, enzyme activity is necessary for initiation and maintenance of the FeSV-transformed phenotype since mutants lacking enzyme activity are nontransforming (Barbacid et al., 1981; Reynolds et al., 1981a).

Analysis of avian and feline viral oncogenes, showed that v-*fps* and v-*fes* were derived from cognate loci of mammals and birds (Shibuya et al., 1981; Shibuya and Hanafusa, 1982; Hampe et al., 1982). As expected, probes prepared to v-*fes* and v-*fps* detect the same cellular proto-oncogene sequences in a variety of mammalian and avian species. In humans, for example, the v-*fes*/v-*fps* sequences have been molecularly cloned and have been mapped to the long arm of human chromosome 15 (Heisterkamp et al., 1982; Franchini et al., 1982; Groffen et al., 1983). More recently, the GR- and HZ-2 strains of FeSV were also found to have acquired previously described transforming genes of the tyrosine kinase gene family (see Table 1). Interestingly, the v-*fgr* gene of GR-FeSV not only contains sequences closely homologous to the v-*src*-2 proto-oncogene but also includes portions of an actin gene (Naharro et al., 1984). The fact that the same proto-oncogene sequences (e.g. v-*fes*/*fps*, v-yes/*fgr*, and v-*abl*) were independently recombined into different viral vectors suggests that the number of genes encoding tyrosine-specific protein kinases in normal cells may be relatively few.

The tyrosine-specific protein kinases encoded by different oncogenes comprise a class of

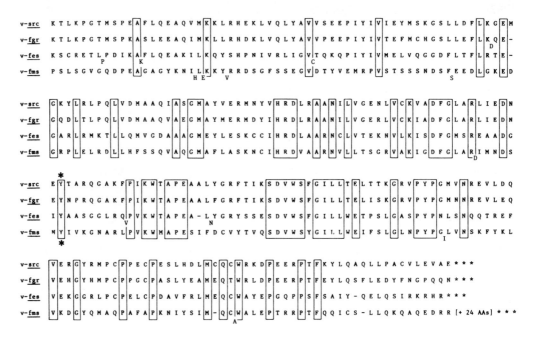

Fig. 2.

proteins with common features. The domain of the protein responsible for enzyme activity is restricted to an evolutionarily conserved portion of each v-*onc* product. For example, a 20 kilodalton carboxylterminal fragment of pp60*src* is active as a kinase *in vitro* (Levinson *et al.*, 1981). Sequence comparisons (see Fig. 2) show that the conserved domain includes both the region for tyrosine autophosphorylation as well as a lysine residue presumed to be the site of ATP binding. Mutations in the conserved regions can affect enzyme activity (Bryant and Parsons, 1983; Even *et al.*, 1983) although in pp60*src* the preferred site for autophosphorylation is not necessary for the transforming function (Cross and Hanafusa, 1983; Snyder *et al.*, 1983). The divergence of other regions of these genes could possibly affect their activities in different tissues by regulating the catalytic efficiencies of the enzymes or by providing ancillary functional domains.

9. AN UNUSUAL MEMBER OF THE KINASE GENE FAMILY (v-*fms*)

The SM-FeSV and HZ-5-FeSV strains are the only retroviruses known to have acquired the oncogene v-*fms*. Molecular cloning of the SM-FeSV provirus (Donner *et al.*, 1982) and nucleotide sequence analysis (Hampe *et al.*, 1984) predicted that the primary translation product would be a 160 kd polyprotein. However, immune precipitation of the metabolically labeled v-*fms* gene product showed that the earliest protein detected *in vivo* was 180 kd (Barbacid *et al.*, 1980; Ruscetti *et al.*, 1980; Van de Ven *et al.*, 1980) due to cotranslational addition of N-linked oligosaccharide chains (Sherr *et al.*, 1980a; Anderson *et al.*, 1982). The glycosylated polyprotein (designated gP180*gag-fms*) is proteolytically cleaved during or soon after synthesis to yield an aminoterminal *gag* gene-coded fragment (p55*gag*) and a glycosylated carboxylterminal polypeptide (gp120*fms*) specified entirely by v-*fms* sequences. Although the v-*fms* gene product is synthesized on membrane-bound polyribosomes and is transported into the cisternae of the endoplasmic reticulum (ER), a hydrophobic 'stop transfer' sequence, located near the middle of the v-*fms*-coded polypeptide immobilizes the protein in the ER membrane (Hampe *et al.*, 1984; Rettenmier *et al.*, 1985b). Assuming that polyribosomes complete translation of the v-*fms*-coded molecules by 'running off' into the cytoplasm, the completed polypeptides were predicted to be integral transmembrane glycoproteins oriented with their aminoterminal portion in the ER and their carboxylterminal domain in the cytoplasm (Sherr et al., 1984). Recent experiments performed with a series of SM-FeSV deletion mutants verified these predictions, and showed that the v-*fms* gene product was transported to the cell surface as an integral transmembrane glycoprotein (Rettenmier et al., 1985b).

Kinetic analyses showed that a small proportion of gp120*fms* molecules was subsequently converted to polypeptides of higher molecular weight designated gp140*fms*. The majority of molecules in transformed cells which remains as gp120*fms* contained N-linked oligosaccharide chains lacking terminal fucose and sialic acid residues. By contrast, gp140*fms* molecules have acquired complex carbohydrate chains and contain terminal sugars, added during passage through the Golgi complex. Although a small proportion of gp140*fms* molecules are detected at the cell surface, most of the glycoproteins remain internally sequestered and appear in juxtanuclear complexes containing membranes and supporting cytoskeletal filaments (Anderson et al., 1984). The basis for the apparent block in transport of v-*fms* coded molecules from the ER through the Golgi system is unclear. Experiments with an engineered viral mutant have indicated, however, that cell surface expression of gp140*fms* correlates with the transformed phenotype (Roussel et al., 1984). The biochemical properties and precursor-product relationships of the v-*fms*-coded molecules are summarized in Fig. 3.

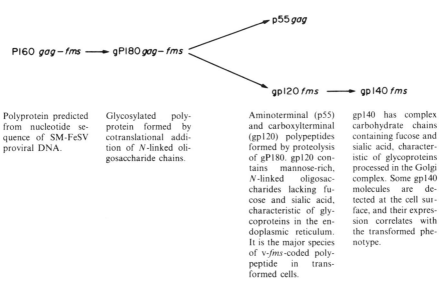

FIG. 3.

Nucleotide sequencing showed that the carboxylterminal domain of the v-*fms*-coded glycoprotein was closely related to the conserved core region of oncogene products with tyrosine-specific protein kinase activity (Hampe et al., 1984, and see Fig. 2). Indeed, *in vitro* assays demonstrated that the v-*fms* gene products were active as protein kinases and catalyzed the transfer of phosphate from ATP into tyrosyl residues of the glycoproteins themselves and to heterologous substrates like immunoglobulin and casein (Barbacid and Lauver, 1981). However, an SM-FeSV mutant that encodes gp120*fms* molecules with *in vitro* kinase activity but lacks the mature form of the glycoproteins was found to be nontransforming, suggesting that enzyme activity in itself was insufficient for transformation (Roussel et al., 1984). Unlike other members of the tyrosine kinase gene family, the major v-*fms* products, are relatively poorly phosphorylated *in vivo* (Sherr et al., 1980a) and lack readily detectable phosphotyrosine (Barbacid and Lauver 1981: Reynolds et al., 1981b). Moreover, cells transformed by SM-FeSV do not show the general elevation in levels of phosphotyrosine observed in cells transformed by most other members of the tyrosine kinase gene family.

The v-*fms* codes for a protein that, in several respects, is similar to the product of the avian erythroblastosis virus oncogene, v-*erb*-B. The latter gene encodes 62–68 kd glycoproteins containing N-linked oligosaccharide chains. These molecules accumulate intracellularly, although a small proportion are detected at the cell surface (Hayman et al., 1983; Privalsky, et al., 1983). The recently published nucleotide sequence of the v-*erb*-B gene (Yamamoto et al., 1983) predicts a product that shares several features with the v-*fms*-coded glycoprotein: namely, (i) the protein contains an internal hydrophobic anchor sequence, (ii) sites for carbohydrate addition cluster in the region aminoterminal to the transmembrane anchor sequence and (iii) the region carboxylterminal to the anchor sequence shows strong homology to pp60*src* even though high levels of enzymatic activity have not been detected. In erythroid precursors, the presence of v-*erb*-B-coded molecules at the cell surface also correlates with expression of the transformed phenotype (Hayman et al., 1983).

Recently, the v-*erb*-B gene was shown to represent a transduced portion of the normal cellular gene which codes for the epidermal growth factor receptor (Downward et al., 1984). The receptors for epidermal growth (EGF), platelet derived growth factor (PDGF) and insulin and insulin-like growth factors (IGFs) each have associated tyrosine-specific protein kinase activities (Ushiro and Cohen, 1980; Kasuga et al., 1982; Ek et al., 1982). In the case of these normal receptor proteins, their enzymatic activity is regulated by binding of ligand to another domain of the receptor. The portion of the EGF receptor gene acquired as v-*erb*-B contains the tyrosine kinase coding function but appears to lack information for ligand binding, suggesting that transformation is mediated by hormone independent receptor signals. Similarities of topology and function suggested that the v-*fms* gene product was also a receptor analog.

10. THE v-*fms* GENE PRODUCT IS RELATED TO THE CSF-1 RECEPTOR

The product of the feline c-*fms* proto-oncogene has been identified and was found to be a 165–170 kd glycoprotein with an associated tyrosine-specific protein kinase activity (Rettenmier et al., 1985a). In adult cat tissues, c-*fms* RNA transcripts were predominantly detected in extracts of spleen, lymph nodes, liver, and bone marrow suggesting that c-*fms* might code for a receptor for one of the interleukins or hematopoietic colony stimulating factors. Using monoclonal antibodies which detect v-*fms*-coded epitopes on the surface of live SM-FeSV transformed cells (Anderson et al., 1982; Roussel et al., 1984; Rettenmier et al., 1985b), the c-*fms* gene product was found to be restricted in its expression to cells of the mononuclear phagocyte series (Sherr et al., 1985).

Antisera raised to a recombinant v-*fms*-coded polypeptide expressed in bacteria cross-reacted with the murine c-*fms*-coded glycoprotein. This glycoprotein was shown to be related to the murine receptor for the mononuclear phagocyte growth factor, CSF-1, by two criteria: first, radiolabeled CSF-1 bound specifically to the murine c-*fms* gene product expressed at the cell surface of murine macrophages. second, membranes prepared from these cells exhibited tyrosine kinase activity *in vitro* in the presence of CSF-1, and phosphorylation of the c-*fms* gene product was greatly enhanced in the presence of the growth factor. Hence, the v-*fms* gene may have arisen by recombination of FeLV with a feline CSF-1 receptor gene (Sherr et al., 1985).

The fact that the v-*fms* gene product (gp140*fms*) retains the major portion of the extracellular, ligand binding domain of its c-*fms* progenitor raised the possibility that the viral protein could function as a competent receptor. Indeed, in recent studies, cells transformed by SM-FeSV, but not other FeSV strains, were found to express receptors for murine CSF-1, and chemical cross-linking experiments revealed that CSF-1 was specifically complexed to gp140*fms*. However, CSF-1 stimulated enhancement of gp140*fms* tyrosine phosphorylation could not be demonstrated (Sacca et al., 1986). Possibly, the truncated viral transforming protein acts as a constitutive kinase, or alternatively, CSF-1 stimulation of receptor kinase activity may only occur in the presence of the homologous feline growth factor.

11. TRANSFORMING GENES AND GROWTH FACTORS (v-*sis*)

The Parodi–Irgens strain of FeSV contains v-*onc* sequences that are related to the v-*sis* sequences of simian sarcoma virus (Besmer *et al.*, 1983a). Unlike the simian sarcoma virus which has transduced v-*sis* sequences in the viral *env* gene (Robbins *et al.*, 1981; Gelman *et al.*, 1981), the v-*sis* gene of PI-FeSV is recombined into the *gag* gene and is expressed in the context of a *gag-sis* fusion protein (see Table 1). The v-*sis* gene of simian sarcoma virus has been sequenced (Devare *et al.*, 1983) and was recently shown to encode one of two protein chains of the platelet derived growth factor (PDGF) (Doolittle *et al.*, 1983; Waterfield *et al.*, 1983). The fact that PDGF itself is a glycoprotein indicates that it segregates into the cisternae of the endoplasmic reticulum during synthesis and can potentially interact with its receptor during intracellular transport. Hence, the persistent production of a growth stimulating polypeptide by a cell that elaborates receptors for the same hormone can provide an autoendocrine mechanism for transformation.

12. CONCLUSIONS

The acutely transforming feline retroviruses have represented a natural source of mobilized oncogenes, recognizable by their unusual properties of rapid tumor induction. In some cases, the viral oncogenes described in FeSV strains have also been acquired by transforming retroviruses of other mammalian and avian species, suggesting that the total number of proto-oncogenes may be relatively few. Genes like v-*abl* and v-*sis* have been much better characterized in murine and primate retroviral isolates, where their functions were originally defined. Conversely, the v-*fms* gene is unique to the feline sarcoma virus system and an understanding of its recently defined function promises to provide new information about mechanisms for transformation and growth control. It seems possible that new FeSV isolates could still be identified, among which novel oncogene sequences might be represented. This is expected because FeLV is a natural vector, ubiquitous in a population of animals routinely subjected to veterinary scrutiny. Conversely, the development of vaccines effective against feline leukemia virus will greatly diminish the probability of obtaining new recombinant sarcoma viruses.

FeLV-induced leukemia is well understood from the perspective of its epidemiology and pathology. It now seems likely that FeLV functions as a tumor-inducing virus through its interaction with host cellular genes. FeLV-induced thymic lymphomas could potentially involve multiple molecular steps, including recombination with endogenous viral sequences and transduction of the feline c-*myc* proto-oncogene. Studies of other forms of feline lymphosarcoma are likely to yield additional insights concerning mechanisms of viral-induced disease.

Clearly, much excitement has been generated by the identification of viral oncogenes and the assignment of functions for their transforming proteins. Work on FeSV has paralleled efforts with other acutely transforming viruses and has provided a route to the characterization of evolutionarily conserved proto-oncogene sequences in different species, including man. Quite naturally, the study of cellular proto-oncogenes has begun to supersede investigations of viral transforming genes although the role of viral transforming proteins and their cellular homologues remains a critical issue. We now recognize that some oncogene products function at the cell surface, either as analogs of hormonal signals or as modified growth factor receptors. As was predicted, the roles of these proteins mimic those of gene products which regulate the growth of normal cells, and describe mechanisms by which cells of different phenotypes intercommunicate. A more detailed understanding of how these proteins interact with cellular target molecules will define how they are intercalated into a metabolism for growth control. It seems likely that the products of other oncogenes may act to regulate gene expression at a level more proximal to the genome itself. The rate of progress in this fundamental area of cell biology owes much to the understanding of the few recombinant tumor viruses that, as accidents of nature, provided the first molecular tools for the study of cancer genes.

REFERENCES

ANDERSON, S. J., GONDA, M. A., RETTENMIER, C. W. and SHERR, C. J. (1984) Subcellular localization of glycoproteins encoded by the viral oncogene, v-*fms J. Virol.* **51**: 730–741.

ANDERSON, S. J., FURTH, M., WOLFF, L., RUSCETTI, S. K. and SHERR, C. J. (1982) Monoclonal antibodies to the transformation-specific glycoprotein encoded by the feline retroviral oncogene v-*fms*. *J. Virol.* **44**: 696–702.

BARBACID, M. and LAUVER, A. V. (1981) The gene products of McDonough feline sarcoma virus have an *in vitro* associated protein kinase that phosphorylates tyrosine residues: lack of detection of this enzymatic activity *in vivo*. *J. Virol.* **40**: 812–828.

BARBACID, M., LAUVER, A. V. and DEVARE, S. G. (1980) Biochemical and immunological characterization of polyproteins coded for by the McDonough, Gardner–Arnstein and Snyder–Theilen strains of feline sarcoma virus. *J. Virol.* **33**: 196–207.

BARBACID, M., DONNER, L., RUSCETTI, S. K. and SHERR, C. J. (1981) Transformation defective mutants of Snyder–Theilen feline sarcoma virus lack tyrosine specific protein kinase activity. *J. Virol.* **39**: 246–254.

BEEMON, K. L., DUESBERG, P. and VOGT, P. K. (1974) Evidence for crossing over between avian tumor viruses based on analysis of viral RNAs. *Proc. natn. Acad. Sci. U.S.A.* **71**: 4254–4258.

BEEMON, K. L., FARAS, A. J., HAASE, P. H., DUESBERG, P. H. and MAISEL, J. E. (1976) Genome complexities of murine leukemia and sarcoma, reticuloendotheliosis and visna viruses. *J. Virol.* **17**: 525–537.

BENVENISTE, R. E., SHERR, C. J. and TODARO G. J. (1975) Evolution of type C viral genes: origin of feline leukemia virus. *Science* **190**: 886–888.

BESMER, P., SNYDER, H. W., MURPHY, J. R., HARDY, W. D., JR and PARODI, A. (1983a) The Parodi–Irgens feline sarcoma virus and simian sarcoma virus have homologous oncogenes. *J. Virol.* **46**: 606–613.

BESMER, P., HARDY, W. D., JR., ZUCKERMAN, E., LEDERMAN, L. and SNYDER, H. W. (1983b) The Hardy–Zuckerman-2 FeSV, a new feline retrovirus with oncogene homology to Abelson-MuLV. *Nature* **303**: 825–828.

BILLETER, M. A., PARSONS, J. T. and COFFIN, J. M. (1974) The nucleotide sequence complexity of avian tumor virus RNA. *Proc. natn. Acad. Sci. U.S.A.* **71**: 3560–3564.

Bishop, J. M. (1983) Cellular oncogenes and retroviruses. *A. Rev. Biochem.* **52**: 301–354.

BRYANT, D. and PARSONS, J. T. (1983) Site directed point mutations in the *src* gene of Rous sarcoma virus results in an inactive *src* gene product. *J. Virol.* **45**: 1211–1216.

CASEY, J. W., ROACH, A., MULLINS, J., BURCK, K. B., NICOLSON, M. O., GARDNER, M. B. and DAVIDSON, N. (1981) The U3 portion of the feline leukemia virus identifies horizontally acquired proviruses in leukemic cats. *Proc. natn. Acad. Sci. U.S.A.* **78**: 7778–7782.

COFFIN, J. (1982) Structure of the retroviral genome. In: RNA *Tumor Viruses*, pp. 261–368, WEISS, R., TEICH, N., VARMUS, H. and COFFIN, F. (eds) Cold Spring Harbor Laboratory, New York.

CRIGHTON, G. W. (1969) Lymphosarcoma in the cat. *Vet. Res.* **84**: 329–331.

CROSS, F. R. and HANAFUSA, H. (1983) Local mutagenesis of Rous sarcoma virus; the major sites of tyrosine and serine phosphorylation of p60*src* are dispensable for transformation. *Cell* **34**: 597–608.

DEVARE, S. G., REDDY, E. P., DORIA-LAW, J., ROBBINS, K. C. and AARONSON, S. A. (1983) Nucleotide sequence of the simian sarcoma virus genome. Demonstration that its acquired cellular sequences encode the transforming gene product p28*sis*. *Proc. natn. Acad. Sci. U.S.A.* **80**: 731–735.

DONNER, L., FEDELE, L. A., GARON, C. F., ANDERSON, S. J. and SHERR, C. J. (1982) McDonough feline sarcoma virus: characterization of the molecularly cloned provirus and its feline oncogene v-*fms*. *J. Virol.* **41**: 489–500.

DOOLITTLE, R. F., HUNKAPILLER, M. W., HOOD, L. E., DEVARE, S. G., ROBBINS, K. C., AARONSON, S. A. and ANTONIADES, H. N. (1983) Simian sarcoma virus *onc* gene, v-*sis*, is derived from the gene (or genes) encoding a platelet-derived growth factor. *Science* **221**: 275–276.

DORN, C. R., TAYLOR, D., SCHNEIDER, R., HIBBARD, H. H. and KLAUBER, M. R. (1968) Survey of animal neoplasms in Alameda and Contra Costa Counties, California, II. Cancer morbidity in dogs and cats from Alameda County. *J. natn. Cancer Inst.* **40**: 307–318.

DOWNWARD, J., YARDEN, Y., MAYES, E., SCRACE, G., TOTTY, N., STOCKWELL, P., ULLRICH, A., SCHLESSINGER, J. and WATERFIELD, M. D. (1984) Close similarity of epidermal growth factor receptor and v-*erb*-B oncogene protein sequences. *Nature* **307**: 521–527.

EK, B., WESTERMARK, B., WASTESON, A. and HELDIN, C. H. (1982) Stimulation of tyrosine-specific phosphorylation by platelet-derived growth factor. *Nature* **295**: 419–420.

ESSEX, M. (1975) Horizontally and vertically transmitted oncornaviruses of cats. *Adv. Cancer Res.* **21**: 175–248.

ESSEX, M., KLEIN, G., SNYDER, S. P. and HARROLD, J. B. (1971) Correlation between humoral antibody and regression of tumors induced by feline sarcoma virus. *Nature* **233**: 195–196.

ESSEX, M., SLISKI, A., COTTER, S. M., JAKOWSKI, R. M. and HARDY, W. D., JR. (1975) Immunosurveillance of naturally occurring feline leukemia. *Science* **190**: 790–792.

EVEN, J., ANDERSON, S. J., HAMPE, A., GALIBERT, F., LOWY, D., KHOURY, G. and SHERR, C. J. (1983) Mutant feline sarcoma provirus containing the viral oncogene (v-*fes*) and either feline or murine control elements. *J. Virol.* **45**: 1004–1016.

FEDELE, L. A., EVEN, J., GARON, C. F., DONNER, L. and SHERR, C. J. (1981) Recombinant bacteriophages containing the integrated transforming provirus of Gardner–Arnstein feline sarcoma virus. *Proc. natn. Acad. Sci. U.S.A.* **78**: 4036–4040.

FRANCHINI, G., GELMANN, E. P., DALLA-FAVERA, R., GALLO, R. C. and WONG-STAAL, F. (1982) Human gene (c-*fes*) related to the *onc* sequences of Snyder–Theilen feline sarcoma virus. *Molec. cell. Biol.* **2**: 1014–1018.

FRANCIS, D. P., ESSEX, M. and HARDY, W. D., JR. (1977) Excretion of feline leukaemia virus by naturally infected pet cats. *Nature* **269**: 252–254.

FRANKEL, A. E., NEUBAUER, R. L. and FISHINGER, P. J. (1976) Fractionation of DNA transcripts from Mononey sarcoma virus and isolation of sarcoma virus-specific complementary DNA. *J. Virol.* **18**: 481–490.

FRANKEL, A. E., GILBERT, J. H., PORZIG, K. J., SCOLNICK, E. M. and AARONSON, S. A. (1979) Nature and distribution of feline sarcoma virus nucleotide sequences. *J. Virol.* **30**: 821–827.

FUNG, Y.-K., LEWIS, W. G., CRITTENDEN, L. B. and KUNG, H.-J. (1983) Activation of the cellular oncogene c-*erb* B by LTR insertion: molecular basis for induction of erythroblastosis by avian leukosis virus. *Cell* **33**: 357–368.

GARDNER, M. B., RONGEY, R. W., JOHNSON, E. Y., DEJOURNETT, R. and HUEBNER, R. J. (1971) C-type virus particles in salivary tissue of domestic cats. *J. natn. Cancer Inst.* **47**: 561–565.

GARDNER, M. B., RONGEY, R. W., ARNSTEIN, P., ESTES, J. D., SARMA, P., HUEBNER, R. J. and RICHARD, C. G. (1970) Experimental transmission of feline fibrosarcoma to cats and dogs. *Nature* **226**: 807–809.

GELMAN, E. P., WONG-STAAL, F., KRAMER, R. and GALLO, R. C. (1981) Molecular cloning and comparative analysis of the genomes of simian sarcoma virus and its associated helper virus. *Proc. natn. Acad. Sci. U.S.A.* **78**: 3373–3377.

GILBOA, E., MITRA, S. W., GOFF, S. and BALTIMORE, D. (1979) A detailed model of reverse transcription and tests of crucial aspects. *Cell* **18**: 93–100.

GROFFEN, J., HEISTERKAMP, N., SHIBUYA, M., HANAFUSA, H. and STEPHENSON, J. R. (1983) Transforming genes of avian (v-*fps*) and mammalian (v-*fes*) retroviruses correspond to a common cellular locus. *Virology* **125**: 480–486.

HAMPE, A., GOBET, M., SHERR, C. J. and GALIBERT, F. (1984) Nucleotide sequence of the feline retroviral oncogene v-*fms* shows unexpected homology with oncogenes encoding tyrosine-specific protein kinases. *Proc. natn. Acad. Sci. U.S.A.* **81**: 85–89.

HAMPE, A., LAPREVOTTE, I., GALIBERT F. and SHERR, C. J. (1982) Nucleotide sequence of feline retroviral oncogenes (v-*fes*) provide evidence for a family of tyrosine-specific protein kinase genes. *Cell* **30**: 775–786.

HARDY, W. D., JR. (1980a) The virology, immunology and epidemiology of the feline leukemia virus. In: *Feline Leukemia Virus*, pp. 33–78, HARDY, W. D. JR., ESSEX, M. and MCCLELLAND, A. J. (eds) Elsevier North Holland, Amsterdam.

HARDY, W. D., JR. (1980b) Feline leukemia virus disease. In: *Feline Leukemia Virus*, pp. 3–31, HARDY, W. D., JR., ESSEX, M. and MCCELLAND, A. J. (eds) Elsevier/North Holland, Amsterdam.

HARDY, W. D., JR. (1981) The feline sarcoma viruses. *J. Am. Hosp. Ass.* **17**: 981–996.

HARDY, W. D., JR. (1984) A new package for an old oncogene. *Nature* **308**: 775.

HARDY, W. D., JR., ZUCKERMAN, E., MARKOVICH, R., BESMER, P. and SNYDER, H. W., JR. (1982) Isolation of feline sarcoma viruses from pet cats with multicentric fibrosarcoma. In: *Advances in Comparative Leukemia Research* pp. 205–206. YOHN, S. and BLAKESLEE, J. R. (eds) Elsevier/North Holland, Amsterdam.

HARDY, W. D., JR., GEERING, G., OLD, L. J., DEHARVEN, E., BRODY, K. S. and MCDONOUGH S. (1969) Feline leukemia virus: occurrence of viral antigen in the tissues of cats with lymphosarcoma and other diseases. *Science* **166**: 1019–1021.

HARDY, W. D., JR., MCCLELLAND, A. J., ZUCKERMAN, E. E., SNYDER, H. W., JR., MACEWEN, E. G., FRANCIS D. and ESSEX, M. (1980) Development of virus nonproducer lymphosarcomas in pet cats exposed to FeLV. *Nature* **288**: 90–92.

HARDY, W. D., JR., HESS, P. W., MACEWEN, E. G., MCCLELLAND, A. J., ZUCKERMAN, E. E., ESSEX, M., COTTER, S. M. and JARRETT, O. (1976) Biology of feline leukemia virus in the natural environment. *Cancer Res.* **36**: 582–588.

HAYMAN, M. J., RAMSEY, G. M., SAVIN, K., KITCHENER, G., GRAF, T. and BEUG, H. (1983) Identification and characterization of the avian erythroblastosis virus *erb* B gene product as a membrane glycoprotein. *Cell* **32**: 579–588.

HAYWARD, W. S. (1977) Size and genetic content of viral RNAs in avian oncovirus-infected cells. *J. Virol.* **24**: 47–63.

HAYWARD, W. S., NEEL, B. G. and ASTRIN, S. M. (1981) Activation of a cellular *onc* gene by promotor insertion in ALV-induced lymphoid leukosis. *Nature* **290**: 475–480.

HEISTERKAMP, N., GROFFEN, J., STEPHENSON, J. R., SPURR, N. K., GOODFELLOW, P. N., SOLOMON, E., CARRITT, B. and BODMER, W. F. (1982) Chromosomal locations of human cellular homologues of two viral oncogenes. *Nature* **299**: 747–749.

HENDERSON, I. C., LIEBER, M. M. and TODARO, G. J. (1974) Mink cell line MvlLu (CCL64): focus-formation and the generation of 'nonproducer' transformed cell lines with murine and feline sarcoma viruses. *Virology* **60**: 282–287.

HSU, T. W., SABRAN, J. L., MARK, G. E., GUNTAKA, R. V. and TAYLOR, J. M. (1978) Analysis of unintegrated avian RNA tumor virus double-stranded DNA intermediates. *J. Virol.* **28**: 810–818.

IRGENS, K., WYER, M., MORAILLON, A., PARODI, A. and FORTUNY, V. (1973) Isolement d'un virus sarcomatogene felin a partir d'un fibrosarcome spontane du chat: etude du pouvoir sarcomatogene *in vivo*. *C.r. Acad. Aci.* **26**: 1783–1786.

JARRETT, O., LAIRD, H. M. and HAY, D. (1973) Determinanats of the host range of feline leukemia viruses. *J. gen. Virol.* **20**: 169–175.

JARRETT, W. F. H., CRAWFORD, E. M., MARTIN, W. B. and DAVEY, F. (1964a) Leukaemia in the cat. A virus-like particle associated with leukaemia (lymphosarcoma). *Nature* **202**: 567–568.

Jarrett, W. F. H., Martin, W. B., Crighton, G. W., Dacton, R. G. and Steward, M. F. (1964b) Transmission experiments with leukaemia (lymphosarcoma). *Nature* **202**: 566–567.

Kasuga, M., Zick, Y., Blithe, D. L., Crettaz, M. and Kahn, R. (1982) Insulin stimulates tyrosine phosphorylation of the insulin receptor. *Nature* **298**: 667–669.

Khoury, G. and Gruss, P. (1983) Enhancer elements. *Cell* **33**: 313–314.

Koshy, R., Gallo, R. C. and Wong-Staal, F. (1980) Characterization of the endogenous feline leukemia virus-related DNA sequences in cats and attempts to identify exogenous viral sequences in tissues of virus-negative leukemic animals. *Virology* **103**: 434–445.

Levinson, A. D., Courtneidge, S. A. and Bishop, J. M. (1981) Structural and functional domains of the Rous sarcoma virus transforming protein (pp60src). *Proc. natn. Acad. Sci. U.S.A.* **78**: 1624–1628.

Levy, L. S., Gardner, M. B. and Casey J. W. (1984) Isolation of a feline leukaemia provirus containing the oncogene *myc* from a feline lymphosarcoma. *Nature* **308**: 853–856,

McCollough, B., Schaller, J., Shadduck, J. A. and Yohn, D. S. (1972) Induction of malignant melanomas associated with fibrosarcomas in gnotobiotic cats inoculated with Gardner feline fibrosarcoma virus. *J. natn. Cancer Inst.* **48**: 1893–1896.

McDonough, S. K., Larsen, S., Brodey, R. S., Stock, N. D. and Hardy, W. D., Jr. (1971) A transmissible feline fibrosarcoma of viral origin. *Cancer Res.* **31**: 953–956.

Mellon, P. and Duesberg, P. H. (1977) Subgenomic, cellular Rous sarcoma virus RNAs contain oligonucleotides from the 3' half and 5' terminus of virion RNA. *Nature* **270**: 631–634.

Mullins, J. I., Brody, D. S., Binari, R. C., Jr. and Cotter, S. (1984) Viral transduction of c-*myc* gene in naturally occurring feline leukaemias: *Nature* **308**: 856–858.

Naharro, G., Robbins, K. C. and Reddy, E. P. (1984) Gene product of a v-*fgr onc*: hybrid protein containing a portion of actin and a tyrosine-specific protein kinase. *Science* **223**: 63–66.

Neil, J. C., Hughes, D., McFarlane, R., Wilkie, N. M., Onions, D. E., Lees, G. and Jarrett, O. (1984) Transduction and rearrangement of the *myc* gene by feline leukaemia virus in naturally occurring T-cell leukaemias. *Nature* **308**: 814–820.

Niederkorn, J. Y., Shadduck, J. A., Albert, D. and Essex, M. (1981) Serum antibodies against feline oncornavirus-associated cell membrane antigen in cats bearing virally-induced uveal melanomas. *Invest. Ophthalmol. vis. Sci.* **20**: 598–605.

Nusse, R. and Varmus, H. E. (1982) Many tumors induced by the mouse mammary tumor virus contain a provirus integrated in the same region of the host genome. *Cell* **31**: 99–109.

Payne, G. S., Bishop, J. M. and Varmus, H. E. (1982) Multiple arrangements of viral DNA and an activated host oncogene in bursal lymphomas. *Nature* **295**: 209–213.

Privalsky, M. L., Sealy, L., Bishop, J. M., McGrath, J. P. and Levinson, A. D. (1983) The product of the avian erythroblastosis virus *erbB* locus is a glycoprotein. *Cell* **32**: 1257–1267.

Quade, K., Smith R. E. and Nichols, J. L. (1974) Evidence for common nucleotide sequences in the RNA subunits comprising Rous sarcoma virus 70S RNA. *Virology* **61**: 287–291.

Quintrell, N., Varmus, H. E., Bishop, J. M., Nicholson, M. O. and McAllister, R. M. (1974) Homologies among the nucleotide sequences of the genomes of C-type viruses. *Virology* **58**: 569–575.

Rettenmier, C. W., Chen, J. H., Roussel, M. F. and Sherr, C. J. (1985a) The product of the c-*fms* proto-oncogene: a glycoprotein with associated tyrosine kinase activity. *Science* **228**: 320–322.

Rettenmier, C. W., Roussel, M. F., Quinn, C. O., Kitchingman, G. R., Look, A. T. and Sherr, C. J. (1985b) Transmembrane orientation of glycoproteins encoded by the v-*fms* oncogene. *Cell* **40**: 971–981.

Reynolds, F. H., Van de Ven, W. J. M., Blomberg, J. and Stephenson J. R. (1981b) Involvement of a high molecular weight protein translational product of Snyder–Theilen feline sarcoma virus in malignant transformation. *J. Virol.* **37**: 643–653.

Reynolds, F. H., Van de Ven, W. J. M., Blomberg, J. and Stephenson, J. R. (1981b) Differences in mechanisms of transformation by independent feline sarcoma virus isolates. *J. Virol.* **38**: 1084–1089.

Rickard, C. G., Post, J. E., Noronha, F. and Barr, L. M. (1969) A transmissible virus-induced lymphocytic leukemia of the cat. *J. natn. Cancer Inst.* **42**: 987–1014.

Robbins, K. C., Devare, S. G., Aaronson, S. A. (1981) Molecular cloning of integrated simian sarcoma virus: genome organization of infectious DNA clones. *Proc. natn. Acad. Sci. U.S.A.* **78**: 2918–2922.

Rojko, J. L., Hoover, E. A., Mathes, L. E., Olsen, R. G. and Schaller, J. P. (1979) Pathogenesis of experimental feline leukemia virus infection. *J. natn. Cancer Inst.* **63**: 759–768.

Rothenberg, E., Donohue, D. J. and Baltimore, D. (1978) Analysis of a 5' leader sequence on murine leukemia virus 21S RNA: Heteroduplex mapping with long reverse transcriptase products. *Cell* **13**: 435–451.

Rous, P. (1911) A sarcoma of the fowl transmissible by an agent separable from the tumor cells. *J. exp. Med.* **13**: 397–411.

Roussel, M. F., Rettenmier, C. W., Look, A. T. and Sherr, C. J. (1984) Cell surface expression of v-*fms*-coded glycoproteins is required for transformation. *Molec. Cell Biol.* **4**: 1999–2009.

Roussel, M., Saule, S., Lagrou, C., Rommens, C., Beug, H., Graf, T. and Stehelin, D. (1979) Three new types of viral oncogene of cellular origin specific for haematopoietic cell transformation. *Nature* **281**: 452–455.

Ruscetti, S. K., Turek, L. P. and Sherr, C. J. (1980) Three independent isolates of feline sarcoma virus code for three distinct gag-x polyproteins. *J. Virol.* **35**: 259–264.

Russell, P. H. and Jarrett, O. (1978a) The specificity of neutralizing antibodies to feline leukaemia viruses. *Int. J. Cancer* **21**: 768–778.

Russell, P. H. and Jarrett, O. (1978b) The occurrence of feline leukemia virus neutralizing antibodies in cats. *Int. J. Cancer* **22**: 351–357.

Sacca, R., Stanley, E. R., Sherr, C. J. and Rettenmier, C. W. (1986). Specific binding of the mononuclear phagocyte colony-stimulating factor CSF-1 to the product of the v-*fans* oncogene, *Proc. Natl. Acad. Sci. USA* **83**: 3331–3335.

Sarma, P. S. and Log, T. (1971) Viral interference in feline leukemia-sarcoma complex. *Virology* **44**: 352–358.

SARMA, P. and LOG, T. (1973) Subgroup classification of feline leukemia and sarcoma viruses by viral interference and neutralization tests. *Virology* **54**: 160–169.

SARMA, P. S., SHARAR, A., WALTERS, V. and GARDNER, M. (1974) A survey of cats and humans for prevalence of feline leukemia-sarcoma virus neutralizing serum antibodies. *Proc. Soc. exp. Biol. Med.* **145**: 560–574.

SCHWARTZ, D. E., TIZARD, R. and GILBERT, W. (1983) Nucleotide sequence of Rous sarcoma virus. *Cell* **32**: 853–869.

SCOLNICK, E. M., RANDS, E., WILLIAMS, D. and PARKS, W. P. (1973) Studies on the nucleic acid sequences of Kirsten sarcoma virus: a model for formation of a mammalian RNA-containing sarcoma virus. *J. Virol.* **12**: 458–463.

SHANK, P. R., HUGHES, S. H., KUNG, H.-J., MAJORS, J. E., QUINTRELL, N., GUNTAKA, R. V., BISHOP, J. M. and VARMUS, H. E. (1978) Mapping unintegrated avian sarcoma virus DNA: termini of linear DNA bear 300 nucleotides present once or twice in two species of circular DNA. *Cell* **15**: 1383–1395.

SHERR, C. J., ANDERSON, S. J., RETTENMIER, C. W. and ROUSSEL, M. F. (1984) The oncogenes *fes* and *fms*. In: *Cancer Cells 2/Oncogenes and Viral Genes* pp. 329–338, VANDE WOUDE, G. F., TOPP, W., LEVINE, A. J. and WATSON, J. D. (eds) Cold Spring Harbor Laboratory, New York.

SHERR, C. J., DONNER, L., FEDELE, L. A., TUREK, L., EVEN, J. and RUSCETTI, S. K. (1980a) Molecular structure and products of feline sarcoma and leukemia viruses: relationship to FOCMA expression. In: *Feline Leukemia Virus*, pp. 293–307, HARDY, W. D., JR., ESSEX, M. and MCCLELLAND, A. J. (eds) Elsevier/North Holland, Amsterdam.

SHERR, C. J., FEDELE, L. A., OSKARSSON, M., MAIZEL, J. and VANDE WOUDE, G. (1980b) Molecular cloning of Snyder–Theilen feline leukemia and sarcoma viruses: comparative studies of feline sarcoma virus with its natural helper virus and with Moloney murine sarcoma virus. *J. Virol.* **34**: 200–212.

SHERR, C. J., RETTENMIER, C. W., SACCA, R., ROUSSEL, M. F., LOOK, A. T. and STANLEY, E. R. (1985) The c-*fms* proto-oncogene product is related to the receptor for the mononuclear phagocyte growth factor, CSF-1. *Cell* **41**: 665–676.

SHERR, C. J., SEN, A., TODARO, G. J., SLISKI, A. and ESSEX, M. (1978) Pseudotypes of feline sarcoma virus contain an 85,000 dalton protein with feline oncornavirus-associated cell membrane antigen (FOCMA) activity. *Proc. natn. Acad. Sci. U.S.A.* **75**: 1505–1509.

SHIBUYUA, M. and HANAFUSA, H. (1982) Nucleotide sequence of Fujinami sarcoma virus: evolutionary relationship of its transforming gene with transforming genes of other sarcoma viruses. *Cell* **30**: 787–795.

SHIBUYA, M., HANAFUSA, T., HANAFUSA, H. and STEPHENSON, J. R. (1980) Homology exists among the transforming sequences of the avian and feline sarcoma viruses. *Proc. natn. Acad. Sci. U.S.A.* **77**: 6536–6540.

SNYDER, M. A., BISHOP, J. M., COLBY, W. W. and LEVINSON, A. D. (1983) Phosphorylation of tyrosine-416 is not required for the transforming properties and kinase activity of pp60 v-*src*. *Cell* **32**: 891–901.

SNYDER, S. P. and THEILEN, G. H. (1969) Transmissible feline fibrosarcoma. *Nature* **221**: 1074–1075.

STEHELIN, D., GUNTAKA, R. V., VARMUS, H. E. and BISHOP, J. M. (1976a) Purification of DNA complementary to nucleotide sequences required for neoplastic transformation of fibroblasts by avian sarcoma viruses. *J. molec. Biol.* **101**: 349–365.

STEHELIN, D., VARMUS, H. E., BISHOP, J. M. and VOGT, P. K. (1976b) DNA related to the transforming gene(s) of avian sarcoma viruses is present in normal avian DNA. *Nature* **260**: 170–173.

STEPHENSON, J. R., KHAN, A. S., SLISKI, A. H. and ESSEX, M. (1977) Feline oncornavirus-associated cell membrane antigen: evidence for an immunologically cross-reactive feline sarcoma virus coded protein. *Proc. natn. Acad. Sci. U.S.A.* **74**: 5608–5612.

TEICH, N. (1982) Taxonomy of retroviruses. In: *RNA Tumor Viruses*, pp. 25–207, WEISS, R., TEICH, N., VARMUS, H. and COFFIN, J. (eds) Cold Spring Harbor Laboratory, New York.

TEICH, N., WYKE, J., MAK, T., BERNSTEIN, A. and HARDY, W. (1982) Pathogenesis of retrovirus-induced disease. In: *RNA Tumor Viruses*, pp. 785–998, WEISS, R., TEICH, N., VARMUS, H. and COFFIN, J. (eds) Cold Spring Harbor Laboratory, New York.

TEMIN, H. M. (1981) Structure, variation, and synthesis of retrovirus long terminal repeat. *Cell* **27**: 1–3.

TEMIN, H. M. and BALTIMORE, D. (1972) RNA-directed DNA synthesis and RNA tumor viruses. *Adv. Virus Res.* **17**: 129–186.

THEILEN, G. H., SNYDER, S. P., WOLFE, L. G. and LANDON, J. L. (1970) Biological studies with viral induced fibrosarcomas in cats, dogs, rabbits, and nonhuman primates. In: *Comparative Leukemia Research*, pp. 393–400, DUTCHER, R. M. (ed) Karger, Basel.

TSICHLIS, P. N. and COFFIN, J. M. (1980) Recombinants between endogenous and exogenous avian tumor viruses: role of the C region and other portions of the genome in the control of replication and transformation. *J. Virol.* **33**: 238–249.

USHIRO, H. and COHEN, S. J. (1980) Identification of phosphotyrosine as a product of epidermal growth factor-activated protein kinase in A-431 cell membranes. *J. biol. Chem.* **255**: 8363–8365.

VAN DE VEN, W. J. M., REYNOLDS, F. H. and STEPHENSON, J. R. (1980) The nonstructural components of polyproteins encoded by transformation-defective mammalian transforming retroviruses are phosphorylated and have associated protein kinase activity. *Virology* **101**: 185–197.

VARMUS, H. E. (1983) Retroviruses. In: *Mobile Genetic Elements*, pp. 411–503, SHAPIRO, J. A. (ed.) Academic Press, New York.

WATERFIELD, M. D., SCRACE, G. T., WHITTLE, N., STROOBANT, P., JOHNSSON, A., WASTESON, A., WESTERMARK, B., HELDIN, C.-H., HUANG, J. S. and DEUEL, T. F (1983) Platelet derived growth factor is structurally related to the putative transforming protein p28*sis* of simian sarcoma virus. *Nature* **304**: 35–39.

WEISS, S. R., VARMUS, H. E and BISHOP, J. M. (1977) The size and genetic composition of virus specific RNAs in the cytoplasm of cells producing avian sarcoma-leukemia viruses. *Cell* **12**: 983–992.

YAMAMOTO, T., NISHIDA, T., MIYAJIMA, N., KAWAI, S., OOI, T. and TOYOSHIMA, K. (1983) The *erb*B gene of avian erythroblastosis virus is a member of the *src* gene family. *Cell* **35**: 71–78.

CHAPTER 11

THE *fos* GENE

Inder M. Verma and W. Robert Graham

Molecular Biology and Virology Laboratory, The Salk Institute, PO Box 85800, San Diego, California 92138, USA

1. HISTORICAL BACKGROUND

Viruses were implicated in the formation of tumors at the turn of the century. In 1908 Ellerman and Bang used a filtrate from a chicken's lymphoma to produce lymphomatosis in chickens (Ellerman and Bang, 1908). Peyton Rous used a similar procedure in 1911 to isolate a different agent (now known as the Rous sarcoma virus) which produced sarcomas in chickens (Rous, 1911). Over the ensuing years filtrates from tumors of many different animal species were found to contain agents that could induce abnormal growths. Mice were added to the list in 1942 when Bittner isolated such an agent from the filtrate of milk from a mouse with a breast tumor. When injected into mice, mammary tumors developed. Subsequent work uncovered many filterable agents which produced tumors in mice (Gross, 1951, 1953a, 1953b, 1958; Friend, 1957). One such agent was isolated by Finkel, Biskis and Jinkins in 1966 from an osteosarcoma that spontaneously arose on the thoracic spine and ribs of a 260-day-old CF1/An1 mouse (Finkel *et al.*, 1966). This strain of mouse had been chosen for study because of its high incidence of spontaneous bone tumors (20–25%). When an extract made from the osteosarcoma was injected subcutaneously into newborn and 33-day-old mice, they developed boney tumors which appeared to arise from the periosteum and grew outward. The pathology of the tumors resembled parosteal sarcomas found in humans (Finkel *et al.*, 1966). Microscopic examination revealed considerable histological variation from region to region within an individual tumor and between tumors. The cellular types included fibroblasts, osteoblasts, osteocytes and giant cells and there was wide variation in the amount of osteoid and the degree of ossification.

2. BIOLOGY OF *fos* VIRUSES

Examination of the tumor tissue by electron microscopy demonstrated viral particles. The virus was designated FBJ after its founders (Finkel *et al.*, 1966). Subsequent pathological studies confirmed the rather wide variation seen in the deposition of osteoid and other interstitial substances (Kellof *et al.*, 1969; Yumoto *et al.*, 1970; Price *et al.*, 1972; Finkel *et al.*, 1972; Ward and Young, 1976). The variation in cellular appearances led Yumoto to subclassify FBJ-induced tumors into osteosarcoma, fibrosarcoma, chondrosarcoma, myxo-fibro-osteosarcoma, osteochondrosarcoma, fibroosteosarcoma and chondrofibrosarcoma (Yumoto *et al.*, 1970). Although such variation was found, when individual osteoblasts were examined by electron miscroscopy no differences were found between those in the tumor and normal osteoblasts, suggesting that the FBJ virus did not inexorably interrupt differentiation. Further characterization of FBJ-induced tumors came through the study of specific markers. It was found that such tumors expressed high levels of alkaline phosphatase, a marker for osteoblast (Price *et al.*, 1972; Ward and Young, 1976).

This finding was consistent with the classification of these tumors as osteosarcomas, which express high levels of this enzyme. Production of type I collagen, which is normally produced by osteoblasts and fibroblasts but not chondroblasts, was also identified (Curran, 1982). These findings support the contention that osteoblasts are primary cellular targets for FBJ. However, it was shown that even those FBJ-induced tumors which arose at sites removed from bone, such as those which arose on the peritoneum, also produced alkaline phosphatase (Ward and Young, 1976). This led them to suggest that a multipotential cell is the target for the FBJ virus. Infection of a multipotential cell could account for the variable histological appearance of the tumor in that once such a cell is infected, it might still be subject to different hormones or differentiation factors and therefore appear as different cellular types. Variations in the synthesis of differentiation-specific products such as type I collagen would also be seen and thus result in a variable histological appearance. However, the predominance of FBJ tumors around bones, the high levels of alkaline phosphatase expression, the production of type I collagen and the deposition of osteoid in these tumors all point to osteoblastic precursors as one major target for viral infection and transformation. Other cellular types which might be the targets of FBJ infection in vivo have not been rigorously determined.

It is important to emphasize that 90–100% of mice infected with the FBJ virus develop tumors that are associated with bone. In that sense it is a highly specific virus. These tumors often arise on several bones and sometimes the peritoneum indicating multiple sites of viral tumor formation, but metastases are not seen (Ward and Young, 1976). Local complications of tumor growth are responsible for the morbidity and mortality associated with the FBJ virus. In many ways, then, it is quite different from what one sees with osteosarcomas in humans, where the tumors develop in the deep boney cortex and metastases are prominent. Parosteal sarcomas that arise in humans are more benign and exhibit a growth pattern which is similar to that seen with FBJ-induced tumors in mice (Van Der Huel and Von Ronnen, 1967). Perhaps the FBJ virus would have been better designated as a parosteal tumor virus.

Early studies with FBJ viral extracts were hampered by the inability to quantitate the viral titer. Consequently, there was considerable variation in latent periods and incidence of tumors; a low titer extract produced a disease with a long latent period or no disease at all whereas with high titer virus 100% of susceptible mice developed parosteal tumors with latent periods as short as three weeks (Finkel et al., 1975). The ability to quantitate the virus came in 1973 when the FBJ viral complex was described (Levy et al., 1973). A helper virus, designated FBJ-MLV, was identified using MLV group specific antisera and XC plaque assays (Levy et al., 1973). The defective virus, FBJ-MSV, was identified by focus formation on rat 208F cells. The helper virus produced much higher titers in NIH-3T3 cells, which carry the (Fv-1^{nn}) genotype, as compared with BALB/c cells, which carry the (Fv-1^{bb}) genotype, indicating that FBJ-MLV is an N-tropic virus (Levy et al., 1973; Hartley et al., 1970; Weiss et al., 1982). The helper virus (FBJ-MLV) did not induce any disease in mice (Levy et al., 1973). When the entire viral complex (FBJ-MLV and FBJ-MSV) was injected, parosteal tumors developed. In addition, when FBJ-MSV was combined with another helper virus, again parosteal tumors were produced. A titer of at least 1000 focus-forming units (FFUs) is needed to produce tumors reliably. As opposed to other sarcomas, the FBJ-induced tumors were found to be transplantable (Levy et al., 1973). It is generally assumed that a tumor which produces virus is not transplantable into an immuno-competent mouse because the viral antigens evoke a vigorous immune response and the tumor is destroyed. The best explanation for these findings is that the FBJ-induced tumors progressed *in vivo* such that they were no longer producing virus. The FBJ viral complex was purported easily to establish primary rat or mouse fibroblasts in cultures, whereas other MSVs had been unable to accomplish this (Bather et al., 1968; Levy, 1971; Levy and Rowe, 1971; Levy et al., 1973). However, others were unable efficiently to transform and establish rat embryo cells in culture using the FBJ viruses (Rhim et al., 1969). Recently another group was unable to establish primary mouse embryo fibroblasts that had been transfected with FBJ-MSV DNA, in spite of morphological transformation (Jenuwein et al., 1985). It is doubtful that FBJ-MSV more efficiently transforms or immortalizes cells when compared with other MSVs. It is probably best to state that, like other defective viruses, the role of FBJ-MSV in the complex scheme of neoplasia remains obscure. This scheme involves a multistep process and some of these steps may be induced by tissue culture conditions or other experimental manipulations, which may explain some of the discrepancies in the studies cited above.

It is clear, however, that FBJ-MSV will mediate the formation of primary tumors in mice and will induce morphological transformation of cells in culture. By analogy with other defective retroviruses, it was assumed that FBJ-MSV harbored an oncogene. The product of this oncogene was identified to be a 55,000 dalton phosphoprotein (Curran and Teich, 1982b), which was designated v-*fos*. This 55 kda protein (p55) was identified by immunoprecipitation using sera from rats that had been injected with FBJ-MSV transformed cells. Such rats developed tumors and their sera were referred to as tumor bearing rat sera (TBRS). This TBRS also precipitated a 39,000 dalton protein of host origin (Curran and Teich, 1982a). p55 could not be precipitated from cells transformed with other oncogenes. v-*fos* thus represents an essential component in the FBJ viral complex and is apparently responsible for the altered (transformed) morphology of cells infected in tissue culture and for the induction of tumors *in vivo*. Further evidence for an important role of v-*fos* in neoplasia came with the study of the FBR virus. This was isolated from an osteosarcoma that developed in a X/GF mouse following treatment with 90Sr (Finkel et al., 1972). These mice were chosen for study because of their resistance to the formation of tumors. The viral complex was shown to contain a defective virus (FBR-MSV) and a B-tropic helper virus (FBR-MLV) (Lee et al., 1979). The FBR-MSV was found to code for a 75,000 dalton *gag-fos* phosphoprotein that had a region homologohus to v-*fos* (Curran and Verma, 1984; Van Beveren et al., 1984; Michiels et al., 1984). Mice injected with the FBR viral complex developed a disease that was identical to that obtained with the FBJ viral complex—i.e., parosteal tumors. As with FBJ, when FBR-MSV DNA was transfected onto cells they exhibited a transformed morphology. FBR-MSV DNA, however, also established primary cells in culture and these cells produced tumors when injected into syngeneic or nude mice (Jenuwein et al., 1985). In certain respects then, the FBR viral complex appears to be more potent when compared with the FBJ viral complex. Nevertheless, these findings again placed v-*fos* sequences in a pivotal role for tumor formation. To gain further insight into the role of the *fos* gene in the induction of tumors, we have undertaken structural and functional analyses.

3. STRUCTURE OF THE *fos* GENE AND PROTEIN

The complete nucleotide sequences of FBJ-MSV, FBR-MSV proviral DNA and the cellular progenitor of the *fos* gene have been determined (Van Beveren et al., 1983, 1984). Fig. 1 is a diagram of the organization of the viral and cellular *fos* genes and their deduced products. The salient features can be summarized as follows:
  1. FBJ-MSV proviral DNA contains 4026 nucleotides, including two long terminal repeats (LTRs) of 617 nucleotides each, 1639 nucleotides of acquired cellular sequences (v-*fos*), and a portion of the envelope (*env*) gene.
  2. Both the initiation and termination codons of the v-*fos* protein are within the acquired sequences that encode a protein of 381 amino acids, having a molecular weight of 49,601.
  3. In cells transformed by FBJ-MSV, a phosphoprotein with an apparent MW of 55,000 (p55) on SDS polyacrylamide gel electrophoresis (SDS-PAGE), has been identified as the transforming protein (Curran et al., 1982). The discrepancy between the observed size and the size predicted by sequence analysis is likely due to the unusual amino acid composition of the *fos* protein (10% proline), since the v-*fos* protein expressed in bacteria has a similar relative mobility (MacConnell and Verma, 1983).
  4. The sequences in the c-*fos* gene that are homologous to those in the v-*fos* gene are

A

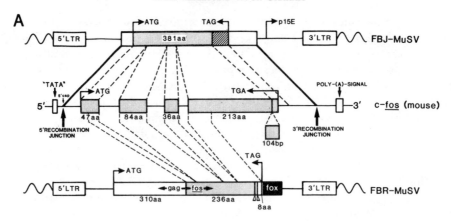

B

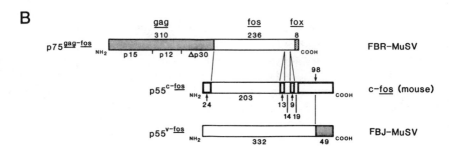

Fig. 1. (A) Molecular architecture of FBJ-MSV (*top*) and FBR-MSV (*bottom*) proviral DNAs and the c-*fos* gene (*middle*). (*Top*) The large open box indicates the acquired cellular sequences; solid, vertical bars indicate the initiation and termination codons of v-*fos* proteins; the hatched region indicates the carboxy-terminal 49 amino acids of the v-*fos* protein encoded in a different reading frame due to deletion of 104 bp of c-*fos* sequences. (*Middle*) The stippled boxes are the exons; the number of amino acids encoded by each exon is given. The 104 bp sequence that has been deleted in the v-*fos* sequence is indicated with a box below the line. Unlike the v-*fos* protein, the c-*fos* protein terminates at a TGA codon. (*Bottom*) Broken lines indicate the portions of the exon acquired from the c-*fos* gene; small, open triangles indicate deletion from FBR-MSV as compared with the c-*fos* gene. (B) A schematic comparison of p75$^{gag\text{-}fos}$ (*top*), p55$^{c\text{-}fos}$ (*middle*), and p55$^{v\text{-}fos}$ (*bottom*) proteins. In p75 the *gag*-encoded portion is indicated with stippled box, and that encoded by v-*fos* is shown by the open box. The regions of p55$^{c\text{-}fos}$ indicated by thickened boxes and vertical arrows are those portions deleted in p75$^{gag\text{-}fos}$. The stippled region in p55$^{v\text{-}fos}$ is the carboxy-terminal portion, which differs from that of p55$^{c\text{-}fos}$. The numbers refer to the number of amino acids encoded by each region. The data in this figure are compiled from Van Beveren *et al.* (1983, 1984).

interrupted by four regions of nonhomology, three of which represent bona fide introns.

5. The 104-nucleotide-long fourth region, which is present in both mouse and human c-*fos* genes, represents sequences that have been deleted during the biogenesis of the v-*fos* gene. (The additional 104 nucleotides in the c-*fos* gene transcripts do not increase the predicted size of the c-*fos* proteins, because of a switch to a different reading frame.)
6. The c-*fos* protein has 380 amino acids, which is remarkably similar to the size of the v-*fos* protein (381 amino acids).
7. In the first 332 amino acids, the v-*fos* and mouse c-*fos* proteins differ at only five residues, whereas the remaining 48 amino acids of the c-*fos* protein are encoded in a different reading frame from that in the v-*fos* protein. Thus, the v-*fos* and c-*fos* proteins, though largely similar, have different carboxyl termini (Fig. 1B).
8. Despite their different caroboxyl termini, both the v-*fos* and c-*fos* proteins are located in the nucleus (Curran *et al.*, 1984).
9. The mouse and human c-*fos* genes share greater than 90% sequence homology,

differing in only 24 residues out of a total of 380 amino acids (van Straaten *et al.*, 1983).

10. The putative parents of FBJ-MSV (namely FBJ-MLV and c-*fos* gene) share a five nucleotide sequence at the 5' end and ten to eleven nucleotides at the 3' end of the v-*fos* region. Sequences involved in re-combination at the 5' end lie in the untranslated region of both FBJ-MLV and mouse c-*fos* gene.

11. FBR-MSV proviral DNA contains 3791 nucleotides (specifically a genome of 3284 bases) and encodes a single *gag-fos* fusion product of 554 amino acids.

12. The *fos* portion of the gene lacks sequences that encode the first 24 and the last 98 amino acids of the 380 amino acids mouse c-*fos* gene product (Fig. 1B). In addition, the coding region has sustained three small in-frame deletions, one in the p30gag portion and two in the *fos* region, as compared with sequences of AKR MLV and the c-*fos* gene, respectively (Van Beveren *et al.*, 1984).

13. The gene product terminates in sequences termed *fox* (Fig. 1A), which are present in normal mouse DNA at loci unrelated to the c-*fos* gene. The c-*fox* gene(s) is expressed as an abundant class of polyadenylated RNA in mouse tissue.

14. A retrovirus, FBJ/R, containing N-terminal sequences of FBJ-MSV and C-terminal sequences of FBR-MSV containing 268 amino acids was generated, which transformed fibroblasts *in vitro* (Miller *et al.*, 1985). Thus, the *gag* moiety of FBR-MSV can be removed without affecting its transforming potential.

15. The transforming *fos* proteins vary in size from 268 amino acids (FBJ/R) to 381 amino acids (FBJ), but they all have nuclear location. Additionally, all *fos* proteins co-immunoprecipitate a 39 K cellular protein whose identity remains obscure.

16. *fos*-specific antisera has been obtained from tumor-bearing rats injected with FBJ-MSV transformed cells (Curran and Teich, 1982a), as well as rabbits injected with specific peptides synthesized from various regions of the *fos* protein (Curran *et al.*, 1985).

17. Both the viral and cellular *fos* proteins are post-translationally modified (Curran *et al.*, 1984). The cellular *fos* protein is more extensively modified with molecular weights ranging from 55 to 72 kD (Curran *et al.*, 1984; Kruijer *et al.*, 1984). The extent and precise nature of the *fos* protein modifications remain unknown but part of the modifications are due to phosphorylation (Barber and Verma, 1987).

4. TRANSFORMATION BY *fos* GENE

Both FBJ-MSV and FBR-MSV containing the *fos* gene can transform established fibroblast cell lines (Curran *et al.*, 1982; Curran and Verma, 1984). Additionally, it has been reported that these viruses can also induce discernible foci in primary mouse embryo fibroblasts (Jenuwein *et al.*, 1985). The cellular *fos* gene can also induce transformation, but requires at least two manipulations: (a) addition of LTR sequences, presumably to increase transcription by enhancer sequences, and (b) removal of sequences downstream of the coding domain (Miller *et al.*, 1984). A number of recombinant constructs (Fig. 2) were generated which contained various portions of the viral and cellular *fos* genes. Briefly, the v-*fos* and c-*fos* genes were split into three parts, namely (i) the promoter region and the first 316 amino acids originating either from the viral or c-*fos* gene; (ii) the C-terminus, 64–65 amino acids of the coding domain of either viral or c-*fos* gene; and (iii) the 3' noncoding domain [including poly(A) addition signal] originating from either v- or c-*fos* gene. Thus, a construct referred to as VVV means that the promoter, the coding domain and the 3' noncoding region all originate from the FBJ-MSV proviral DNA, while MMM signifies that complete c-*fos* (mouse) gene was used. A construct VMV would indicate that the C-terminus of the *fos* protein is cellular, while MVM would contain the viral C-terminus. The results of transformation by various constructs unequivocally demonstrate that both the v-*fos* and the c-*fos* proteins can induce cellular transformation.

Constructs like VMM which do not induce transformation are efficiently transcribed but are unable to synthesize sufficient p55 *fos* protein (Miller *et al.*, 1984). In comparison when the 3′ noncoding sequences are removed as in the transforming construct VM(A)$_n$, at least ten times more *fos* protein is synthesized (F. Meijlink, T. Curran and I. M. Verma, unpublished results). It is worth noting that the only difference between transforming VVM and non-transforming VMM is the altered C-terminus. Thus it would appear that the noncoding sequences interact with the C-terminus to prevent transformation. The nature of the sequences in the noncoding domain that is involved in 3′ interactions has been extensively analyzed and localized to an A-T rich 67 bp region located some 500 nucleotides downstream from the end of the coding domain and about 120 nucleotides upstream of the poly(a) addition signal sequence (Meijlink *et al.*, 1985). The precise nature of the 67 bp region in abolishing the transforming potential of c-*fos* gene is not understood. Two possible mechanisms can be advanced: (i) autoregulation of c-*fos* protein synthesis by interaction of the c-*fos* protein with the 67 bp region and/or the C-terminus, or (ii) the presence of the 67 bp region influences the stability of the c-*fos* mRNA.

An underlying basic principle of tumor induction is the assumption that a number of events conspire to acquire the malignant phenotype. Support of this multistep carcinogenesis was advanced by the observation that more than one oncogene is required to induce transformation of primary embryo fibroblasts (Land *et al.*, 1983; Ruley, 1983). It is now a general consensus that nuclear oncoproteins like *myc, myb* polyoma large T antigens, etc., collaborate with other oncogenes (cytoplasmic or plasma membrane) to induce transformation of primary embryo fibroblasts. The *fos* gene product, despite being a nuclear protein, defies any such categorization because it can induce transformation of both primary cultures as well as those of established cell lines (Jenuwein *et al.*, 1985). It appears that FBR-MSV may be more adept at extending self-renewal of non-established mouse cells than FBJ-MSV (Jenuwein *et al.*, 1985). The FBR-MSV has undergone several structural alterations as compared to FBJ-MSV, which may in part account for its higher transforming potential.

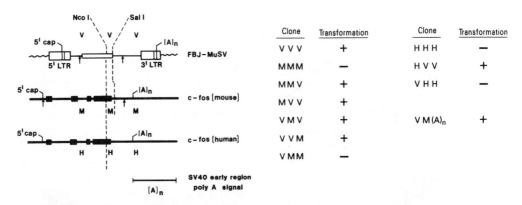

FIG. 2. Transforming potential of c-*fos* gene: The top line depicts the FBJ-MuSV provirus. The v-*fos* coding region is shown by an open box in the middle of the proviral DNA. Rat DNA sequences surrounding the provirus are shown by wavy lines. The middle and lower lines depict the mouse and human c-*fos* genes, respectively, with the c-*fos* coding regions shown by closed boxes. The coding regions of each of the c-*fos* genes are separated by three introns. The restriction endonucleases Nco I and Sal I divide the v-*fos* and mouse c-*fos* genes into three regions, and the human c-*fos* into two regions, as shown. RNA 5′ cap and polyadenylation signals are shown. The origins of these plasmids are described (Miller *et al.*, 1984). The arrows in the FBJ-MuSV provirus and in the c-*fos* gene indicate the positions of recombination between the mouse gene and the helper retrovirus that generated FBJ-MuSV. Clones were tested for transforming ability by transfection onto rat 208F cells. (+) indicates transformation efficiency of about 200 foci/μg DNA. (−) indicates transforming efficiency of <10 foci/μg DNA.

5. EXPRESSION OF THE c-fos GENE

Proto-oncogene *fos* is a highly inducible gene. It is expressed in a wide variety of cell types during development, growth and differentiation, often in response to a variety of mitogens or differentiation-specific inducers. Table 1 shows a variety of cell types where *fos* gene is induced (Verma, 1986). Below we illustrate some specific examples.

TABLE 1

c-*fos* induction

1) Promonocyte/Monomyelocyte $\frac{TPA}{Vit\ D_3}$ macrophages
2) PC12 $\frac{NGF}{cAMP,\ K^+}$ neurites
3) Partial hepatectomy
4) Spleen cells stimulated with ConA or LPS
5) Resting BALB/C or NIH 3T3 cells stimulated with PDGF, serum or TPA
6) Hepatocytes stimulated with growth factors
7) A431 + EGF
8) Primary rat pituitary cells + GRF

No c-*fos* induction

1) Monomyelocyte $\xrightarrow{DMSO}$ granulocytes
2) PC12 $\xrightarrow{DEX}$ chromaffin cells

5.1 Expression during prenatal development

During mouse prenatal development, the highest accumulation of c-*fos* transcripts was detected in late gestational extraembryonal membranes (amnion, yolk sac) (Müller et al., 1982; Müller et al., 1983a; Müller and Verma, 1984). Small amounts of *fos* transcript are also detected in placenta and mid-gestation fetal liver (Muller et al., 1984b). The levels of C-*fos* transcripts are low during day 10–11, but increase by day 17–18. At day 18 nearly all cells in mouse amnion contain c-*fos* transcripts as judged by *in situ* hybridizaton (Deschamps et al., 1985b). Not only is the *fos* gene transcribed but also the *fos* proteins can be identified in day 17–18 amniotic membranes (Curran et al., 1984). c-*fos* expression was also witnessed in human amnion and placenta (Müller et al., 1983b). Expression of c-*fos* transcripts is generally detected following induction with mitogenic or differentiation-specific agents, but in amnion cells there appears to be a constitutive synthesis of the *fos* transcripts. It was previously postulated that perhaps the amnion cells are continuously stimulated with growth factors from placenta (Verma et al., 1985). Evidence for this assumption is provided by the experiment that amnion cells in an *in vitro* culture do not synthesize c-*fos* after plating but can be stimulated to make c-*fos* transcripts if dialyzed placenta or embryo-conditioned medium is added (Müller et al., 1986). It is not clear, however, if c-*fos* plays any role during prenatal development. It doesn't appear to be involved in proliferation of amnion cells because in vitro cultures of amnion cells synthesize DNA without making *fos* transcripts (Müller et al., 1986).

5.2. Expression during cell growth

When quiescent mouse fibroblasts are treated with serum or growth factors like PDGF, EGF or TPA, the proto-oncogene *fos* is rapidly induced (Cochran et al., 1984; Greenberg and Ziff, 1984; Kruijer et al., 1984; Müller et al., 1984a; Bravo et al., 1985). The salient features of these observations can be summarized as follows: (1) Within 2–3 min of stimulation of growth, c-*fos* transcripts can be detected as measured by hybridization with ^{32}P-labeled cRNA (Kruijer et al., 1984). (2) Maximal levels of induction occur within 20 min (20-fold induction) of the exposure of cells to 0.83 nM purified PDGF. The levels declined by 60 min and by 240 min little or no c-*fos* transcripts could be detected. (3) Addition of cycloheximide resulted in a 50-fold induction. Addition of anisomycin, another inhibitor of protein synthesis, also results in 'superinduction' of the c-*fos* gene (Greenberg *et al.*, 1986). We assume that this 'superinduction' represents stabilization of the mRNA since little or no *de novo* transcription is observed. (4) We estimate that after 20 min of exposure to PDGF, 0.0001% of NIH-3T3 cell RNA (0.0005% of mRNA) is c-*fos* mRNA. Assuming a cellular RNA content of 6 pg, this corresponds to about 5–10 copies of *fos* mRNA per cell. (5) Exposure to PDGF for as short as 30 min induces the synthesis of *fos* protein which can be detected by immunoprecipitation with *fos*-specific peptide antisera. (6) At least 6–8 polypeptides are identified by immune precipitation, most of which represent modified forms of *fos* protein; however, some non-*fos* polypeptides are also precipitated. One possibility is that some of them may be related to *fos* (R-*fos*) and may react with peptide antisera (Cochran *et al.*, 1984). (7) c-*fos* protein synthesis was maximal with PDGF concentrations that saturate PDGF binding sites at 37°C (1.0 nM) and half-maximal at 0.3–0.5 nM.

Induction of the c-*fos* gene when cells transit from the G_0 to the G_1 state suggests that it may have a role in the cell cycle. Another proto-oncogene, c-*myc*, has also been shown to be induced during the G_0–G_1 transition (Kelly *et al.*, 1983). The role of proto-oncogenes *fos* and *myc* during the cell cycle is difficult to reconcile in view of the results of their invariant amounts during the cell cycle (Thompson *et al.*, 1985; Haan *et al.*, 1985; Bravo *et al.*, 1986).

Rapid and transient induction of the *fos* gene was observed in regenerating liver 10–60 min following partial hepatectomy (Kruijer *et al.*, 1986). Addition of cycloheximide stabilized c-*fos* RNA. Modified forms of the c-*fos* protein can be identified (Kruijer *et al.*, 1986). The in vivo results can be simulated by exposing quiescent adult rat hepatocytes in primary cultures to hepatotrophic factors like EGF or serum. Recently it has been observed that when primary pituitary cells are treated with growth hormone release factor (GRF), c-*fos* transcripts are rapidly and transiently induced (R. Mitchell and I. Verma unpublished material). The function of the *fos* protein during cell growth at present remains elusive and largely correlative.

5.3. Expression during differentiation

Proto-oncogene *fos* is expressed during the differentiation of a variety of cell lines. Depending on the nature of the inducer, in some cases c-*fos* expression appears to be lineage specific (Table 1). Below we describe two systems:

5.3.1 *Hematopoietic differentiation*

The earliest hint that c-*fos* expression may be modulated during hematopoiesis came from the work of Gonda and Metcalf (1984). They observed c-*fos* transcripts when WEH1-3B murine myeloid leukemia cell line was induced to macrophage-like differentiation with granulocytic colony stimulating factors. Subsequently it was observed that c-*fos* transcripts can be found in bone marrow and parietal exudates containing macrophages (Müller *et al.*, 1984b, 1985; Mitchell *et al.*, 1985). When human monomyelocytic cell line, HL-60, or promonocytic cell line U-937 are treated with phorbol esters (TPA), c-*fos* gene is rapidly induced. Within minutes, the c-*fos* transcripts can be observed. They accumulate to maximal levels in 30–60 min followed by a decline of 4- to 5-fold and then the levels remain unchanged for the next 109 hours (Mitchell *et al.*, 1985). c-*fos* protein can, however, be detected for only 60–120 min post-induction. In contrast, when HL-60 cells are induced to differentiate to granulocytes by addition of DMSO, no c-*fos* transcripts can be observed (Mitchell *et al.*, 1985). Thus, it is tantalizing to propose some role of c-*fos* protein during the monocytic differentiation pathway. Recent results from our laboratory, however, show that c-*fos* expression is neither sufficient nor obligatory for macrophage differentiation (Mitchell *et al.*, 1986). Two lines of evidence advance these inferences: (i) expression of high levels of c-*fos* by serum or diacylglycerol (DAG) does not commit U-937 cells to differentiate to macrophages, and (ii) TPA-resistant HL-60 cell lines can be induced to differentiate to macrophages with 1,25-dihydroxy vitamin D_3 without eliciting c-*fos* expression. At present the role of c-*fos* during myeloid differentiation remains uncertain.

5.3.2. *Neuronal differentiation*

A rat pheochromocytoma cell line (PC12) can be induced to differentiate to neurites upon addition of nerve growth factor (NGF), dibutryl cyclic AMP and 50 mM K^+ ions. One of the first molecular events to take place upon addition of the differentiation-specific inducers is the rapid but transient expression of the c-*fos* gene (Curran and Morgan, 1985; Kruijer *et al.*, 1986; Greenberg *et al.*, 1985). Within 5 min of post-induction, c-*fos* transcripts can be observed which reach maximal levels by 30–60 min. No c-*fos* specific transcripts can be observed by 120 min. Modified forms of the c-*fos* protein can also be detected during this period. When PC12 cells are treated with dexamethasone, they differentiate to become chromaffin-like cells. No c-*fos* expression is observed during this differentiation, again suggesting some lineage-specific expression of the c-*fos* gene.

It appears that most inducers of c-*fos* gene are also activators of protein kinase C. Thus one could tentatively assume that c-kinase may be involved in c-*fos* induction. In the absence of any specific inhibitors of c-kinase, it is difficult to undertake definitive experiments. Similarly, the role of the c-*fos* gene during differentiation remains conjectural and correlative. A definitive answer to the role of c-*fos* protein during differentiation will have to await experiments where its synthesis is selectively blocked or can be constitutively turned on without addition of inducing agents.

5.3.3. c-fos *induced differentiation*

Introduction of c-*fos* gene into undifferentiated F9 teratocarcinoma cell line leads to endoderm-like differentiation. When normal c-*fos* gene or metallothionein promoter linked to c-*fos* gene were introduced by DNA transfection into F9 stem cells, colonies of morphologically altered cells were obtained (Müller and Wagner, 1984). However, the c-*fos* induced differentiation was different from chemically induced differentiation because only a part of the expected morphological and biochemical changes were registered. Furthermore, many cell clones were isolated which showed high expression of the c-*fos* protein without any morphological or biochemical alterations (Rüther *et al.*, 1985). It would thus appear that expression of exogenously added *fos* gene may not be sufficient to induce complete differentiation of F9 stem cells. Other genes or factors may be required.

6. TRANSCRIPTION OF THE c-*fos* GENE

The rapid and transient induction of the c-*fos* gene transcription lends itself to a search for transcriptional enhancers and inducible sequences in the c-*fos* gene. Several reports have recently characterized the identification and location of transcriptional enhancer elements of the c-*fos* promotor (Treisman, 1985; Deschamps *et al.*, 1985a,b; Renz *et al.*, 1985). An element essential for transcriptional activation and inducibility in response to serum is located between nucleotides -276 and -332, relative to the 5' cap site. However, when this sequence is linked to a heterologous promoter, the extent of induction with either serum or TPA is only 3 to 5-fold (Treisman, 1985; Deschamps *et al.*, 1985b). Furthermore, the transcripts are more stable with increased constitutive levels (Treisman, 1985). By making fusion genes between human c-*fos* and mouse β-globin genes it has been shown that in addition to the 5' activating element, transient accumulation of c-*fos* RNA following induction with serum also requires sequences at the 3' end of the c-*fos* gene. The precise nature of these sequences remains unknown but they are located downstream of the 3' coding domain and include the interacting 67 nucleotide sequence discussed above. Two points are worth noting: (1) The 5' upstream sequences essential for transcriptional activation are conserved between human and mouse c-*fos* genes; these sequences also contain one of the two DNase I hypersensitive sites (Deschamps et al., 1985a). An additional DNase I hypersensitive site is found intragenically in the c-*fos* mouse gene but not the human (Renz et al., 1985). (2) The large variety of cell types where *fos* gene expression can be induced show that, as expected, the *fos* enhancer is not tissue-specific (Deschamps et al., 1985a).

7. REGULATION OF *fos* EXPRESSION

The gene is versatile; the gene product may play a role during development, cellular differentiation and cell growth. Since the c-*fos* protein can induce transformation of at least fibroblasts *in vitro*, it is puzzling that cells expressing *fos* genes in response to inducers do not succumb to transformation. Perhaps fibroblasts and other cells susceptible to transformation by c-*fos* protein are not transformed because the expression of the *fos* protein is transient. It is possible that some cell types, such as peritoneal macrophages or macrophages in culture are refractory to transformation by *fos* proteins. The synthesis of the *fos* gene product may be regulated post-transcriptionally or even more likely, at the translational level. As mentioned before, c-*fos* can induce cellular transformation if an AT-rich stretch of 67 base pairs located downstream of the termination codon is removed. We have no firm idea of the manner in which the 67 base pair sequences influence the synthesis of the c-*fos* protein, but it could either affect the stability of the mRNA or alter the translational efficiency of the c-*fos* mRNA. Little or no c-*fos* protein is detected in cells transfected with *fos* recombinant DNA constructs containing the 67 base pairs. Promonocytic or monomyelocytic cell lines induced to differentiate into macrophages continue to express *fos* mRNA for at least 10 days, but the *fos* protein is detected only for up to 120 minutes following induction. It is possible that the *fos* antiserum is unable to detect *fos* proteins because they are extensively modified. It is also difficult to comprehend how a protein found exclusively in the nucleus can influence translation. Apparently post-transcriptional or translational control of the expression of *fos* gene product is abrogated in mouse amnion cells where both the RNA and protein can be detected during prenatal development. It is possible that due to the presence of growth factors in the placenta, c-*fos* gene is continuously stimulated in amnion cells. Regardless of the molecular mechanism influencing *fos* expression, we believe that the natural expression of the c-*fos* protein does not transform cells because it is synthesized only transiently. In contrast, the v-*fos* gene escapes this regulation, due to an altered carboxyl terminus, and its sustained synthesis leads to cellular transformation.

Delineation of the function of the *fos* protein is a top priority for researchers in the field and elucidation of the molecular mechanism by which non-coding sequences influence the transforming potential of the c-*fos* gene remains a challenge. Why does *fos* protein cause only bone tumors? Are rearrangements or chromosomal translocation of the *fos* gene involved in any human tumors? With such marvellous prospects, no wonder the *fos* gene will continue to generate excitement.

Acknowledgments—We thank our many colleagues for providing unpublished data and Carolyn Goller for typing the manuscript. We extend special appreciation to Marguerite Vogt for her advice and interest.
    This work was supported by grants from the National Institutes of Health and the American Cancer Society to IMV and a grant from the National Cancer Institute to WRG.

REFERENCES

Bather, R., Leonard, A. and Yang, J. (1968) Characteristics of the *in vitro* assay of murine sarcoma virus (Moloney) and virus-infected cells. *J. Natl. Cancer Inst.* **40**: 551–560.

Bittner, J. J. (1942) Milk-influence of breast tumors in mice. *Science* **45**: 462–463.

Barber, J. & Verma, I. M. (1987) Modification of Fus proteins: phosphorylation of c. *fos*, but not v. *fos* is stimulated by TPA *mol. cell. biol.* (in press).

Bravo, R., Burckhardt, J., Curran, T. and Müller, R. (1985) Stimulation of growth by EGF in different A431 cells is accompanied by rapid induction of c-*fos* and c-*myc* genes. *EMBO J.* **4**: 1193–1198.

Bravo, R., Burckhardt, J., Curran, T. and Müller, R. (1986) Expression of c-*fos* in NIH3T3 cells is very low but inducible throughout the cell cycle. *EMBO J.* **5** 695–700.

Cochran, B. M., Zullu, J., Verma, I. M. and Stiles, C. D. (1984) Expression of c-*fos* oncogene and of a *fos*-related gene as stimulated by platelet-derived growth. *Science* **226**: 10–80.

Curran, T. (1982) FBJ murine osteosarcoma virus. Ph.D. Thesis, Imperial Cancer Research Fund Labs., London, England.

Curran, T. and Morgan, J. P. (1985) Superinduction of the c-*fos* by nerve growth factor in the presence of peripherally active benzodiazepines. *Science* **229**: 1265–1268.

Curran, T. and Teich, N. M. (1982a) Identification of a 39,000 dalton protein in cells transformed by the FBJ murine osteosarcoma virus. *Virology* **116**: 221–235.

Curran, T. and Teich, N. M. (1982b) Product of the FBJ murine osteosarcoma virus oncogene: characterization of a 55,000 dalton phosphoprotein. *J. Virol.* **42**: 114–122.

Curran, T. and Verma, I. M. (1984) FBR murine osteosarcoma virus. I. Molecular analysis and characterization of a 75,000 Da *gag-fos* fusion product. *Virology* **135**: 218–228.

Curran, T., Miller, A. D., Zokas, L. and Verma, I. M. (1984) Viral and cellular *fos* proteins: a comparative analysis. *Cell* **36**: 259–268.

Curran, T., Peters, G., Van Beveren, C., Teich, N. and Verma, I. M. (1982) FBJ murine osteosarcoma virus: identification and molecular cloning of biologically active proviral DNA. *J. Virol.* **44**: 674–682.

Curran, T., Van Beveren, C., Ling, N. and Verma, I. M. (1985) Viral and cellular *fos* proteins are complexed with a 39,000 dalton cellular protein. *Mol. Cell. Biol.* **5**: 107–112.

Deschamps, J., Meijlink, F. and Verma, I. M. (1985a) Identification of a transcriptional enhancer element upstream from the proto-oncogene *fos*. *Science* **230**: 1174–1177.

Deschamps, J., Mitchell, R. L., Meijlink, F., Kruijer, W., Schubert, D. and Verma, I. M. (1985b) Proto-oncogene *fos* is expressed during development, differentiation, and growth. *Cold Spring Harbor Symp. Quant. Biol.* **50**: 733–745.

Ellerman, V. and Bang, O. (1908) Experimentelle leukamie bei huhnern centralb. *F. Bact. Abt.* I, **46**: 595–609.

Finkel, M. P., Biskis, B. O. and Jinkins, P. B. (1966) Virus induction of osteosarcomas in mice. *Science* **151**: 698–701.

Finkel, M. P., Reilly, C. A., Biskis, B. O. and Greco, J. L. (1972) Bone Tumor Viruses. In Colson Papers, *Proceedings of the 24th Symp. of the Colson Research Society*, pp. 353–366.

Finkel, M. P., Reilly, C. A. and Biskis, B. O. (1975) Viral etiology of bone cancer. *Front. Radiat. Ther. Onc.* **10**: 29–39.

Friend, C. (1957) Cell-free transmission in adult Swiss mice of a disease having the character of a leukemia. *J. Exp. Med.* **105**: 304–318.

Gonda, T. J. and Metcalf, D. (1984) Expression of *myb*, *myc* and *fos* genes during the differentiation of murine myeloid leukemia. *Nature* **310**: 249–251.

Greenberg, M. E. and Ziff, E. B. (1984) Stimulation of 3T3 cells induces transcription of the c-*fos* proto-oncogene. *Nature* **311**: 433–438.

Greenberg, M. E., Greene, L. A. and Ziff, E. B. (1985) Nerve growth factor and epidermal growth factor induce rapid transient changes in proto-oncogene transcription in PC12 cells. *J. Biol. Chem.* **260**: 14101–14110.

Greenberg, M. E., Hermanowski, A. L. and Ziff, E. B. (1986) Effect of protein synthesis inhibitors on growth factor activation of c-*fos*, c-*myc* and actin gene transcription. *Mol. Cell. Biol.* **6**: 1050–1057.

Gross, L. (1951) 'Spontaneous' leukemia developing in C3H mice following inoculation in infancy with AK leukemia extracts or AK embryos. *Proc. Soc. Exp. Biol. N.Y.* **76**: 27–32.

Gross, L. (1953a) Filterable agent recovered from AK leukemia extracts causing salivary gland carcinomas in C3H mice. *Proc. Soc. Exp. Biol. Med.* **83**: 414–421.

Gross, L. (1953b) Neck tumors of leukemia developing in adult C3H mice following inoculation in early infancy with fitered or centrifuged AK leukemic extracts. *Cancer* **6**: 948–957.

Gross, L. (1958) Attempts to recover filterable agent from X-ray induced leukemia. *Acta Haemat.* **19**: 353–361.

Haan, S. R., Thompson, C. B. and Eisenmann, R. (1985) c-*myc* oncogene protein synthesis is independent of the cell cycle in human and avian cells. *Nature* **314**: 369–371.

Hartley, J. W., Rowe, W. P. and Huebner, R. J. (1970) Host range restrictions of murine leukemia virus or mouse embryo cultures. *J. Virol.* **5**: 221–225.

Jenuwein, T., Müller, D., Curran, T. and Müller, R. (1985) Extended life span and tumorigenicity of nonestablished mouse connective tissue cells transformed by the *fos* oncogene of FBR-MuLV. *Cell* **41**: 629–637.

Kellof, G. S., Lane, W. T., Turner, A. C. and Huebner, R. J. (1969) *In vivo* studies of the FBJ murine osteosarcoma virus. *Nature* **223**: 1379–1380.

Kelly, K., Cochran, B. H., Stiles, D. and Leder, P. (1983) Cell-specific regulation of the c-*myc* gene by lymphocyte mitogens and platelet-derived growth factor. *Cell* **35**: 603–610.

Kruijer, W., Cooper, J. A., Hunter, T. and Verma, I. M. (1984) Platelet-derived growth factor induces rapid but transient expression of the c-*fos* gene and protein. *Nature* **312**: 711–716.

KRUIJER, W., SKELLY, H., BOTTERI, F., V.D. PUTTEN, H., BARBER, J., VERMA, I. M. and LEFFERT, H. (1986) Proto-oncogene expression in regenerating liver is simulated in cultures of primary rat hepatocytes. *J. Biol. Chem.* **261**: 7929–7933.

LAND, H., PARADA, L. F. and WEINBERG, R. A. (1983) Tumorigenic conversion of primary embryo fibroblasts requires at least two cooperating oncogenes. *Nature* **304**: 596–598.

LEE, C. K., CHAN, E. W., REILLY, C. A., PAHNKE, V. A., ROCKUS, G. and FINKEL, M. P. (1979) *In vitro* properties of FBR murine osteosarcoma virus. *Proc. Soc. Exp. Biol, and Med.* **162**: 214–220.

LEVY, J. A. (1971) Demonstration of difference in murine sarcoma virus foci formed in mouse and rat cells under soft agar overlay. *J. Natl. Cancer Inst.* **46**: 1001–1007.

LEVY, J. A. and ROWE, W. P. (1971) Lack of requirement of murine leukemia virus for early steps in infection of mouse embryo cells by murine sarcoma virus. *Virology* **45**: 844–847.

LEVY, J. A., HARTLEY, J. W., ROWE, W. P. and HUEBNER, R. J. (1973) Studies of FBJ osteosarcoma virus in tissue culture. I. Biological characteristics of 'C' type viruses. *J. Natl. Cancer Inst.* **51**: 525–539.

MACCONNELL, W. P. and VERMA, I. M. (1983) Expression of FBJ-MSV oncogene (*fos*) product of bacteria. *Virology* **131**: 367–372.

MEIJLINK, F., CURRAN, T., MILLER, A. D. and VERMA, I. M. (1985) Removal of a 67 base pair sequence in the non-coding region of proto-oncogene *fos* converts it to transforming gene. *Proc. Natl. Acad. Sci. USA* **82**: 4987–4991.

MICHIELS, L., PEDERSEN, S. and MERREGAERT, J. (1984) Characterization of the FBR murine osteosarcoma virus complex: FBR-MuSV encodes a *fos*-derived oncogene. *Int. J. Cancer* **33**: 511–517.

MILLER, A. D., CURRAN, T. and VERMA, I. M. (1984) c-*fos* protein can induce cellular transformation: a novel mechanism of activation of a cellular oncogene. *Cell* **36**: 51–60.

MILLER, A. D., VERMA, I. M. and CURRAN, T. (1985) Deletion of the *gag* region from FBR murine osteosarcoma virus does not affect its enhanced transforming activity. *J. Virol.* **55**: 521–526.

MITCHELL, R. L., HENNING-CHUBB, C., HUBERMAN, E. and VERMA, I. M. (1986) c-*fos* expression is neither sufficient nor obligatory for differentiation of monomyelocytes to macrophages. *Cell* **45**: 497–504.

MITCHELL, R. L., ZOKAS, L., SCHREIBER, R. D. and VERMA, I. M. (1985) Rapid induction of the expression of proto-oncogene *fos* during human monocytic differentiation. *Cell* **40**: 209–217.

MÜLLER, R. and VERMA, I. M. (1984) Expression of cellular oncogenes. *Curr. Topics Microbiol. Immunol.* **112**: 73–115.

MÜLLER, R., BRAVO, R., BURCKHARDT, J. and CURRAN, T. (1984a) Induction of c-*fos* gene and protein by growth factors precedes activation of c-*myc*. *Nature* **312**: 716–720.

MÜLLER, R., CURRAN, T., MÜLLER, D. and GUILBERT, L. (1985) Induction of c-*fos* during myelomonocytic differentiation and macrophage proliferation. *Nature* **314**: 546–548.

MÜLLER, R., MÜLLER, D. and GUILBERT, L. (1984b) Differential expression of c-*fos* during myelomonocytic differentiation and macrophage proliferation. *EMBO J.* **3**: 1887–1890.

MÜLLER, R., MÜLLER, D., VERRIER, B., BRAVO, R. and HERBST, H. (1986) Evidence that expression of c-*fos* protein in amnion cells is regulated by external signals. *EMBO J.* **5**: 311–316.

MÜLLER, R., SLAMON, D. J., ADAMSON, E. D., TREMBLAY, J. M., MÜLLER, D. J., CLINE, M. T. and VERMA, I. M. (1983a) Expression of c-*onc* genes c-*fos*^{Ki} and c-*fms* during mouse development. *Mol. Cell. Biol.* **8**: 1062–1069.

MÜLLER, R., SLAMON, D. J., TREMBLAY, J. M., CLINE, M. J. and VERMA, I. M. (1982) Differential expression of cellular oncogenes during pre- and postnatal development of the mouse. *Nature* **299**: 640–644.

MÜLLER, R., TREMBLAY, J. M., ADAMSON, E. D. and VERMA, I. M. (1983b) Tissue and cell type-specific expression of two human c-*onc* genes. *Nature* **304**: 454–456.

MÜLLER, R. and WAGNER, E. F. (1984) Differention of F9 teratocarcinoma stem cells after transfer of c-*fos* proto-oncogenes. *Nature* **311**: 438–442.

PRICE, C. H. G., MOORE, M. and JONES, D. B. (1972) FBJ virus-induced tumors in mice. *Brt. J. Cancer* **26**: 15–27.

RENZ, M., NEUBERG, M., KURZ, C., BRAVO, R. and MÜLLER, R. (1985) Regulation of c-*fos* transcription in mouse fibroblasts: identification of DNase I-hypersensitive sites and regulatory upstream sequences. *EMBO J.* **4**: 3711–3716.

RHIM, J. S., HUEBNER, R. J., LANE, W. J., TURNER, A. C. and RABSTEIN, L. (1969) Neoplastic transformation and derivation of a focus-forming sarcoma virus in cultures of rat embryo cells infected with osteosarcoma (FBJ) virus. *Proc. Soc. Exp. Biol. Med.* **132**: 1091–1098.

ROUS, P. (1911) Transmission of a malignant new growth by means of a cell-free filtrate. *JAMA* **56**: 198.

RULEY, H. E. (1983) Adenovirus early region 1A enables viral and cellular transforming genes to transform primary cells in culture. *Nature* **304**: 602–606.

RÜTHER, U., WAGNER, E. F. and MÜLLER, R. (1985) Analysis of the differentiation-promoting potential of inducible c-*fos* genes introduced into embryonal cells. *EMBO J.* **4**: 1775–1781.

THOMPSON, C. B., CHALLONER, P. B., NEIMAN, P. E. and GROUDINE, M. (1985) Levels of c-*myc* oncogene mRNA are invariant throughout the cell cycle. *Nature* **314**: 363–366.

TREISMAN, R. (1985) Transient accumulation of c-*fos* RNA following serum stimulation requires a conserved 5' element and c-*fos* 3' sequences. *Cell* **42**: 889–902.

VAN BEVEREN, C., ENAMI, S., CURRAN, T. and VERMA, I. M. (1984) FBR murine osteosarcoma virus. II. Nucleotide sequence of the provirus reveals that the genome contains sequences acquired from two cellular genes. *Virology* **135**: 229–243.

VAN BEVEREN, C., VAN STRAATEN, F., CURRAN, T., MÜLLER, R. and VERMA, I. M. (1983) Analysis of FBJ-MuSV provirus and c-*fos* (mouse) gene reveals that viral and cellular *fos* gene products have different carboxy termini. *Cell* **32**: 1241–1255.

VAN DER HEUL, R. O. and VON RONNEN, J. R. (1967) Juxtacortical osteosarcoma. *The Journal of Bone and Joint Surgery* **99-A**: 415–439.

VAN STRAATEN, F., MÜLLER, R., CURRAN, T., VAN BEVEREN, C. and VERMA, I. M. (1983) Complete nucleotide sequence of a human c-*fos* gene: deduced amino acid sequence of the human c-*fos* protein. *Proc. Natl. Acad. Sci. USA* **80**: 3183–3187.

VERMA, I. M. (1986) Proto-oncogene *fos*: a multifaceted gene. *Trends in Genetics* **2**(4):93–96.
VERMA, I. M., MITCHELL, R. L., KRUIJER, W., VAN BEVEREN, C., ZOKAS, L., HUNTER, T. and COOPER, J. A. (1985) Proto-oncogene *fos*: induction and regulation during growth and differentiation. In *Cancer Cells 3/Growth Factors and Transformation*, pp. 275–287. Cold Spring Harbor, New York.
WARD, J. M. and YOUNG D. M. (1976) Histogenesis and morphology of periosteal sarcomas induced by FBJ virus in NIH Swiss mice. *Cancer Res.* **36**: 3985–3992.
WEISS, R., TEICH, N., VARMUS, H. and COFFIN, J. (1982) RNA tumor viruses. *Molecular Biology of Tumor Viruses* (2nd ed.), pp. 69–75.
YUMOTO, T., POEL, W. E., KODAMA, T. and DMOCHOWSKI, L. (1970) Studies on FBJ virus-induced bone tumors in mice. *Texas. Rep. Biol. Med.* **28**: 145–165.

CHAPTER 12

CELLULAR TRANSFORMATION BY AVIAN VIRUSES

DIANE R. MAKOWSKI, PAUL G. ROTHBERG and SUSAN M. ASTRIN

Institute for Cancer Research, Fox Chase Cancer Center, Philadelphia, Pennsylvania 19111, U.S.A.

1. INTRODUCTION

We owe much of our current understanding of the mechanisms of transformation by retroviruses and other agents to the study of the avian RNA tumor viruses. The first of these viruses to be described was avian leukosis virus, an infectious agent that induced bursal lymphomas in chickens (Ellerman and Bang, 1908, 1909). Three years later, Peyton Rous described Rous sarcoma virus, a filterable agent that induced sarcomas in inoculated birds (Rous, 1911). We now know that these discoveries represent truly signal findings: the first long latent period tumor virus and the first acute tumor virus. Elucidation of the means by which these viruses transform cells, not achieved until many decades later, gave us our first real insights into the molecular events in the neoplastic process.

The gene responsible for the tumor-inducing capabilities of Rous sarcoma virus was isolated in the 70's (Stehelin *et al.*, 1976). Surprisingly, it was shown that this gene, named *src*, was highly homologous to a gene present in the normal cells of all known vertebrates. The transforming genes of other acute transforming retroviruses were described and isolated shortly thereafter; we now have identified a total of almost 20 oncogenes, each of which is carried by an RNA tumor virus and each of which is homologous to a normal cellular gene. What is the function of these cellular 'oncogenes'? What is their relationship to oncogenic transformation and the neoplastic process? These questions are just now being broached experimentally.

Much recent effort has focused on the role of cellular oncogenes in malignancy. This work is based on the idea that cancer is due to a genetic change and that many of these changes will affect oncogene loci. The concept that cancer is a disease induced by genetic change was first put forth by Theodore Boveri early in this century (Boveri, 1914). However, it was not until the 1980's that an alteration in a human tumor was actually pinpointed to a known genetic locus. This change was ultimately determined to be a mutation in the protein coding sequence of the H-*ras* oncogene locus. Such a mutation results in the synthesis of a qualitatively altered oncogene protein; however, there is also precedent for quantitative changes in oncogene products figuring in tumorigenesis.

The first example of activation of a cellular oncogene leading to elevation of its level of expression and ultimately to neoplasia came from work by Hayward and Astrin on avian leukosis virus. This virus, which has no oncogene of its own, apparently transforms by activating the chicken c-*myc* oncogene via an insertion of the viral promotor for RNA synthesis adjacent to that gene (Neel *et al.*, 1981; Hayward *et al.*, 1981). The promoter insertion mechanism, as it was called, has served as a model for quantitative activation of cellular oncogenes in human cancer. It is now clear that such activation of human c-*myc* can be achieved by a variety of mechanisms, including gene amplification and rearrangement (Astrin and Rothberg, 1983; Perry, 1983).

Thus, the study of the mechanisms of induction of neoplasia has come a long way since the isolation of the first oncogenic tumor virus. The viral oncogenes, especially those of the avian viruses, have figured prominently in this process. It is the aim of this review to describe these viruses and their relationship to oncogenesis.

2. ROUS SARCOMA VIRUS

2.1 HISTORY AND PATHOGENESIS

Rous sarcoma virus (RSV) is the best characterized oncovirus out of all the RNA tumor viruses which have been studied to date. Not only was it one of the first tumor-inducing virus strains isolated (from a tumor filtrate; Rous, 1911), but it is the only nondefective acute transforming virus. RSV is an extremely efficient transforming virus, both *in vivo* and in cell culture. The various strains of RSV cover a host range from birds to rodents to primates (for review, see Purchase and Burmester, 1978; Beard, 1980) and possess a broad pathogenic spectrum. RSV-induced tumors mainly consist of cells of mesodermal origin such as fibroblasts, their predominant target *in vitro*. It has also been shown that RSV will transform myoblasts (Kaighn *et al.*, 1966; Fiszman and Fuchs, 1975; Tato *et al.*, 1983), chondroblasts (Pacifici *et al.*, 1977; Boettiger *et al.*, 1983) and epithelial cells from chick iris (Ephrussi and Temin, 1960) and retina (Pessac and Calothy, 1974; Boettiger *et al.*, 1977) with a resulting blockage of their respective differentiation programs.

Transformation of cells in culture by RSV results in numerous changes in cell phenotype, which have been reviewed extensively (Hanafusa, 1977). The earliest alterations induced by the transforming protein of RSV involve both cellular morphology and the cytoskeleton. Temperature downshifted chick embryo fibroblasts (CEF) infected by RSV mutants with a thermolabile transforming protein rapidly begin to develop plasma membrane 'ruffles', termed 'flowers' (Ambros *et al.*, 1975; Boschek *et al.*, 1979, 1981; Boschek, 1982). Later changes in cytoskeletal structure include a loss of microfilament anchorage within cellular adhesion plaques (Boschek, 1982) and a generalized depolymerization of actin with a concomitant disorganization of stress fibers (Ash *et al.*, 1976; Edelman and Yahara, 1976; Wang and Goldberg, 1976; Boschek *et al.*, 1981; Boschek, 1982). It is this latter series of events which result in the alteration of cell shape (rounding up) commonly observed in cellular transformation. A simultaneous loss of the cell surface protein fibronectin has been observed (Hynes, 1974; Vaheri and Ruoslahti, 1975). All these alterations may in turn contribute to the ability of transformed fibroblasts to grow in soft agar, whereas their normal counterparts cannot. Another phenotypic marker of RSV transformation of CEF cells which involves the plasma membrane is an increase in hexose transport (Kletzien and Perdue, 1975). However, results from Shiu *et al.* (1977) demonstrate that the levels of two normal membrane proteins decrease as a result of glucose starvation in the rapidly depleted cell culture medium. Thus, the increased glucose uptake may be a secondary effect to the increased growth rate characteristic of most tumor cells, including RSV-transformed fibroblasts *in vitro*. The possible relationship between the function of the RSV transforming protein and this enhanced glycolysis will be looked at in Section 2.4 of this review.

2.2 GENOME STRUCTURE AND REPLICATION

A composite diagram of the RSV genome derived from the Prague (reviewed in Weiss *et al.*, 1982; Schwartz *et al.*, 1983) and Schmidt-Ruppin strains (Czernilofsky *et al.*, 1980a,b; Swanstrom *et al.*, 1982) is presented in Fig. 1. A recent comprehensive review of the viral genes and their products can be found in Weiss *et al.* (1982); only a brief outline will be presented here.

As mentioned above, the structure of the RSV genome is unique among the replication-competent retroviruses, as it is the only one which carries a transforming oncogene. At the 5′ and 3′ ends of the RNA genome is a short repeated sequence (R) which is used during replication for nascent DNA chain transfer. The U_5 and U_3 regions possess promoter elements and polyadenylation signals respectively. The *gag* region codes for a large polyprotein (Pr76) which is post-translationally cleaved to yield mature proteins of molecular weights ranging from 12 kd to 27 kd in size. These smaller proteins comprise the internal structure of the virion. The viral RNA-dependent DNA polymerase or reverse transcriptase protein is encoded by the *pol* region of the genome. It is originally synthesized as a 180 kd readthrough product of the entire *gag-pol* region, and is later processed into

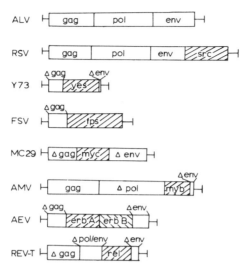

Fig. 1. Depicted here are the genomes of seven different avian sarcoma viruses and one lymphoid leukosis virus (ALV). Replicative functions are indicated by labelled boxes, with partially deleted genes denoted by a Δ prefix. Transforming sequences are presented as hatched boxes. Abbreviations are as follows: ALV = avian leukosis virus; RSV = Rous sarcoma virus; Y73 = Yamaguchi 73 sarcoma virus; FSV = Fujinami sarcoma virus; MC29 = myelocytomatosis 29 virus; AMV = avian myeloblastosis virus; AEV = avian erythroblastosis virus; REV-T = reticuloendotheliosis virus (turkey).

three lower MW polypeptides. The *env* region encodes the surface glycoproteins of the virion envelope, which are originally translated from a spliced subgenomic-size mRNA to yield Pr57, a nonglycosylated precursor which is subsequently cleaved. The 5' end of the *env* region has been found to overlap for 113 b.p. with the 3' end of the *pol* region, with each protein being read from a different reading frame (Weiss et al., 1982). The *src* region will be discussed in Section 2.3.

The molecular events which take place during RSV replication have primarily been studied in CEF cells *in vitro*. In general, mammalian cells appear to be nonpermissive for replication of RSV (Altaner and Temin, 1970; Boettiger et al., 1975) with the exception of certain RSV stocks which can enter and get as far as the integrated proviral DNA stage (Varmus et al., 1973). The restrictions on viral host range appear to be imposed by entry into the cell: cells which lack appropriate receptors for the viral *env* gene product are not penetrated by the virions. The initial step in RSV replication involves the synthesis of a double-stranded DNA copy of the viral genome. This DNA is then integrated into the host cell genome, apparently with little or no site specificity (Hughes et al., 1978; Quintrell et al., 1980). Integration is followed by the synthesis of viral RNAs, which may make up as much as 20% of the polyadenylated RNA in an infected permissive cell (see Bishop, 1978 for a review). The quantity of RSV-specific RNA in a nonpermissive mammalian cell is orders of magnitude lower than in a permissive avian cell (Coffin and Temin, 1972; Bishop et al., 1976; Quintrell et al., 1980). Three species of viral RNA can be detected in chicken cells infected with wild-type RSV—about 9 kb, 5.4 kb and 3.3 kb in length. The smallest contains only *src* sequences, the intermediate both *src* and *env* sequences, and the 9 kb RNA hybridizes to all four genomic regions (Brugge et al., 1977; Hayward, 1977; Weiss et al., 1977; Krzyzek et al., 1978; Parsons et al., 1978; Martin et al., 1979). Although the same three major RNA species are present in nonpermissive (mammalian) cells, the relative concentrations of each type are altered. In avian cells the predominant species is the genomic-length 9 kb RNA, whereas the 3.3 kb *src*-specific RNA is the highest represented in nonpermissive cells (Quintrell et al., 1980). *In vitro* translation experiments confirm the above hybridization data, with 9 kb RNA directing the synthesis of *gag* and *pol* precursor proteins (Von der Helm and Dusberg, 1975; Pawson et al., 1976; Purchio et al., 1977; McGinnis et al., 1978; Weiss et al., 1978; Katz et al., 1979), 5.4 kb viral RNA directing *env* precursor protein synthesis (Pawson *et al.*, 1977, 1980b), and 3.3 kb RSV RNA promoting synthesis of the *src* protein (Yamamoto *et al.*, 1980; Weiss *et al.*, 1981).

2.3 IDENTIFICATION OF THE RSV TRANSFORMING REGION

The task of discriminating between regions necessary for essential viral functions and oncogenic regions has been greatly facilitated by the isolation and characterization in the early 1970's of a large number of RSV mutants (for a recent listing of these mutants, see Weiss *et al.*, 1982). RSV mutants fall into two classes. Conditional mutants are those whose mutant gene or genes function under some conditions but not under other conditions. To date, all known conditional mutants of RSV are temperature-sensitive (i.e. have permissive and nonpermissive temperatures for gene activity). Nonconditional mutants on the other hand possess structural defects in genes essential for replication and can only be propagated in the proviral DNA state or with the aid of wild-type or complementary helper viruses. A further breakdown of viral mutants can be made based on which class of gene or genes are affected. Those with mutations in the oncogenic region of the genome are termed transformation defective (td); those with mutations in essential viral structural and/or replicative regions are termed replication-defective (rd).

Early genetic mapping experiments and later sequence analysis utilizing td mutants has shown that it is the 3' end of the RSV genome which is responsible for fibroblast transformation *in vivo* and *in vitro*. This region codes for a 60 kd polypeptide termed pp60$^{v\text{-}src}$ as shown in Fig. 2. This protein will be discussed in detail in Section 2.4. The nucleotide region coding for the viral *src* gene has been sequenced in its entirety from both the Schmidt-Ruppin (SR) strain of RSV (Czernilofsky *et al.*, 1980a) and the Prague-C (Pr-C) strain (reviewed in Weiss *et al.*, 1982). The SR strain possesses a pair of direct repeat sequences which flank the v-*src* gene (shown in Fig. 2 as 'dr'-labeled boxes). It has been hypothesized that homologous crossing-over through these repeat sequences can serve to explain the uniformity in genome size for a large fraction of the numerous td RSV isolates characterized to date (Duesberg and Vogt, 1970; Bernstein *et al.*, 1976; reviewed in Weiss *et al.*, 1982). The incorporation of the v-*src* sequence into a murine retrovirus genome and the ability of these recombinant viruses to induce splenic foci in Swiss mice further corroborates the identification of the *src* gene as the oncogenic region of the RSV genome (Anderson and Scolnick, 1983). As a final confirmation, Copeland *et al.* (1980) have shown that DNA fragments bearing only the viral *src* gene are capable of transforming fibroblasts *in vitro* which are tumorigenic in syngeneic animals.

Molecular hybridization analysis has revealed that the v-*src* gene has a homologue in the genome of normal uninfected vertebrate cells, termed c-*src* (cellular *src*) (Stehelin *et al.*, 1976). The presence of c-*src* sequences in poly(A) RNA of normal cells indicates that the c-*src* gene codes for a normal cellular protein (Wang *et al.*, 1977; Spector *et al.*, 1978a,b). The level of activity of the c-*src* gene is 100-fold lower in normal cells than that of the v-*src* gene in transformed cells (Bishop, 1978) leading to the theory that RSV-induced transformation is accomplished by an *src* protein dosage effect. An alternative possibility is that subtle differences exist between the viral gene and the cellular gene which account for the oncogenic effect of the former and not the latter. Molecular cloning of the chicken c-*src* locus and nucleotide sequencing of regions within the gene have shown that the terminal 19 amino acids coded for in the c-*src* sequence were replaced by a new set of 12 amino acids in v-*src* (Takeya *et al.*, 1982; Takeya and Hanafusa, 1982; 1983). It has been proposed that viral oncogenes originally arose through transduction of their corresponding cellular homologues during the course of infection (Bishop, 1978; Takeya and Hanafusa, 1983; Swanstrom *et al.*, 1983). If the above differences between the viral and cellular *src* sequences are responsible for the transforming ability, one prediction would be that the c-*src* gene could not complement a td-RSV strain via homologous recombination. This is not the case. Several partial td deletion mutants of RSV have been found to induce tumors in chickens and quails (Hanafusa *et al.*, 1977; Wang *et al.*, 1978; Halpern *et al.*, 1979; Karess *et al.*, 1979; Vigne *et al.*, 1979; Enrietto *et al.*, 1983c). During passage through the animal each of the recovered viruses has reacquired an intact, functional *src* gene, parts of

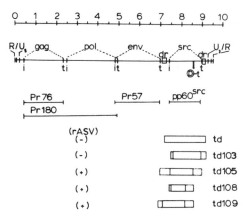

FIG. 2. Depicted in this figure is the genome of Rous sarcoma virus as derived from nucleotide sequence and RNA and protein mapping (see text for details). The top line represents the scale in kilobases. The following symbols represent: R = repeat sequence; U_5 = promotor region; *gag* = group-specific antigen gene; *pol* = reverse transcriptase gene; *env* = envelope glycoprotein gene; dr = direct repeat sequence; *src* = transforming gene; ⓟ—t = phosphotyrosine location on *src* protein, U_3 = polyadenylation region; i = initiation of translation; t = termination of translation. Below the genome structure are the four major polypeptides whose synthesis is directed by the RSV genome with their respective molecular weights in kilodaltons. The bottom half of this figure shows the structure of a number of transformation-defective mutants of RSV, with the boxes delineating the extent of the genomic deletion. Dotted areas demonstrate the degree of uncertainty in the boundaries of these regions. The column at the left indicates the ability of these mutants to produce recovered ASV (rASV) after injection into host animals (modified from Weiss *et al.*, 1982).

which were obtained from the cellular *src* sequence. Not all td deletion mutants can give rise to recovered ASV (rASV) however. Of the viruses shown at the bottom of Fig. 2, only those that retain a short region near the 3' end of *src* are capable of this event (Hanafusa *et al.*, 1980a; Wang *et al.*, 1980). Whether this 3' region is required to mediate the recombinational event or contains some important sequence necessary for transformation cannot be answered by the studies performed to date.

As mentioned in Section 2.1, there are many biochemical and morphological changes associated with cellular transformation, including density-independent growth, growth in agar and increased glucose uptake. Studies with ts-mutant infected cells have not only shown that the *src* gene is responsible for these events, but in addition that some mutants can dissociate these parameters of transformation from one another. For example, cells transformed by mutants GI251–253 were able to grow in soft agar and to high density without forming foci at 41°C (Becker *et al.*, 1977; Weber and Friis, 1979). A recently isolated pair of RSV mutants (PA101 and PA104) have the capability of inducing the growth of both neuroretinal cells (Poirier *et al.*, 1982, 1984) and bone marrow stem cells (Boettiger *et al.*, 1984) without producing a morphologic transformation of the cells. In the case of neuroretinal cell proliferation, the mitogenic capacity of RSV does not require elevated levels of *src* gene activity. The combined data from analysis of these mutants would seem to indicate that there exist multiple targets for the *src* protein within the cell, as opposed to a single target protein which initiates a cascade of biochemical and morphological changes.

Experiments carried out with either avian or mammalian cells infected with a ts RSV mutant have shown that conversion between the transformed and normal states is reversible within a few hours after the appropriate temperature shift. The switch from a transformed to a normal appearance is unaffected by protein synthesis inhibitors (Ash *et al.*, 1976). This might indicate that the components necessary for normal cellular function are present in the ts mutant-transformed cell, but are modified in a readily reversible manner.

2.4 TRANSFORMING GENE PRODUCT, pp60^{src}

2.4.1. *Structure and Intracelular Location of the Protein*

The product of the RSV *src* gene was originally identified by immunoprecipitation of

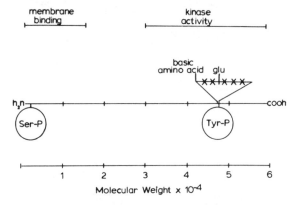

FIG. 3. Represented in this figure is the structure of the RSV *src* protein as derived from a number of separate studies. At the bottom is a scale in daltons. The location of phosphorylated serine and tyrosine residues is indicated by Ser-P and Tyr-P. A consensus sequence for tyrosine phosphorylation is indicated directly above the protein, with X = one amino acid residue. At the top of the figure the assignment of functions of the protein to sub-regions of the polypeptide is indicated.

radiolabeled proteins from RSV-transformed avian and mammalian cells using serum from RSV-induced tumor-bearing rabbits. This immunoprecipitation, as well as cell-free translation of the 3.3 Kb RNA found in infected cells led to the identification of a 60 Kd protein product (Brugge and Erikson, 1977; Brugge et al., 1978; Levinson et al., 1978; Sefton et al., 1978; Beemon and Hunter, 1978; Purchio et al., 1978; Weiss et al., 1981). The nucleotide sequence of *src* has confirmed this data, revealing an open reading frame capable of generating a 60,000 dalton protein (Czernilofsky et al., 1980a; reviewed in Weiss et al., 1982). Partial proteolytic hydrolysis has shown pp60^{v-src} to consist of two distinct domains. The enzymatic activity of the protein (a kinase activity, described in next section) resides in the carboxy-terminal 30 kd portion of the molecule (Levinson et al., 1981). A domain of about 8 kd at the amino-terminal end of the protein anchors it to the plasma membrane (Krueger et al., 1982; Levinson et al., 1981; see Fig. 3). Site-specific mutagenesis at various restriction endonuclease sites within the *src* gene confirm these results (Bryant and Parsons, 1982; 1983).

The pp60^{v-src} protein, as its name indicates, exists as a phosphoprotein within the cell. There are probably two major sites of serine phosphorylation, both of which occur in the amino-terminus of the protein (Sefton et al., 1982a; reviewed in Weiss et al., 1982). The single phosphotyrosine residue lies at position 419 in the SR strain of RSV (position 416 in the PR-RSV strain), and is surrounded by the amino acid sequence indicated in Fig. 3 (Patschinsky et al., 1982; reviewed in Weiss et al., 1982). Similar structural features, as related to proteolytic sensitivity and phosphorylation sites, have been seen in the *fps* and *yes* (Fujinami and Y73 sarcoma virus transforming sequences, respectively) proteins (Neil et al., 1982).

Analyses to date suggest that the pp60^{v-src} protein lies in close proximity to the plasma membrane in infected cells (Willingham et al., 1979; Courtneidge et al., 1980; Krueger et al., 1980b) possibly anchored to the cytoskeleton (Burr et al., 1980). The protein seems to be concentrated within focal adhesion plaques (Rohrschneider, 1980; Nigg et al., 1982) and at tight junctions between adjacent cells (Willingham et al., 1979; Levinson et al., 1981; Nigg et al., 1982). Mutant RSV isolates with alterations in the amino-terminal portion of the *src* molecule show altered subcellular localization with respect to the original RSV strain. In one case, mutant *src* protein was found to be associated with the nuclear envelope (Garber et al., 1982), whereas with another two mutant isolates the *src* protein was found to fractionate largely as soluble, cytoplasmic protein (Krueger et al., 1982). The mutant RSV coding for cytoplasmic protein showed reduced *in vivo* tumorigenicity. Studies done by a number of researchers have shown that the v-*src* protein is synthesized on free ribosomes rather than on membrane-bound polyribosomes, as are most membrane-bound proteins (Lee et al., 1979; Purchio et al., 1980; Levinson et al., 1981; Courtneidge and Bishop, 1982). The newly synthesized pp60^{v-src} molecule then forms a complex with two cellular proteins of MW 50 kd and 89. This complex is located entirely in the cytoplasm (Courtneidge and Bishop, 1982). pp60^{v-src} reaches the plasma membrane within 5 to 15 min of synthesis (Levinson et al., 1981). While in the complex, pp60^{v-src} is phosphorylated predominantly on serine residues. Only after transfer from the complex to the plasma membrane is the tyrosine residue phosphorylated (Courtneidge and Bishop, 1982). A ts-mutant of RSV was found that showed diminished translocation of pp60^{v-src} to the plasma membrane and accumulated instead as a complex within the cytoplasm. The ts-mutant-encoded pp60^{v-src} already established within the membrane at the permissive temperature was found to be released to the cytoplasm and returned to the cellular protein-complex form upon shifting to the restrictive temperature (Courtneidge and Bishop, 1982). It has also been found that newly synthesized pp60^{v-src} lacks, but mature membrane-bound pp60^{v-src} possesses, tightly bound lipid in its NH$_2$-terminal domain (Sefton et al., 1982b; Garber et al., 1983). Furthermore, amino-terminal variant RSV isolates that behave as soluble, cytoplasmic proteins (Krueger et al., 1982) with *in vivo* reduced tumorigenicity contain no detectable lipid, and lipid association is temperature sensitive in a mutant whose membrane association is temperature sensitive (Garber et al., 1983). These data support the roles of lipid binding and cytoplasmic protein association in the rapid transport of the v-src molecule to the plasma membrane.

As mentioned above, the RSV *src* gene is derived from normal avian cells. The normal cellular gene, c-*src*, has been shown to produce a protein similar in size and structure to the v-*src* gene product by cross-reactive sera from RSV-induced tumor-bearing animals (Collett et al., 1978; Oppermann et al., 1979; Rohrschneider et al., 1979; Sefton et al., 1980a; Schartl and Barnekow, 1982). The expression of the c-*src* protein has been looked at in both chicken and human tissues. The highest levels of protein in both species are found in neural tissues, including brain, retina and spinal ganglia (Jacobs and Rübsamen, 1983; Cotton and Brugge, 1983; Sorge et al., 1984). pp60^{c-src} is developmentally regulated in both chick embryo brain tissue (Cotton and Brugge, 1983) and neural retina (Sorge et al., 1984). Enhancements of 4- to 20-fold in pp60^{c-src} activity were found in a number of human sarcomas and mammary carcinomas as compared to normal control tissue levels (Jacobs and Rübsamen, 1983).

2.4.2. *Tyrosine Kinase Activity of pp60^{src}*

Early studies on the phosphoprotein pp60^{v-src} have revealed that the serine residue(s) near the amino-terminus is phosphorylated by a cyclic AMP(c-AMP)-stimulated protein kinase activity within the infected cell and in cell-free extracts (Collett et al., 1979a). Phosphorylation of the tyrosine residue, however, is accomplished by a cAMP-independent kinase activity (Collett and Erikson, 1978; Levinson et al., 1978). Phosphotyrosine is a relatively rare amino acid in normal cells (Hunter and Sefton, 1980) and has been found in a number of viral oncogene products (see next section). As previously mentioned, analysis of RSV mutants leads to the belief that the *src* gene product acts by regulating diverse cellular functions in a readily reversible manner. One means of mediating such a pleiotropic effect is through phosphorylation of cellular proteins (Greengard, 1978). Therefore, initial studies directed at identifying an activity associated with the v-*src* gene product involved assays for protein kinase activity. Unexpectedly, it was found that although pp60^{v-src} failed to catalyze the transfer of radiolabeled phosphate from ATP to a number of common substrates of protein kinases, it did phosphorylate the IgG antibody used to immunoprecipitate the protein from cellular extracts (Collett and Erikson, 1978; Levinson et al., 1978; Rübsamen et al., 1979). Cell-free translation of 3.3 Kb RNA from infected cells generated a pp60^{src} molecule possessing similar activity (Erikson et al., 1978). Continued research revealed the identity of other target proteins for the pp60^{v-src} kinase activity, all becoming phosphorylated on tyrosine residues (see Section 2.4.3).

An early observation which is still under question involves the apparent autophosphorylation of pp60$^{v\text{-}src}$. One study showed that the rate and extent of tyrosine phosphorylation of pp60$^{v\text{-}src}$ was independent of the concentration of the protein in the cellular extracts (Levinson et al., 1978). Preparations of the protein which differ in extent of purification have demonstrated the ability to phosphorylate itself with variable success (Erikson et al., 1979a; 1979b; Levinson et al., 1980; Purchio, 1982). Although the thermolability of both the kinase activity and the self-phosphorylation activity of ts mutants of RSV are consistent with the autophosphorylation hypothesis (Collett et al., 1979a; Erikson and Purchio, 1982), the data present at hand are not conclusive.

Comparison of the amino acid sequence of the SR-RSV *src* gene and that of the catalytic chain of bovine cAMP-dependent protein kinase shows a strong relationship, with a key lysine residue being conserved between the sequences (Barker and Dayhoff, 1982). The strongest homology lay in the carboxyl-terminal half, where the protein kinase activity is localized. The enzymic activity of pp60$^{v\text{-}src}$ can use not only ATP or GTP as phosphate donors, a predicted property for a cyclic-nucleotide-independent protein kinase (Rubin and Rosen, 1975), but a variety of nucleoside triphosphates (Collett and Erikson, 1978; Levinson et al., 1978; Richert et al., 1979; Graziani et al., 1983a; Collett et al., 1983). Incubation with GTP as the phosphate donor results in phosphorylation of the same tyrosine residue of pp60^{src} as that seen in infected cells. In contrast, the use of ATP resulted in the modification of several more tyrosine residues on both the amino-terminal and carboxy-terminal portions of the pp60$^{v\text{-}src}$ molecule (Graziani et al., 1983a; Collett et al., 1983). Comparison of the sequence of amino acids surrounding sites of tyrosine phosphorylation catalyzed by the *src* kinase, including IgG and *src* itself, has revealed similarities (see Fig. 3) between the sites. A basic amino acid (frequently arginine) lies 7 residues to the amino terminal side of the targeted tyrosines and a glutamic acid is found 4 residues on the amino terminal side (Neil et al., 1981; Patschinsky et al., 1982).

The function of phosphoproteins within the cell is frequently influenced by their phosphorylation state (Krebs and Beavo, 1979). Is this also true for the v-*src* gene product? A conclusive answer to this question has not been arrived at. Studies done with a ts-mutant of RSV have shown that the phosphotyrosine residue on the carboxyterminal domain of v-*src* is not significantly phosphorylated when the cells are being grown at the nonpermissive temperature. Shifting to the permissive temperature results in phosphorylation at levels comparable to that of pp60^{src} encoded by wild-type virus (Collett et al., 1979a). However, mutant RSV isolates which code for p60^{src} proteins lacking the major site of either serine or tyrosine phosphorylation have been shown to be competent in the induction of morphological transformation and soft agar colony formation (Cross and Hanafusa, 1983). These RSV mutants possess wild-type levels of tyrosine kinase activity, as measured both *in vivo* and *in vitro*.

The recovered virus experiments mentioned above in Section 2.3 as well as the close structural relationship between the viral and cellular p60 molecules predicted that pp60$^{c\text{-}src}$ would similarly possess a protein kinase activity. This was found to be the case when the immune complex assay (phosphorylation of IgG) was carried out with the product encoded by the normal cellular gene (Collett et al., 1980). This protein kinase activity is similarly specific for tyrosine residues. After extensive purification and functional analysis, it has been concluded that pp60$^{c\text{-}src}$ is indistinguishable from pp60$^{v\text{-}src}$ in enzymatic activity (Purchio et al., 1981).

2.4.3. Target Proteins for the pp60^{src} Protein Kinase

The ultimate goal of the studies described here is to gain insights into the action of the RSV *src* gene product, pp60^{src}, and the mechanism of RSV-induced cellular transformation. Although the modulation of a function of any protein *in vivo* or *in vitro* by a tyrosine-specific protein kinase has yet to be discovered, the fact that many retroviral transforming proteins and some cellular receptors possess this activity may be an indication that tyrosine phosphorylation is an important mediator of growth regulatory mechanisms. To date, seven retroviral oncogenes have been found to have a tyrosine kinase activity associated with their viral proteins, including *src*, *yes*, *fps* and *ros* in avian systems and *abl*, *fes* and *fgr* in mammalian retroviruses (see next section). In addition, it has been found that the binding of epidermal growth factor (EGF) (Carpenter et al., 1979; Cohen et al., 1980; Ushiro and Cohen, 1980; Buhrow et al., 1982; Cohen et al., 1982), platelet-derived growth factor (PDGF) (Ek et al., 1982; Nishimura et al., 1982; Ek and Heldin, 1982) and insulin (Kasuga et al., 1981; Avruch et al., 1982; Petruzzelli et al., 1982) to their respective cellular receptor proteins results in the phosphorylation of the receptor on tyrosine residues. Preliminary evidence now exists which indicates that transforming growth factor (TGF) produced by a human melanoma cell line may operate in a manner analogous to EGF by stimulating a tyrosine kinase activity within the cell membrane (Pike et al., 1982). Interestingly enough, the combined actions of EGF and insulin can induce chicken heart mesenchymal cells to proliferate at a rate comparable to the same cells infected by RSV (Balk et al., 1982).

One piece of circumstantial evidence implicating the protein kinase activity of $pp60^{src}$ in the transformation process is the time scale of events following the downshifting of ts RSV mutant-infected cells to the permissive temperature. Within 15 min, protein kinase activity appears and $pp60^{src}$ becomes phosphorylated (Friis et al., 1980; Ziemiecki and Friis, 1980). The formation of dorsal surface ruffles or flowers can be seen within 30 min of temperature shifting. The reactivation of kinase activity, flower formation and loss of stress fibers all take place even in the presence of the protein synthesis inhibitor cycloheximide (Friis et al., 1980). In addition to this data, it has also been shown that the amount of phosphotyrosine in cellular proteins increases in cells transformed by *src* by about one order of magnitude whereas in cells transformed by chemical carcinogens or non-kinase-associated viral oncogenes this is not seen (Sefton et al., 1980b).

Although the substrate specificity of the purified $pp60^{src}$ enzyme has been examined *in vitro*, few clues have been obtained as to the relevant cellular substrates within the infected cell. Many proteins which serve as substrates for the tyrosine kinase activity of the *src* protein in the test tube, such as tubulin, casein and actin, do not exist as tyrosine phosphoproteins *in vivo* (Erikson et al., 1979; Levinson et al., 1980). Recent evidence of *in vitro* phosphorylation of analogs of the peptide hormone angiotensin has been shown (Wong and Goldberg, 1983).

Early experiments have demonstrated the coprecipitation of two cellular proteins with antibody directed against $pp60^{v-src}$, of MWs 50 kd and 90 kd (pp50 and pp90; Sefton et al., 1978). These two proteins are both phosphoproteins *in vivo*; however, only pp50 possesses a phosphorylated tyrosine residue (Hunter and Sefton, 1980; Oppermann et al., 1981). A potential role in transformation for this phosphotyrosine form of pp50 might be concluded as it exists in RSV-infected, but not in normal cells. However, it has also been shown that pp50 complexed with ts-mutant-encoded $pp60^{src}$ proteins is phosphorylated at tyrosine equally at the permissive and nonpermissive temperatures (Brugge and Darrow, 1982). This paradox has been attributed to the tight binding between the two molecules. Although the function of pp90 within the cell is unknown, further studies have shown that the protein is down-regulated in glucose-deprived murine cells and is induced in heat-shocked cells (Lanks et al., 1982). Both pp50 and pp90 have been demonstrated to form a complex with newly synthesized, non-tyrosine phosphorylated $pp60^{src}$ within the cytoplasm (as discussed in Section 2.4.1 above; Courtneidge and Bishop, 1982; Brugge et al., 1983).

Electrophoretic analysis of cellular proteins which become phosphorylated in RSV-transformed CEF cells has identified a number of putative targets for the $pp60^{src}$ protein. Possibly the most abundant phosphotyrosine-containing protein in an RSV-transformed cell is a 36 kd basic protein (Radke and Martin, 1979) which has been detected by a number of investigators who have reported MWs varying between 34 kd and 39 kd (Erikson and Erikson, 1980; Kobayashi and Kaji, 1980; Cooper and Hunter, 1981a; Cheng and Chen, 1981; Nakamura and Weber, 1982; Martinez et al., 1982; Decker, 1982; Courtneidge et al., 1983; Greenberg and Edelman, 1983). A rather large body of information has accumulated on the structure, location and expression of this protein (for a review, see Cooper and Hunter, 1983a). The nonphosphorylated form of pp36 is a major protein in chick fibroblasts, although an essential role in the cell is contraindicated by its low expression in some lymphoid cell lines and various animal tissues (Radke et al., 1980; Sefton et al., 1983; reviewed in Cooper and Hunter, 1983a,b). Both the phosphorylated and nonphosphorylated forms of the protein are found at the inner face of the plasma membrane, possibly involved in membrane–cytoskeletal interactions (Cheng and Chen; Cooper and Hunter, 1982; Courtneidge et al., 1983; Greenberg and Edelman, 1983). Preliminary evidence from one laboratory has shown that pp36 is associated with cytosolic malic dehydrogenase activity (Rübsamen et al., 1982). Interestingly enough, Graziani et al. (1983b) have recently demonstrated that the $pp60^{src}$ protein is associated with glycerol kinase activity. Both of these activities are key reactions essential in the shuttling of reducing equivalents from the cytosol to the mitochondria. A recent report has demonstrated that $pp60^{v-src}$ can phosphorylate phosphatidylinositol and 1,2-diacylglycerol both in vitro and in vivo (Sugimoto et al., 1984). It has been proposed that this may account for the increased turnover of phosphatidylinositol in RSV-transformed cells (Diringer and Friis, 1977).

Because cytoskeletal alterations are among the earliest effects of cellular transformation by RSV, a survey of cytoskeletal proteins was undertaken by Sefton et al. (1981) in an attempt to identify which, if any, of the cytoskeletal components became phosphorylated upon infection. To date, only vinculin, myosin heavy chain and filamin have been found to contain phosphotyrosine residues in vivo, and only vinculin shows an increase in the level of phosphotyrosine in transformed cells (Hunter and Sefton, 1980; Sefton et al., 1981). This finding caused much speculation as cellular vinculin is concentrated in focal adhesion plaques, where it is believed to be involved in the anchorage of stress fibers, composed of actin microfilaments, to the plasma membrane (Geiger, 1979; Burridge and Feramisco, 1980). Another early target of RSV-induced transformation is the cellular microfilament system and the adhesion plaques (Ash et al., 1976; Edelman and Yahara, 1976; Wang and Goldberg, 1976). As mentioned earlier, within three hours of temperature shifting in ts-RSV mutant-infected CEF cells, there is a reduction in adhesion plaques and a co-accumulation of vinculin and $pp60^{src}$ within the reduced numbers of adhesion plaques remaining (David-Pfeuty and Singer, 1980; Rohrschneider, 1980; Shriver and Rohrschneider, 1981; Sefton et al., 1982b; Nigg et al., 1982). Vinculin has been reported to have the ability to initiate in vitro bundling of actin microfilaments (Jockusch and Isenberg, 1981). Theoretically, if phosphorylation of the tyrosine residue in vinculin interferes with its performance in these critical architectural functions within the cell, this could provide the mechanism by which RSV infection leads to the morphological changes observed in transformed cells. As appealing as this hypothesis may seem, the phosphorylation of vinculin may not be the only basis by which cellular shape is altered. Although there is a ten-fold increase in phosphotyrosine content in the vinculin of transformed cells, this still only comprises approximately 1% of the vinculin molecules in an RSV-transformed cell (Sefton et al., 1981). Furthermore, no correlation can be found between the phosphorylation state of vinculin and subcellular localization of $pp60^{v-src}$ (Rohrschneider and Rosok, 1983). Thus vinculin phosphorylation cannot fully explain the RSV-induced changes in the cytoskeleton.

Two-dimensional protein gel electrophoretic analysis of radiolabeled proteins from RSV-transformed CEF cells by Cooper and Hunter (1981a; 1983b) showed the phosphorylation of tyrosine residues in three cellular proteins of MW 46 kd, 35 kd and 28 kd respectively. A more extended analysis revealed that these same proteins were similarly phosphorylated in a number of virally transformed cells, both in chicken and mouse 3T3 cells (Cooper and Hunter, 1981b; Hunter and Cooper, 1983). These three proteins have now been identified as the glycolytic enzymes enolase (46 kd), lactate dehydrogenase (35 kd) and phosphoglycerate mutase (28 to 29 kd) (Cooper et al., 1983). In view of the increased rate of anaerobic glycolysis in tumor cells first noted by Warburg (1930), the identification of $pp60^{v-src}$-modified proteins as glycolytic enzymes may explain this phenomenon. However, as with the case of vinculin, the phosphorylated forms of these enzymes constitute only a small part (5–10%) of the cellular population of each protein in the transformed cell (Cooper and Hunter, 1983b). Thus, phosphorylation by pp60^{src} could only be effective if there existed a critical subpopulation of enzyme molecules within the cell that is actively involved in glycolysis. This is a distinct possibility as there are theories that glycolytic enzymes are spatially organized within the cell (Sigel and Pette, 1969; Clarke and Masters, 1974; Fulton, 1982). The *in vivo* activity of the enzymes seems unchanged as the difference in substrate/product ratios for all three enzymes between normal and RSV-transformed CEF cells was found to be minimal (Bissell *et al.*, 1973; Singh *et al.*, 1974). As a final note, it should be mentioned that although many types of transformed and tumor cells exhibit the Warburg effect, phosphorylation of these glycolytic enzymes has only been seen in cells transformed by some of the protein kinase-encoding tumor viruses (Cooper and Hunter, 1981a; 1981b). Studies of target proteins for the pp60^{v-src} tyrosine kinase has not yet led to an understanding of the mechanism of cellular transformation by RSV, but further work in this area along similar experimental lines may yet make the connection between an enzymatic activity of pp60^{v-src} and the measurable consequences of RSV transformation. A clearer understanding of normal and aberrant growth control will undoubtedly result from this work.

3. DEFECTIVE AVIAN RETROVIRUSES

In contrast to RSV, all other characterized acutely transforming retroviruses are defective. They lack viral functions required for forming infectious virions. The missing functions are supplied by helper viruses which allow the defective acutely oncogenic particles to be propagated.

The transforming component of acutely transforming retroviruses contains a piece of DNA closely related to a host cell gene. It is presumed, and in many cases proved to one degree or another, that the captured host gene (oncogene) is responsible for the oncogenicity of the virus. The strongest evidence for a host cellular sequence endowing a virus with oncogenic properties is found in RSV and was discussed in the previous section.

The defective avian acute transforming retroviruses can be divided into a sarcoma group and a leukemia group depending on the predominant neoplasm caused by the virus. The individual isolates are further divided into subgroups corresponding to the oncogene carried by the virus (Table 1). In this section we will discuss the oncogenic spectrum, genome structure and the properties of the host derived oncogenes of defective avian retroviruses.

3.1 DEFECTIVE SARCOMA VIRUSES

3.1.1. *fps*

The prototype virus in this group is Fujinami sarcoma virus (FSV) (Fujinami and Inamoto, 1914). The presumed transforming gene, designated *fps*, was shown to be different from the *src* gene (Lee *et al.*, 1980; Hanafusa *et al.*, 1980b). The *fps* gene takes up about 2.7 kb of the 4.4 kb genome displacing most of the viral replicative genes as shown in Fig. 1 (Shibuya and Hanafusa, 1982). The protein product of FSV, FSV p130, is a fusion protein containing the remaining part of the *gag* structural protein fused with the *fps* gene information (Hanafusa *et al.*, 1980b). FSV p130 is a tyrosine kinase as is the *src* gene product, and is located in the cytoplasm of infected cells (Feldman *et al.*, 1980; Mathey-Prevot *et al.*, 1982).

Although *fps* is nonhomologous to *src* by molecular hybridization, a comparison of nucleic acid sequences shows 40% homology in the carboxy-terminal 280 amino acids (Shibuya and Hanafusa, 1982). Another indication of similarity between *fps* and *src* is the production of monoclonal antibodies against *src* that also recognize *fps* (Lipsich *et al.*, 1983). This shows that *src* and *fps*, although distinct genes, may have a common ancestor. Differences have been noted between FSV p130 and the *src* gene product including

TABLE 1.

Oncogene	Viral Isolates
fps	<u>FSV</u> PRC11 PRC11-p UR1 PRCIV
yes	<u>Y73</u> ESV
ros	UR2
ski	SK 770 SK 780 SK 790
lil	B77 variant
myc	<u>MC29</u> CM II OK 10 MH2
mil (mht)	MH2
myb	<u>AMV</u> E26
ets	E26
erbA	<u>AEV</u> strains R and ES4
erbB	<u>AEV</u> strains R and ES4 AEV-H
rel	<u>REV-T</u>

Listed in this figure are the oncogenes originally identified in avian viruses and the viral isolates in which they are found. The prototype viruses are underlined for each oncogene except for ski, lil, mil and ets, which are less well characterized.

substrate specificity (FSV p130 uses only ATP, not GTP, as a phosphate donor while pp60$^{v\text{-}src}$ uses both ATP and GTP; Feldman et al., 1980) and binding to subcellular components (FSV p130 binding is salt sensitive while the pp60$^{v\text{-}src}$ protein binds in a salt resistant manner; Feldman et al., 1983).

Fps is related by nucleic acid homology to fes, the oncogene of feline sarcoma virus (Shibuya and Hanafusa, 1982; Hampe et al., 1982). Later work showed that these two genes represent a common cellular locus that has been highly conserved during vertebrate evolution (Groffen et al., 1983).

There is strong evidence that fps is responsible for the ability of FSV to cause tumors. It is the only gene expressed from the transforming virus and its 130,000 dalton mass (130 kd) nearly exhausts the coding capacity of the genome (Lee et al., 1980, Hanafusa et al., 1980b). A temperature sensitive mutant of FSV has been found that induces transformation of infected cells at the permissive temperature (37°C). At 41.5°C these cells revert to an apparently normal phenotype. The transformed state correlated with phosphorylation of FSV p130 on tyrosine residues and an increase in the total level of tyrosine phosphorylation in cellular proteins (Lee et al., 1981; Pawson et al., 1980a). This correlation of transformation and tyrosine phosphorylation is reminiscent of RSV-induced transformation.

An interesting issue was resolved when the gag structural protein part of FSV P130 was shown not to be required for transformation. The fps portion of FSV was detached from the gag determinants, cloned into another retroviral vector, and shown to be capable of inducing transformation of avian cells (Foster and Hanafusa, 1983).

A host cellular protein has been identified by immunoprecipitation that is presumably the product of c-fps, the cellular homologue of v-*fps* (Mathey-Prevot et al., 1982). Its MW is 98,000 daltons and it is a phosphoprotein with an associated protein kinase. c-*fps* is very similar in many respects to v-*fps* (Mathey-Prevot et al., 1982). The chicken c-*fps* gene contains sequences not found in v-*fps* (Lee et al., 1983; Seeburg et al., 1984). Most of the differences are due to the presence of introns in c-*fps* that are absent from v-*fps*, but at least some amount of c-*fps* 5'-end coding sequence is missing from v-*fps*.

Other viruses containing *fps* include three isolates discovered at the Poultry Research Center, Edinburgh, Scotland (PRCII, PRCII-p, and PRCIV; Carr and Campbell, 1958) and one named after the University of Rochester (UR1; Balduzzi et al., 1981; Wang et al., 1981). These viruses differ from each other in the size of their oncogene, and the amount and nature of helper virus genome remaining in the transforming virus. PRCII-p and PRCIV have identical 2.9 kb v-*fps* sequences but differ in their helper virus related sections (Wong et al., 1982). UR1 has 3.3 kb of *fps* sequence and differs from the others in helper sequences as well (Wang et al., 1981). PRCII on the other hand has only 1.7 kb of *fps* sequence. It is missing 1020 bases found near the 5' end of the FSV v-*fps* gene (Wong et al., 1982, Duesberg et al., 1983; Huang et al., 1984). This deficiency of *fps* genetic information may account for the lower tumorigenicity of PRCII compared to the other *fps* containing viruses (Duesberg et al., 1983).

All five viral isolates described here are probably the result of independent events in which a replication competent avian retrovirus captured the host c-*fps* gene and deleted replicative genes; in the process becoming transformation-competent and replication-deficient.

3.1.2. yes

Two viral isolates have been found in this group. Y73, the prototype, was isolated from the 33rd passage of a spontaneous tumor obtained from a White Leghorn hen in Yamaguchi, Japan in 1973 (Itohara et al., 1978). Esh sarcoma virus (ESV) was isolated at the New Bolton Center, Pennsylvania from a tumor in a White Leghorn chicken brought to their attention by a farmer named Esh (Wallbank et al., 1966). The Y73 genome (Fig. 1) is 3.7 kb and is expressed as a 90,000 dalton fusion protein containing *gag* structural information attached to the v-*yes* gene (Kawai et al., 1980; Kitamura et al., 1982). The ESV genome is probably less than 5 kb, although its size is not accurately known, and is also expressed as a *gag-yes* fusion protein with a MW of 80,000 daltons (Ghysdael et al., 1981). A study of the phenotypic characteristics of chick embryo fibroblasts transformed by RSV (*src*), FSV (*fps*) and Y73 (*yes*) showed that although *fps* and *yes*-transformed cells are qualitatively similar to *src*-transformed cells, the extent of change was quantitatively less for *fps* and *yes* as compared to *src* (Guyden and Martin, 1982).

The *yes* sequence of Y73 was shown to be different from the oncogenes of RSV, FSV, AEV, AMV and MC29 by nucleic acid hybridization (Yoshida et al., 1980; Kawai, et al., 1980). A c-*yes* gene was detected in the chicken genome and is a distinct locus from the c-*src* gene (Yoshida et al., 1980). c-*yes* RNA is expressed in all tissues analyzed, but was particularly high (26 copies per cell) in chicken kidney cells (Shibuya et al., 1982a).

The complete sequence of Y73 provided an unexpected result: the *yes* gene is closely related to *src* in amino acid sequence; in a patch of 436 amino acids *yes* and *src* are 82% homologous (Kitamura et al., 1982). The lack of hybridization noted earlier (Yoshida et al., 1980; Kawai et al., 1980) was due to divergence in nucleotide sequence, but most of the nucleotide differences are silent at the amino acid coding level (Kitamura et al., 1982). The similarity between *src* and *yes* is seen in the ability of polyclonal and certain monoclonal antibodies raised against denatured pp60^{v-src} to react with the v-*yes* gene product (Lipsich et al., 1983; Erikson and Erikson, 1983). Similar to pp60^{v-src}, the v-*yes* gene products of Y73 and Esh are both tyrosine specific protein kinases that accept ATP and GTP as phosphate donors (Kawai et al., 1980; Ghysdael et al., 1981; Feldman et al., 1982).

3.1.3. ros

The only viral strain containing the *ros* oncogene is UR2, named after the University of Rochester (Balduzzi *et al.*, 1981). The genome RNA consists of approximately 3,300 nucleotides of which ca. 1,200 are unique to the transforming component (Wang *et al.*, 1982). v-*ros* was demonstrated to be a new oncogene because it does not share obvious structural homology with any other known oncogene (Wang *et al.*, 1982). Homology between *src*, *fps* and *yes* was revealed by study of their nucleotide sequences. It will be interesting to see if there is some as yet undetected homology between *ros* and the other sarcoma inducing oncogenes when the UR2 nucleotide sequence is obtained. One indication of possible homology is the existence of an anti-*src* monoclonal antibody that also reacts with the *ros* product (Lipsich *et al.*, 1983).

The protein product of UR2 is a *gag–ros* fusion protein with a MW of 68,000 daltons (Feldman *et al.*, 1982). It has a tyrosine specific protein kinase activity which differs from the RSV (*src*) and Y73 (*yes*) products in using only ATP as a phosphate donor (Feldman *et al.*, 1982). Both RSV-pp60$^{v\text{-}src}$ and Y73-p90 can use GTP and ATP as phosphate donors. The UR2-p68 kinase activity differs from the FSV (*fps*) product in its stability in alkaline conditions which will inhibit the FSV protein kinase (Feldman *et al.*, 1982). The *ros* gene product is located on cytoplasmic membranes in infected cells (Land *et al.*, 1983b).

3.1.4. ski and lil

Two potential new oncogenes have been described in the literature, but little is known about them. They were generated in the laboratory by passing B77 avian sarcoma virus (a close relative of RSV) in tissue culture.

One group passaged a transformation-defective strain of B77 that lacked the *src* gene and obtained three replication-defective viruses (SK770, SK780, SK790) that were capable of transforming fibroblasts (Stavnezer *et al.*, 1981). These new viruses lacked some viral replicative genes and contained new information unrelated to the other known viral oncogenes. The creators of these new transforming viruses were unable to find evidence for a viral protein kinase function in the protein products or the transformed cells (Stavnezer *et al.*, 1981). The putative oncogene will probably be called *ski*.

A new cellular sequence was found in a stock of B77 avian sarcoma virus and named *lil* (Boccara *et al.*, 1982). c-*lil* was detected in chickens of the genus *Gallus*, but was not detectable in other species. This distinguishes c-*lil* from the avian oncogenes, which have been found to be highly conserved during evolution. It will be interesting to see if v-*lil* is an oncogenic sequence.

3.1.5. Defective Avian Sarcoma Viruses: Unifying concepts

The viruses described in this section containing *fps*, *yes* and *ros* are transcribed and then translated as fusion proteins containing *gag* determinants on their amino termini. In no case, however, has the *gag* information been shown to be necessary for transformation. All are associated with or contain tyrosine specific protein kinase activity as is pp60$^{v\text{-}src}$ described earlier. Some homology has been shown between the *src*, *fps* and *yes* amino acid sequences and is predicted between this group and *ros* when the *ros* sequence is available. Despite this homology, they are distinct genes whose protein products show variation in the properties of their enzymatic activities. Although a great deal of research has been devoted to determining the target protein(s) of v-*onc* protein kinases (discussed above for v-*src*) a mechanism for transformation has not been uncovered. Thus, it is not possible at this time to conclude that the sarcoma virus oncogenes transform cells using either common or disparate pathways.

3.2 DEFECTIVE LEUKEMIA VIRUSES

3.2.1. myc, mil (mht) and raf

Unlike the sarcoma viruses which transform fibroblasts in tissue culture and cause sarcomas *in vivo*, the viruses of the *myc* group can transform fibroblasts, epithelial cells and bone marrow cells in culture and cause sarcomas, carcinomas and myelocytomas *in vivo* (reviewed in Graf and Beug, 1978). This is of great clinical interest in view of the fact that in humans the incidence of carcinoma is much greater than sarcoma (Cairns, 1978).

The prototype virus of this group is MC29 (Fig. 1) whose ca. 5700 base genome contains ca. 1600 bases of *myc* specific sequences (Reddy *et al.*, 1983). The v-*myc* gene is distinct from the other known oncogenes (Roussel *et al.*, 1979). MC29's v-*myc* protein product is synthesized as a fusion protein with *gag*, a theme reminiscent of the defective avian sarcoma viruses (Bister *et al.*, 1977). The 110 kd protein has no detectable protein kinase activity (Bister *et al.*, 1980), but has been found to bind to DNA (Donner *et al.*, 1982). It is located in the nucleus of infected cells (Abrams *et al.*, 1982; Donner *et al.*, 1982; Hann *et al.*, 1983).

There is ample evidence that v-*myc* sequences are responsible for the ability of MC29 to cause tumors. Three mutants of MC29 have been isolated that are capable of transforming fibroblasts *in vitro*, but are severely deficient in transforming macrophages, as compared to wild type virus (Ramsay *et al.*, 1980). The mutants also show lower pathogenic potential *in vivo* (Enrietto *et al.*, 1983a). Analysis of the RNA of these mutants shows overlapping deletions between 200 and 600 bases in length (Bister *et al.*, 1982b). The protein products of the mutants are smaller than the wild type 110 kd *gag*-*myc* fusion protein, with MWs of 100 kd, 95 kd and 90 kd (Ramsay and Hayman, 1982). Studies of both viral RNA and protein demonstrated that the deletions occurred in *myc*-specific sequences and did not involve the *gag* domain (Bister *et al.*, 1982; Ramsay and Hayman, 1982). The protein products produced by the mutants showed decreased DNA binding compared to wild type protein (Donner *et al.*, 1983). Several conclusions can be reached from these studies: (1) v-*myc* sequences are involved in transformation; (3) separate functional domains of v-*myc* may be involved in fibroblast and macrophage transformation; and (4) the binding of v-*myc* protein to DNA may be involved in the ability of MC29 to transform macrophages and cause tumors *in vivo*.

A partial revertant of td 10H, one of the deletion mutants described above, was obtained by passage through macrophages (Ramsay *et al.*, 1982). The recovered virus has regained the ability to transform macrophages efficiently. Analysis of viral proteins and proviral DNA in the revertant-infected cells showed a restoration of *myc* information very similar, but not identical, to that deleted in the original mutant. The replaced *myc* sequences were presumably derived via recombination with the host cell c-*myc* gene. The revertant virus, when inoculated into chickens, caused lymphoma, a disease that is not caused by the wild type MC29 virus (Enrietto *et al.*, 1983b). The ability to cause lymphomas is presumably the result of having c-*myc* sequences in a critical section of the viral gene. It is appropriate at this point to mention that c-*myc* has been implicated in avian leukosis virus-induced bursal lymphomas (discussed in Section 4).

A comparison of the nucleotide sequences of MC29 v-*myc* and chicken c-*myc* shows two major differences: (1) the 5'-end of both genes are unrelated but still code for amino acids; and (2) c-*myc* has a 971 base pair intron in its 3' end that is not found in v-*myc* (Watson *et al.*, 1983). A study of murine and human c-*myc* sequences show that the gene is split into three exons the first of which does not code for any amino acids (Bernard *et al.*, 1983; Battey *et al.*, 1983). The v-*myc* sequence apparently resulted from the capture of only a portion of the c-*myc* sequence representing the second and third exons.

The function of c-*myc* is unknown, but recent experiments have shed some light on a possible role in cell division. Growth stimulation of quiescent mouse tissue culture cells resulted in increased expression of c-*myc* RNA (Campisi *et al.*, 1984). On the other hand, terminal differentiation of mouse teratocarcinoma stem cells into nonproliferating endoderm correlated with a reduction in c-*myc* RNA. Chemically transformed mouse fibroblasts, however, had relatively high levels of c-*myc* RNA whether growing or not growing. These experiments suggest that *myc* may be involved in cell proliferation. In another series of experiments, also using mouse cells, agents that initiate cellular proliferation, like lipopolysaccharide or concanavalin A in lymphocytes and platelet-derived growth factor in fibroblasts, caused an increase in c-*myc* RNA expression (Kelly *et al.*, 1983). Cell proliferation in HL60, a human promyelocytic leukemia cell line, also correlates with elevated c-*myc* RNA expression. When HL60 is induced to mature to nonproliferating granulocytes by exposure to DMSO or to nonproliferating monocytes by exposure to 1,25 $(OH)_2D_3$ (a vitamin D metabolite) the *myc* RNA level drops substantially (Westin *et al.*, 1982; Reitsma *et al.*, 1983).

The cellular *myc* gene has been implicated in several nonviral neoplasms. In murine plasmacytoma and human Burkitt's lymphoma characteristic chromosome translocations unite the *myc* gene with immunoglobulin genes (for review: Perry, 1983). In addition, several cell lines derived from human tumors have an amplification of c-*myc* DNA (Dalla Favera *et al.*, 1982; Collins and Groudine, 1982; Alitalo *et al.*, 1983). Finally, fresh tumor material from adenocarcinoma of the colon and a variety of human hematopoietic cancers, particularly acute leukemia, exhibit several examples of an elevation in c-*myc* RNA content (Rothberg *et al.*, 1984; Erisman *et al.*, in preparation).

Other viral isolates containing *myc* include CMII, OK10 and MH2 (Graf and Beug, 1978; Roussel *et al.*, 1979). CMII is similar to MC29 but contains less *gag* sequence (Hayman *et al.*, 1979). Its protein product is a 90 kd *gag-myc* fusion protein that, like MC29-p100, is a DNA binding protein that is located in the nucleus of infected cells (Bunte *et al.*, 1983).

Mill Hill 2 (MH2) virus was isolated in 1927 from an endothelioma in the ovarian region of a hen (reviewed Graf and Beug, 1978). Its pathogenicity differs from MC29 in Japanese quails where MH2 is much more oncogenic than MC29 (Linial, 1982). The genome is ca. 5.2 kb with ca. 1.3 kb of v-*myc* sequences (Kan *et al.*, 1983).

MH2 codes for a 100 kd protein that contains *gag* determinants and was originally thought to be a *gag-myc* fusion protein as in MC29 and CMII. This, however, is not the case. The MH2-p100 protein does not have any detectable *myc* determinants (Jansen *et al.*, 1983). MH2 infected cells contain a 2.6 kb subgenomic mRNA that codes for a 57 Kd *myc* protein without any *gag* information (Pachl *et al.*, 1983; Saule *et al.*, 1983). This indicates that v-*myc* does not have to be expressed as a *gag-myc* fusion protein. A highly oncogenic strain of MH2 has been found that does not produce p100 but does produce p57 (Linial, 1982). This strain shows that transformation correlates with v-*myc* expression while expression of the remaining portion of the *gag* gene is dispensible.

Analysis of MH2 proviral DNA from infected cells shows that a non-*myc* nonviral sequence of ca- 1.2 kb is present between *gag* and *myc* (Saule *et al.*, 1983). This sequence, named *mht* by one group (Kan *et al.*, 1983) and *mil* by another (Coll *et al.*, 1983) is derived from the host genome. The 100 kd protein found in MH2-infected cells is presumably a *gag-mil* fusion protein. This protein is located in the cytoplasm of infected cells and binds to DNA and RNA *in vitro* in contrast to MC29 *gag-myc* p110 which is located in the nucleus and binds to DNA but not RNA (Bunte *et al.*, 1983).

The consequences of the expression of the *mil* (*mht*) sequence is not known. As mentioned earlier there is a strain of MH2 that does not appear to express these sequences, but is still oncogenic (Linial, 1982). The *mil* determinants, therefore, are probably not responsible for MH2 being more oncogenic than MC29 in Japanese quails.

Mil is homologous to *raf*, a viral oncogene from a murine retrovirus (murine sarcoma virus 3611) (Jansen *et al.*, 1984). The viral genes were derived from homologous cellular genes in their respective species. This is similar to the relationship between avian *fps* and feline *fes* mentioned above. The amino acid sequence of v-*raf* shows slight similarity to the sequence of *src* (Mark and Rapp, 1984). *Mil*, therefore, can also be considered a distant relative of *src*.

The cellular gene c-*mil*/c-*raf* is detectable in several species, including man, indicating that it is phylogenetically conserved, a characteristic common to all oncogenes (Coll *et al.*, 1983). The gene is expressed at the RNA level in several different species with an RNA differing in size from c-*myc* RNA (Coll *et al.*, 1983). Human *mil*/*raf* sequences are located on chromosomes 3 and 4 (Bonner *et al.*, 1984) while human c-*myc* is located on chromosome 8 (Taub *et al.*, 1982); the two genes are not linked in the human genome as they are in MH2. The virus probably captured the cellular *myc* and *mil* sequences in two independent events.

OK10 has a much larger genome (7.5 kb) than MC29 (5.7 kb). It contains a complete *gag* gene unlike any of the defective avian viruses discussed so far (Ramsay and Hayman, 1980; Bister *et al.*, 1980; Pfeifer *et al.*, 1983). The *myc* sequence is inserted between partial sequences of the *pol* and *env* genes. The pattern of gene expression in OK10 transformants has not been completely clarified. There are two RNA transcripts consistently observed, 8.0–8.6 kb and 3.5 kb in length. Both transcripts contain *myc* sequences (Chiswell *et al.*, 1981; Saule *et al.*, 1982). Two primary protein products have been detected to date, a 76 Kd *gag* precursor protein and a 200 Kd *gag-pol-myc* fusion protein (Chiswell *et al.*, 1981). Both proteins can be synthesized by *in vitro* translation of the 8.0–8.6 kb genomic RNA. A product has not yet been identified for the 3.5 kb RNA, but it may code for a *myc* protein that is free of *gag* determinants (Saule *et al.*, 1982). Thus in OK10 two types of viral *myc* expression may be present: (1) a polyprotein containing both *myc* and *gag* information as in MC29 and CMII and (2) a *myc* protein free of *gag* sequences as in MH2.

3.2.2. *myb and ets*

The prototype virus in this group is avian myeloblastosis virus (AMV). It was isolated from a chicken with neurolymphomatosis (reviewed in Graf and Beug, 1978). After several *in vivo* passages the virus produced only a myeloblastic leukemia. *In vitro* the virus can transform growing macrophages and cells committed to the macrophage lineage (Boettiger and Durban, 1984) but not fibroblasts (Graf and Beug, 1978).

The genome of AMV is 7.2–7.5 kb with 1.2 kb of *myb* sequence, as shown in Fig. 1 (Duesberg *et al.*, 1980; Souza *et al.*, 1980; Klempnauer *et al.*, 1982). V-*myb* is inserted between the *pol* and *env* genes, both of which are partially deleted (Klempnauer *et al.*, 1982). A complete functional *gag* gene is present in AMV (Duesberg *et al.*, 1980). The v-*myb* putative transforming protein is translated from a spliced subgenomic RNA (Chen *et al.*, 1981a; Gonda *et al.*, 1981). It has a MW of 45,000 daltons. The first six amino terminal amino acids are apparently coded by the *gag* gene and the final 11 carboxy-terminal amino acids by the *env* gene (Klempnauer *et al.*, 1982; Klempnauer *et al.*, 1983). As a result the transforming protein could correctly be called a *gag-myb-env* fusion protein, but the contribution of viral structural gene information in the final product is very small. *Myb* appears to be a distant relative of *myc* because of a low amount of homology in their amino acid sequences (Ralston and Bishop, 1983). Like *myc*, the *myb* gene product has an intranuclear location (Land *et al.*, 1983b).

A temperature-sensitive mutation of AMV has been found (Moscovici and Moscovici, 1983). Cells transformed with the mutant at 35.5°C were no longer fully transformed when shifted to 41°C. Although presumably v-*myb* is the mutated gene, genetic mapping of this mutant on the viral genome will be necessary in delineating the sequences essential for transformation.

The c-*myb* locus has been identified and portions of its nucleotide sequence determined (Klempnauer *et al.*, 1982). It differs from v-*myb* in two major aspects: (1) c-*myb* contains seven introns which are spliced out in the production of mature messenger RNA, whereas in v-*myb* all of the intron sequences have been removed except for part of one intron which still remains in the AMV genome (Klempnauer *et al.*, 1982 and 1983); and (2) c-*myb* produces a 75,000 dalton protein which is considerably larger than the 45,000 dalton v-*myb* protein. The v-*myb* protein product represents a truncation of the c-*myb* product (Klempnauer *et al.*, 1983). A possibility is that the oncogenicity of AMV is due to its possession of a truncated version of c-*myb* as opposed to possession of the complete gene.

The function of the normal c-*myb* homologue is unknown, but its pattern of expression suggests that it is involved in hematopoiesis (Gonda *et al.*, 1982). c-*myb* RNA is present at a high level in thymus, yolk sac and bone marrow which are all centers of hematopoiesis in the chicken. Avian erythroblastosis virus (AEV; discussed below) -transformed erythroblasts have a high level of c-*myb* RNA whereas fibroblasts transformed by AEV are much lower in c-*myb* RNA content. This suggests that the AEV target cell in the erythroid lineage normally contains large amounts of c-*myb* RNA where perhaps the *myb* protein has some function in cell division. If the c-*myb* RNA elevation were due to AEV transformation then AEV-transformed fibroblasts would also contain a high level of c-*myb* RNA, which as mentioned above is not the case.

E26 was discovered in Bulgaria in 1962 from a chicken with erythroblastosis (reviewed in Graf and Beug, 1978). It was shown to contain *myb*, making it taxonomically related to AMV (Roussel et al., 1979). E26 differs in pathogenicity from AMV by causing mostly erythroblastosis *in vivo* (Radke et al., 1982; Moscovici et al., 1983).

E26 has a 5.7 kb genome and produces a 135,000 dalton protein which contains *gag* information (Bister et al., 1982a; Klempnauer and Bishop, 1984). Besides *myb* and some viral replicative genetic information E26 contains in addition a new sequence called *ets* (Nunn et al., 1983; Leprince et al., 1983). The structure of the 135kd protein is the result of translation of a *gag* (1.2 kb)-*myb* (0.8 kb)-*ets* (1.5 kb) RNA (Nunn et al., 1983). V-*ets* is distinct from the other oncogenes and is homologous to a cellular gene c-*ets* that has been phylogenetically conserved (Leprince et al., 1983). V-*ets* may be a new oncogene which is responsible for the ability of E26 to transform erythroid cells or it may alter the transforming ability of the *myb* gene information to which it is joined in the 135 kd protein.

3.2.3. erb

There are two indistinguishable strains of avian erythroblastosis virus (AEV), R and ES4 (reviewed in Graf and Beug, 1978). They cause mainly erythroblastosis, but can cause sarcoma after intramuscular injection. This disease spectrum corresponds to the ability of AEV to transform fibroblasts and cells of the erythroid lineage (Graf et al., 1981; reviewed in Graf and Neug, 1978).

The fact that differentiation in the erythroid lineage is well characterized has allowed for a fine dissection of both the hematopoietic target cells for AEV and the maturational events that take place after infection. AEV infects cells at the burst forming unit (familiarly: BFU-E) stage which is an early progenitor of the erythroid lineage (Gazzolo et al., 1980). The transformed cells are at the colony forming unit (familiarly: CFU-E) stage, which is a later progenitor of the erythroid lineage (Samarut and Gazzolo, 1982; Beug et al., 1982). This shows that BFU-E cells infected with AEV go through further differentiation to the CFU-E stage, at which their further progress towards hemoglobin-containing mature erythroid cells is halted. One avian hemoglobin gene, α^A, is transcribed, but the transcript is not properly processed and never leaves the nucleus, whereas the α^D and β globin genes are not transcribed at all (Therwath et al., 1984).

The genome of AEV (Fig. 1) is ca. 5.5 kb with 3 kb of host-derived sequences in between what remains of the viral replicative genes *gag* and *env* (Bister and Duesberg, 1979). The host derived sequences were identified as the *erb* oncogene (Roussel et al., 1979; Saule et al., 1981) which we now know represents two different cellular loci: *erb*A and *erb*B (Vennstrom and Bishop, 1982). AEV-infected cells have two different RNAs which contain v-*erb* sequences. (Anderson et al., 1980; Sheiness et al., 1981). The larger, genome sized RNA codes for a 75,000 dalton protein which contains *gag* information as well as a portion (*erb*A) of the host derived oncogene. The smaller 3.5 kb subgenomic RNA codes for a ca. 68,000 dalton protein which contains the *erb*B information (Privalsky and Bishop, 1982; Hayman et al., 1983; Privalsky et al., 1983). The *erb*A protein has been shown to be phosphorylated, but no evidence could be found for a protein kinase activity (Anderson and Hanafusa, 1982). It is found in the cytoplasm of infected cells (Graf and Beug, 1983).

The significance of the v-*erb*A gene remains an area of active investigation. A deletion mutant of AEV, unable to synthesize the *erb*A gene product, was able to transform fibroblasts in culture but had decreased leukemogenic potential *in vivo* (Graf and Beug, 1983; Sealy et al., 1983b). In contrast to this, a newly discovered avian erythroblastosis virus, AEV-H, was found to carry only *erb*B yet was able to cause erythroblastosis and sarcoma (Yamamoto et al., 1983a). Although *erb* B appears to be the key oncogene in causing both sarcoma and erythroblastosis, *erb* A may enhance erythroid cell transformation (Sealy et al., 1983b; Frykberg et al., 1983; Graf and Beug, 1983).

Deletion mutants in *erb* B show that it is essential for cellular transformation and *in vivo* oncogenicity (Frykberg et al., 1983; Sealy et al., 1983a). The *erb* B protein product is phosphorylated, glycosylated and located on the plasma membrane of infected cells (Privalsky et al., 1983; Hayman et al., 1983). It has some homology to the *src* gene product, but no protein kinase activity has yet been detected (Yamamoto et al., 1983b; Privalsky et al., 1984). The *erb* B sequence is very close to that of a portion of human epidermal growth factor (EGF) receptor (Downward et al., 1984). Whether or not the EGF receptor gene is c-*erb* B or just a close relative remains to be determined. The similarity of the two genes allows the formulation of a very provocative hypothesis for the mode of action of v-*erb* B. EGF receptor can be divided into three domains; (1) a cell surface EGF binding domain, (2) a transmembrane domain and (3) a cytoplasmic section with tyrosine specific protein kinase activity. The latter two domains transmit a cellular growth signal when EGF binds to the external EGF binding domain. The signal is propagated to the interior of the cell where it presumably activates the tyrosine kinase activity. The v-*erb* B gene contains the information for only the last two domains; the external EGF binding domain is missing from v-*erb* B. The theory is that the v-*erb* B protein inserts into the membrane and provides a continuous signal to divide, equivalent to the signal produced when normal EGF receptor is stimulated by binding of EGF (Downward et al., 1984).

The structures of the c-*erb* genes in chickens (Vennstrom and Bishop, 1982; Sergeant et al., 1982) and humans (Jansson et al., 1983) have been investigated. In chickens c-*erb* A contains at least three introns and c-*erb* B contains at least 11 introns in contrast to the viral oncogenes which are not interrupted by any introns. The full extents of the cellular *erb* loci are not known yet. *erb* A and *erb* B are separated by at least 15 kb, and are probably totally independent genes (Vennstrom and Bishop, 1982). In support of this the human homologue of *erb* B is located on chromosome 7 while the closest human homologue to *erb* A is on chromosome 17 (Spurr et al., 1984). In the human genome, besides the *erb* B homologue, three loci were detected with homology to *erb* A; however, two of these were only distantly related to the v-*erb* A sequence (Jansson et al., 1983).

3.2.4. *rel*

The reticuloendotheliosis viruses form a group that is unrelated to all the other oncogenic viruses discussed in this review (Kang and Temin, 1973; Purchase et al., 1973). The members of this group are duck infectious anemia virus, Trager duck spleen necrosis virus, chick syncytial virus, reticuloendotheliosis-associated virus and the prototypical reticuloendotheliosis virus strain T (REV-T). In this group only REV-T is a defective retrovirus which contains a host derived oncogene, a description that qualifies it for treatment in this section. The other viruses of this group are replication-competent and presumably initiate their associated pathology through some mechanism which does not involve a viral oncogene. They will not be dealt with further in this section.

REV-T was isolated from a sick adult turkey with reticuloendotheliosis (Theilen et al., 1966). It causes a proliferation of reticuloendothelial cells resulting in death in turkeys, chickens and Japanese quail. It can transform chick embryo fibroblasts (Franklin et al., 1977) and splenic lymphocytes (Hoelzer et al., 1980) *in vitro*. An early B-lymphocyte precursor transformed by REV-T *in vitro* caused reticuloendotheliosis *in vivo* (Lewis et al., 1981). This suggests that the *in vivo* target cell is a lymphocyte of the B-cell lineage.

The genome of REV-T (Fig. 1) is ca. 5.5 kb with 1.4 kb of sequence unrelated to the helper virus (Chen et al., 1981b; Wong and Lai, 1981; Rice et al., 1982). This sequence, named *rel*, is not closely related to the other oncogenes. The *rel* gene is inserted into a partially deleted *env* gene (Stephens et al., 1983). Portions of the *gag* and *pol* genes have also been deleted. The nucleotide sequence shows that translation of *rel* probably begins in the *env* gene, continues through *rel*, and then concludes in 3′-terminal *env* sequences. The *env-rel-env* fusion protein has a predicted MW of 56 kd for an unmodified protein.

A protein of 64 kd has been detected in infected cells and is probably post-translationally modified. A somewhat tenuous relationship has been detected between *src* and *rel*. Whether this relationship is a genuine homology, an example of convergent evolution because of functional similarities, or merely a chance event remains to be determined.

Besides the presence of the v-*rel* gene, other structural elements are necessary for REV-T to be oncogenic (Chen and Temin, 1982). When helper virus sequences were inserted into REV-T to replace missing viral genetic information, the resultant virus was non-transforming. The additional helper virus sequences suppress v-*rel*-induced transformation. Thus, the acquisition of *rel* information was only one event involved in creating REV-T; deletion of viral replicative and structural gene information was also important in making REV-T transformation-competent.

The *rel* sequence, like the other oncogenes, was derived from a host cell gene. In this case, however, the closest v-*rel* homologue is found in the turkey, not the chicken as for all the other viral oncogenes described in this review (Wong and Lai, 1981). The capture of *rel* sequences by a replication-competent virus probably occurred in turkey. Turkey c-*rel* has at least seven introns and covers at least 23 kb of the turkey genome including both introns and exons (Wilhelmsen and Temin, 1984). This contrasts with the 1.4 kb intronless v-*rel* gene. A study of chicken c-*rel* shows that at least part of a 3′-c-*rel* exon is missing from v-*rel* (Chen *et al.*, 1983). The chicken c-*rel* mRNA is 4 kb, which also points out that not all of c-*rel* is contained in v-*rel*. The importance of the c-*rel* sequences not maintained in v-*rel* is unknown.

3.2.5. *Defective Avian Leukemia Viruses: Unifying Concepts*

The defective leukemia viruses (DLV) contain their acquired oncogenes as either *gag* gene inserts (*myc* in MC29 and CMII; *myb* in E26; *erb* A in AEV; *mil* in MH2) or *env* gene inserts (*myc* in MH2 and OK 10; *myb* in AMV; *rel* in REV-T). This situation differs from that found in the defective sarcoma viruses, which have their oncogenes in the *gag* position. It is not known if the position of the oncogene in the virus has any bearing on its oncogenicity. Future experiments will determine if the sarcoma-inducing oncogenes can cause disease when inserted in the *env* gene of a defective retrovirus.

The defectiveness of the DLVs is caused by a deletion of genetic information coding for structural and replicative proteins. In the case of REV-T this deletion is important for oncogenicity. In the other DLVs the importance of viral gene deletion is unknown.

The oncogenes of the DLVs form a more diverse lot than the sarcoma oncogenes. *erb* B, *mil* and *rel* have some structural similarity to the *src* group which suggests that a tyrosine specific protein kinase may be an intrinsic activity of their respective protein products, although such an activity has not been detected yet in any members of the DLV group. *myc* and *myb* are distantly related to each other and to the E1A protein of adenovirus 12 but are distinct from the other oncogenes.

The theme of multiple oncogenic events leading to cancer is being renewed in the current research environment. Three examples of multiple oncogenes in one virus are found in the DLVs: AEV contains *erb* A and *erb* B which are separately expressed, MH2 contains *myc* and *mil* also expressed on separate protein products and E26 contains *myb* and *ets* which are expressed as a *gag-myb-ets* fusion protein. Oncogenic viruses containing only *mil/raf* (murine sarcoma virus 3611; discussed more fully in the Chapter on murine transforming viruses), *erb* B (AEV-H), *myc* (MC29) or *myb* (AMV) are found in nature. The additional oncogenes in AEV, MH2 and E26 appear to supplement the transforming potential of their constituent viruses. Transformation of primary tissue culture cells by two oncogenes, where either one alone would not suffice, has been shown (Land *et al.*, 1983a,b; Ruley, 1983).

4. LYMPHOID LEUKOSIS VIRUS

4.1 History and Pathogenesis

Lymphoid leukosis disease in chickens was first seen over a century ago, and has been the topic of a number of comprehensive reviews (see Hanafusa, 1975; Weiss, 1975). Since the evidence for the transmission of this disease with cell-free filtrates was provided (Burmester et al., 1946), a great deal of work has been done on both the structure and mode of action of lymphoid leukosis viruses. This section will briefly review the pathogenesis of nondefective leukosis viruses (or lymphoid leukosis viruses, LLVs) and their genomic structure, and end with a review of current theories as to their mode of action.

As can be inferred from the name, LLV most commonly induces a proliferation of B lymphoblasts accompanied by an enlarged liver and spleen when injected into susceptible hosts, which include chickens, Japanese quail, pheasant and a number of other avian species (see review by Purchase and Burmester, 1978). The origin of these tumor cells is the bursa of Fabricius, where lymphoblast accumulation can be seen as early as one month after infection (Cooper et al., 1968; Neiman et al., 1980a). This accumulation regresses in about half of chronically infected chickens, whereas the others slowly grow into lymphoma nodules which eventually metastasize to the liver and spleen, killing the bird in an average of 20 weeks post-inoculation (reviewed in Graf and Beug, 1978; Neiman et al., 1980a). Clone-purified strains of LLV have also been known to induce osteopetrosis, nephroblastoma, erythroblastosis, fibrosarcoma and hemangiocarcinoma (Purchase et al., 1977; Okazaki et al., 1982). The development of these neoplasms all require the same long latency period (as compared to acute transforming viruses); however, their spectrum and incidence is influenced by the strains of LLV used for the inoculation and the host strain (reviewed in Graf and Beug, 1978). Bursal lymphomatosis by LLV, as compared to sarcoma virus tumorigenesis (see previous sections), is a relatively rare event, and the tumors and subsequent metastasis produced are clonal, originating from a single transformed cell (Neiman et al., 1980b; Neel et al., 1981; Payne et al., 1981; Fung et al., 1981).

LLVs do not induce morphological transformation at detectable frequency in tissue culture cells. Several strains are capable of inducing cytopathic alterations and plaques in CEF cultures; however, these cells are not tumorigenic and cannot develop colonies in soft agar (reviewed in Graf and Beug, 1978). The identity of the major target cell of LLV has been shown by *in vivo* experiments, where susceptibility to virus-induced lymphatic leukemia in chicks was removed via bursectomy and returned by reimplantation of bursa cells (Purchase and Gilmour, 1975). The lymphoid target cells are arrested at a developmental stage just prior to the switch from IgM- to IgG- or IgA-producing cells (Cooper et al., 1974).

4.2 GENOME STRUCTURE AND CLASSIFICATION

A prototype genome structure for the leukosis viruses is presented in Fig. 1. The viruses are fully competent for replication, possessing both structural (*gag*, *env*) and enzymatic (*pol*) sequences. They are classified into seven subgroups which are defined by their viral envelope antigen. Subgroups A through E are chicken isolates, whereas subgroups F and G are from pheasants (Vogt and Ishizaki, 1966; Duff and Vogt, 1969; Hanafusa et al., 1970; Fujita et al., 1974; Hanafusa, 1975). There are at least four major genetic loci in chickens which determine cellular resistance to infection or viral host range, one of them associated with the R_1 erythrocyte antigen (Crittenden, 1975).

The LLVs can occur as independent oncogenic agents or associated with replication defective strains of leukemia or sarcoma viruses. The defect in virus replication of these rd-strains can be complemented by the avian leukosis viruses, also termed 'helper viruses'. Missing or defective structural or enzymatic functions are supplied by the helper virus by coinfecting the same cell with both strains of virus. Both sarcoma virus and helper virus are produced from the same infected cell and in cases where the sarcoma virus is env⁻ both viruses have envelope antigens identical to those of the helper virus. It can be seen that the helper virus can thus determine the range of infectivity and the rate of virus maturation for both itself and the acutely transforming virus it is associated with.

4.3. PROMOTER INSERTION MECHANISM OF ONCOGENESIS

The leukosis viruses differ from the acute transforming viruses discussed in the earlier sections of this review in three major properties: (1) their latent period for tumor induction is 4–12 months as compared with a 2–3 week period for the acute viruses; (2) they have no *in vitro* transformation assay which is related to their *in vivo* effect; (3) they do not possess a distinct oncogene. The oncogenic potential of avian leukosis virus (or ALV) has been shown to reside within the 3' region of the viral genome in the LTR region, within approximately 500 bases of the poly(A) tract (Robinson *et al.*, 1979; Crittenden *et al.*, 1980; Tsichlis and Coffin, 1980). This region does not seem to encode a viral protein. It has also been demonstrated that viral gene expression is not necessary for maintenance of the transformed phenotype. Many of the B cell lymphomas induced by ALV do not contain viral 21S and/or 35S mRNAs (Neel *et al.*, 1981; Payne *et al.*, 1981). In addition, many of the integrated proviruses in these tumors are defective, with deletions ranging from one small 5' region to almost the entire genome (Neel *et al.*, 1981; Payne *et al.*, 1981; Fung *et al.*, 1981). Although evidence for specificity of integration for retroviral provirus DNA has never been demonstrated (Hughes *et al.*, 1978; Steffen and Weinberg, 1978; Ringold *et al.*, 1979; Quintrell *et al.*, 1980), tumors induced by ALV in a number of different birds had integration sites in common (Neel *et al.*, 1981; Payne *et al.*, 1981; Fung *et al.*, 1981). The tumor cells were shown to synthesize discrete new poly(A) RNAs consisting of viral sequences covalently linked to cellular sequences.

All this evidence put together pointed towards the conclusion that ALV induced neoplastic transformation by activating a normal cellular gene(s). In 1981, Hayward, Neel and Astrin revealed that in 85% of the ALV-induced B-cell lymphomas studied, proviral integration had occurred upstream from the chicken c-*myc* gene, the cellular counterpart of the transforming gene of MC29 virus (see Figs 1 and 4). Levels of *myc*-specific RNA in these tumors were 30–100 fold higher than in normal uninfected tissue (Hayward *et al.*, 1981). This mechanism of tumorigenesis was termed 'promoter insertion' and a model of its action is as follows. Although ALV does not possess an oncogene as do the acute sarcoma viruses, both types possess strong transcriptional promotors repeated at both ends of their integrated proviral DNA form, the LTRs (see Section 2). These sequences are responsible for the efficient transcription of the provirus subsequent to integration into the host cell genome, with transcription initiating within the left LTR (Fig. 5) and proceeding through the viral genome, terminating in the 5' end of the right LTR.

In the case of an acute transforming virus, transcription initiated in the left LTR and terminated in the right LTR would result in expression of the viral oncogene encoded

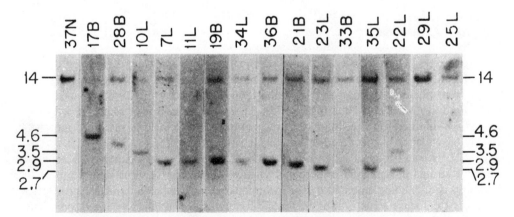

FIG. 4. This figure depicts the pattern of hybridization of [^{32}P]-labeled v-*myc* DNA to Eco RI-digested genomic DNA from normal and tumor tissue derived from ALV-infected chickens, transferred to nitrocellulose and hybridized by the method of Southern (1975). In 12/15 cases a new tumor-specific c-*myc* band has appeared. Sample 22 represents a biclonal tumor, whereas samples 25 and 29 have not had integration events in the vicinity of the c-*myc* gene. N = normal tissue; B = bursal lymphoma; L = liver metastasis.

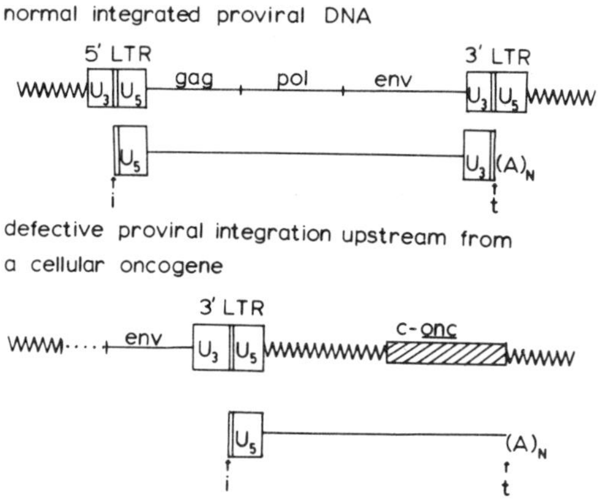

FIG. 5. *Top:* the structure of proviral DNA integrated into the cellular genome is shown. Right below this is the RNA transcript whose synthesis would be promoted by the 5' LTR region. *Bottom:* a situation is indicated in which the 5' LTR sequence has been deleted and the proviral DNA has integrated upstream from a cellular oncogene sequence. The predicted hybrid mRNA sequence is presented below. ⋯—cellular sequence; *gag* = group-specific antigen; *pol* = reverse transcriptase; *env* = envelope protein; i = initiation of transcription; t = termination of transcription; $(A)_N$ = poly(A) sequence.

between the two repeat sequences (see Fig. 1). However, since the right LTR sequence possesses the same promoter elements as the left LTR, transcripts could also initiate within this region, resulting in expression of the flanking cellular sequences. If these sequences contain an oncogene, expression of the gene could result in tumor formation (Fig. 5). As proviral integration adjacent to a cellular oncogene would be a relatively rare event in the infected tissue, this model explains both the latency time characteristic of ALV-induced bursal lymphomas and their clonal nature.

Results obtained from cloning provirus–host cell tumor-specific junction fragments from two independent bursal lymphomas supported the promoter insertion model (Neel *et al.*, 1982). Analysis of the cloned tumor DNA demonstrated two facts: (1) that the proviral DNA had integrated without grossly rearranging the c-*myc* gene; and (2) that the integration event had situated the ALV provirus in such a way that the right hand viral LTR could promote transcription of c-*myc*. This same orientation was found in all tumors where integration had occurred in the vicinity of the c-*myc* locus. Recent work has provided evidence to support the hypothesis that deletion of the 5' LTR is required in order for the promoting activity of the 3' proviral LTR to be effective. It was found that transcription of a marker gene downstream from the 3' LTR was elevated five-fold upon mutation of the 5' LTR sequence, a phenomenon termed 'transcriptional overlap interference' (Cullen *et al.*, 1984).

Similar work done in the same system has revealed defective ALV proviral integration in bursal tumors not only upstream and in the same orientation as the c-*myc* gene, but downstream from the gene and also upstream in the opposite orientation as well (Payne *et al.*, 1982). As both of these latter arrangements preclude direct utilization of the 3' LTR as a promoter for the c-*myc* locus it has been proposed that the inverted or downstream provirus is providing 'enhancement' functions similar to those seen in a number of DNA tumor viruses (reviewed in Khoury and Gruss, 1983 and Gluzman and Shenk, 1983). One such position-independent transcriptional enhancement sequence has already been identified in the Abelson murine leukemia virus LTR (Srinivasan *et al.*, 1984).

Lymphomatosis induction via promoter insertion is a mechanism by no means limited to avian leukosis virus. A nondefective strain of reticuloendotheliosis virus, chick syncytial virus (CSV), is distinct from ALV, but causes a similar pathology. Noori-Daloii *et al.* (1981) demonstrated that in over 90% of CSV-induced chicken lymphomas analyzed by their group the CSV provirus had integrated next to the c-*myc* gene. Yet another example of this phenomenon is shown by lymphomas induced by one of the helper viruses of avian myeloblastosis virus (myeloblastosis-associated virus-1, or MAV-1), a virus differing from ALV predominantly in the LTR domain. There it was found that in MAV-1-induced lymphomas in quail, MAV-specific DNA could be mapped adjacent to the quail c-*myc* gene (Varmus, 1983).

Promoter insertion in the vicinity of an endogenous oncogene other than the c-*myc* gene has been seen in a study of the induction of erythroblastosis in chickens by ALV. As mentioned in Section 4.1. the type of neoplasm induced by avian leukosis virus is influenced by the strain of ALV inoculated and the host strain. Six cases of erythroblastosis were induced by infecting the appropriate strain of chickens (Fung et al., 1983). Analysis of isolated erythroblast DNA revealed that in four of the six samples ALV proviral integration had occurred adjacent to the c-*erb* B gene (the cellular homologue to the viral *erb* B gene in avian erythroblastosis virus, AEV; see Fig. 1). Increased *erb* B-specific RNAs were detected in all six tumors. DNA mapping and sequence analysis of a cloned viral-erythroblast cell junction fragment showed that the orientation of insertion of the ALV genome was such that the right-hand LTR could promote downstream transcription of the c-*erb* B gene (Fung et al., 1983). A great deal of work in non-avian as well as avian systems has accumulated delineating the role played by insertional mutagenesis in and around cellular oncogenes in the induction of cellular transformation (for a recent review, see Makowski et al., 1984). Recent work by Steffen (1984) has shown that murine leukemia virus is similarly capable of initiating the formation of a lymphoma by inserting a proviral copy adjacent to the c-*myc* gene.

Acknowledgements—The authors gratefully acknowledge the help of Dr W. S. Mason and J. M. Spandorfer in the preparation of this manuscript. We also thank M. D. Erisman for disclosing experimental results prior to publication. D.R.M. is a Hoffman-LaRoche Fellow of the Life Sciences Research Foundation. P.G.R. was supported by CA-09035 from the National Institutes of Health. S.M.A. was supported by CD-174 from the American Cancer Society and CA-06927 and RR-05539 from the National Institutes of Health, and an appropriation from the Commonwealth of Pennsylvania.

REFERENCES

ABRAMS, H. D., ROHRSCHNEIDER, L. R. and EISENMAN, R. N. (1982) Nuclear location of the putative transforming protein of avian myelocytomatosis virus. *Cell* **29**: 427–439.

ALITALO, K., SCHWAB, M., LIN, C. C., VARMUS, H. E. and BISHOP, J. M. (1983) Homogeneously staining chromosomal regions contain amplified copies of an abundantly expressed cellular oncogene (c-*myc*) in malignant neuroendocrine cells from a human colon carcinoma. *Proc. natn. Acad. Sci. U.S.A.* **80**: 1707–1711.

ALTANER, C. and TEMIN, H. M. (1970) Carcinogenesis by RNA sarcoma viruses. XII. A quantitative study of infection of rat cells *in vitro* by avian sarcoma viruses. *Virology* **40**: 118–134.

AMBROS, V. R., CHEN, L. B. and BUCHANAN, J. M. (1975) Surface ruffles as markers for studies of cell transformation by Rous sarcoma virus. *Proc. natn. Acad. Sci. U.S.A.* **72**: 3144–3148.

ANDERSON, S. and HANAFUSA, H. (1982) Characterization of avian erythroblastosis virus p75. *Virology* **121**: 32–50.

ANDERSON, S. M., HAYWARD, W. S., NEEL, B. G and HANAFUSA, H. (1980) Avian erythroblastosis virus produces two mRNAs. *J. Virol.* **36**: 676–683.

ANDERSON, S. M. and SCOLNICK, E. M. (1983) Construction and isolation of a transforming murine retrovirus containing the *src* gene of Rous sarcoma virus. *J. Virol.* **36**: 594–605.

ASH, J. F., VOGT, P. K. and SINGER, S. J. (1976) Reversion from transformed to normal phenotype by inhibition of protein synthesis in rat kidney cells infected with a temperature-sensitive mutant of Rous sarcoma virus. *Proc. natn. Acad. Sci. U.S.A.* **73**: 3603–3607.

ASTRIN, S. M. and ROTHBERG, P. G. (1983) Oncogenes and cancer. *Cancer Invest.* **1**: 355–364.

AVRUCH, J., NEMENOFF, R. A., BLACKSHEAR, P. J., PIERCE, M. W. and OSATHANODH, R. (1982) Insulin-stimulated tyrosine phosphorylation of the insulin receptor in detergent extracts of human placental membranes. *J. biol. Chem.* **257**: 15162–15166.

BALDUZZI, P. C., NOTTER, M. F. D., MORGAN, H. R. and SHIBUYA, M. (1981) Some biological properties of two new avian sarcoma viruses. *J. Virol.* **40**: 268–275.

BALK, S. D., SHIU, R. P. C., LAFLEUR, M. M. and YOUNG L. L. (1982) Epidermal growth factor and insulin cause normal chicken heart mesenchymal cells to proliferate like their Rous sarcoma virus-infected counterparts. *Proc. natn. Acad. Sci. U.S.A.* **79**: 1154–1157.

BARKER, W. C. and DAYHOFF, M. O. (1982) Viral *src* gene products are related to the catalytic chain of mammalian cAMP-dependent protein kinase. *Proc. natn. Acad. Sci. U.S.A.* **79**: 2836–2839.

BATTEY, J., MOULDING, C., TAUB, R., MURPHY, W., STEWART, T., POTTER, H., LENOIR, G. and LEDER, P. (1983) The human c-*myc* oncogene: structural consequences of translocation into the Igh Locus in Burkitt Lymphoma. *Cell* **34**: 779–787.

BEARD, J. W. (1980) Biology of avian oncorna viruses. In: *Viral Oncology*, pp. 55–87, KLEIN, G. (ed.) Raven Press, New York.

BECKER, D., KURTH, R., CRITCHLEY, D., FRIIS, R. and BAUER, H. (1977) Distinguishable transformation-defective phenotypes among temperature-sensitive mutants of Rous sarcoma virus. *J. Virol.* **21**: 1042–1055.

BEEMON, K. and HUNTER, T. (1978) Characterization of Rous sarcoma virus *src* gene products synthesized *in vitro*. *J. Virol.* **28**: 551–566.

BERNARD, O., CORY, S., GERONDAKIS, S., WEBB, E. and ADAMS, J. M. (1983) Sequence of the murine and human cellular *myc* oncogenes and two modes of *myc* transcription resulting from chromosome translocation in B lymphoid tumours. *EMBO J.* **2**: 2375–2383.

BERNSTEIN, A., MACCORMICK, R. and MARTIN, G. S. (1976) Transformation defective mutants of avian sarcoma viruses; the genetic relationship between conditional and non-conditional mutants. *Virology* **70**: 206–209.

BEUG, H., PALMIERI, S., FRUEDENSTEIN, C., ZENTGRAF, H. and GRAF, J. (1982) Hormone-dependent terminal differentiation *in vitro* of chicken erythroleukemia cells transformed by ts mutants of avian erythroblastosis virus. *Cell* **28**: 907–919.

BISHOP, J. M. (1978) Retroviruses. *A. Rev. Biochem.* **47**: 35–88.

BISHOP, J. M., DENG, C.-T., MAHY, B. W. J., QUINTRELL, N., STRAVNEZER, E. and VARMUS, H. E. (1976) Synthesis of viral RNA in cells infected by avian sarcoma viruses. In: *Animal Virology*, p. 1–20, BALTIMORE, D. (ed) Academic Press, New York.

BISSELL, M. J., WHITE, R. C., HATIE, C. and BASSHAM, J. A. (1973) Dynamics of metabolism of normal and virus-transformed chick cells in culture. *Proc. natn. Acad. Sci. U.S.A.* **70**: 2951–2955.

BISTER, K. and DUESBERG, P. H. (1979) Structure and specific sequences of avian erythroblastosis virus RNA: evidence for multiple classes of transforming genes among avian tumor viruses. *Proc. natn. Acad. Sci. U.S.A.* **76**: 5023–5027.

BISTER, K., HAYMAN, M. J. and VOGT, P. K. (1977) Defectiveness of avian myelocytomatosis virus MC29: isolation of long-term nonproducer cultures and analysis of virus-specific polypeptide synthesis. *Virology* **82**: 431–448.

BISTER, K., LEE, W. H. and DUESBERG, P. H. (1980) Phosphorylation of the nonstructural proteins encoded by three avian acute leukemia viruses and by avian Fujinami sarcoma virus. *J. Virol.* **36**: 617–621.

BISTER, K., NUNN, M., MOSCOVICI, C., PERBAL, B., BALUDA, M. A. and DUESBERG, P. H. (1982a) Acute leukemia viruses E26 and avian myeloblastosis virus have related transformation-specific RNA sequences but different genetic structures, gene products, and oncogenic properties. *Proc. natn. Acad. Sci. U.S.A.* **79**: 3677–3681.

BISTER, K., RAMSAY, G. M. and HAYMAN, M. J. (1982b) Deletions within the transformation-specific RNA sequences of acute leukemia virus MC29 give rise to partially transformation-defective mutants. *J. Virol.* **41**: 754–766.

BISTER, K., RAMSAY, G., HAYMAN, M. J. and DUESBERG, P. H. (1980) OK10, an avian acute leukemia virus of the MC29 subgroup with a unique genetic structure. *Proc. natn. Acad. Sci. U.S.A.* **77**: 7142–7146.

BOCCARA, M., PLUQUET, N., COLL, J., ROMMENS, C. and STEHELIN, D. (1982) Characterization of c-*lil*, a chicken cellular sequence associated with a stock of B77 avian sarcoma virus. *J. Virol.* **43**: 925–934.

BOETTIGER, D., ANDERSON, G. and DEXTER, T. M. (1984) Effect of src infection on long-term marrow cultures: increased self-renewal of hemopoietic progenitor cells without leukemia. *Cell* **36**: 763–773.

BOETTIGER, D. and DURBAN, E. (1984) Target cells for avian myeloblastosis virus in embryonic yolk sac and relationship of cell differentiation to cell transformation. *J. Virol.* **49**: 841–847.

BOETTIGER, D., LOVE, D. N. and WEISS, R. A. (1975) Virus envelope markers in mammalian tropism of avian RNA tumor viruses. *J. Virol.* **15**: 108–114.

BOETTIGER, D., ROBY, K., BRUMBAUGH, J., BIEHL, J. and HOLTZER, H. (1977) Transformation of chicken embryo retinal melanoblasts by a temperature-sensitive mutant of Rous sarcoma virus. *Cell* **11**: 881–890.

BOETTIGER, D., SOLTESZ, R., HOLTZER, H. and PACIFICI, M. (1983) Infection of chick limb bud presumptive chondroblasts by a temperature-sensitive mutant of Rous sarcoma virus and the reversible inhibition of their terminal differentiation in culture. *Molec. cell. Biol.* **3(8)**: 1518–1526.

BONNER, T., O'BRIEN, S. J., NASH, W. G., RAPP, U. R., MORTON, C. C. and LEDER, P. (1984) The human homologs of the *raf*(*mil*) oncogene are located on human chromosome 3 and 4. *Science* **223**: 71–74.

BOSCHEK, C. B. (1982) Organizational changes of cytoskeletal proteins during cell transformation. In: *Advances in Viral Oncology*, pp. 173–187, KLEIN, G. (ed.) Raven Press, New York.

BOSCHEK, C. B., JOCKUSCH, B. M., FRUS, R. R., BACK, R. and BAUER, H. (1979) Morphological alterations to the cell surface and cytoskeleton in neoplastic transformation. *Beitr. Elektronen-mikroskop. Direktabb. Oberfl.* **12**: 47–54.

BOSCHEK, C. B., JOSKUSCH, B. M., FRUS, R. R., BACK, R., GRUNDMANN, E. and BAUER, H. (1981) Early changes in the distribution and organization of microfilament proteins during cell transformation. *Cell*, **24**: 175–184.

BOVERI, T. (1914) *Zur Frage der Entstehung maligner Tumoren*, Gustav Fischer, Jena.

BRUGGE, J. S. and DARROW, D. (1982) Rous sarcoma virus-induced phosphorylation of a 50,000-molecular weight cellular protein. *Nature* **295**: 250–253.

BRUGGE, J. S. and ERIKSON, R. L. (1977) Identification of a transformation-specific antigen induced by an avian sarcoma virus. *Nature* **269**: 346–348.

BRUGGE, J. S., ERIKSON, E., COLLETT, M. S. and ERIKSON, R. L. (1978) Peptide analysis of the transformation-specific antigen from avian sarcoma virus-transformed cells. *J. Virol.* **26**: 773–782.

BRUGGE, J. S., PURCHIO, A. F. and ERIKSON, R. L. (1977) Virus-specific RNA species present in the cytoplasm of Rous sarcoma virus-infected chicken cells. *Virology* **83**: 16–26.

BRUGGE, J. S., YONEMOTO, M. and DARROW, D. (1983) Interaction between the Rous sarcoma virus transforming protein and two cellular phosphoproteins: analysis of the turnover and distribution of this complex. *Molec. cell. Biol.* **3**: 9–19.

BRYANT, D. and PARSONS, J. T. (1982) Site-directed mutagenesis of the *src* gene of Rous sarcoma virus: construction and characterization of a deletion mutant temperature-sensitive for transformation. *J. Virol.* **44**: 683–691.

BRYANT, D. and PARSONS, J. T. (1983) Site-directed point mutation in the *src* gene of Rous sarcoma virus results in an inactive *src* gene product. *J. Virol.* **45**: 1211–1216.

BUHROW, S. A., COHEN, S. and STAVROS. J. V. (1982) Affinity labeling of the protein kinase associated with the epidermal growth factor receptor in membrane vesicles from A431 cells. *J. biol. Chem.* **257**: 4019–4022.

BUNTE, T, GREISER-WITHE, I. and MOELLING, K. (1983) The transforming protein of the MC29-related virus CMI is a nuclear DNA-binding protein whereas MH2 codes for a cytoplasmic RNA–DNA binding polyprotein. *EMBO J.* **2**: 1087–1092.

BURMESTER, B. R., PRICKETT, C. O. and BELDING, T. C. (1946) A filtrable agent producing lymphoid tumors and osteopetrosis in chickens. *Cancer Res.* **6**: 189–196.

BURR, J. G., DREYFUSS, G., PENMAN, S. and BUCHANAN, J. M. (1980) Association of the *src* gene product of Rous sarcoma virus with cytoskeletal structures of chicken embryo fibroblasts. *Proc. natn. Acad. Sci. U.S.A.* **77**: 3484–3488.

BURRIDGE, K. M. and FERAMISCO, J. (1980) Micro-injection and localization of a 130 K protein in living fibroblasts: a relationship to actin and fibronectin. *Cell* **19**: 587–595.

CAIRNS, J. (1978) *Cancer: Science and Society*, W. H. Freeman, San Francisco.

CAMPISI, J., GRAY, H. E., PARDEE, A. B., DEAN, M. and SONENSHEIM, G. E. (1984) Cell-cycle control of c-*myc* but not c-*ras* expression is lost following chemical transformation. *Cell* **36**: 241–247.

CARPENTER, G., KING, L., JR. and COHEN, S. (1979) Rapid enhancement of protein phosphorylation in A-431 cell membrane preparations by epidermal growth factor. *J. biol. Chem.* **254**: 4884–4891.

CARR, J. G. and CAMPBELL, J. G. (1958) Three new virus-induced fowl sarcomata. *Br. J. Cancer* **12**: 631–635.

CHEN, J. H., HAYWARD, W. S. and MOSCOVICI, C. (1981a) Size and genetic content of virus-specific RNA in myeloblasts transformed by avian myeloblastosis virus (AMV). *Virology* **110**: 128–136.

CHEN, I. S. Y., KAH, T. W., O'REAR, J. J. and TEMIN, H. M. (1981b) Characterization of reticuloendotheliosis virus strain T DNA and the isolation of a novel variant of reticuloendotheliosis virus strain T by molecular cloning. *J. Virol.* **40**: 880–811.

CHEN, I. S. Y. and TEMIN, H. M. (1982) Substitution of 5' helper virus sequences into non-*rel* portion of reticuloendotheliosis virus strain T suppresses transformation of chicken spleen cells. *Cell* **31**: 111–120.

CHEN, I. S. Y., WILHELMSEN, K. C. and TEMIN, H. M. (1983) Structure and expression of c-*rel*, the cellular homolog to the oncogene of reticuloendotheliosis Virus Strain T. *J. Virol.* **45**: 104–113.

CHENG, Y. S. E. and CHEN, L. B. (1981) Detection of phosphotyrosine-containing 34,000 dalton protein in the framework of cells transformed with Rous sarcoma virus. *Proc. natn. Acad. Sci. U.S.A.* **78**: 2388–2392.

CHISWELL, D. J., RAMSAY, G. and HAYMAN, M. J. (1981) Two virus-specific RNA species are present in cells transformed by defective leukemia virus OK10. *J. Virol.* **40**: 301–304.

CLARKE, F. M. and MASTERS, C. J. (1974) On the association of glycolytic components in skeletal muscle extracts. *Biochem. biophys. Acta* **358**: 193–207.

COFFIN, J. M. and TEMIN, H. M. (1972) Hybridization of Rous sarcoma virus deoxyribonucleic acid polymerase product and ribonucleic acids from chicken and rat cells infected with Rous sarcoma virus. *J. Virol.* **9**: 766–775.

COHEN, S., CARPENTER, G. and KING, L. R. (1980) Epidermal growth factor-receptor-protein interactions. *J. biol. Chem.* **255**: 4834–4842.

COHEN, D., USHIRO, H., STOSCHEK, C and CHINKERS, M. (1982) A native 170,000 epidermal growth factor receptor–kinase complex from shed plasma membrane vesicles. *J. biol. Chem.* **257**: 1523–1531.

COLL, J., RIGHI, M., DETAISNE, C., DISSOUS, C., GEGONNE, A. and STEHELIN, D. (1983) Molecular cloning of the avian acute transforming retrovirus MH2 reveals a novel cell-derived sequence (v-*mil*) in addition to the *myc* oncogene. *EMBO J.* **2**: 2189–2194.

COLLETT, M. S., BRUGGE, J. S. and ERIKSON, R. L. (1978) Characterization of a normal avian cell protein related to the avian sarcoma virus transforming gene product. *Cell* **15**: 1363–1369.

COLLETT, M. S. and ERIKSON, R. L. (1978) Protein kinase activity associated with the avian sarcoma virus *src* gene product. *Proc. natn. Acad. Sci. U.S.A.* **75**: 2021–2024.

COLLETT, M. S., ERIKSON, E. and ERIKSON, R. L. 1979a) Structural analysis of the avian sarcoma virus transforming protein: sites of phosphorylation. *J. Virol.* **29**: 770–781.

COLLETT, M. S., PURCHIO, A. F. and ERIKSON, R. L. (1980) Avian sarcoma virus-transforming protein, pp60^{src}, shows protein kinase activity specific for tyrosine. *Nature* **285**: 167–169.

COLLETT, M. S., WELLS, S. K. and PURCHIO, A. F. (1983) Physical modification of purified Rous sarcoma virus pp 60$^{v\text{-}src}$ protein after incubation with ATP/Mg^{2+}. *Virology* **128**: 285–297.

COLLINS, S. and GROUDINE, M. (1982) Amplication of endogenous *myc*-related DNA sequences in a human myeloid leukemia cell line. *Nature* **298**: 679–681.

COOPER, J. A. and HUNTER, T. (1981a) Changes in protein phosphorylation in Rous sarcoma virus-transformed chicken embryo cells. *Molec. cell. Biol.* **1**: 165–178.

COOPER, J. A. and HUNTER, T. (1981b) Four different classes of retroviruses induce phosphorylation of tyrosines present in similar cellular proteins. *Molec. cell. Biol.* **1**: 394–407.

COOPER, J. A. and HUNTER, T. (1982) Discrete primary locations of a tyrosine protein kinase and of three proteins that contain phosphotyrosine in virally-transformed chick fibroblasts. *J. Cell. Biol.* **94**: 287–296.

COOPER, J. A. and HUNTER, T. (1983a) Regulation of cell growth and transformation of tyrosine-specific protein kinases: the search for important cellular substrate proteins. *Curr. Top. Microbiol. Immunol.* **107**: 125–161.

COOPER, J. A. and HUNTER, T. (1983b) Identification and characterization of cellular targets for tyrosine protein kinase. *J. biol. Chem.*, **258**: 1108–1115.

COOPER, J. A., REISS, N. A., SCHWARTZ, R. J. and HUNTER, T. (1983) Three glycolytic enzymes are phosphorylated at tyrosine in cells transformed by Rous sarcoma virus. *Nature* **302**: 218–223.

COOPER, M. D., PAYNE, L. N., DENT, P. B., BURMESTER, B. R. and GOOD, R. A. (1968) Pathogenesis in avian lymphoid leukosis. I. Histogenesis. *J. natn. Cancer Inst.* **41**: 373–389.

COOPER, M. D., PURCHASE, H. G. BOCKMAN, D. E. and GATHINGS, W. E. (1974) Studies on the nature of the abnormailty of B cell differentiation in avian lymphoid leukosis: production of heterogeneous IgM by tumor cells. *J. Immunol.* **113**: 1210–1222.

COPELAND, N. G., ZELENETZ, A. D. and COOPER, G. M. (1980) Transformation by subgenomic fragments of Rous sarcoma virus DNA. *Cell* **19**: 863-870.
COTTON, P. C. and BRUGGE, J. S. (1983) Neural tissues express high levels of the cellular *src* gene product pp60$^{c\text{-}src}$. *Molec. cell. Biol.* **3**: 1157-1162.
COURTNEIDGE, S. A. and BISHOP, J. M. (1982) Transit of pp60$^{v\text{-}src}$ to the plasma membrane. *Proc. natn. Acad. Sci. U.S.A.* **79**: 7117-7121.
COURTNEIDGE, S. A., LEVINSON, A. D. and BISHOP, J. M. (1980) The protein encoded by the transforming gene of avian sarcoma virus (pp60^{src}) and a homologous protein in normal cells (pp60$^{proto\text{-}src}$) are associated with the plasma membrane. *Proc. natn. Acad. Sci. U.S.A.* **77**: 3783-3787.
COURTNEIDGE, S., RALSTON, R., ALITALO, K. and BISHOP, J. M. (1983) The subcellular location of an abundant substrate (p 36) for tyrosine-specific protein kinases. *Molec. cell. Biol.* **3**: 340-350.
CRITTENDEN, L. B. (1975) Two levels of genetic resistance to lymphoid leukosis. *Avian Dis.* **19**: 281-292.
CRITTENDEN, L. B., HAYWARD, W. S., HANAFUSA, H. and FADLY, A. M. (1980) Induction of neoplasms by subgroup E recombinants of exogenous and endogenous avian retroviruses (Rous-associated virus type 60). *J. Virol.* **33**: 915-919.
CROSS, F. R. and HANAFUSA, H. (1983) Local mutagenesis of Rous sarcoma virus: the major sites of tyrosine and serine phosphorylation of pp60^{src} are dispensable for transformation. *Cell* **34**: 597-607.
CULLEN, B. R., LOMEDICO, P. T. and JU, G. (1984) Transcriptional interference in avian retroviruses—implications for the promoter insertion model of leukaemogenesis. *Nature* **307**: 241-245.
CZERNILOFSKY, A. P., DELORBE, W., SWANSTROM, R., VARMUS, H. E., BISHOP, J. M., TISCHER, E. and GOODMAN, H. M. (1980b) The nucleotide sequence of an untranslated but conserved domain at the 3' end of the avian sarcoma virus genome. *Nucleic Acids Res.* **8**: 2967-2984.
CZERNILOFSKY, A. P., LEVINSON, A. D., VARMUS, H. E., BISHOP, J. M., TISCHER, E. and GOODMAN, H. M. (1980a) Nucleotide sequence of an avian virus oncogene (src) and proposed amino acid sequence for gene product. *Nature* **287**: 198-203.
DALLA FAVERA, R., WONG-STAAL, F. and GALLO, R. C. (1982) *Onc* gene amplification in promyelocytic leukaemia cell line HL-60 and primary leukaemia cells of the same patient. *Nature* **299**: 61-63.
DAVID-PFEUTY, T. and SINGER, S. J. (1980) Altered distribution of the cytoskeletal proteins vinculin and α-actinin in cultured fibroblasts transformed by Rous sarcoma virus. *Proc. natn. Acad. Sci. U.S.A.* **77**: 6687-6691.
DECKER, S. (1982) Phosphorylation of the $M_r = 34,000$ protein in normal and Rous sarcoma virus-transformed rat fibroblasts. *Biochem. biophys. Res. Commun.* **109**: 434-441.
DIRINGER, H. and FRIIS, R. R. (1977) Changes in phosphatidylinositol metabolism correlated to growth state of normal and Rous sarcoma virus transformed Japanese quail cells. *Cancer Res.* **37**: 2979-2984.
DONNER, P., BUNTE, T., GREISER-WILKE, I. and MOELLING, K. (1983) Decreased DNA-binding ability of purified transformation-specific proteins from deletion mutants of the acute avian leukemia virus MC29. *Proc. natn. Acad. Sci. U.S.A.* **80**: 2861-2865.
DONNER, P., GREISER-WILKE, I. and MOELLING, K. (1982) Nuclear localization and DNA binding of the transforming gene product of avian myelocytomatosis virus. *Nature* **296**: 262-266.
DOWNWARD, J., YARDEN, Y., MAYES, E., SCRACE, G., TOTTY, N., STOCKWELL, P., ULLRICH, A., SCHLESSINGER, J. and WATERFIELD, M. D. (1984) Close similarity of epidermal growth factor receptor and v-*erb* B oncogene protein sequences. *Nature* **307**: 521-526.
DUESBERG, P. H., BISTER, K. and MOSCOVICI, C. (1980) Genetic structure of avian myeloblastosis virus released from transformed myeloblasts as a defective virus particle. *Proc. natn. Acad. Sci. U.S.A.* **77**: 5120-5124.
DUESBERG, P. H., PHARES, W. and LEE, W. (1983) The low tumorigenic potential of PRCII, among viruses of the Fujinami sarcoma virus subgroup, corresponds to an internal (*fps*) deletion of the transforming gene. *Virology* **131**: 144-158.
DUESBERG, P. H. and VOGT, P. K. (1970) Differences between ribonucleic acids of transforming and non-transforming avian tumor viruses. *Proc. natn. Acad. Sci. U.S.A.* **67**: 1673-1680.
DUFF, R. G. and VOGT, P. K. (1969) Characteristics of two new avian tumor virus subgroups. *Virology* **39**: 18-30.
EDELMAN, G. M. and YAHARA, I. (1976) Temperature-sensitive changes in surface modulating assemblies of fibroblasts transformed by mutants of Rous sarcoma virus. *Proc. natn. Acad. Sci. U.S.A.* **73**: 2047-2051.
EK, B. and HELDIN, C. H. (1982) Characterization of a tyrosine-specific kinase activity in human fibroblast membranes stimulated by platelet-derived growth factor. *J. biol. Chem.* **257**: 10486-10492.
EK, B., WESTERMARK, B., WASTESON, A. and HELDIN, C. H. (1982) Stimulation of tyrosine-specific phosphorylation by platelet-derived growth factor. *Nature* **295**: 419-420.
ELLERMANN, V. and BANG, O. (1908) Experimentelle Leukämie bei hühnern. *Zentralbl. Bakteriol.* **46**: 595-609.
ELLERMANN, V. and BANG, O. (1909) Experimentelle Leukämie bei hühnern. *Z. Hyg. Infektionskr.* **63**: 231-272.
ENRIETTO, P. J., HAYMAN, M. J., RAMSAY, G. M., WYKE, J. A. and PAYNE, L. N. (1983a) Altered pathogenicity of avian myelocytomatosis (MC29) viruses with mutations in the v-*myc* gene. *Virology* **124**: 164-172.
ENRIETTO, P. J., PAYNE, L. N. and HAYMAN, M. J. (1983b) A recovered avian myelocytomatosis virus that induces lymphomas in chickens: pathogenic properties and their molecular basis. *Cell* **35**: 369-379.
ENRIETTO, P. J., PAYNE, L. N. and WYKE, J. A. (1983c) Analysis of the pathogenicity of transformation-defective partial deletion mutants of avian sarcoma virus: characterization of recovered viruses which encode novel *src* specific proteins. *Virology* **127**: 397-411.
EPHRUSSI, B. and TEMIN, H. M. (1960) Infection of chick iris epithelium with the Rous sarcoma virus *in vitro*. *Virology* **11**: 547-552.
ERIKSON, E, COLLETT, M. S. and ERIKSON, R. L. (1978) *In vitro* synthesis of a functional avian sarcoma virus transforming-gene product. *Nature* **274**: 919-921.
ERIKSON, R. L., COLLETT, M. S., ERIKSON, E. and PURCHIO, A. F. (1979) Evidence that the avian sarcoma virus transforming gene product is a cyclic AMP-independent protein kinase. *Proc. natn. Acad. Sci. U.S.A.* **76**: 6260-6264.
ERIKSON, R. L., COLLETT, M. S., ERIKSON, E., PURCHIO, A. F. and BRUGGE, J. S. (1979) Protein phosphorylation

mediated by partially purified avian sarcoma virus transforming-gene product. *Cold Spring Harb. Symp. Q. Biol.* **44**: 907-917.

ERIKSON, E. and ERIKSON, R. L. (1980) Identification of a cellular protein substrate phosphorylated by the avian sarcoma virus transforming gene product. *Cell* **21**: 829-836.

ERIKSON, E. and ERIKSON, R. L. (1983) Antigenic and structural studies on the transforming proteins of Rous sarcoma virus and Yamaguchi 73 avian sarcoma virus. *Virology* **130**: 221-226.

ERIKSON, R. L. and PURCHIO, A. F. (1982) Avian sarcoma viruses, protein kinases, and cell transformation. In: *Advances in Viral Oncology*, pp. 43-57, KLEIN, G. (ed.) Raven Press, New York.

FELDMAN, R. A., HANAFUSA, T. and HANAFUSA, H. (1980) Characterization of protein kinase activity associated with the transforming gene product of Fujinami sarcoma virus. *Cell* **22**: 757-765.

FELDMAN, R. A., WANG, E. and HANAFUSA, H. (1983) Cytoplasmic localization of the transforming protein of Fujinami sarcoma virus: salt-sensitive association with subcellular components. *J. Virol.* **45**: 782-791.

FELDMAN, R. A., WANG, L. H., HANAFUSA, H. and BALDUZZI, P. C. (1982) Avian sarcoma Virus UR2 encodes a transforming protein which is associated with a unique protein kinase activity. *J. Virol.* **42**: 228-236.

FISZMAN, M. Y. and FUCHS, P. (1975) Temperature-sensitive expression of differentiation in transformed myoblasts. *Nature* **254**: 429-431.

FOSTER, D. A. and HANAFUSA, H. (1983) A *fps* gene without *gag* sequences transforms cells in culture and induces tumors in chickens. *J. Virol.* **48**: 744-751.

FRANKLIN, R. B., KANG, C. Y., WAN, K. M. M. and BOSE, H. R., Jr (1977) Transformation of chick embryo fibroblasts by reticuloendotheliosis virus. *Virology* **83**: 313-321.

FRIIS, R. R., JOCKUSCH, B. M., BOSCHEK, C. B., ZIEMIECKI, A., RÜBSAMEN, H. and BAUER, H. (1980) Transformation defective temperature-sensitive mutants of Rous sarcoma virus have a reversibly-defective *src* gene product. *Cold Spring Harb. Symp. Q. Biol.* **44**: 1007-1012.

FRYKBERG, L., PALMIERI, S., BEUG, H., GRAF, T., HAYMAN, M. J. and VENNSTROM, B. (1983) Transforming capacities of avian erythroblastosis virus mutants deleted in the *erb*A or *erb*B oncogenes. *Cell* **32**: 227-238.

FUJINAMI, A. and INAMOTO, K. (1914) Veber Geschwulste bei japanischen Haushuhnern insbesondere uber einen transplantablen Tumor. *Z. Krebsforsch* **14**: 94-119.

FUJITA, D. J., CHEN, Y. C., FRIIS, R. R. and VOGT, P. K. (1974) RNA tumor viruses of pheasants: characterization of avian leukosis subgroups F and G. *Virology* **60**: 558-571.

FULTON, A. B. (1982) How crowded is the cytoplasm? *Cell* **30**: 345-347.

FUNG, Y. T., FADLY, A. M., CRITTENDEN, L. B. and KUNG, H. J. (1981) On the mechanism of retrovirus-induced avian lymphoid leukosis: deletion and integration of the proviruses. *Proc. natn. Acad. Sci. U.S.A.* **78**: 3418-3422.

FUNG, Y. T., LEWIS, W. G., CRITTENDEN, L. B. and KUNG, H. J. (1983) Activation of the cellular oncogene c-*erb*B by LTR insertion: molecular basis for induction of erythroblastosis by avian leukosis virus. *Cell* **33**: 357-368.

GARBER, E. A., KRUEGER, J. G. and GOLDBERG, A. R. (1982) Novel localization of pp60^{src} in Rous sarcoma virus-transformed rat and goat cells and in chicken cells transformed by viruses rescued from these mammalian cells. *Virology* **118**: 419-429.

GARBER, E. A., KRUEGER, J. G., HANAFUSA, H. and GOLDBERG, A. R. (1983) Only membrane-associated RSV^{src} proteins have amino-terminally bound lipid. *Nature* **302**: 161-163.

GAZZOLO, L., SAMARUT, J., BOUABDELLI, M. and BLANCHET, J. P. (1980) Early precursors in the erythroid lineage are the specific target cells of avian erythroblastosis virus *in vitro*. *Cell* **22**: 683-691.

GEIGER, B. (1979) A 130kd protein from chicken gizzard: its location at the termini of microfilament bundles in cultured chicken cells. *Cell* **18**: 193-205.

GHYSDAEL, J., NEILL, J. C., WALLBANK, A. M. and VOGT, P. K. (1981) *Esh* avian sarcoma virus codes for a *gag*-linked transformation-specific protein with an associated protein kinase activity. *Virology* **111**: 386-400.

GILMORE, T. D., RADKE, K. and MARTIN, G. S. (1982) Tyrosine phosphorylation of a 50kd cellular polypeptide associated with the Rous sarcoma virus transforming protein pp60^{src}. *Molec. cell. Biol.* **2**: 199-206.

GLUZMAN, Y. and SHENK, T. (1983) *Enhancers and Eukaryotic Gene Expression*. Cold Spring Harb. Lab., Cold Spring Harbor, New York.

GONDA, T. J. and BISHOP, J. M. (1983) Structure and transcription of the cellular homolog (c-*myb*) of the avian myeloblastosis virus transforming gene (v-*myb*). *J. Virol.* **46**: 212-220.

GONDA, T. J., SHEINESS, D. K. and BISHOP, J. M. (1982) Transcripts from the cellular homologs of retroviral oncogenes: distribution among chicken tissues. *Molec. cell. Biol.* **2**: 617-624.

GONDA, T. J., SHEINESS, D. K., FANSHIER, L., BISHOP, J. M., MOSCOVICI, C. and MOSCOVICI, M. G. (1981) The genome and the intracellular RNAs of avian myeloblastosis virus. *Cell* **23**: 279-290.

GRAF, T. and BEUG, H. (1978) Avian leukemia viruses: interaction with their target cells *in vivo* and *in vitro*. *Biochim. biophys. Acta* **516**: 269-299.

GRAF, T. and BEUG, H. (1983) Role of the v-*erb*A and v-*erb*B oncogenes of avian erythroblastosis virus in erythroid cell transformation. *Cell* **34**: 7-9.

GRAF, T., KIRCHBACH, A. V. and BEUG, H. (1981) Characterization of the hematopoietic target cells of AEV, MC29 and AMV avian leukemia viruses. *Exp. cell. Res.* **131**: 331-343.

GRAZIANI, Y., ERIKSON, E. and ERIKSON, R. L. (1983a) Characterization of the Rous sarcoma virus transforming gene product: *in vitro* phosphorylation with ATP and GTP as phosphate donors. *J. biol. Chem.* **258**: 6344-6351.

GRAZIANI, Y., ERIKSON, E. and ERIKSON, R. L. (1983b) Evidence that the Rous sarcoma virus transforming gene product is associated with glycerol kinase activity. *J. biol. Chem.* **258**: 2126-2129.

GREENBERG, M. E. and EDELMAN, G. M. (1983) The 34kd pp60^{src} substrate is located at the inner face of the plasma membrane. *Cell*, **33**: 767-779.

GREENGARD, P. (1978) Phosphorylated proteins as physiological effectors. *Science* **199**: 146-152.

GROFFEN, J., HEISTERKAMP, N., SHIBUYA, M., HANAFUSA, H. and STEPHENSON, J. R. (1983) Transforming genes of avian (v-fps) and mammalian (v-fes) retroviruses correspond to a common cellular locus. *Virology* **125**: 480-486.

GUYDEN, J. C. and MARTIN, G. S. (1982) Transformation parameters of chick embryo fibroblasts transformed by Fujinami, PRC11, PRC11-p, and Y73 avian sarcoma viruses. *Virology* **122**: 71–83.

HALPERN, C. C., HAYWARD, W. S. and HANAFUSA, H. (1979) Characterization of some isolates of newly recovered avian sarcoma virus. *J. Virol.* **29**: 91–101.

HAMPE, A., LAPREVOTTE, I., GALIBERT, F., FEDELE, L. A. and SHERR, C. J. (1982) Nucleotide sequences of feline retroviral oncogenes (v-*fes*) provide evidence for a family of tyrosine-specific protein kinase genes. *Cell* **30**: 775–785.

HANAFUSA, H. (1975) Avian RNA tumor virus. In: *Cancer: A Comprehensive Treatise*, pp. 49–90, BECKER, F. F. (ed.) Plenum, New York.

HANAFUSA, H. (1977) Cell transformation by RNA tumor viruses. In: *Comprehensive Virology*, pp. 401–483, FRAENKEL-CONRAT, H. and WAGNER, R. P. (eds) Plenum Press, New York.

HANAFUSA, H., HALPERN, C. C., BUCHHAGEN, D. L. and KAWAI, S. (1977) Recovery of avian sarcoma virus from tumors induced by transformation-defective mutants. *J. exp. Med.* **146**: 1735–1747.

HANAFUSA, H., WANG, L. H., ANDERSON, S. M., KARESS, R. E. and HAYWARD, W. S. (1980a) The nature and origin of the transforming gene of avian sarcoma virus. In: *Animal Virus Genetics*, pp. 483–497, FIELDS, B. et al. (eds) Academic Press, New York.

HANAFUSA, T., HANAFUSA, H. and MIYAMOTO, T. (1970) Recovery of a new virus from apparently normal chick cells by infection with avian tumor viruses. *Proc. natn. Acad. Sci. U.S.A.* **67**: 1797–1803.

HANAFUSA, T., WANG, L. H., ANDERSON, R. G., KARESS, R. E., HAYWARD, W. S. and HANAFUSA, H. (1980b) Characterization of the transforming gene of Fujinami sarcoma virus. *Proc. natn. Acad. Sci. U.S.A.* **77**: 3009–3013.

HANN, S. R., ABRAMS, H. D., ROHRSCHNEIDER, L. R. and EISENMAN, R. N. (1983) Proteins encoded by v-*myc* and c-*myc* oncogenes: identification and localization in acute leukemia virus transformants and bursal lymphoma cell lines. *Cell* **34**: 789–798.

HAYMAN, M. J., RAMSAY, G. M., SAVIN, K., KITCHNER, G., GRAF, T. and BEUG, H. (1983) Identification and characterization of the avian erythroblastosis virus *erb*B gene product as a membrance glycoprotein. *Cell* **32**: 579–588.

HAYMAN, M. J., KITCHNER, G. and GRAF, T. (1979) Cells transformed by avian myelocytomatosis virus strain CM11 contain a 90kd *gag*-related protein. *Virology* **98**: 191–199.

HAYWARD, W. S. (1977) Size and genetic content of viral RNAs in avian oncovirus-infected cells. *J. Virol.* **24**: 47–63.

HAYWARD, W. S., NEEL, B. G. and ASTRIN, S. M. (1981) Activation of a cellular *onc* gene by promoter insertion in ALV-induced lymphoid leukosis. *Nature* **290**: 475–480.

HOELZER, J. D., LEWIS, R. B., WASMUTH, C. R. and BOSE, H. R., Jr. (1980) Hematopoietic cell transformation by reticuloendotheliosis virus: characterization of the genetic defect. *Virology* **100**: 462–474.

HUANG, C. C., HAMMOND, C. and BISHOP, J. M. (1984) Nucleotide sequence of v-*fps* in the PRCII strain of avian sarcoma virus. *J. Virol.* **50**: 125–131.

HUGHES, S. H., SHANK, P. R., SPECTOR, D. H., KUNG, H. J., BISHOP, J. M., VARMUS, H. E., VOGT, P. K. and BREITMAN, M. L. (1978) Proviruses of avian sarcoma virus are terminally redundant, co-extensive with unintegrated linear DNA and integrated at many sites. *Cell* **15**: 1397–1410.

HUNTER, T. and COOPER, J. A. (1983) The role of tyrosine phosphorylation in malignant transformation and in cellular growth control. In: *Prog. Nucleic Acid Res. Molec. Biol.* pp. 221–233, COHN, W. (ed.) Academic Press, New York.

HUNTER, T. and SEFTON, B. M. (1980) Transforming gene product of Rous sarcoma virus phosphorylates tyrosine. *Proc. natn. Acad. Sci. U.S.A.* **77**: 1311–1315.

HYNES, R. O. (1974) Role of surface alterations in cell transformation: the importance of proteases and surface proteins. *Cell* **1**: 147–156.

ITOHARA, S., KIRATA, K., INOVE, M., HATSUOKA, M. and SATO, A. (1978) Isolation of a sarcoma virus from a spontaneous chicken tumor. *Gann* **69**: 825–830.

JACOBS, C. and RÜBSAMEN, H. (1983) Expression of pp60^{c-src} protein kinase in adult and fetal human tissue: high activities in some sarcomas and mammary carcinomas. *Cancer Res.* **43**: 1696–1702.

JANSEN, H. W., LURZ, R., BISTER, K., BONNER, T. I., MARK, G. E. and RAPP, U. R. (1984) Homologous cell-derived oncogenes in avian carcinoma virus MH2 and murine sarcoma virus 3611. *Nature* **307**: 281–284.

JANSEN, H. W., PATSCHINSKY, T. and BISTER, K. (1983) Avian oncovirus MH2: molecular cloning of proviral DNA and structural analysis of viral RNA and protein. *J. Virol.* **48**: 61–73.

JANSSON, M., PHILIPSON, L. and VENNSTROM, B. (1983) Isolation and characterization of multiple human genes homologous to the oncogenes of avian erythroblastosis virus. *EMBO J.* **2**: 461–465.

JOCKUSCH, B. M. and ISENBERG, G. (1981) Interaction of α-actinin and vinculin with actin: opposite effects on filament network formation. *Proc. natn. Acad. Sci. U.S.A.* **78**: 3005–3009.

KAIGHN, M. E., EBERT, J. D. and STOTT, P. M. (1966) The susceptibility of differentiating muscle clones to Rous sarcoma virus. *Proc. natn. Acad. Sci. U.S.A.* **56**: 133–140.

KAN, N., FLORDELLIS, C., GARON, C., DUESBERG, P. and PAPAS, T. (1983) Avian carcinoma virus MH2 contains a transformation-specific sequence, *mht*, and shares the *myc* sequence with MC29, CMII, and OK10 viruses. *Proc. natn. Acad. Sci. U.S.A.* **80**: 6566–6570.

KANG, C. Y. and TEMIN, M. H. (1973) Lack of sequence homology among RNAs of avian leukosis-sarcoma viruses, reticuloendotheliosis viruses, and chicken endogenous RNA-directed DNA polymerase activity. *J. Virol.* **12**: 1314–1324.

KARESS, R. E., HAYWARD, W. S. and HANAFUSA, H. (1979) Cellular information in the genome of recovered avian sarcoma virus directs the synthesis of transforming protein. *Proc. natn. Acad. Sci. U.S.A.* **76**: 3154–3158.

KASUGA, M., KARLSSON, F. A. and KAHN, C. R. (1981) Insulin stimulates the phosphorylation of the 95,000-dalton subunit of its own receptor. *Science* **215**: 185–187.

KATZ, R. A., MANIATIS, G. M. and GUNTAKA, R. V. (1979) Translation of avian sarcoma virus RNA in *Xenopus laevis* öocytes. *Biochem. biophys. Res. Commun.* **86**: 447–453.

KAWAI, S., YOSHIDA, M., SEGAWA, K. SUGIYAMA, H., ISHIZAKI, R. and TOYOSHIMA, K. (1980) Characterization of Y73, an avian sarcoma virus: a unique transforming gene and its product, a phosphopolyprotein with protein kinase activity. *Proc. natn. Acad. Sci. U.S.A.* **77**: 6199–6203.

KELLY, K., COCHRAN, B. H., STILES, C. D. and LEDER, P. (1983) Cell-specific regulation of the c-*myc* gene by lymphocyte mitogens and platelet-derived growth factor. *Cell* **35**: 603–610.

KHOURY, G. and GRUSS, P. (1983) Enhancer elements. *Cell* **33**: 313–314.

KITAMURA, N., KITAMURA, A., TOYOSHIMA, K., HIRAYAMA, Y. and YOSHIDA, M. (1982) Avian sarcoma virus Y73 genome sequence and structural similarity of its transforming gene product to that of Rous sarcoma virus. *Nature* **297**: 205–208.

KLEMPNAUER, K. H. and BISHOP, J. M. (1984) Neoplastic transformation by E26 leukemia virus is mediated by a single protein containing domains of *gag* and *myb* genes. *J. Virol.* **50**: 280–283.

KLEMPNAUER, K. H., GONDA, T. J. and BISHOP, J. M. (1982) Nucleotide sequence of the retroviral leukemia gene v-*myb* and its cellular progenitor c-*myb*: the architecture of a transduced oncogene. *Cell* **31**: 453–463.

KLEMPNAUER, K. H., RAMSAY, G., BISHOP, J. M., MOSCOVICI, M. G., MOSCOVICI, C., MCGRATH, J. P. and LEVINSON, A. D. (1983) The product of the retroviral transforming gene v-*myb* is a truncated version of the protein encoded by the cellular oncogene c-*myb*. *Cell* **33**: 345–355.

KLETZIEN, R. F. and PERDUE, J. F. (1975) Regulation of sugar transport in chick embryo fibroblasts infected with a temperature-sensitive mutant of RSV. *Cell* **6**: 513–520.

KOBAYASHI, N. and KAJI, A. (1980) Phosphoprotein associated with activation of the *src* gene product in myogenic cells. *Biochem. biophys. Res. Commun.* **93**: 278–284.

KREBS, E. G. and BEAVO, J. A. (1979) Phosphorylation-dephosphorylation of enzymes. *A. Rev. Biochem.* **48**: 923–959.

KRUEGER, J. G., GARBER, E. A., GOLDBERG, A. R. and HANAFUSA, H. (1982) Changes in amino-terminal sequences of pp60^{src} lead to decreased membrane association and decreased *in vivo* tumorigenicity. *Cell* **28**: 889–896.

KRUEGER, J. G., WANG, E. and GOLDBERG, A. R. (1980b) Evidence that the *src* gene product of Rous sarcoma virus is membrane associated. *Virology* **101**: 25–40.

KRUEGER, J. G., WANG, E., GARBER, E. A. and GOLDBERG, A. R. (1980a) Differences in intracellular location of pp60^{src} in rat and chicken cells transformed by Rous sarcoma virus. *Proc. natn. Acad. Sci. U.S.A.* **77**: 4142–4146.

KRZYZEK, R. A., COLLETT, M. S., LAU, A. F., PERDUE, M. L., LEIS, J. P. and FARAS, A. J. (1978) Evidence for splicing of avian sarcoma virus 5'-terminal genomic sequences onto viral-specific RNA in infected cells. *Proc. natn. Acad. Sci. U.S.A.* **75**: 1284–1288.

LAND, H., PARADA, L. F. and WEINBERG, R. A. (1983) Tumorigenic conversion of primary embryo fibroblasts requires at least two cooperating oncogenes. *Nature* **304**: 596–602.

LAND, H., PARADA, L. F. and WEINBERG, R. A. (1983) Cellular oncogenes and multi-step carcinogenesis. *Science* **222**: 771–778.

LANKS, K. W., KASAMBALIDES, E. J., CHINKERS, M. and BRUGGE, J. S. (1982) A major cytoplasmic glucose-regulated protein is associated with the Rous sarcoma virus pp60^{src} protein. *J. biol. Chem.* **257**: 8604–8607.

LEE, J. S., VARMUS, H. E. and BISHOP, J. M. (1979) Virus-specific messenger RNAs in permissive cells infected by avian sarcoma virus. *J. biol. Chem.* **254**: 8015–8022.

LEE, W. H., BISTER, K., MOSCOVICI, C. and DUESBERG, P. H. (1981) Temperature-sensitive mutants of Fujinami sarcoma virus: tumorigenicity and reversible phosphorylation of the transforming p140 protein. *J. Virol.* **38**: 1064–1976.

LEE, W. H., BISTER, K., PAWSON, A., ROBINS, T., MOSCOVICI, C. and DUESBERG, P. H. (1980) Fujinami sarcoma virus: an avian RNA tumor virus with a unique transforming gene. *Proc. natn. Acad. Sci. U.S.A.* **77**: 2018–2022.

LEE, W. H., PHARES, W. and DUESBERG, P. H. (1983) Structural relationship between the chicken DNA locus, proto-*fps*, and the transforming gene of Fujinami sarcoma virus, *gag-fps*. *Virology* **129**: 79–93.

LEPRINCE, D., GEGONNE, A., COLL, J., DETAISNE, C., SCHNEEBERGER, A., LAGROU, C. and STEHELIN, D. (1983) A putative second cell-derived oncogene of the avian leukemia retrovirus E26. *Nature* **306**: 395–397.

LEVINSON, A. D., COURTNEIDGE, S. A. and BISHOP, J. M. (1981) Structural and functional domains of the Rous sarcoma virus transforming protein (pp60^{src}). *Proc. natn. Acad. Sci. U.S.A.* **78**: 1624–1628.

LEVINSON, A. D., OPPERMANN, H., LEVINTOW, L., VARMUS, H. E. and BISHOP, J. M. (1978) Evidence that the transforming gene of avian sarcoma virus encodes a protein kinase associated with a phosphoprotein. *Cell* **15**: 561–572.

LEVINSON, A. D., OPPERMANN, H., VARMUS, H. E. and BISHOP, J. M. (1980) The purified product of the transforming gene of avian sarcoma virus phosphorylates tyrosine. *J. biol. Chem.* **255**: 11973–11980.

LEWIS, R. B., MCCLURE, J., RUP, B., NIESEL, D. W., GARRY, R. F., HOELZER, J. D., NAZERIAN, K. and BOSE, H. R. (1981) Avian reticuloendotheliosis virus: identification of the hematopoietic target cell for transformation. *Cell* **25**: 421–431.

LINIAL, M. (1982) Two retroviruses with similar transforming genes exhibit differences in transforming potential. *Virology*, **119**: 382–391.

LIPSICH, L. A., LEWIS, A. J. and BRUGGE, J. S. (1983) Isolation of monoclonal antibodies that recognize the transforming proteins of avian sarcoma viruses. *J. Virol.* **48**: 352–360.

MAKOWSKI, D. R., ROTHBERG, P. G. and ASTRIN, S. M. (1984) The role of promoter insertion in the induction of neoplasia. *Surv. Synth. Path. Res.* **3**: 342–349.

MARK, G. E. and RAPP, U. R. (1984) Primary structure of v-*raf*: relatedness to the *src* family of oncogenes. *Science* **224**: 285–289.

MARTIN, G. S., RADKE, K., HUGHES, S., QUINTRELL, N., BISHOP, J. M. and VARMUS, H. E. (1979) Mutants of Rous sarcoma virus with extensive deletions of the viral genome. *Virology* **96**: 530–546.

MARTINEZ, R., NAKAMURA, K. D. and WEBER, M. J. (1982) Identification of phosphotyrosine-containing proteins in untransformed and Rous sarcoma virus-transformed chicken embryo fibroblasts. *Molec. cell. Biol.* **2**: 653–665.

MATHEY-PREVOT, B., HANAFUSA, H. and KAWAI, S. (1982) A cellular protein is immunologically cross reactive with and functionally homologous to the Fujinami sarcoma virus transforming protein. *Cell* **28**: 897–906.

MCGINNIS, J., HIZI, A., SMITH, R. E. and LEIS, J. P. (1978) *In vitro* translation of a 180,000 dalton Rous sarcoma virus precursor polypeptide containing both the DNA polymerase and the group-specific antigens. *Virology* **84**: 518–522.

MOSCOVICI, M. G., JURDIC, P., SAMARUT, J., GAZZOLO, L., MURA, C. V. and MOSCOVICI, C. (1983) Characterization of the hemopoietic target cells for the avian leukemia virus E26. *Virology* **129**: 65–78.

NAKAMURA, K. D. and WEBER, M. J. (1982) Phosphorylation of a 36,000 M_r cellular protein in cells infected with partial transformation mutants of Rous sarcoma virus. *Molec. cell. Biol.* **2**: 147–153.

NEEL, B. G., GASIC, G. P., ROGLER, C. E., SKALKA, A. M., JU, G., HISHINUMA, F., PAPAS, T., ASTRIN, S. M. and HAYWARD, W. S. (1982) Molecular analysis of the c-*myc* locus in normal tissue and in avian leukosis virus-induced lymphomas. *J. Virol.* **44**: 158–166.

NEEL, B. G., HAYWARD, W. S., ROBINSON, H. L., FANG, J. M. and ASTRIN, S. M. (1981) Avian leukosis virus-induced tumors have common proviral integration sites and synthesize discrete new RNAs: oncogenesis by promoter insertion. *Cell* **23**: 323–334.

NEIL, J. C., GHYSDAEL, J., SMART, J. E. and VOGT, P. K. (1982) Structural similarities of proteins encoded by three classes of avian sarcoma viruses. *Virology* **121**: 274–287.

NEIL, J. C., GHYSDAEL, J., VOGT, P. K. and SMART, J. E. (1981) Homologous tyrosine phosphorylation sites in transformation-specific gene products of distinct avian sarcoma viruses. *Nature* **291**: 675–677.

NEIMAN, P. E., JORDAN, L., WEISS, R. A. and PAYNE, L. N. (1980a) Malignant lymphoma of the bursa of Fabricius: analysis of early transformation. *Cold Spring Harb. Conf. Cell Proliferation* **8**: 519–528.

NEIMAN, P., PAYNE, L. N. and WEISS, R. A. (1980b) Viral DNA in bursal lymphomas induced by avian leukosis viruses. *J. Virol.* **34**: 178–186.

NIGG, E. A., SEFTON, B. M., HUNTER, T., WALTER, G. and SINGER, S. J. (1982) Immunofluorescent localization of the transforming protein of Rous sarcoma virus with antibodies against a synthetic *src* peptide. *Proc. natn. Acad. Sci. U.S.A.* **79**: 5322–5326.

NISHIMURA, J., HUANG, J. S. and DEVEL, T. F. (1982) Platelet-derived growth factor stimulates tyrosine-specific protein kinase activity in Swiss mouse 3T3 cell membranes. *Proc. natn. Acad. Sci. U.S.A.* **79**: 4303–4307.

NOORI-DALOII, M. R., SWIFT, R. A. and KING, H. J. (1981) Specific integration of REV proviruses in avian bursal lymphomas. *Nature* **294**: 574–576.

NUNN, M. F., SEEBURG, P. H., MOSCOVICI, C. and DUESBERG, P. H. (1983) Tripartite structure of the avian erythroblastosis virus E26 transforming gene. *Nature* **306**: 391–395.

OKAZAKI, W., PURCHASE, H. G. and CRITTENDEN, L. B. (1982) Pathogenicity of avian leukosis viruses. *Avian Dis.* **26**: 553–559.

OPPERMANN, H., LEVINSON, A. D., LEVINTOW, L., VARMUS, H. E., BISHOP, J. M. and KAWAI, S. (1981) Two cellular proteins that immunoprecipitate with the transforming protein of Rous sarcoma virus. *Virology* **113**: 736–751.

OPPERMANN, H., LEVINSON, A. D., VARMUS, H. E., LEVINTOW, L. and BISHOP, J. M. (1979) Uninfected vertebrate cells contain a protein that is closely related to the product of the avian sarcoma virus transforming gene (*src*). *Proc. natn. Acad. Sci. U.S.A.* **76**: 1804–1808.

PACHL, C., BIEGALKE, B. and LINIAL, M. (1983) RNA and protein encoded by MH2 virus: evidence for subgenomic expression of v-*myc*. *J. Virol.* **45**: 133–139.

PACIFICI, M., BOETTIGER, D., ROBY, K. and HOLTZER, H. (1977) Transformation of chondroblasts by Rous sarcoma virus and synthesis of the sulfated proteoglycan matrix. *Cell* **11**: 891–899.

PARSONS, J. T., LEWIS, P. and DIERKS, P. (1978) Purification of virus specific RNA from chicken cells infected with avian sarcoma virus: identification of genome-length and subgenomic-length viral RNAs. *J. Virol.* **27**: 227–238.

PATSCHINSKY, T., HUNTER, T., ESCH, F. S., COOPER, J. A. and SEFTON, B. M. (1982) Analysis of the sequence of amino acids surrounding sites of tyrosine phosphorylation. *Proc. natn. Acad. Sci. U.S.A.* **79**: 973–977.

PAWSON, T., GUYDEN, J., KUNG, T. H., RADKE, K., GILMORE, T. and MARTIN, G. S. (1980a) A strain of Fujinami sarcoma virus which is temperature-sensitive in protein phosphorylation and cellular transformation. *Cell* **22**: 767–775.

PAWSON, T., HARVEY, R. and SMITH, A. E. (1977) The size of Rous sarcoma virus mRNAs active in cell-free translation. *Nature* **268**: 416–420.

PAWSON, T., MARTIN, G. S. and SMITH, A. E. (1976) Cell-free translation of virion RNA from nondefective and transformation-defective Rous sarcoma viruses. *J. Virol.* **19**: 950–967.

PAWSON, T., MELLON, P., DUESBERG, P. H. and MARTIN, G. S. (1980b) *env* gene of Rous sarcoma virus: identification of the gene product by cell-free translation. *J. Virol.* **33**: 993–1003.

PAYNE, G. S., BISHOP, J. M. and VARMUS, H. E. (1982) Multiple arrangements of viral DNA and an activated host oncogene in bursal lymphomas. *Nature* **295**: 209–214.

PAYNE, G. S., COURTNEIDGE, S. A., CRITTENDEN, L. B., FADLY, A. M., BISHOP, J. M. and VARMUS, H. E. (1981) Analysis of avian leukosis virus DNA and RNA in bursal tumors: viral gene expression is not required for maintenance of the tumor state. *Cell* **23**: 311–322.

PERRY, R. P. (1983) Consequences of *myc* invasion of immunoglobulin loci: facts and speculation. *Cell* **33**: 647–649.

PESSAC, B. and CALOTHY, G. (1974) Transformation of chick-embryo neuroretinal cells by Rous-sarcoma virus, *in vitro* induction of cell-proliferation. *Science* **185**: 709–710.

PETRUZZELLI, L. M., GANGULY, S., SMITH, C. J., COBB, M. H., RUBIN, C. H. and ROSEN, O. M. (1982) Insulin activates a tyrosine-specific protein kinase in extracts of 3T3-L1 adipocytes and human placenta. *Proc. natn. Acad. Sci. U.S.A.* **79**: 6792–6796.

PFEIFER, S., ZABIELSKI, J., OHLSSON, R., FRYKBERG, L., KNOWLES, J., PETTERSSON, R., OKER-BLOM, N., PHILIPSON,

L., VAHERI, A. and VENNSTROM, B. (1983) Avian acute leukemia virus OK10: analysis of its *myc* oncogene by molecular cloning. *J. Virol.* **46**: 347–354.

PIKE, L. J., MARQUARDT, H., TODARO, G. J., GALLIS, B., CASHELLIE, J. E., BORNSTEIN, P. and KREBS, E. G. (1982) Transforming growth factor and epidermal growth factor stimulate the phosphorylation of a synthetic, tyrosine-containing peptide in a similar manner. *J. biol. Chem.* **257**: 14628–14631.

POIRIER, F., CALOTHY, G., KARESS, R. E., ERIKSON, E. and HANAFUSA, H. (1982) Role of p60^{src} kinase activity in the induction of neuroretinal cell proliferation by Rous sarcoma virus. *J. Virol.* **42**: 780–789.

POIRIER, F., JULLIEN, P., DEZELEE, P., DAMBRINE, G., ESNAULT, E., BENATRE, A. and CALOTHY, G. (1984) Role of the mitogenic property and kinase activity of p60^{src} in tumor formation by Rous sarcoma virus. *J. Virol.* **49**: 325–332.

PRIVALSKY, M. L. and BISHOP, J. M. (1982) Proteins specified by avian erythroblastosis virus: coding region localization and identification of a previously undetected erbB polypeptide. *Proc. natn. Acad. Sci. U.S.A.* **79**: 3958–3962.

PRIVALSKY, M. L., RALSTON, R. and BISHOP, J. M. (1984) The membrane glycoprotein encoded by the retroviral oncogene v-*erb*B is structurally related to tyrosine-specific protein kinases. *Proc. natn. Acad. Sci. U.S.A.* **81**: 704–707.

PRIVALSKY, M. L., SEALY, L., BISHOP, J. M., MCGRATH, J. P. and LEVINSON, A. D. (1983) The product of the avian erythroblastosis virus *erb*B locus is a glycoprotein. *Cell* **32**: 1257–1267.

PURCHASE, H. G. and BURMESTER, B. R. (1978) Neoplastic disease. Leukosis/sarcoma group. In: *Diseases of Poultry* pp. 418–468, M. S. HOFSTAD, *et al.* (Eds) Iowa State University Press, Ames, Iowa.

PURCHASE, H. G. and GILMOUR, D. G. (1975) Lymphoid leukosis in chickens chemically bursectomized and subsequently inoculated with bursal cells. *J. natn. Cancer Inst.* **55**: 851–855.

PURCHASE, H. G., LUDFORD, C., NAZERIAN, K. and COX, H. W. (1973) A new group of oncogenic viruses: reticuloendotheliosis, chick syncytial, duck infectious anemia, and spleen necrosis viruses. *J. natn. Cancer Inst.* **51**: 489–499.

PURCHASE, H. G., OKAZAKI, W., VOGT, P. K., HANAFUSA, H., BURMESTER, B. R. and CRITTENDEN, L. B. (1977) Oncogenicity of avian leukosis viruses of different subgroups and of mutants of sarcoma viruses. *Infect. Immun.* **15**: 423–428.

PURCHIO, A. F. (1982) Evidence that pp60^{src} the product of the Rous sarcoma virus src gene, undergoes autophosphorylation. *J. Virol.* **41**: 1–7.

PURCHIO, A. F., ERIKSON, E., BRUGGE, J. S. and ERIKSON, R. L. (1978) Identification of a polypeptide encoded by the avian sarcoma virus *src* gene. *Proc. natn. Acad. Sci. U.S.A.* **75**: 1567–1571.

PURCHIO, A. F., ERIKSON, E., COLLETT, M. S. and ERIKSON, R. L. (1981) Partial purification and characterization of pp60^{c-src} a normal cellular protein structurally and functionally related to the avian sarcoma virus *src* gene product. In: *Cold Spring Harbor Conferences on Cell Proliferation—Protein Phosphorylation*, pp. 1203–1215, ROSEN, O. M. and KREBS, E. G. (eds) Cold Spring Harbor.

PURCHIO, A. F., ERIKSON, E. and ERIKSON, R. L. (1977) Translation of 35S of and of subgenomic regions of avian sarcoma virus RNA. *Proc. natn. Acad. Sci. U.S.A.* **74**: 4661–4665.

PURCHIO, A. F., JOVANOVICH, S. and ERIKSON, R. E. (1980) Sites of synthesis of viral proteins in avian sarcoma virus-infected chicken cells. *J. Virol.* **35**: 629–636.

QUINTRELL, N., HUGHES, S. H., VARMUS, H. E. and BISHOP, J. M. (1980) Structure of viral DNA and RNA in mammalian cells infected with avian sarcoma virus. *J. molec. Biol.* **143**: 363–393.

RADKE, K., BEUG, H., KORNFIELD, S. and GRAF, T. (1982) Transformation of both erythroid and myeloid cells by E26, an avian leukemia virus that contains the *myb* gene. *Cell* **31**: 643–653.

RADKE, K., GILMORE, T. and MARTIN, G. S. (1980) Transformation by Rous sarcoma virus: a cellular substrate for transformation-specific protein phosphorylation contains phosphotyrosine. *Cell* **21**: 821–828.

RADKE, K. and MARTIN, G. S. (1979) Transformation by Rous sarcoma virus: effects of *src* gene expression on the synthesis and phosphorylation of cellular polypeptides. *Proc. natn. Acad. Sci. U.S.A.* **76**: 5212–5216.

RALSTON, R. and BISHOP, J. M. (1983) The protein products of the *myc* and *myb* oncogenes and adenovirus El$_a$ are structurally related. *Nature* **306**: 803–806.

RAMSAY, G. M., ENRIETTO, P. S., GRAF, T. and HAYMAN, M. J. (1982) Recovery of myc-specific sequences by a partially transformation defective mutant of avian myelocytomatosis virus, MC29, correlates with the restoration of transforming activity. *Proc. natn. Acad. Sci. U.S.A.* **79**: 6885–6889.

RAMSAY, G., GRAF, T. and HAYMAN, M. J. (1980) Mutants of avian myelocytomatosis virus with smaller *gag* gene-related proteins have an altered transforming ability. *Nature* **288**: 170–172.

RAMSAY, G. and HAYMAN, M. J. (1980) Analysis of cells transformed by defective leukemia virus OK10: production of noninfectious particles and synthesis of Pr76^{gag} and an additional 200,000-dalton protein. *Virology* **106**: 71–81.

RAMSAY, G. M. and HAYMAN, M. J. (1982) Isolation and biochemical characterization of partially transformation defective mutants of avian myelocytomatosis virus strain MC29: localization of the mutation to the *myc* domain of the 110,000-Dalton *gag-myc* polyprotein. *J. Virol.* **41**: 745–753.

REDDY, E. P., REYNOLDS, R. K., WATSON, D. K., SCHULTZ, R. A., LAUTENBERGER, J. and PAPAS, T. S. (1983) Nucleotide sequence analysis of the proviral genome of avian myelocytomatosis virus (MC29). *Proc. natn. Acad. Sci. U.S.A.* **80**: 2500–2504.

RICHERT, N. D., DAVIES, P. J. A., JAY, G. and PASTAN, I. H. (1979) Characterization of an immune complex kinase in immunoprecipitates of avian sarcoma virus-transformed fibroblasts. *J. Virol.* **31**: 695–706.

REITSMA, P. H., ROTHBERG, P. G., ASTRIN, S. M., TRIAL, J., BAR-SHAVIT, Z., HALL, A., TEITELBAUM, S. L. and KAHN, A. J. (1983) Regulation of *myc* gene expression in HL60 leukaemia cells by a vitamin D metabolite. *Nature* **306**: 492–494.

RICE, N. R., HIEBSCH, R. R., GONDA, M. A., BOSE, H. R., Jr. and GILDEN, R. V. (1982) Genome of reticuloendotheliosis virus: characterization by use of cloned proviral DNA. *J. Virol.* **42**: 237–252.

RINGOLD, G. M., SHANK, P. R., VARMUS, H. E., RING, J. and YAMAMOTO, K. R. (1979) Integration and transcription of mouse mammary tumor virus DNA in rat hepatoma cells. *Proc. natn. Acad. Sci. U.S.A.* **76**: 665–669.

ROBINSON, H. L., PEARSON, M. N., DESIMONE, D. W., TSICHLIS, P. N. and COFFIN, J. M. (1979) Subgroup-E avian-leukosis-virus-associated disease in chickens. *Cold Spring Harb. Symp. Q. Biol.* **44**: 1133–1142.

ROHRSCHNEIDER, L. R. (1980) Adhesion plaques of Rous sarcoma virus-transformed cells contain the *src* gene product. *Proc. natn. Acad. Sci. U.S.A.* **77**: 3514–3518.

ROHRSCHNEIDER, L. R., EISENMAN, R. N. and LEITCH, C. R. (1979) Identification of a Rous sarcoma virus transformation-related protein in normal avian and mammalian cells. *Proc. natn. Acad. Sci. U.S.A.* **76**: 4479–4483.

ROHRSCHNEIDER, L. R. and ROSOK, M. (1983) Transformation parameters and pp60^{src} localization in cells infected with partial transformation mutants of Rous sarcoma virus. *Molec. cell. Biol.* **3**: 731–746.

ROHRSCHNEIDER, L. R., ROSOK, M. and SHRIVER, K. (1981) Mechanism of transformation by Rous sarcoma virus: events within adhesion plaques. *Cold Spring Harb. Symp. Q. Biol.* **46**: 953–965.

ROTHBERG, P. G., ERISMAN, M. D., DIEHL R. E., ROVIGATTI, U. G. and ASTRIN, S. M. (1984) Structure and expression on the oncogene c-*myc* in fresh tumor material from patients with hematopoietic malignancies. *Molec. cell. Biol.* **4**: 1096–1103.

ROUS, P. (1911) A sarcoma of the fowl transmissible by an agent separable from the tumor cells. *J. exp. Med.* **13**: 397–411.

ROUSSEL, M., SAULE, S., LAGROU, C., ROMMENS, C., BEUG, H., GRAF, T. and STEHELIN, D. (1979) Three new types of viral oncogene of cellular origin specific for hematopoietic cell transformation. *Nature* **281**: 452–455.

RUBIN, C. S. and ROSEN, O. M. (1975) Protein phosphorylation. *A. Rev. Biochem.* **44**: 831–887.

RÜBSAMEN, H., FRIIS, R. R. and BAUER, H. (1979) *Src* gene product from different strains of avian sarcoma virus: kinetics and possible mechanism of heat inactivation of protein kinase activity from cells infected by transformation defective temperature-sensitive mutant and wild-type virus. *Proc. natn. Acad. Sci. U.S.A.* **76**: 967–971.

RÜBSAMEN, H., SALTENBERGER, K., FRIIS, R. R. and EIGENBRODT, E. (1982) Cytosolic malic dehydrogenase activity is associated with a putative substrate for the transforming gene product for Rous sarcoma virus. *Proc. natn. Acad. Sci. U.S.A.* **79**: 228–232.

RULEY, H. E. (1983) Adenovirus early region 1A enables viral and cellular transforming genes to transform primary cells in culture. *Nature* **304**: 602–606.

SAMARUT, J. and GAZZOLO, L. (1982) Target cells infected by avian erythroblastosis virus differentiate and become transformed. *Cell* **28**: 921–929.

SAULE, S., COLL, J., RIGHI, M., LAGROU, C., RAES, M. B. and STEHELIN, D. (1983) Two different types of transcription for the myelocytomatosis viruses MH2 and CM11. *EMBO J.* **2**: 805–809.

SAULE, S., ROUSSEL, M., LAGROU, C. and STEHELIN, D. (1981) Characterization of the oncogene (*erb*) of avian erythroblastosis virus and its cellular progenitor. *J. Virol.* **38**: 409–419.

SAULE, S., SERGEANT, A., TORPIER, G., RAES, M. B., PFEIFER, S. and STEHELIN, D. (1982) Subgenomic mRNA in OK10 defective leukemia virus-transformed cells. *J. Virol.* **42**: 71–82.

SCHARTL, M. and BARNEKOW, A. (1982) The expression in eukaryotes of a tyrosine kinase which is reactive with pp60^{v-src} antibodies. *Differentiation* **23**: 109–114.

SCHWARTZ, D. E., TIZARD, R. and GILBERT, W. (1983) Nucleotide sequence of Rous sarcoma virus. *Cell* **32**: 853–869.

SEALY, L., MOSCOVICI, G., MOSCOVICI, C. and BISHOP, J. M. (1983b) Site-specific mutagenesis of avian erythroblastosis virus: v-*erb*A is not required for transformation of fibroblasts. *Virology* **130**: 179–194.

SEALY, L., PRIVALSKY, M. L., MOSCOVICI, G., MOSCOVICI, C. and BISHOP, J. M. (1983a) Site-specific mutagenesis of avian erythroblastosis virus: *erb*B is required for oncogenicity. *Virology* **130**: 155–178.

SEEBURG, P. H., LEE, W. H., NUNN, M. F. and DUESBERG, P. H. (1984) The 5' ends of the transforming gene of Fujinami sarcoma virus and of the cellular proto-*fps* gene are not colinear. *Virology* **133**: 460–463.

SEFTON, B. M., BEEMON, K. and HUNTER, T. (1978) Comparison of the expression of the *src* gene of Rous sarcoma virus *in vitro* and *in vivo*. *J. Virol.* **28**: 957–971.

SEFTON, B. M., HUNTER, T., BALL, E. H. and SINGER, S. J. (1981) Vinculin: a cytoskeletal substrate of the transforming protein of Rous sarcoma virus. *Cell* **24**: 165–174.

SEFTON, B. M., HUNTER, T. and BEEMON, K. (1980a) Relationship of polypeptide products of the transforming gene of Rous sarcoma virus and the homologous gene of vertebrates. *Proc. natn. Acad. Sci. U.S.A.* **77**: 2059–2063.

SEFTON, B. M., HUNTER, T., BEEMON, K. and ECKHART, W. (1980b) Evidence that the phosphorylation of tyrosine is essential for cellular transformation by Rous sarcoma virus. *Cell* **20**: 807–816.

SEFTON, B. M., HUNTER, T. and COOPER, J. A. (1983) Some lymphoid cell lines transformed by Abelson murine leukemia virus lack a major 36,000-dalton tyrosine protein kinase substrate. *Molec. cell. Biol.* **3**: 56–63.

SEFTON, B. M., PATSCHINSKY, T., BERDOT, C., HUNTER, T. and ELLIOT, T. (1982a) Phosphorylation and metabolism of the transforming protein of Rous sarcoma virus. *J. Virol.* **41**: 813–820.

SEFTON, B. M., TROWBRIDGE, I. S., COOPER, J. A. and SCOLNICK, E. M. (1982b) The transforming proteins of Rous sarcoma virus, Harvey sarcoma virus and Abelson virus contain tightly bound lipid. *Cell* **31**: 465–474.

SERGEANT, A., SAULE, S., LEPRINCE, D., BEGUE, A., ROMMENS, C. and STEHELIN, D. (1982) Molecular cloning and characterization of the chicken DNA locus related to the oncogene *erb* B of avian erythroblastosis virus. *EMBO J.* **1**: 237–242.

SHEINESS, D., VENNSTROM, B. and BISHOP, J. M. (1981) Virus-specific RNAs in cells infected by avian myelocytomatosis virus and avian erythroblastosis virus: modes of oncogene expression. *Cell* **23**: 291–300.

SHIBUYA, M. and HANAFUSA, H. (1982) Nucleotide sequence of Fujinami sarcoma virus: evolutionary relationship of its transforming gene with transforming genes of other sarcoma viruses. *Cell* **30**: 787–795.

SHIBUYA, M., HANAFUSA, H. and BALDUZZI, P. C. (1982a) Cellular sequences related to three new *onc* genes of avian sarcoma virus (*fps*, *yes* and *ros*) and their expression in normal and transformed cells. *J. Virol.* **42**: 143–152.

Shibuya, M., Wang, L. H. and Hanafusa, H. (1982b) Molecular cloning of the Fujinami sarcoma virus genome and its comparison with sequences of other related transforming viruses. *J. Virol.* **42**: 1007–1016.

Shih, T. T. and Weeks, M. O. (1984) Oncogenes and cancer: the p21 *ras* genes. *Cancer Invest.* **2**: 109–123.

Shiu, R. P. C., Pouyssegur, J. and Pastan, I. (1977) Glucose depletion accounts for the induction of two transformation-sensitive membrane proteins in Rous sarcoma virus-transformed chick embryo fibroblasts. *Proc. natn. Acad. Sci. U.S.A.* **74**: 3840–3844.

Shriver, K. and Rohrschneider, L. R. (1981) Organization of pp60^{src} and selected cytoskeletal proteins within adhesion plaques and junctions of Rous sarcoma virus-transformed rat cells. *J. cell. Biol.* **89**: 525–535.

Sigel, P. and Pette, D. (1969) Intracellular localization of glycogenolytic and glycolytic enzymes in white and red rabbit skeletal muscle. *J. Histochem. Cytochem.* **17**: 225–237.

Singh, V. N., Singh, M., August, J. T. and Horecker, B. L. (1974) Alterations in glucose metabolism in chick-embryo cells transformed by Rous sarcoma virus: intracellular levels of glycolytic intermediates. *Proc. natn. Acad. Sci. U.S.A.* **71**: 4129–4132.

Sorge, L. K., Levy, B. T. and Maness, P. F. (1984) pp60^{c-src} is developmentally regulated in the neural retina. *Cell* **36**: 249–257.

Southern, E. M. (1975) Detection of specific sequences among DNA fragments separated by gel electrophoresis. *J. molec. Biol.* **98**: 503–517.

Souza, L. M., Strommer, J. N., Hillyard, R. L., Komaromy, M. C. and Baluda, M. A. (1980) Cellular sequences are present in the presumptive avian myeloblastosis virus genome. *Proc. natn. Acad. Sci. U.S.A.* **77**: 5177–5181.

Spector, D. H., Baker, B., Varmus, H. E. and Bishop, J. M. (1978b) Characteristics of cellular RNA related to the transforming gene of avian sarcoma viruses. *Cell* **13**: 381–386.

Spector, D. H., Smith, K., Padgett, T., McCombe, P., Roulland-Dussoix, D., Moscovici, C., Varmus, H. E. and Bishop, J. M. (1978a) Uninfected avian cells contain RNA related to the transforming gene of avian sarcoma virus. *Cell* **13**: 371–379.

Spurr, N. K., Solomon, E., Jansson, M., Sheer, D., Goodfellow, P. N., Bodmer, W. F. and Vennstrom, B. (1984) Chromosomal localisation of the human homologues to the oncogenes *erb* A and B. *EMBO J.* **3**: 159–163.

Srinivasan, A., Reddy, E. P., Dunn, C. Y. and Aaronson, S. A. (1984) Molecular dissection of transcriptional control elements within the long terminal repeat of the retrovirus. *Science* **223**: 286–189.

Stavnezer, E., Gerhard, D. S., Binari, R. C. and Balazs, I. (1981) Generation of transforming viruses in cultures of chicken fibroblasts infected with an avian leukosis virus. *J. Virol.* **39**: 920–934.

Steffen, D. (1984) Proviruses are adjacent to c-*myc* in some murine leukemia virus-induced lymphomas. *Proc. natn. Acad. Sci. U.S.A.* **81**: 2097–2101.

Steffen, D. and Weinberg, R. A. (1978) The integrated genome of murine leukemia virus. *Cell* **15**: 1003–1010.

Stehelin, D., Varmus, H. E., Bishop, J. M. and Vogt, P. K. (1976) DNA related to the transforming gene(s) of avian sarcoma viruses is present in normal avian DNA. *Nature* **260**: 170–173.

Stephens, R. M., Rice, N. R., Hiebsch, R. R., Bose, H. R. and Gilden, R. V. (1983) Nucleotide sequence of v-*rel*: the oncogene of reticuloendotheliosis virus. *Proc. natn. Acad. Sci. U.S.A.* **80**: 6229–6233.

Sugimoto, Y., Whitman, M., Cantley, L. C. and Erikson, R. L. (1984) Evidence that the Rous sarcoma virus transforming gene product phosphorylates phosphatidylinositol and diacylglycerol. *Proc. natn. Acad. Sci. U.S.A.* **81**: 2117–2121.

Swanstrom, R., Parker, R. C., Varmus, H. E. and Bishop, J. M. (1983) Transduction of a cellular oncogene: the genesis of Rous sarcoma virus. *Proc. natn. Acad. Sci. U.S.A.* **80**: 2519–2523.

Swanstrom, R., Varmus, H. E. and Bishop, J. M. (1982) Nucleotide sequence of 5' noncoding region and part of the gag gene of Rous sarcoma virus. *J. Virol.* **41**: 535–541.

Takeya, T., Feldman, R. A. and Hanafusa, H. (1982) DNA sequence of the viral and cellular *src* gene of chickens I. Complete nucleotide sequence of an EcoRI fragment of recovered avian sarcoma virus which codes for gp37 and pp60^{src}. *J. Virol.* **44**: 1–11.

Takeya, T. and Hanafusa, H. (1982) DNA sequence of the viral and cellular *src* of chickens II. Comparison of the *src* genes of two strains of avian sarcoma virus and of the cellular homolog. *J. Virol.* **44**: 12–18.

Takeya, T. and Hanafusa, H. (1983) Structure and sequence of the cellular gene homologous to the RSV *src* gene and the mechanism for generating the transforming virus. *Cell* **32**: 881–890.

Tato, F., Alema, S., Dlugosz, A., Boettiger, D., Holtzer, D., Cossu, G. and Pacifici, M. (1983) Development of "revertant" myotubes in cultures of Rous sarcoma virus-transformed avian myogenic cells. *Differentiation* **24**: 131–139.

Taub, R., Kirsch, I., Morton, C., Lenoir, G., Swan D., Tronick, S., Aaronson, S. and Leder, P. (1982) Translocation of the c-*myc* gene into the immunoglobulin heavy chain locus in human Burkitt lymphoma and murine plasmacytoma cells. *Proc. natn. Acad. Sci. U.S.A.* **79**: 7838–7841.

Theilen, G. H., Zeigel, R. F. and Twiehaus, M. J. (1966) Biological studies with RE virus (strain T) that induces reticuloendotheliosis in turkeys, chickens, and Japanese quail. *J. natn. Can. Inst.* **37**: 731–744.

Therwath, A., Mengod, G. and Sherrer, K. (1984) Altered globin gene transcription pattern and the presence of a 7–8 kb α globin gene transcript in avian erythroblastosis virus-transformed cells. *EMBO J.* **3**: 491–495.

Tsichlis, P. N. and Coffin, J. M. (1980) Recombinants between endogenous and exogenous avian tumor viruses: role of the C region and other portions of the genome in the control of replication and transformation. *J. Virol.* **33**: 238–249.

Ushiro, H. and Cohen, S. (1980) Identification of phosphotyrosine as a product of epidermal growth factor-activated protein kinase in A-431 cell membranes. *J. biol. Chem.* **255**: 8363–8365.

Vaheri, A. and Ruoslahti, E. (1975) Fibroblast surface antigen molecules and their loss from virus-transformed cells: a major alteration in cell surface. *Cold Spring Harb. Conf. Cell Proliferation* **2**: 967–975.

Varmus, H. E. (1983) *Oncogenes and Retroviruses: Evaluation of Basic Findings and Clinical Potential*, Alan R. Liss, Inc. New York, NY.

Varmus, H. E., Vogt, P. K. and Bishop, J. M. (1973) Integration of deoxyribonucleic acid specific for Rous

sarcoma virus after infection of permissive and non-permissive hosts. *Proc. natn. Acad. Sci. U.S.A.* **70**: 3067–3071.
VENNSTROM, B. and BISHOP, J. M. (1982) Isolation and characterization of chicken DNA homologous to the two putative oncogenes of avian erythroblastosis virus. *Cell* **28**: 135–143.
VIGNE, R., BREITMAN, M. L., MOSCOVICI, C. and VOGT, P. K. (1979) Restitution of fibroblast-transforming ability in *src* deletion mutants of avian sarcoma virus during animal passage. *Virology* **93**: 413–426.
VOGT, P. K. and ISHIZAKI, R. (1966) Criteria for the classification of avian tumor viruses in: *Viruses Inducing Cancer*, pp. 71–90, BURDETT, W. J. (ed.) Univ. Utah Press, Salt Lake City.
VON DER HELM, K. and DUESBERG, P. H. (1975) Translation of Rous sarcoma virus RNA in cell-free systems from ascites Krebs II cells. *Proc. natn. Acad. Sci. U.S.A.* **72**: 614–618.
WALLBANK, A. M., SPERLING, F. G., HUBBEN, K. and STUBBS, E. L. (1966) Isolation of a tumor virus from a chicken submitted to a Poultry Diagnostic Laboratory—Esh sarcoma virus. *Nature* **209**: 1265.
WANG, E. and GOLDBERG, A. R. (1976) Changes in microfilament organization and surface topography upon transformation of chick embryo fibroblasts with Rous sarcoma virus. *Proc. natn. Acad. Sci. U.S.A.* **73**: 4065–4069.
WANG, L. H., FELDMAN, R., SHIBUYA, M., HANAFUSA, H., NOTTER, M. F. P. and BALDUZZI, P. C. (1981) Genetic structure, transforming sequence and gene product of avian sarcoma virus, UR1. *J. Virol.* **40**: 258–267.
WANG, L. H., HALPERN, C. C., NADEL, M. and HANAFUSA, H. (1978) Recombination between viral and cellular sequences generates transforming sarcoma virus. *Proc. natn. Acad. Sci. U.S.A.* **75**: 5812–5816.
WANG, L., HANAFUSA, H., NOTTER, M. F. D. and BALDUZZI, P. C. (1982) Genetic structure and transforming sequence of avian sarcoma virus UR2. *J. Virol.* **41**: 833–841.
WANG, L. H., SNYDER, P., HANAFUSA, T., MOSCOVICI, C. and HANAFUSA, H. (1980) Comparative analysis of cellular and viral sequences related to sarcoma genic cell transformation. *Cold Spring Harb. Symp. Q. Biol.* **44**: 755–764.
WANG, S. Y., HAYWARD, W. S. and HANAFUSA, H. (1977) Genetic variation in the RNA transcripts of endogenous virus genes in uninfected chicken cells. *J. Virol.* **24**: 64–73.
WARBURG, O. (1930) *The Metabolism of Tumors*. DICKENS, E. (trans.) Constable, London.
WATSON, D., REDDY, E., DUESBERG, P. and PAPAS, T. (1983) Nucleotide sequence analysis of the chicken c-*myc* gene reveals homologous and unique coding regions by comparison with the transforming gene of avian myelocytomatosis virus MC29, Δ*gag-myc*. *Proc. natn. Acad. Sci. U.S.A.* **80**: 2146–2150.
WEBER, M. J. and FRIIS, R. R. (1979) Dissociation of transformation parameters using temperature-conditional mutants of Rous sarcoma virus. *Cell* **16**: 25–32.
WEISS, R. A. (1975) Genetic transmission of RNA tumor viruses. *Perspect. Virol.* **9**: 165–205.
WEISS, R., TEICH, N., VARMUS, H. and COFFIN, J. (1982) *RNA Tumor Viruses* (2nd edn.) Cold Spring Harbor Laboratory, Cold Spring Harbor, New York.
WEISS, S. R., HACKETT, P. B., OPPERMANN, H., ULLRICH, A., LEVINTOW, L. and BISHOP, J. M. (1978) Cell-free translation of avian sarcoma virus RNA: suppression of the *gag* termination codon does not augment synthesis of the joint *gag/pol* product. *Cell* **15**: 607–614.
WEISS, S. R., VARMUS, H. E. and BISHOP, J. M. (1977) The size and genetic composition of virus-specific RNAs in the cytoplasm of cells producing avian sarcoma-leukemia viruses. *Cell* **12**: 983–992.
WEISS, S. R., VARMUS, H. E. and BISHOP, J. M. (1981) Cell-free translation of purified avian sarcoma virus *src* mRNA. *Virology* **110**: 476–478.
WESTIN, E., WONG-STAAL, F., GELMANN, E. P., DALLA FAVERA, R., PAPAS, T. S., LAUTENBERGER, J. A., EVA, A., REDDY, E. P., TRONICK, S. R., AARONSON, S. A. and GALLO, R. C. (1982) Expression of cellular homologues of retroviral *onc* genes in human hematopoietic cells. *Proc. natn. Acad. Sci. U.S.A.* **79**: 2490–2494.
WILHELMSEN, K. C. and TEMIN, H. M. (1984) Structure and dimorphism of c-*rel* (turkey), the cellular homolog to the oncogene of reticuloendotheliosis virus strain T. *J. Virol.* **49**: 521–529.
WILLINGHAM, M. C., JAY, G. and PASTAN, I. (1979) Localization of the ASV *src* gene product to the plasma membrane of transformed cells by electron microscopic immunocytochemistry. *Cell* **18**: 125–134.
WONG, T. C., LAI, M. M. C., HU, S. S. F., HIRANO, A. and VOGT, P. K. (1982) Class II defective avian sarcoma viruses: comparative analysis of genome structure. *Virology* **120**: 453–464.
WONG, T. W. and GOLDBERG, A. R. (1983) *In vitro* phosphorylation of angiotension analogs by tyrosyl protein kinases. *J. biol. Chem.* **258**: 1022–1025.
WONG, T. C. and LAI, M. M. C. (1981) Avian reticuloendotheliosis virus contains a new class of oncogene of turkey origin. *Virology*, **111**: 289–293.
YAMAMOTO, T., DE CROMBRUGGHE, B. and PASTAN, I. (1980) Identification of a functional promoter in the long terminal repeat of Rous sarcoma virus. *Cell* **22**: 787–797.
YAMAMOTO, T., HIHARA, H., NISHIDA, T., KAWAI, S. and TOYOSHIMA, K. (1983a) A new avian erythroblastosis virus, AEV-H carries *erb*B gene responsible for the induction of both erythroblastosis and sarcomas. *Cell* **34**: 225–232.
YAMAMOTO, T., NISHIDA, T., MIYAJIMA, N., KAWAI, S., OOI, T. and TOYOSHIMA, K. (1983b) The *erb*B gene of avian erythroblastosis virus is a member of the *src* gene family. *Cell* **35**: 71–78.
YOSHIDA, M., KAWAI, S. and TOYOSHIMA, K. (1980) Uninfected avian cells contain structurally unrelated progenitors of viral sarcoma genes. *Nature* **287**: 653–654.
ZIEMIECKI, A. and FRIIS, R. R. (1980) Phosphorylation of pp60^{src} and the cycloheximide insensitive activation of pp60^{src} associated kinase activity of transformation-defective temperature-sensitive mutants of Rous sarcoma virus. *Virology* **106**: 391–394.

CHAPTER 13

CELLULAR TRANSFORMATION BY HUMAN T LYMPHOTROPIC RETROVIRUSES

Prem S. Sarin

Laboratory of Tumor Cell Biology, National Cancer Institute, Bethesda, Maryland 20892, U.S.A.

Abstract—Human T lymphotropic retroviruses (HTLV) have recently been isolated from patients with adult leukemias and immune deficiency syndrome. Four subgroups of HTLV family of retroviruses have so far been identified. HTLV-1 is the etiological agent of adult T cell leukemia (ATLL), a disease endemic in Japan, and HTLV-3 has been identified as the etiological agent of acquired immune deficiency syndrome (AIDS). HTLV-3 belongs to the family of human immune deficiency viruses called HIV. HTLV-2 has been isolated from a few patients with hairy cell leukemia, and HTLV-4 has been obtained from some healthy donors from Africa whereas HIV-2 has been obtained from West African AIDS patients. HTLV-2 is related to HTLV-1, whereas HTLV-4 and HIV-2 show relatedness to STLV-3. All the human retroviruses are T cell tropic and preferentially infect T-helper cells. HTLV-1 and HTLV-2 are transforming viruses whereas HTLV-3 and HIV-2 are cytopathic to the cells they infect.

1. INTRODUCTION

Retroviruses have been shown to be involved in the pathogenesis of leukemias and lymphomas in several animal species (Gallo, 1984; Sarin and Gallo, 1983). However, evidence for their role in human neoplasias has not been available until recently. The induction of leukemia by retroviruses in the feline and murine systems is usually associated with viremia and abundant virus production (Gross, 1983), whereas in human leukemia/lymphoma, a retrovirus could not be detected in fresh tumor cells. This lack of virus detection in human leukemia/lymphoma was considered as evidence against the involvement of retroviruses in the pathogenesis of this disease. Two examples from the animal systems provide interesting parallels for the humn disease. In the case of feline leukemia, the isolation of the virus (FeLV), epidemiological role of FeLV in cat leukemia, and the induction of leukemia in cats by inoculation of FeLV clearly established FeLV as the cause of the disease (Essex, 1975; Hardy *et al.*, 1976; Jarrett *et al.*, 1964). Although FeLV is not detectable in 30% of the cats with leukemia, it has been detected in the normal bone marrow cells (Rojko *et al.*, 1981). In the case of bovine leukosis, the involvement of bovine leukemia virus (BLV) was suspected to be the cause of the disease long before BLV was isolated. BLV is not detectable in primary leukemic cells, but it is expressed in cells after short-term culture. BLV has since been isolated and transmitted to permanently growing cell lines (Burny *et al.*, 1980). These examples of retrovirus isolation from the animal system have been very useful in the subsequent isolation of human T-cell tropic retroviruses from patients with adult T-cell malignancies and from patients with acquired immune deficiency syndrome (AIDS) and AIDS related complex (ARC) (Sarin and Gallo, 1984b, 1986; Wong-Staal and Gallo, 1985)

2. ISOLATION OF HTLV

The isolation of T-cell growth factor (TCGF, IL-2) from the conditioned media of lectin stimulated peripheral blood lymphocytes allowed, for the first time, the growth of T cells in culture (Morgan et al., 1976; Sarin and Gallo, 1984a). A number of human T-cell lines have been established from patients with these T-cell malignancies with the help of purified IL-2. These cultured T-cell lines release a type-C retrovirus which has been named human T-cell leukemia/lymphoma virus (HTLV-1) (Poiesz et al., 1980b; Popovic et al., 1983b; Sarin and Gallo, 1983, 1984b, Gallo, 1986).

The clinical features of patients with adult T-cell leukemia/lymphoma (ATLL), a disease endemic in Japan, include high leukocyte counts, skin lesions, hepatosplenomegaly, lymphadenopathy, hypercalcemia and an acute clinical course (Uchiyama et al., 1977). Hypercalcemia with lytic bone lesions and diffuse and generalized bone resorption is highly characteristic of this disease. Response to conventional therapy is generally poor. The ATLL cells circulating in peripheral blood are pleomorphic, of variable size, have lobulated nuclei, coarsely clumped nuclear chromatin, and have mature T-cell surface markers. The T cells belong to the helper cell phenotype as shown by the monoclonal antibody OKT4, although functional assays indicate that these cells act as suppressor cells (Waldmann et al., 1984; Yamada, 1983). The fresh primary cells from ATLL patients also possess IL-2 receptors (IL-2R) as detected with a monoclonal antibody termed anti-TAC (Waldmann et al., 1984).

HTLV-1, like other animal retroviruses, contains a reverse transcriptase (100,000 daltons), a high molecular weight RNA genome (9 Kb, 70S) and viral core *gag* proteins, composed of units similar to other mammalian retroviruses (Kalyanaraman et al., 1981b; Reitz et al., 1981; Rho et al., 1981). HTLV-1 is an exogenous human virus belonging to the group of chronic leukemia viruses and does not contain an *onc* gene (Reitz et al., 1981; Reitz et al., 1983). In addition, HTLV-1 sequences were not found in the normal Epstein-Barr virus-infected B cells from an HTLV-1-positive T-cell lymphoma patient. Only his neoplastic T cells contained HTLV-1-specific sequences (Gallo et al., 1982; Reitz et al., 1983). Since the first isolation of HTLV-1 (Poiesz et al., 1980a, 1981), approximately 200 HTLV-1 isolates have been obtained from cell lines established from patients with mature T-cell malignancies (Gallo et al., 1982, 1983a; Kalyanaraman et al., 1982b; Popovic et al., 1983b; Sarin and Gallo, 1983, 1984b; Sarin et al., 1983). Recently, HTLV-3 (HIV-1 and HIV-2 have also been isolated from patients with the acquired immune deficiency syndrome (AIDS) (Gallo, 1987, Gallo et al., 1983b, 1984; Popovic et al., 1984b; Sarin and Gallo, 1984b, 1986, Clavel et al., 1986). A comparison of the properties of HTLV isolates belonging to the three subgroups is given in Table 1. The cell lines derived from patients

TABLE 1. *Relatedness of HTLV-3 to HTLV-1, HTLV-2 and HTLV-4*

Property	HTLV-1	HTLV-2	HTLV-3	HTLV-4 (HIV-2)
1a. Cell specificity for infection	Lymphocytes	Lymphocytes	Lymphocytes	Lymphocytes
b. T cell (OKT4$^+$, T helper cell)	+	+	+	+
2. Presence of giant multi-nucleated cells	+	+	+	+
3. Major core protein	p24	p24	p24	p26
4. Common p24 epitope	+	+	+	−
5. Size of reverse transcriptase (RT)	~100K	~100K	~100K	ND*
6. RT divalent cation preference	Mg^{++}	Mg^{++}	Mg^{++}	ND
7. Nucleic acid homology to HTLV-1 (stringent conditions)	+++	±	−	−
8. Nucleic acid homology to HTLV-1 (moderate stringency)	+++	++	−	−
9. Homology to other retro-viruses except PTLV	−	−	−	−
10. Presence of pX region	+	+	+	+
11. Transacting transcriptional (TAT) activation of viral LTR	+	+	+	+
12. Mode of transmission. Sexual, blood, congenital	+	+	+	+
13. Presence of related viruses in old world monkeys	STLV-1		STLV-3	STLV-3
14. Syncytia formation	+	+	+	−HTLV-4 / +HIV-2

*ND = not determined.

TABLE 2. *HTLV-1 Antibodies in Sera of Patients with Leukemia/ Lymphoma from the Caribbean, Japan and the United States (Gallo et al., 1984; Robert-Guroff et al., 1984; Sarin and Gallo, 1983, 1984b, 1986)*

Disease Category*	# Positive/# Tested	% Positive
ATLL/T-LCL	55/62	89
T-ALL	0/18	0
T-CLL	0/40	0
CTCL	6/260	2
T-NHL	25/68	37
NHL (non-T)	0/54	0
ALL (non-T)	6/167	4
CLL (non-T)	4/28	1
Hodgkins Lymphoma	0/120	0

*ATLL/T-LCL, adult T cell leukemia, T lymphosarcoma cell leukemia; T-NHL, non-Hodgkins lymphoma; T-CLL, T chronic lymphocytic leukemia; T-ALL, adult lymphocytic leukemia T cell type; CTCL, cutaneous T cell lymphoma. The incidence of HTLV-1 antibodies in healthy individuals in the endemic regions of Japan (Kyushu and Shikoku) is 10–16%; in the Caribbean the incidence is 6–12%; in the United States, especially Georgia and Florida, the incidence is 1–2% whereas in West Germany, Netherlands and the United Kingdom no HTLV-1 infection was detectable in healthy individuals. HTLV-1 infection was also observed in Surinam immigrants to Holland (12%), different parts of Venezuela (14%), and Africa (2–8%).

with mature T-cell malignancies express HTLV belonging to subgroup 1 (HTLV-1). The virus isolated from a cell line established from a patient (MO) with a T-cell variant of hairy cell leukemia belongs to subgroup 2 (HTLV-2). The virus isolated from patients with ARC and AIDS belong to subgroup 3 (HTLV-3). Recently a retrovirus belonging to subgroup 4 has been isolated from healthy donors from West Africa (Kanki et al., 1986). This virus does not show cytopathic effect. A similar virus has been isolated from West African AIDS patients and is called HIV-2 (Clavel et al., 1986). In contrast to HTLV-4, HIV-2 shows cytopathic effect in culture. The presence of HTLV-1 was detected by competition radioimmunoprecipitation assay (RIPA) for the major core protein p24 (Kalyanaraman et al., 1981b), indirect immunofluorescence assay (IFA) using highly specific monoclonal antibody for p19 (Robert-Guroff et al., 1981) and reverse transcriptase activity in culture fluids. As shown in Table 1, HTLV-1 was fully expressed in all established T-cell lines.

HTLV-1 has also been isolated independently by a number of other laboratories (Haynes et al., 1983; Miyoshi et al., 1982a; Vyth-Dreese and DeVries, 1982; Yoshida et al., 1982). Virus isolates obtained by Japanese workers from adult T-cell leukemia/lymphoma (ATLL) have been named adult T-cell leukemia virus (ATLV), although biochemical and molecular biological studies show that the ATLV and HTLV-1 are closely related or identical (Popovic et al., 1982).

All the HTLV isolates belonging to subgroup 1 (HTLV-1) are closely related or identical in their serologic cross-reactions of the viral core proteins and homology of the viral RNA (Kalyanaraman et al., 1981b; Kalyanaraman et al., 1982b; Reitz et al., 1983; Sarin and Gallo, 1983, 1984b, 1986). Examination of the leukemic cells for the presence of HTLV-1 shows that all HTLV-1 isolates, except $HTLV_{MO}$ (HTLV-2), are closely related and have highly conserved genomes as determined by restriction enzyme mapping. $HTLV_{MO}$, a member of HTLV subgroup 2, competes poorly in the p24 assays (Kalyanaraman et al., 1982b), and nucleic acid sequence homology of this virus with HTLV-1 was detectable only under very nonstringent hybridization conditions (Gelmann et al., 1984). Using two different biological assays (syncytia induction and vesicular stomatitis virus pseudotypes), it was shown that the $HTLV-2_{MO}$ isolate has markedly different envelope antigens compared to other HTLV-1 isolates (Nagy et al., 1983, 1984). Recently, HTLV isolates belonging to subgroup 3 have been isolated from bone marrow, peripheral blood, plasma, semen, saliva and tears of patients with acquired immune deficiency syndrome (AIDS) and ARC (Gallo, 1987, Gallo et al., 1984b; Fujikawa et al., 1985 Groopman et al., 1984; Popovic et al., 1984b; Zagury et al., 1984, 1985a, b). Over 100 isolates of HTLV-3 have so far been obtained and analysed.

3. OTHER RETROVIRUSES

HTLV-1 does not show any immunological relatedness to other avian or mammalian retroviruses (Kalyanaraman et al., 1981a,b; Reitz et al., 1981; Sarin and Gallo, 1983, 1984b; 1986). However, HTLV-1 and BLV show a significant amino acid sequence homology in the gag proteins (Oroszlan et al., 1982). A comparison of the amino acid sequence of p24 of HTLV-1 and BLV show a 40% homology in the sequence from the amino terminal end and 60% homology in the sequence from the carboxyl terminal end of the two proteins (Oroszlan et al., 1984). HTLV-1 p15 and BLV p12 are basic linear polypeptides composed of 85 amino acids and 69 amino acids, respectively. Amino acid sequence analyses of the two proteins show a significant relatedness between the two, although HTLV-1 p15 has 16 more amino acid residues than BLV p12 (Oroszlan et al., 1984). The amino acid sequence of the HTLV-1 core proteins match the recently published nucleotide sequence (9032 bases) of the cloned HTLV-1 (ATLV) provirus (Seiki et al., 1983). Analysis of the nucleotide sequence suggests that the provirus DNA is composed of two long terminal repeats (LTRs), each consisting of 755 bases, which could be arranged into a unique secondary structure making possible the transcriptional termination within the 3' LTR but not in the 5' LTR. The nucleotide sequence of the provirus contains three open reading frames, which are capable of coding for proteins of 48,000, 99,000 and 54,000 daltons from the 5' end of the viral genome and presumably code for gag, pol and env genes, respectively. HTLV-3 has recently been cloned in our laboratory (Hahn et al., 1984; Shaw et al., 1984) and the complete nucleotide sequence determined (Ratner et al., 1985). The sequence shows the presence of four long open reading frames. The first open reading frame codes for the gag gene, the second for the pol gene, the third for the sor (short open reading frame) and the fourth for the env gene. Both restriction enzyme mapping and nucleotide sequence analysis show polymorphism in the different HTLV-3 isolates obtained, from different patients. HIV-2 has also been recently cloned and complete nucleotide sequence determined (Guyader et al, 1987, Clavel et al, 1986). Molecular cloning and sequencing of HTLV-4 and STLV-3 show them to be very closely related (Hirsch et al, 1986, Kornfeld et al., 1987).

The human T-lymphotropic retrovirus (HTLV-1, HTLV-2) and BLV differ from other leukemia and sarcoma viruses in that they do not utilize a conserved site of proviral integration for transformation suggesting the possible involvement of a transacting transcriptional regulation (TAT) element in mediation of their biological activity (Sodroski et al., 1984; Haseltine et al., 1985). The transfection of cells with the TAT gene has led to the direct demonstration of transcriptional activation of an LTR linked CAT gene. TAT gene activation has also been demonstrated by HTLV-3, lenti-viruses, BLV and simian T-lymphotropic retrovirus (STLV-1). The heterogeneity observed in the env region of HTLV-3 isolates (Wong-Staal et al., 1985) may present problems in the development of a vaccine to combat the disease. However, other approaches such as chemotherapy are currently being explored in several laboratories. Several chemotherapeutic agents that are being examined for possible treatment of AIDS include suramin (Mitsuya et al., 1984), foscarnet (Sarin et al., 1985), AL721 (Sarin et al., 1985), ribarvarin (McCormick et al., 1984), azidothymidine (Mitsuya et al., 1985), HPA-23 (Rosenbaum et al., 1985), amphotericias (Shafner et al, 1986), avarol and averone (Sarin et al, 1987) and dideoxy-cytidine. The development of drugs which can inhibit virus replication without showing undue toxicity will be extremely useful in controlling the disease in AIDS and ARC patients and could be potentially useful in ATLL patients infected with HTLV-1.

4. EPIDEMIOLOGY

Seroepidemiologic studies were carried out on sera of patients with T-cell malignancies, their relatives and normal healthy donors for the presence of HTLV-1 antibodies in order to determine HTLV infection in these populations. A large number of patients with a variety of T-cell malignancies were found to be positive for HTLV-1 (Essex et al., 1983; Gallo et al., 1984a; Reitz et al., 1983; Sarin and Gallo, 1983, 1984b), and the results are summarized in Table 2. As shown in this table, HTLV-1-positive T-cell malignancies occur sporadically in the U.S., but adult T-cell leukemia/lymphoma (ATLL) is more common in Japan and in the Caribbean. Clinical studies on Japanese ATLL indicate a geographic clustering of this disease in the southwestern islands of Kyushu and Shikoku (Sarin and Gallo, 1983; Takatsuki et al., 1979). Seroepidemiological studies indicate that nearly 90% of the sera of Japanese ATLL patients have antibodies to HTLV-1, thus indicating that HTLV-1 is endemic in the southwestern parts of Japan (Kalyanaraman et al., 1982a; Robert-Guroff et al., 1984). Another region endemic for HTLV-1 is the Caribbean (Blattner et al., 1982, 1985; Catovsky et al., 1982). All the West Indian patients with T-cell lymphosarcoma cell leukemia (T-LCL), a disease similar to Japanese ATLL with an aggressive course, were found to be positive for HTLV-1 antibodies. HTLV-1 associated T-cell malignancies have also been found in the southeastern U.S., Boston, Alaska, Central and South America, Africa and Israel (Gallo et al., 1984; Sarin and Gallo, 1983, 1984b).

A considerable variation in the prevalence of HTLV-1 antibodies in the normal Japanese population has been observed in different parts of Japan and correlates well with the incidence of ATLL in these regions (Hinuma et al., 1981; Kalyanaraman et al., 1982a; Robert-Guroff et al., 1982). The prevalence of HTLV-1 antibodies in the normal healthy donors from Nagasaki and Kagoshima areas is 16% whereas in the Uwajima area of Shikoku Island and the Honshu Island, the incidence of HTLV-1 antibodies is 9% and 2% respectively. It is even lower in the northern parts of Japan (Hokkaido Island) (Kalyanaraman et al., 1982a; Robert-Guroff et al., 1982; Sarin and Gallo, 1983, 1984b, 1986). The Caribbean region, which has also been identified as endemic for the HTLV-1 infection, shows a higher proportion of normal healthy donors with antibodies to HTLV-1 (Blattner et al., 1982, 1985).

Analysis of the sera of family members of ATLL patients indicates that a high percentage of family members contain HTLV-1 antibodies (Robert-Guroff et al., 1984). The highest incidence was found among relatives of Japanese ATLL patients (48%), whereas 17–19% of the relatives of U.S. and Caribbean patients with T-cell malignancies had HTLV-1 antibodies. The presence of HTLV-1 infection in the family members of patients with T-cell malignancies was confirmed by the expression of HTLV-1 in the cultured T cells of family members of a patient from Japan (Sarin et al., 1983). Recently, HTLV-1 antibodies have been detected in Jamaican and Colombian patients with tropical spastic paraperisis (TSP) (Johnson et al, 1985) and HTLV-1 has been isolated from the cerebrospinal fluid of one of the patients. This HTLV-1 isolate is being characterized to determine if this isolate is identical or different from use brown HTLV-1 isolates (Sarai et al, unpublished results).

Examination of sera from patients with AIDS, ARC and intravenous drug users show that 70–95% of these patients have antibodies to HTLV-3 (Goedert et al., 1984; Safai et al., 1984; Sarngadharan et al., 1984). Twenty-five to thirty percent of the healthy homosexuals have also been found to have HTLV-3 antibodies.

5. IN VITRO TRANSMISSION

The in vitro infectivity of HTLV-1 was first observed by Miyoshi et al. (1982a) when the virus was isolated from cord blood cells used as a feeder layer during attempts to establish a cell line (MT-2) from an ATLL patient in Japan. Virus isolated from MT-2 has also been transmitted into peripheral blood leukocyte (Yamamoto et al., 1982b). HTLV-1 isolated from patients with T cell malignancies has been transmitted into human T cells from cord blood, bone marrow, or peripheral blood by either co-cultivation of the HTLV-1-positive cells with recipient cells or by infection of the recipient cells with cell-free virus. In the co-cultivation experiments, the HTLV-1-positive donor cells are X-irradiated or treated with mitomycin-C and co-cultivated with HTLV-1-negative recipient T cells of the opposite sex (male × female and vice versa). After two to four weeks of co-cultivation, the cultures are analyzed for expression of HTLV antigens, extracellular virus, karyotype and HLA profiles.

To firmly establish the transmission of HTLV, two different approaches were used. In the first series of experiments, transmission of the virus from an HTLV-1 positive T-cell line (HTLV-1_{MJ}) into human umbilical cord blood T cells from four different newborns was carried out. On each successive transmission, the infected recipient cord blood T cells became the donor for infection of the new recipient cord blood T cells. Analysis of the cell cultures for HTLV-1 p19, p24, RT and virus particles by electron microscopy showed that HTLV-1 was fully expressed in all the infected cord blood T-cell lines (Table 3), and the cell lines exhibited distinct HLA profiles. When C3/MJ cord blood T cells with a female karyotype were co-cultured with C5 cord blood cells from a male, the recipient cells showed male karyotype, and HLA profiles matched the recipient cord blood cells (C5). Cord blood T cells co-cultured with HTLV-negative T cells from peripheral blood of a normal donor, or PHA-stimulated cell cultures of the cord blood recipient cells were consistently negative for HTLV-1 p24 and p19. HTLV-1 has also been transmitted into thymus T cells from a fetus with Klinefelter's syndrome, and a number of other HTLV-1 isolates (Table 3) and an HTLV-2 isolate from a patient with a hairy cell leukemia (HTLV-2_{MO}) have been transmitted into fresh human cord blood or bone marrow T cells resulting in productive infection (Popovic et al., 1983b, 1984a; Sarin and Gallo, 1983, 1984b). The results from transmission studies clearly demonstrate that HTLV isolates of both subgroups 1 and 2 can infect and replicate in human T cells. Transmission studies using only concentrated (cell-free) virus of HTLV-1 (Ruscetti et al., 1983; Markham et al., 1983) or HTLV-2 (Popovic et al., 1984a) for infection of T cells have also been successful. Cell-free virus transmission of HTLV-3 into T cells from peripheral blood, bone marrow or cord blood usually results in cell death, although a T cell line (H9) productively infected with HTLV-3 has recently been obtained (Popovic et al., 1984b) and this cell line has been the source of large quantities of the virus for use in the manufacture of test kits for the screening of blood supply by Elisa and Western blot techniques.

6. T-CELL TROPISM

Detection of HTLV-1 in neoplastic T cells and not in B cells of the same patient (CR) suggests that HTLV-1 is an exogenous virus and T-cell tropic (Gallo et al., 1982, 1984; Miyoshi et al., 1982a; Poiesz et al., 1980a; Popovic et al., 1983b; Sarin and Gallo, 1983, 1984b, 1986). This is supported by seroepidemiological and nucleic acid studies and by establishment of HTLV-1-positive T-cell lines from a number of patients with mature T-cell malignancies and their close relatives. The establishment of a B-cell line from an ATLL patient positive for both HTLV-1 and EBV has been reported (Yamamoto et al., 1982a). The results of our HTLV-1 transmission studies into cells derived from fetal thymus and spleen; cord blood of newborns; bone marrow, peripheral blood, liver and spleen of adults; and nasopharyngeal tonsils of an adolescent are summarized in Table 4. In all cases, the HTLV-1-positive T cells were initially partially dependent upon exogenous IL-2, and possessed mature T cell markers (OKT4^+ and Leu-3^+) (Mann et al., 1983a; Popovic et al., 1983b; Sarin and Gallo, 1983, 1984b). In co-infection experiments (HTLV + EBV) of cells derived from nasopharyngeal lymphoid tissues in the presence of IL-2, only HTLV-positive T cells with T-cell markers grew. In contrast, EBV-positive cells with B-cell markers grew in cell cultures simultaneously infected with HTLV-1 and EBV and cultured in the absence of IL-2. Thus, the HTLV-1 positivity of fetal thymus T cells (OKT6^+) suggests that immature T cells can also be infected with HTLV-1. T cells from subhuman primates and rodents have also been shown to be susceptible to infection by HTLV-1 (Miyoshi et al., 1982b, 1983; Popovic et al., 1984a). It is possible that a precursor cell of B and T cell lineage can be infected by HTLV-1, which after maturation becomes susceptible to EBV infection, and hence may explain the establishment of a cell line infected with both HTLV-1 and EBV as described by Yamamoto et al. (1982a).

TABLE 3. *Representative examples of Transmission of HTLV-1 and HTLV-2 into Human Cord Blood or Bone Marrow T Cells (Popovic et al., 1984a; Sarin and Gallo 1983, 1984b)*

Cocultured Cells* (Recipient/Donor)	HTLV-1 Expression			
	p19	p24	RT	EM
1. *HTLV-1*				
C5/MJ	++	++	++	+
C10/MJ	++	++	++	+
C4/UK	++	+++	++	+
C21/MI	++	++	++	+
C91/PL	++	++	++	+
C8/SK	++	+++	+	+
C7/TK	++	++	++	+
C90/HK	++	++	++	+
2. *HTLV-2*				
BM/MO	++	+	++	+
C344/MO	++	+	++	+
C218/MO	++	+	++	+
C346/MO	++	+	++	+
C446/MO	++	+	++	+

*HTLV-1 transmission into T cells from cord blood (C) or bone marrow (BM) was carried out by co-cultivation of T cells with X-irradiated (6,000–10,000 R) HTLV positive donor T cells.

TABLE 4. *T-Cell Tropism of HTLV-1 and HTLV-2 (Popovic et al., 1984a; Sarin and Gallo, 1983, 1984b, 1986)*

Source of Cells	HTLV-1*	HTLV-2†
Human T Cells:		
Fetal thymus	+	+
Fetal spleen	+	+
Newborn cord blood	+	+
Adolescent nasopharyngeal tonsils‡	+	+
Adult peripheral blood	+	+
Adult bone marrow	+	+
Adult spleen	+	ND
Adult liver	+	ND
Human B Cells:		
Newborn cord blood	−	−
Adolescent nasopharyngeal tonsils‡	−	−
Adult peripheral blood	−	−
Marmoset T Cells From:		
Peripheral blood	+	+

*HTLV-1$_{TK}$ and HTLV-1$_{MJ}$ isolates were used for infection.

†In the case of HTLV-2, MO and MO-F isolates from the same patient (JM) were used. +, positive for p19, p24 and RT; −, negative for p19 and RT; ND, not done.

‡In co-infection (HTLV-1 + EBV) experiments, HTLV-1 positive T cells were obtained only in the presence of IL-2 and EBV-positive B cells in the absence of IL-2.

7. CELL TRANSFORMATION

HTLV-1-infected T cells show several characteristic features which distinguish them from uninfected T cells and show similarities to T-cell lines established from patients with T-cell malignancies (Gallo *et al.*, 1983a; Miyoshi *et al.*, 1982a; Popovic *et al.*, 1983a; Sarin and Gallo, 1983, 1986). A comparison of the properties of HTLV-infected T cells with normal T cells and HTLV-1-positive neoplastic T cells derived from patients with mature T-cell malignancies (Table 5) clearly demonstrates that the *in vitro* infected T cells with

TABLE 5. *Characteristics of HTLV-1 Positive T Cells and Mitogen-Stimulated Human Cord Blood T Cells (Popovic et al., 1984a; Sarin and Gallo, 1983, 1984b, 1986)*

Characteristic	HTLV-1 Positive Neoplastic T Cell Lines	HTLV-1 Infected and Transformed T Cells		Lectin Stimulated Cord Blood T Cells
		Cord Blood	Bone Marrow	
1. Cell morphology:				
(a) Presence of multinucleated giant cells	++	++	±	−
(b) Presence of lobulated nuclei	++	++	±	−
2. Cell phenotype: (% positive cells)				
(a) Inducer/helper (OKT4)	50–95	70–95	0–90	65–95
(b) Suppressor/cytotoxic (OKT8)	0–30	0–20	0–40	0–40
3. E-Rosette	+++	+++	+++	+++
4. S-IgG*, EBNA†, TdT‡	−	−	−	−
5. HLA 'Modification'				
(a) Expression of additional HLA antigens	+	+	+	−
(b) Expression of HLA-DR§	0–5%	0–5%	ND	10–12%
6. Requirement for exogenous IL-2 (v/v)	Indefinite	Indefinite	Indefinite	Limited
7. *In vitro* growth	+++	+++	+++	+
8. IL-2 receptor (IL-2R)	+	+	+	−
9. Lymphokine production	+	+	+	−
10. HTLV-1 expression as detected by p19, p24, RT and type C virus particles (EM)				

*S-IgG, cell surface immunoglobulins; †EBNA, Epstein–Barr virus nuclear antigen; ‡TdT, terminal deoxynucleotidyl transferase; §Determined by cell sorter using monoclonal antibodies.

HTLV-1 are transformed. Like primary neoplastic T cells, the HTLV-1-transformed T cells can grow indefinitely, show a decreased or complete independence of requirement for IL-2, have helper/inducer (OKT4$^+$/Leu-3$^+$) phenotype, become constitutive producers of lymphokines, and show morphological and cell surface alterations (Gallo et al., 1983a; Popovic et al., 1983b; Sarin and Gallo, 1983, 1984b, 1986; Salahuddin et al., 1984).

Lectin-stimulated T cells from cord blood grow as single cell suspensions with small clumps, whereas the HTLV-1-infected T cells from cord blood grow predominantly as large clumps. The growth in large clumps is a characteristic of HTLV-1-infected T cells which become less dependent or independent of exogenous IL-2. The HTLV-1-infected T cells grow as mutinucleated giant cells. The growth pattern and morphology of the HTLV-1-infected cord blood T cells are very similar to the neoplastic T-cell lines derived from patients with mature T-cell malignancies (Gallo et al., 1983a; Poiesz et al., 1980b; Popovic et al., 1983a; Sarin and Gallo, 1983, 1984b, 1986). Electron microscopic examination of HTLV-1-infected and mitogen-stimulated cord blood T cells shows the presence of lobulated nuclei, a feature observed in the fresh neoplastic cells of the patients with ATLL.

HTLV-1-infected T cells can grow indefinitely in tissue culture (Gallo et al., 1983a; Miyoshi et al., 1982a; Popovic et al., 1983a; Sarin and Gallo, 1983, 1984b), whereas the lectin-stimulated T cells from peripheral blood, bone marrow, or cord blood reach a growth crisis around 45 days (Gallo et al., 1983a; Popovic et al., 1983a; Yamamoto et al., 1982b). In the absence of HTLV-1 infection, it is extremely difficult to obtain a IL-2-dependent or IL-2-independent T-cell line from human cord blood, peripheral blood, or bone marrow, whereas HTLV-1-infected T-cell lines can grow indefinitely in the presence or absence of IL-2.

Cells from patients with HTLV-1-positive T-cell leukemia/lymphoma contain IL-2 receptors and, therefore, they do not require lectin activation to respond to IL-2 (Gallo et al., 1983a; Poiesz et al., 1980b; Popovic et al., 1983b), whereas the T cells from normal healthy donors acquire IL-2 receptors only after mitogen stimulation. A comparison of the IL-2 requirement of HTLV-1-infected cord blood T cells and HTLV-1-positive neoplastic T cells and normal uninfected T cells shows that the uninfected T cells require two- to tenfold more IL-2 than the HTLV-1-positive T cells. Recent studies show that HTLV-1 can also induce direct activation of resting lymphocytes (Gazzdo and Dodon, 1987). HTLV-3 infection of T cells from peripheral blood, bone marrow or cord blood results in cell death due to the cytopathic effect of the virus.

8. PHENOTYPE OF HTLV-1-TRANSFORMED T CELLS

The presence of receptors for IL-2 on HTLV-1-positive T cells has been demonstrated by a monoclonal antibody (anti-TAC) (Waldmann et al., 1984), and 70% of the T cells of HTLV-1-positive cell lines react with this antibody (Popovic et al., 1983b). Another characteristic feature of the cultured T cells is the expression of HLA-Dr (Metzgar et al., 1979), a determinant not expressed by mitogen-stimulated normal cord blood T cells (Yodoi et al., 1982). It was of interest, therefore, to examine whether HTLV-1-infected cord blood T cells express IL-2 receptors and HLA-Dr determinants. Examination of HTLV-1-transformed cord blood T cells and mitogen-stimulated T cells from the same donor by fluorescence-activated cell sorter (FACS) analysis using the anti-TAC monoclonal antibody and a monomorphic anti-HLA-Dr antibody (3.1) showed that HTLV-1-infected T cells were highly positive for the expression of both IL-2 receptors and HLA-Dr. The expression of IL-2 receptors and HLA-Dr on HTLV-1-infected T cells was similar to those observed on the neoplastic T cells from the patients with T-cell malignancies and was approximately 50-fold greater than observed in mitogen-stimulated normal human cord blood T cells. The expression of IL-2 receptors and HLA-Dr determinants on HTLV-1-infected T cells may be important in the control of T cell proliferation. The density of transferrin receptors on the HTLV-1-transformed T cells is also relatively high.

Examination of the peripheral blood T cells and HTLV-1-positive cultured T cells from patients with T-cell malignancies for HLA determinants suggested the expression of additional HLA-A and -B locus antigens on the HTLV-1-positive cultured T cells. These HLA determinants were not expressed on the EBV-transformed B cells or on the fresh peripheral blood lymphocytes from the same patient. Cultured T cells positive for HTLV-1 derived from patients or transformed *in vitro* showed parallel expression of both HTLV-1 and altered HLA alloantigen (Gallo *et al.*, 1982; Mann *et al.*, 1983b).

9. RELEASE OF LYMPHOKINES

The role of T cells in humoral and cell-mediated immunity is considered to be both as an effector and as modulator cells. These T-cell functions, along with cell proliferation, are mediated by the release of biologically active factors termed lymphokines. A number of active factors released by T cells have been reported (Salahuddin *et al.*, 1984). Examination of the conditioned media from a number of HTLV-1-positive T-cell cultures established from patients with T-cell malignancies or obtained after HTLV-1 infection of T cells from peripheral blood, cord blood or bone marrow showed that all HLTV-1-transformed T-cell lines became constitutive producers of several lymphokines (Table 6). These include macrophage migration-inhibitory factor (MIF), leukocyte-inhibitory factor (LIF), migration-enhancement factor (MEF), macrophage-activating factor (MAF), differentiation-inducing factor (DIF), colony-stimulating factor (CSF), eosinophil growth-maturation activity (eos.GMA), interleukin 3 (IL-3), fibroblast-activating factor (FAF), B-cell growth factor(s) (BCGF), and gamma interferon (γ-interferon). Low levels of IL-2 were detectable

TABLE 6. *Lymphokine Production by HTLV-1 Transformed Human T Cells* (Salahuddin et al., *1984*)

Source of HTLV-Positive T Cells*	Biological Activities						
	MIF	LIF/MEF	MAF	DIF	CSF	eos.GMA	FAF
A. *Leukemic*							
CR	+	+	−	+	+ +	+ + +	+
MJ		+	−	+	+	+ + +	+
UK		+	−		+	+ + +	−
B. *Cord Blood*							
C10/MJ-2	+	−	+	+ + +	+		+
C5/MJ	+	−	+	+	+		−
C91/PL	+	+	−	+	+		+
C43/UK	+	+	+	+ +	+	+ + +	−
C63/CR	+	+	+	+ + +	+	+ + +	+
C. *Bone Marrow*							
B1/MJ	+	−	−	+ + +	+	+	+
B2a/MJ	+	+	+	+	+	+	+
B2/UK			+	+	+	+	+
B2/CR			−	+ + +	+	+ + +	+
B9a/C10UK	+	+	−	+ +	+	−	−
B9/C10MJ			−	+	+ +		
B9b/C10UK†		+			+	+	−
B10/C10UK†	+	+	−	+	+ +	+	+

*Cultured leukemic T cells initially established in suspension culture from peripheral blood of an adult T-cell leukemia patient in the presence of added IL-2 were lethally irradiated and used as a source of virus to infect umbilical cord blood T cells and adult bone marrow T cells by co-cultivation procedures. Virus donor cells are represented by the initials of the donor leukemia patient. Other cell lines used labelled C for cord blood or B for bone marrow followed by sample number in the numerator and the cell line used as a sounce of HTLV-1 in the denominator. All cell lines shown except UK grew independently of added IL-2. Cell lines C43/UK, C63/CR, B2/CR and B1/MJ were nonproductively infected by HTLV-1. Biological activities are: macrophage migration-inhibiting factor (MIF), leukocyte migration-inhibitory (LIF) or migration-enhancing (MEF) factor, macrophage-activating factor (MAF), differentiation-inducing factor (DIF), colony-stimulating factor (CSF), eosinophil growth-maturation activities (eos.GMA), and fibroblast-activity factor (FAF).

†Cell lines B9b/C10UK and B10/C10UK also produce high levels of γ-interferon.

in only a few of the HTLV-1-transformed T-cell lines. Thus, in the case of HTLV-1-transformed T cells, a decreased requirement for IL-2, cell surface alterations, and constitutive production of various lymphokines may influence their growth characteristics. It will be of interest to examine the lymphokines produced by HTLV-3, HTLV-4 and HIV-2 infected T cells.

10. MECHANISM OF TRANSFORMATION

HTLV-1-infected T cells offer an excellent model system for studying the mechanism of transformation by retroviruses. HTLV-1-infected T cells express multiple species of viral mRNA coding for *gag, pol* region (9 Kb mRNA); *env* (4Kb); and TAT, LTR sequences (2 Kb), suggesting the possible importance of the TAT region and the expression of other viral proteins in the cell transformation by HTLV-1. Analyses of cloned complete HTLV-1 genomes indicate that HTLV-1 does not contain a cell-derived *onc* gene (Seiki *et al.*, 1983), suggesting that HTLV-1 is a chronic leukemia virus despite its *in vitro* transforming activity. Several chronic leukemia viruses are known to induce leukemia by activating cellular *onc* genes (e.g. *myc* in B-cell lymphomas) (Hayward *et al.*, 1981) by integration in the proximity of these genes. Examination of the HTLV-1-infected neoplastic T cells and cord blood T cells for possible *onc* gene activation with cloned *onc* gene probes showed no evidence for expression of any *onc* gene, with the possible exception of low-level *sis* expression in some HTLV-1-infected cells. The c-*sis* gene has been shown to code for the gene of the platelet-derived growth factor (PDGF), which normally acts on fibroblasts, smooth muscle, and glial cells (Doolittle *et al.*, 1983; Waterfield *et al.*, 1983). Activation of growth factor genes and c-*onc* genes as a possible mechanism of leukemogenesis by HTLV-1 was examined in the HTLV-1-infected cells. All HTLV-1-infected cells express IL-2 receptors. Examination of the HTLV-1-infected T cells for the expression of IL-2 mRNA with cloned IL-2 gene probes (Clark *et al.*, 1984) showed the expression of very low levels of TCGF mRNA expression in a few cases. Although high levels of IL-2 receptors are expressed in HTLV-1-infected T cells, a simple IL-2 receptor autostimulation mechanism is probably not operative in the initiation or maintenance of the leukemic state. In the case of HTLV-3, the cytopathic effect of HTLV-3 on T cells may be due to the alteration of cellular transcriptional control by an HTLV-3 TAT gene product (Ratner *et al.*, 1985).

REFERENCES

BLATTNER, W. A., KALYANARAMAN, V. S., ROBERT-GUROFF, M., LISTER, T. A., GALTON, D. A. G., SARIN, P. S., CRAWFORD, M. H., CATOVSKY, D., GREAVES, M. and GALLO, R. C. (1982) The human type-C retrovirus, HTLV, in Blacks from the Caribbean region and relationship to adult T-cell leukemia/lymphoma. *Int. J. Cancer* **39**: 257–264.

BLATTNER, W. A., BIGGAR, R. J., WEISS, S.H., CLARK, J. W. and GOEDERT, J. J. (1985) Epidemiology of human lymphotropic retroviruses: an overview. *Cancer Res. (Suppl.)* **45**: 4558–4601.

BURNY, A., BRUCK, C., CHANTRENNE, H., CLEUTER, Y., DEKEZEL, D., GHYSDAEL, J., KETTMANN, R., LECLERCQ, M., LEUNEN, J., MAMMERICKX, M. and PORTELLE, D. (1980) Bovine leukemia virus: molecular biology and epidemiology. In: *Viral Oncology,* pp. 231–280, KLEIN, G. (ed.). Raven Press, New York.

CATOVSKY, D., GREAVES, M. F., ROSE, M., GALTON, D. A. G., GOOLDEN, A. W. G., MCCLUSKEY, D. R., WHITE, J. M., LAMPERT, I., BOURIKAS, G., IRELAND, R., BROWNELL, A. I., BRIDGES, J. M., BLATTNER, W. A. and GALLO, R. C. (1982) Adult T-cell lymphoma-leukaemia in Blacks from the West Indies. *Lancet* i (#8372): 639–642.

CLARK, S. C., ARYA, S. K., WONG-STAAL, F., MATSUMOTO-KOBAYASHI, M., KAY, R. M., KAUFMAN, R. J., BROWN E. L., SHOEMAKER, C., COPELAND, T., OROSZLAN, S., SMITH, K., SARNGADHARAN, M. G., LINDNER, S. G. and GALLO, R. C. (1984) Human T cell growth factor: partial amino acid sequence, CDNA cloning and organization and expression in normal and leukemic cells. *Proc. natn. Acad. Sci. U.S.A.* **81**: 2543–2547.

CLAVEL, F., GUYADER, M., GUETARD, D., SALLE, M., MONTAGNIER, L. and ALIZON, M. (1986) Molecular Cloning and Polymorphism of Human Immunedeficiency Virus Type 2. *Nature* **324**: 691–695.

DOOLITTLE, R. F., HUNKAPILLAR, M. W., HOOD, L. E., DEVARE, S. G., ROBBINS, K. C., AARONSON, S. A. and ANTOINIADES, H. N. (1983) Simian sarcoma virus *onc* gene, v-*sis*, is derived from the gene (or genes) encoding a platelet derived growth factor. *Science* **221**: 275–277.

ESSEX, M. (1975) Horizontally and vertically transmitted oncornaviruses of cats. *Adv. Cancer Res.* **21**: 175–264.

Essex, M., McLane, M. F., Lee, T. H., Falk, L., Howe, C. W. S., Mullins, J. I., Cabradilla, C. and Francis, D. P. (1983) Antibodies to cell membrane antigens associated with human T-cell leukemia virus in patients with AIDS. *Science* **220**: 859–862.

Fujikawa, L. S., Salahuddin, S. Z., Palestine, A. G., Masur, H., Nussenblatt, R. B. and Gallo, R. C. (1985) Isolation of human T lymphotropic retrovirus type III from the tears of a patient with the acquired immune deficiency syndrome. *Lancet*, **2**: 529–530.

Gallo, R. C (1984) Human T cell leukemia-lymphoma virus and T cell malignancies in adults. In: *Cancer Surveys*, Vol. 3, pp. 113–159, Franks, L. M., Wyke, J. and Weiss, R. A. (eds). Oxford University Press, Oxford.

Gallo, R. C (1986) The first human retrovirus. *Sci. Amer.* **254**: 88–98.

Gallo, R. C. (1987) The AIDS Virus. *Sci. Amer.* **256**: 46–56.

Gallo, R. C., Mann, D., Broder, S., Ruscetti, F. W., Maeda, M., Kalyanaraman, V. S., Robert-Guroff, M. and Reitz, M. S. (1982) Human T-cell leukemia-lymphoma virus (HTLV) is in T- but not B-lymphocytes from a patient with cutaneous T-cell lymphoma. *Proc. natn. Acad. Sci. U.S.A.* **79**: 4680–4683.

Gallo, R. C., Popovic, M., Lange-Wantzin, G., Wong-Staal, F. and Sarin, P.S. (1983a) Stem cells, leukemia viruses, and leukemia of man. In: *Haemopoietic Stem Cells*, pp. 155–170, Killmann, Sv.-Aa., Cronkite, E. P. and Muller-Berat, C. H. (eds). Munksgaard, Copenhagen.

Gallo, R. C., Salahuddin, S. Z., Popovic, M., Shearer, G. M., Kaplan, M., Haynes, B. F., Palker, T. J., Redfield, R., Oleske, J., Safai, B., White, G., Foster, P. and Markham, P. D. (1984) Frequent detection and isolation of cytopathic retroviruses (HTLV-III) from patients with AIDS and at risk for AIDS. *Science* **224**: 500–503.

Gallo, R. C., Sarin, P. S., Gelmann, E. P., Robert-Guroff, M., Richardson, E., Kalyanaraman, V. S., Mann, D., Sidhu, G. D., Stahl, R. E., Zolla-Pazner, S., Leibowitch, J. and Popovic, M. (1983b) Isolation of human T-cell leukemia virus in acquired immune deficiency syndrome (AIDS). *Science* **220**: 865–867.

Gazzolo, L. and Dodon, M. D. (1987) Direct activation of resting T lymphocytes by human T-lymphotropic virus type I. *Nature* **326**: 714–717.

Gelmann, E. P., Franchini, G., Manzari, V., Wong-Staal, F., and Gallo, R. C. (1984) Molecular cloning of a new unique human T cell leukemia virus (HTLV-2_{MO}). In: *Human T Cell Leukemia/Lymphoma Virus*, pp. 184–195, Gallo, R. C., Essex, M. E. and Gross, L. (eds). Cold Spring Harbor Laboratory, Cold Spring Harbor, New York.

Goedert, J., Sarngadharan, M. G., Biggar, R., Weiss, S., Winn, D., Grossman, R., Greene, M., Bodner, A., Mann, D., Strong, D., Gallo, R. and Blattner, W. (1984) Determinants of retrovirus (HTLV-III) antibody and immunodeficiency conditions in homosexual men. *Lancet* **ii**: 711–716.

Groopman, J. E., Salahuddin, S. Z., Sarngadharan, M. G., Markham, P. D., Gonda, M., Sliski, A. and Gallo, R. C. (1984) HTLV-III in saliva of people with AIDS related complex and healthy homosexual men at risk of AIDS. *Science* **226**: 447–449.

Gross, L. (1983) Onogenic viruses, Vol. I and II 3rd edition. Pergamon Press, New York.

Guyaderm, M., Emerman, M., Sonigo, P., Clavel, F., Montagnier, L. and Alizon, M. (1987) Genome organization and transaction of the human immune deficiency virus type 2. *Nature* **326**: 662–669.

Hahn, B. H., Shaw, G. M., Arya, S. K., Popovic, M., Gallo, R. C. and Wong-Stall, F. (1984) Molecular cloning and characterization of the HTLV-III virus associated with AIDS. *Nature* **312**: 166–169.

Hardy, W. D., Jr., Hess, P. W., MacEwen, E. G., McClelland, A. J., Zuckerman, E. E., Essex, M. and Cotter, S. M. (1976) The biology of feline leukemia virus in the natural environment. *Cancer Res.* **36**: 582–588.

Haseltine, W. A., Sodroski, J. and Rosen, C. (1985) The lor gene and the pathogenesis of HTLV-I, II and III. *Cancer Res. (Suppl).* **45**: 4545–4549.

Haynes, B. F., Miller, S. E., Moore, T. O., Dunn, P. H., Bolognesi, D. P. and Metzgar, R. S. (1983) Identification of human T cell leukemia virus in a Japanese patient with adult T cell leukemia and cutaneous lymphomatous vasculitis. *Proc. natn. Acad. Sci. U.S.A.* **80**: 2054–2058.

Hayward, W. S., Neel, B. G. and Astrin, S. M. (1981) Activation of a cellular *onc* gene by promoter insertion in ALV-induced lymphoid leukosis. *Nature* **209**: 475–480.

Hinuma, Y., Nagata, K., Misoka, M., Nakai, M., Matsumoto, T., Kinoshita, K. I., Shirakawa, S. and Miyoshi, I. (1981) Adult T-cell leukemia: antigen in an ATL cell line and detection of antibodies to the antigen in human sera. *Proc. natn. Acad. Sci. U.S.A.* **78**: 6476–6480.

Hirsch, V., Riedel, N., Kornfield, H., Kanki, P., Essex, M. and Mullins, J. L. (1986) Cross reactivity to human T lymphotropic virus type III from African green monkeys. *Proc. Natl. Acad. Sci. U.S.A.* **83**: 9754–9758.

Jarrett, W. F. H., Crawford, E. M., Martin, W. B. and Davie, F. (1964) A virus-like particle associated with leukemia (lymphosarcoma). *Nature* **202**: 567–570.

Johnson, P. R., Gajdusek, D. C., Morgan, D. C., Zaninovic, V., Sarin, P. S. and Graham, D. S. (1985) HTLV-I and HTLV-III antibodies and tropical spastic paraperesis. *Lancet* **2**: 1247–1248.

Kalyanaraman, V. S., Sarngadharan, M. G., Bunn, P. A., Minna, J. D. and Gallo, R. C. (1981a) Antibodies in human sera reactive against an internal structural protein of human T-cell lymphoma virus. *Nature* **294**: 271–273.

Kalyanaraman, V. S., Sarngadharan, M. G., Nakao, Y., Ito, Y., Aoki, T. and Gallo, R. C. (1982a) Natural antibodies to the structural core protein (p24) of the human T-cell leukemia (lymphoma) retrovirus (HTLV) found in sera of leukemic patients in Japan. *Proc. natn. Acad. Sci. U.S.A.* **79**: 1653–1657.

Kalyanaraman, V. S., Sarngadharan, M. G., Poiesz, B. J., Ruscetti, F. W. and Gallo, R. C. (1981b) Immunological properties of a type C retrovirus isolated from cultured human T-lymphoma cells and comparison to other mammalian retroviruses. *J. Virol.* **38**: 906–913.

Kalyanaraman, V. S., Sarngadharan, M. G., Robert-Guroff, M., Blayney, D., Golde, D. and Gallo, R. C. (1982b) A new subtype of human T-cell leukemia virus (HTLV-II) associated with a T-cell variant of Hairy cell leukemia. *Science* **218**: 571–573.

Kanki, P. J., Barin, F., M'boup, S., Allan, J. S., Rometlemone, J. L., Marlink, R., McClane, M. F., Lee,

T. H., ARBEILLE, B., DENIS, F. and ESSEX, M. (1986) New human T-lymphotropic retrovirus related to simian T-lymphotropic virus type III (STLV-III AGM). *Science* **232**: 238–243.

MCCORMICK, J. B., MITCHELL, S. W., GETCHELL, J. P. and HICKS, D. R. (1984) Ribavarin suppresses relication of lymphadenopathy associated virus in cultures of human adult T-lymphocytes. *Lancet,* **2**: 1367–1369.

MANN, D. L., POPOVIC, M., MURRAY, C., NEULAND, C., STRONG, D. M., SARIN, P. S., GALLO, R. C. and BLATTNER, W. A. (1983a) Cell surface antigen expression in newborn cord blood lymphocytes infected with HTLV. *J. Immunol.* **131**: 2621–2624.

MANN, D. L., POPOVIC, M., SARIN, P., MURRAY, C., NEWLAND, C., STRONG, D. M., HAYNES, B. F., GALLO, R. C. and BLATTNER, W. A. (1983b) Cell lines producing human T-cell lymphoma virus (HTLV) have altered HLA expression. *Nature* **305**: 58–60.

MARKHAM, P., SALAHUDDIN, Z., KALYANARAMAN, V. S., POPOVIC, M., SARIN, P. and GALLO, R. C. (1983) Infection and transformation of fresh human umbilical cord blood cells by multiple sources of human T-cell leukemia/lymphoma virus (HTLV) *Int. J. Cancer* **31**: 413–420.

METZGAR, R. S., BERTOGLIO, I., ANDERSON, J. K., BONNARD, G. B. and RUSCETTI, F. W. (1979) Detection of HLA-DRw (Ia-like) antigens on human T lymphocytes grown in tissue culture. *J. Immunol.* **122**: 949–953.

MITSUYA, H., POPOVIC, M., YARCHOAN, R., MATSUSHITA, S., GALLO, R. C. and BRODER, S. (1984) Suramin protection of T cells *in vitro* against infectivity and cytopathic effect of HTLV-III. *Science,* **226**: 172–174.

MITSUYA, H., WEINHOLD, K. J., FURMAN, P. A., ST. CLAIR, M. H., LEHRMAN, S. N., GALLO, R. C., BOLOGNESI, D., BARRY, D. W. and BRODER, S. (1985) 3′-Azido-3′-deoxythymidine (BWA 509U): an antiviral agent that inhibits the infectivity and cytopathic effect of HTLV-III/LAV *in vitro. Proc. natn. Acad. Sci. U.S.A.* **82**: 7096–7100.

MIYOSHI, I., KUBONISHI, I., YOSHIMOTO, S., AKAGI, T., OHTSUKI, Y., SHIRAISHI, Y., NAGATA, K. and HINUMA, Y. (1982a) Type C virus particles in a cord T-cell line derived by co-cultivating normal human cord leukocytes and human leukemic T-cells. *Nature* **294**: 770–771.

MIYOSHI, I., TAGUCHI, H., FUJISHITA, M., YOSHIMOTO, S., KUBONISHI, I., OHTSUKI, Y., SHIRAISHI, Y. and AKAGI, T. (1982b) Transformation of monkey lymphocytes with adult T-cell leukemia virus. *Lancet* **i**: 1016.

MIYOSHI, I., YOSHIMOTO, S., TAGUCHI, H., KOBONISHI, I., FUJISHITA, M., OHTSUKI, Y. and SHIRAISHI, Y. (1983) Transformation of rabbit lymphocytes with T-cell leukemia virus. *Gann* **74**: 1–4.

MORGAN, D. A., RUSCETTI, F. W. and GALLO, R. C. (1976) Selective *in vitro* growth of T-lymphocytes from normal human bone marrow. *Science* **193**: 1007–1008.

NAGY, K., CLAPHAM, P., CHEINSONG-POPOV, R. and WEISS, R. A. (1983) Human T cell leukemia virus type 1. Induction of syncytia and inhibition by patient's sera. *Int. J. Cancer* **32**: 321–328.

NAGY, K., WEISS, R. A., CLAPHAM, P. and CHEINGSONG-POPOV, R. (1984) Human T-cell leukemia/lymphoma virus envelope antigens. In: *Human T Cell Leukemia/Lymphoma Virus,* pp. 121–131, GALLO, R C., ESSEX, M. E. and GROSS, L. (eds). Cold Spring Harbor Laboratory, Cold Spring Harbor, New York.

OROSZLAN, S., COPELAND, T. D., KALYANARAMAN, V. S., SARNGADHARAN, M. G., SCHULTZ, A. M. and GALLO, R. C. (1984) Chemical analysis of human T cell leukemia virus structural proteins. In: *Human T Cell Leukemia/Lymphoma Virus,* pp. 101–110, GALLO, R. C., ESSEX, M. E. and GROSS, L. (eds). Cold Spring Harbor Laboratory, Cold Spring Harbor, New York.

OROSZLAN, S., SARNGADHARAN, M. G., COPELAND, T. D., KALYANARAMAN, V. S., GILDEN, R. V. and GALLO, R. C. (1982) Primary structure analysis of the major internal protein p24 of human type-C T cell leukemia virus. *Proc. natn. Acad. Sci. U.S.A.* **79**: 1291–1294.

POIESZ, B. J., RUSCETTI, F. W., GAZDAR, A. F., BUNN, P. A., MINNA, J. D. and GALLO, R. C. (1980a) Isolation of type-C retrovirus particles from cultured and fresh lymphocytes of a patient with cutaneous T-cell lymphoma. *Proc. natn. Acad. Sci. U.S.A.* **77**: 7415–7519.

POIESZ, B. J., RUSCETTI, F. W., MIER, J. W., WOODS, A. M. and GALLO, R. C. (1980b) T-cell lines established from human T-lymphocytic neoplasias by direct response to T-cell growth factor. *Proc. natn. Acad. Sci. U.S.A.* **77**: 6815–6819.

POIESZ, B. J., RUSCETTI, F. W., REITZ, M. S., KALYANARAMAN, V. S. and GALLO, R. C. (1981) Isolation of a new type C retrovirus (HTLV) in primary uncultured cells of a patient with Sezary T-cell leukaemia. *Nature* **294**: 268–271.

POPOVIC, M., KALYANARAMAN, V. S., MANN, D. L., RICHARDSON, E., SARIN, P. S. and GALLO, R. C. (1984a) Infection and transformation of T cells by human T cell leukemia/lymphoma virus of subgroups 1 and 2 (HTLV-1 and HTLV-2). In: *Human T Cell Leukemia/Lymphoma Virus,* pp. 217–227, GALLO, R. C., ESSEX, M. E. and GROSS, L. (eds). Cold Spring Harbor Laboratory, Cold Spring Harbor, New York.

POPOVIC, M., LANGE-WANTZIN, G., SARIN, P. S., MANN, D. and GALLO, R. C. (1983a) Transformation of human umbilical cord blood T-cell leukemia/lymphoma virus (HTLV). *Proc. natn. Acad. Sci. U.S.A.* **80**: 5402–5406.

POPOVIC, M., REITZ, M. S., JR., SARNGADHARAN, M. G., ROBERT-GUROFF, M., KALYANARAMAN, V. S., NAKAO, Y., MIYOSHI, I., MINOWADA, J., YOSHIDA, M., ITO, Y. and GALLO, R. C. (1982) The virus of Japanese adult T-cell leukaemia is a member of the human T-cell leukaemia virus group. *Nature* **300**: 63–66.

POPOVIC, M., SARIN, P., ROBERT-GUROFF, M., KALYANARAMAN, V. S., MANN, D., MINOWDA, J. and GALLO, R. C. (1983b) Isolation and transmission of human retrovirus (human T-cell leukemia virus). *Science* **219**: 856–859.

POPOVIC, M., SARNGADHARAN, M. G., READ, E. and GALLO, R. C. (1984b) Detection, isolation, and continuous production of cytopathic retroviruses (HTLV-III) from patients with AIDS and pre-AIDS. *Science* **224**: 497–500.

RATNER, L., HASELTINE, W., PATAREA, R., LIVAK, K. J., STARCICH, B., JOSEPHS, S. J., DORAN, E. R., RAFALSKI, J. A., WHITEHORN, E. A., BAUMEISTER, K., IVANOFF, L., PETTEWAY, S. R., PEARSON, M. L., LAUTENBERGER, J. A., PAPAS, T. A., GHRYEB, J., CHANG, N. T., GALLO, R. C. and WONG-STAAL, F. (1985) Complete nucleotide sequence of the AIDS virus, HTLV-III. *Nature,* in press.

REITZ, M. S., KALYANARAMAN, V. S., ROBERT-GUROFF, M., POPOVIC, M., SARNGADHARAN, M. G., SARIN, P. S. and GALLO, R. C. (1983) Human T-cell leukemia/lymphoma virus: the retrovirus of adult T-cell leukemia/lymphoma. *J. Infect. Dis.* **147**: 399–405.

REITZ, M. S., POIESZ, B. J., RUSCETTI, F. W. and GALLO, R. C. (1981) Characterization and distribution of nucleic acid sequences of a novel type C retrovirus isolated from neoplastic human T lymphocytes. *Proc. natn. Acad. Sci. U.S.A.* **78**: 1887–1891.

RHO, H. M., POIESZ, B. J., RUSCETTI, F. W. and GALLO, R. C. (1981) Characterization of the reverse transcriptase from a new retrovirus (HTLV) produced by a human cutaneous T-cell lymphoma cell line. *Virology* **112**: 355–358.

ROBERT-GUROFF, M., NAKAO, Y., NOTAKE, K., ITO, Y., SLISKI, A. and GALLO, R. C. (1982) Natural antibodies to human retrovirus HTLV in a cluster of Japanese patients with adult T-cell leukemia. *Science* **215**: 975–978.

ROBERT-GUROFF, M., RUSCETTI, F. W., POSNER, L. E., POIESZ, B. J. and GALLO, R. C. (1981) Detection of the human T-cell lymphoma virus p19 in cells of some patients with cutaneous T-cell lymphoma and leukemia using a monoclonal antibody. *J. exp. Med.* **154**: 1957–1965.

ROBERT-GUROFF, M., SCHÜPBACH, J., BLAYNEY, D. W., KALYANARAMAN, V., MERINO, F., LANIER, A., SARNGADHARAN, M. G., CLARK, J., SAXINGER, W. C., BLATTNER, W., A. and GALLO, R. C. (1984) Seroepidemiologic studies on HTLV-1. In: *Human T Cell Leukemia/Lymphoma Viruses*, pp. 285–295, GALLO, R. C., ESSEX, M. E. and GROSS, L. (eds). Cold Spring Harbor Laboratory, Cold Spring Harbor, New York.

ROJKO, J. L., HOOVER, E. A., FINN, B. L. and OLSEN, G. R. (1981) Determinants of susceptibility and resistance to feline leukemia virus infection. II. Susceptibility of feline lymphocytes to productive feline leukemia. *JNCI* **67**: 899–909.

ROSENBAUM, W. D., DORMONT, P., SPIRE, B., VILMER, E., GENTILINI, M., GRISELLI, C., MONTAGNIER, L., BARRE-SINOUSSI, F. and CHERMANN, J. C. (1985) Antimoniotung state (HPA 23) treatment of three patients with AIDS and one with prodrome. *Lancet*, **1**: 450–451.

RUSCETTI, F. W., ROBERT-GUROFF, M., CECCHERINI-NELLI, L., MINOWADA, J., POPOVIC, M. and GALLO, R. C. (1983) Persistent *in vitro* infection by human T-cell leukemia-lymphoma virus (HTLV) of normal human T-lymphocytes from blood relatives of patients with HTLV-associated mature T-cell neoplasms. *Int. J. Cancer* **31**: 171–183.

SAFAI, B., SARNGADHARAN, M. G., GROOPMAN, J. E., ARNETT, K., POPOVIC, M., SLISKI, A., SCHUPBACH, J. and GALLO, R. C. (1984) Seroepidemiologic studies of human T lymphotropic retrovirus type III in acquired immune deficiency syndrome. *Lancet* i; 1438–1440.

SALAHUDDIN, S. Z., MARKHAM, P. D., LINDNER, S. G., GOOTENBERG, J., POPOVIC, M., HEMMI, H., SARIN, P. S. and GALLO, R. C. (1984) Lymphokine production by cultured human T-cells transformed by human T-cell leukemia-lymphoma virus-I. *Science* **223**: 703–707.

SARIN, P. S. and GALLO, R. C. (1983) Human T cell leukemia virus (HTLV). In: *Progress in Hematology*, pp. 149–161, BROWN, E. B. (ed.). Grune and Stratton, New York.

SARIN, P. S. and GALLO, R. C. (1984a) Human T cell growth factor. *CRC crit. Rev. Immunol.* **4**: 279–305.

SARIN, P. S. and GALLO, R. C. (1984b) Human T lymphotropic retroviruses in adult T cell leukemia-lymphoma and acquired immune deficiency syndrome. *J. clin. Immunol.* **4**: 415–423.

SARIN, P. S. and GALLO, R. C. (1986) The involvement of human T-lymphotropic retroviruses in T cell leukemia and immune deficiency. *Cancer Rev.*, in press.

SARIN, P., AOKI, T., SHIBATA, A., OHNISHI, Y., AOYAGI, Y., MIYAKOSHI, H., EMURA, I., KALYANARAMAN, V. S., ROBERT-GUROFF, M., POPOVIC, M., SARNGADHARAN, M. G., NOWELL, P. C. and GALLO, R. C. (1983) High incidence of human type-C retrovirus (HTLV) in family members of an HLTV-positive Japanese T-cell leukemia patient. *Proc. natn. Acad. Sci. U.S.A.* **80**: 2370–2374.

SARIN, P. S., GALLO, R. C., SCHEER, D. I., CREWS, F. and LIPPA, A. (1985) Effects of a novel compound (AL 721) on HTLV-III infectivity *in vitro*. *New Engl. J. Med.* **313**: 1289–1290.

SARIN, P. S., SUN, D. K., THORNTON, A. H., TAGUCHI, Y. and MUELLER, W.E.G. (1987) Inhibition of the replication of the etiologic agent of AIDS (HTLU-III/lau) by avarol and avarone. *J. Natl. Cancer Inst.* **78**: 663–666.

SARIN, P. S., TAGUCHI, Y., SUN, D., THOENTON, A., GALLO, R. C. and OBERG, B. (1985) Inhibition of HTLV-III/LAV replication by foscarnet. *Biochem. Pharmacol.* **34**: 4075–4078.

SARNGADHARAN, M. G., POPOVIC, M., BRUNCH, L., SCHUPBACH, J. and GALLO, R. C. (1984) Antibodies reactive with human T-lymphotropic retroviruses (HTLV-III) in the serum of patients with AIDS. *Science* **224**: 506–508.

SEIKI, M., HATTORI, S., HIRAYAMA, Y. and YOSHIDA, M. (1983) Human adult T-cell leukemia virus: complete nucleotide sequence of the provirus genome integrated in leukemia cell DNA. *Proc. natn. Acad. Sci. U.S.A.* **88**: 3618–3622.

SHAFFNER, C. P., PLESCIA, O. J., PONTANI, D., SUN, D., THORNTON, A., PANDEY, R. C. and SARIN, P. S. (1986) Antiviral activity of amphotericin-B-methyl ester. Inhibition of HTLV-III replication in cell culture. *Biochem Pharmacol.* **35**: 4110–4113.

SHAW, G. M., HAHN, B. H., ARYA, S. K., GROOPMAN, J. E., GALLO, R. C. and WONG-STAAL, F. (1984) Molecular characterization of human T cell leukemia (lymphotropic) virus type III in the acquired immune deficiency syndrome. *Science* **226**: 1165–1171.

SODROSKI, J. G., ROSEN, C. A. and HASELTINE, W. A. (1984) Transacting transcriptional activation of the long terminal repeat of human T-lymphotropic viruses in infected cells. *Science*, **225**: 381–385.

TAKATSUKI, K., UCHIYAMA, T., UESCHIMA, Y. and HATTORI, T. (1979) Adult T cell leukemia: further clinical observations and cytogenetic and functional studies of leukemic cells. *Jap. J. clin. Oncol.* **9**: 317–324.

UCHIYAMA, T., YODOI, J., SAGAWA, K., TAKATSUKI, K. and UCHINO, H. (1977) Adult T-cell leukemia: clinical and hematological features of 16 cases. *Blood* **50**: 481–503.

VYTH-DREESE, F. A. and DEVRIES, J. E. (1982) Human T-cell leukemia virus in lymphocytes from T-cell leukemia patient originating from Surinam. *Lancet* i: 993.

WALDMANN, T., BRODER, S., GREENE, W., SARIN, P. S., SAXINGER, C., BLAYNEY, D. W., BLATTNER, W. A., GOLDMAN, C., FROST, K., SHARROW, S., DEPPER, J., LEONARD, W., UCHIYAMA, T. and GALLO, R. C. (1984) A functional and phenotropic comparison of human T-cell leukemia/lymphoma virus (HTLV) positive adult T-cell leukemia with HTLV negative Sezary leukemia and their distinction using anti-TAC, a monoclinal antibody identifying the human receptor for T-cell growth factor. *J. clin. Invest.* **73**: 1711–1718.

WATERFIELD, M. D., SCRACE, G. T., WHITTLE, N., STROBANT, P., JOHNSON, A., WATESON, A., WESTERMARK, B., HELDIN, C. H., HUANG, J. S. and DEUEL, J. F. (1983) Platelet derived growth factor is structurally related to the putative transforming protein p24 of simian sarcoma virus. *Nature* **304**: 35–39.

WONG-STAAL, F. and GALLO, R. C. (1985) Human T-lymphotropic retroviruses (HTLV). *Nature,* **317**: 395–403.

WONG-STAAL, F., SHAW, G., HAHN, B. H., SALAHUDDIN, S. Z., POPOVIC, M., MARKHAM, P., REDFIELD, R. and GALLO, R. C. (1985) Genomic diversity of human T-lymphotropic virus Type III (HTLV-III). *Science,* **229**: 759–762.

YAMADA, Y. (1983) Phenotypic and functional analysis of leukemic cells from 16 patients with adult T-cell leukemia/lymphoma. *Blood* **61**: 192–199.

YAMAMOTO, N., MATSUMOTO, T., KOYANAGI, T., TANAKA, Y. and HINUMA, Y. (1982a) Unique cell lines harboring both Epstein-Barr virus and adult T-cell leukemia virus established from leukemia patients. *Nature* **249**: 367–369.

YAMAMOTO, N., OKADA, M., KOYANAGI, Y., KANAGI, M. and HINUMA, Y. (1982b) Transformation of human leukocytes by cocultivation with an adult T-cell leukemia virus. *Science* **217**: 737–739.

YODOI, T., MIYAWAKI, T., YACHIE, A., OHZEKI, S. and TANIGUCHI, N. (1982) Discrepancy in expression ability of TAC antigen and Ia determinants defined by monoclonal antibodies on activated or cultured cord blood T lymphocytes. *J. Immunol.* **129**: 1441–1445.

YOSHIDA, M., MIYOSHI, I. and HINUMA, Y. (1982) Isolation and characterization of retrovirus from cell lines of human adult T-cell leukemia and its implication in the disease. *Proc. natn. Acad. Sci. U.S.A.* **79**: 2031–2035.

ZAGURY, D., BERNARD, J., LEIBOWITCH, J., SAFAI, B., GROOPMAN, J. E., FELDMAN, M., SARNGADHARAN, M. G. and GALLO, R. C. (1984) HTLV-III in cells cultured from semen of two patients with AIDS. *Science* **126**: 449–451.

ZAGURY, D., FOUCHARD, M., CHEYNIER, R., BERNARD, J. K., CATTAN, A., SALAHUDDIN, S. Z. and SARIN, P. S. (1985a) Evidence for HTLV-III in T-cells from semen of AIDS patients. Expression in primary cell culture, long term mitogen stimulated cell cultures and cocultures with a permissive T cell line. *Cancer Res.* (Suppl) **45**: 4595–4597.

ZAGURY, D., FOUCHARD, M., VOL, J. C., CATTAN, A., LEIBOWITCH, J., FELDMAN, M., SARIN, P. S. and GALLO, R. C. (1985) Detection of infectious HTLV-III/LAV virus in cell-free plasma from AIDS patients. *Lancet,* **2**: 505–506.

CHAPTER 14

THE ROLE OF CELLULAR ONCOGENES IN CANCERS OF NON-VIRAL ETIOLOGY

MITCHELL P. GOLDFARB

Department of Biochemistry and Molecular Biophysics, Columbia University, 630 West 168th Street, New York, NY 10032, U.S.A.

The oncogenic potential of acutely tumorigenic retroviruses derive from viral-encoded oncogenes (v-*onc*). All retroviral oncogenes are transduced and frequently mutated cellular genes (c-*onc*). Do the c-*onc* genes play a role in cancers of non-viral etiology? Extensive research in the last four years shows that, in many cases, the answer is yes. In this paper, current knowledge of altered cellular gene function in non-viral malignancies will briefly be summarized. In addition to listing cancers that have been associated with misfunction of particular genes, the nature of these genetic misfunctions and their role in the overall multistep process of cancer, where it is known, will be emphasized. Table 1 summarizes the identified cellular oncogenes and their mode of oncogenic activation.

1. HUMAN AND MOUSE B-CELL MALIGNANCIES HAVE ALTERED c-*myc* EXPRESSION

The human and murine cellular *myc* genes (c-*myc*) encode proteins of 439 amino acids (Colby *et al.*, 1983; Stanton *et al.*, 1984a; Watt *et al.*, 1983). The protein is located in the cell nucleus and has a DNA binding capacity *in vitro* (Donner *et al.*, 1982; Persson and Leder, 1984). c-*myc* is expressed at radically higher levels in growing as opposed to resting cells (Goyette *et al.*, 1984; Kelly *et al.*, 1983), and in G_0 resting mouse fibroblasts, the mitogenic platelet-derived growth factor (PDGF) rapidly increases the level of c-*myc* transcripts (Kelly *et al.*, 1983). Furthermore, fibroblasts in which c-*myc* expression is rendered constitutive by an artificial promoter have a reduced PDGF requirement for growth (Armelin *et al.*, 1984). The precise function of c-*myc* is unknown, but given the above findings together with the known tumor-inducing properties of v-*myc* (Weiss, 1982), it appears that *myc* protein is a signal for cell division, and appropriate growth control requires precise regulation of c-*myc* gene expression.

Chicken bursal lymphomas (B-cell tumors) are usually associated with the insertion of an avian leukosis retroviral DNA sequence near the chicken c-*myc* gene. The transcriptional enhancing elements in the long terminal repeated sequences of the leukosis provirus serve to elevate the transcription of c-*myc*. The frequency of these events strongly implicates elevated *myc* gene expression in the genesis of chicken B-cell neoplasia. It is now clear that altered c-*myc* expression occurs in mouse and human B-cell lymphomas. Here the mechanism of altered expression is frequently chromosomal translocation.

1.1 TRANSLOCATION BRINGS c-*myc* INTO PROXIMITY WITH IMMUNOGLOBULIN GENES

Mouse plasmacytomas usually bear a translocation between chromosomes 12 and 15 (Klein, 1981) and chromosome 12 contains the immunoglobulin heavy chain genes (Ohno *et al.*, 1979). Human Burkitts lymphomas are characterized by one of three chromosomal

TABLE 1. c-onc Activation in Non-Viral Cancers

Oncogene	Activated c-onc in:	Frequency	Mode of Activation
c-myc	Burkitt lymphoma	often	translocation
	murine plasmacytoma	often	translocation
	human small cell lung cancer	25%	amplification
	human promyelocytic leukemia	one case	amplification
N-myc	human neuroblastoma	often	amplification
	human retinoblastoma	often	amplification
	human small cell lung cancer	20%	amplification
L-myc	human small cell lung cancer	20%	amplification
c-myb	human colon carcinoma	one case	amplification
	human leukemia (AML)	one case	amplification
c-abl	human leukemia (CML)	often	translocation
c-mos	murine plasmacytoma	two cases	insertion element
c-H-ras	human bladder carcinoma	three cases	structural mutation
	human lung carcinoma	two cases	structural mutation
	murine epidermal carcinoma	often	ND
	rat mammary carcinoma	often	structural mutation
c-K-ras	human lung carcinoma	10–20%	structural mutation
	human colon carcinoma	10–20%	structural mutation
	human pancreatic cancer	two cases	structural mutation
	human T-cell leukemia	one case	ND
	murine thymoma	often	ND
	murine adrenal tumor	one case	amplification
c-N-ras	human T-cell leukemia	often	ND
	human leukemia (AML)	often	structural mutation
	human neuroblastoma	one case	structural mutation
	human promyelocytic leukemia	one case	structural mutation
	human gastric carcinoma	one case	structural mutation
	human lung carcinoma	one case	structural mutation
	murine thymoma	often	structural mutation

The table summarizes the collective data which is reviewed in this paper. The precise frequencies of c-*onc* activation in specific forms of cancer are, in most cases, not yet known. The term 'often' is used to imply activation frequencies of greater than 35%. Where samplings are still too small to give any credible activation frequency, the number of positive samples is shown. ND, not determined.

translocations, T8;14, T8;22, T2;8 (Rowley, 1982). The three chromosomes with which the long arm of chromosome 8 becomes joined contain the immunoglobulin heavy chain genes (chromosome 14) (Croce *et al.*, 1979; Hobart *et al.*, 1981; Kirsch *et al.*, 1982), the κ light chain genes (chromosome 2) (Malcolm *et al.*, 1982), and λ light chain genes (chromosome 22) (Erikson *et al.*, 1981). The frequency of these rearrangements led to the speculation that a critical step leading to B-cell neoplasia is the translocation of a cellular oncogene into the proximity of immunoglobulin genes, thereby deregulating the expression of the oncogene (Klein, 1981; Rowley, 1982). Consistent with this idea, many murine B-cell tumors were shown to harbor a non-productive immunoglobulin gene rearrangement between the heavy chain constant region and a specific segment of DNA mapping outside the heavy chain gene cluster in germ line DNA (Adams *et al.*, 1982; Calame *et al.*, 1982; Harris *et al.*, 1982; Kirsch *et al.*, 1981). The non-immunoglobulin-associated rearranged DNA segment was mapped to chromosome 15 (Calame *et al.*, 1982) and later identified as murine c-*myc* (Marcu *et al.*, 1983; Shen-Ong *et al.*, 1982; Taub *et al.*, 1982). The human c-*myc* gene has been mapped to the distal end of chromosome 8, q24—qter (Dalla-Favera *et al.*, 1982a; Taub *et al.*, 1982), the chromosomal segment involved in all Burkitt lymphoma translocations. In some cases, the translocation breakpoint is very close to c-*myc* (see below).

1.2. Relationship of Translocation Breakpoints to c-*myc* and Immunoglobulin Genes

A description of the translocation breakpoints in B-cells tumors requires a short discussion of c-*myc* and immunoglobulin gene organizations. The murine and human c-*myc* gene structures are highly conserved. Both genes consistent of two coding exons (Colby *et al.*, 1983; Shen-Ong *et al.*, 1982; Stanton *et al.*, 1983, 1984a; Watt *et al.*, 1983) which are predicted to encode proteins of 439 amino acid residues. Additionally, both genes contain a 5' non-coding exon (Battey et al., 1983; Stanton et al., 1984a,b; Watt et al., 1983) which bear substantial sequence homology to one another (Battey et al., 1983). Analysis of normal human c-*myc* transcripts has revealed two promoters 150 base pairs apart at the 5' end of the non-encoding exon (Battey et al., 1983). Murine c-*myc* dual promoters are suggested by the presence of two consensus sequences for transcription promotion and initiation (Stanton et al., 1984b).

The immunoglobulin locus contains a linear array of potential coding segments which undergo a choice of site and region-specific recombination events to generate a diverse set of functional heavy chain-encoding genes (Leder, 1982; Tonegawa, 1983). In the first phase of functional rearrangement, V, D, and J segments are brought together, looping out the interposing sequences, generating the variable domain of a μ heavy chain gene. Subsequent recombination between heavy chain switch regions can generate genes expressing proteins with different functional domains (e.g. γ, α) but with the same antigenic specificity. The light chain gene clusters are simpler, lacking D segments and undergoing only V–J rearrangement.

Plasmacytomas induced with pristane in Balb/c mice usually contain rearrange *myc* genes (Adams et al., 1982; Calame et al., 1982; Harris et al., 1982; Marcu et al., 1983; Shen-Ong et al., 1982; Taub et al., 1982). The chromosomal breakpoints in these tumors vary with respect to c-*myc* (Harris et al., 1982; Shen-Ong et al., 1982), occurring either within the 5' non-coding exon or the first intron (Stanton et al., 1983). In plasmacytomas which secrete IgA antibodies, the 5' end of c-*myc* has recombined into the α switch region of the heavy chain locus (Calame et al., 1982; Harris et al., 1982; Shen-Ong et al., 1982). The rearrangement is head to head (5'–5') with respect to the transcriptional orientation of the two loci, and α switch region has no known transcriptional promoter. Therefore, transcription of the rearranged c-*myc* gene is likely mediated by unmasked cryptic promoters. In two examples of IgG secreting tumors, the 5' end of c-*myc* rearranged with either the γ2b switch region or a fused μ–γ2b switch region in head to head fashion (Dunnick et al., 1983; Stanton et al., 1984b). In other plasmacytomas, the sequences with which c-*myc* recombine have not been identified.

In human Burkitts lymphoma, the translocation breakpoints are somewhat different. In only about half of the T8;14 translocations is the breakpoint in chromosome 8 close enough to the 5' end of c-*myc* to be detected by molecular cloning and restriction enzyme analysis. The breakpoint has occurred in several instances within the first intron (ar-Rushdi et al., 1983; Leder et al., 1983; Saito et al., 1983) and sometimes the breakpoint is 1–5 kb pairs 5' to the non-coding exon (Battey et al., 1983; Rabbitts et al., 1983; Taub et al., 1984). In these cases where the translocation breakpoint is very near c-*myc*, the gene is rearranged head to head into the μ region of the immunoglobulin heavy chain locus, either within or 5' to the μ switch region (Battey et al., 1983; Marcu et al., 1983; Saito et al., 1983; Taub et al., 1984). In one example of a Burkitt lymphoma with the T2;8 translocation, chromosome 2 was severed in the κ variable region gene cluster, transferring variable and constant regions to a region of chromosome 8 at least 5 kb pairs 3' to c-*myc* (Erikson et al., 1983b). In an example of a lymphoma with a T8.22 translocation chromosome 22 breaks within the λ gene cluster, with the arm transferring to chromosome 8 at a position at least 5 kb pairs from c-*myc* on its 3' side (Croce et al., 1983).

1.3 Consequences of translocations involving c-*myc*

There has been debate as to the change in structure and expression of c-*myc* which is crucial for neoplastic progression. Alternative models for derangement include: (a) enhanced and/or constitutive transcription mediated by loss of negative control elements flanking c-*myc* or positive control elements brought fortuitously close to c-*myc*; (b) altered translation efficiency of *myc* RNA mediated by different promoter utilization; or (c) structural mutations in c-*myc* precipitated by an earlier translocation event.

Sixteen structural mutations were reported in the c-*myc* gene of the Burkitt lymphoma cell line Raji (Rabbitts *et al.*, 1983). However, this cell line has been maintained in culture for many years, during which time some or all of the detected mutations may have been accumulated. In the other Burkitts lymphomas, the translocated c-*myc* has no structural lesions (Battey *et al.*, 1983; Rabbits *et al.*, 1984). Furthermore, no structural mutations were observed in the rearranged c-*myc* gene of a murine plasmacytoma (Stanton *et al.*, 1984a). Therefore, structural mutation plays, at best, an occasional role in the progression of B-cell tumors.

The translocation-mediated translational activation model of Tonegawa and colleagues (Saito *et al.*, 1983) is intriguing. They have detected substantial sequence homology between the 5' non-coding exon and the first coding exon of human c-*myc* and speculate that normal c-*myc* transcripts could assume a stem-loop structure, blocking ribosome translation and allowing for regulation of *myc* protein synthesis via factors which modify this RNA secondary structure. (This method of activation could only apply in those cases where translocation breakpoints occur 3' to the natural c-*myc* promoters.) Arguing against this model is the finding that RNA synthesized off c-*myc* genes deleted in the portion of the non-coding exon which mediates stem-loop structure is translated no more efficiently *in vitro* than in normal c-*myc* RNA (Persson *et al.*, 1984).

The general feature of B-cell tumors is the deregulated transcription of the rearranged c-*myc* gene. *Myc* RNA is elevated to a greater or lesser extent in murine plasmacytomas compared to levels in normal tissue, mouse fibroblasts, and several other types of tumors, including pre-B-cell tumors (Marcu *et al.*, 1983; Mushinski *et al.*, 1983; Shen-Ong *et al.*, 1982). In human Burkitts lymphoma, c-*myc* RNA concentration is 2–20-fold higher than in Epstein–Barr virus-immortalized B lymphoblasts (Nishikura *et al.*, 1983; Taub *et al.*, 1984) or other types of tumors (Saito *et al.*, 1983). More strikingly, when comparison allows, the translocated murine or human c-*myc* gene is transcribed at far higher levels than the non-rearranged *myc* allele (Marcu *et al.*, 1983; Mushinski *et al.*, 1983; Shen-Ong *et al.*, 1982; Taub *et al.*, 1984). The expression of the human c-*myc* alleles on rearranged or normal chromosomes has been further studied by Croce and colleagues through the use of somatic cell hybrids between Burkitts cells and mouse plasmacytoma cells or human lymphoblastoid cells. These experiments showed that: (a) human c-*myc* on a translocated chromosome is transcribed in mouse plasmacytoma hybrids while normal human c-*myc* is not (Croce *et al.*, 1983; Erikson *et al.*, 1983a,b; Nishikura *et al.*, 1983), and (b) normal human c-*myc* is transcribed in lymphoblastoid cells while translocated c-*myc* is often not (Croce *et al.*, 1984).

These results help paint a likely picture for the role of c-*myc* in B-cell tumorigenesis: c-*myc* is expressed in normal proliferating lymphoid cells and, presumably, this expression is a signal for continued growth. In terminally differentiating B-cells, c-*myc* transcription is arrested, resulting in a non-proliferating plasma cell. However, when chromosomal rearrangements bring c-*myc* into the vicinity of an immunoglobulin gene locus, c-*myc* transcription can be positively driven by that locus to a greater or lesser extent, thus providing continued proliferative capacity to the plasma cell. How the immunoglobulin locus can activate c-*myc* at sometimes long chromosomal distances (Croce *et al.*, 1983; Erikson *et al.*, 1983b) is still obscure.

2. ALTERED c-*abl* GENE IN HUMAN CHRONIC MYELOGENOUS LEUKEMIA

Human chronic myelogenous leukemias (CMLs) are almost always cytogenetically characterized by the 'Philadelphia' translocation between chromosomes 22 and 9, generating $22q^-$ and $9q^+$ chromosomes (Lawler, 1977; Rowley, 1973). It has now been shown that this translocation is reciprocal, and the small piece of chromosome 9 (q34—qter) transferred to $22q^-$ contains the c-*abl* gene (deKlein *et al.*, 1982).

c-*abl* is the cellular homologue to the transforming gene (v-*abl*) of Abelson murine leukemia virus (Shields et al., 1979; Witte et al., 1979) a virus which can transform fibroblasts and pre-B lymphoid cells (Abelson and Rabstein, 1970; Rosenberg and Baltimore, 1976; Scher and Siegler, 1975). v-abl encodes a fusion protein consisting of retroviral structural and c-abl-specific protein sequences (Witte et al., 1978). v-abl protein is a tyrosine-specific protein kinase (Witte et al., 1980a) and kinase activity appears essential for the transforming capacity of the protein (Reynolds et al., 1980; Rosenberg et al., 1980; Witte et al., 1980b). The v-abl kinase is mediated by the abl-specific domain of the protein (Prywes et al., 1983), but original studies on the 150 kD c-abl protein failed to detect kinase activity (Ponticelli et al., 1982; Konopka et al., 1984). Recently, c-abl tyrosine kinase activity has been detected using modified assay conditions (Konopka and Witte, 1985). Presumably, the c-abl p150 kinase activity, which is expressed in many tissues (Müller et al., 1982), has different substrate specificity than v-abl kinase or is controlled by regulatory mechanisms which cannot influence the altered v-abl kinase.

The translocation breakpoints in some CMLs have been characterized. The breakpoint in chromosome 22, while not unique, is confined to within a 5 kb pair 'cluster region' termed bcr (Groffen et al., 1984). The breaks on chromosome 9 are more widely scattered, varying over a minimum of 20 kb pairs (Groffen et al., 1984; Heisterkamp et al., 1983) toward the 5' side of c-abl. All CML cell lines tested express a 210 kD abl-specific protein (p210) (Konopka et al., 1984, 1985) which is structurally overlapping with c-abl p150 and which has tyrosine kinase activity indistinguishable from v-abl kinase (Konopka et al., 1984). Expression of CML p210 is the direct consequence of the Philadelphia translocation. All CMLs express a novel 8.2 kb abl-specific RNA (Collins et al., 1984) which contains bcr-specific sequences (Shtivelman et al., 1985). This RNA encodes a protein with bcr-derived N-terminal amino acid sequence replacing the normal N-terminus of c-abl p150 (Shtivelman et al., 1985). The bcr translocation region is part of a gene of unknown function (Heisterkamp et al., 1985). The Philadelphia translocation joins introns of the bcr and c-abl genes, generating a rearranged gene encoding the p210 fusion protein.

The Philadelphia chromosome (and, hence, the bcr-abl fused gene) is detectable in the early stages of CML (Collins et al., 1984). What molecular feature distinguishes early CML from the acute blast crisis stage of the disease? One possibility is that blast crisis results from increase in the amount of p210 kinase synthesized. In support of this model, blast crisis cells have much higher levels of c-abl RNA than early CML (Collins et al., 1984). In one case, elevated expression results from amplification of the translocated c-abl

3. CERTAIN TUMORS HAVE AMPLIFIED c-onc GENES

Tumor cells sometimes contain double minute chromosomes (DMs), which are small chromosomal segments lacking centromeres, and homogenously staining regions of centromere-containing chromosomes (HSRs) (Barker, 1982). These karyotypic abnormalities have been shown to contain amplified cellular genes (Cowell, 1982; Schimke et al., 1978). It has been hypothesized that amplification and concommitant increased expression of certain genes can have malignant consequences; naturally, reasonable candidate genes for such amplification are the c-onc genes. Amplied c-onc genes have now been detected in certain tumors, with these genes often residing within HSRs or DMs.

A consistent correlation between a type of tumor and specific gene amplification occurs in neuroectodermal tumors. Neuroblastomas frequently contain an amplified gene with limited homology to c-myc, which has been termed N-myc (Brodeur et al., 1984; Kohl et al., 1983; Schwab et al., 1983a). N-myc has been shown to encode a protein of 464 amino acids with approximately 50% homology to c-myc protein (Kohl et al., 1986). Whereas c-myc is expressed in virtually all growing tissues, N-myc expression is more limited, being detected in developing neuroectoderm, kidney, intestine, lung, and heart, as well as in pre-B lymphoid cells (Kohl et al., 1984; Zimmerman et al., 1986).

N-myc is amplified in virtually all neuroblastomas harboring HSRs or DMs (Kohl et al., 1983; Schwab et al., 1983a) and these cells have high levels of N-myc transcripts.

Amplification has only been observed in neuroblastomas clinically classified as late-stage and highly malignant (Brodeur et al., 1984). The early phases of the disease may involve other alterations in N-*myc* or in different genes. N-*myc* amplification has also been detected in two of ten human primary retinoblastomas and in a retinoblastoma cell line (Lee et al., 1984). Amplification was precisely correlated with the presence of DMs or HRs in these tumors. In several of the retinoblastomas examined lacking gene amplification, N-*myc* transcription was none the less elevated, indicating alternative mechanisms for N-*myc* overexpression.

Another class of tumors which shows frequent amplification of specific c-onc genes is small 'oat' cell lung carcinoma (SCLC). Seventy per cent of these tumors have multiple copies of either c-*myc* (Little et al., 1983), N-*myc* (Nau et al., 1986), or a third *myc*-related gene termed L-*myc* (Nau et al., 1985). The more malignant variants of SCLC consistently show far greater *myc* gene amplification along with karyotypic abnormalities (Little et al., 1983). It is possible that continued *myc* gene amplification mediates progression of SCLC malignancies.

Other examples of c-*onc* amplification in tumors do not follow consistent patterns with respect to specific forms of neoplasia. These examples include:

c-*myc* amplification in a human promyelocytic leukemia cell line (Collins and Groudine, 1982; Dalla-Favera et al., 1982b), a human neuroblastoma cell line (Kohl et al., 1983) and a human colon carcinoma cell line (Alitalo et al., 1983);

c-*myb* amplification in a human colon carcinoma cell line (Alitalo et al., 1984) and a human myelogenous leukemia (Pelicci et al., 1984);

c-K-*ras* amplification in a mouse adrenocortical tumor cell line (Schwab et al., 1983b).

4. ACTIVATION OF c-*mos* IN MURINE MYELOMA BY RETROVIRUS-RELATED ELEMENT

The c-*mos* gene is the cellular homologue of the transforming gene of Moloney murine sarcoma virus (v-*mos*). The v-*mos* 37,000 dalton gene product mediates transformation of fibroblasts *in vitro* and sarcomas in mice (Papkoff et al., 1982). When a molecularly cloned c-*mos* gene is linked to a strong retroviral transcriptional promoter and introduced into a fibroblast cell line by gene transfer techniques (to be described in the next section), the c-*mos* gene is expressed and the cells are rendered tumorigenic (Blair et al., 1981). Transfer of the c-*mos* gene alone has no effect on fibroblasts (Blair et al., 1981). In a wide survey of normal and malignant cell types, c-*mos* was found to be unexpressed (Gattoni et al., 1982; Müller et al., 1982). Recently, c-*mos* transcripts have been detected in early embryos as well as in testes and ovaries of adult mice (Propst and Vande Woude, 1985).

The potential role of c-*mos* in neoplasia has been examined by screening tumors for c-*mos* transcription and mutation. While an early study proved negative (Gattoni et al., 1982; Müller et al., 1982) two independently derived mouse myeloma tumors were later found to have rearranged and active c-*mos* genes (Cohen et al., 1983; Gatoni-Celli et al., 1983; Rechavi et al., 1982). The rearranged c-*mos* can transform fibroblast cultures upon gene transfer (Rechavi et al., 1982). In both myelomas, rearrangement resulted from the insertion of a DNA sequence which is homologous to a murine endogenous retroviral genome (Canaani et al., 1983; Cohen et al., 1983). The insertions occurred within the c-*mos* coding sequence, at 89 and 30 condons within the initiator ATG triplet in the two tumors (Cohen et al., 1983; Rechavi et al., 1982) and insertion allowed for c-*mos* transcription (Gattoni-Celli et al., 1983; Rechavi et al., 1982).

One of the two c-*mos* activated myelomas, XRPC24, has been assayed for c-*myc* function, and found to contain a rearranged and overexpressed c-*myc* gene (Mushinski et al., 1983). These two oncogenetic abnormalities might combine to generate a highly malignant myeloma. Broad screenings of murine and human myelomas are required to assess the general importance of c-*mos* activation in these tumors.

5. THE TRANSFORMING GENES IN MANY TYPES OF TUMORS ARE c-*ras*

Previous sections of this paper described oncogenes identified on the basis of screening known c-*onc* genes for altered structure or expression in tumors. The use of DNA-mediated gene transfer to characterize oncogenes in tumor cells has been an alternative and fruitful approach to the problem of altered gene expression in neoplasia. The basis of bioassays for oncogenes evolved from *in vitro* tumor virus transformation assays. DNA and RNA tumor viruses which induce solid tumors *in vivo* were found to transform morphologically and physiologically embryo fibroblast cells derived from the susceptible host species. At limiting dilutions, these viruses would generate foci of transformed cells on a background of normal fibroblasts within a short time after infection. The advent of immortal fibroblast cell lines with growth-restrictive properties comparable to primary cell cultures facilitated the use of focus assays to detect transforming agents, whether viral or chemical (see Tooze, 1973 for review).

With the development of techniques for DNA-mediated gene transfer (Graham and van der Eb, 1973; Hill and Hillova, 1972) came the ability to detect and localize transforming genes in viral genomes (Andersson *et al.*, 1979; Chang *et al.*, 1980; Graham and van der Eb, 1973; Lai and Nathans, 1974). When sufficiently high efficiencies of gene transfer were eventually obtained by the calcium phosphate DNA transfer method, it became feasible to ask whether cellular oncogenes in tumors or transformed cells could be detected by gene transfer and focus assays. C. Shih *et al.* (1979) demonstrated that DNA from chemically transformed rodent cells could transform the growth-restricted NIH 3T3 mouse fibroblast cell line. DNA from a focus could again induce foci, and the sensitivity of the transforming activity to certain restriction endonucleases was the same for both the original donor DNA and DNA from the primary foci. Therefore, the chemically transformed cell had a unique transforming gene. The 3T3 focus assay was later used to show that human tumor cell lines contain transforming genes (Krontiris and Cooper, 1981; Lane *et al.*, 1981, 1982; Perucho *et al.*, 1981; Shih *et al.*, 1981). Surprisingly, hybridization studies between viral oncogenes and cloned transforming genes (Goldfarb *et al.*, 1982; Pulciani *et al.*, 1982b; Shih and Weinberg, 1982; Shimizu *et al.*, 1983b) or DNA from 3T3 transformants revealed that many of the human transforming genes are homologous to two related retroviral oncogenes, v-H-*ras* and v-K-*ras* (Der *et al.*, 1982; Parada *et al.*, 1982; Pulciani *et al.*, 1982a; Santos *et al.*, 1982; Shimizu *et al.*, 1983c).

The two viral *ras* genes encode serologically related, but distinct, 21,000 dalton proteins (p21) (T. Shih *et al.*, 1979, 1980). *Ras* p21 is synthesized as a cytosolic precursor, which matures to the inner side of the plasma membrane (Shih *et al.*, 1982; Willingham *et al.*, 1980) where it acquires covalently bound lipid (Sefton *et al.*, 1982) most likely flanked to a cysteine residue near the C-terminus (Willumsen *et al.*, 1984). Mature p21 can bind GTP and GDP (Scolnick *et al.*, 1979). The amino acid sequences of H- and K-*ras* p21s, as predicted from DNA sequence data (Dhar *et al.*, 1982; Tsuchida *et al.*, 1982), differ substantially only near their C-termini (see Fig. 1). Mammalian genomes contain at least three genetic homologues to the *ras* genes (Chang *et al.*, 1982; Ellis *et al.*, 1981; Hall *et al.*, 1983; Shimizu *et al.*, 1983c). One of these genes is c-H-*ras*, whose sequence is highly conserved between rat and human genomes, specifying proteins with only one amino acid difference (Capon *et al.*, 1983; Dhar *et al.*, 1982; Fasano *et al.*, 1983). The c-K-*ras* gene is again highly conserved (McGrath *et al.*, 1983; Shimizu *et al.*, 1983a; Tsuchida *et al.*, 1982). The third cellular *ras* gene is N-*ras*, which was originally isolated as a transforming gene in a human neuroblastoma (Shimizu *et al.*, 1983b) and fibrosarcoma (Hall *et al.*, 1983). Though only weakly homologous to v-H-*ras* and v-K-*ras* sequences, N-*ras* specifies a 21 kilodalton protein serologically related to H- and K-*ras* p21 (Shimizu *et al.*, 1983c) and substantially differing in amino acid sequence from those p21s only near its C-terminus (Taparowsky *et al.*, 1983) (see Fig. 1). Interestingly, all three *ras* genes are split into four coding exons at identical positions, arguing for a spliced, ancestral *ras* gene (Capon *et al.*, 1983; Fasano *et al.*, 1983; McGrath *et al.*, 1983; Shimizu *et al.*, 1983a; Taparowsky *et al.*,

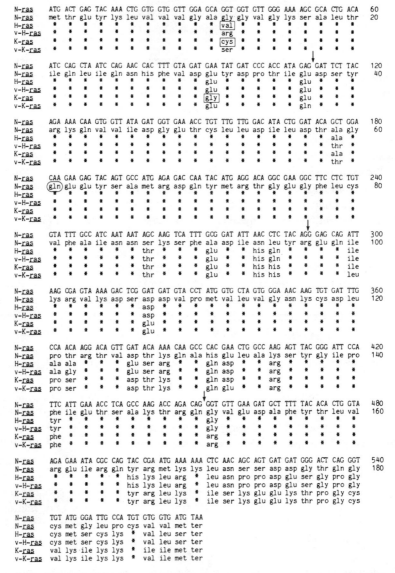

FIG. 1. Comparison of *ras* protein sequences. The nucleotide sequence is that of the nontransforming N-*ras* gene. Below this are the amino acid sequences of normal N-*ras* (with the circled residue at position 61 being the only change in a transforming N-*ras* protein), transforming human H-*ras* from T24 bladder carcinoma cells (Capon *et al.*, 1983; Fasano *et al.*, 1983) viral H-*ras* (Dhar *et al.*, 1982), transforming human K-*ras* from Calu-1 lung carcinoma cells (McGrath *et al.*, 1983; Shimizu *et al.*, 1983a), and viral D-*ras* (Tsuchida *et al.*, 1982). Asterisks under amino acid residues of the latter four sequences indicate positions of identity for all five *ras* proteins.

1983). The c-K-*ras* gene has two alternative fourth exons, allowing for the synthesis of two K-*ras* p21s differing only at their C-termini (McGrath *et al.*, 1983; Shimizu *et al.*, 1983a). The c-H-*ras* and c-K-*ras* genes are transcribed at low levels in all tissues that have been examined (Müller *et al.*, 1983, 1982), and are expressed in resting as well as growing fibroblast cultures (M. Goldfarb, unpublished data). c-K-*ras*, but not c-H-*ras*, transcription increases dramatically during rat liver regeneration (Goyette *et al.*, 1984).

Whereas c-*ras* genes cloned from normal human DNA cannot transform NIH 3T3 cells (Reddy *et al.*, 1982; Tabin *et al.*, 1982; Taparowsky *et al.*, 1983, 1982), transforming *ras* genes have been detected in roughly 20% of human tumors and tumor cell lines which have been examined. Activated H-*ras* has been observed in urinary tract carcinomas (Der *et al.*, 1982; Fujita *et al.*, 1984; Parada *et al.*, 1982; Santos *et al.*, 1982), and a lung carcinoma cell line (Yuasa *et al.*, 1983). c-K-*ras* transforming genes have been detected in lung carcinoma cell lines and tumors (Nakano *et al.*, 1984; Pulciani *et al.*, 1982a; Shimizu *et al.*, 1983a), colon carcinoma cell lines and tumors (Der *et al.*, 1982; Pulciani *et al.*, 1982a; Shimizu *et al.*, 1983c), a pancreatic carcinoma cell line (Cooper *et al.*, 1984), a sarcoma (Pulciani *et al.*, 1982a) and T-cell leukemia cell line (Eva *et al.*, 1983). The transforming N-*ras* gene has been observed in nearly half of T-cell leukemia cell lines and tumors (Eva *et al.*, 1983; Souryi and Fleissner, 1983) and in several human acute myelogenous leukemias (Bos *et al.*, 1985), as well as in a neuroblastoma cell line (Shimizu *et al.*, 1983c), fibrosarcoma cell lines (Hall *et al.*, 1983), a lung carcinoma cell line (Yuasa *et al.*, 1984), and a teratocarcinoma cell line (Cooper *et al.*, 1984). Transforming *ras* are dominant acting, meaning that they are oncogenic in cells despite the expression of normal *ras* genes; such is the case in recipient NIH 3T3 cells and, presumably, in tumor cells which have a normal as well as a transforming *ras* allele (Fujita *et al.*, 1984; Taparowsky *et al.*, 1983).

In all cases examined, transforming *ras* genes are distinguished from their normal counterparts by structural point mutations. These mutations have always mapped to amino acid codons 12, 13, or 61, and several different amino acid substitutions at these positions have transforming consequences (Bos *et al.*, 1985; McCoy *et al.*, 1984; McGrath *et al.*, 1983; Shimizu *et al.*, 1983a; Taparowsky *et al.*, 1982, 1983; Yuassa *et al.*, 1983, 1984). Where normal tissue has been available from cancer patients, it has been demonstrated that transforming *ras* genes arose by somatic mutation (Bos *et al.*, 1985; Fujita *et al.*, 1984; Santos *et al.*, 1984).

In vitro mutagenesis of a normal c-H-*ras* gene has shown that virtually any amino acid substitution for glycine at position 12 generates a transforming p21 (Fasano *et al.*, 1984; Seeburg *et al.*, 1984). This suggests that glycine$_{12}$ is vital to the regulation of p21 activity, with mutation resulting in a constitutive biochemical process. Glycine$_{13}$ and glutamine$_{61}$ may also be vital for functional regulation. *In vitro* mutagenesis has also shown that substitutions at codons 59 and 63 generate transforming genes as determined by the 3T3 focus assay (Fasano *et al.*, 1984). The failure to detect mutations in these positions in tumors is subject to various interpretations.

What are the normal and transforming functions of *ras* proteins? The recent capability of expressing *ras* genes in bacteria to generate and purify these proteins in abundance (McGrath *et al.*, 1984; Stacey and Kung, 1984; Sweet *et al.*, 1984) is providing insights. Normal human c-H-*ras* p21 purified from *E. coli* can bind GTP and GDP, and also has a GTPase activity. Transforming p21 (valine$_{12}$) has virtually identical nucleotide binding properties, but is deficient in GTPase (McGrath *et al.*, 1984; Sweet *et al.*, 1984) Since the transforming, but not normal p21 made in bacteria can induce growth of resting fibroblasts upon microinjection, this observed biochemical difference is likely not an artifact of *E. coli* synthesized protein (Feramisco *et al.*, 1984; Stacey and Kung, 1984). It is possible that p21–GTP complex is the biochemically active form of the protein. In normal cells, a growth signal (such as a hormonal growth factor) could promote this complex. The H-, K- and N-*ras* p21s could be responsive to different signals, due to the different C-termini. Whereas growth stimulation triggered by normal *ras* p21 is transient due to its GTPase activity, a transforming p21 complexed with GTP could be locked into the growth-promoting state because of the absence of GTPase. This biochemical model is similar to those of other enzymatic systems involved in information transfer and amplification across the plasma membrane. Hormone stimulation or inhibition of adenylate cyclase in many responsive cell types and light-induced stimulation of cyclic GMP phosphodiesterase in photoreceptor cells are examples of enzymatic complexes that possess a GTP/GDP binding regulatory subunit (Bokoch *et al.*, 1984; Northup *et al.*, 1982; Stryer *et al.*, 1981). However, unlike these rather well understood enzyme systems, we still know neither the biochemical effects that *ras* proteins mediate nor the presumed *ras* effector molecules in mammalian cells.

How does *ras* gene structural mutation fit into the presumed universal principle of multistep neoplastic progression? Although NIH 3T3 cells are malignantly transformed by mutated *ras* genes, this does not imply that such mutation is the only event required for tumor development. The NIH 3T3 cells used for transforming assays are not normal, but growth-restricted immortal cells which arose as rare variants of normal fibroblasts in senescence. The transfer of transforming *ras* genes into primary embryo fibroblasts causes transient transformation, but does not generate immortal, malignant cells. Neoplastic transformation of these cells requires mutated *ras* plus a second gene, either constitutively expressed v-*myc* or the early IA gene of adenovirus (Land et al., 1983; Ruley, 1983). Since it has been shown that v-*myc* and EIA can immortalize fibroblast and epithelial cells (Houweling et al., 1980; Ruley et al., 1984), it has been speculated that mutation of a *ras* gene is an event in tumor progression secondary to earlier, immortalizing events. The stage in tumor development when *ras* mutation can occur will, hopefully, be elucidated by studies in animal model tumor systems. Several laboratory groups have shown that certain types of tumors resulting from chemical or radiation carcinogenesis consistently have transforming *ras* genes. Mouse skin carcinomas induced by chemical carcinogen and tumor promoter and rat mammary carcinomas induced by single-dose chemical carcinogen consistently have transforming H-*ras* genes (Balmain and Pragnell, 1983; Sukumar et al., 1983). Mouse thymomas induced by chemical treatment have a transforming N-*ras* gene, while thymomas induced by γ irradiation have activated c-K-*ras* in four of seven cases examined (Guerrero et al., 1984). If *ras* mutation is an early event, possibly caused directly by carcinogenesis, then preneoplastic lesions will be found to have the mutated *ras* allele and, perhaps each carcinogen will repeatedly induce precisely the same mutation. Recently, Zarble et al. (1985) have shown that all rat mammary tumors induced by methylnitrosourea have transforming c-H-*ras* genes with the same nucleotide substitution, namely a guanine to adenine transition in codon 12, while dimethylbenzanthracene induced mammary tumors in the same animal never showed this codon 12 mutation. This finding is consistent with mutation mediated by direct carcinogen-DNA binding, thereby placing *ras* mutation as a very early step in tumor progression.

6. DIRECTIONS FOR FUTURE RESEARCH

The last several years' research have elucidated some of the genetic alterations which accompany, and likely contribute to certain forms of cancer. But the full spectrum of oncogenetic lesions is far from known. In those cancers where altered c-*onc* expression has been found, what other genetic changes may have occurred in the multistep development of the tumor? Filling in the huge gaps in our knowledge shall require new approaches to identify oncogene activity. The NIH 3T3 focus assay cannot detect genes which only manifest malignant potential in specifically differentiated cells, nor identify genes which immortalize cells. The number and chromosomal location of genetic changes conferring immortality to cells could be studied by somatic cell hybridization studies; hybrids between human tumor cells and rodent primary culture cells segregating human chomosomes could determine the location of gene(s) required for the immortal phenotype. Identification of such genes will require gene transfer experiments in various types of non-established cultured cells. For example, higher than currently obtained efficiencies of gene transfer could allow for detection of immortalizing genes in tumors which, together with a transforming *ras* gene, can transform primary rodent fibroblast cells.

Another limitation of the NIH 3T3 focus assay is that it can only detect genes which morphologically alter the cells; more subtle physiological changes which, none the less, represent a progression towards malignancy would go unnoticed. An alternative 3T3 transformation assay could be based upon a selection for cells that can grow in defined culture media lacking one or more of the required hormonal growth factors. Hormone-independent growth could be mediated by altered expression or structure of any gene product along the physiological circuit between hormone binding and cell entry into DNA synthesis. For example, platelet-derived growth factor induction of c-*myc* gene expression (Kelly et al., 1983) suggests that a 3T3 transformation assay based on PDGF-independent growth could detect activated c-*myc* genes in certain tumors.

Beyond the determination of cellular oncogene misfunction in tumors lies the need to determine the biochemical functions of normal and mutated c-*onc* gene products and the physiological consequences of these activities. The potentially long road towards this understanding will hopefully be shortened by a couple of means. First, proteins that are identified as directly or indirectly interacting with c-*onc* gene products may be part of well-studied biochemical systems in the cell. These include ion gates and pumps, hormone receptors and their dependent enzymatic activities, and associations among membrane and cyto-skeleton components. Second, genetic approaches towards elucidating oncogene function is hampered by the lack of genetic tools (e.g. homologous DNA recombination between exogenous and chromosomal sequences) and cumbersome cell culture methods, problems which distinguish studies on higher organisms from more maleable simple eukaryotes. Recent discoveries of homologues to c-*onc* genes in yeast (DeFeo-Jones *et al.*, 1983; Kataoka *et al.*, 1984; Lorincz and Reed, 1984; Peterson *et al.*, 1984; Powers *et al.*, 1984; Tachell *et al.*, 1984) hold promise for a more rapid genetic and biochemical analysis of ancestral c-*onc* function. Recent findings in Michael Wigler's laboratory demonstrate that a yeast homologue of c-*ras* genes encodes a guanine nucleotide-binding protein which regulates the activity of yeast adenylate cyclase (Toda *et al.*, 1985). Although a role for *ras* proteins in mammalian adenylate cyclase is not yet evident, continued findings in yeast should, in part, be applicable to our understanding of the mammalian *ras* genes.

REFERENCES

ABELSON, H. T. and RABSTEIN, L. S. (1970) Lymphosarcoma: virus-induced thymic independent disease in mice. *Cancer Res.* **30**: 2213–2222.

ADAMS, J. M., GERONDAKIS, S., WEBB, E., MITCHELL, J., BERNARD, O. and CORY, S. (1982) Transcriptionally active DNA that rearrange frequently in murine lymphoid tumors. *Proc. natn. Acad. Sci. U.S.A.* **79**: 6966–6970.

ALITALO, K., SCHWAB, M., LIN, C. C., VARMUS, H. E. and BISHOP, J. M. (1983) Homogeneously staining chromosomal regions contain amplified copies of an abundantly expressed cellular oncogene (c-*myc*) in malignant neuroendocrine cells from a human colon carcinoma. *Proc. natn. Acad. Sci. U.S.A.* **80**: 1707–1711.

ALITALO, K., WINQVIST, R., LIN, C. C., DE LA CHAPELLE, A., SCHWAB, M. and BISHOP, J. M. (1984) Aberrant expression of an amplified c-*myb* oncogene in two cell lines from a colon carcinoma. *Proc. natn. Acad. Sci. U.S.A.* **81**: 4534–4538.

ANDERSSON, P., GOLDFARB, M. and WEINBERG, R. A. (1979) A defined subgenomic fragment of *in vitro* synthesized *Moloney sarcoma* virus DNA can induce cell transformation upon transvection. *Cell* **16**: 63–75.

ARMELIN, H. A., ARMELIN, M. C. S., KELLY, K., STEWART, T., LEDER, P., COCHRAN, B. H. and STILES, C. D. (1984) Functional role for c-*myc* in mitogenic response to platelet-derived growth factor. *Nature* **310**: 655–660.

AR-RUSHDI, A., NISHIKURA, K., ERIKSON, J., WATT, R., ROVERA, G. and CROCE, C. M. (1983) Differential expression of the translocated and the untranslocated c-*myc* oncogene in Burkitt lymphoma. *Science* **222**: 390–393.

BALMAIN, A. and PRAGNELL, I. B. (1983) Mouse skin carcinomas induced *in vivo* by chemical carcinogens have a transforming Harvey-*ras* oncogene. *Nature* **303**: 72–74.

BARKER, P. E. (1982) Double minutes in human tumor cells. *Cancer Genet. Cytogenet.* **5**: 81–94.

BATTEY, J., MOULDING, C., TAUB, R., MURPHY, W., STEWART, T., POTTER, H., LENOIR, G. and LEDER, P. (1983) The human c-*myc* oncogene: structural consequences of translocation into the IgH locus in Burkitt lymphoma. *Cell* **34**: 779–787.

BLAIR, D. G., OSKARSSON, M., WOOD, T. G., MCCLEMENTS, W. L., FISCHINGER, P. J. and VANDEWOUDE, G. G. (1981) Activation of the transforming potential of a normal cell sequence: a molecular model for oncogenesis. *Science* **212**: 941–943.

BOKOCH, G. M., KATADA, T., NORTHUP, J. K., UI, M. and GILMAN, A. G. (1984) Purification and properties of the inhibitory guanine nucleotide-binding regulatory component of adenylate cyclase. *J. biol. Chem.* **259**: 3560–3567.

BOS, J. L., TOKSOZ, D., MARSHALL, C. J., DE VRIES, M. V., VEENEMAN, G. H., VAN DER EB, A. J., VAN BOOM, J.H., JANSSEN, J. W .G. and STEENVORDEN, A.C.M. (1985) Amino acid substitutions at codon 13 of the N-*ras* oncogene in human acute myelogenous leukemia. *Nature* **315**: 726–730.

BRODEUR, G. M., SEEGER, R. C., SCHWAB, M., VARMUS, H. E. and BISHOP, J. M. (1984) Amplification of N-*myc* in untreated human neuroblastomas correlates with advanced disease stage. *Science* **224**: 1121–1124.

CALAME, K., KIM, S., LALLEY, P., HILL, R., DAVIS, M. and HOOD, L. (1982) Molecular cloning of translocations involving chromosome 15 and the immunoglobulin C alpha gene from chromosome 12 in two murine plasmacytomas. *Proc. natn. Acad. Sci. U.S.A.* **79**: 6994–6998.

CANAANI, E., DREAZEN, O., KLAR, A., RECHAVI, G., RAM, D., COHEN, J. B. and GIVOL, D. (1983) Activation of the c-*mos* oncogene in a mouse plasmocytoma by insertion of an endogenous intracisternal A-particle genome. *Proc. natn. Acad. Sci. U.S.A.* **80**: 7118–7122.

CAPON, D., ELLSON, Y., LEVINSON, A., SEEBURG, P. and GOEDDEL, D. (1983) Complete nucleotide sequences of the T24 human bladder carcinoma oncogene and its normal cellular homologue. *Nature* **302**: 33–37.

CHANG, E. H., GONDA, M. A., ELLIS, R. W., SCOLNICK, E. M. and LOWY, D. R. (1982) Human genome contains four genes homologous to transforming genes of Harvey and Kirsten murine sarcoma viruses. *Proc. natn. Acad. Sci. U.S.A.* **79**: 4848–4852.

CHANG, E. H., MARYAK, J. M., WEI, C.-M., SHIH, T. Y., SHOBER, R., CHEUNG, H. L., ELLIS, R. W., HAGER, G. L., SCHOLNICK, E. M. and LOWY, D. R. (1980) Functional organization of the Harvey murine sarcoma virus genome. *J. Virology* **35**: 76–92.

COHEN, J. B., UNGER, T., RECHAVI, G., CANAANI, E. and GIVOL, D. (1983) Rearrangement of the oncogene c-*mos* in mouse myeloma NSI and hybridomas. *Nature* **306**: 797–799.

COLBY, W. C., CHEN, E. Y., SMITH, D. H. and LEVINSON, A. D. (1983) Identification and nucleotide sequence of a human locus homologous to the v-*myc* oncogene of avian myelocytomatosis virus MC29. *Nature* **301**: 722–725.

COLLINS, S. J. and GROUDINE, M. T. (1982) Amplification of endogenous *myc*-related DNA sequences in a human myeloid leukemia cell line. *Nature* **298**: 679–681.

COLLINS, S. J. and GROUDINE, M. T. (1983) Rearrangement and amplification of c-*abl* sequences in the human chronic mylogenous leukemia cell line K562. *Proc. natn. Acad. Sci. U.S.A.* **80**: 4813–4817.

COLLINS, S. J., KUBONISHI, I., MIYOSHI, I. and GROUDINE, M. T. (1984) Altered transcription of the c-*abl* oncogene in K-562 and other chronic myelogenous leukemia cells. *Science* **225**: 72–74.

COOPER, C. S., BLAIR, D. G., OSKARSSON, M. K., TAINSKY, M. A., EADER, L. A. and VANDEWOUDE, G. F. (1984) Characterization of human transforming genes from chemically transformed, teratocarcinoma and pancreatic carcinoma cell lines. *Cancer Res.* **44**: 1–10.

COWELL, J. K. (1982) Double minutes and homogeneously staining regions: gene amplification in mammalian cells. *Ann. Rev. Genet.* **16**: 21–59.

CROCE, C. M., ERIKSON, J., AR-RUSHDI, A., ADEN, D. and KISHIKURA, K. (1984) Translocated c-*myc* oncogene of Burkitt lymphoma is transcribed in plasma cells and repressed in lymphoblastoid cells. *Proc. natn. Acad. Sci. U.S.A.* **81**: 3170–3174.

CROCE, C. M., SHANDER, M., MARTINIS, J., CICUREL, L., D'ANCONA, G. G., DOLBY, T. W. and KOPROWSKI, H. (1979) Chromosomal location of the genes for human immunoglobulin heavy chains. *Proc. natn. Acad. Sci. U.S.A.* **76**: 3416–4319.

CROCE, C. M., THIERFELDER, W., ERIKSON, J., NISHIKURA, K., FINAN, J., LENOIR, G. M. and NOWELL, P. C. (1983) Transcriptional activation of an unrearranged and untranslocated c-*myc* oncogene by translocation of a C lambda locus in Burkitt lymphoma cells. *Proc. natn. Acad. Sci. U.S.A.* **80**: 6922–6926.

DALLA-FAVERA, R., BREGNI, M., ERIKSON, J., PATTERSON, D., GALLO, R. C. and CROCE, C. M. (1982a) Human c-*myc* oncogene is located on the region of chromosome 8 that is translocated in Burkitt lymphoma cells. *Proc. natn. Acad. Sci. U.S.A.* **79**: 7824–7827.

DALLA-FAVERA, R., WONG-STAAL, F. and GALLO, R. C. (1982b) *Onc* gene amplification in promyelocytic leukaemia cell line HL60 and primary leukaemic cells from the same patient. *Nature* **299**: 61–63.

DEFEO-JONES, D., SCOLNICK, E., KOLLER, R. and DHAR, R. (1983) *Ras*-related gene sequences identified and isolated from *Saccharomyces cerevisiae*. *Nature* **306**: 707–709.

DEKLEIN, A., VAN KESSEL, A. G., GROSVELD, G., BARTRAM, C. R., HAGEMEIJER, A., BOOTSMA, D., SPURR, N. K., HEISTERKAMP, N., GROFFEN, J. and STEPHENSON, J. R. (1982) A cellular oncogene is translocated to the Philadelphia chromosome in chronic myelogenous leukemia. *Nature* **300**: 765–767.

DER, C. J., KRONTIRIS, T. G. and COOPER, G. M. (1982) Transforming genes of human bladder and lung carcinoma cell lines are homologous to the *ras* genes of Harvey and Kirsten sarcoma viruses. *Proc. natn. Acad. Sci. U.S.A.* **79**: 3637–3640.

DHAR, R., ELLIS, R. W., SHIH, T. Y., OROSZLAN, S., SHAPIRO, B., MAIZEL, J. LOWY, D. and SCHOLNICK, E. (1982) Nucleotide sequence of the p21 transforming protein of Harvey murine sarcoma virus. *Science* **217**: 934–937.

DONNER, P., GREISER-WILKE, I. and MOELLING, K. (1982) Nuclear localization and DNA binding of the transforming gene product of avian myelocytomatosis virus. *Nature* **296**: 262–266.

DUNNICK, W., SHELL, B. E. and DERG, C. (1983) DNA sequences near the site of reciprocal recombination between a c-*myc* oncogene and an immunoglobulin switch region. *Proc. natn. Acad. Sci. U.S.A.* **80**: 7269–7273.

ELLIS, R. W., DEFEO, D., SHIH, T. Y., GONDA, M. A., YOUNG, H. A., TSUCHIDA, T., LOWY, D. R., COLLINS, S. J. and SCOLNICK, E. (1981) The p21 src genes of Harvey and Kirsten sarcoma viruses originate from divergent members of a family of normal vertebrate genes. *Nature* **292**: 506–511.

ERIKSON, J., AR-RUSHDI, A., DRWINGA, H. L., NOWELL, P. C. and CROCE, C. M. (1983a) Transcriptional activation of the translocated c-*myc* oncogene in Burkitt lymphoma. *Proc. natn. Acad. Sci. U.S.A.* **80**: 820–824.

ERIKSON, J., MARTINIS, J. and CROCE, C. M. (1981) Assignment of the genes for human lambda immunoglobulin chains to chromosome 22. *Nature* **294**: 173–175.

ERIKSON, J., NISHIURA, K., AR-RUSHDI, A., FINAN, J., EMANUEL, B., LENIOR, G., NOWEL, P. C. and CROCE, C. M. (1983b) Translocation of an immunoglobulin K locus to a region 3' of an unrearranged c-*myc* oncogene enhances c-*myc* transcription. *Proc. natn. Acad. Sci. U.S.A.* **80**: 7581–7585.

EVA, A., TRONICK, S. R., GOL, R. A., PIERCE, J. H. and AARONSON, S. A. (1983) Transforming genes of human hematopoietic tumors; frequent detection of *ras*-related oncogenes whose activation appears to be independent of tumor phenotype. *Proc. natn. Acad. Sci. U.S.A.* **80**: 4926–4930.

FASANO, O., ALDRICH, T., TAMANOI, F., TAPAROWSKY, E., FURTH, M. and WIGLER, M. (1984) Analysis of the transforming potential of the human H-*ras* gene by random mutagenesis. *Proc. natn. Acad. Sci. U.S.A.* **81**: 4008–4012.

FASANO, O., TAPAROWSKY, E., FIDDES, J., WIGLER, M. and GOLDFARB, M. (1983) Sequence and structure of the coding region of the human H-*ras*-1 gene from T24 bladder carcinoma cells. *J. Molec. appl. Genet.* **2**: 173–180.

FERAMISCO, J. R., GROSS, M., KAMATA, T., ROSENBERG, M. and SWEET, R. W. (1984) Microinjection of the oncogene form of the human H-*ras* (T24) protein results in rapid proliferation of quiescent cells. *Cell* **38**: 109–117.

FUJITA, J., YOSHIDA, O., RHIM, J. S., HATANAKA, M. and AARONSON, S. A. (1984) Ha-*ras* oncogenes are activated by somatic alterations in human urinary tract tumors. *Nature* **306**: 464–466.

GATTONI-CELLI, S., HSIANO, W.-L. W. and WEINSTEIN, I. B. (1983) Rearranged c-*mos* locus in a MOPC21 murine myeloma cell line and its persistence in hybridomas. *Nature* **309**: 795–796.
GATTONI, S., KIRSCHMEIER, P., WEINSTEIN, I. B., ESCOBEDO, J. and DINA, D. (1982) Cellular Moloney murine sarcoma (c-*mos*) sequences are hypermethylated and transcriptionally silent in normal and transformed rodent cells. *Molec. cell. Biol.* **2**: 42–51.
GOLDFARB, M., SHIMIZU, K., PERUCHO, M. and WIGLER, M. (1982) Isolation and preliminary characterization of a human transforming gene from T24 bladder carcinoma cells. *Nature* **296**: 404–409.
GOYETTE, M., PETROPOULOS, C. J., SHANK, P. R. and FAUSTO, N. (1984) Regulated transcription of C-Ki-*ras* and c-*myc* during compensatory growth of rat liver. *Molec. cell. Biol.* **4**: 1493–1498.
GRAHAM, F. L. and VAN DER EB, A. J. (1973) A new technique for the assay of infectivity of human adenovirus 5 DNA. *Virology* **52**: 456–467.
GROFFEN, J., STEPHENSON, J. R., HEISTERKAMP, N., DEKLEIN, A., BARTRAM, C. R. and GROVSVELD, G. (1984) Philadelphi chromosomal breakpoints are clustered within a limited region, bcr, on chromosome 22. *Cell* **36**: 93–99.
GUERRERO, I., CALZADA, P., MAYER, A. and PELLICER, A. (1984) A molecular approach to leukemogenesis: mouse lymphomas contain an activated c-*ras* oncogene. *Proc. natn. Acad. Sci. U.S.A.* **81**: 202–205.
HALL, A., MARSHALL, C., SPURR, N. and WEISS, R. (1983) Identification of the transforming gene in two human sarcoma cell lines as a new member of the *ras* gene family located on chromosome 1. *Nature* **303**: 396–400.
HARRIS, L. J., LANG, R. B. and MARCU, K. B. (1982) Non-immunoglobulin-associated DNA rearrangements in mouse plasmocytomas. *Proc. natn. Acad. Sci. U.S.A.* **79**: 4175–4179.
HEISTERKAMP, N., KEES, S., GROFFEN, J., DEKLEIN, A. and GROSVELD, G. (1985) Structural organization of the bcr gene and its role in the Ph' translocation. *Nature* **315**: 758–761.
HEISTERKAMP, N., STEPHENSON, J. R., GROFFEN, J., HANSEN, P. F., DEKLEIN, A., BARTRAM, C. R. and GROSVELD, G. (1983) Localization of the c-*abl* oncogene adjacent to a translocation breakpoint in chronic myelocytic leukemia. *Nature* **306**: 239–242.
HILL, M. and HILLOVA, J. (1972) Recovery of the temperature-sensitive mutant of *Rous sarcoma* virus from chicken cells exposed to DNA from hamster cells transformed by the mutant. *Virology* **49**: 309.
HOBART, M. J., RABBITS, T. H., GOODFELLOW, P. N., SOLOMON, E., CHAMBERS, S., SPURE, N. and POVEY, S. (1981) Immunoglobulin heavy chain genes in humans are located on chromosome 14. *Ann. hum. Genet.* **45**: 331–335.
HOUWELING, A., VAN DE ELSEN, D. and VAN DER EB, A. (1980) Partial transformation of primary rat cells by the leftmost 4.5% fragment of adenovirus 5 DNA. *Virology* **105**: 537–550.
KATAOKA, T., POWERS, S., MCGILL, C., FASANO, O., STRATHERN, J., BROACH, J. and WIGLER, M. (1984) Genetic analysis of yeast *ras* 1 and *ras* 2 genes. *Cell* **37**: 437–445.
KELLY, K., COCHRAN, B. H., STILES, C. D. and LEDER, P. (1983) Cell-specific regulation of the c-*myc* gene by lymphocyte mitogens and platelet-derived growth factor. *Cell* **35**: 603–610.
KIRSCH, I. R., MORTON, C. C., NAKAHARA, K. and LEDER, P. (1982) Human immunoglobulin heavy chain map to a region of translocations in malignant B lymphocytes. *Science* **216**: 301–303.
KIRSCH, I. R., RAVETCH, J. V., KWAN, S.-P., MAX, E. E., NEY, R. L. and LEDER, P. (1981) Multiple immunoglobulin switch region homologies outside the heavy chain constant region locus. *Nature* **293**: 585–587.
KLEIN, G. (1981) The role of gene dosage and genetic transpositions in carcinogenesis. *Nature* **294**: 313–318.
KOHL, N. E., GEE, C. E. and ALT, F. W. (1984) Activated expression of the N-*myc* gene in human neuroblastomas and related tumors. *Science* **226**: 1335–1336.
KOHL, N. E., KANDA, N., SCHRECK, R. R., BRUNS, G., LATT, S. A., GILBERT, F. and ALT, F. (1983) Transposition and amplification of oncogene-related sequences in human neuroblastomas. *Cell* **35**: 359–367.
KOHL, N. E., LEGUOY, E., DEPINHO, R. A., NISEN, P. D., SMITH, R. K., GEE, C. E. and ALT, F. W. (1986) Human N-*myc* is closely related in organization and nucleotide sequence to c-*myc*. *Nature* **319**: 73–77.
KONOPKA, J. B., WATANABE, S. M., SINGER, J. W., COLLINS, S. J. and WITTE, O. N. (1985) Cell lines and clinical isolates derived from Ph-positive chronic myelogenous leukemic patients express c-*abl* proteins with a common structural alteration. *Proc. natn. Acad. Sci. U.S.A.* **82**: 1810–1814.
KONOPKA, J. B., WATANABE, S. M. and WITTE, O. N. (1984) An alteration of the human c-*abl* protein in K562 leukemia cell unmasks associated tyrosine kinase activity. *Cell* **37**: 1035–1042.
KONOPKA, J. B. and WITTE, O. N. (1985) Detection of c-*abl* tyrosine kinase activity in vitro permits direct comparison of normal and altered abl gene products. *Molec. Cell. Biol.* **5**: 3116–3123.
KRONTIRIS, T. and COOPER, G. M. (1981) Transforming activity of human tumor DNAs. *Proc. natn. Acad. Sci. U.S.A.* **78**: 1181–1184.
LAI, C.-J. and NATHANS, D. (1974) Mapping temperature-sensitive mutants of simian virus 40: rescue of mutants by fragments of viral DNA. *Virology* **60**: 466–475.
LAND, H., PARADA, L. F. and WEINBERG, R. A. (1983) Tumorigenic conversion of primary embryo fibroblasts requires at least two cooperating oncogenes. *Nature* **304**: 596–602.
LANE, M. A., SAINTEN, A. and COOPER, G. M. (1981) Activation of related transforming genes in mouse and human mammary carcinomas. *Proc. natn. Acad. Sci. U.S.A.* **78**: 5185–5189.
LANE, M. A., SAINTEN, A. and COOPER, G. M. (1982) Stage-specific transforming genes of human and mouse B- and T-lymphoctye neoplasia. *Cell* **28**: 873–880.
LAWLER, S. D. (1977) *Clin. Haemat.* **6**: 55–75.
LEDER, P. (1982) The genetics of antibody diversity. *Scient. Am.* **246**(5): 102–115.
LEDER, P., BATTEY, J., LENOIR, G., MOULDING, C., MURPHY, W., POTTER, H., STEWART, T. and TAUB, R. (1983) Translocations among antibody genes in human cancer. *Science* **222**: 765–771.
LEE, W.-H., MURPHREE, A. L. and BENEDICT, W. F. (1984) Expression and amplification of the N-*myc* gene in primary retinoblastoma. *Nature* **309**: 458–460.
LITTLE, C. D., NAU, M. M., CARNEY, D. N., GAZDAR, A. F. and MINNA, J. D. (1983) Amplification and expression of the c-*myc* oncogene in human lung cancer cell lines. *Nature* **306**: 194–196.
LORINCZ, A. T. and REED, I. S. (1984) Primary structure homology between the product of yeast cell division control gene CDC28 and vertebrate oncogenes. *Nature* **307**: 183–185.

Malcolm, S., Barton, P., Murphy, C., Smith, M. A. F., Bentley, D. L. and Rabbits, T. H. (1982) Localization of human immunoglobulin kappa light chain variable region genes to the short arm of chromosome 2 by *in situ* hybridization. *Proc. natn. Acad. Sci. U.S.A.* **79**: 4957–4661.

Marcu, K. B., Harris, L. J., Stanton, L. W., Erikson, J., Watt, R. and Croce, C. M. (1983) Transcriptionally active c-*myc* oncogene is contained within NIARD, a DNA sequence associated with chromosome translocations in B-cell neoplasia. *Proc. natn. Acad. Sci. U.S.A.* **80**: 519–523.

McCoy, M. S., Bargmann, C. I. and Weinberg, R. A. (1984) Human colon carcinoma Ki-*ras* 2 oncogene and its corresponding proto-oncogene. *Molec. cell. Biol.* **4**: 1577–1582.

McGrath, J. P., Capon, D. J., Chen, E. Y., Seeburg, P. H., Goeddel, D. V. and Levinson, A. D. (1983) Structure and organization of the human Ki-*ras* proto-oncogene and a related processed pseudogene. *Nature* **304**: 501–506.

McGrath, J. P., Capon, D. J., Goeddel, D. V. and Levinson, A. D. (1984) Comparative biochemical properties of normal and activated human *ras* p21 protein. *Nature* **310**: 644–649.

Muller, R., Slamon, D. J., Adamson, E. D., Tremblay, J. M., Muller, D., Cline, M. J. and Verma, I. M. (1983) Transcription of c-*onc* genes c-*ras* and c-*fms* during mouse development. *Molec. cell. Biol.* **3**: 1062–1069.

Muller, R., Slamon, D. J., Tremblay, J. M., Cline, M. J. and Verma, I. M. (1982) Differential expression of cellular oncogenes during pre- and post-natal development of the mouse. *Nature* **299**: 640–644.

Mushinski, J. F., Bauer, S. R., Potter, M. and Reddy, E. P. (1983) Increased expression of *myc*-related oncogene mRNA characterizes most Balb/c plasmacytomas induced by pristane or Abelson murine leukemia virus. *Proc. natn. Acad. Sci. U.S.A.* **80**: 1073–1077.

Nakano, H., Yamamoto, F., Neville, C., Evans, D., Mizuno, T. and Perucho, M. (1984) Isolation of transforming sequences of two human lung carcinomas: structural and functional analysis of the activated c-K-*ras* oncogenes. *Proc. natn. Acad. Sci. U.S.A.* **81**: 71–75.

Nau, M. M., Brooks, B. J., Battey, J., Sausville, E., Gazdar, A. F., Kirsch, I. R., McBride, O. W., Bertness, V., Hollis, G. F. and Minna, J. D. (1985) L-*myc*, a new *myc*-related gene amplified and expressed in human small cell lung cancer. *Nature* **318**: 69–73.

Nau, M. M., Brooks, B. J., Carney, D. N., Gazdar, A. F., Battey, J. F., Sausville, E. A. and Minna, J. D. (1986) Human small-cell lung cancers show amplification and expression of the N-*myc* gene. *Proc. Natn. Acad. Sci. U.S.A.* **83**: 1092–1096.

Nishikuri, K., ar-Rushdi, A., Erikson, J., Watt, R., Rovera, G. and Croce, C. M. (1983) Differential expression of the normal and the translocated human c-*myc* oncogenes in B cells. *Proc. natn. Acad. Sci. U.S.A.* **80**: 4822–4826.

Northup, J. K., Smigel, M. D. and Gilman, A. G. (1982) The guanine nucleotide activating site of the regulatory component of adenylkate cyclase. *J. biol. Chem.* **257**: 11416–11423.

Ohno, S., Babonits, M., Wiener, F., Spira, J., Klein, G. and Potter, M. (1979) Non-random chromosome changes involving the Ig gene-carrying chromosomes 12 and 6 in pristane-induced mouse plasmacytomas. *Cell* **18**: 1001–1007.

Papkoff, S., Verma, I. M. and Hunter, T. (1982) Detection of a transforming gene product in cells transformed by Moloney murine sarcoma virus. *Cell* **29**: 417–426.

Parada, L. F., Tabin, C. J., Shih, C. and Weinberg, R. A. (1982) Human EJ bladder carcinoma oncogene is homologue of Harvey sarcoma virus *ras* gene. *Nature* **297**: 474–478.

Pelicci, P.-G., Lanfrancone, L., Brathwaite, M. D., Wolman, S. R. and Dalla-Favera, R. (1984) Amplification of the c-*myb* oncogene in a case of human acute myelogenous leukemia. *Science* **224**: 1117–1121.

Persson, H. and Leder, P. (1984) Nuclear localization and DNA binding properties of a protein expressed by human c-*myc* oncogene. *Science* **225**: 718–720.

Persson, H., Hennighauser, L., Taub, R., DeGrado, W. and Leder, P. (1984) Antibodies to human c-*myc* oncogene product is evidence of an evolutionarily conserved protein induced during cell proliferation. *Science* **225**: 687–693.

Perucho, M., Goldfarb, M., Shimizu, K., Lama, C., Fogh, J. and Wigler, M. (1981) Human tumor-derived cell lines contain common and different transforming genes. *Cell* **27**: 467–476.

Peterson, T. A., Yochem, J., Byers, B., Nunn, M. F., Deusberg, P. H., Doolittle, R. F. and Reed, I. S. (1984) A relationship between the yeast cell cycle genes CDC4 and CDC36 and the *ets* sequence of oncogenic virus E26. *Nature* **309**: 556–558.

Ponticelli, A. S., Whitlock, C. A., Rosenberg, N. and Witte, O. N. (1982) *In vivo* tyrosine phosphorylations of the Abelson virus transforming protein are absent in its normal cellular homolog. *Cell* **29**: 953–960.

Powers, S., Kataoka, T., Fasano, O., Goldfarb, M., Strathern, J., Broach, J. and Wigler, M. (1984) Genes in *S. cerevisiae* encoding proteins with domains homologous to the mammalian *ras* proteins. *Cell* **36**: 607–612.

Propst, F. and Vande Woude, G. F. (1985) Expression of c-*mos* proto-oncogene transcripts in mouse tissues. *Nature* **315**: 516–518.

Prywes, R., Foulkes, J. G., Rosenberg, N. and Baltimore, D. (1983) Sequences of the A-MuLV protein needed for fibroblast and lymphoid cell transformation. *Cell* **34**: 569–579.

Pulciani, S., Santos, E., Lauver, A. V., Long, L. K., Aaronson, S. A. and Barbacid, M. (1982a) Oncogenes in solid human tumors. *Nature* **300**: 539–542.

Pulciani, S., Santos, E., Lauver, A. V., Long, L. K., Robbins, K. C. and Barbacid, M. (1982b) Oncogenes in human tumor cell lines: molecular cloning of a transforming gene from human bladder carcinoma cells. *Proc. natn. Acad. Sci. U.S.A.* **79**: 2845–2849.

Rabbits, T. H., Foster, A., Hamlyn, D. and Baer, R. (1984) Effect of somatic mutation within translocated c-*myc* genes in Burkitt lymphoma. *Nature* **309**: 592–597.

Rabbitts, T. H., Hamlyn, D. and Baer, R. (1983) Altered nucleotide sequences of a translocated c-*myc* gene in Burkitt lymphoma. *Nature* **306**: 706–765.

Rechavi, G., Givol, D. and Canaani, E. (1982) Activation of a cellular oncogene by DNA rearrangement: possible involvement of an IS-like element. *Nature* **300**: 607–611.

REDDY, E. P., REYNOLDS, R. K., SANTOS, E. and BARBACID, M. (1982) A point mutation is responsible for the acquisition of transforming properties by the T24 human bladder carcinoma oncogene. *Nature* **300**: 149–152.

REYNOLDS, R. K., VAN DE VEN, W. J. M. and STEPHENSON, J. R. (1980) Abelson murine leukemia virus transformation defective mutants with impaired P120 associated protein kinase activity. *J. Virol.* **36**: 374–386.

ROSENBERG, N. and BALTIMORE, D. (1976) A quantitative assay for transformation of bone marrow cells by Abelson murine leukemia virus. *J. exp. Med.* **143**: 1453–1463.

ROSENBERG, N., CLARK, D. R. and WITTE, O. N. (1980) Abelson murine leukemia virus mutants deficient in kinase activity and lymphoid cell transformation. *J. Virol.* **36**: 766–774.

ROWLEY, J. D. (1973) A new consistent chromosomal abnormality in chronic myelogenous leukemia identified by quinacrine fluorescence and giemsa staining. *Nature* **243**: 290–293.

ROWLEY, J. D. (1982) Identification of the constant chromosome regions involved in human hematologic malignant disease. *Science* **216**: 749–751.

RULEY, H. E. (1983) Adenovirus early region 1A enables viral and cellular transforming genes to transform primary cells in culture. *Nature* **304**: 602–606.

RULEY, H. E., MOOMAW, J. F. and MARUGAMA, R. (1984) Avian myelocytomatosis virus *myc* and adenovirus early region 1A promote the *in vitro* establishment of cultured primary cells. In: *Cancer Cell*, Vol. 2, VANDEWOUDE, G., LEVINE, A., TOPP, W. and WATSON, J. (eds) Cold Spring Harbor, New York.

SAITO, H., HAYDAY, A. C., WIMAN, K., HAYWARD, W. S. and TONEGAWA, S. (1983) Activation of the c-*myc* gene by translocation: a model for translational control. *Proc. natn Acad. Sci. U.S.A.* **80**: 7476–7480.

SANTOS, E., MARTIN-ZANCA, D., REDDY, E. P., PIEROTT, M. A., DELLAPORTA, G. and BARBACID, M. (1984) Malignant activation of a K-*ras* oncogene in lung carcinoma but not in normal tissue of the same patient. *Science* **223**: 661–664.

SANTOS, E., TRONICK, S. R., AARONSON, S. A., PULCIANI, S. and BARRICID, M. (1982) T24 human bladder carcinoma oncogene is an activated form of the normal human homologue of Balb- and Harvey-MSV transforming genes. *Nature* **298**: 343–347.

SCHER, C. D. and SIEGLER, R. (1975) Direct transformation of 3T3 cells by Abelson murine leukemia virus. *Nature* **253**: 729–731.

SCHIMKE, R. T., KAUFMAN, R. J., ALT, F. and KELLEMS, R. F. (1978) Gene amplification and drug resistance in cultured murine cells. *Science* **202**: 1051–1055.

SCHWAB, M., ALITALO, K., KLEMPNAUER, K.-H., VARMUS, H. E., BISHOP, J. M., GILBERT, F., BRODEUR, G., GOLDSTEIN, M. and TRENT, J. (1983a) Amplified DNA with limited homology to *myc* cellular oncogene is shared by human neuroblastoma cell lines and a neuroblastoma tumor. *Nature* **305**: 245–248.

SCHWAB, M., ALITALO, K., VARMUS, H. E., BISHOP, J. M. and GEORGE, D. (1983b) A cellular oncogene (C-Ki-*ras*) is amplified, overexpressed, and located within karyotypic abnormalities in mouse adrenocortical tumor cells. *Nature* **303**: 497–411.

SCOLNICK, E. M., PAPAGEORGE, A. G. and SHI, T. Y. (1979) Guanine nucleotide-binding activity as an assay for *src* protein of rat-derived murine sarcoma viruses. *Proc. natn. Acad. Sci. U.S.A.* **76**: 5355–5359.

SEEBURG, P. H., COLBY, W. W., CAPON, D. J., GOEDDEL, D. V. and LEVINSON, A. D. (1984) Biological properties of human c-Ha-*ras* 1 gene initiated at codon 12. *Nature* **312**: 71–73.

SEFTON, B. M., TROWBRIDGE, I. S., COOPER, J. A. and SCOLNICK, E. M. (1982) The transforming proteins of Rous sarcoma virus, Harvey sarcoma virus, and Abelson leukemia virus contain tightly bound lipid. *Cell* **31**: 465–474.

SHEN-ONG, G. L. C., KEATH, E. J., PICCOLI, S. P. and COLE, M. D. (1982) Novel *myc* oncogene RNA from abortive immunoglobulin-gene recombination in mouse plasmacytomas. *Cell* **31**: 443–452.

SHIELDS, A., GOFF, S., PASKIND, M., OTTO, G. and BALTIMORE, D. (1979) Structure of the Abelson murine leukemia virus genome. *Cell* **18**: 955–962.

SHIH, C. and WEINBERG, R. A. (1982) Isolation of a transforming sequence from a human bladder carcinoma cell line. *Cell* **29**: 161–169.

SHIH, C., PADHY, L. C., MURRAY, M. and WEINBERG, R. A. (1981) Transforming genes of carcinomas and neuroblastomas introduced into mouse fibroblasts. *Nature* **290**: 261–264.

SHIH, C., SHILO, B.-Z., GOLDFARB, M., DANNENBERG, A. and WEINBERG, R. A. (1979) Passage of phenotypes of chemically transformed cells via transfection of DNA and chromatin. *Proc. natn. Acad. Sci. U.S.A.* **76**: 5714–5718.

SHIH, T. Y., WEEKS, M. O., GRUSS, P., KHAR, R., OROSZLAN, S. and SCOLNICK, E. M. (1982) Identification of a precursor in the biosynthesis of the p21 transforming protein of Harvey murine sarcoma virus. *J. Virol.* **43**: 253–261.

SHIH, T. Y., WEEKS, M. O., YOUNG, H. A. and SCOLNICK, E. M. (1979) Identification of a sarcoma virus-coded phosphoprotein in nonproducer cells transformed by Kirsten or Harvey murine sarcoma viruses. *Virology* **96**: 64–79.

SHIH, T. Y., WEEKS, M. O., YOUNG, H. A. and SCOLNICK, E. M. (1980) P21 of Kirsten Murine sarcoma virus is thermolabile in a viral mutant temperature sensitive for the maintenance of transformation. *J. Virol.* **31**: 546–556.

SHIMIZU, K., BIRNBAUM, D., RULEY, M. A., FASANO, O., SUARD, Y., EDLUND, L., TAPAROWSKY, E., GOLDFARB, M. and WIGLER, M. (1983a) Structure of the K-*ras* gene of the human lung carcinoma cell line Calu-1. *Nature* **304**: 497–500.

SHIMIZU, K., GOLDBARB, M., PERUCHO, M. and WIGLER, M. (1983b) Isolation and preliminary characterization of the transforming gene of a human neuroblastoma cell line. *Proc. natn. Acad. Sci. U.S.A.* **80**: 383–387.

SHIMIZU, K., GOLDFARB, M., SUARD, Y., PERUCHO, M., LI, Y., KAMATA, T., FERAMISCO, J., STAVNEZER, E., FOGH, J. and WIGLER, M. H. (1983c) Three human transforming genes are related to the viral *ras* oncogenes. *Proc. natn. Acad. Sci. U.S.A.* **80**: 2112–2116.

SHTIVELMAN, E., LIFSHITZ, B., GALE, R. P. and CANAANI, E. (1985) Fused transcript of abl and bcr genes in chronic myelogenous leukaemia. *Nature* **315**: 550–554.

SOURYI, M. and FLEISSNER, E. (1983) Identification by transfection of transforming sequences in DNA of human T-cell leukemias. *Proc. natn. Acad. Sci. U.S.A.* **80**: 6676–6679.

STACEY, D. W. and KUNG, H.-F. (1984) Transformation of NIH 3T3 cells by microinjection of Ha-*ras* p21 protein. *Nature* **310**: 508–511.

STANTON, L. W., FAHRLANDER, P. D., TESSER, P. M. and MARCU, K. B. (1984a) Nucleotide sequence comparison of normal and translocated murine c-*myc* genes. *Nature* **310**: 423–425.

STANTON, L. W., WATT, R. and MARCU, K. B. (1983) Translocation, breakage, and truncated transcripts of c-*myc* oncogene in murine plasmacytomas. *Nature* **303**: 401–406.

STANTON, L. W., YANG, J.-Q., ECKHARDT, L. A., HARRIS, L. J., BIRSHTEIN, B. K. and MARCU, K. B. (1984b) Products of a reciprocal translocation involving the c-*myc* gene in a murine plasmacytoma. *Proc. natn. Acad. Sci. U.S.A.* **81**: 829–833.

STRYER, L., HURLEY, J. B. and FUNG, B. K.-K. (1981) First stage of amplification in the cyclic-nucleotide cascade of vision. *Curr. Top. membr. Transp.* **15**: 93–108.

SUKUMAR, S., NOTARIO, V., MARTIN-ZANCA, D. and BARBACID, M. (1983) Induction of mammary carcinomas in rats by nitroso-methylurea involves malignant activation of H-*ras*-1 locus by single point mutations. *Nature* **306**: 658–661.

SWEET, R., YOKOYAMA, S., KAMATA, T., FERAMISCO, J. R., ROSENBERG, M. and GROSS, M. (1984) The product of *ras* is a GTPase and the T24 oncogenic mutant is defective in this activity. *Nature* **311**: 273–275.

TABIN, C., BRADLEY, S., BARGMANN, C., WEINBERG, R. A., PAPAGEORGE, A., SCOLNICK, E. M., DHAR, R., LOWY, D. and CHANG, E. (1982) Mechanism of activation of a human oncogene. *Nature* **300**: 143–148.

TAPAROWSKY, E., SHIMIZU, K., GOLDFARB, M. and WIGLER, M. (1983) Structure and activation of the human N-*ras* gene. *Cell* **34**: 581–586.

TAPAROWSKY, E., SUARD, Y., FASANO, O., SHIMIZU, K., GOLDFARB, M. and WIGLER, M. (1982) Activation of the T24 bladder carcinoma transforming gene is linked to a single amino acid change. *Nature* **300**: 762–765.

TATCHELL, K., CHALEFF, D. T., DEFEO-JONES, D. and SCOLNICK, E. M. (1984) Requirement of either of a pair of *ras*-related genes of *Saccharomyes cerevisiae* for spore viability. *Nature* **309**: 523–527.

TAUB, R., KIRSCH, I., MORTON, C., LENOIR, G., SWAN, D., TRONICK, S., AARONSON, S. and LEDER, P. (1982) Translocation of the c-*myc* gene into the immunoglobulin heavy chain locus in human Burkitt lymphoma and murine plasmacytoma cells. *Proc. natn. Acad. Sci. U.S.A.* **79**: 7837–7841.

TAUB, R., MOULDING, C., BATTEY, J., MURPHY, W., VASICEK, T., LENOIR, G. M. and LEDER, P. (1984) Activation and somatic mutation of the translocated c-*myc* gene in Burkitt lymphoma cells. *Cell* **36**: 339–348.

TODA, T., UNO, I., ISHIKAWA, T., POWERS, S., KATAOKA, T., BROEK, D., CAMERON, S., BROACH, J., MATSUMOTO, K. and WIGLER, M. (1985) In yeast, RAS proteins are controlling elements of adenylate cyclase. *Cell* **40**: 27–36.

TONEGAWA, S. (1983) Somatic generation of antibody diversity. *Nature* **302**: 575–581.

TOOZE, J. (ed.) (1973) *Molecular Biology of Tumor Viruses*. Cold Spring Harbor, New York.

TSUCHIDA, N., RYDER, T. and OHTSUBO E. (1982) Nucleotide sequence of the oncogene encoding the p21 transforming protein of Kirsten murine sarcoma virus. *Science* **217**: 937–939.

WATT, R., STANTON, L. W., MARCU, K. B., GALLO, R. C., CROCE, C. M. and ROVERA, G. (1983) Nucleotide sequence of cloned cDNA of human c-*myc* oncogene. *Nature* **303**: 725–728.

WEISS, R. (1982) The *myc* oncogene in man and birds. *Nature* **299**: 9–10.

WILLINGHAM, M. C., PASTAN, I., SHIH, T. Y. and SCOLNICK, E. M. (1980) Localization of the *src* gene product of the Harvey strain of MSV to plasma membrane of transformed cells by electron microscopic immunocytochemistry. *Cell* **19**: 1005–1014.

WILLUMSEN, B. M., CHRISTENSEN, A., HUBBERT, N. L., PAPAGEORGE, A. G. and LOWY, D. R. (1984) The p21 *ras* C-terminus is required for transformation and membrane association. *Nature* **310**: 583–586.

WITTE, O. N., DASGUPTA, A. and BALTIMORE, D. (1980a) Abelson murine leukemia virus protein is phosphorylated *in vitro* to form phosphotyrosine. *Nature* **283**: 826–831.

WITTE, O. N., GOFF, S. P., ROSENBERG, N. and BALTIMORE, D. (1980b) A transformation defective mutant of Abelson murine leukemia virus lacks protein kinase activity. *Proc. natn. Acad. Sci. U.S.A.* **77**: 4993–4997.

WITTE, O. N., ROSENBERG, N. and BALTIMORE, D. (1979) A normal cell protein cross-reactive to the major Abelson and murine leukemia virus gene product. *Nature* **281**: 396–398.

WITTE, O. N., ROSENBERG, N., PASKIND, M., SHIELDS, A. and BALTIMORE, D. (1978) Identification of an Abelson murine leukemia virus encoded protein present in transformed fibroblast and lymphoid cells. *Proc. natn. Acad. Sci. U.S.A.* **75**: 2488–2492.

YUASA, Y., GOL, R. A., CHANGE, A., CHIU, I.-M., REDDY, E. P., TRONICK, S. R. and AARONSON, S. A. (1984) Mechanism of activation of an N-*ras* oncogene of SW-1271 human lung carcinoma cells. *Proc. natn. Acad. Sci. U.S.A.* **81**: 3670–3674.

YUASA, Y., SRIVASTAVA, S. K., DUNN, C. Y., RHIM, J. S., REDDY, E. P. and AARONSON, S. A. (1983) Acquisition of transforming properties by alternative point mutations with c-*has/bas* human proto-oncogene. *Nature* **303**: 775–779.

ZARBL, H., SUKUMAR, S., ARTHUR, A. V., MARTIN-ZANCA, D. and BARBACID, M. (1985) Direct mutagenesis of H-*ras*-1 oncogenes by N-nitroso-N-methylurea during initiation of mammary carcinogenesis in rats. *Nature* **315**: 382–385.

ZIMMERMAN, K. A., YANCOPOULOUS, G. D., COLLUM, R. G., SMITH, R. K., KOHL, N. E., DENIS, K. A., NAU, M. M., WITTE, O. N., TOREN-ALLERAND, GEE, C. E., MINNA, J. D. and ALT, F. W. (1986) Differential expression of *myc* family genes during murine development. *Nature* **319**: 780–783.

INDEX

Acyclovir
  efficacy 283
  and herpes virus 274, 281, 286
    latent infections 287
  phosphorylated 284, 285
  structure and prodrug 282
  virus inhibition and reversibility 283, 285
Acyltransferase, metabolic activation 9
Adult T-cell leukemia/lymphoma (ATLL) 392
  clustering, Japan 395
  HTLV antibodies 395
Aflatoxin B 5, 25
  DNA adducts 38, 52, 53
  fluorescence 52
  metabolic activation 23
  potency 23
  structure 3
Agar in dose-response analyses 172
Agar Suspension assay 162, 163
AIDS
  antibodies 395
  Epstein-Barr infections 278, 279
  HTLV 391
  therapeutic agents 394
  viruses isolated 393
AIDS-related complex (ARC) 391
Aliphatic halides 4
Alkene oxides, DNA modification 42, 43
Alkylating agents
  bifunctional 40, 41
  chemical carcinogen properties 39–42
  DNA adduct formation 39–42
  metabolic activation 6, 7, 41
Alkyl nitrosamines 40
Ames assay 35
Amino-azo dyes 3, 4
  metabolic activation 7–10
  tumor induction 4
Aminofluorene, DNA modification 49
Anchorage independence
  fibroblasts 144
  phorbol diester effects 86
Antigens
  hepatitis 301
    hepatocellular carcinoma 313, 314
    hepatocyte distribution 308
  Woodchuck hepatitis virus 301
Antipain
  anticarcinogenic 177
  and radiogenic transformation 177, 178
Antiviral drugs 273, 279–85 (*see also* individual drugs)
  AIDS 394
  halogenated nucleoside 280, 281
    efficacy 283
    herpes virus 274
  herpes virus effects 274
  mechanism of action 283–5
  reversal 287
  target 283, 284
    enzymes 283, 284
  toxicity 283
Aplysiatoxin structure 81, 82
Ara-A (9-β-D-arabinofuranosyladenine) 279
  target 284
Ara-C (1β-D-arabinofuranosylcytosine) 279
  target 284
Aromatic amines 3
  carcinogens 51
  DNA adducts 50, 51
  excision repair and 138, 140
  metabolic activation 7–10

Bay region theory 45, 46
Benzamides, radiogenic transformation 181, 182
Benz[a]anthracene
  DNA modification 47
  methyl substitute 47
Benzidine, DNA adducts 50
Benzo[a]pyrene
  activation steps 43
  Bay region theory 45, 46
  dihydrodiol epoxide 11
    potency 13
    stereoisomers 11, 12
  DNA adducts 44, 45
    structure 44
  isomers 43, 44
  non-Bay region adducts 46, 47
  properties 43
  reactive metabolites 45
Bladder cancer 4
Bloom syndrome
  cells 188
    growth 163
Burkitt's lymphoma, viral antibodies 277

Calcium dependence and transformation 160
Cancer (*see also* individual organs)
  herpes virus links 290
  viral etiology 272, 273
Carcinogenesis (*see also* Chemical, Mouse skin)
  action 136
  in utero data 168
  two-stage 74–7
    initiation and promotion 75
    models 79, 80
Carcinogens (*see also* individual chemicals and classes)
  cytotoxicity and repair 137–9
Catalase, radiogenic transformation 179, 180
CC-1065, DNA adducts 55
Cell communication, phorbol diester effects 86–8
Cell culture, radiogenic transformation 152, 184

Cell differentiation
　calcium regulation   96
　malignant hematopoietic and phorbol
　　diesters   87, 88
　phorbol diester effects   87, 95
　　hematopoietic cells   97, 98
　　nonhematopoietic cells   95–7
　TPA-interferon synergism   96
Cell lines
　C3H/10T1/2, features   99
　　transformation   99–101
　　transformation assay   157
　　X-ray induced transformation   100
　epithelial, properties   154
　fibroblast, properties   154
　hamster embryo   102–4, 156
　　assay   155
　　cocarcinogens   103
　　culture   102
　　phorbol diesters   102, 103
　　radiogenic transformation   172
　　X-ray transformed   156
　human cell systems   107, 108
　　embryo, radiogenic transformation   172
　　KD, X-ray transformation   171, 172
　　phenotypic markers   107
　　senescence   108
　　XP   136–47
　mouse epidermis
　　growth   104
　　phorbol diesters   104, 105
　mouse submandibular gland   107
　mouse 3T3 and phorbol diesters   101, 102
　radiogenic transformation   154–7
　　BALB-3T3   167
　　hamster cell features   161
　rat trachea epithelium   105–7
　　phorbol diesters   106
　　transformation assay   106
Cell morphology, phorbol diester effects   85
Cell proliferation, phorbol diester effects   92–4
Cell transformation
　assay   157
　avian viruses   355–78
　　cytoskeleton   364
　criteria   159–64
　DNA excision repair role   145
　DNA lesion and repair role   135–47
　excision repair and DNA synthesis interval   145
　inhibitors   101
　rate   158
　two-stage model   75
　in vitro   98–108
　　mouse C3H-10T1/2 cells   98–101
　　radiation-induced   100
　viral mechanism   197–227
Cervical carcinoma   290, 291
Chemical carcinogenesis, initiation and
　promotion   33, 73
Chemical carcinogen (see also individual
　compounds and classes)
　classes   39–56
　cytotoxicity and excision repair   139
　DNA interactions   34
　mechanism of action   5–7
　metabolic activation   7–23
　natural products   23, 24
　structures   2, 3
Chromosomal aberrations, phorbol diester
　effects   94
Chromosomes
　G-banding   171
　phorbol diester effects   86, 94

sister chromatid exchanges   94
　transformation criteria   159, 160
Chrysene, DNA adducts   48
Co-carcinogens (see also Tumor promoters)
　radiogenic transformation   175
Contact inhibition, transformation   160
Crotonaldehyde   41
Croton oil   80
Croton tiglium, phorbol diesters   80, 81
Cycacin
　DNA modification   53
　structure   3
Cytoskeleton, phorbol diester effects   85

Debromoplysiatoxin structure   81, 82
Deoxyguanosine, adduct structure   41
2′-Deoxyguanosine, structure and activity   280
DHPG, 9-(1,3-dihydroxy-2-propoxymethylguanine)
　efficacy   283
　and herpes virus   274, 280, 281
　protein synthesis   286, 287
　virus inhibition and reversibility   282
Dibenz[a,h]acridine   3
Dibenz[a,h]anthracene   1
　structure   2
Diet
　cancer risk   174
　protective   188
Diethylstilbestrol (DES) and keratinocyte
　growth   97
Dihydrodiolepoxide   10, 11
　anti or trans   11
　stereoselectivities   12
　syn or cis   11
　tumor potency   12
4-Dimethylaminoazobenzene
　DNA adducts   51
　structure   2, 3
Dimethylbenzanthracene (DMBA)
　Bay region activation   38
　DNA adduct formation   47, 48
　fluorescent DNA   38
　K-region oxide   48
　structure   2, 3
1,2-Dimethylhydrazine, metabolic scheme   22
DNA
　hepadnavirus replication   308, 309
　postlabelling technique   36
　premutagenic lesions   142
　synthesis and phorbol diesters   92, 93
　transformation criteria   159, 160
DNA adducts   34–57
　assays   34, 35
　and carcinogenesis   56, 57
　cis platinum   56
　detection   34–9
　　fluorescence   37, 38
　　immunological   36, 37
　direct analysis   35
　human population   35
　photochemical   55, 56
　radiolabelled carcinogens   35, 36
DNA excision repair
　carcinogen cytotoxicity   137–9
　human cell transformation   135–47
　methyltransferase   138
　mutation frequency   139–41
　XP cells and UV   137, 138
DNA lesions
　human cell transformation   135–47
　transformation process   145, 146
DNA modification   33–57

carcinogen interactions 34
  nucleophilic sites 34
DNA transfection 163, 164
  X-ray transformed cells 183, 184

ELISA (enzyme-linked immunosorbent assay)
  DNA adduct detection 37
  sensitivity 37
Epoxides, DNA modification 42, 43
β-Estradiol, X-ray induced transformation 176
Estragole 24
  structure 3
Ethionine, DNA modification 41

Fibroblasts 136
  neoplastic transformation 143–5, 186
    agents 144
  6-thioguanine resistance 145
2-Fluorenylacetamide
  activation reaction 7–10
  detoxification reactions 7
  hydroxylation 10
  N-O bond formation 9
  ultimate carcinogens 9
Fluorescence
  DNA adduct detection 37, 38
  line narrowing 38
Fluorescence-activated cell sorter, analysis 399
Free radical scavengers 179, 180

Gene amplification, phorbol diester effects 94, 95
Glucose metabolism, phorbol diester effects 90
Glycoprotein, surface and phorbol diesters 89, 90
Guanosine fluorescence 37

Hepatocarcinogenesis assay 80
Hepatocellular carcinoma
  demography 313
  hepadna viruses 299–318
    infection 313–15
  hepatitis antigens 313–15
  integrated viral DNA 315, 316
    multiple sites 316
  virus state 315–18
Heterocyclic aromatics 3
Hydrazines 5
  carcinogenicity 21
  metabolic activation 21, 22
  naturally occurring 21
  structures 2, 3
N-Hydroxy-2-fluorenylacetamide
  carcinogens derived 8, 9
  hepatotoxicity 9
N-Hydroxy metabolites
  routes 8
    oxidative 9
β-Hydroxynitrosamines, sulfate conjugation 20

Immunocompromised hosts 289
Indomethacin, tumor promotion modification 78
Initiation 73, 74
  speed 75
Interferon, phorbol diester effects 92
5-Iodo-2′-deoxyuridine, antiviral 279

Kaposi's sarcoma
  herpes virus links 290
  virus in allografts 278

Ligand binding, phorbol diesters 88, 89
Lung cancer, asbestos and cigarettes 73
Lymphocyte
  characteristics of HTLV-positive 397–9
  T-cell growth factor 391
  T-cell tropism in HTLV 396, 397
Lymphokines
  HTLV-1 transformed T-cells 400, 401
  phorbol diester effects 92

Membrane enzymes, normal and transformed
    fibroblasts 178, 179
Membrane structure and transformation 161, 162
Metabolic activation 6–24
  and cancer site 25
Methapyrilene 25
Methotrexate resistance 94, 95
3-Methyladenine 35
3-Methylcholanthrene, DNA adducts 48
N-Methyl-N′-nitro-N-nitrosoguanidine (MNNG)
  cytotoxic and mutagenic effects 140, 142, 143
Mezerein 77
Mitomycin C
  DNA adduct structure 53, 54
  DNA modification 53
Model systems, carcinogenesis 74 (see also Mouse
    skin)
Monocrotaline, deoxyguanosine adduct 54
Mouse skin model, two-stage carcinogenesis 75–7
  initiators 76
  malignant tumors 76
  modification 78, 79
  promotion biology 76, 77
  putrescine 78
Mutagenic action
  DNA replication 141
  excision repair and carcinogens 139–43

Naphthylamines
  DNA adduct structure 50
  DNA modification 49, 50
  structure 2, 3
Nasopharyngeal carcinoma and Epstein-Barr
    virus 277, 293
  episomes 288
  reactivation 289
Neutrons, rodent cell transformation 170
Nitroaromatics, DNA adducts 51, 52
Nitrogen release 15
4-Nitroquinoline oxide
  DNA adducts 52
  structure 3
N-Nitrosamines
  carcinogenic activation 16
  cyclic, metabolic activation 18, 19
  α-ester chemistry 16
  α-hydroxylation pathway 14, 15
    nitrogen yield 15
    stability 15, 16
  metabolic activation 14–21
  ω-oxidation 17
  β-oxidized
    metabolic activation 19–21
    pancreatic carcinogens 19
  structures 2, 3
N-Nitroso-(acetoxymethyl)methylamine
  reactions 16
N-Nitrosoarylalkylamines, metabolic activation
N-Nitrosobis-(2-oxopropyl)amine, Baeyer-Villiger
  oxidation 20

N-Nitroso compounds, 4, 5
  structures 2, 3
Nitrosodialkylamines, metabolic activation 14–17
N-Nitrosodiethanolamine, liver carcinogen 20
N-Nitroso-2,6-dimethylmorpholine
    hydroxylation 19
  liver carcinogen 18
N-Nitrosodi-n-butylamine, ω-oxidation 17
N-Nitrosodipropyl system, β-oxidised, redox
    interconversion 20
N-Nitrosomethylalkylamines
  even and odd, cancers 17
  metabolic products 17
  structure 2, 3
N-Nitrosomethylaniline, oxidative methylation 21
N'-Nitrosonornicotine (NNN), metabolic
    pathways 18, 19
N-nitrosopyrollidine
  α-hydroxylation 18
  structure 18
N-substituted aromatics, DNA adduct
    formation 49, 52

Oncogenes
  in chemical carcinogenesis 5, 6
  defective avian retroviruses 365–74
  erb 366, 372, 373
    deletion mutants 372, 373
    genome 372
  ets 366, 371, 372
  feline sarcoma virus 329
  lil 366, 368
  malignancy 355
  mil 366, 368–71
  mutation 33
  myb 366, 371, 372
    nucleotide sequence 371
    protein coded 372
  myc 366, 368, 369
    nonviral neoplasms 370
    protein coded 370
    sequences 369
  c-myc
    activation 355
    regulation 111
  v-myc sequences 369
  phorbol diester effects 109, 111, 112
  radiogenic transformation role 182–4
  raf 368
  rel 366, 373, 374
    sequence 374
  ros 366, 368
  ski 366, 368
  transcript identification 184
  viral Harvey-ras 104
  yes 366, 367
Oncogenesis and latency 293
Ornithine decarboxylase, phorbol diester
    effects 91, 92
Oxides, DNA modification 42, 43

Phenobarbital, liver tumorigenesis 79
1-Phenyl-3,3-dimethyltriazene, metabolic
    pathway 23
Phorbol diesters
  binding assay 82
  binding sites 82, 83
  intracellular effects 90–2
  membrane effects 84–90
  metabolic effects 85–90
  mouse skin effects 77

  oncogene expression 109, 111, 112
  protein kinase binding 83
  skin tumor promotion 80, 81
  structure 80
  virus activation 110, 111
  virus-induced transformation 108–12
Phorbol structure 80
Phospholipid metabolism and phorbol diesters 89
Phosphonoacetic acid (PAA), structure and
    activity 280
Phosphonoformic acid (PFA)
  herpes virus 274
  structure and activity 280
Platelet aggregation, phorbol diesters 86
Platelet-derived growth factor (PDGF)
  simian sarcoma virus code 335
Poly(APP-ribose)
  and 3-aminobenzamide 182
  cell role 181
  radiogenic transformation 181, 182
Polycyclic aromatic hydrocarbons 2
  Bay regions 46
  DNA-bound forms 12
  DNA modification 43
  environmental 2
  epoxide double bonds 13
  metabolic activation 10–14, 25
    Bay region dihydrodiol epoxide 10, 11
    epoxide deposition 13
    reactions 13
  oxide formation sites 45
  ricinal dihydrodiol epoxide 13
  tumor initiation bond 14
Procarbazine, metabolic scheme 22
Promoter-insertion mechanism 316, 355
  avian leukosis virus 376–8
Promotion 73, 74 (see also Tumor promoters)
  inhibitor 76
  modifier 78, 79
β-Propiolactone 40
Protease secretion, phorbol diester effects 90
Protein phosphorylation, phorbol diester
    effects 90, 91
Proteolytic enzymes, transformed cell 162
Psoralens
  DNA adducts 55, 56
  uses 55
Pyriolizidine alkaloids, DNA adducts 54
Pyrollo[1,4]benzodiazepine antibiotics, DNA
    adducts 54

Radiation
  cancer risk 174, 184
  linear energy transfer 151
  public interest 184
  relative biological efficiencies 151
  types 151
Radiation-induced oncogenic transformation, assay
    systems 185, 186
Radiogenic transformation 151–89 (see also UV,
    X-ray)
  agar growth 162, 163
  agents affecting 158
  benzamide effects 181, 182
  chemicals affecting 175
  carcinogens 175
  criteria 159–64
  culture systems used 154–7
  diploid cells in vitro 172
  DNA target 182
  DNA transfection 183, 184
  dose-response relationship 165, 166

frequency 186
hamster cell characteristics 161
high-LET 168, 169
hormones affecting 176, 177
inhibitors 178–80
initiation 157
in vitro 164–73
low-LET, oncogenicity 164–7
methods used 158, 159
modulators 174–82, 187
oncogene role 182–4, 188
permissive factors 188
promoters 175, 176
protective factors 180, 181, 188
rates 158, 159
sequence of events 157, 158
split dose effects 167, 168
surface expression 187
Radiolabelling, carcinogens 35, 36
Reactive electrophilic metabolites 26
Receptor, CSF-1, v-*fms* gene product 334, 335
Receptor, T-cell growth factor 399
Receptor, transferrin 399
Restriction enzyme analysis 184
Retinoids and radiogenic transformation 178, 179, 188
Retinyl acetate 101
Reverse transcripticase 392
mRNA, viral transformed cell transcripts 214
Rodent cell radiation transformation 169, 170

Safrole
　carcinogenicity 53
　DNA adducts 53
　metabolic activation 24
　structure 3
Selenium, radiogenic transformation 180, 181, 188
Serum dependence and transformation 160, 161
Small molecular transport, phorbol diester effects 88
Soot and scrotal cancer 1
Southern blot analysis 184
Streptozotocin, DNA interaction 54
Structure-activity studies 6
Superoxide dismutase, radiogenic transformation 179, 180, 188
Swain and Scott parameters 34

Telecidins 81, 82
12-*O*-Tetradecanoylphorbol-13-acetate (TPA) (*see also* Phorbol esters)
　biochemical effects 84–111
　skin tumor promotion 80
Thyroid hormones and radiogenic transformation 176, 177, 186, 188
Tobacco-specific carcinogen 18, 19, 51
Transfection and viral DNA 317
Transformation (*see also* Cell, radiogenic)
　assay 136
Transforming growth factors 205, 224
Triazenes
　DNA modification 41
　metabolic activation 22, 23
　structures 3
Tumorigenesis and carcinogenesis 76
Tumorigenicity and transformation 163
Tumor promoters 73–113 (*see also* Phorbol diesters)
　models *in vivo* 75–80 (*see also* Mouse skin)
　modifiers 78, 79

mouse skin 74, 82
　biochemical effects 77, 78
rat liver 79
skin 80–3
Tyrosine kinase, mouse sarcoma virus transformation protein 361, 362

Ultraviolet light
　cytotoxicity, mutagenicity 143
　excision and DNA synthesis 141, 142, 144
　rodent cell transformation 169, 170
　sunlamps 174
　and transformation 33, 173
　　human cells 174
　　rodent cells 173, 174

Vinyl chloride
　cancers 42
　DNA modification 42
　structure 2, 3
Viral integration and transformation 316
Viral oncogenes 325
Viral transformation (*see also* individual viruses)
　enhancement 175
　mechanism 197–227
Virus, adenovirus
　cellular transformation 237–61
　　inefficiency 238
　discovery 237
　DNA homology 238
　EIA region 241–3
　　amino acid sequences 242
　　function of products 246
　　genetic analysis, protein role 251
　　hybrid plasmid transformation 248, 249
　　mutations, subgroup C effects 247, 248
　　polypeptide properties 258, 259
　　product scheme 241
　　products and transformation 246
　　mRNA species 242, 243
　　super transformation 250
　　temperature dependent phenotypes 249
　　eranscription 241
　　transformed cell phenotype 248, 249
　　transformation assay 249, 250
　EIB region 243
　　gene product function 251–5
　　19kd protein, properties 260, 261
　　19kd protein tumorigenicity 256
　　55kd protein properties 259, 260
　　mutant expression 252
　　mutants and defects 253
　　polypeptide properties 243–5
　　product scheme 243, 244
　　products in transformation 246
　　protein lesion effects 254
　　proteins and roles 252–253
　　mRNA species 245
　　N-terminal end 243
　　transformed cells 245
　　host defence evasion 257
　　oncogenic properties 243
　　transformation assay 249
　　transformed cell tumorigenicity 255–8
　　　EIB 19kd protein 256
　　　MHC antigen expression 257
　　transforming genes 238–41
　　　group N mutations 240
　　　mutants and defects 239
　　　permissiveness and temperature 239
　　　products 240
　　　purified DNA fragments 240, 241

Virus, adenovirus—*continued*
  tumorigenicity 255–8
    A and C subgroups 257
    determinants 255
Virus, avian
  defective 365–74
  cellular transformation 355–78
Virus, avian erythroblastosis
  defective 372, 373
  genome 357
Virus, avian leukemia
  defective 368–74
  multiple oncogeneic events 374
Virus, avian leucosis
  classification 375
  genome 257, 275
    structure 375
  and leukemia 328
  lymphoid leukosis disease 374, 375
  lymphomatosis induction 377
  oncogenesis 374
  promoter insertion mechanisms 376–8
  proviral integration 376
  strains and erythroblastosis 378
Virus, avian myeloblastosis
  defective 371
  genome 357, 371
Virus, avian sarcoma, defective 368
Virus, bovine leukemia 391
  HTLV relatedness 394
Virus, cytomegalovirus 272
  cancer links 290, 292
  cellular transformation 290
  fibroblast transformation 290, 291
Virus, duck hepatitis
  distribution 314, 315
  replication 309
Virus, Epstein-Barr
  acyclovir mechanism 274, 281, 286
    reversal 287
  cancers associated 272
  cell systems 285, 286
  genome sequence 275
  infection in AIDS 278, 279
  infective states 286
  latent infections 287
  lymphocyte transformation 288
  lymphoproliferative syndromes 277, 278
  nasopharyngeal carcinoma 277, 288, 289, 293
  oncogenic 292, 293
  pathogenesis 288, 289
  protein synthesis effects 286, 287
  therapy prediction 288, 289
  transformation evidence 290
  treatment 285–9
Virus, FJB-induced tumors 341, 342
  bone 342
  fos, expression regulation 350, 351
  fos gene 343
    nucleotide sequence 343
    product 343
  c-*fos* gene
    differentiation 348, 349
    expression 347–50
    fos-related R-fos 345
    growth 348
    hematopoietic 349
    induction 347
    neonatal 349
    prenatal 347, 348
    properties 344–6
    protein features 344
    proviral DNA 344
    sequence homology 344, 345
    transcription 350
    transformation 345, 346
    transforming potential 346
  v-*fos* gene properties 343–5
    protein 343
  helper virus 342
  isolation 341
  -MSV 342, 343
    molecular architecture 344
Virus, feline leukemia 325, 335
  disease induced 327, 328
  features 391
  genetic structure 326
  insertional mutagen 328
  replication 326
  spread 326, 327, 391
  wild cats 327
Virus, feline retroviruses 325–36
  effects 325, 326
Virus, feline sarcoma 325
  v-*fes* 331
  v-*fms* 331–5
    nucleotide sequences 333
    polyproteins 333
    product and CSF-1 receptor 334, 335
    protein coded 334
    translation product 332
  gene order 330
  isolation 328, 329
  oncogenes 329
  platelet-derived growth factor 335
  strains 329, 330
  structure and replication 329, 330
  transcription regulation 330, 331
  transforming function 329
  transforming genes 335
  tyrosine-specific protein kinases 331, 332
    oncogene sequences 332
Virus, *fos* 341–3 (*see also* Virus, FBJ)
Virus, Fujinami sarcoma
  defective 365
  fps gene 365
  genome 357
Virus, ground squirrel hepatitis 299
  disease response 314
  polypeptides 301, 302
Virus, hepadna
  antigenicity 301
  C gene 304, 306
  codon preferences 307
  DNA replication scheme 312
    mechanism 308–11
  genome structure 302–5
  hepatocellular carcinoma role 299–318
  infection and hepatocellular carcinoma 313–15
  morphology 300
  name 300
  polymerase amino acids 306
  polypeptides 301, 302
  protein DNA primer 309
  retrovirus homology 305–7
  retrovirus relationship 311–13
  reverse transcripticase 300
  RNA template 310, 312
  viral integration 310, 311
  X-gene 307
Virus, hepatitis B 271
  DNA digestion sequences 310
  genetic map 303
    5′ ends 304
    nucleotide sequences 304

hepatitis 299
integrated DNA 316
polypeptides 301, 302
  genes 304
prevalence 299
Virus, herpes (*see also* Virus, Epstein-Barr)
  antigenicity 276
  antiviral drug effects 273, 274
  appearance 274, 275
  cancer associated types 272
  cellular transformation 271–93
  genome features 275
  human neoplasia 289–93
  latency 291
    oncogenesis 293
Virus, human T-cell leukemia (HTLV-1) 271
  isolation 271
Virus, human T-lymphotropic retroviruses
    (HTLV) 391–401
  cell transformation 397–9
  epidemiology 394, 395
  isolates, relatedness 392, 393
  isolation 391, 392
  oncogenes 401
  patient antibodies 393
  properties of HTLV-1, -2 and -3 392
  reverse transcripticase 392
  T-cell positive, characteristics 397–9
  T-cell transformed
    lymphokine release 400, 401
    mechanism 401
    phenotype 399, 400
  T-cell tropism 396, 397
  transmission *in vitro* 395–7
    bone marrow 397
    cord blood 397
Virus, murine leukemia; genetic map 303, 305
Virus, myelomatosis 29, genome 357
Virus, oncogenic properties 273
Virus, phorbol diester effects 108–12
  expression 110, 111
  transformation 108–10
Virus, polyoma
  cell interactions 197
  DNA map 199
  features 197
  host control of transformation 224, 225
  host metabolism 204
  infection process 201, 202
  isolation 197, 205
  oncogenic potential 205, 206
    cell culture type 208
  permissiveness 198
    transformation 207, 208
  regulatory protein 200
  mRNA features 199
  RNA synthesis induction 204, 205
  T-antigen 200, 201
    fibroblast transfection 223
    function 203
    molecular weight 203
    mutants 222
    primary cell immortalization 223, 224
    transformation role 222–4
  transcription 200
  transformation aspects 198
  transformation models 226, 227
  transformed cell properties 206, 207
    early protein role 224
    evolution 217, 218
    excision and amplification mechanisms 212, 213
    gene integration scheme 209, 210

insertion site 209
integrated genome expression 213–17
integrated genome instability 210–13
persistence 208
transcription regulation 215
viral free DNA 211
viral genome 208–13
viral proteins 216, 217
viral transcripts 214–17
transforming genes 226, 227
Virus, reticuloendotheliosis, genome 357, 373
Virus, retrovirus (*see also* individual retroviruses)
  DNA effects 317
  hepadnavirus genomes homology 305–7
  hepadnavirus relationships 311–13
Virus, Rous sarcoma
  cell transformation 356
  characteristics 356
  genome 356, 357, 359
  replication 357
  RNA species 357
  transforming gene product 359
    coprecipitation 363
    location 360, 361
    pp60 360
    structure 359, 360
    target proteins 362–5
    tyrosine kinase 361, 362
    tyrosine phosphorylation 364
  transforming protein 358, 359
  ts mutant 358, 359
  v-*src* gene 358, 359, 361
    amino acid sequence 362
Virus, SV40
  cell interactions 197
  DNA map 200, 201
  features 197, 272
  gpt gene activation 216
  host control of transformation 224, 225
  host metabolism 264
  infection process 201, 202
  oncogenic potential 206
    cell culture type 208
  permissiveness 198
    transformation 207, 208
  RNA synthesis induction 204, 205
  T-antigen 201
    functional map 220
    multifunction 221
    properties 202, 203, 220, 221
    replication-defective 220
    transformation competent, replication
      defective 220
    transformation role 218–21
  transcription 200, 201
  transformation aspects 198
    viral protein role 218–21
  transformation model 226, 227
  transformed cell 206, 207
    evolution 217, 218
    excision and amplification 213
    gene integration 209
    insertion site 209
    integrated genome expression 213–17
    integrated genome stability 210–13
    transcription regulation 215
    viral genome 208–13
    viral protein 216, 217
    viral transcripts 214–17
  transforming genes 227
  ts-SV40 transferred 3T3 cells 225
Virus, varicella-zoster features 272, 276
Virus, Woodchuck hepatitis 299

Virus, Woodchuck hepatitis—*continued*
  distribution 314
  polypeptides 301, 302
  tumor DNA 316
Virus, Yamaguchi 73 Sarcoma genome 357
Vitamin E, radiogenic transformation 180, 181

Xeroderma pigmentosum (*see also* Cells, human)
  cell transformation 188

  excision repair-deficient cells 136
  mutation assay 136, 137
X-rays
  carcinogenic 152
  human cell transformation 169, 171–3
  induced transformation 164–73
  neoplastic transformation, hamster embryo 152, 153
  rodent cell transformation 169, 170